ICALT 2001

IEEE International Conference on Advanced Learning Technologies (ICALT 2001)

Proceedings

IEEE International Conference on Advanced Learning Technologies

6-8 August 2001

Madison, Wisconsin, USA

Sponsored by
IEEE Computer Society
IEEE Computer Society Learning Technology Task Force
University of Wisconsin-Madison

Hosted by
University of Wisconsin-Madison, USA

Supported by
Microsoft Research
IEEE Learning Technology Standards Committee
Army Reserve Readiness Training Center

Edited by
Toshio Okamoto
Roger Hartley
Kinshuk
John P. Klus

Los Alamitos, California
Washington   Brussels   Tokyo

Copyright © 2001 by The Institute of Electrical and Electronics Engineers, Inc.
All rights reserved

Copyright and Reprint Permissions: Abstracting is permitted with credit to the source. Libraries may photocopy beyond the limits of US copyright law, for private use of patrons, those articles in this volume that carry a code at the bottom of the first page, provided that the per-copy fee indicated in the code is paid through the Copyright Clearance Center, 222 Rosewood Drive, Danvers, MA 01923.

Other copying, reprint, or republication requests should be addressed to: IEEE Copyrights Manager, IEEE Service Center, 445 Hoes Lane, P.O. Box 133, Piscataway, NJ 08855-1331.

The papers in this book comprise the proceedings of the meeting mentioned on the cover and title page. They reflect the authors' opinions and, in the interests of timely dissemination, are published as presented and without change. Their inclusion in this publication does not necessarily constitute endorsement by the editors, the IEEE Computer Society, or the Institute of Electrical and Electronics Engineers, Inc.

IEEE Computer Society Order Number PR01013
ISBN 0-7695-1013-2
ISBN 0-7695-1014-0 (case)
ISBN 0-7695-1015-9 (microfiche)
Library of Congress Number 2001087975

Additional copies may be ordered from:

IEEE Computer Society	IEEE Service Center	IEEE Computer Society
Customer Service Center	445 Hoes Lane	Asia/Pacific Office
10662 Los Vaqueros Circle	P.O. Box 1331	Watanabe Bldg., 1-4-2
P.O. Box 3014	Piscataway, NJ 08855-1331	Minami-Aoyama
Los Alamitos, CA 90720-1314	Tel: + 1 732 981 0060	Minato-ku, Tokyo 107-0062
Tel: + 1 714 821 8380	Fax: + 1 732 981 9667	JAPAN
Fax: + 1 714 821 4641	http://shop.ieee.org/store/	Tel: + 81 3 3408 3118
http://computer.org/	customer-service@ieee.org	Fax: + 81 3 3408 3553
csbooks@computer.org		tokyo.ofc@computer.org

Editorial production by Danielle C. Young

Cover art production by Joe Daigle/Studio Productions

Printed in the United States of America by The Printing House

Table of Contents

IEEE International Conference on Advanced Learning Technologies (ICALT 2001)

Preface xviii
Conference Organization xx
Reviewers xxii

Full and Short Papers

Theme 1: Architectures

A Standards-Driven Open Architecture for Learning Systems 3
    L. Anido, M. Llamas, M. Fernández, J. Rodríguez, M. Caeiro, and J. Santos

The Educational Digital Entities as a Component of the New Hybrid and
Learning School Environment: Expectations and Speculations 5
    P. Anastasiades

Virtual Adaptive Learning Architecture (VALA) 7
    A. Metcalfe, M. Snitzer, and J. Austin

Managing Large Scale On-Line Systems 11
    G. Aldridge

Architectures Supporting e-Learning through Collaborative Virtual Environments:
The Case of INVITE 13
    Ch. Bouras, G. Hornig, V. Triantafillou, and Th. Tsiatsos

Developing an Architecture for the Software Subsystem of a Learning
Technology System — An Engineering Approach 17
    A. Paris, R. Simos, P. Andreas, and S. Manolis

An Architecture for Defining Re-usable Adaptive Educational Content 21
    C. Karagiannidis, D. Sampson, and F. Cardinali

Architectures for Computer Supported Collaborative Learning 25
    D. Suthers

The Architecture of a Framework for Building Distributed Learning
Environments 29
    C. Allison, D. McKechan, A. Ruddle, and R. Michaelson

Theme 2: Theories and Formalisms

Constructivist Values for Web-Based Instruction: University and NSA Collaboration 35
    P. Semrau and B. Boyer

Research on Role-Based Learning Technologies 37
    B. Slator, J. Clark, P. Juell, P. McClean, B. Saini-Eidukat, D. Schwert, and A. White

An Ontology for Modeling Ill-Structured Domains in Intelligent Educational Systems 41
    C. Papaterpos, N. Georgantis, and T. Papatheodorou

An Affective Model of Interplay between Emotions and Learning: Reengineering Educational Pedagogy—Building a Learning Companion 43
    B. Kort, R. Reilly, and R. Picard

Theme 3: Instructional Design (Including 3D)

Instructional Design in a Technological World: Fitting Learning Activities into the Larger Picture 49
    R. Soine

The Design of a Lexical Difficulty Filter for Language Learning on the Internet 53
    C.-H. Kuo, D. Wible, C.-C. Wang, and F.-Y. Chien

An Engineering Process for Constructing Scaffolded Work Environments to Support Student Inquiry: A Case Study in History 55
    K. Luchini, P. Oehler, C. Quintana, and E. Soloway

The Authority Structure Problem of Computer Supported Collaborative Concept Mapping System for Elementary Students 57
    C.-H. Chiu

Decompressing and Aligning the Structures of CBI Design 61
    J. Duffin and A. Gibbons

A Design Framework for Interaction in 3D Real-Time Learning Environments 63
    F. Klett

The RSVP Unibrowser: Bringing Humanness and Interactivity to e-learning 67
    E. Cupp, M. Danchak, K. Foster, A. Johnson, C. Kim, and D. Sarlin

Prototype of *Cyber Teaching Assistant* 70
    Y. Shindo and H. Matsuda

Carpe Diem: Models and Methodologies for Designing Engaging and Interactive
e-learning Discourse 74
    A. Ravenscroft and M. Matheson

SAC: A Self-Paced and Adaptive Courseware System 78
    A. Chan, S. Chan, and J. Cao

Collaborative Analysis and Tutoring: The FACT Framework 82
    M. Mora and R. Moriyón

Implementing Collaborative Learning Research in Web-Based Course Design and
Management Systems 86
    K. Wijekumar

Structure of Training Cases in Web-Based Case-Oriented Training Systems 90
    T. Illmann, A. Seitz, A. Martens, and M. Weber

Stream-Based Lecturing System and Its Instructional Design 94
    N.-S. Chen and Y.-C. Shih

Learning through Ad-hoc Formative Paths 96
    V. Carchiolo, A. Longheu, and M. Malgeri

Theme 4: Teaching/Learning Strategies (Including Concept Mapping)

The Internet and the "Learning by Doing" Strategy in the Educational Processes:
A Case of Study 103
    N. Sala

Effective Learning Strategies for the On-Line Learning Environment: Including
the Lost Learner 105
    M. Axmann

Analyzing Middle School Students' Use of the ARTEMIS Digital Library 107
    J. Abbas, C. Norris, and E. Soloway

DIAL: Serendipitous DIAlectic Learning 109
    G. da Nóbrega, S. Cerri, and J. Sallantin

Learning Management in Integrated Learning Environments 111
    I. Galeev, S. Sosnovsky, and V. Chepegin

How Undergraduate Students' Learning Strategy and Culture Effects Algorithm Animation Use and Interpretation 113
    T. Hubscher-Younger and N. Narayanan

Theme 5: Specific Applications

Domain-Expert Repository Management for Adaptive Hypermedia Learning System 119
    N. Yusof and P. Samsuri

Application of Web-Based Learning in Sculpture Curves and Surfaces 121
    W.-T. Sung and S.-C. Ou

Making Design Accessible 123
    C. Lawrence and R. Baird

English Assistant: A Support Strategy for On-Line Second Language Learning 125
    G. Weir and G. Lepouras

Automating Repeated Exposure to Target Vocabulary for Second Language Learners 127
    D. Wible, C.-H. Kuo, F.-Y. Chien, and N. Taso

Using Speech Analysis Techniques for Language Learning 129
    V. Petrushin

Web Passive Voice Tutor: An Intelligent Computer Assisted Language Learning System over the WWW 131
    M. Virvou and V. Tsiriga

The Development of CALL Environment on the WWW for Teaching Academic English 135
    J. Chen, H. Inoue, T. Okamoto, S. Belkada, and A. Cristea

OWLs in Flight: Online Writing Labs for Distance Learning 137
    M. Spore

Bridging Gaps in Computerised Assessment of Texts 139
    D. Callear, J. Jerrams-Smith, and V. Soh

From Concrete Experiences to Abstract Formalisms: Learning with Interactive Simulations that Combine Physical and Computational Media 141
    M. Milrad

The Design and Evaluation of Simulations for the Development of Complex Decision-Making Skills 145
    R. Hartley and G. Varley

Evaluation of a Learning System – and Learning to Evaluate 149
    H. Keegan

The Effects of Simulation Participation on the Perception of Threatening
Cultural Dynamics in a Collaborative Virtual Learning Environment 151
    E. Raybourn

Computerised Problem-Based Scenarios in Practice — A Decade of DIAGNOSIS 153
    T. Stewart, R. Kemp, and P. Batrum

Asynchronous Distributed Problem-Based Learning 157
    F. King and H. Mayall

Software Support for Creative Problem Solving 160
    A. Aurum, J. Cross, M. Handzic, and C. Van Toorn

Effect of Using Computer Graphics Animation in Programming Education 164
    H. Matsuda and Y. Shindo

X-Compiler: Yet Another Integrated Novice Programming Environment 166
    G. Evangelidis, V. Dagdilelis, M. Satratzemi, and V. Efopoulos

How Students Learn to Program: Observations of Practical Tasks Completed 170
    P. Thomas and C. Paine

C-VIBE: A Virtual Interactive Business Environment Addressing Change
Management Learning 174
    A. Angehrn and T. Nabeth

How to Design Web-Based Counseling Systems 178
    M. Kuittinen, S. Pöntinen, and E. Sutinen

Application of Multipoint Desktop Video Conferencing System (MDVC) for
Enhancement of Trainee-Supervisor Discourse in Teacher Education 180
    M. Khine, L. Sharpe, H. Chun, L. Crawford, S. Gopinathan, M. Ngoh
    and A. Wong

Theme 6: Authoring and Development of Systems

Authoring and Delivering Adaptive Web-Based Textbooks Using WEAR 185
    M. Moundridou and M. Virvou

Using the REDEEM ITS Authoring Environment in Naval Training 189
    S. Ainsworth, B. Williams, and D. Wood

Adaptive Presentation of Evolving Information Using XML 193
    M. Bieliková

A Collaborative Courseware Generating System Based on WebDAV, XML, and JSP 197
    C. Qu, J. Gamper, and W. Nejdl

Use of RDF for Content Re-purposing on the ARKive Project 199
    A. Dingley and P. Shabajee

From CD-ROM to Web-Served: Reverse Engineering of an Interactive Multimedia
Course 203
    D. Maraschi, S. Cerri, and G. Martinengo

A Matter of Life or Death: Re-engineering Competency-Based Education through
the Use of a Multimedia CD-ROM 205
    J. Wilkinson

From Educational Meta-data Authoring to Educational Meta-data Management 209
    V. Papaioannou, P. Karadimitriou, A. Papageorgiou, C. Karagiannidis,
    and D. Sampson

Design and Implementation of WBT System Components and Test Tools for WBT
Content Standards 213
    K. Nakabayashi, Y. Kubota, H. Yoshida, and T. Shinohara

UNIVERSAL - Design and Implementation of a Highly Flexible E-Market-Place for
Learning Resources 215
    S. Brantner, T. Enzi, S. Guth, G. Neumann, and B. Simon

Theme 7: ITS and Agent Based Methods

Investigation of Learning Object Metadata and Application to a Search Engine
for K-12 Schools in Japan 221
    M. Okamoto, M. Shinohara, Y. Okui, S. Terashima,
    and M. Hashimoto

K-InCA: Using Artificial Agents to Help People Learn and Adopt New Behaviours 225
    A. Angehrn, T. Nabeth, L. Razmerita, and C. Roda

Human Teacher in Intelligent Tutoring System: A Forgotten Entity! 227
    Kinshuk, A. Tretiakov, H. Hong, and A. Patel

Multi-media as a Cognitive Tool: Towards a Multi-media ITS 231
    F. Albalooshi and E. Alkhalifa

A Case-Based Agent Framework for Adaptive Learning 235
    C.-S. Lee and Y. Singh

A Web-Based ITS Controlled by a Hybrid Expert System 239
  J. Prentzas, I. Hatzilygeroudis, and C. Koutsojannis

E-learning from Expertize: A Computational Approach to a Non-textual Culture of Learning 241
  G. Albano and F. Formato

An Adapted Virtual Class Based on Intelligent Tutoring System and Agent Approaches 243
  A. Hernández-Domínguez and A. do Socorro da Silva

Theme 8: CMC, Dialogue, Collaborative Learning

An Analysis of Critically Reflective Teacher Dialogue in Asynchronous Computer-Mediated Communication 247
  M. Hawkes

Facilitating Entry into Professional Life with CMC Support 251
  D. Walker

Facilitating Computer-Mediated Discussion Classes: Exploring Some Teacher Intervention Strategies 253
  S. Walker and R. Pilkington

How to Learn the Many Unwritten "Rules of the Game" of the Academic Discourse: A Hybrid Approach Based on Critiques and Cases to Support Scientific Writing 257
  S. Aluísio, I. Barcelos, J. Sampaio, and O. Oliveira, Jr.

Helping the Tutor Facilitate Debate to Improve Literacy Using CMC 261
  P. Kuminek and R. Pilkington

A Study of Social-Learning Networks of Students Studying an On-Line Programme 263
  G. Singh and J. O'Donoghue

Proposal of a Collaborative Learning Standardization 267
  T. Okamoto, M. Kayama, and A. Cristea

Cosar: Collaborative Writing of Argumentative Texts 269
  J. Jaspers, G. Erkens, and G. Kanselaar

The Knowledge Management for Collaborative Learning Support in the INTERNET Learning Space 273
  M. Kayama and T. Okamoto

Theme 9: Virtual Learning Environments

Virtual Classroom 279
    A. Krukowski and I. Kale

A Virtual Learning Environment for Short Age Children 283
    P. González, F. Montero, V. López, A. Fernández-Caballero, J. Montañés, and T. Sánchez

Using a Virtual Learning Environment in Higher Education to Support Independent and Collaborative Learning 285
    C. Bennett and R. Pilkington

Applying Interactive Mechanism to Virtual Experiment Environment on WWW with Experiment Action Language 289
    R. Kuo, M. Chang, and J.-S. Heh

A Criterion-Referenced Approach to Assessing Perioperative Skills in a VR Environment 291
    W. Witzke, D. Witzke, J. Hoskins, M. Mastrangelo, U. Chu, I. George, and A. Park

Theme 10: Support and Tools Systems

Student Adaptivity in TILE: A Client-Server Approach 297
    Kinshuk, B. Han, H. Hong, and A. Patel

MachineShop: Steps toward Exploring Novel I/O Devices for Computational Craftwork 301
    G. Blauvelt and M. Eisenberg

Supporting Knowledge Communities with Online Distance Learning System Platform 305
    J. Correa, D. Fink, C. Moraes, and A. Sonntag

Support Tools for Graphs in Computer Science Education 307
    V. Kasyanov

A Template-Based Concept Mapping Tool for Computer-Aided Learning 309
    A. Arruarte, J. Elorriaga, and U. Rueda

Educational Web Portal Based on Personalized and Collaborative Services 313
    C. Martel and L. Vignollet

Using WebCT to Support Team Teaching 315
    A. Fuller, G. Awyzio, and P. McFarlane

WhiteboardVCR – A Presentation Tool Using Text-to-Speech Agents 319
    N. Chong, P. Tosukhowong, and M. Sakauchi

Effective Use of WebBoard for Distance Learning 323
    H. Wang

Decision-Making Resources for Embedding Theory into Practice 327
    G. Conole and M. Oliver

Theme 11: Evaluation/Monitoring/Student Models/Information Needs

Student Models Construction by Using Information Criteria 331
    M. Ueno

The Effectiveness of an Intelligent Tutoring System for Rocket Training 335
    R. Wisher, L. Abramson, and J. Dees

Evaluation of the Advice Generator of an Intelligent Learning Environment 339
    M. Virvou and K. Katerina

The Success of Advanced Learning Technologies for Instruction: Research and
Evaluation of Human Factors Issues 343
    D. O Coldeway

Mapping Information Needs 345
    G. Conole

On Workflow Enabled e-Learning Services 349
    J. Lin, C. Ho, W. Sadiq, and M. Orlowska

Issues and Methods for Evaluating Learner-Centered Scaffolding 353
    C. Quintana, J. Krajcik, and E. Soloway

Towards Evaluating Learners' Behaviour in a Web-Based Distance Learning
Environment 357
    O. Zaïane and J. Luo

On Monitoring Study Progress with Time-Based Course Planning 361
    I. Fung

What's in a Prerequisite 365
    R. Hübscher

Exploring Learning Problems of Cyber University 369
  K. Lin and N. Chen

Evaluating the Learning Object Metadata for K-12 Educational Resources 371
  D. Suthers

Measuring Knowledge Transfer Skills by Using Constrained-Student Modeler Autonomous Agent 375
  S. Belkada, A. Cristea, and T. Okamoto

Studying the Learning Practice: Implications for the Design of a Lifelong Learning Support System 379
  G. Vavoula and M. Sharples

Courseware Accessibility: Recommendations for Inclusive Design 381
  R. Luke

Design for Web-Based On-Demand Multiple Choice Exams Using XML 383
  R. Lister and P. Jerram

Theme 12: Linkages/Networks for Linkages and Lifelong Learning/ Distance Education/Internet Resources

Resource Manager for Distance Education Systems 387
  G. Kimovski, V. Trajkovic, and D. Davcev

The Tuneup and Integration of Resources in Web-Based Learning 391
  L. Xiao, L. Jianguo, and Z. Xiaozhen

Interactivity and Integration in Virtual Courses 395
  C. Pahl

Networked Facilitated Open and Distance Learning in Continuing Engineering Education 397
  S. Payr, F. Reichl, G. Csanyi, and U. Vierlinger

Linking Experiences: Issues Raised Developing Linkservices for Resource Based Learning and Teaching 401
  H. Davis and S. White

Distance Learning Technologies and an Interactive Multimedia Educational System 405
  G. Jiang, J. Lan, and X. Zhuang

The Continuous Education Solution for a Country Wide Telecommunication Company 409
  M. Leifheit, J. Correa, and D. Fink

The Practice in the Web-Based Teaching and Learning for Three Years 411
    F. Lin and X. Xie

Running a European Internet School – OTIS at Work 413
    M. Beer, G. Armitt, J. van Bruggen, R. Daniels, L. Ghyselen, S. Green,
    J. Sandqvist, and A. Sixsmith

Considering Automatic Educational Validation of Computerized Educational
Systems 415
    A. Cristea and T. Okamoto

The Distance Ecological Model to Support Self/Collaborative-Learning in the
Internet Environment 418
    T. Okamoto, M. Kayama, A. Cristea, and K. Seki

A Cooperative Linkage between University and Industry via an Internet Distance
Education System 422
    T. Okamoto, H. Inoue, A. Cristea, M. Kayama, T. Matsui, and K. Seki

Internet Based Course Delivery: Technology and Implementation 426
    M. Sun

Posters

ITS Design Technology for the Broad Class of Domains 431
    I. Galeev, S. Sosnovsky, and V. Chepegin

Constructivist Learning Systems: A New Paradigm 433
    W. Li

Promoting the Introduction of Lifelong Learning Related Concepts in the
Description of Information Resources Using Metadata Technology 435
    R. Garcia Robles

A Framework for Constructing Adaptive Web-Based Educational Systems 437
    M. Obitko, L. Kurz, and I. Glücksmann

Domain Instruction Server (DIS) 439
    J. Gilbert and D.-M. Wilson

A Web-Enabled Exam Preparation and Evaluation Service: Providing Real-Time
Personalized Tests for Academic Enhancement 441
    S. Abidi and A. Goh

An Evolving Instructional Design Model for Designing Web-Based Courses 443
    S.-W. Ling, C.-W. Khong, and C.-S. Lee

Web-Based Educational Tools in Buraydah College of Technology 445
  J. Al-Herbish and A. Arbaoui

The University of Bristol DataHub - A Prerequisite for an Integrated Learning
Environment? 447
  P. Browning and G. Conole

Displaying Mental Powers: The Case of VirtualMente 449
  M. del Carmen Malbrán and C. Mariela Villar

An Evolutionary Distribution System for Web-Based Teaching Materials 451
  T. Ishikawa, H. Matsuda, and H. Takase

ACME Project, Internet-Based Systems That Advocate Academic Credit for Military
Experience and Analyze Options for Veterans in Career Transition 453
  D. Wenger, M. Rufflo, and F. Bertalan

Panels

Panel 1: Agents, Believability, and Embodiment in Advanced Learning Environments

Agents, Believability and Embodiment in Advanced Learning Environments:
Introduction to a Panel Discussion 457
  A. Nijholt

Teaching with the Help of Talking Heads 460
  A. Graesser, X. Hu, and N. Person

Cognitive Requirements for Agent-Based Learning Environments 462
  A. Baylor

Contributions to Learning in an Agent-Based Multimedia Environment:
A Methods-Media Distinction 464
  R. Moreno

Panel 2: The Integration of Television and the Internet

Integration of TV and the Internet: Design Implications and Issues 469
  M. Ally

Genres, User Attitude and Prospects for Learning through Video on the WWW 471
  P. Kommers

Prefetch Agent: Virtual Internet Based on CATV 473
    H. Lu, Z. Lu, and Y. Li

The Integration of Television and the Internet 475
    C. O'Hagan

Edutainment – The Integration of Education and Interactive Television 478
    S. Skelton

Surgeon Training via Video Conferencing and the Internet 480
    A. Vranch and A. Kingsnorth

Panel 3: Collaborative Context-Mediated Experiential Learning through Asynchronous Learning

Collaborative Learning through Versatile Representations in Asynchronous Learning Transactions via the WWW 485
    E. McKay

Collaborative Knowledge Management Requirements for Experiential Learning (CKM) 488
    B. Garner

Collaborative Learning Support Knowledge Management for Asynchronous Learning Networks 490
    T. Okamoto, M. Kayama, and A. Cristea

Sharing Learning Experiences through Correspondence on the WWW 492
    P. Kommers

Peer-to-Peer and Learning Objects: The New Potential for Collaborative Constructivist Learning Online 494
    D. Wiley

Tutorials

Learner-Centered Design: Developing Software That Scaffolds Learning 499
    C. Quintana, E. Soloway, and C. Norris

Animated Pedagogical Agents for Education Training and Edutainment 501
    W. Johnson

Adaptive Educational Environments for Cognitive Skills Acquisition 502
    A. Patel and Kinshuk

Author Index 505

Preface

The International Conference on Advanced Learning Technologies (ICALT 2001) brings together researchers, academics, and industry practitioners who are involved or interested in the design and development of advanced and emerging learning technologies. Understanding the challenges faced in providing technology tools to support the learning process and ease the creation of instructional material will help build a valuable direction for further research and implementation.

The rapidly increasing interest in advanced learning technologies provides many challenges to those engaged in such research and development enterprises. On the one hand, the capabilities of digital technologies in providing and contributing to learning environments are opening up new approaches that utilise, for example, multimedia, virtuality, and collaborative methods of knowledge management. On the other hand, the changing and increasing demands of education in this technological age require practical techniques and applications that benefit a wider range of abilities, learning styles, and organisations.

Interesting and important issues being addressed by the Conference include:

Where should the computer be placed in these developments and what roles should it undertake in learning environments?

What theories and representations should underpin research and what are the fruitful directions to follow and exploit?

How should the adaptive intelligences of computing systems and teachers/students interact and collaborate?

What pedagogies are appropriate and useful to guide applications, and what tools and media are required for developers, teachers, and students?

Evaluation is an important but often neglected issue; what methods are appropriate to provide guidance and empowerment to these advances in learning technologies and their implementations?

Under these themes, several papers focus on the architectures of systems that link pedagogic approaches to digital technologies in order to provide a variety of learning environments that may be virtual, large scale, adaptive, and able to support collaborative interactions. Issues of standards and the re-usability of software and materials are also discussed.

The learning theories and formalisms underpinning instructional designs within these environments include affective and attitudinal factors, and link to a variety of teaching/learning strategies dealing with constructivism and scaffolding, activity and problem based methods, and learning management. Specific issues address structuring, hypermedia, and concept mapping, interactivity and engagement, collaborative techniques, and case-based training. These applications are set in contexts that span CAD, Computer Programming, Business Studies, Language Learning, Computer Based Assessment, and Student Counselling.

The papers focusing on Authoring and System Development are largely Web oriented but feature techniques of reverse-and re-engineering of materials, and systems that provide management structures as well as adaptive interactions. These relate to Tools and Systems to support knowledge management and collaborative working, and techniques of concept mapping, audio-graphics devices and the use of WebCT. Applications using Virtual Learning Environments and Intelligent Tutoring Systems are also well represented, tackling issues such as the role of the human teacher in relation to ITSs, multi-modalities as cognitive tools, the application of hybrid expert systems, and Text to Speech Agent techniques.

Within interactive and cooperative learning, systems support for asynchronous and synchronous collaborative discourse is becoming increasingly important. Several papers consider the analysis of discourse, the roles of tutors in facilitating and managing computer based discourse, the development of argumentation skills, the social networks that can be established, and the fostering of collaborative writing.

Evaluation has not been neglected within the Conference agenda. Studies that are being reported have examined projects that use advanced learning technologies and intelligent tutoring systems, learner centred scaffolding, student modelling and learning practices, and Web based environments. The papers also comment on techniques of determining pre-requisites, mapping student information needs, and monitoring.

Finally, on a larger scale, some papers discuss the establishment and management of learning networks, and the interlinking of resources within and between institutions that feature Distance Learning, and have a concern for widening access within a concept of Lifelong Learning.

There is no doubt that, in the future, e-learning, in response to the requirements of society will transform education and training in comprehensive ways. From an institutional and administrative standpoint, e-Universities and Colleges will expand their businesses via distance learning; the production of material will essentially involve companies that utilise the instructional intelligence and capital of teachers who are effective communicators in the new media. The future is hard to read, with its increasing diversity and volatility, and with an increasing emphasis on marketability and cost-effectiveness set against political and social pressures for widening access under a lifelong learning concept. Most predictions prove wrong, but the success of e-learning depends critically on developing pedagogies and teaching/learning methods that are able to adapt to changing circumstances. For this to be manageable and affordable, research needs to provide a framework to guide the pragmatics of progress. Hopefully ICALT2001, launched by Professor O'Shea's keynote address can make its distinctive contribution to this objective. The Conference presents a rich variety of theories, applications and expertise relating to Advanced Learning Technologies which should stimulate an active and positive response from Conference participants.

We are delighted to report that IEEE-ICALT2001 attracted almost 200 submissions, from 34 countries. All submissions were reviewed by an international panel of expert referees, resulting in about 30% acceptance rate in the form of full papers. Some other high standard submissions were selected for presentation and publication as short papers and as posters. We regard all of them as having the higher academic value, and expect all authors will contribute much to improvement of education in each country as well as international fields. Finally, we are pleased to acknowledge the invaluable assistance of the international referees, who are named separately in this proceedings.

J. R. Hartley, Toshio Okamoto, Kinshuk and John P. Klus
Editors

Conference Organization

Program Co-Chairs
Roger Hartley, *University of Leeds, United Kingdom*
Toshio Okamoto, *University of Electro-Communications, Japan*

General Chair
Kinshuk, *Massey University, New Zealand*

Organizing Chair
John P. Klus, *UW-Madison, USA*

Panel Chair
Piet Kommers, *Twente University, The Netherlands*

Tutorial Chair
Clark Quinn, *KnowledgePlanet, USA*

Workshop Chair
Chul-Hwan Lee, *Inchon National University of Education, Korea*

Steering Committee
Michael Freeman, *Army Programs Training, Computer Sciences Corporation, USA*
Ildar Galeev, *Kazan State Technological University, Russia*
Roger Hartley, *University of Leeds, United Kingdom*
Randy Hinrichs, *Microsoft Research, USA*
Chris Jesshope, *Massey University, New Zealand*
Kinshuk, *Massey University, New Zealand*
Piet Kommers, *University of Twente, The Netherlands*
Toshio Okamoto, *University of Electro-Communications, Japan*
Ashok Patel, *De Montfort University, United Kingdom*
Katherine Sinitsa, *Int'l Res. and Trng Ctr of Info. Tech. and Systems, Kiev, Ukraine*

Program Committee
Rosa Maria Bottino, *Consiglio Nazionale delle Ricerche, Genova, Italy*
Stefano Cerri, *University of Montpellier II, France*
Betty Collis, *Twente University, The Netherlands*
Grainne Conole, *University of Bristol, United Kingdom*
Jonathan Darby, *Oxford University, United Kingdom*
Chris Dede, *Harvard Graduate School of Education, USA*
Darina Dicheva, *Winston-Salem State University, USA*
Ben du Boulay, *University of Sussex, United Kingdom*
Gerhard Fischer, *University of Colorado, USA*
Ildar Galeev, *Kazan State Technological University, Russia*

Peter Goodyear, *Lancaster University, United Kingdom*
Randy Hinrichs, *Microsoft Research, USA*
Ulrich Hoppe, *University of Duisburg, Germany*
Kouji Itoh, *Tokyo Science University, Japan*
Chris Jesshope, *Massey University, New Zealand*
W. Lewis Johnson, *University of Southern California, USA*
Judy Kay, *University of Sydney, Australia*
Badrul Khan, *The George Washington University, USA*
Ray Kemp, *Massey University, New Zealand*
Alfred Kobsa, *University of California, Irvine, USA*
Chul-Hwan Lee, *Inchon National University of Education, Korea*
Ruddy Lelouche, *University of Laval, Canada*
Tanja Mitrovic, *University of Canterbury, New Zealand*
Riichiro Mizoguchi, *Osaka University, Japan*
G.U. Matushansky, *Kazan State Technological University, Russia*
Steve Molyneux, *University of Wolverhampton, United Kingdom*
Som Naidu, *University of Melbourne, Australia*
Toshio Okamoto, *University of Electro-Communications, Japan*
Ron Oliver, *Edith Cowan University, Australia*
Reinhard Oppermann, *GMD, Germany*
Valery A. Petrushin, *Andersen Consulting, USA*
Clark Quinn, *KnowledgePlanet, USA*
Roy Rada, *University of Maryland, Baltimore County, USA*
Thomas C. Reeves, *University of Georgia, USA*
Robby Robson, *Saba Software, USA*
Jeremy Roschelle, *SRI International, USA*
Alex Shafarenko, *University of Herfordshire, United Kingdom*
Elliot Soloway, *University of Michigan, USA*
J. Michael Spector, *Syracuse University, USA*
Barbara Wasson, *University of Bergen, Norway*
Beverly Park Woolf, *University of Massachusetts, USA*

Organising Committee

John P. Klus, *University of Wisconsin, Madison, USA*
Kinshuk, *Massey University, Palmerston North, New Zealand*
James Goff, *ARRTC-Fort McCoy, Wisconsin, USA*
Patrick J. Koehler, *ARRTC-Fort McCoy, Wisconsin, USA*

Reviewers

Akiko Inaba
Andrew Ravenscroft
Antonija Mitrovic
Baltasar Fernandez-Manjon
Barbara Wasson
Benedict du Boulay
Bert Zwaneveld
Bottino Rosa Maria
Brent Muirhead
Carolyn Dowling
Catherine McLoughlin
Cecilia Baranauskas
Chris DiGiano
Chris Quintana
Clark Quinn
David Shaffer
David Wiley
Deborah Tatar
Debra Sprague
Denise Whitelock
Ray Kemp
Alexandra Cristea
Robert Atkinson
Robert Watson
Errol Thompson
Garlatti Serge
Gerald Edmonds
Giuliana Dettori
Golitsina I.N.
H. Ulrich Hoppe

Hong Hong
ITOH Kohji
J. Michael Spector
Jenny Zhang
Jonathan Darby
Jorge Adolfo Ramirez Uresti
Kalina Yacef
Kevin Ruess
Kinshuk
Lev Gordon
Loia
M. Loomes
Marcus Specht
Maria Augusta Silveira Netto Nunes
Mark Chung
Martin Muehlenbrock
Martyn Wild
Michael Mayo
Michael Szabo
Mike Chorost
Mikhail Morozov
Mitsuru Ikeda
Mizue KAYAMA
Mohamed Ally
Monique Grandbastien
Muhammad Betz
Nelson Baloian
Nicola Henze
Oktay V. Ibrahimov

Paul Brna
Chul-Hwan Lee
Rachel Pilkington
Randall Hill
Richard Joiner
Riichiro Mizoguchi
Robby Robson
Roger Hartley
Rory McGreal
Roy Rada
Safia Belkada
Sergey Sosnovsky
Shaaron Ainsworth
St. Trausan-Matu
Stacy Marsella
Susan Bull
Tatsunori Matsui
Ted Smith
Thomas C. Reeves
Tom Conlon
Vadim Chepegin
Valery A. Petrushin
Vladan Devedzic
Vladimir Godkovsky
Young Woo Sohn
Yukari Kato

Full and Short Papers

Theme 1: Architectures

A Standards-driven Open Architecture for Learning Systems

L. Anido, M. Llamas, M.J. Fernández, J. Rodríguez, M. Caeiro and J. Santos
Depto. Tecnologías de las Telecomunicaciones, University of Vigo, SPAIN
{lanido,martin,manolo,jestevez,mcaeiro,jsgago}@ait.uvigo.es

Abstract

The learning technology standardization process is taking one of the leading roles in the research efforts into computer-based education. Institutions like the IEEE, the US Department of Defense and the European Commission have set up committees to deliver recommendations and specifications on this area. Their goal is to provide interoperability among heterogeneous systems and learning objects reuse. From this standardization work, we propose an open and distributed architecture that identifies some common software services for the e-learning domain. Our aim is to contribute to this standardization process from the software perspective.

1. Introduction

Advances in Information and Communication technologies and specifically in Multimedia, Networking, and Software Engineering enable the apparition of a new generation of computer-based training systems. Internet is today the ubiquitous supporting environment for virtual and distributed learning environments. As a consequence, many institutions take advantage of new technologies to offer training products and services at all levels.

In this scene, educational systems and resources proliferate, and a need for standardization becomes apparent. Like in other standard-driven initiatives, standardization applied to learning technologies will enable reuse and interoperation among heterogeneous software systems. To achieve this, a consensus is needed on architectures, services, data models and open interfaces.

So far, most efforts have been focused on the definition of an adequate information model that enables reutilization and interoperability, see section 2 for an introduction on this. Our contribution is centered on the definition of commonly needed learning software services over the defined standardized data models. For this, we define a reference model (section 3) and an open reference architecture that identifies every needed software component (section 4). The main aim of this work is to serve as a starting point towards standardized software services for distributed e-learning.

2. The learning technology standardization

Initiatives like the IEEE's LTSC [1], IMS project [2], Aviation Industry's AICC [3], US DoD's ADL initiative [4], ARIADNE [5], GESTALT project [6], PROMETEUS [7], CEN/ISSS/LT [8], and many others are contributing to the learning technology standardization process. The IEEE LTSC is the institution that is actually gathering recommendations and proposals from other learning standardization institutions and projects. ISO/IEC JTC1 created in November 1999, the Standards Committee for Learning Technologies, SC36 [9], who is continuing this work to create ANSI or ISO standards.

The main results obtained so far are in the fields of educational metadata, learner's profiles, course structure formats, content packaging and runtime environments.

No open reference architecture or service definitions have been produced so far. Although the IEEE has a group focused on architectural issues, its outcome, the LTSA, is more a conceptual model than a software reference architecture. The work that we present in this paper is a proposal towards such open architecture. We took into account the works enumerated in this section. Over their standardized information models, a set of services is offered to the domain community of distributed e-learning systems developers. Eventually, these services together with the underlying architecture will support our proposal for a new domain facility on e-learning.

3. Reference Model

A reference model is a division of the functionality together with data flow between the pieces. We derive our reference model from a set of use cases.

We identified five functional modules in our reference model. (1) Educational Content Providers (ECP) are responsible for developing educational contents and offering them, perhaps under payment, to the (2) Educational Service Providers (ESP). ESP are the modules learners interact with along their learning process: from course enrollment to graduation. The learning process must be guided by the competency definition goals established at the (3) Education Bureau

(EB). Learners find training resources through (4) learning Brokers, which help them to find and locate those ESP that offer the courses they look for. Brokers must maintain metadata information about courses offered by associated ESP. Finally, employers look for experts into the (5) Profile Repository (PR), which maintain learner's high-level performance data and overall records.

4. The reference architecture

The reference architecture is a reference model mapped onto software components. Whereas a reference model divides the functionality, a reference architecture is the mapping of that functionality onto a system decomposition. Mature domains are supposed to have reference architectures that guide software developers to build their specific systems. The reference architecture definition process is an iterative and incremental work. We derived a reference architecture from each of the modules above, taking into account their relationship.

Because of space restrictions we just outline here the reference architecture for ESP and Brokers, see Figure 1. We encourage the interested reader to contact the authors to get the whole architecture specification.

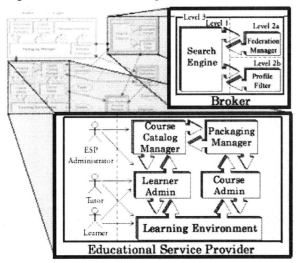

Figure 1: Partial view of reference architecture

Broker modules are composed of three different software components: Search Engine (SE), Federation Manager (FM) and Profile Filter (PF). SE deals with the metadata repository to search for a learning object. Search results may be filtered according to user preferences. The PF is responsible for this. While at the reference model the broker is a single module, at the reference architecture its functionality may have to be implemented by the collaboration of a group of federated brokers. The FM software component manages federations according to whichever topology. To be compliant with this architecture, only the SE component must be implemented. The Brokerage architecture defines several service levels. According the level the broker is compliant with, PF and/or FM have to be implemented. However, this architecture is fully scalable, a single SE-broker may add PF or FM components in the future, since automatic component configuration mechanisms are defined. The standardized information model each component deals with drove our design criteria. Updating the data model that defines learning metadata or user preference information would only affect the SE or PF component respectively.

The Educational Service Provider reference architecture is composed of five software modules (c.f. Figure 1): Learning Environment (LE), Learner Admin (LA), Courses Admin (CA), Packaging Manager (PM) and Course Catalogue Manager (CCM). The LE component is responsible for student routing through a course and student tracking. Learner Admin manages course enrollment, and payment, if applicable, user authentication and student curriculum records and profiles updating. CA module manages the actual learning resources each course is composed of. It delivers contents to the LE and receives new resources from ECP modules through the PM. CCM maintains up-to-date information about the actual courses being offered by the institution and publicizes them through the Brokers.

5. Summary and Future Work

This paper outlines our contribution to the learning technology standardization process: the definition of a reference architecture and a set of software services oriented to the e-learning domain. A domain CORBA [10] facility proposal will support formalization of these services.

6. Acknowledgements

This works was funded by FEDER 1FD097-0100 and PGIDTOOTIC32203PR.

7. References

[1] IEEE's LTSC web site at http://ltsc.ieee.org
[2] IMS web site at http://www.imsproject.org
[3] AICC web site at http://www.aicc.org
[4] DoD's ADL web site at http://www.adlnet.org
[5] ARIADNE web site at http://ariadne.unil.ch
[6] GESTALT web site at http://www.fdgroup.co.uk/gestalt
[7] PROMETEUS web site at http://prometeus.org
[8] CEN/ISSS/LT. http://www.cenorm.be/isss/Workshop/lt/
[9] ISO-IEC JTC1 SC36 web site at http://www.jtc1sc36.org
[10] CORBA technology web site at http://www.omg.org

The Educational Digital Entities as a Component of the New Hybrid and Learning School Environment: Expectations and Speculations

Panagiotes Anastasiades Ph.d in Social Informatics
*Visiting Lecturer, University Of Cyprus, Department of Computer Science
75 Kallipoleos Street, P.O.Box.20537CY-1678 Nicosia, CYPRUS
Tel.: +357-2-892277 Fax.: +357-2-339062 panas@ucy.ac.cy*
Department of Computer Engineering and Information Systems , Polytechnic School, University of Patras, GR26500 Patra, GREECE, *panas@ceid.upatras.gr*
Chairman of the Standing Committee on Social Issues of the Greek Computer Society

Abstract

This work puts under the judgment and speculation of the educational community, the concept of the educational digital entity, that is the digital representation of the teacher or as a virtual figure with human characteristics. With the help of three dimensional graphics and advanced animation techniques, the digital entity will have (?) the possibility to imitate or to substitute the a teacher's human existence itself mainly in regards to the lecturing of the courses to be taught. Moreover, as part of this work the methodology of the development of a digital entity is determined, its introduction as a pilot in the educational sector is described.

1. Introduction

The emerging Information Society [1], is marked by a reversal of the basic relationship between man and information [2], meaning that in our days the digital information – in the form of products and services – is flowing through to the people looking for it through the screens of their computers, without having to leave their place, gaining precious time, improving the conditions of our everyday life. The area of education is the most important priority of the Information Society, one of the most important mainstays of the new digital age [3].

2. The basic characteristics of the hybrid learning environment

The basic characteristic of the traditional model of education, was the simultaneous presence of teachers and students in the same space (the course room) and at the same time (schedule). The best possible use of the basic services of teleconference and electronic exchange of messages and data [4] in combination with the Internet makes possible the implementation of open and at a distance education [6]. Forming a completely different learning environment, within which the basic relationship between the teacher and the student is reversed. The simultaneous presence of the two mainstays of school education (teacher – student) in the course rooms is not anymore a precondition, since the teaching – lecturing procedures, which are of the highest importance, can take place through a PC [7].

As hybrid environment is meant that learning environment, within which the best possible combination of modern applications of technologies is attempted – which liberate the educational process from its space and time bonds – and of the respective pedagogic reports based upon the traditional school approach of the educational process [8].

3. The educational digital entities

With the term digital entity, we refer to a virtual figure with human characteristics, which tries to imitate or substitute human existence itself in its various manifestations [9]. With the help of three dimensional graphics and advanced animation techniques, digital entities are part of an unknown up to now environment, as part of which the virtual aspect of their component parts flows through to the tangible and intangible elements which make up our everyday life. An educational digital entity, will be in essence the virtual representation of a teacher, and will undertake in the course rooms a part of the teaching of the scheduled courses.

As part of a comprehensive hybrid environment [10], the students will be connected –from the place of their choice through the Internet- with the digital entity at the time they want to. The whole endeavor will be taking place under the supervision of the Ministry of Education and the pedagogical institution, while it is going to be applied as a pilot in pre-selected schools based on geographical, educational and other criteria. The steps for the realization of the project inductively can be described as following:

1st Step: Working out – development of the form and the characteristics of the Digital Educational Entity. Description: The characteristics of the entity, which must be approaching the average characteristics of the teacher of a course, are defined in detail. Involved Parties: All groups. Result: the form of the entity.

2nd Step: Transformation of the courses to be taught to teachings for the digital educational entity. Description: The courses to be taught are divided in sections based on a fixed number of teachings. The major points on which the digital entity must focus are underlined so that they are mentioned in the teachings. Texts are prepared which will comprise the content of the teaching and which will have for the sake of our sample the following form: Text, Text, Text, Text, Text, (the voice of the digital teacher is normal). **Text, Text, Text, Text, Text,** (the voice of the digital teacher slows down emphasizing the selected text).

Text, Text, Text, Text, Text, (The speed of lecturing by the digital teacher increases). Text, Text, Text, Text, Text, (the underlined text signifies a question or a query).
Involved Parties: The group of authors and educators. The group of psychologists pedagogues has a consulting role. Result: The teachings (lectures) are formatted into electronic form (word processor) based upon the above rules.

3rd Step: Development of voice techniques and text – speech systems. Description: The formatted texts are transformed into voice based on the specifications that have been set. The voice is entered into the text and speech processing system, so that the entity's face and mouth motions are in harmony with the content of the teachings. Involved Parties: Animation experts group, text and speech processing system experts group. Result: The educational entity acquires a voice.

4th Step: Testing period. Description: The digital entity teaches its first courses with the work parties (groups) as the audience, which make the necessary corrections, modifications and changes. Involved Parties: All groups. Result: The educational entity acquires "existence".

5th Step: Indicative pilot operation. Description: The educational entity gives its first lecture to selected student classes, who have been selected with such criteria as their performance, geographical distribution etc. After the end of the lecturing the instructors clarify various points, answer possible questions. At the end of the pilot operation period the progress of the children is evaluated, while at the same time they express their views about the new system. A statistical data processing follows in order to draw conclusions, which afterwards are going to be useful.
Involved Parties: All groups, selected classes of students. Result: A first package of evaluation data about the educational entity. Depending on the evaluation of the educational entity and the correlation with the other statistical data, the modifications continue until the work party judges that the digital instructor can appear in school classes. The future of the experimental educational entity depends upon the degree of its acceptance by the educational and student community, as well as by the children's parents.

4. Conclusion

The development of the educational digital entities, is considered by many as a completion of the digital environment of the new form of education, since it ensures pedagogic and tutorial uniformity which will be materialized by work parties, which will consist of teachers, child psychologists, pedagogues etc. However, for some others it is the beginning of an Orwellean scene in the fragile area of learning. Expectations, worries, fears and speculations make up a double-meaning puzzle in the beginnings of the third millennium.

5. References

[1] M. Bangemann, "The Europe and the global information society" Recomendtions to the E.U Brussels 1995.
[2] Anastasiades, S.P.,. *In the Information Age,* A.A Libanis, Greece, Athens, 2000.
[3] European Commission, COM (96) 471 final.
[4] BergeZ.,M.Collins,"Computer mediated Communication and the Online Classroom in Higher education", *Computer – Mediated Communication Magazine,* 1995a, vol.2, no.2.
[5] DeLoach A., "A starting point of Educators on the internet", *CMC Magazine,* 1995, vol.2, no.6.
[6] European Commission, COM (96) 471 final.
[7] Retalis S., Papaspyrou N., Markakis V., Skordalakis E., "An enriched classroom model based on the WWW technology", *Active Learning*, July 1998, no.8.
[8] Anastasiades P., 'The basic characteristics of the new hybrid learning environment on the emerging information society', Proceedings of the 2nd Panhellenic Conference on Computer and Telecommunications Technology in Education, 13-15 October 2000, Patra, Greece.
[9] Anastasiades P., 'The basic characteristics of the digital educational entity', Proceedings of the Panhellenic Conference on 'education and Informatics', 11-12 January 2000, Thessaloniki, Greece.
[10] Anastasiades P., 'The school in the information society: the new role of secondary level teachers'. Proceedings of the Annual Seminar on 'Computing in secondary education', February 2000, Athens, Greece.

Virtual Adaptive Learning Architecture (VALA)

Amy Metcalfe	Marcy Snitzer	James Austin
University of Arizona	University of Arizona	University of Arizona
amysm@u.arizona.edu	marcys@u.arizona.edu	jaustin@u.arizona.edu

Abstract

The Virtual Adaptive Learning Architecture (VALA) project has developed a learning architecture with user interface adaptability that provides a personalized learning environment as well as a repository of reusable learning objects. VALA is designed to be customized easily while allowing for future expansion. Instructors from multiple disciplines can easily find, develop, edit and publish multimedia content in support of their instructional endeavors with VALA. In addition, research questions regarding how the delivery of a learning experience affects learning outcomes can be explored through the system.

1. Introduction

The Virtual Adaptive Learning Architecture (VALA) project represents a dramatic shift from the current practice in which the learner is subject to a predetermined presentation style, to one in which the learner can select the presentation of content to suit his or her individual learning preferences. Although past computer-mediated instructional environments typically delivered learning objects in one style, more recent web-based instructional environments allow for more interactivity. However, some learners are excluded because their learning styles differ from the common approaches used by today's instructional media. If students are to succeed using web-based options, they must be able to engage in the learning process using their own identified learning strengths. From there, they can be stimulated to develop critical thinking skills and to integrate other styles of learning. The need for learning environments that allow for greater learner variability requires the management of vast amounts of academic information. Standardization, systems of classification, retrieval methods that can identify commonalties in information, and preservation of original formats are necessary. In addition, convenient methods for identifying learner preferences and mapping them to learning objects must be developed. The Virtual Adaptive Learning Architecture (VALA) project seeks to resolve these issues by creating an interface that allows selection of content presentation based on a student's preferred learning style, a repository of reusable learning objects standardized to national initiatives, and an "expert" learning assessment tool capable of evaluating both how and what students are learning. VALA's design and components are based on the Learning Technology Systems Architecture (LTSA) Specification [1].

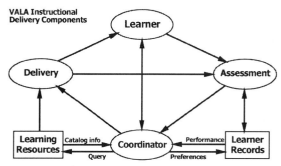

Figure 1. VALA instructional delivery components

VALA is a three-year grant project sponsored by the U.S. Department of Education's Fund for the Improvement of Post-Secondary Education (FIPSE) Learning Anytime, Anywhere Partnerships (LAAP) competition. The University of Arizona is the lead institution in a partnership comprised of Pima Community College, Sun Microsystems, Silicon Graphics, and Oracle. The VALA partnership provides a scalable adaptive learning architecture that can be readily packaged, with or without content, for transfer to other universities, colleges, and communities across multiple academic platforms. Now in the second year of development, total funding is comprised of $1.4 million in grant funds and $5.6 million in partnership match.

2. Learner Inventory

Upon entering VALA for the first time, the learner will be asked to complete a twenty eight question inventory known as the VALA Inventory of Personal Educational Resources (VIPER), designed to assess preferences relative to learning style and modality. The results from VIPER will then be used to define the learning architecture and to create a customized user interface that provides a personalized learning environment conducive to the learner's profile. The VALA team developed VIPER based on available research on learning styles and modalities. Learning style is defined as "the different ways in which children and adults think and learn" [2] while learning modality refers to the learners' preferred manner of sensory intake of information—visual, auditory, or kinesthetic. By addressing the learning style and learning modality, VALA determines an appropriate electronic learning environment.

A variety of instruments to measure either learning style or learning modality are available. The VALA team surveyed existing instruments but quickly discovered that none measured both learning style and learning modality. Since both the cognitive aspects of learning style and sensory aspects of learning modality are critical to VALA's ability to accurately type learners, the VALA team determined that it would be necessary to develop a new instrument that combines the needed constructs. Of the literally dozens of constructs which contribute to learning style, the constructs selected to be identified by the initial learner inventory interface in VALA are based on Sternberg's model of mental self government which includes local/global and internal/external categorizations [3]. The local/global categorization indicates a learner's preference relative to details and abstract, general thinking. Since most learning tasks require learners to process information both ways, the local/global categorization is used as an indication of which type of information the learner prefers to process first, representing, in essence, convergent versus divergent thinking. The internal/external categorization indicates a learner's preference for working alone or individually.

Additionally, due to the interactive, multimedia nature of the instructional platform created for the VALA project, it is necessary to determine the user's preferred learning modality—that is whether they prefer to intake information visually, aurally, or kinesthetically. The instrument used for obtaining learning modality is the University of Arizona Learning Center's Learner Inventory. The above instruments were selected based on the available literature and an examination of the instruments themselves, including measures of reliability and validity. Using a sample of 300 students from the University of Arizona and Pima Community College, VIPER yielded extremely high reliability and validity rates. The instrument has an overall reliability index of .72 and item-total correlations range from .20 to .52, making the reliability and validity higher than the majority of instruments surveyed [4].

3. Reusable Learning Objects

The ability to match a learner's preferences to his or her experience within the learning environment is dependent upon having content available in a variety of modular formats that will appeal to different types of learners. For example, a lesson on the abacus might include a video showing an abacus in use, an audio file describing the abacus while a picture is shown, and interactive abacus for the learner to manipulate. Each of the formats would achieve the same learning objective, but appeal to different types of learners. In order to reduce duplication of effort and the expense of creating and digitizing learning objects in a variety of formats for a single lesson, vast repositories of learning objects must be available to content developers. Content developers will then have the option of creating their own objects or searching the index of existing objects to create a lesson. While keeping costs down is certainly an issue in any discussion of interactive distance learning, a searchable repository of reusable learning objects also plays the important role of increasing interdisciplinary connections. Objects created for one academic discipline may easily meet the needs of faculty in other disciplines, creating interdisciplinary connections that would not be feasible in traditional classroom environments. The content developer who found an object on Celtic calendars while developing a lesson on Celtic culture may create a link so that learners can jump into the calendar lesson if they choose to pursue that topic. Likewise, the student may choose to follow the calendar lesson and discover that it connects to a larger unit on Astronomy. Such interdisciplinary connections enrich the learning experience and acknowledge learners' tendency to be nomadic rather than linear. In order for the learning objects to be reusable and for interdisciplinary connections to be facilitated, learning objects must be supported by an extensive system of metadata.

4. VALACat

VALACat is the cataloging utility that is used by VALA content developers to index and classify their online learning material. VALACat was developed using industry standards for data management. The metadata specifications developed by the Dublin Core Metadata Initiative [5] and the IMS Global Learning Consortium [6] were used as the basis of design for the data structure that is used by VALACat. The data classification scheme that has been developed is broad based and is designed to be both accurate and applicable to all types of online learning objects. By using the industry standards for classification, interoperability between VALA and other systems will be possible. This will enable VALA objects to be discovered from outside systems and allow VALA to search other systems to locate learning objects.

The functionality of VALACat was designed to be as easy to use as possible. Through a web interface, basic information describing the resource is entered (title, creator, description, etc). Information is also collected on goals and objectives of the learning experience and teaching methods used in designing the learning resource. The classification of teaching methods is a key to matching the learning resource to the correct learner after the learner has been classified by VIPER (see Fig. 2). Using the URL of the resource, the VALACat parser is 'pointed' at the learning resource and VALACat then classifies the entire learning resource, and the learning objects that are used to comprise the entire resource. Granularity of description is achieved by having each object inherit the description for the entire resource. Each learning object can then be given a more detailed description if necessary, but this is not required. By classifying the learning resources at the object level, the ability is then in place to search the VALA database at the object level (jpegs, sound files, video files) independent of the object's origin.

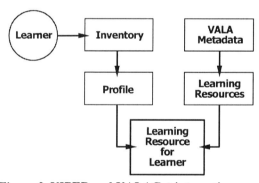

Figure 2. VIPER and VALACat interactions

5. Assessment

VALA is a continually evolving project that will change and improve over time. As part of the process of improvement that will take place in VALA, three forms of assessment are being developed: system, object, and learner assessment. System assessment will consist of evaluations of the functionality of the various aspects of VALA: user interface, developer interface, metadata functions and overall design usability. The evaluation will not only assess the usability of these aspects, but will look at ways of enhancing functionality and adapting the system to specific types of users and their specific needs. Object assessment will evaluate how instructional objects are being used and have been used by the VALA system to deliver learning to users. The object assessment will track how many times instructional objects were accessed and used, in what context they were used, how the learners performed on expected outcomes for goals and objectives that are associated with objects, and how the learner's VIPER classification has affected the usability and accessibility of the objects.

Learner assessment of knowledge gained from a particular lesson is critical to helping both the learner and the instructor identify areas of confusion. VALA's developer interface will provide content developers with the opportunity to insert assessments at any point during the instructional process. The VALA team is currently analyzing programs to determine if a program that allows for web based development and delivery of assessments already exists. The final component of assessment is the ability to adjust the learner's profile based on actual performance within the system. The VIPER tool will set the "baseline" for the learner's style and modality preferences. Based on this profile, VALA will direct learners through the program. As the learner works through lessons in VALA, he or she will be free to choose other alternative content presentations based on learning style and modality preferences. VALA will track this route, cataloging the learner's behavior. The learner's path will be stored along with his or her VIPER results, providing an indication of how effectively VIPER matched the learner's actual preferences. When the learner returns to VALA to complete subsequent lessons, he or she will have the option to use either the learner profile established by VIPER or the profile that emerged through the history of interactions.

6. Project Evaluation

Two questions guide VALA development: "How is learning different when learning style is taken into account and content is adapted interactively?" and, "What combination of learning path and activities will maximize learning outcomes?" Embedded in each learning module, assessment objects provide ongoing, real-time information about how and what students are learning that can be immediately utilized. Formative evaluation will focus on developing systematic feedback for project developers regarding the effectiveness of various adaptive instructional strategies embedded within prototype materials. It will also seek to determine which learner variables (learning style, modality, and perceived learner control) are most productive as a basis for determining adaptive instructional strategies. Summative evaluation will focus on the key project goals, demonstrating the effectiveness of those adaptive instructional strategies (for both instructors and learners) ultimately implemented in instructional modules, the cost-effectiveness of modularized instructional materials, "recyclability" and transferability of materials, overall usability and value of materials in peer institutions, and increased accessibility and utilization to institutional research materials.

7. References

[1] F. Farance and J. Tonbel. LTSA Specification: Learning Technology Systems Architecture, www.edutool.com/ltsa, 2001.

[2] M.E. Litzinger and B. Osif, "Accommodating diverse learning styles: Designing instruction for electronic information sources," in *What is Good Instruction Now? Library Instruction for the 90s.* Linda Shirato, L. ed., Pierian Press, Ann Arbor, MI, 1993.

[3] R.J. Sternberg, *Thinking Styles,* Cambridge University Press, New York, NY, 1999.

[4] See the VALA web site at www.vala.arizona.edu for more information.

[5] Dublin Core Metadata Initiative, http://dublincore.org/

[6] IMS Global Learning Consortium, http://www.imsproject.org/metadata/index.html

Managing Large Scale On-Line Systems

Geoff Aldridge BA (Hons), AMBCS
The Open University, UK
G.Aldridge@open.ac.uk

Abstract

This paper discusses the implications and difficulties of running large-scale on-line applications, drawing on the extensive experience of the On-Line Applications group at The Open University. The history of on-line services at The Open University is given, with the main advantages the University has reaped through its use of Computer Mediated Communication. A series of success factors are offered that are applicable to managing large-scale on-line systems.

1. Introduction

The Open University (OU) is a world leader in the use of computer conferencing and on-line services for distance education. This paper discusses the issues involved in maintaining large-scale on-line services. Some success factors for large-scale system management proposed are: Empowerment, Moderation, Training and Automation.

2. The History

The OU has around 200,000 students studying from around the world each year on over 350 courses. It has been using Computer Mediated Conferencing (CMC) for course development since 1984. The first OU course to utilise CMC for teaching and support was in 1988. FirstClass™ is a Canadian system developed by Centrinity Inc. that was swiftly integrated with the OU's Post Graduate Certificate in Education (PGCE) programme in 1994. Since then the growth of conferencing within the university has been exponential, going to over 140,000 users.

3. The academic use of CMC

The OU FirstClass™ system is used extensively in academic arenas in terms of project-based teams, tutor support areas, course topic discussions areas and peer-to-peer self-help areas. These would generally be classified as a *community of interest* [1], namely their interaction is on specific topics, with a common interest although there is also a significant high level of interpersonal communication.

The most distinct advantage through the use of CMC in distance learning is allowing the students to more actively and collectively interact with the course material. The OU course "T171: You, Your Computer and the Net" is taught entirely on-line with Web based materials and fully integrated on-line CMC; this has proved to be a huge success with 12,000 students a year enrolled to study [2].

4. On-line communities

In additional to the academic discussion forum on the OU system, there are a large number of social and student-run self-help conference areas. These are administered by the OU Students Association (OUSA) and vary from 'unofficial' course-focused conferences to fully social discussion areas.

The boundaries between the academic use of CMC and the social aspects are diminishing, and although this can make conference moderation a more difficult prospect, this does enhance the learning process, and research [3] has shown that this can dramatically improve both the learning experience and also the gains involved as, "learning is fundamentally experiential and fundamentally social" [3].

The use of CMC has enabled students on courses to form invaluable university-wide bonds with others, reducing the feeling of isolation and allowing for courses of much increased size. Various self-help groups form outside the tutor area, reducing the burden on the tutor.

5. Success Factors

From the accumulation of experience in CMC within the OU, below are the key success factors for large-scale system management of education or training discussion-based systems: Empowerment, Training, Moderation, and Automation.

5.1. Empowerment

The FirstClass™ system is highly scalable and reliable and allows for thousands of users on a single machine. As on-line systems become larger and more complex, there is an increasing requirement for system administrators to be in a viable position to empower selected users, namely academics and course teams, to run their own conference areas.

5.2 Training

Salmon [4] has identified a five-stage model for describing the processes that students (and lecturing staff) go through in order to be able to use the on-line medium for productive learning.

The five stages that on-line learners go through are as follows; Access and motivation; On-line socialization; Information exchange; Knowledge construction and Development.

The main premise behind this model is that these stages are inherent in on-line development and cannot be ignored or rushed; the real skill for an on-line moderator is to ensure that progress is made through these stages, in order for the students to develop as on-line learners. Also, not all on-line groups will develop through all five stages. The T171 experience showed that most students were at stage 4, with some at stage 5 at the end of the course [2].

5.3 Moderation

Moderation is important in two key areas. Firstly, with the changes in legislation throughout the UK, there is an importance to ensure that personal liabilities are not violated in such a public forum.

The second, and far more difficult area, is the moderating function that is designed to encourage students to participate and share ideas and ultimately knowledge that has been constructed within the group. This type of moderation tends to require full support from the course team, and with incentives for conferencing built into the course assessment framework. Courses without this inbuilt incentive tend to find that conferencing participation levels are lower, and the conference moderators themselves have a more difficult job to achieve the same levels of personal development.

5.4 Automation

Through the growth of the on-line systems at the OU the support units have had to develop more scalable, reliable and complex automatic registration systems. The systems now in existence within the OU handle all aspects of students' on-line activities. These systems, have permitted much of the growth of the OU's conferencing and on-line systems, whilst maintaining the existing levels of services, and also offering additional, more complex services.

6. Conclusions

The most pertinent success factor for managing large-scale on-line systems is a need for distributing power to the right level of user. In the majority of cases, the power should lie within the course teams or Tutor, not central staff. This is for a number of reasons: increasing the involvement of course teams in the development of CMC activities, increased responsiveness to changes to student numbers or work patterns and greater ability to cope with more courses and more complicated on-line courses.

The boundaries between social and academic use are diminishing over time and this needs to be incorporated into the design of on-line systems and also to be taken into consideration when training on-line moderators to manage the conferencing environments.

6. References

[1] Armstrong, A. and Hagel, J. "The real value of on-line communities" *Harvard Business Review*, May – June 1996, pp. 134-7.

[2] Weller, M. J. "Creating a Large-Scale, Third Generation, Distance Education Course" *Open Learning*, Vol. 15, No. 3, November 2000, pp. 243-251.

[3] Wenger, E. *Communities of Practice* Cambridge University Press, Cambridge, 1998.

[4] Salmon, G. *E-Moderating: The Key to Teaching and Learning Online* Kogan Page Limited, London, 2000.

Architectures supporting e-Learning through Collaborative Virtual Environments: The case of INVITE

Ch. Bouras	G. Hornig	V. Triantafillou	Th. Tsiatsos
Computer Technology Institute, Greece	*blaxxun interactive AG, Germany*	*Computer Technology Institute, Greece*	*Computer Technology Institute, Greece*
bouras@cti.gr	guido.hornig@blaxxun.de	triantaf@cti.gr	tsiatsos@cti.gr

Abstract

The growing need for communication, visualization and organization technologies in the field of e-learning environments has led to the application of virtual reality and the use of collaborative virtual environments. This paper presents a system architecture to support such environments, defined by user needs and using state of the art technologies.

1. Introduction

New types of applications have been developed in order to incorporate information technology and its advances in the learning environment. Many systems already support e-learning, using different types of learning methods as well as different technologies to implement these methods; many of which have good potential but are either too difficult or expensive to implement. On the other hand, many technologies are applied in systems offering small functionality or complexity to the system. A lot of research work has been developed in the last few years concerning e-learning systems. The need for communication, visualization and organization is a strong prerequisite for providing efficient e-learning platforms. Many research projects have been engaged in this area during the last few years [1]. Although research projects try to fit in technological advances to produce up to date technological e-learning systems, commercial products available seem to focus more on the already available technological solutions to provide efficient e-learning solutions. Products like InterWise Millenium 3.2 [2], Centra Symposium 4.0 [3] and LearnLinc 4.5 [4] are promising and leading software solutions for synchronous teaching environments. All these tools enable large groups of dispersed individuals to interact, collaborate and learn in real-time over intranets, extranets and the Internet. Most of the commercial web-based training solutions lack sufficient realization of real-time communication features, meaning a shared sense of space and presence. In general, third party applications refer to features such as application sharing and video conferencing. So there is a definite need for integrated solutions that can offer a much higher degree of usability and that do not demand additional technical requirements as in the case of most commercial systems.

In this paper we present a system architecture to support collaborative e-learning using collaborative virtual environments. This system applies various technologies, such as collaborative virtual environments, on-line translation, real-time audio and intelligent agents, integrating them in a flexible and efficient way to support various training models in virtual environments. A community of learners and tutors is possible. In addition this system makes it possible for the participants to use the communication space for their own needs. Interaction between all members is not dependent only on the lesson or the role itself. In the next section we describe a European project in the area of the collaborative learning environments for tele-training. We then present our methodology for designing the system and our vision for its system architecture.

2. Intelligent Distributed Virtual Training Environment project

INVITE (Intelligent Distributed Virtual Training Environment) [1] is a project in the framework of the Information Society Technology (IST) Program of the European Commission. It started in February 2000, and it will run for almost 3 years.

The main aim of the project is to build a platform for synchronous tele-learning which can be interfaced with standardized content management and/or instructional management systems. In order to reach this aim the following objectives have been set: (a) Identification of the relevant cognitive and social processes in the collaborative learning situation and extraction of those factors into user requirements. (b) Development of an integrated system based on distributed virtual environment technologies, including intelligent agents' real-time translation facilities, realistic avatar representation and enhanced interactivity of avatars. (c) Evaluation of the prototype within different learning contexts. (d) Research results on social learning processes within virtual environments.

In the following paragraphs we present both the first steps for the development of the INVITE system: design and system architecture. To present these steps in a uniform and understandable way, we present a summary of the user requirements and the basic learning models.

3. User Requirements

Before the development of the system we gather a first set of user requirements, in order to include this in our scenarios and then to define the functional specifications of the system. According to the users' opinions, the proposed e-learning system should conform to the following set of requirements:

- To be easy to use
- To offer user-friendly help
- To easily integrate existing digital materials
- To support audio communication
- To give the lecturer the capability to administer her/his own courses and to monitor the learners' progress and participation
- To support multi-modal interaction between the users through visual communication, realistic user representation, and real-time display of users' movements
- To support application sharing and text communication
- To offer tools for recording the communication in learning sessions as well as whole learning sessions
- To visualize the learning environment as realistically as possible
- To offer an interactive and shared whiteboard
- To support audio and text translation into other languages
- To leave certain degrees of freedom for the learners giving them the option of self control in order to enable them to work autonomously

In short, users want an e-learning system that can support three types of training: **synchronous training** (on-line lectures from a trainer on a specific theme), **asynchronous training** (autonomous training using educational material and notes from previous lectures or minutes from collaboration), **collaborative training** (on-line communication and collaboration between the members of a usergroup on a specific theme).

4. System Design

During the system design phase we "compile" the user requirements and training types into functional specifications of the system and then (in the implementation of the system architecture) we will choose the available technological solution that will apply in the INVITE system. The main issue here is which technology we should use.

4.1 Functional Specifications

In order to satisfy the user requirements, our proposed INVITE system should support a collaborative virtual environment which is a set of different virtual worlds that offers the users the ability to navigate and interact in a 3D shared space. The main functionality of the system is the following:

3D visualization of a virtual learning environment: The realistic visualization of the classrooms can only be accomplished by a 3D model of a virtual learning environment [5]. There are three main categories into which all of the virtual worlds created can be classified: Lectures' Virtual Rooms-LVR (for synchronous training), Subject-specific Collaborative Virtual Rooms-CVR (for collaborative training), Private Virtual Rooms-PVR (for asynchronous training). Also an entry point (Entry Virtual Room-EVR) should supplement the above structure of the INVITE virtual community, where the users should enter in order to inform themselves about the specific lectures, meetings and news.

User representation by avatars: Full-body, photo-realistic avatars will be used for the visual representation of the users in a realistic way. Photo-realistic avatars are more effective when used in a collaborative environment offering multi-modal interaction between the learners via both gestures and mimics (like waving, nodding, bowing, disagreeing, etc.) as well as real-time movements of the users [6].

Autonomous/asynchronous learning services: In order to meet the need of offering autonomous learning, INVITE will integrate available asynchronous learning systems and it will support audio/video streaming of available content. Additionally, in order to manage documents and other educational material, a document repository will be implemented, to facilitate data visualization and implementation of structured search engines. Tools for the recording of lectures and storage in the repository will also support autonomous learning activities.

Chat, audio, and multilingual collaboration: In order to meet the need for learner/learner and learner/teacher interaction, we will offer chat and audio communication. For data communication, both application sharing and whiteboard functionality will be used. Furthermore to facilitate the collaboration between multilingual users, we plan to offer audio and text translation capabilities.

Lecture administration: The teacher/trainer can use administration and moderation tools in order to administer the on-line lecture and to moderate the users respectively.

Intelligent help: The INVITE system will provide the users with many capabilities and useful functionality (as described above). In order for the users to use this functionality effectively and easily, they could be supported by intelligent agents, which will help, trigger, and guide them, according to their profile [7], [8].

In the following paragraph we present a range of technologies and standards which are useful for the implementation of the above-described functionality.

4.2 Technologies and Standards

The range of technologies available for developing e-learning oriented virtual environments is more varied than ever. According to the above-described functionality, the main components of such a system are the following: streaming video, document repository, avatars, DVE functionality/user interfaces, 3D community, translation system, real-time conferencing, and intelligent agents. Each component, their functionality and the useful technology and standards for its implementation are depicted in Figure 1.

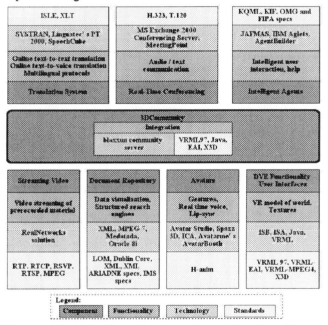

Figure 1. Basic Components, Functionality, technologies and standards for a Virtual Collaborative Learning environment

5. System Architecture

According to the above criteria, the system architecture is defined by some basic principles. The system should be based on a variety of communication protocols, be scalable, be platform independent and be based on open standards. Our proposal is based on several components that provide all the needed functionality. These components are the following: the Virtual Worlds Platform, the language server, the voice server, the agent server, the document repository, the streaming server and the avatar server.

The Virtual Worlds Platform includes all virtual community features and acts on server side as an integration platform through the extended API. For multi-user language translation services, there is a language server connected to the API of the virtual worlds server. Other services are the voice server for voice communication between clients or groups of clients, and an agent server that works as support for client agents and can provide services. A document repository allows managing several kinds of documents and works as a document archive and supports versioning.

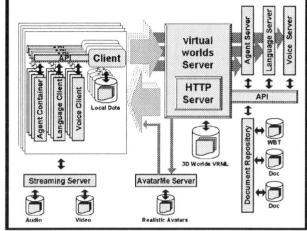

Figure 2. INVITE Architecture

The AvatarMe Server [9] provides user-specific access to personal avatars, which can be accessed from outside the system. The client side has the ability to store data locally. The open architecture of the client software is synchronized with the virtual worlds server and allows the adding of client functionality through an client-API. This is used to integrate local agent containers, language support clients and a voice client connected to 3D events and avatar functionality, like gestures or lip synchronization. In the following paragraphs the above modules are described in more detail.

Content Management Module-Document Repository: The content management module stores, archives and retrieves documents. Meta-Data support the retrieval of content. MPEG-7 will allow the retrieval of information within video data. Existing content from CBT and WBT can be integrated, if the content was created according to standards like XML. Other learning - content standards like LOM or Dublin Core will be supported.

Virtual Worlds Platform Module: The Virtual Worlds Platform [10] provides server and client technology for synchronous and asynchronous communication. There are text chat, message boards, user registration, user profiling and tracking, encryption, instant messaging, list of friends, question and answer modules, 2D and 3D user home space, public meeting places, 2D and 3D shared objects. Objects can be interactive web content, like 3D-VRML [11] or HTML and other data formats specified in MPEG4.

Agent Module: The agent module consists of the agent container on the client side and the agent server. Functions for information retrieval in documents and pre-recorded sessions include: order translations, converting into other formats, synchronizing local with central data and storing recorded sessions for reflection. The technology that will be used to implement this module is blaxxun server agents and IBM aglets [12].

Translation Module: The translation module can translate online chat and stored documents. It can translate on demand, just in time, or as a background service for later use. Translation and look-up is possible for several languages and also for specific use of terminology. For this module we will use Linguatec's solution [13].

3D Worlds Module: The client software is able to visualize dynamic 3D data in high quality and performance. The 3D Worlds module is part of the blaxxun's Virtual Worlds Platform and provides functions for representing relations in 3D space, like participant groups or any data that is useful to be viewed in 3D. The functionality is easily adaptable due the client API [10].

Avatar Module: For the purpose of immersive creation of learners, tutors and other roles participating in the learning environment, there is a photo-realistic representation available on the Internet, hosted by an Avatar server. The H-Anim standard is a starting point for Avatar functionality. 3D visualisation of participants as avatars is required for better self-creation in collaborative learning situations [5].

Streaming Video Module: The 3D Worlds module offers the possibility for streaming media content. Depending on other system requirements, the streaming servers from Real networks and Microsoft are supported, and content can be viewed on a screen in the 3D environment or in 2D with the client plug-ins from the server vendor. The technology for streaming content will change with the implementation of MPEG-4. The 3D client supports also the MPEG-4 standard.

6. Conclusion-Future Work

In this paper we have presented the INVITE project approach for the architecture of a system which targets the offering of collaborative e-learning services using collaborative virtual environments. The design of this architecture is based on the requirements of the target groups, who are professionals of certain organizations. To achieve this we have designed a system that applies various technologies, integrating them in a flexible and efficient way to support various training models in virtual environments.

Our next step is to implement the first functional prototype and to engage the end-users in its evaluation. Re-usability of learning content will also be one of the most challenging parts of the future. The improvement of standards, as well as the dissemination of standards and meta-data, is the key to efficient implementation of learning systems. Another goal is soft skill training, which will be more in the foreground in the future. Rich multimedia communication will help to give all participants the possibility to use it for more purposes than they can imagine today. Users on both user sides may need templates for organizing their daily learning. The present situation, where one tutor supports many learners needs to be enhanced by the possibility to have many tutors supporting one learner.

However, beyond all techical augmentation we must bear in mind that only good teachers can be good learning facilitators - a rule for real world and for virtual learning environments.

7. Acknowledgements

We would like to thank all the INVITE partners for their collaboration and contribution to our work.

8. References

[1] INVITE: Intelligent Distributed Virtual Training Environment project, http://invite.fh-joanneum.at/
[2] InterWise Millenium 3.2, http://www.interwise.com/
[3] Centra Symposium 4.0, http://www.centra.com/
[4] LearnLinc 4.5, http://www.learnlinc.com/
[5] M.D Dickey," 3D Virtual Worlds and Learning: An Analysis of the Impact of Design Affordances and Limitations in Active Worlds, blaxxun interactive, and OnLive!Traveler; and A Study of the Implementation of Active Worlds for Formal and Informal Education.", *Dissertation*, The Ohio State University, 1999.
[6] D.Thalmann, The Role of Virtual Humans in Virtual Environment Technology and Interfaces, In *Proceedings* of Joint EC-NSF Advanced Research Workshop, Bonas, France, 1999.
[7] Shaw, E., Johnson, W.L., & Ganeshan, R. Pedagogical agents on the Web. In *Proceedings* of the Third Annual Conference on Autonomous Agents (Seattle, WA, May 1999), ACM Press, 283-290.
[8] Fabri M., Gerhard M., Hobbs J.D., Moore J.D., "Agents for Networked Virtual Learning Environments." In proceedings of Neties '99 Conference, 18-19 March, 1999, Krems, Austria.
[9] AvatarMe, http://www.avatarme.com
[10] Blaxxun Interactive http://www.blaxxun.de
[11] VRML97 Specification, http://www.web3d.org/
[12] IBM Aglets, http://www.trl.ibm.co.jp/aglets/about.html,
[13] Linguatec, http://www.linguatec.de

Developing an Architecture for the Software Subsystem of a Learning Technology System – an Engineering Approach

Avgeriou Paris	Retalis Simos	Papasalouros Andreas	Skordalakis Manolis
National Technical University of Athens Dept. of Electrical and Computer Engineering pavger@softlab.ntua.gr	*Department of Computer Science University of Cyprus retal@softlab.ntua.gr*	*National Technical University of Athens Dept. of Electrical and Computer Engineering {andpapas, skordala}@softlab.ntua.gr*	

Abstract

There exists an urgent demand on defining architectures for Learning Technology Systems (LTS), so that high-level frameworks for understanding these systems can be discovered, portability, interoperability and reusability can be achieved and adaptability over time can be accomplished. In this paper we propose an architecting process for only the software subsystem of an LTS. We base our work upon the LTSA working standard of IEEE LTSC, which serves as a business model and on the practices of a well-established software engineering process. Special emphasis is granted on imposing a component-based nature on the produced architecture.

1. Introduction

Learning Technology Systems (LTS) are learning, education and training systems that are supported by the Information Technology. Examples of such systems are computer-based training systems, intelligent tutoring systems, web-based distance learning systems and so on.

It is common knowledge that the application of Learning Technologies does not comprise a panacea to the problem of accomplishing knowledge-driven education and training and performing the "educational shift" from teacher to learner-centered [1]. Even though LTS are quite promising in aiding to the accomplishment of this cause, undoubtedly a vast amount of research needs to be conducted in order to move from promise to practice [2]. Much of this research effort is focused on developing system architectures for LTS.

In this paper we profess the numerous advantages of introducing a component-based architecture for the software subsystem of LTS, seen from a software engineering point of view. The added value of our work is the proposal of a component-based architecting process for the software part of Learning Technology Systems, i.e. a software architecting process. This process has three important key aspects: it is founded on the higher-level architecture of IEEE P1484.1 Learning Technology Systems Architecture [http://ltsc.ieee.org/]; it adopts and customizes a big part of the well-established, widely-adopted, industry-leading software engineering process, the Unified Software Development Process (USDP) [3]; and it is fundamentally and inherently component-based.

The structure of the paper is as follows: In section 2 we provide the theoretical background of the process, which derives both from the software engineering discipline and the LTS standardization efforts. Section 3 deals with the description of the process itself, focusing also on the fact that it receives input from LTS working standards and that special care is taken to produce an inherently component-based architecture. Section 4 contains conclusions about the added value of our approach and future plans.

2. Theoretical background

We consider Learning Technology Systems, to be comprised of a human subsystem (learners, tutors, administrators etc.), a software subsystem, and a subsystem of miscellaneous non-software resources (workstations, computer networks, printed material etc.). In this section we will present the theoretical background of this paper, by focusing on two different concepts: 1) the holistic architecture of LTS, that contains all the aforementioned subsystems and is an interdisciplinary subject of study from engineering, instructional theory and design etc.; 2) the specific architecture of the software subsystem, which is a subject of study of the software engineering discipline. The first concept is being discussed because the holistic architecture of an LTS can actually provide the business model for the software subsystem of the LTS. The second concept is naturally being discussed because it will briefly outline the software engineering process and other concepts, needed to comprehend the proposed architecting process.

The largest effort on developing an LTS architecture has been carried out in the IEEE P1484.1 Learning Technology Systems Architecture (LTSA) workgroup. The LTSA deals with the Learning Technology System as a whole, encompassing a software system, human resources and other non-software resources and their interactions. The LTSA describes a high-level system architecture and layering for learning technology systems, and identifies the objectives of human activities and computer processes and their involved categories of knowledge. These are all encompassed into the 5 layers, where each layer is a refinement of the concepts in the above layer: "Learner and Environment Interactions", "Human-Centered and Pervasive Features", "System Components", "Stakeholder Perspectives and Priorities", and "Operational Components and Interoperability - codings, APIs, protocols". Similar work of defining abstraction-implementation levels has recently commenced within the ISO/IEC JTC1 SC36 [http://jtc1sc36.org].

To pinpoint the exact relation of the architecting process under study with LTSA, we must clarify that an architecture produced by this process cannot be straightforwardly matched into a layer of the LTSA. On the other hand, an architecture produced this way, defines software components that derive from Layer 3 components; it also takes under consideration several of the stakeholder perspectives of layer 4 and deals with some of the low-level issues of layer 5. A final point is that the LTSA does not deal with specific details of implementation technologies necessary to create the system components, while our approach suggests technologies of this kind, because they comprise a fundamental aspect of a software architecture.

As far as the involvement of software engineering to the proposed architecting process is concerned, we have chosen to adopt the USDP, an architecture-centric, use-case-driven, iterative and incremental process. The USDP incorporates the views, i.e. the most significant modeling elements, of five different models: the use-case model, the analysis model, the design model, the deployment model, and the implementation model. This set of views corresponds with the classic *4+1 views* described in [4]. Except for the five architectural views, the architecture also contains some non-functional requirements, platform decisions, architecture patterns contained and other generic features.

The notation used to describe the architecture is the Unified Modeling Language [5], a widely adopted visual modeling language in the software industry and an Object Management Group [http://www.omg.org] standard.

Another concept that we adopt from the software engineering discipline is the component-based nature of the architecting process. A software component can be deployed independently and is subject to composition by third parties [6]. Components can be plugged together, according to certain rules, and constitute greater components, also referred to as **component frameworks**.

The component-based nature of the proposed architecting process derives from the fifth and final view of the architecture description, that is the implementation model view. Together with the provision of USDP to promote a component-based architecture, our approach further enforces this by proposing binding and implementation technologies for the development of system components.

3. The process of architecting

The architecting process for the software system of an LTS combines the issues discussed in the previous section into a simple process model depicted in Figure 1.

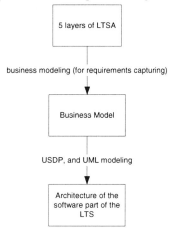

Figure 1- The macroscopic view of the architecting process for the software subsystem of an LTS

The first step produces a business model from the first 4 layers of the LTSA, so that the context of the software subsystem will be firmly grasped and all requirements will be captured. In our case the context of the software subsystem is the LTS itself, as it is particularly seen for the purposes of the system under development, e.g. a web-based distance learning system, an intelligent tutoring system etc. In other words at this stage, particular LTSA stakeholder perspectives must be chosen in order to define the business model. Next, the human activities and computer processes incorporated in these perspectives, serve as business use cases, which are the business processes involved in an LTS. Also the people and other non-human entities, which interoperate with the system, serve as business actors. The business use cases and the business actors together form the business use case model, which is the first part of a business model. The second part comprises of a business object model, which depicts how the business use cases, i.e. the system's functionality, is realized. The result of business modeling is a complete set of the LTS's processes, fully analyzed, from an Information Technology point of view, as the LTSA does not encompass a theory of learning. In order for the requirements capturing to be completed from the pedagogical point of view, an instructional or learning theory needs to be taken under account [7].

After the business model is specified, the USDP puts into effect the workflows and builds the software architecture. Our aim though, is to produce an inherently component-based architecture with the help of the USDP. How can that be achieved? As stated in [6], a software system architecture in the component-based paradigm consists of a set of component frameworks, an interoperation design for the component frameworks, and a set of platform decisions. This statement corresponds with the architecture description given in the USDP, where the architectural views of the models describe the component frameworks and the interoperation design between them, from five different viewpoints, while platform decisions are matched with the rest of the architecture description, as described earlier. We shall follow this pattern in order to enforce the component-based nature in the proposed architecting process. We shall first analyze the system into component frameworks or as we simply call them *subsystems*, describe their interaction and lastly make platform decisions.

The business processes defined in the business model are transformed into the use-case model by refining the business model, and elaborating on those business use cases that relate with the software system to be developed. This results into capturing all the functional and non-functional requirements that are specific to individual use-cases. In the next workflow, that generates the analysis model, every use case will be realized in-depth, and a first-level decomposition of the system into analysis packages will be performed, also showing their dependencies and their contents, which will be used as an input to the design model.

The decomposition of the Learning Technology System is continued during the design model, by specifying the very coarse-grained discrete subsystems, as they have derived from the use case and analysis model. Especially for the purpose of identifying subsystems, the analysis packages, together with their dependencies and contents are being used as a starting point.

These subsystems, that are in essence component frameworks, are meant to be further processed by identifying their contents and specifying their interfaces. The process then continues by building the deployment model of the system, which actually maps the software components into hardware components. Finally, the last workflow of this process produces the implementation model, which defines the executable components and their dependencies on each other.

After the five models of the USDP have been completed, all the component frameworks and interoperations between them have been identified. At the last part of the component-based architecting process, we make platform and implementation decisions, that we consider to be the most suitable for a component-based system. These technologies embodied in a component development model are depicted in Figure 2.

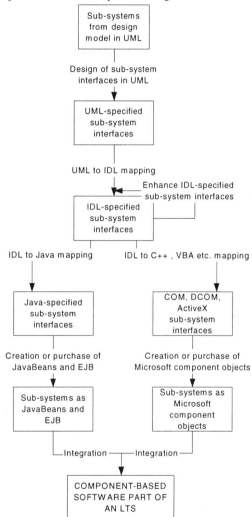

Figure 2- Component development model

The artifacts from the design model, that is sub-systems with textually described interfaces are provided as an input to the above development model. These interfaces are then designed with concrete UML notation and then mapped into the Interface Definition Language (IDL), which is an ISO standard for formally defining interfaces. Because the UML to IDL mapping is incomplete, the produced IDL interfaces need to be elaborated, so that a more accurate specification can be achieved. The next step is to transform the IDL interfaces into the implementation platform, in our case Java or Microsoft technologies, through the Java IDL API, or the Microsoft IDL APIs. The components now have concretely defined interfaces in the programming language, and they can either be constructed from scratch, or acquired from existing implementations and possibly modified to exactly fit the interfaces. The result is the implementation of the sub-systems as JavaBeans or Enterprise JavaBeans (EJB), which is the Java form of components, or, as Microsoft component objects (COM/DCOM objects, ActiveX controls etc.). The final step is to integrate the components through an integration and testing process into the final outcome: the component-based software part of an LTS.

4. Conclusions and future work

Each one of these three key concepts of the proposed process adds special value to the proposed architecture.

To start with, the proposed architecting process professes the same principles as the LTSA, namely [8]: it provides a framework for understanding existing and future systems; it promotes interoperability, portability and reusability by identifying critical system interfaces; and it remains adaptable to new technologies and learning technology systems.

An architecture that is built with the aid of the Unified Software Development Process [3]: helps all concerned stakeholders (e.g. developers, managers, customers) to understand the system through a common language; organizes the development effort, eliminating the communications overhead; fosters reuse of system components; and helps the maintenance and evolvement of the system through development iterations and product lifecycles, thus making the system change-tolerant. Moreover, Software Engineering is unique in that it is heavily driven by risk, and architecture-based development is the primary successful approach in risk-driven engineering [9].

Last but not least, as far as the enforced component-based paradigm is concerned, it is claimed in [6] that component-based architectures are inherently modular and as such have significant software engineering advantages: good modular architectures make dependencies explicit and help to reduce and control these dependencies; are naturally layered, leading to a natural distribution of responsibilities; and it is easier to migrate part of a system by adopting relevant component interface standards.

Based on these points, it is concluded that an inherently component-based software architecture is the right step towards bringing the economies of scale, needed to build affordable, interoperable as well as effective software subsystems of Learning Technology Systems.

We are currently investigating the use of this process into real LTS implementations and the subsequent evaluation of this process. This will raise several issues such as: whether the LTSA is able to provide a full, well-documented business model; how can a learning theory be combined with the business model in order to provide a full set of system requirements; whether the USDP, which is a generic software engineering process, works well in this type of applications; whether the binding technologies and platforms proposed, will efficiently help in the software system implementation; and whether the proposed process indeed leads to a pure component-based system.

References

[1] J. M. Spector, "Trends and Issues in Educational Technology: How Far We Have Not Come", *ERIC-IT Newsletter,* Sep. 2000.
[2] The web-based education commission, "The power of the Internet for learning: moving from promise to practice", Washington DC, December 2000, available on-line at [http://www.ed.gov/offices/AC/WBEC/FinalReport/WBECReport.pdf].
[3] I. Jacobson, G. Booch and J. Rumbaugh, *The Unified Software Development Process*, Addison-Wesley, 1999.
[4] P.B Kruchten, "The 4+1 view model of architecture", *IEEE Software,* November 1995.
[5] G. Booch, J. Rumbaugh and I. Jacobson, *The UML User Guide*, Addison-Wesley, 1999
[6] C. Szyperski, *Component Software – Beyond Object-Oriented Programming*, ACM Press, 1999.
[7] C. McCormack and J. D. Jones, *Building a Web-based Education System*, Wiley Computer Publishing, 1997.
[8] IEEE Learning Technology Standards Committee, "Draft Standard for Learning Technology Systems Architecture (LTSA)", November 2000.
[9] T. Mowbray and W. Ruh, *Inside CORBA – Distributed Object Standards and Applications*, Addison-Wesley, 1997.

An Architecture for Defining Re-usable Adaptive Educational Content

Charalampos Karagiannidis and Demetrios Sampson
Informatics and Telematics Institute (I.T.I.)
Centre for Research and Technology – Hellas (CE.R.T.H.)
1, Kyvernidou Street, Thessaloniki, GR-54639 Greece
Tel: +30-31-868324, 868785, 868580, internal 105
Fax: +30-31-868324, 868785, 868580, internal 213
E-mail: karagian@iti.gr, sampson@iti.gr
www.iti.gr

Fabrizio Cardinali
GIUNTI Interactive Labs S.r.l.
Via al Ponte Calvi 3/15
16124 Genova, Italy
Tel.:+39-010-2465178
Fax: +39-010-2465179
E-mail: f.cardinali@giuntilabs.it
www.giuntilabs.it

Abstract

This paper addresses re-usability in personalized learning environments. The paper presents the work of the European Project KOD "Knowledge on Demand", towards the development of an architecture for defining re-usable adaptive educational content which can be easily interchanged and re-used across different personalized learning applications and services.

1. Introduction and Background – The Need and the current Practice

The KOD project aims to address the needs of the different categories of users (market players) involved in the *e*-Learning arena, including: *e-Learning assets publishers*, aiming to version their learning assets for different online learning solutions in a re-usable and interoperable way; *e-Learning platform providers*, mainly interested in providing architectural solutions for *e*-Learning at different levels, e.g. learning management systems, assessment systems, performance support systems, etc; and *e-Learning service providers*, mainly using existing *e*-Learning platforms in conjunction with their own, or third-party learning content, to support service provision in the *e*-Learning arena.

In general, these users need to be able to publish to (or access from) public knowledge repositories learning material (either "single" learning assets, or "learning packages"), so that it can be easily interchanged across different applications and services. This, in turn, requires that learning material is described and published in a *common format* [1].

This need has resulted in a number of international standardization activities aiming to define common learning technologies specifications and standards that can ensure interoperability in the *e*-Learning arena. The main initiatives in the area are the IEEE LTSC (Learning Technologies Standards Committee, ltsc.ieee.org), the European CEN/ISSS Learning Technologies Workshop (www.cenorm.be/isss/Workshop/lt/), the IMS (Instructional Management Systems) Global Learning Consortium Inc (www.imsproject.org) and the US ADLnet (Advanced Distributed Learning Network, www. adlnet.org) [2].

These efforts have already resulted in a number of specifications for *e*-Learning applications and services. In fact, current specifications already encompass (or will, in the next future) almost all aspects of a standard *e*-Learning architecture (e.g. the IEEE LTSA, Learning Technologies Standard Architecture) from the description of learning objects meta-data based on shareable XML-based data structures (e.g. through the IEEE LOM, Learning Objects Metadata Schemas) to the assessment of user performances (e.g. through the IMS QTI, Question and Testing Interoperability Schemas). That is, existing specifications enable the common description of learning units, questions and tests, learner profiles, etc, so that they can be easily interchanged between different applications.

Moreover, existing specifications enable users (publishers, platform providers, service providers, etc) to "package" and publish content structures which are built on content building blocks in a standard way, and in particular through the IMS Content Packaging (CP) Specification. The CP XML schema reports a meta-data header describing the packaged resource itself, an *organization* field describing content structure and content packages included in the proposed learning path, and the list of meta-data of all referenced resources together with links to the encapsulated resources themselves. The schema allows for iterative descriptions, i.e. one can build onion like structures encapsulating resources within resources; every new resource has a new general meta-data describing it and an organization field describing its navigational structure. Using the CP specification, a publisher may describe a course as a "packaged pathway" through existing modular resources own by either the publisher or third parties. These can be described and chained in the standard description of the flat file based on a XML notation implemented in the CP format.

However, existing standards do not adequately support the definition and interchange of reusable adaptive and flexible learning methods which are beyond the "rigid" approach of directive, curricular-based, linear learning, as enabled by the envisaged hierarchical structure description in the Content Packaging standard. In particular, the organization field in the XML schema enabling the IMS content packaging manifest, although open to any notational description of navigation, mainly considers "rigid" hierarchical tree-based content structures description; no standard declarative notation, toolkit and viewer is available for conditional branching navigation or for path redirection.

In this context, the KOD project has identified a procedure for defining adaptive educational content in a way that is can be easily interchanged and re-used across different *e*-Learning applications and services [3]. The steps of this procedure include the definition of:

1. the concept ontology of the learning material to be presented,
2. the learning resources, i.e. the learning units that are to be communicated to the learner,
3. the competencies which are related to each node of the concept ontology,
4. the questions and tests that define when a learner has acquired a specific competency,
5. the different user profiles of the user groups that are expected to be interacting with the system, and finally
6. the navigation rules which define how different learning objects are selected for different learners; these "rules" specify the "matching" between the learner profiles and the learning content, that is the "learning path" that is appropriate for each different learner profile.

This paper focuses on the architecture that has been developed in the KOD project for defining adaptive educational content in a common format, following the above procedure. The architecture is shown in Fig. 1, and described in the following section.

2. The Proposed Architecture

The architecture includes a *Learner Interface*, which facilitates access to the learners' functionality. The Learner Interface supports all the features that are available in existing state-of-the-art *e*-Learning, *e*-Publishing and *e*-Knowledge tools. In addition, it facilitates access to PL services, i.e. enabling the user to authenticate himself, and subsequently to define, review and modify his/her user profile, so that the PL environment is adapted to the user's requirements, preferences, interests, goals, etc.

In the following paragraphs, the "KOD Factory" is described, i.e. the components of the architecture which enable the definition of adaptive educational material. The description is given in terms of a scenario of use, describing the steps that need to be performed by an "editor", i.e. a person responsible for defining adaptive content so that is can be interchanged across different PL environments.

2.1 Ontology toolkit

First of all, editors need to define the content to be presented to the learners. To this end, the first step in the definition of a "knowledge route" (i.e. the output of the PL environment) is the definition of the *ontology* that describes the concepts to be communicated. It should be noted that this ontology contains a factual description of the learning concepts (e.g. computer science is composed of operating systems, programming languages and databases), which is *independent* from the learning process. This ontology can be provided by a content expert, who is being interviewed by an editor. In this context, the Editor Factory includes an *Ontology Toolkit*, which assists editors in this process. The Ontology Toolkit employs a specific *knowledge representation* technique for storing the ontology, and provides as output XML files (*Ontology Profiles*), which are maintained in the *XML database* of the PL environment.

2.2 Meta-Data Toolkit

Following the above process, editors need to define the learning material / resources that are available for each "atom" of the ontology (i.e. the leafs in the ontology structure). In this context, the Editor Factory includes a *Meta-Data Toolkit*, which assist editors in defining the meta-data of the learning resources that are available. The Meta-Data Toolkit stores the meta-data as XML files (*Meta-Data Profiles*), which are also included in the *XML database* of the PL environment.

2.3 Questions & Tests Toolkit

In addition, the ontology should be "enriched" with questions and tests, which can specify when the learner has mastered a concept in the ontology hierarchy, and can therefore proceed with the next concept. The process of defining these tests is assisted by a *Question & Test Toolkit*, which enables the editor (with the help of the content expert) to define the questions and tests that are related for each concept in the ontology hierarchy. The

Question & Test Toolkit stores ***Question & Test Information*** this information into XML files, maintained into the ***XML database***.

2.4 Competencies Toolkit

Based on the ontology, editors then need to define the competencies that are related to each node of the ontology. In this context, the Editor factory includes a ***Competency Toolkit***, assisting editors interviewing content experts, and storing this information into XML files (***Competency Profiles***), maintained in the ***XML Database***.

2.5 User Profiles Toolkit

Then editors need to define which are different profiles of the users (learners) who are expected, foreseen, etc, to interact with the system. The definition of the user profiles (which is assisted by a content expert) is assisted by a ***User Profiles Toolkit***, with similar functionality with the above tools. The User Profiles Toolkit stores ***User Profiles*** in XML format, maintained within the ***XML database***.

2.6 Rule Toolkit

Finally, editors need to define how the learners will navigate the concepts of the ontology (i.e. viewing the learning resources, questions and tests, etc) that have been specified for each concept in the ontology, based on the user profile of the learner, as well as the competency level. Therefore, a ***Navigation Rules Toolkit*** is included in the Editor Factory, which assists editors to define the rules that determine the learning paths in the ontology that should be followed, and their matching to user profiles and competencies. It should be noted that these rules (***Navigation Rules***) are *dependent on the specific learning content* to be communicated, and are again stored in XML files within the ***XML database***.

The Rule Toolkit also enables the editor to define ***general rules***, which are applicable to every learning context, and stored in the SQL database of the PL environment.

All the above information form the basis for the definition of the different *knowledge routes* of the PL environment. Knowledge routes contain the ontology of the learning material, the learning objects and questions and tests that are related to each node in the ontology, the different user profiles and competency profiles, as well as the *navigation rules* which determine how the ontology is navigated for different learner profiles.

2.7 Agents Toolkit

The proposed architecture is based on software agents, which are the knowledge analyzing, monitoring, generating, adapting and delivering "pulse" of the PL environment. In this context, the Editor Factory includes also an ***Agents Toolkit***, which enables the editor to construct new agents, destruct existing agents, modify agent parameters, etc.

Agents are capable of presenting knowledge routes to learners, i.e. processing knowledge routes files. In particular, agents extract the information contained in the knowledge routes, and present learning material according to the learner profile. At each node of the ontology, agents are capable of identifying whether the user can understand the respective concept, or whether there is a need for presenting some "pre-requisite" concepts before. In case that the user cannot understand these concepts based on the learning material that is available (i.e. the physical learning resources indicated in the knowledge routes file), the agent automatically searches for additional information, both from specific repositories (defined by the editor), and from the Internet.

Also, agents provide assistance for the "verification" of the information that is encapsulated in the knowledge routes. For example, agents can notify the editor that the users that have accessed a specific knowledge route can be classified into more (or less) user profiles (employing data mining techniques).

3. Discussion and Conclusions

The different types of information that are defined through the procedure described in the previous section are necessary for the provision of adaptive content. In order for this information to be transferred across different applications, it needs to be maintained in *a common format*, i.e. following the existing and emerging *e-*Learning standards and specifications. For example, the description of the learning objects needs to be based on the Learning Objects Metadata (LOM) standard of the IEEE LTSC P1484.12 Working Group; the competencies need to be represented following the specification of the IEEE LTSC P1484.20 Competency Definitions Working Group; etc.

On the other hand, *all* the above types of information need to be represented *together* following a single common specification. To this end, the KOD project is currently working on the proposition of an extension of the IMS Content Packaging Specification [4], to include the above information. In particular, as it has been described in the introductory section, the current version of the specification enables (through the "organization" element) the description of "rigid" hierarchical tree-based content structures description (e.g. simple table of content).

The KOD project aims to propose an extension of this specification, where all the information presented in the previous section is also included in the content packaging description. This will facilitate the description of adaptive educational content in a common format, thus enabling users and publishers to share not only content and content routes, but also navigation algorithms (i.e. conditional branching based on user performances). As a result, adaptive educational content can be interchanged and re-used across different personalized learning applications and services, and thus re-usability in personalized learning can be promoted.

References

[1] Sampson, D., & Karagiannidis, C. (eds.) (2000). *Report on User Needs Analysis and the KOD Model Definition*. KOD Project Deliverable D1.1 (available from the authors).
[2] Bacsich, P., Heath, A., Lefrere, P., Miller, P., & Riley K. (1999). *The Standards Fora for Online Education*. D-Lib Magazine, 5(12).
[3] Karagiannidis, C., Sampson, D. & Cardinali, F. (2001). *Integrating Adaptive Educational Content into Different Courses and Curricula*. Educational Society and Technology, Special Issue on Learning, Instruction, Curriculum and the Internet, 4(3), In Press.
[4] IMS Project (2000). *Content Packaging Information Model*. Version 1.0 – Final Specification.

Acknowledgements

The KOD "Knowledge on Demand" Project (www.kodweb.org, kod.iti.gr) is partially funded by the European Commission Information Society Technologies (IST) Programme.

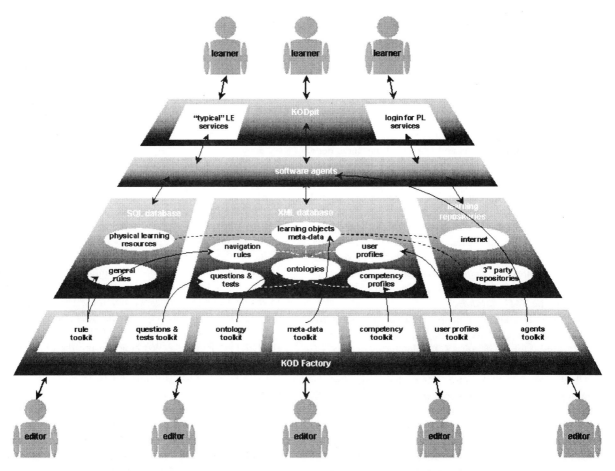

Fig. 1. The KOD Architecture for the Definition of Re-usable Adaptive Educational Content

Architectures for Computer Supported Collaborative Learning

Daniel D. Suthers
Department of Information and Computer Sciences
University of Hawai`i at Manoa
suthers@hawaii.edu

Abstract

Four architectures for computer supported collaborative learning systems are analyzed using the model-view-controller design pattern and compared from the standpoints of coupling between activities of the users and suitability for educational use, as well as network load and ease of implementation. The architectures are illustrated with examples from the developmental history of Belvedere, an environment for collaborative construction of knowledge representations during problem solving. A hybrid architecture that supports model-level coupling is shown to provide the best design tradeoffs.

1. Introduction

In the last decade we have witnessed a dramatic rise in popularity of computer-mediated learning, as evidenced by the growth in conferences on learning technologies, the emergence of online universities and the efforts of traditional universities to develop online degree programs. Consistent with research that demonstrates the value of collaboration in learning, computer support for collaborative learning has become of greater interest, and various architectures for synchronous and asynchronous collaboration have been explored. In this paper I discuss the suitability of four such architectures from the standpoint of the types of coupling or decoupling between the activities of learners that they can support, the ease of converting existing applications, and considerations of network load. I illustrate the architectures with examples from the developmental history of Belvedere [6]. Belvedere is an evolving environment for student-construction of explicit models (knowledge representations) while learning in science and other domains requiring critical inquiry (evaluation of alternatives with respect to evidence and criteria). In developing Belvedere, we explored all four of the architectures described herein. I conclude that a hybrid architecture offers the most flexibility for collaborative learning applications.

1.1. MVC

This paper uses the design pattern known as Model-View-Controller (MVC) to analyze the architectures [4]. The *Model* is an internal

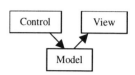

Figure 1. Model-View-Controller

representation of a semantic model of the problem of interest. The *View* displays the model in some visual representation. Software components implementing a View are registered as observers of the Model, so that changes in the Model will automatically result in an update to the View display. A *Controller* enables the user or the environment to modify the state of the Model. (A controller can be a human-computer interface widget, or it can be software reading from a physical sensor.) Software implementing the Model is registered as an observer of the Controllers, so that actions on the Controllers automatically result in an update to the Model state (and hence of the View).

1.2. Coupling

The architectures to be discussed differ on the degree of coupling that they support (or require) between the activities of different users and the state of applications used by those users. I define three levels of coupling.

Strict "what you see is what I see" or WYSIWIS ("whiz-e-whiz"), provides all users with exactly the same view and controller states. Strict WYSIWIS can support the collaboration of two or at most three users whose activities are *tightly* coupled. Strict WYSIWIS is problematic for larger groups or loosely coupled interactions because everyone sees the same cursor or set of cursors, and view states such as scroll position are the same for everyone. NetMeeting is an example of strict WYSIWIS.

Relaxed WYSIWS does not insist that the state of the view be exactly the same, provided the same view is being presented. Different users can scroll to different viewports on this view, and perform their own operations such as editing or moving objects without distracting the others, at least until a model change forces an update in the view. An example of relaxed WYSIWIS is Teamwave (www.teamwave.com/).

Model level coupling, which I call "what you model is what I model" or WYMIWIM ("whim-e-whim"), guarantees only that users will see the same semantic state of a shared model. The views may be entirely different, even to the extent of using different representations. For example, in Belvedere 3.x one person can view the model as a graph and another as a matrix.

I now analyze the architectures in terms of the distribution of MVC components and the type of coupling between them.

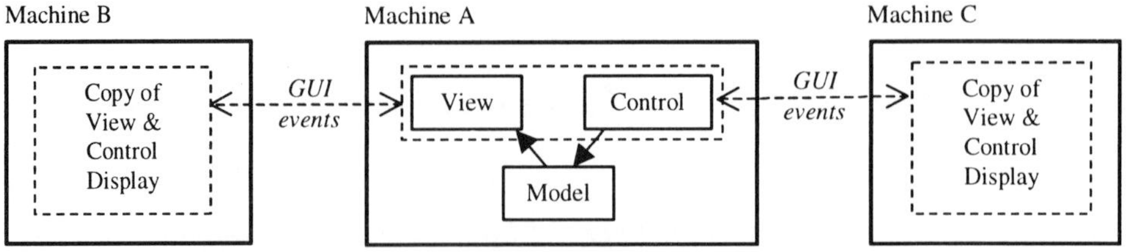

Figure 2. Centralized architecture

2. Centralized

A *centralized* architecture provides only one application, and distributes *copies* of the GUI (view and controller) by sending window system events to all participating client machines. The actual model, view and controller all remain on one host machine (Figure 2).

A well known example is NetMeeting, in which the applications run on one Wintel machine and other participating Wintel machines see these applications with strict WYSIWIS. Only one person can use the mouse or keyboard at a time. This architectures' primary advantage is that it is possible to take arbitrary applications and make them collaborative *at run time* by capturing and broadcasting window system events: developers need not know in advance which applications will be shared.

The Centralized architecture's transmission of complete display information and interface events over the network does not make efficient use of bandwidth. In some implementations (exemplified by NetMeeting) it also enforces too strict a form of WYSIWIS for learning applications. Although tight coupling may be appropriate for one-on-one training such as demonstrating the use of a software system, in collaborative learning applications it is more appropriate to allow learners to shift freely between working in parallel and working together.

Belvedere 1.x was implemented in Common Lisp in the Common Lisp Interface Manager (CLIM) using a modified centralized architecture, which addressed the latter problem. The application ran as a single process on one machine, and hence is classified as centralized because the model and all of the views and controllers remained on one machine. However, we achieved relaxed WYSIWIS by generating multiple view-controller instances, each view-controller instance being a CLIM application frame associated with a client IP number. These frames were displayed on the remote clients via X-windows. Yet this architecture still suffered from bandwidth issues and the need for persons at all machines to coordinate setting up the displays (e.g., provide xhost permission at B, then send display from A to B), which was too complex for classroom use.

3. Replicated

In a *replicated* architecture [1] the entire application is installed and run (i.e., replicated) on each client machine; and some means of synchronization between them is provided. This architecture is characterized by having all three MVC components – model, view and controller – replicated on each client machine (Figure 3). Examples of replicated collaborative systems include E-Slate (E-Slate.cti.gr), Habanero [2], and MatchMaker [7]. All of these systems require that applications be written with collaboration in mind, using an API for event sharing. In contrast, JAMM (Java Applications Made Multiuser, [1]), provides a way to convert existing single user applications to collaborative use without modifying the code: Java Swing interface classes are modified to broadcast the events on each copy of the application to the other copies.

Replicated architectures improve on use of network resources because display data is not transmitted over the network: only Controller events need to be distributed. Also, the client can be used without the network. Replicated architectures based on the automatic broadcast of controller events (as in JAMM) have the disadvantage that they are most naturally suited for strict rather than

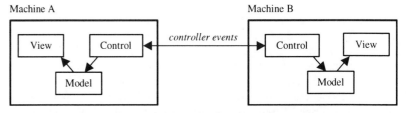

Figure 3. Replicated architecture

relaxed WYSIWIS. This disadvantage can be avoided by selecting relevant events when manually building a collaborative application. Yet synchronization via controller events may be at the wrong level of abstraction for many learning applications. Learners will be more interested in each others' semantic changes (model updates) rather than in the manipulations of the GUI by which other learners achieve these semantic changes. MatchMaker's synchronization is at the model level: one can select which objects are to be synchronized, and even turn synchronization on and off at runtime.

The Belvedere 2.x series used a replicated architecture more complex than that pictured. Version 2.1 is described in [5]. We reimplemented Belvedere as a stand-alone Java application with a self-contained MVC architecture. We then provided the application with a *Listener* on a dedicated port to listen for events from the other clients. Each client also had a component that informed the outside world of changes to its model. However, rather than informing the other clients directly, this component, known as the *Belvedere Object Request Broker Interface* or BORBI, communicated with a server providing persistent storage of the model. We shall see that in this respect Belvedere 2.x represented a hybrid of Replicated and Distributed architectures. BORBI updated the remote database for each change, and also informed a Java process on the server. This Connection Manager kept a table of all active clients and the workspaces they had opened, and would broadcast change events to the Listeners of clients that had opened the workspace being changed. Belvedere 2.x also provided a simple Chat facility: users of any given workspace received messages typed into Chat by others working on that workspace. Belvedere 2.x's replicated architecture transmitted model change events rather than controller events. This reduces network traffic and opens up the possibility of model-based coupling or WYMISIM. The shared persistent store is a step towards supporting asynchronous collaboration. The architecture of Belvedere 2.2 also forms the basis for a coached collaborative distance learning system known as COLER [3].

4. Distributed

A distributed architecture is characterized by the distribution of the MVC components across multiple hosts. Typically, the Model lives on a shared server and each client has its own View and Controller (Figure 4).

The most familiar example of the distributed architecture may be database-backed web sites such as airline reservation systems and other e-commerce systems. The user's web browser provides the view and controller and the server stores the data. This type of distributed

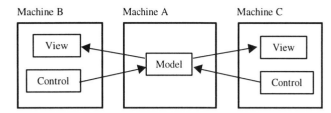

Figure 4. Distributed architecture

architecture shares some features with Centralized in that specifications of the View and Controller are actually constructed on a server and sent to the client. Hence some of the problems of ineffective use of bandwidth apply.

A somewhat more network-efficient implementation is exemplified by Enterprise Java Beans (EJB). Simply stated, EJB enables one to run the Model as a Java Bean on a server, and have this bean shared by multiple clients consisting of View-Controller software. The View and Controller originate on the clients and are not sent over the network. One can program the View-Controller as if the model were running on the client machine. EJB handles the distribution of the model on the network and WYMIWIM updating. Other services such as transactional behavior are provided automatically. During the past two years, my students Hongli Xiang and Bo Yang experimented with an EJB architecture for Belvedere 3.0. We found that EJB provides a high initial learning curve, yet once this is overcome one can program a distributed application quickly.

Properly implemented, a distributed architecture requires only model update events to be sent over the network, making this architecture more efficient in terms of network resources. Since coupling is at the level of the model, the distributed architecture can support WYMIWIM: users can collaborate on the same model while using entirely different visual representations of the model. This motivated our experimentation with an EJB-based distributed architecture for Belvedere 3.0.

From a user's perspective, the primary disadvantage of a truly distributed architecture is the reliance on the network. The ability to run as a stand-alone application has important advantages, particularly in a classroom environment where the network may be unreliable and the teacher must be able to continue class activities after discovery of an outage, with no more than a minute's delay before chaos ensues! This motivates our current hybrid architecture.

5. Hybrid

Belvedere 2.x introduced a hybrid between Replicated and Distributed architectures (Figure 5). In this architecture, synchronization is at the model level via a persistent model. Applications can run standalone with their own models, saving state to the local file system, or can connect to a persistent store that provides WYMIWIM updating between active clients.

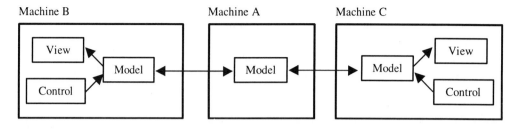

Figure 5. Hybrid architecture

Belvedere 3.0 differs from the previous versions of Belvedere in one important respect: it provides *multiple views* on a given model. One can construct an evidence model using any of Graph, Matrix or Tree visual representations. Updates in one view are immediately available in the others, and one can switch between views freely as one works. A collaborative version of Belvedere 3.0 *requires* model-level coupling, as the views may be entirely different. We are implementing Belvedere 3.0 using the Hybrid architecture to achieve model-level coupling along with the flexibility of running either networked or stand-alone.

6. Conclusions

I defined three architectures for collaborative learning systems in terms of the location of model, view and controller components and the means of coupling between applications, and identified advantages and disadvantages for each. I described a hybrid architecture that endows each client application with its own model/view/controller components, yet couples these via a shared model on a server. While slightly more complex to implement, this architecture addresses the tradeoff between independence and coupling of applications. More importantly, coupling at the level of model state enables applications to use different visual representations for their views on this model, enabling learners to work within the view that best meets their current needs while still being able to collaborate with others. The architectures were illustrated with a series of implementations of Belvedere. Ongoing work is exploring the design of coupling between shared knowledge representations and computer mediated communication media such as threaded discussion. Future work may be needed to understand how collaboration may be affected by the use of multiple views.

7. Acknowledgements

Thanks to Bin Ma, Bo Yang and Hongli Xiang for their work on Belvedere 3.0, and to Dan Jones for his work on the Belvedere 2.x versions. This work was partially supported by a grant from the National Science Foundation's Learning and Intelligent Systems program.

8. References

[1] Begole, J., C. A. Struble, et al. (1997). Transparent Sharing of Java Applets: A Replicated Approach. 1997 Symposium on User Interface Software and Technology, New York, NY, ACM Press.

[2] Chabert, A., Grossman, E., Jackson, L., Pietrowicz, S., and Seguin, C. (1998). Java Object-Sharing in Habanero. *Communications of the ACM*, Vol. 41 # 6, June 1998, pp 69-76.

[3] Constantino-González, M.A. and Suthers, D.D. (2000). A Coached Collaborative Learning Environment for Entity-Relationship Modeling. In Gauthier, G., Frasson, C. and VanLehn, K. (Eds.) *Intelligent Tutoring Systems Proceedings of the 5th International Conference, ITS 2000,* Montreal, June, pp 324-333. Available: http://lilt.ics.hawaii.edu/lilt/papers/COLER-ITS00.pdf

[4] Erich Gamma, Richard Helm, Ralph Johnson, and John Vlissides. Design Patterns: Elements of Reusable Object-Oriented Software. Addison Wesley, 1995.

[5] Suthers, D.D. and Jones, D. (1997, August). An Architecture for Intelligent Collaborative Educational Systems. In B. du Boulay, R. Mizoguchi (Eds.) *8th World Conference on Artificial Intelligence in Education (AIED'97), pp. 55-62.*

[6] Suthers, D.D. Toth, E. E. and Weiner, A. (1997). An integrated approach to implementing collaborative inquiry in the classroom. In *Proc. of the 2nd International Conference on Computer Supported Collaborative Learning (CSCL'97)* (pp. 272-279). Toronto.

[7] Tewissen, F., Baloian, N., Hoppe, U., and Reimberg, E. (2000). "MatchMaker": Synchronising Objects in Replicated Software-Architectures. *Proc. 6th Int. Workshop on Groupware, CRIWG 2000,* Madeira, Portugal, 18 - 20 October 2000, IEEE CS Press.

The Architecture of A Framework for Building Distributed Learning Environments[†]

Colin Allison   David McKechan   Alan Ruddle

School of Computer Science
University of St Andrews, Scotland
{ca,dm,alanr}@dcs.st-and.ac.uk

Rosa Michaelson

Department of Accountancy and Business Finance
University of Dundee, Scotland
r.michaelson@dundee.ac.uk

Abstract

Distributed Learning Environments (DLEs) are being cited as solutions to the ambitious political goals of better education, wider access and lower costs. This paper outlines the architecture of the TAGS framework for building DLEs. The main features that the architecture facilitates are strong support for group work, anytime/anywhere access, the input of real world data, event monitoring facilities, authentication, authorisation and protection, and the controlling of delay between users actions and system responses. These features have required the development of an expanded view of Quality of Service (QoS) parameters. The architecture has been refined over a number of years as the result of collaborative development between users and programmers. In particular, we have found that the bringing together of systems programmers, computer science researchers and subject-specific educationalists has allowed needs to be identified and addressed effectively. We believe that TAGS makes it possible for educationalists to easily build DLEs that meet their needs.

1. Introduction

Distributed Learning Environments (DLEs), built from communication and information technologies are being cited as solutions to the ambitious goals of better education, wider access and lower costs in the education and training sector [1]. This paper discusses system support issues not addressed by other learning environments [2].

Who are the key users of a DLE? End users typically fall into one of three major roles: student, tutor and lecturer. These users are predominantly non-computer specialists. Tutors and Lecturers in particular must be provided with suitable abstractions of online classes. It is also important that DLEs support developers and maintainers of educational resources.

From the perspective of good educational practice a DLE should provide certain features for students:

- interactive, engaging
- group based, not just single-user mode
- real world input, not only simulations
- student centered, with individual needs addressed
- anytime/anywhere, allowing flexible access.

These pedagogical goals imply certain technical requirements:

Interactivity implies that online working is much more than simply browsing lecture notes placed on the Web. Responsiveness is essential for creating an interactive feel when working on the web.

Support for group working means that the architecture needs to facilitate the bringing together of groups and students with their tutors and educational materials. In addition multi-user awareness is important for teamwork.

Real world input implies a direct Internet connection. For example the Finesse portfolio management facility [3] allows the management of a portfolio of shares by a student group. The groups maintain their portfolios by buying and selling shares chosen from a database of live (every five minutes) data for approximately 1100 companies quoted in the London Stock Exchange.

Student Centered means understanding individual user needs. It implies strong monitoring capabilities, be they for manual inspection or automatic adaptation.

Anytime/anywhere implies a resilient, highly available web-based service. Availability means coping with faults. Providing a service based upon multiple servers

---

[†] This work was supported by the SHEFC C&IT programme.

can negate periods of non-availability caused by server crashes.

2. The TAGS Framework for DLEs

TAGS [4, 5] is a framework for DLE construction and maintenance, which has been made increasingly QoS aware [6], through our experience in its development and deployment. At present TAGS is used by tutors and students at six Scottish Universities in eight different subject areas. The project has addressed issues in usability, security, responsiveness, concurrency control, availability and infrastructure limitations that would have been unlikely to arise in a pure research environment.

Figure 1 gives an overview of the current TAGS framework. TAGS presents four simple abstractions

abstractions for tutors	support for developers
users *groups* *resources* *events*	*generic facilities*
User Interface	DAPSI

systems aspects
Middleware, QoS

concurrency, availability, responsiveness, scalability, replication, distribution, coherence, security, portability......

Figure 1 The TAGS framework

for end users: users, groups, resources and events. It also provides a developer and application programmer service interface (DAPSI), which allows educational resource developers to concentrate on content, by supplying commonly needed system services.

3. System Interactions

Figure 2 shows the main interactions in the system during the lifetime of a user session. When the user logs on their username and password are authenticated. The portal generator retrieves information about the user from the TAGS database and generates a personalized portal to the system. The user may then request access to one of their resources. The status of the user is checked and control passed to resource

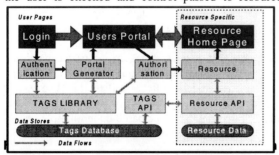

specific code, which may in turn access the TAGS database through a safe API and retrieve resource specific data through its own API. The view of the resource that is created depends upon the status of the user.

There are also special system resources, which have direct access to the TAGS library. These provide a number of utilities for System Administrators, which among other functions facilitate the allocation of users and resources to groups.

4. Users Groups and Resources

The user model for TAGS is based around the three simple components of users, groups and resources (see figure 3). Sets of users are brought together with the resources that support their collective work by the TAGS group mechanism.

Users and groups are unique by name; resources are unique by name and type. Groups form the basis of privileges and access, they provide the building blocks for the user's home page, and they can also act as dissemination channels when a resource is involved. There is no restriction on the nature of a resource. It can be a simple timetable, an interactive multi-user spreadsheet or a live data feed. All of a user's allocated resources appear on their home page, which is generated from their group memberships. Access rights can be specified when a resource is allocated to a group.

When a resource is edited the changes are passed on to all members of the group(s) it has been allocated to. If it is deleted, it is removed from all the groups it was associated with. When a group is deleted, the mappings it has formed between individuals and resources are removed. If a group's membership is changed then only the updated membership will have access to the resources allocated to that group.

There are three main levels of privilege in TAGS appropriate for students, tutors and lecturers respectively. Membership of a privilege group defines the privilege level a user has, which in turn defines the view that a user receives. For example, in the Register

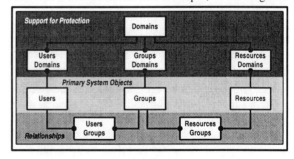

Figure 3 TAGS database schema

resource students can only view their own individual

marks. A tutor is able to view and edit the marks for all members of their group, and a lecturer is allowed to view and edit all marks.

5. Collaborative Learning and Protection

Each instance of a TAGS server may service multiple courses. This reduces the cost to each course but brings with it the need for protection so that administrative rights granted to a user on one course are not global throughout the system. This is particularly important for activities such as user creation, and the allocation of users and resources to groups.

From the point of view of a course administrator the requirement for protection is two-fold. On the one hand they do not want to have to deal with users, groups and resources that belong to other courses. On the other they require that tutors and system administrators who are not associated with their course should not be able to gain access to their work or educational materials. The TAGS architecture resolves this by constraining the rights of an administrator within the boundaries of an Administrative Domain of Protection (see Figure 3).

When we consider the possibility of a user being associated with more than one course, there are further requirements: the need for a single user to have access to multiple domains, and that a single log in and password should allow access to all of the domains that a user has the right to access.

The protection that Domains provide should not prevent members of separate administrative domains from engaging in collaborative learning. It follows that a mechanism to allow inter-domain communication is needed.

The problem of protection is one that has been explored in the context of operating system design. Silberschatz and Galvin [7] make the following observations:
1) A computer system can be thought of as a collection of processes and objects. The operations that are possible may depend upon the object.
2) A process should be allowed to access only those resources it has been authorised to access. Furthermore, at any time it should be able to access only those resources that it requires to complete its task. This "need-to-know" principle limits the damage that can be done by a faulty process.

These observations provided the starting point for the design of TAGS domains. In the TAGS system a user plays a role analogous to that of a process in an operating system and Users, Groups and Resources that of operating system objects.

A domain is an area of protection, which maps to an administrative area of control. A user, resource or group may belong to one or to many domains of protection. Normally a user will only be active in one domain at a time. If they wish to manage objects in another domain to which they have management privileges, they must explicitly change the domain in which they are active. This helps support a weak version of the need-to-know principle. The user only has access to those objects they need for the tasks they are undertaking at a particular point in time.

As discussed above, group membership is used to structure collaborative learning. TAGS treats group membership as a tighter binding than domain membership. Consequently, assigning members of different domains to the same group will facilitate inter-domain communication. The group has to belong to multiple domains but the users may be in separate domains. Groups may also be used as a mechanism for allowing resources in one domain to be made available to users in other domains.

In this way domains provide the administrative protection that is required without unnecessarily constraining the freedom of communication that collaborative group work requires.

6. TAGS Resources

Resources provide the workspace utilised by students and tutors and facilitate cooperation between members of a group. There currently exist a substantial suite of resources for TAGS. These resources fall broadly into three categories:

The first category is made up of subject specific resources such as the Finesse Portfolio for Finance students, the foreign language-writing tool Ecrire and a tool for medical students called The Model Patient.

The second consists of generic resources that support a range of functions. There is a Notebook and Question and Answer tool, which facilitate communication between students and tutors. There are separate tools that support online submission (DAT), evaluation of teaching and learning (Marksheet) and performance and attendance tracking (Register).

The third includes those that allow educational providers to take advantage of the TAGS allocation, authentication and authorisation mechanisms in order to make educational materials available to specific groups of students. These include the Page Set, and simple URL.

The TAGS framework also eases the construction of new resource types by providing a data access API, the TAGS Resource API, and the mechanisms of group and resource allocation (see figure 3). This means that developers can concentrate on the educational needs of their resource and not have to worry about low-level code and allocation mechanisms.

7. Architecture of Interactive Resources

The delay between requesting a page and receiving it

can make web browsing a frustrating experience. It has been shown that the length of delay directly impacts on the users perception of the quality of the page displayed and that the tolerance of delay is relative to users expectations [8]. Interactivity requirements are particularly important in influencing the architecture of TAGS resources. A user can be expected to perform many interactions during a single session; a tutor may need to fill in all of the class marks for an assignment. Thus, high interactivity is combined with locality of reference. These characteristics lead to the observation that the minimisation of delay is a critical element of delivering QoS to users of the system [9].

As users expectations are important in determining acceptable levels of delay it is relevant to consider the behaviour of non-distributed applications. User application interactions can be divided into two phases. In the first phase the user issues a command to load the application. For this to be completed libraries of code need to be loaded from disk into memory before the user interface can be presented to the user. It is not unusual for this process to take several seconds. By analogy, with TAGS an experienced user of

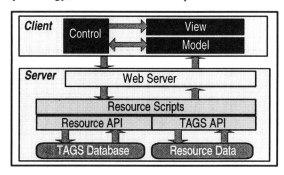

Figure 4 TAGS resource architecture

computer applications will be less tolerant of delays once a resource has been loaded.

This expectation has been met by the architecture illustrated in Figure 4. When a resource is in use all the data that the resource or resource component needs is transferred to the client. A model of the relevant data is then constructed in the browser. At this point, most interactions with the user do not require network communication, but cause the local copy of the model to be modified. When the users work has been completed the changes to the model can be committed to the server in one batch and the database updated. Facilities are also provided to allow the user to commit changes at intermediary points.

This approach leads to a high proportion of users requests being resolved within the client browser, thereby reducing the number of requests that result in interactions with the server and the TAGS database; hence meeting the interactivity requirement. The distribution of computation from the server to the client increases scalability and the control that end users have over performance. Upgrades in the client systems will result in a perceived improvement in performance.

8. Conclusion

This paper has outlined the architecture of the TAGS framework for building Distributed Learning Environments. The starting point for the design of TAGS was the identification of the pedagogical needs of educationalists. The initial architecture has been developed and refined over several years, during which time TAGS DLEs have been deployed and used in a number of higher educational institutes.

We have found that the feedback from the user base has been overwhelmingly positive and has been an important input into the evolution of the system. In particular the need to provide inter-domain protection and to improve the interactivity of the system arose out of the deployment of the system. Addressing these and other issues arising from the ongoing collaboration between educationalists and developers has required a fuller understanding of the importance of QoS in distributed learning environments.

References

1. Finlay, R.M., *Tele-Learning: The "Killer App"* ? IEEE Communications, 1999. **37**(3): p. 80-118.
2. Britain, S. and O. Liber, *A framework for pedagogical evaluation of Virtual Learning Environments*, . 1999, JTAP.
3. Helliar, C.V., R. Michaelson, D.M. Power, and C.D. Sinclair, *Using a Portfolio Management Game (FINESSE) to Teach Finance*. Accounting Education, 2000. **9**(1): p. 37-51.
4. Allison, C., M. Bramley, R. Michaelson, and J. Serrano. *An Integrated Framework for Distributed Learning Environments*. in *"Advances in Concurrent Engineering", 6th ISPE International Conference on Concurrent Engineering*. 1999. Bath.
5. Allison, C., D. McKechan, H. Lawson, and R. Michaelson. **The TAGS Framework for Web-Based Learning Environments**. in *Web Based Learning Environments*. 2000. Portugal: University of Porto, FEUP Editions.
6. Allison, C., D. McKechan, H. Lawson, and A. Ruddle, *An Holistic View of Quality of Service*. Interactive Learning Environments, 2001. **9**(1).
7. Silbershatz, A. and P. Galvin, *Operating System Concepts*. 4th ed. 1994: Addison-Wesley.
8. Ramsay, J., B. Alessandro, and J. Preece, *A psychological investigation of long retrieval times on the World Wide Web*. Interacting with Computers, 1998. **1998**(10): p. 77-86.
9. Allison, C., M. Bramley, and J. Serrano. *The World Wide Wait: Where Does the Time Go?* in *Engineering Systems and Software for the Next Decade*. 1998. Vasteras, Sweden: IEEE Press.

Theme 2: Theories and Formalisms

Constructivist Values for Web-based Instruction: University and NSA Collaboration

Penelope Semrau, Ph.D.
Email. psemrau@calstatela.edu
Educational Foundations & Interdivisional Studies

Barbara A. Boyer, Ph.D.
Email. bboyer@calstatela.edu
Art Department

California State University, Los Angeles
5151 State University Drive
Los Angeles, CA 90032 U.S.A.
Tel. (323) 343-5506
Fax. (323) 343-5336

Abstract

CSULA was charged with assisting the NSA in the research, development, documentation and the necessary expertise to convert a face-to-face course to a multimedia web-based course called "EEO 100: Equal Employment Opportunity Law and Diversity Training for Selection Boards." The focus of our work was on ways to encourage interactive learning and the use of motivational strategies online.

CONSTRUCTIVIST APPROACHES TO LEARNING AND RESEARCH

In July 1998 CSULA began a 3-year research and courseware conversion project sponsored by the National Security Agency (NSA) to convert a face-to-face course to a multimedia web-based course using "constructivist" approaches as the basic foundation for organizing and developing the project. It involved two professors and an interdisciplinary team of graduate and undergraduate students from Instructional Technology, Art Education, Computer Science, and Graphic Design. Recent studies in education have promoted the constructivist approach to learning "...where students develop their knowledge through team collaboration, discuss different interpretations of a problem, and negotiate and synthesize ideas drawing from various disciplines" (Boyer & Semrau, 1995, 14). Ultimately, students will be engaged in their learning, see relevance in it, and take on an ownership in it. To meet the needs and interests of students, education must emphasize problem solving and applications of information, instead of memorization of facts.

YEAR ONE OF PROJECT: INITIATING THE PROJECT

The NSA provided one of their platform-based courses called "EEO 100: Equal Employment Opportunity Law and Diversity Training for Selection Boards" for the CSULA group to convert. NSA included course material and a ½" VHS videotape of an instructor answering questions and discussing material related to the course content. EEO 100 is directed to employees of the NSA who serve or who may serve on boards, panels, or committees charged with the responsibility of rating workers for selection purposes.

Training was provided to the CSULA student team on the basics of html coding, Adobe Photoshop, Premiere, Pagemill and WebCT. WebCT is an authoring and course management tool for web-based courses. The students were instructed regarding the overall "look & feel" of the website layout and screen designs. For a foundation in working collaboratively in building a website the students first designed a demonstration site on streaming media using RealAudio and RealVideo.

YEAR TWO OF PROJECT: CONVERSION STAGE

During the fall, 1999 we moved onto converting the EEO 100 course content into a web-based course. The EEO 100 content was subdivided into the following main topic areas: Current requirements, additional requirements, definitions of discrimination, disparate impact, harassment, affirmative action review, and tools. It was necessary to rewrite and revise the content, as well as integrate multimedia to accommodate a web-based learning environment.

A focus was to keep the web course interactive with an emphasis on learner control. It was also important that there be a balance between text and the use of graphics, diagrams, video and audio to reinforce the learning. Another significant area was the focus on culturally diverse distance learners. Sanchez and Gunawardena (1998) identified the challenge of recognizing both the diversity of culture and individual learning styles with diversity in the learners, cultures, and learning styles – variety in learning may be the solution (p. 61).

In the spring of 2000 and into early summer, we synthesized various course components (content, video, test items) into a final course website and designed visual learning aids (diagrams, charts, and pictures) and tools (glossary, law books, and court cases) to enhance learning of the course content.

CONCLUSION

As Boettcher and Cartwright (1997) emphasized, web course designers must realize that the "individual learner must be viewed as the key design element." The focus of our work was on ways to encourage interactive learning and the use of motivational strategies online.

The new web-based version of this course will make the content accessible to more NSA employees worldwide. Plus, the online version includes interactive pop quizzes with immediate scoring and feedback as well as streaming media lectures to complement the course content, which is presented as text on the web pages.

Through this project the students became constructively involved in their own learning and have acquired in-depth experiences in collaborative learning and team approaches. The students became empowered to be developers of their own curricular materials and course websites instead of being passive viewers of others'. All of Bloom's higher-level taxonomies were implemented in this project – the students analyzed websites, synthesized criteria that they collected, compared and contrasted their criteria, designed and produced their own web pages with a focus on sound educational practices. As Gibson (1998) noted "…we, as distance educators, need to be learner-centered reflective practitioners" (p. 143).

REFERENCES

Boettcher, J. & Cartwright, G. P. (Sept/Oct 1997). Designing and supporting courses on the web. [On-line]. Available:
http://www.kentinfoworks.com/change/articles/sepoct97.html

Boyer, B. A. & Semrau, P. (January/February, 1995). A constructivist approach to social studies integrating technology, *Social studies & the young learner,* 7(3), 14-16.

Gibson, C. C. (1998). In retrospect, *Distance learners in higher education: Institutional responses for quality outcomes.* Chere Campbell Gibson, editor. Madison, WI: Atwood Publishing.

Sanchez, I. & Gunawardena, C. N. (1998). Understanding and supporting the culturally diverse distance learner, *Distance learners in higher education: Institutional responses for quality outcomes.* Chere Campbell Gibson, editor. Madison, WI: Atwood Publishing.

Research on Role-based Learning Technologies

Brian M. Slator(a), Jeffrey Clark(b), Paul Juell(a), Phil McClean(c), Bernhardt Saini-Eidukat(d), Donald P. Schwert(d), Alan R. White(e)
Departments of a. Computer Science, b. Anthropology, c. Plant Sciences,
d. Geosciences, e. Biological Sciences
slator@cs.ndsu.edu, Jeffrey_Clark@ndsu.nodak.edu, Paul_Juell@ndsu.nodak.edu, Phillip_Mcclean@ndsu.nodak.edu, Bernhardt_Saini-eidukat@ndsu.nodak.edu, Donald_Schwert@ndsu.nodak.edu, Alan_White@ndsu.nodak.edu

Abstract

One of the goals of science education is to familiarize students with an intellectual framework based on established scientific principles and general approaches that can later be used to solve science-based problems. Science is also content-based, and students must master the content of a discipline in order to succeed. The challenge for science educators is to develop educational tools and methods that deliver the principles but at the same time teach the important content material, but in a meaningful way. This paper describes research based on experimental virtual role-based environments built to explore the following beliefs:

- educational technology should capitalize on the natural human propensity for role-playing;

- students will be willing to assume roles if the environment makes it easy to do, and if the environment reinforces role-playing through careful crafting of explicit tutorial components;

- that educational software should be engaging, entertaining, attractive, interactive, and flexible: in short, game-like.

The experiences provided to the student within these virtual worlds can be both meaningful and authentic, although some trade-offs are required to make them fun, challenging, and occasionally unpredictable.

1. Introduction

The NORTH DAKOTA STATE UNIVERSITY World Wide Web Instructional Committee (WWWIC; McClean et al., 1999; Slator et al., 1999b) is engaged in several virtual/visual research and development projects: the Geology Explorer (Schwert et al., 1999), the Virtual Cell (White et al., 1999), and the ProgrammingLand MOOseum of Computer Science (Slator and Hill, 1999). Each has shared and individual goals. Shared goals include the mission to teach science structure and process: the scientific method, scientific problem solving, diagnosis, hypothesis formation and testing, and experimental design. The individual goals are to teach the content of specific scientific disciplines: Geology, Cell Biology, Computer Science.

In addition, WWWIC is applying what has been learned in Science education to new domains: history, microeconomics, and anthropology. Further, WWWIC has active research projects in three highly related areas: 1) qualitative assessment of student learning, 2) tools for building virtual educational environments, and 3) intelligent software tutoring agents (Slator, 1999).

The WWWIC program for designing and developing educational media implements a coherent strategy for all its efforts: to deploy teaching systems that share critical assumptions and technologies (e.g. LambdaMOO; Curtis 1992, 1997), in order to leverage from each other's efforts. In particular, systems are designed to employ consistent elements across disciplines and, as a consequence, foster the potential for intersecting development plans and common tools for that development. Our simulations are implemented by building objects and interfaces onto a MOO ("MUD, Object-Oriented", where MUD stands for "Multi-User Domain"). MUDs are typically text-based electronic meeting places where players build societies and fantasy environments, and interact with each other. Technically, a MUD is a networked multi-user database and messaging system. The basic components are "rooms" with "exits", "containers" and "players". MUDs support the object management and inter-player messaging that is required for multi-player games, and at the same time provide a programming language for writing the simulations and customizing the environments.

2. Role-based Environments

The theory of role-based environments (Slator and Chaput, 1996), is both simple to explain and complex to implement. An apprentice watches their master, learning techniques and practicing their craft; they observe the master's actions and internalize them (Brown et al., 1989; McGee et al., 1998). When confronted with a problem, the apprentice asks, "what would the master do in this situation?" And then the apprentice models the expertise of the master in the pursuit of their goals. This is a common experience shared by silversmiths, doctors, Ph.D. candidates, and anyone else enculturating themselves into what they want to be. When John Houseman says, in the *Paper Chase* that "We are not teaching you the law, we are teaching you to think like a lawyer," this is what he means. Similarly, there is little argument that immersive foreign language learning is most effective; to learn French, go to France. At some point, it is widely reported, you begin to "think in French", and that is what you want.

Our idea is to put students into authentic situations that will challenge them to think through problems and act like scientific problem solvers. Students in our role-based environments are presented with a series of high-level goals, which are designed to promote science reasoning. For example, "locate a kimberlite deposit" is designed to raise inquiries like "what is kimberlite?", "what properties does it have?", "what experiments relating to these properties will confirm an identification?", "what instruments are needed for these experiments?", and "how are these instruments used?". These questions exemplify science reasoning in terms we are attempting to promote: the "doing" of hypothesis formation and experimentation, and the acquisition of conceptual knowledge in a problem-solving context.

3. Developing Role-based Environments

Developing such systems is difficult, expensive, and inherently collaborative, and requires three components: 1) content expertise by subject matter experts and the associated (non-standard) pedagogical design; 2) software design, development, and project management; and 3) fundamental Computer Science research in the areas of distributed systems, software agents and intelligent tutoring, and virtual environments. In practice this means implementing authentic simulated environments where students can learn-by-doing. This entails implementing a vast range of interacting locations, artifacts, and instruments to support 1) authentic problem solving situations, 2) research questions and content-relevant goals, and 3) online help and tutorial advice in context.

4. Major WWWIC Team Efforts

Space does not permit full description of the WWWIC projects. This paper describes only two: the Geology Explorer and the Virtual Cell.

4.1 Example: the Geology Explorer

The Geology Explorer project (Saini-Eidukat et al., 1999; Slator et al., 1999a), implements a virtual world where learners assume the role of a geologist on an expedition to explore the geology of a mythical planet. Learners participate in field-oriented expedition planning, sample collection, and "hands on" scientific problem solving. The Geology Explorer world, Planet Oit, is simulated on an Object Oriented Multi-user Domain. A text-based version of Geology Explorer was tested in an introductory geology class during the Summer 1998. Results of that test were used to prepare for a much larger test during the next semester. A graphical user interface to the Geology Explorer is under development.

To play the game, students are transported to the planet's surface and acquire a standard set of field instruments. Students are issued an "electronic log book" to record their findings and, most importantly, are assigned a sequence of exploratory goals. These goals are intended to motivate the students to view their surroundings with a critical eye, as a geologist would. Goals are assigned from a principled set, in order to leverage the role-based elements of the game. The students make their field observations, conduct small experiments, take note of the environment, and generally act like geologists as they work towards their goal of, say, locating a kimberlite or a graphite deposit. A scoring system has been developed, so students can compete with each other and with themselves. The Geology Explorer prototype can be visited at http://oit.cs.ndsu.nodak.edu/.

4.2 Example: the Virtual Cell

The Virtual Cell (VCell) is an interactive, 3-dimensional visualization of a eukaryotic cell. VCell has been prototyped using the Virtual Reality Modeling Language (VRML), and is available via the Internet. To the student, the Virtual Cell looks like an enormous navigable space populated with 3D organelles. In this environment, experimental goals in the form of question-based assignments promote diagnostic reasoning and problem-solving in an authentic visualized context.

The initial point of entry for the Virtual Cell is a VRML-based laboratory. Here the learner encounters a scientific mentor (a software agent), and receives specific assignments. In this laboratory, the student performs simple experiments and learns the basic physical and chemical features of the cell and its components. More notably, our laboratory procedures are crafted such that they necessitate a voyage into the Virtual Cell where experimental Science meets virtual reality. As the student progresses, they revisit the laboratory to receive more assignments. Periodically, the student will bring cellular samples back to the virtual lab for experimentation. The Virtual Cell prototype can be visited at http://vcell.ndsu.edu/.

5. Background: Authentic Assessment

All WWWIC projects are based on the idea of authentic assessment (Bell, Bareiss, and Beckwith, 1994), within authentic contexts, where the goal is to determine the benefit students derive from their "learn by doing" experience using our virtual environments (Duffy and Jonassen, 1992; Edelson et al., 1996; Lave and Wenger, 1991; Reid, 1994). Our scenario-based assessment protocol is a qualitative one that seeks to measure how student thinking has improved.

When learners join the synthetic environment they are assigned goals, selected by content matter experts to be appropriate to the learner's experience. Goals are assigned point values, and learners accumulate objectively measured scores as they achieve their goals. The goals are taken from a principled set, where easier goals are followed by more advanced ones. Similarly, certain goals in a set are required while others are optional. In this way, we can insure that highly important concepts are thoroughly covered while allowing the maximum flexibility to the learner. Subject matter experts identify teaching objectives in more-or-less traditional ways, while learner outcomes are assessed in terms of the performance of specific and authentic tasks. This is the particular strength of learn-by-doing immersive environments, that a learner's success in achieving their goals provides an automatic measure of their progress.

In addition to these outcome-based measures, all students are asked to answer open-ended scenario-based questions before and after the experiment. These scenario questions are word problems that present the student with a situation that a scientist might confront (the complete set of Planet Oit scenarios can be viewed at http:// www.cs.ndsu.nodak.edu/ ~slator/ html/ PLANET/ assess-scen.html). Students respond to the question with a narrative answer, which is evaluated according to an established protocol.

6. Next Steps

The Blackwood Project (http:// lions. cs. ndsu. nodak. edu/ ~blackwood/), is the first attempt at the "next generation" of role-based virtual environments for education where the pedagogical simulation will support cross disciplinary content and a choice of varied and specific roles to promote player interaction and potential collaborations. Blackwood combines microeconomics with Western history and implements a virtual reconstruction of a 19th Century Western town populated with intelligent software agents to simulate an economic environment representative of the times.

The Virtual Polynesia project will develop an educational software system for presenting, in an intellectually engaging and stimulating way, a humanities perspective on human culture. The system will be used to teach and present research methods and critical analysis for the often-overlooked humanities tradition within cultural anthropology. Most funding for anthropology comes from social science sources, hence the humanities aspect of anthropology has become neglected. Yet it is that very dimension of anthropology that is crucial today for understanding globalization and the interaction of diverse groups.

Our goal is to give students a tool for understanding other peoples by teaching ethnographic techniques and methodology through an interactive, on-line, game environment for stimulating learning about human culture and social behavior. Specifically, the application will teach ethnography techniques to focus on (1) global awareness, (2) cultural diversity and interaction, (3) decision-making in a cultural environment, and (4) anthropology in general.

7. Conclusion

Role-based instruction allows the student to participate in the practices of a discipline within a naturalistic context. These environments allow the students access to techniques in the context in which they actually work; such techniques, as noted by Lave and Wegner (1991), embody the practices of the culture. Our goal is to extend this pedagogical approach from Science to the domain of socio-cultural systems.

8. References

Bell, B. R. Bareiss, and R. Beckwith (1994). The Role of Anchored Instruction in the Design of a Hypermedia Museum Exhibit. Journal of the Learning Sciences, 2(2).

Brown, J. S., Collins, A., & Duguid, P. (1989). Situated Cognition and the Culture of Learning, Ed. Researcher, 18(1), 32-42.

Curtis, P. (1997). Not Just a Game: How LambdaMOO Came to Exist and What It Did to Get Back at Me. in Cynthia Haynes and Jan Rune Holmevik, Editors: High Wired: On the Design, Use, and Theory of Educational MOOs. Ann Arbor: U. of Michigan Press.

Duffy, T.M. & Jonassen, D.H. (1992). Constructivism: new implications for instructional technology. In Duffy and Jonassen (eds.), *Constructivism and the Technology of Instruction*. Hillsdale: Lawrence Erlbaum.

Edelson, D., Pea, R., & Gomez, L. (1996). Constructivism in the Collaboratory. In B. G. Wilson (Ed.), *Constructivist learning environments: Case studies in instructional design*, (pp. 151-164). Englewood Cliffs, NJ: Educational Technology Publications.

Lave, Jean, and Wenger, Etienne. (1991). Situated Learning. Cambridge: Cambridge University Press..

McClean, P.E., Schwert, D.P., Juell, P., Saini-Eidukat, B., Slator, B.M., &White, A. (1999). Cooperative Development of Visually-Oriented, Problem-Solving Science Courseware. Proc. Intl. Conf. on Mathematics/Science Ed. & Technology (M/SET-99), San Antonio, TX.

McGee, S., Howard, B., & Hong, N. (1998, April). Cognitive Apprenticeship, Activity Structures, and Scientific Inquiry. In S. McGee (Chair), Changing the game: Activity structures for reforming education. Symposium conducted at the annual meeting of the American Educational Research Association, San Diego, CA.

Reid, T. A. (1994). Perspectives on computers in education: the promise, the pain, the prospect. Active Learning, 1(1). Oxford, UK: CTI Support Service.

Saini-Eidukat, B., Schwert, D. & Slator, B.M. (1999). Designing, Building, and Assessing a Virtual World for Science Education. Proc. 14th Intl. Conf. on Computers and Their Applications (CATA-99), April 7-9, Cancun.

Schwert, D.P., Slator, B.M., & Saini-Eidukat, B. (1999). A virtual world for earth science education in secondary and post-secondary environments: The Geology Explorer. Intl. Conf. on Mathematics/Science Ed. & Technology (M/SET-99), March 1-4, San Antonio, TX.

Slator, B.M. (1999). Intelligent Tutors in Virtual Worlds. Proc. 8th Intl. Conf. on Intelligent Systems (ICIS-99). June 24-26. Denver, CO, pp. 124-127.

Slator, B. M. & Hill, C. (1999). Mixing Media For Distance Learning: Using IVN And Moo in Comp372. World Conf. on Educational Media, Hypermedia and Telecommunications (ED-MEDIA 99), June 19-24, Seattle, WA.

Slator, B.M., Schwert, D.P., & Saini-Eidukat, B. (1999). Phased Development of a Multi-Modal Virtual Educational World. Proc. Intl. Conf. on Computers and Advanced Technology in Ed. (CATE'99), Cherry Hill, NJ, May 6-8.

Slator, B.M., Juell, P., McClean, P., Saini-Eidukat, B., Schwert, D.P., White, A., & Hill, C. (1999). Virtual Worlds for Education. Journal of Network and Computer Applications, 22 (4), 161-174.

Slator, B. M. & Chaput, H. "C". (1996). Learning by Learning Roles: a virtual role-playing environment for tutoring. Proc. Third Intl. Conf. on Intelligent Tutoring Systems (ITS'96). In C. Frasson, G. Gauthier, & A. Lesgold (Eds.) Lecture Notes in Computer Science. pp. 668-676. Montreal: Springer-Verlag,

White, A. R., McClean, P.E., & Slator, B.M. (1999). The Virtual Cell: An Interactive, Virtual Environment for Cell Biology. World Conf. on Educational Media, Hypermedia and Telecommunications (ED-MEDIA 99), June 19-24, Seattle, WA.

Acknowledgements

WWWIC research is supported by the National Science Foundation under grants DUE-9752548, EAR-9809761, DUE-9981094, and EAI-0086142, and US. Department of Education FIPSE #P116B000734. We also acknowledge the large team of dedicated undergraduate and graduate students who have helped to make these projects possible.

An ontology for modeling ill-structured domains in intelligent educational systems

Christos M. Papaterpos, Nektarios P. Georgantis, Theodore S. Papatheodorou
High Perfomance Computing Lab, University of Patras, Greece
{cmp,npg,tsp}@hpclab.ceid.upatras.gr

Abstract

We introduce a generic and flexible ontology for modeling ill-structured knowledge domains within intelligent educational systems. The ontology is based on the Theory of Cognitive Flexibility. We have prototyped the ontology on a deductive object oriented database system. We provide simple and concrete examples that demonstrate benefits of the ontology, such as extensibility, flexibility and support for implementing authoring environments.

1. Introduction and background

The objective of this work is the definition and prototyping of a generic and flexible ontology for modeling ill-structured knowledge domains (ISDs) in intelligent educational systems (IES) for advanced learning. Besides introduction and implementation of the ontology, we substantiate and demonstrate its utility through concrete examples pertaining to important merits, namely extensibility, support for individualized sequencing and support for an authoring environment. Modeling constructs are derived from the Theory of Cognitive Flexibility (CFT). As far as we know, this is the first generic implementation of CFT principles. Although usable in many fields, our application class of choice is Web Based Adaptive Educational Systems (AES) [1]; hypermedia-based learning environments on the web, adapting instruction to the learner's skills, needs and goals.

Knowledge domains are considered ill-structured [7] if they demonstrate interaction of complex conceptual structures (cases and abstract concepts) in an across-case irregular fashion. Benefits of using ontologies are presented in [13]. Use of domain and task ontologies within IES is demonstrated in [10][9]. However, emphasis is placed on task ontologies and little provision is made for ISDs. Application in ISDs has been demonstrated for systems founded on "Goal Based Scenarios". In principle, the proposed ontology is complementary to such approaches, with regard to domain modeling (cf. Section 2). Sophisticated domain models in AES do exploit explicit ontologies, as in KBS-Hyperbook [4], to support mechanisms such as indexing of learning goals with the concepts documented in the educational material. Nevertheless, these AES focus on teaching abstract concepts (atomic or composite), with cases (examples or projects) having secondary roles.

The Theory of Cognitive Flexibility [7] provides a good basis for modeling ISDs. It targets advanced learning for ISDs, evangelizes case-based learning and dismisses the incremental complexity approach followed by most AES to tackle flaws in teaching that hinder increased knowledge transfer, such as overgeneralization and oversimplification. Indications of CFT efficiency are given in [6], where it is remarked that "students who received the CFTH performed significantly better than students who worked in a hierarchical hypertext or textbook learning environments, especially in conceptually complex tasks like comparison, analysis and synthesis". CFT mandates are transformed into a set of hypertext design decisions in [7]. From these, we define a simple ontology to provide a generic implementation of CFT, unlike existing implementations ([6], [7]) which demonstrate solutions crafted against specific problems. The fundamental units of instruction are **mini-cases**, smaller parts of larger cases that retain complexity to a smaller extent. Abstract concepts are termed **themes and** the way they are applied in mini-cases is typed by **sub-themes**, a mechanism that allows the definition of themes as complex conceptual structures. Theme application on a specific mini-case is substantiated in a **theme-instance**. A mini-case's **thematic commentary** integrates all theme-instances applicable in the mini-case, as well as relationships among these with other theme instances in the domain. The basic operation in the domain is the **theme-based exploration**, which for a particular theme or theme combination returns all associated mini-cases. All material returned should be **sequenced to produce cognitive structures with woven interconnectedness**. According to the "intermediates principle", "mini-cases should be sequenced for presentation in such a manner that two representational extremes will be avoided". This should help learners "avoid overgeneralization" and "avoid the perception that each case is unique" [7]. To this end, we define similarity metrics among mini-cases and explorations with regard to themes or to theme combinations.

2. The ontology and its utility

Sub-themes are defined as *subclasses* of themes and their *instances* describe how they are applied on mini-cases. Each mini-case is interpreted by a thematic commentary that aggregates all instances of themes applied on the mini-case and integrates relationships among theme instances that may be defined by the model author or inferred. These are modeled through an instance_relationship class. The frame below contains these classes in F-logic notation [2]. Square brackets enclose properties, and =>> denotes multiple values.

```
theme.
mini_case[isDefinedBy=>URI;inpterpretedBy=>
    thematic_commentary].
thematic_commentary[interpretsThemes=>>theme,
    relates_instances=>>instance_relationship].
instance_relationship[instance1*=>theme;
instance2*=>theme].
```

Exploration operations are modeled as F-logic predicates. We currently support (possibly negated) themes, theme conjunctions and disjunctions (as series of evaluations of conjunctions). To **sequence mini-cases**, according to the "intermediates principle", we introduce a metric termed l-equivalence that indicates how equivalent two mini-cases are with regard to a theme or a theme combination. During presentation, the system can sequence mini-cases with intermediate l-equivalence values close to each other. Calculation of this metric is based on theme hierarchy and similarities identified among theme instances through the *instance_relationship* class. Distance in hierarchies has been previously used in case-based reasoning systems for the definition of similarity metrics. It is reasonable to assume that two mini-cases are more similar with regard to two themes if the primary classes of the corresponding theme instances are close in the theme hierarchy. This distance also moderates the effect of declared or inferred instance relationships on the calculation of l-equivalence. Combining these two factors – "instance_similarity" and distance in the hierarchy – we can calculate l-equivalence. Averaging these values on all instances of a combination on two mini-cases we estimate similarity of mini-cases with regard to the combination. This is similarly extended to compare explorations.

The classes and operations introduced here have been implemented in a prototype, built over a deductive object oriented database system (DOODS). The implementation platform comprises an F-Logic compiler (FLORA [3]) on top of XSB [8]. F-logic's combination of object-oriented definitions and inferencing offers an elegant way to add **extensibility and flexibility**. A simple example of this is the inclusion in the ontology of standard axioms, such a transitivity and declaration of a sub-class of instance_relationship as transitive. A more complicated example is the introduction of weights in a subclass of instance_relationship to denote importance of theme instances on the particular mini-cases and define a variant algorithm to exploit this information. The original algorithm can still use the transitive and "quantified" subclasses of *instance_relationship*. Based on the l-equivalence metric, we can built **sequencing algorithms to promote adaptive learning** by taking into account evidence of knowledge-transfer learner capabilities for specific theme combinations, maintained in a User Model. This can be accomplished by integrating such information with our metric through mechanisms formalized in [11], namely metric Product and Prioritization. A final merit of our proposal is its **support for authoring** the domain model. Ontology elements such as explorations and similarity can provide immediate feedback to model and teaching material authors during authoring. Following [9], we have formalized these operations into a task ontology. This is used in the implementation of a working prototype for a simple authoring system [12] that extends Protégé [5] interfaced with the XSB prototype to support: inspecting mini-cases in an embedded browser, simulation of explorations and calculation of similarity metrics. Currently, we are validating similarity metrics in History, with help from domain experts. Models and teaching material produced shall be tested with students to evaluate educational efficiency of our approach.

3. References

[1] Brusilovsky P. "Adaptive Educational Systems on the World-Wide-Web: A Review of Available Technologies". ITS'98,
[2] Kifer M. et al. "Logical Foundations of Object-Oriented and Frame-Based Languages", Journal of the ACM 1995, vol. 42
[3] Guizhen Yang, Michael Kifer, "FLORA: Implementing an Efficient DOOD System Using a Tabling Logic Engine", DOOD'2000: 6th Int. Conf. on Rules and Objects in Databases.
[4] N. Henze et al, "Modeling Constructivist Teaching Functionality and Structure in the KBS Hyperbook System" AIED'99.
[5] http://www.smi.stanford.edu/projects/protege
[6] Sojoong J. "Effects of Cognitive Flexibility Theory-based instruction on Korean high school history teaching". "Distance Education an International Journal", v21-n1, 2000
[7] Spiro, R. J., & Jehng, J. "Cognitive flexibility theory: Theory and technology for the nonlinear and multidimensional traversal of complex subject matter". "Cognition, Education, and Multimedia", Lawrence Erlbaum Associates (1990).
[8] http://xsb.sourceforge.net/
[9] J Lai, et al, "An Ontology-Aware Authoring Tool", AIED99
[10] Bourdaugh J., Mizoguchi R. "Ontology Engineering of Instruction: a Perspective" AIED99.
[11] H. Osborne et alt., "Similarity Metrics: A Formal Unification of Cardinal and Non-Cardinal Similarity Measures", ICCBR99.
[12] Papaterpos C. "A CFT inspired authoring system", Technical Report HPCLAB-TR-010203.
[13] R. Mizoguchi. "A Step towards ontology engineering", 12th National Conference on AI of JSAI, June, 1998

An Affective Model of Interplay Between Emotions and Learning: Reengineering Educational Pedagogy—Building a Learning Companion

Barry Kort, Rob Reilly, Rosalind W. Picard
M.I.T. Media, Laboratory
{bkort, reilly, picard}@media.mit.edu

Abstract

There is an interplay between emotions and learning, but this interaction is far more complex than previous theories have articulated. This article proffers a novel model by which to: 1). regard the interplay of emotions upon learning for, 2). the larger practical aim of crafting computer-based models that will recognize a learner's affective state and respond appropriately to it so that learning will proceed at an optimal pace.

1. Looking around then moving forward

> The extent to which emotional upsets can interfere with mental life is no news to teachers. Students who are anxious, angry, or depressed don't learn; people who are caught in these states do not take in information efficiently or deal with it well.
> - Daniel Goleman, *Emotional Intelligence*

Educators have emphasized conveying information and facts; rarely have they modeled the learning process. When teachers present material to the class, it is usually in a polished form that omits the natural steps of making mistakes (e.g., feeling confused), recovering from them (e.g., overcoming frustration), deconstructing what went wrong (e.g., not becoming dispirited), and starting over again (with hope and enthusiasm). Those who work in science, math, engineering, and technology (SMET) as professions know that learning naturally involves failure and a host of associated affective responses. Yet, educators of SMET learners have rarely illuminated these natural concomitants of the learning experience. The result is that when students see that they are not getting the facts right (on quizzes, exams, etc.), then they tend to believe that they are either 'not good at this,' 'can't do it,' or that they are simply 'stupid' when it comes to these subjects. What we fail to teach them is that all these feelings associated with various levels of failure are normal parts of learning, and that they can actually be helpful signals for *how* to learn better.

Expert teachers are very adept at recognizing and addressing the emotional state of learners and, based upon their observation they take some action that positively impacts learning. But what do these expert teachers 'see' and how do they decide upon a course of action? How do students who have strayed from learning return to a productive path, such as the one that Csikszentmihalyi [1990] refers to as his "zone of flow"?

Skilled humans can assess emotional signals with varying degrees of accuracy, and researchers are beginning to make progress giving computers similar abilities at recognizing affective expressions. We believe that accurately identifying a learner's cognitive-emotional state is a critical mentoring skill. Although computers perform as well as or better than people in selected domains, they do not yet rise to human levels of mentoring. We envision that computers will soon become capable of recognizing human behaviors indicative of the user's affective state.

We have begun research that will lead to our building of a computerized Learning Companion that will track the affective state of a learner through their learning journey. It will recognize cognitive-emotive state (affective state), and respond appropriately. We believe that the first task is to evolve new pedagogical models, which assess whether or not learning is proceeding at a healthy rate and intervene appropriately; then these pedagogical models will be integrated into a computerized environment. Two issues face us, one is to research new educational pedagogy, and the other is a matter of building computerized mechanisms that will accurately and immediately recognize a learner's state by some ubiquitous method and activate an appropriate response.

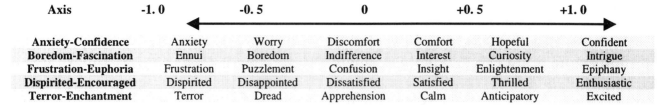

Figure 1 – Emotion sets possibly relevant to learning

2. Two sets of research results

This research project will have two sets of results. This paper offers the first set of results, which consists of our model and a research method to investigate the issue. A future paper will contain the results of the empirical research—the second set of results.

This paper will address two aspects of our current research. Section 3 will outline our theoretical frameworks and define our model (Figures 1 and 2). Section 4 will describe our empirical research methods.

3. Guiding theoretical frameworks: An ideal model of learning process

Before describing the model's dynamics, we should say something about the space of emotions it names. Previous emotion theories have proposed that there are from two to twenty basic or prototype emotions (see for example, Plutchik, 1980; Leidelmeijer, 1991). The four most common emotions appearing on the many theorists' lists are fear, anger, sadness, and joy. Plutchik [1980] distinguished among eight basic emotions: fear, anger, sorrow, joy, disgust, acceptance, anticipation, and surprise. Ekman [1992] has focused on a set of from six to eight basic emotions that have associated facial expressions. However, none of the existing frameworks address emotions commonly seen in SMET learning experiences, some of which we have noted in Figure 1. Whether all of these are important, and whether the axes shown in Figure 1 are the "right" ones remains to be evaluated, and it will no doubt take many investigations before a "basic emotion set for learning" can be established. Such a set may be culturally different and will likely vary with developmental age as well. For example, it has been argued that infants come into this world only expressing interest, distress, and pleasure [Lewis, 1993] and that these three states provide sufficiently rich initial cues to the caregiver that she or he can scaffold the learning experience appropriately in response. We believe that skilled observant human tutors and mentors (teachers) react to assist students based on a few 'least common denominators' of affect as opposed to a large number of complex factors; thus, we expect that the space of emotions presented here might be simplified and refined further as we tease out which states are most important for shaping the companion's responses.

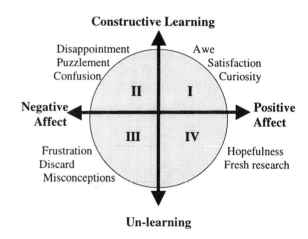

Figure 2 – Proposed model relating phases of learning to emotions in Figure 1

Nonetheless, we know that the labels we attach to human emotions are complex and can contain mixtures of the words here, as well as many words not shown here. The challenge, at least initially, is to see how our model and its hypothesis can do initially with a very small space of possibilities, since the smaller the set, the more likely we are to have greater classification success by the computer.

Figures 2 attempts to interweave the emotion axes shown in Figure 1 with the cognitive dynamics of the learning process. The horizontal axis is an Emotion Axis. It could be one of the specific axes from Figure 1, or it could symbolize the n-vector of all relevant emotion axes (thus allowing multi-dimensional combinations of emotions). The positive valence (more pleasurable) emotions are on the right; the negative valence (more unpleasant) emotions are on the left. The vertical axis is what we call the Learning Axis, and symbolizes the construction of knowledge upward, and the discarding of misconceptions downward. (Note: we do not view learning as being simply a process of constructing/deconstructing information or simply a process of adding/subtracting information; this terminology is merely a projection of one aspect of how people can think about learning. Other aspects could be similarly included along the Learning Axis.)

The student ideally begins in Quadrant I or II: they might be curious and fascinated about a new topic of interest (Quadrant I) or they might be puzzled and motivated to reduce confusion (Quadrant II). In either case, they are in the top half of the space, if their focus is on constructing or testing knowledge. Movement happens in this space as learning proceeds. For example, when solving a puzzle in *The Incredible Machine*, a student gets an idea how to implement a solution and then builds its simulation. When she runs the simulation and it fails, she sees that her idea has some part that doesn't work – that needs to be deconstructed. At this point it is not uncommon for the student to move down into the lower half of the diagram (Quadrant III) where emotions may be negative and the cognitive focus changes to eliminating some misconception. As she consolidates her knowledge—what works and what does not—with awareness of a sense of making progress, she may move to Quadrant IV. Getting a fresh idea propels the student back into the upper half of the space, most likely Quadrant I. Thus, a typical learning experience involves a range of emotions, moving the student around the space as they learn. Typically, movement would be in a counter-clockwise direction

If one visualizes a version of Figure 2 for each axis in Figure 1, then at any given instant, the student might be in multiple Quadrants with respect to different axes. They might be in Quadrant II with respect to feeling frustrated; and simultaneously in Quadrant I with respect to interest level. It is important to recognize that a range of emotions occurs naturally in a real learning process, and it is not simply the case that the positive emotions are the good ones. We do not foresee trying to keep the student in Quadrant I, but rather to help them see that the cyclic nature is natural in SMET learning, and that when they land in the negative half, it is only part of the cycle. Our aim is to help them to keep orbiting the loop, teaching them how to propel themselves especially after a setback.

A third axis (not shown), can be visualized as extending out of the plane of the page—the Knowledge Axis. If one visualizes the above dynamics of moving from Quadrant I to II to III to IV as an orbit, then when this third dimension is added, one obtains an excelsior spiral when evolving/developing knowledge. In this diagram (which is know as a phase plane plot in systems theory), time is parametric as the orbit is traversed in a counterclockwise direction. In Quadrant I, anticipation and expectation are high, as the learner builds ideas and concepts and tries them out. Emotional mood decays over time, either from boredom or from disappointment. In Quadrant II, the rate of construction of new concepts diminishes, and negative emotions emerge as progress flags. In Quadrant III, the learner discards misconceptions and ideas that didn't pan out, as the negative affect runs its course. In Quadrant IV, the learner recovers hopefulness and positive attitude as the knowledge set is now cleared of unworkable and unproductive concepts, and the cycle begins anew. In building a complete and correct mental model associated with a learning opportunity, the learner may experience multiple cycles around the phase plane until completion of the learning exercise. Each orbit represents the time evolution of the learning cycle. (Note: the orbit doesn't close on itself, but gradually moves up the knowledge axis.)

4. Empirical research to validate the model

The results of this part of the research will provide data that will validate our model and control the action of the automated Learning Companion.

A number of 6-11 year old subjects will be video taped while individually playing the *Incredible Machine* or *Gizmos and Gadgets*. There are two video cameras and a posture sensing device gathering data. One camera is a version of IBM's *Blue Eyes Camera* eye-tracking device (see URL http://www.almaden.ibm.com/cs/blueeyes). The other camera, which is a conventional camcorder, provides a split-screen view of the subject's upper body and the other part of the split-screen will show the computer display as the subject sees it. The posture sensing device uses an array of force sensitive resistors similar to the *SmartChair* employed by Tan *et al* (1997).

Blue Eyes and *SmartChair* data will be gathered and synchronized with the data from the split-screen video tapes and will be coded based upon three observable factors: 1) surface level behavior (e.g., facial expression, body language), 2) inferred emotional state, and 3) task/game-state.

Part of our research currently involves developing and testing appropriate interventions strategies when the learner is found to be stuck. In general the Learning Companion might intervene when a learner is not focused on a relevant part of the computer screen, or is focused completely outside the task area for a certain period of time, or their eye gaze is sufficiently quick/jerky for a given period of time.

In Quadrant I a learner is happily engaged in exploratory learning and/or discovery learning, there needs to be little or no intervention (short of ensuring that all the resources that the learner will need are present and accessible as they are needed).

In Quadrant II, where a learner is beginning to encounter difficulties arising from a misconception or an incomplete understanding, the intervention must serve the purpose of helping the learner recognize and identify the gaps and errors in his or her mental model. The method of intervention, ranging from subtle hints in the form of Socratic questioning to direct diagnosis and give-away hints, depends on the learning orientation of the individual. At the same time, the Learning Companion must guard against the possibility that the learner might become overly crestfallen in the process.

In Quadrant III, where the learner has recognized and acknowledged that they had been working from an erroneous or incomplete model, the intervention focuses on providing the emotional support required to survive and emerge from the disappointment, chagrin, anger, anguish, self-doubt, or whatever other dispiritedness may arise during the retreat and recovery phase of the learning cycle. Again, some learners require more emotional support and spiritual coaching than others. Quadrant III intervention is arguably the most challenging and uncertain. The point of Quadrant III intervention is to successfully grieve the loss and get on with life.

In Quadrant IV, where the learner has 'gone back to the drawing board' to construct an improved understanding of the subject at hand, requires the kind of scaffolding we find in current theories for the support of model-based learning [See e.g., Soloway, 1999]. Again, Socratic inquiry methods, hints, and direct teaching may all be appropriate, depending again on the learning orientation of the student [See e.g., Jones and Martinez, 2001].

Finally, when the student makes the breakthrough back to Quadrant I with a fresh insight and a new idea, an acknowledgement ritual may be in order to celebrate progress or success. Here we want to reinforce and celebrate the feelings of pleasure and delight that accompany successful learning, so as to fuel and recharge the spirit for the next travail around the loop.

5. Assessing and Applying Our Results

The timing and the nature of the intervention strategies will depend upon a valid assessment of the learner's cognitive-emotive state and the state of the learner's progress in the underlying learning task.

We are presently testing and revising appropriate intervention strategies. We also expect to make use of dovetailing theories of intervention that consider individual idiosyncratic styles of learning; in particular we are impressed with the research of Martinez and Bunderson [2000] and Jones and Martinez [2001] relating to the theory of learning orientation, which carries on the work begun by Chronbach and Snow [1977].

Acknowledgements: We are indebted to Ashish Kapoor and Selene Mota, respectively, for their efforts in constructing and adapting an IBM *Blue Eyes Camera* and a *Smart Chair* for our research. We are also indebted to the National Science Foundation for providing the funds to conduct this project (NSF grant #0088776).

6. References

[1] Csikszentmihalyi, M. (1990). Flow: The Psychology of Optimal Experience, Harper-Row: NY.

[2] Cronbach, L. & Snow, R. (1977). Aptitudes and Instructional Methods: A Handbook for Research on Interactions. New York: Irvington. Note: for a summary of Chronbach and Snow see URL http://tip.psychology.org/chronbach.html

[3] Ekman, Paul. (1992). Are there basic emotions?, Psychological Review, 99(3): 550-553.

[4] Goleman, D., (1995). Emotional Intelligence. Bantam Books: New York.

[5] Jones, E. R. and Martinez, M. (2001) Learning Orientations in University Web-Based Courses - a paper submitted for publication in the Proceedings of WebNet 2001, Oct 23-27, Orlando, Florida.

[6] Leidelmeijer, K. (1991). Emotions: An Experimental Approach. Tilburg University Press.

[7] Lewis M., (1993). Ch. 16: The emergence of human emotions. In M. Lewis and J. Haviland, (Eds.), Handbook of Emotions, pages 223-235, New York, NY. Guilford Press.

[8] Martinez, Margaret and C. Victor Bunderson. (2000). Foundations for Personalized Web Learning Environments, ALN Magazine, vol. 4 no. 2.

[9] Plutchik, R. 'A general psychoevolutionary theory of emotion,' in Emotion Theory, Research, and Experience (R. Plutchik and H. Kellerman, eds.), vol. 1, Theories of Emotion, Academic Press, 1980.

[10] Soloway, Eliot. (1999). Scaffolded Technology Tools to Promote Teaching and Learning in Science, Available at: http://hi-ce.eecs.umich.edu/hiceinformation/papers/index.html

[11] Tan H.Z., Ifung Lu and Pentland A. (1997). The Chair as a Novel Haptic User Interface. In Proceedings of the Workshop on Perceptual User Interfaces, Banff, Alberta, Canada, Oct. 1997.

Theme 3: Instructional Design (Including 3D)

Instructional Design in a Technological World: Fitting Learning Activities Into the Larger Picture

Robin Soine
Wisconsin Technical College System Foundation, Inc.
soiner@wids.org

Abstract

With the prevalence and pressure to learn and use new technology, teachers may rush through or even avoid instructional design. This results in learners not clearly knowing what they're going to do, how to go about doing it, or when they'll know they've finished it. In designing and planning instruction, teachers must avoid the temptation to become so engaged in using technology that they neglect to design effective learning activities. This temptation may also cause teachers to neglect connecting the learning activities to a larger outcome. These outcomes, along with clear performance expectations--stated in advance of instruction--become the backbone of learning in any environment.

1. Introduction

Designing quality instruction today is taking on fresh importance as technology creeps, or sometimes sweeps, its way into classrooms around the country. With most of society (teachers, school boards, government, parents, and the workplace) [9] showing such enthusiasm for using technology in the classroom, it is easy to get caught up in the glamour of designing lessons or learning activities using technology, *just because they involve technology*.

Teachers may rush through or even avoid instructional design, because of the pressure to learn and use new technology. This results in learners *not* clearly knowing what they're going to do, how to go about doing it, or when they'll know they've finished it. In designing and planning instruction, teachers must avoid the temptation to become so engaged in using technology that they neglect to design effective learning activities.

The temptation of technology may also cause teachers to neglect connecting the learning activities to a larger outcome. These outcomes, along with clear performance expectations--stated in advance of instruction--become the backbone of learning in any environment. Well-designed learning activities have meaning for learners and connect them to a larger learning outcome. And finally, well-designed learning activities guide learners to achievement of an outcome, which sets the stage for assessment. Strong instructional design and planning, therefore, becomes the critical piece to a successful learning experience, *especially* in a distance learning environment.

2. Model of Instructional Design

Instructional design, in general, is a "procedure which begins with a determination of instructional goals based upon the requirements of the job and the needs of the learner and the organization" [10]. These goals are then used to construct "objectives and criterion-referenced assessment procedures" followed by designing a delivery plan, trying out and revising the instruction, and evaluating the effectiveness of the instruction [10].

Borrowing from this idea and the instructional systems design (ISD) model, the Wisconsin Technical College System set out to develop a statewide model of instructional design—a common language for all 16 college districts. With input from instructional designers, practitioners, researchers, business representatives, and K-12 educators, a model informed by theory and best practice developed. From 1992-1994, this statewide committee spearheaded the design of the model, and still meets today to continuously improve and develop the model. Application of the models of several prominent researchers/practitioners such Bloom, McCarthy, Gardner, and Clark, were initially used in the development of this model [8] and the instructional design software tool that was eventually released.

This model, called the WIDS model (Wisconsin Instructional Design System) is currently the only statewide model of instructional design in use—developed by theory and practitioner input. The model was released in 1994, and has been used at the college and statewide level in Wisconsin for over 7 years.

3. Designing Effective Learning Activities in the WIDS Model

Learning activities are strategies to help learners master an outcome or group of related outcomes. They direct the learner along a path to achievement of a higher level skill, one that has more than likely been established collaboratively, either through a curriculum committee, the local school district, or in some cases, the state.

Given the increasing use of "computer-mediated instructional materials" [5] one must consider instructional design before using technology to any extent. The need for clear activities is even greater in a location-independent environment, as the student is typically not face-to-face with the instructor or other class members. A clear map is needed to lead the learner to the desired destination: achievement of a major skill or larger outcome. For example, with poorly designed learning activities, students end up finding huge amounts of information electronically, but then aren't directed to evaluate its usefulness or value.

Four major principles govern the design of effective learning activities, according to the WIDS model [8].

Ensure Validity and Value

Learning activities should address target outcomes and broad transferable skills like communication, critical thinking, and problem solving. For example, if oral communication is a broad transferable skill to be covered in a course, it can be embedded into the learning activities by building oral presentations into the methods.

Support the Thinking Process

By following brain-based theories on learning when designing activities, teachers support the thinking process. Choosing learner-centered activities, breaking instruction into "chunks", and providing several opportunities for elaborative rehearsal and frequent practice all support the way in which we know how learners learn. These points reflect the work of Dr. Ruth Clark, educational psychologist [3].

Accommodate Varied Learning Styles

Not all learners learn the same way. By varying the methods (strategies which cause learning) and media (format that delivers methods) in learning activities, more learner styles can be addressed. Well-written activities progress through all stages of the learning cycle created by Neill: motivation, comprehension, practice, application. Dave Meier, founder of the Center for Accelerated Learning uses a similar cycle of preparation, presentation, practice, and performance [8].

Provide Learning Plans

Learning plans help learners navigate through materials. These plans list the larger outcomes, the benchmark skills, the performance standards, and any related core ability. Assessment activities are noted, as well as the list of learning activities. Instructional materials can be attached or linked to a learning plan, but the plan itself is the navigational tool leading to success for learners.

Well thought-out learning activities extend the learner from listener and gatherer to analyst, researcher, collaborator, and synthesizer. They become actively involved in learning and are held accountable for achieving the larger outcome. The teacher becomes an inspiration, mentor, guide, and facilitator. Neill says "careful instructional design, based on what we know about how people learn, provides a map that guides (teachers) and learners toward successful completion of a learning experience" [8].

4. Connecting Learning Activities to Larger Outcomes

Once learning activities are soundly built, how do they relate to the larger outcomes or learning results? It's easy to skip this design question, as students may be motivated simply by the computer itself. This is dangerous, as the computer medium alone does not cause learning. Adela Najarro, a third grade bilingual teacher in San Francisco, is a "believer" that students "do more work for you and do better work with a computer." She says, "Just because it's on a monitor, kids pay more attention. There's this magic to the screen" [9]. This magic could be constructed into a valuable motivational tool, but unless there is up-front organization, clear communication about performance expectations, outcome driven assessment, and imaginative learning strategies, the magic is wasted.

Learning activities must be connected to a larger, higher level skill, observable and measurable, and they must be built into the class design. They follow the assessment design and precede the development of instructional materials.

Hardy agrees that outcomes, or "What you want the students to learn" must come before technology inclusion and the planning for delivery of the course. Technology decisions come *after* decisions about *what* teachers want learners to learn, and why students enroll in a particular course in the first place [1].

4.1. Connection Comes from Broadening the Teacher's Role

In connecting teaching to learning, the teacher becomes learning facilitator, life long learning model, guide, designer, and developer of activities that enable learners to learn. Neill and Mashburn say broadening the teacher's role is crucial to the learning process because it "requires that you know about your learners, be an expert in your discipline and skill areas, apply what you know about how people learn, and apply what you know about your discipline or skill areas to the design and implementation of instruction" [8].

In designing an online course, these concepts still hold true. In "Online Teaching: Moving From Risk to

Challenge" Cini says the "new" role of the online teacher is to:
- design experiences and activities to facilitate student learning
- encourage students to become active learners and constructors of knowledge
- guide the process of learning while encouraging student initiative, and
- teach, as well as learn from, students [2].

4.2. Connecting Experience and Learning

New brain research adds a new dimension to our knowledge about the learning connection:

This research provides growing evidence that learning is about making connections—whether the connections are established by firing synapses in the brain, the "ah ha" experience of seeing the connection between two formerly isolated concepts, or the satisfaction of seeing the connection between an abstraction and a 'hands-on' concrete application [4].

Connection has strong support in research, Cross says, and closely examining these connections is not likely to be just another passing fad. Besides presenting connections in neurological, cognitive, and social terms, Cross goes on to say that connecting between experience and learning is important because it improves learning and performance. Teachers must help learners make the connections that constitute learning.

How specifically are these connections made? It's easy to send a student off on Internet research hunts or collaborative chats. Taking this experience and connecting it with a major skill or outcome specified in the course is done by providing clear navigational tools such as learning plans or course schedules broken down into "chunks" with linked or listed readings and well-designed activities.

5. Matching Outcomes and Assessment

As mentioned earlier, up-front organization and clear communication about performance expectations and outcome driven assessment are even more important when the learner may not be face-to-face with the teacher, or the learner is working independently within a traditional classroom. Learners want to know up-front what is expected of them and what they must do to succeed in the class. They should never have to ask, "Is this right? Is this good enough? How am I doing?" Designing assessments that state expectations in advance and reliably measure learner performance can become burdensome. The initial time investment of the teacher can greatly impede completion. Designing assessments come right before the design of the learning activities in Neill's Instructional Design Flowchart [8].

Typically, assessments are designed at the course level, rather than the school or national level, and they take on two levels of purpose. The first involves reporting learning results (accountability assessment) at the *end* of instruction. Examples of these include written/oral tests, performance tests, portfolios, simulations, and projects. These measure larger outcomes and provide credentials for documenting a learner's competence.

The second purpose of assessment involves evaluating the learner's progress toward achieving the instructional outcomes, or in other words, giving feedback to learners (continual improvement assessment) *when they have an option to do something about it*. This type of assessment also gives teachers feedback on teaching strategies [8]. Examples include self-checks, peer assessments, guided practice checks, feedback on drafts, and practice activities. These occur during the course of instruction, and are often used to indicate if learners can perform the lower-level (benchmark) skills that make up an outcome, or to check on how learners are progressing with achievement of the outcome [8].

Assessments can be thought of as indicators or answers to the question, "What will indicate learner success?" Evaluation *must* be linked to the initial course outcomes. They must relate. They must match. Carolyn Jarmon, in "Testing and Assessment at a Distance" [1] outlines basic evaluation questions and categories, and states that an "integral part of the distance learning course development process is the incorporation of measures of success" [1]. In sound instructional design, performance assessment is closely tied to the performance standards of each major skill or ability needed to perform a task effectively and efficiently. Performance assessment is the process of determining that learners can perform these skills according to specified performance criteria and conditions. Jarmon parallels Neill's statements about continual improvement assessment when she says, "This evaluation process should be integrated throughout the course, not seen as a single measure at the end" [1]. By including multiple methods of evaluation and communication, teachers can identify work that is not consistent with learner work or conversations. The teacher should know that the work belongs to the learner claiming it whether this work is submitted in a 10th grade English class or a graduate class in education.

Assessment Tasks

Cross and Neill both suggest considering several assessment options, including accountability and continual improvement assessments, for each learning outcome. Assessment can be set up as a task--an application of a skill practiced beforehand, one in which the learner has been given timely feedback on prior

attempts. According to Neill, assessments should reflect the outcome in six areas. The assessment should:
- squarely match the content
- match the process or product
- match the domain and level of the competency (outcomes may call for higher level domains, such as application, so a multiple choice test will not match)
- be clear on whether or not the learner met all the performance criteria (specifications by which performance of an outcome is evaluated)
- be clear on whether or not the learner met all the performance *conditions* (situation in which performance will be assessed). For example, if learners are asked to critique a role play, but instead critique a video, there is a mismatch in assessment and performance condition.
- engage learners in applying knowledge and skills in the ways they are used in the "real world." Because such skills are transferable, they induce learner motivation.

Punctuating courses with assessments will offer valuable feedback on both affective success (learners' feelings about the experience) and cognitive success (intellectual success, by both teacher and learner).

5.1. Assessment in Location-independent Classes

Success in distance education involves effectively using techniques not typically used in the traditional classroom. Learners use a vast array of resources and may approach activities differently because of the resources available, or unavailable, to them. Frequent assessments give teachers feedback, and the chance to trying something new, if necessary. They become useful to both teachers and learners in this environment.

Examples of assessments for location-independent classes include developing a web page or template, developing a multi-media package, or submitting writing.

In order not to waste the magic referred to in Oppenheimer's article, teachers play a crucial role in guiding learners through the learning and technology maze. Building effective learning activities is just a start. Showing them exactly how they will achieve a larger outcome and having them practice doing it along the way is the next step. And finally, assessing learning based on closely tied performance expectations for each outcome gives the student a fair chance at learning what was intended to be taught.

6. Instructional Design in the Future

Technology in learning and instructional design will evolve into *systems* using a blend of instructional design models[6]. Canada's open Learning Agency, for example, employs a blend of constructivism and cognitivism in instructional design. Working with industry-defined competencies and clearly specified learner paths and assessments, the Agency uses schema theory to take data structures representing the "generic concepts stored in human memory" and categorizes them into systems of parsing processes for storage, retrieval, and analysis [6]. Information is structured around learner processes for the ongoing design of learning.

It is not yet known as to whether the inclusion of technology improves learning. But recent work seems to point in the same direction: give students clear expectations in advance of instruction, learning plans with effective learning activities, and outcome driven assessments. The quality of this design will compliment and outlast any technology or system.

References

[1] Boaz, Mary, Betty Elliott, Don Foshee, Darcy Hardy, Carolyn Jarmon, and Don Jr. Olcott, *Teaching at a Distance: A Handbook for Instructors*, League for Innovation in the Community College Archipelago, Mission Viejo, 1999.

[2] Cini, Marie, and Boris Vilic, "Online teaching: Moving from risk to challenge", *Syllabus*, June 1999, pp. 38-40.

[3] Clark, Ruth, *Mind Tools: How to Apply Human Learning Principles to Training Design*, Center for Performance Technology.

[4] Cross, K. Patricia, "Learning is About Making Connections", T*he Cross papers*, League for Innovation in the Community College, Mission Viejo, 1999.

[5] Gifford, Bernard, *Computer-mediated Instruction and Learning in Higher Education Bridging the Gap Between Promise and Practice* [WWW], North Central Association, 1999 [cited June 25 1999]. Available from http://www.ncacihe.org/conference/99material/gifford.pdf.

[6] Laks, Alex. *Designing Choice Through Structure*, Paper presented at the Learning Paradigm Conference (San Diego, CA, March 12, 2001.)

[7] Meier, Dave, Accelerated Learning Training Methods, The Center for Accelerated Learning, Lake Geneva, 1999.

[8] Neill, Judith, and Deborah Mashburn, *Instructional Design and Planning Study Guide*, Wisconsin Technical College System Foundation, Inc., Waunakee, 1999.

[9] Oppenheimer, Todd, *The Computer Delusion* [WWW], 1997 [cited June 24 1999]. Available from http://www2. the Atlantic.com/issues/97jul/computer.htm.

[10] Richey, Rita. "Instructional design can make a difference in staff development", *Journal of Staff Development*, 1993, pp. 44-47.

The Design of a Lexical Difficulty Filter for Language Learning on the Internet

Chin-Hwa Kuo[*], David Wible[**], Chih-Chiang Wang[*], and Feng-yi Chien[**]

[*]Computers and Networking (CAN) Laboratory, Department of Computer Science and Information Engineering, Tamkang University, chkuo@mail.tku.edu.tw

[**]Graduate Institute of Western Languages and Literature, Tamkang University, dwible@mail.tku.edu.tw

Abstract

In this paper, we describe the design and implementation of a tool called the Lexical Difficulty Filter (LDF) intended to help learners or teachers in selecting sentences appropriate to the level of students for English vocabulary learning. The LDF serves as a bridge between learner and information resource bank. It selects suitable sentences from the output of standard English corpus concordancers. The core technology in the design of LDF is a fuzzy expert system. The results obtained are very encouraging in this pioneering work. We achieve 94.33% accuracy rate in imitating the judgments of a human expert in determining the degree of difficulty of a sentence for a given learner level.

1. Introduction

The Internet is an important resource for information acquisition. This information can benefit to teachers as well as learns. However, the information obtained from the Internet directly may not be suitable to learners, and further processing is often required. In this paper, a *Lexical Difficulty Filter* (LDF) for language learning on the Internet is described. The main purpose of the designed LDF is to determine the *degree of difficulty* of a given English sentence or paragraph. After this filtering process, the learning materials that are suitable to learner's level are delivered. As a result, learners are in a better-prepared learning environment.

Text processing, such as retrieval, understanding, summarization, clustering and classification, has been a lively field of research for several decades. The design concept of the LDF in our work is similar to text classification. The core technology used in this paper is based on a fuzzy rule generation system, an extended development based on the work by Hong and Lee in [1]. The LDF is able to achieve 94.33% accuracy in determining the degree of difficulty of the corresponding text.

2. System Overview

Using the Internet and/or electronic documents is an important trend in delivering learning materials. However, the materials obtained directly from such an environment may not be suitable for learners. Therefore, pre-processing may be required. As shown in Fig. 1, the LDF determines the degree of the difficulty of a given electronic text. This can be applied to offer learners numerous sentences that contain a new vocabulary word in context, but with the difficult sentences filtered out by the LDF. Or it can be used by teachers to check the level of lexical difficulty of a certain text for his or her students.

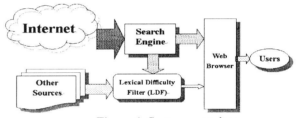

Figure 1. System overview

3. Design of LDF

Sentence examples provided by British National Corpus Online Concordancing are used as our sample data set. Each sentence is translated into four representative parameters:

1. Threshold Value
British National Corpus maintains a database called "vocabulary frequency list". Each user determines a limit value based on this database. Words that are within this limit value are assumed to be included in the user's vocabulary.

2. Words
The number of words contained in the sentence.

3. Percentage (PofW)
The percentage of words that are within the given threshold.

4. Difficulty

The degree of difficulty should be provided by a human expert for each sentence in the training sample. After training, the LDF will then determine the difficulty levels for sentences in the test sample.

The system architecture of the LDF is illustrated in figure 2. The functions of each component are as follows.

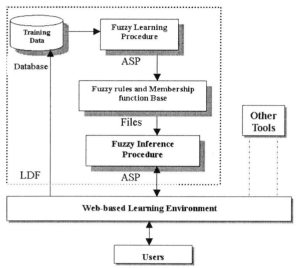

Figure 2. Architecture of the LDF

- Web-Based Learning Environment
  This is a collection of web-based GUI. It includes many tools to facilitate the use of the learning environment. The LDF can be viewed as one of these tools.
- Training Data
  A database contains training sentences with user dependent degree of difficulty judged by a human expert.
- Fuzzy Learning Procedure
  After the inputs to the learning algorithm are the training examples. And the corresponding outputs are the fuzzy rules, and membership functions.
- Fuzzy Rules and Membership Functions Base
  Files that encapsulate fuzzy rules and membership functions are generated by the fuzzy learning procedure.
- Fuzzy Inference Procedure
  This procedure classifies sentences into difficulty levels by applying fuzzy rules and membership functions on parameters of the sentences.

4. Results

A large data set of 260 sentences was analyzed to ascertain the practicality of our system. Professional English teachers were asked to classify the data. Half of the data set was then used as the training sample for our algorithm. The rest was used as the testing sample. The fuzzy learning algorithm generates 167 rules. The computations of the degree of difficulty are based on the concept of a contingency table [2].

The statistical results are shown in Table 1. The overall accuracy achieves 94.33%.

Table 1. The error rate of the LDF

	Error Rate
Easy	8.46 %
Normal	0.1 %
Hard	8.46 %
Avg.	**5.67 %**

The average error rate of our algorithm on the testing sample is 5.67%. Furthermore, a scrutiny of the inaccurately classified sentences shows that the errors are never off by more than a single level. Thus, an "easy" classification by the system will at most be a "normal" classification and not a "hard" even if the classification is incorrect.

5. Conclusions

An integration of the lexical difficulty filter into web-based environments will allow teachers to provide students with materials of appropriate difficulty levels from the web without spending precious human capital to ascertain the difficulty levels of the materials [3].

The core technology in the designing of LDF is based on fuzzy expert systems. The results obtained are very encouraging in this pioneer work. We achieve a 94.33% accuracy rate in the determination of the degree of difficulty for a given learner level. There are several areas of potential extension to this research. The analysis of vocabulary only includes each word's value in the vocabulary frequency list. No information about the actual word is used. The additional information may enable the algorithm to classify difficulty levels more accurately. Another extension is the analysis of optimal sample sizes of the training data set.

6. References

[1] T. P. Hong and C. Y. Lee, "Learning fuzzy knowledge from training examples", In Proceedings of *ACM-CIKM98*, page 161-166, 1998.

[2] D. D. Lewis, "Evaluation and optimizing autonomous text classification systems", In Proceedings of *SIGIR*, page 246-254, 1995.

[3] IWiLL group, http://www.iwillnow.org, 2001.

An Engineering Process for Constructing Scaffolded Work Environments to Support Student Inquiry: A Case Study in History

Kathleen Luchini, Paul Oehler, Chris Quintana, Elliot Soloway
University of Michigan
kluchini@umich.edu, poehler@umich.edu, quintana@umich.edu, soloway@umich.edu

Abstract

Inquiry-based curricula encourage students to develop research and collaboration skills by working with evidence and exploring real problems. Students engaging in inquiry projects must be supported in learning the content and work practices of a domain. SWEets, or scaffolded work environments, are one method of providing this support. We present a new software engineering process for designing SWEets for any educational domain and describe the application of this process to the development of Clio, a web-based SWEet for high school history students. The Clio SWEet was piloted in two classroom studies and preliminary data indicate that students were able to use the software to successfully gather and analyze data as part of an inquiry-based classroom curriculum.

1. Introduction: The Need to Support Inquiry

The National Standards for History Education [1] call for students to develop historical thinking skills by engaging in activities such as defining historical questions, gathering evidence, analyzing data, and working with their peers to generate and critique solutions. An inquiry-based curriculum can provide students with opportunities to participate in such investigations, but studies have shown that students need a great deal of support in order to conduct historical inquiry effectively [2]. One way to provide this support is through the use of scaffolded work environments (SWEets), which are suites of software tools designed to support student inquiry within an educational domain. SWEets like Symphony [3] have been successful in supporting inquiry in science classrooms, and this paper describes work to extend the benefits of SWEet technology to the domain of history. Specifically, we present a software engineering methodology for the design of educational SWEets and describe the application of this design process to the creation of Clio, a web-based SWEet for history. Clio was developed in close collaboration with Dr. R. Bain and his group at the University of Michigan School of Education (SoE), who created the curriculum and classroom materials that make use of the Clio SWEet software.

Preliminary results from two classroom trials indicate that students were able to use the Clio software to gather and analyze historical evidence as part of their classroom inquiry projects. Not all of the tools provided by the Clio software were fully utilized during these first trials, and future work will focus on better integration of the SWEet with the classroom curriculum and on providing more professional development for teachers.

2. An Engineering Process for SWEets

Building on the work of Quintana [4], we are developing the software engineering process described here to guide the construction of SWEets to support student inquiry in any domain. This process is informed by the theory of Learner-Centered Design [5], which recognizes learners as novices with unique needs that must be addressed during the software development process, and was used to develop the Clio software. Although we present these steps as an ordered list, they are not intended to be carried out in a linear fashion. Indeed, like many design techniques this software engineering methodology is iterative and involves revisions throughout the lifecycle of the SWEet.

1. *Identify the Learning Objectives*. We first identify what students will learn, what instructional methods will be used, what tasks and activities students will undertake, and what skills students will develop. Clio's pedagogy and curriculum were developed by our SoE colleagues based on Bain's previous work [6] and focus on research problems relevant to students' own experiences.

2. *Identify Expert Strategies*. The next step is to determine what strategies and techniques a professional would employ to complete the various learning objectives. Example inquiry tasks from the Clio curriculum include evaluating the relevance and credibility of sources and corroborating different accounts of the same event. Bain's curriculum team identified specific expert strategies applicable to each of these tasks in Clio.

3. *Identify Student Needs*. Next we evaluate each task and strategy and identify areas where students are likely to have difficulties. The SoE curriculum team identified aspects of the historical inquiry process where students needed support – for example, in order for students to determine the significance and relevance of a source, they needed access to background information about the artifact.

4. *Identify Teacher Needs*. It is also critical to work with teachers to identify where they need assistance to integrate the SWEet into their classrooms. To meet these needs for Clio, the SoE team developed lesson plans and professional development, and all of the developers (both curriculum and software) went into the classroom to demonstrate and troubleshoot the Clio software.

5. *Identify Classroom Needs*. When developing SWEets for education, it is critical to understand the classroom limitations that may affect the use of the software. For Clio, these considerations included limited computer hardware and Internet bandwidth and were resolved by developing the software using html and scripts and by minimizing the number of graphics in the web interface.

6. *Identify Scaffolds and Supports*. The next step is to determine which needs can be met with educational scaffolds and learner supports (tools assist students in completing tasks that would otherwise be too difficult or complex for novices [7]). To determine how the Clio software could best support these needs, we looked at Bain's experiences with historical inquiry [6] and at SWEets designed to support similar investigations in science, including Model-It [7] and Symphony [3]. The most prominent scaffolds in Clio are the multi-level Strategy Prompts, which support students in analyzing primary source materials. Other supportive tools provided by the Clio software include: Research Problem Folders, which provide persistent workspaces for students to collect information and take notes; Additional Information Sources, which supply additional background and context information for individual artifacts; and the Discussion Forum for online, threaded discussions.

7. *Design and Implementation*. The Clio database maintains information for the artifacts, scaffolds, and individual students, and the SWEet interface is built using scripted webpages that format information from the database dynamically in response to users' actions. The flexibility of this database architecture and dynamically controlled interface content allows the Clio software to be quickly modified as the associated curriculum evolves.

8. *Classroom Trials*. Early prototyping and classroom testing can help identify unanticipated needs; for example, during the first classroom trials of the Clio SWEet we discovered that students needed more flexible tools for navigating through the historical artifacts in the database.

9. *Curriculum & Software Revisions*. Both the Clio curriculum and the SWEet software have been revised based on data from the classroom pilot studies. Examples include adding a Search tool to the interface to help students navigate through the artifact database and creating a process map that hangs in the classroom to remind students of the different steps of the historical inquiry process. This engineering process for SWEets is non-linear, so the redesign and revision of the software should take place throughout the development process.

3. Conclusions and Future Work

Preliminary data gathered from two pilot trials of Clio in high school history classrooms indicate that the software supports students in successfully gathering and analyzing primary source materials and conducting historical investigations. During these first trials, the Clio software was used primarily to gather and analyze historical data. Future trials will explore how the software and curriculum can be more tightly integrated to take advantage of other inquiry tools provided in the SWEet and how we can better support teachers in using the software in their classrooms.

4. Acknowledgements

The authors would like to gratefully acknowledge the work of the Clio curriculum development team, particularly Bob Bain, Stephen Mucher, Mimi Lee and their colleagues at Henry Ford Academy and Museum.

This material is based upon work supported by the National Science Foundation under both NSF REC 9980055 and a Graduate Research Fellowship. Any opinions, findings, and conclusions or recommendations expressed in this material are those of the authors and do not necessarily reflect the views of the National Science Foundation.

5. References

[1] G. B. Nash and C. A. Crabtree, *National Standards for History: Basic Edition*, National Center for History in the Schools, Los Angeles, 1996.

[2] J. Bransford, A. Brown, and R. Cocking, *How People Learn: Brain, Mind, Experience, and School (Expanded Edition)*, National Academy Press, Washington, D.C., 2000.

[3] C. Quintana, J. Eng, A. Carra, H.-K. Wu, and E. Soloway, "Symphony: A Case Study in Extending Learner-Centered Design Through Process Space Analysis," *Proceedings of SIGCHI '99*, Pittsburgh, PA, 1999.

[4] C. Quintana, "Symphony: A Case Study for Exploring and Describing Design Methods and Guidelines for Learner-Centered Design," Ph.D. Dissertation, Ann Arbor: University of Michigan, 2001.

[5] E. Soloway, M. Guzdial, and K. E. Hay, "Learner-Centered Design: The Challenge for HCI in the 21st Century," *Interactions*, vol. 1, 1994, pp. 36-48.

[6] R. B. Bain, "Into the Breach: Using Research and Theory to Shape History Instruction," in *Knowing, Teaching and Learning History*, P. Stearns, P. Seixas, and S. Weinburg, Eds., New York University Press, NYC, 2000, pp. 331-352.

[7] S. L. Jackson, J. Krajcik, and E. Soloway, "The Design of Guided Learner-Adaptable Scaffolding in Interactive Learning Environments," *Proceedings of SIGCHI '98*, Los Angeles, 1998.

The Authority Structure Problem of Computer Supported Collaborative Concept Mapping System for Elementary Students

Chiung-Hui Chiu

Graduate Institute of Computer and Information Education, National Tainan Teachers College,
cchui@ipx.ntntc.edu.tw

Abstract

The purpose of this study was to examine the importance of the authority structure in the design of a computer supported collaborative concept mapping system for elementary students. An experiment was implemented using ninety-six fifth- and sixth-grade Taiwanese students. The independent variable was the authority structure, which was assign, rotate, give, or open, and designed into the system. The dependent variables included the student performance in concept mapping, the student attitude toward collaborative concept mapping, and the quantity of knowledge related interactions by students. The findings suggest that the authority structure design results in different effects on elementary students.

1. Introduction

Concept mapping is a technique developed by Prof. Joseph D. Novak at Cornell University in the 1960s for representing knowledge in a graph. The graph is composed of nodes and links. The nodes represent concepts, while the links represent the relationships between the concepts. Since the efforts of Novak and his colleagues ([1]), concept mapping has established its utility in learning and evaluation. Researchers agree that concept mapping, which includes integrating new related concepts, establishing new links, or re-arranging existing concepts and links, can assist and bring meaningful learning ([2]). A meta-analysis of 19 studies did reveal that concept mapping generally had a positive effect on both knowledge attainment and attitude [3]. Concept mapping is often incorporated into cooperative learning activities in a traditional classroom. Students are arranged into three to five person groups to collaboratively construct group concept maps. Although not many studies have investigated the effects, most have found that collaborative concept mapping can lead to effective discussions concerning concepts, and thus enhance meaningful learning (e.g., [4], [5]).

Traditionally, concept mapping was carried out by hand, using a pen or pencil on paper. To make a paper-and-pencil concept map usually required a significant amount of effort. Any deletions, additions, or changes to the map could be very frustrating. With the advancements and popularization of the personal computer, several computer applications (e.g., Inspiration, Axon Idea Processor) have been developed to support concept mapping. These applications make generating, modifying, or manipulating a concept map relatively easy ([6]). With the advent of ubiquitous computer networking, concept mapping might be moved to network platforms to support distributed collaborative concept mapping. Individuals then might connect, exchange ideas, and generate joint concept maps with others regardless of their individual locations or the time ([7]).

To allow separated individuals to collaboratively participate in a concept mapping activity, a network application is required. Because collaborators cannot directly converse face to face, the application must support mutual communication. Currently, technology has made it feasible to support text, audio, and video interfaces. In addition, the application must support constructing and viewing a common concept map. This will involve the problem of resolving mapping conflicts. To prevent or manage these conflicts, a scheme involving restricting the mapping control to one student at a given time could be adopted. This scheme will bring about three alternatives as follows: (1) *assign* design, in which the mapping control is designated to a particular member; (2) *rotate* design, in which the mapping control is rotated among the group members; and (3) *give* design, in which the mapping controller may decide whom to relinquish control. Another scheme involves allowing every group member to control the mapping all of the time. That is, all group members can manipulate joint concept maps simultaneously. We describe this as an *open* design. Considering an inadequate design can lead to an inadequate system, this study was designed to investigate if the authority structure design in a collaborative concept mapping system creates differences, and further which authority structure design results in a better influence upon student learning.

2. Methodology

An experiment was set up to compare the effects of assign, rotate, give and open authority structure conditions on elementary student interactions, attitudes, and performance. These four authority structures were supported by a web-based collaborative concept mapping system. The independent variable was the authority structure, while the dependent variables were student interaction, attitude and performance. ANOVA analysis was used to determine if the four authority structure conditions caused any difference.

2.1. Subjects

This study involved 96 students drawn from one fifth- and three sixth-grade classes at two elementary schools located in Southern Taiwan. These subjects were about average compared to all students in the same-grades regarding academic performance. They had about two-years of formal education in computer basics and applications. They were able to use one word processing software, one painter software and an Internet browser.

2.2. Collaborative Concept Mapping System

This system adopted a centralized architecture instead of a replicated architecture because of its simplicity at handling concurrency ([8], [9]). The architecture was web-based, so that a group of geographically dispersed users could utilize a common web browser to interact with one another and generate a joint concept map. It included six main components: mapping, communication, tracing, management, authority, and awareness. Figure 1 presents this architecture, while Figure 2 shows the user interface. The mapping component provided the functions to generate, modify, or manipulate a concept map. The communication component supported text-based synchronous communication. The tracing component could trace the entire mapping process and conversations within each group and record these data into the system database. Through the management component, an approved teacher could create, modify, delete, and manage a collaborative concept mapping activity. The authority component supported the design, rotate, give, open authority structures. The awareness component supported each group member maintaining up-to-the-moment awareness of one another concerning whether a member was absent and who now had mapping control(s). Regardless who had mapping control, every member's computer screen was updated accordingly.

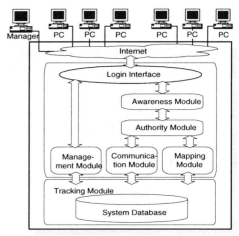

Figure 1. Architecture of the collaborative concept mapping system

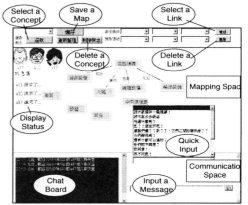

Figure 2. User interface for the collaborative concept mapping system

2.3. Treatments

The experiment was run in the computer laboratories of the respective schools. Both laboratories had IBM-compatible computers. An NT server with the collaborative concept mapping system installed was brought into the laboratories to support this experiment. Prior to the formal experiment, the subjects were introduced to the paper-and-pencil concept mapping method, the collaborative concept mapping system, and how to implement the various authority structure designs to generate a group concept map. Subjects in the same class were randomly arranged into the assign, rotate, give, and open structures. In each authority structure, the subjects were arbitrarily organized into groups of three members. Eight groups were formed for each authority structure. There were four distinct treatments. In the assign structure, one member always had mapping control and the other members did not. In the rotate structure, group members followed one another in three-minute rotations for mapping control. In the give structure, one group member with mapping control could give control to another member. In the open structure all group members had mapping control at the same time. These treatments were carried out class by class (only the subject students attended). Groups were required to generate concept maps on the hereditary features of humans, atmospheric layers of the Earth and the food chain. The first two mapping activities lasted 40 minutes and the third one lasted 60 minutes.

2.4. Measurements

The communication contents of the four treatment conditions were coded and analyzed. Because the amount and time spent explaining correlated highly with the amount learned ([10], [11]), only communication used to discuss and relate to the knowledge covered by the concept mapping activities was used. Social contacts, vague critique, or procedure-related messages were put aside (such as "What's your name?" "You make up the map terribly." or "Please quick."). The discussion messages on the concept, proposition, or hierarchy (such as "What's the relationship between gene and human features?" "It should be the atmosphere contains troposphere." or "Place consumer below producer.") from each student was counted to reflect student cooperative interactions.

The Attitude toward Collaborative Concept Mapping (ACCM) was designed to measure student attitudes and perceptions towards using concept mapping and collaboratively using concept mapping. This instrument consisted of 18 Likert-scale statements, 7 for concept mapping usage (e.g., Concept mapping can help me to understand the relationships among concepts.), and 11 for collaborative concept mapping (e.g., During collaboratively constructing a joint concept map, we can clear up many confusing matters together.). Students can indicate their feelings by selecting one of five choices from agree strongly to disagree strongly. Cronbach's alpha estimates of reliability for the two subscales were 0.80 and 0.83 in a pilot test, and 0.82 and 0.84 in the formal study.

Student performance was measured using (a) in-treatment group concept maps, which represented group collaborative achievement; and (b) subsequent individual concept maps two-weeks later, which reflected the knowledge retained by individuals. For each group, three joint concept maps were generated. Three individual concept maps were completed for each subject. All subjects in a group received the same score for the group concept maps. The individual concept maps were scored individually. Three elementary natural science teachers worked together to score the group and individual concept maps according to the following rules: (a) Every valid proposition (concept-link-concept) scored 1 point; (b) Every valid level of hierarchy scored 5 points; and (c) Each cross link scored 10 points if significant, or 1 point if not significant (modified from [1]). To eliminate the influence of variability in the original scores caused by unequal difficulty among the three mapping themes, T-scores were computed ([12]) for each map and summed up for the in-treatment group maps and the subsequent individual maps.

3. Results

Table 1 summarizes the mean and standard deviations for the student knowledge related interactions, attitudes toward concept mapping and collaborative concept mapping, and performance in terms of in-treatment T-scores and subsequent T-scores under the four authority structures.

Table 1. Means on the dependent variables

		Assign Mean (N, SD)	Rotate Mean (N, SD)	Give Mean (N, SD)	Open Mean (N, SD)
Interaction	Message counts	9.5000 (24, 15.243)	.5417 (24, .7211)	1.2917 (24, 2.1362)	.6250 (24, .9696)
Attitude	Concept Mapping	29.2917 (24, 4.7501)	27.2500 (24, 4.8026)	26.6522 (23, 4.3339)	27.0000 (24, 4.6625)
	Collaborative Concept Mapping	46.0417 (24, 5.9306)	41.2500 (24, 7.2427)	41.5652 (23, 6.1852)	41.8333 (24, 5.9539)
Performance	In-treatment T-scores	164.1742 (24, 22.6774)	132.8773 (24, 24.3361)	156.9290 (24, 22.7251)	154.3266 (24, 16.4224)
	Subsequent T-scores	160.2647 (24, 26.2989)	138.7955 (24, 22.0488)	146.7639 (24, 24.2452)	150.8199 (24, 21.3768)

There were many more interaction messages in the assign structure than in the rotate, give and open structures. The ANOVA analysis of the amount of interaction indicated a significant difference among the four structures, $F(3, 92) = 7.632$, $p = .000$). The Scheffé test showed that the amount of interaction in the assign structure was significantly greater than in the rotate, give, and open structures ($p = .002, .005, .002$). There was no significant difference among the rotate, give, and open structures.

The ANOVA analysis of attitude toward concept mapping indicated no significant difference among the four structures, while the ANOVA analysis of attitude toward collaborative concept mapping did reveal a significant difference among the four authority structures, $F(3, 91) = 3.024$, $p = .034$. The Scheffé test was not able to determine which two structures were different because the test is quite conservative ([13]).

The ANOVA analysis of the in-treatment T-scores indicated a significant difference among the four structures, $F(3, 92) = 9.191$, $p = .000$. The Scheffé test showed that the in-treatment T-scores for the assign, give, open structures were significantly higher than the rotate structure (p = .000, .003, .011). There was no significant difference among the assign, give and open structures. The ANOVA analysis of the subsequent T-scores indicated a significant difference among the four structures, F(3, 92) = 3.444, p = .020. The Scheffé test showed that the subsequent T-scores for the assign structure were significantly higher than the rotate structure (p = .023). There was no significant difference among the rotate, transfer and open structures.

4. Discussions and conclusions

Significant differences were found between the assign, rotate, give, and open authority structures in terms of student knowledge related interaction, attitude toward collaborative concept mapping and in-treatment performance and subsequent performance. These results suggest that the authority structure design be considered while developing collaborative concept mapping systems. In addition, the results show that students in the assign structure would tend to cooperatively interact more with their group members than students in the rotate, give, and open structures. Students in the assign structure demonstrated superiority over students in the rotate structure in the in-treatment performance and subsequent performance. This would suggest that developers of collaborative concept mapping systems could just select an easy scheme, like restricting the mapping control to a designated member. They do not have to deliberate long and deeply upon simultaneous requirements or concurrency control problems. The findings of this study may be inferred to the development of systems to support general collaborative tasks, not just concept mapping. If a collaborative system is to be developed with the intention to benefit student learning, selecting an overly powerful conflict management scheme may not be necessary. However, to determine an optimal design or find principles to follow for designing effective general-purpose collaborative learning systems, further research is required. In addition, this study was conducted with elementary school students. Different students may be suited to different authority structure designs. This study therefore recommends that future research investigate the influence on students different in gender, ability or learning styles.

5. Acknowledgements

This research was supported by the National Science Council of the Republic of China (ROC) under grant number NSC 89-2511-S-024-007.

6. References

[1] J.D. Novak and D.B. Gowin. *Learning How to Learn*, Cambridge University Press, Cambridge, London, 1984.
[2] J. A. Heinze-Fry and J.D. Novak. Concept Mapping Brings Long-term Movement toward Meaningful Learning. *Science Education*, 74(4), 461-472, 1990.
[3] P.B. Horton, A.A. McConney, M. Gallo, A.L. Woods, G.J. Senn, and D. Hamelin. An Investigation of the Effectiveness of Concept Mapping as an Instructional Tool. *Science Education*, 77, 95-111, 1993.
[4] W.M. Roth. Student Views of Collaborative Concept Mapping: An Emancipatory Research Project. *Science Education*, 78, 1-34, 1994.
[5] W.M. Roth and A. Roychoudhury. The Concept Map as a Tool for the Collaborative Construction of Knowledge: A Microanalysis of High School Physics Students. *Journal of Research in Science Teaching*, 30, 503-534, 1993.
[6] L. Anderson-Inman and M. Horney. Computer-based Concept Mapping: Enhancing Literacy with Tools for Visual Thinking. *Journal of Adolescent and Adult Literacy*, 40(4), 302-306, 1996/1997
[7] W.K. Chung, F. O'Neil, E. Herl, and A. Dennis. *Use of Networked Collaborative Concept Mapping to Measure Team Processes and Team Outcomes*. Paper presented at the annual meeting of the American Educational Research Association, Chicago, IL, 1997. (ERIC Document Reproduction Service No. ED 412 225)
[8] S. Greenberg and D. Marwood. Real Time Groupware as a Distributed System: Concurrency Control and Its Effect on the Interface. In *Proceedings of the ACM Conference on Computer Supported Cooperative Work*, pages 207-217, 1994.
[9] J.F. Patterson, R.D. Hill, S.L. Rohall, and W.S. Meeks. Rendezvous: An Architecture for Synchronous Multi-user Applications. In *Proceedings of the ACM Conference on Computer Supported Cooperative Work*, 1990.
[10] G. Hill. Group versus Individual Performance: Are N+1 Heads Better than One? *Psychological Bulletin*, 91, 517-539, 1982.
[11] D.W. Johnson, and R.T. Johnson. Positive Interdependency: Key to Effective Cooperation. In R. Hertz-Lazarowitz and N. Miller (Eds.), *Interaction in Cooperative Groups: The Theoretical Anatomy of Group Learning*, pages 174-199, Cambridge University Press, New York, 1992.
[12] L.H. Cross. Grading Students (ERIC/AE Digest Series EDO-TM-95-5). In *ERIC Clearinghouse on Assessment and Evaluation*, Washington, DC, 1995. (ERIC Document Reproduction Service No. ED 398 239)
[13] L.E. Toothacker. *Multiple Comparisons Procedures*, Sage Publications, Thousand Oaks, CA, 1993.

Decompressing and Aligning the Structures of CBI Design[1]

Joel W. Duffin
Utah State University
jwduffin@yahoo.com

Andrew S. Gibbons
Utah State University
gibbons@cc.usu.edu

Abstract

CBI researchers have long pursued goals of adaptivity, generativity, and scalability. The architecture of a CBI product has a large influence on how well it achieves these goals. This paper introduces principles called "decompression" and "alignment" that can be used to guide the design of a CBI product's architecture.

1. CBI designers often compress their designs

CBI designers tend to think of their designs in monolithic terms because of the tools they use. However, thinking in terms of structures allows "decompression" of the designs into structural layers. When this happens it is clear that structures at one layer place requirements on structures at other layers. This suggests that there should not only be alignment of components within the structures of a given layer, but also alignment of components and structures between layers. This principle guides the appropriate modularization of program modules.

2. Decompressing and aligning the structures of a CBI product can help make the product adaptive, generative, and scalable

SRI researchers observe "individual projects tend to consider one structure dominant and cast other structures, when possible in terms of the dominant structure. Often, but not always, the dominant structure is the module structure." One reason for this is that the module structure tends to mirror the project (work assignment) structure. In CBI this is not the case, division of labor typically falls along areas of expertise such as subject matter expert, instructional designer, computer programmer, and graphic designer. Nevertheless, discussions about CBI architectures are often carried out only in terms of the structures used by the computer programmer. We believe that in order to accomplish CBI goals of adaptivity, generativity, and scalability, architecture discussions should include representations of those structures important to stakeholders other than the computer programmer. In addition, the connections between non-programming structures and programming structures should be defined in ways that are traceable through their alignments.

2.1. Decompression involves considering the structures of a CBI product independently before collectively

When developing stand-alone CBI products, all structures used to represent the design are eventually combined and embodied in constructs offered by the media-logic tools used to create the CBI product. The process of combining the structures of a design is referred to as "compression".

In a decompressed design process, decisions at each layer are considered first independently and then with respect to their influence on each other in light of priorities of the project. This allows designers to give attention to the details at each layer of a design.

2.2. Alignment involves creating modularizations at each layer that do not conflict with modularizations at interdependent layers

The alignment of CBI design layers refers to how well the components at each layer of design match up with and support the requirements established by the constructs used at the other layers of design. Design layers are aligned through linkages or articulation points analogous to where one structure of a building joins with another. Most designers compress their consideration of content, strategy, and representation by giving first priority to decisions about media-logic.

3. The Mystery Box simulation: A case study demonstrating decompression and alignment

---

[1] For the 4 page paper submitted (has references), see http://cc.usu.edu/~sllcd/Papers/ICALT2001/ICALT2001.doc

For the purpose of research on instructional feedback we created a "Mystery Box" simulation environment in which learners solve complex problems situated in the context of an electric laboratory. To accompany the simulation environment we developed an intelligent agent that delivered detailed feedback and commented demonstrations of expert performance. The program logic for the intelligent agent is separate and independent of the program logic for the simulation environment. This is an example of program modularization along the lines of layers of a design. It also demonstrates aligning the modularizations at non-program logic layers with program logic modules. Figure 1 shows the Mystery Box user interface:

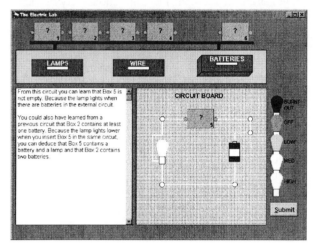

Figure 1: Mystery box user interface

The Mystery Box simulation presents an electrical laboratory in which learners are given the task of discovering what is contained in six mystery boxes.

3.1. The Mystery Box aligns modularization across several design layers

Figure 2 shows the Mystery Box program modules. The simulation environment program module embodies structures from the model and representation layers. The pedagogical planner (strategy layer) uses the problem generator (problem layer) to create a problem to pose to the learner in the simulation environment. As problem solvers work in the simulation environment, their actions produce event descriptions that are passed to the event interpreter (expert performance model layer). The event interpreter translates event descriptions into data that the pedagogical planner uses to decide and carry out responses to problem solving steps. The pedagogical planner determines the type of instructional message to deliver and activates the logic needed to deliver it.

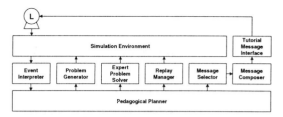

Figure 2: Mystery box program modules

The system is built in such a way that the source of a circuit can be any of the following: (1) a learner acting on the simulation environment, (2) a circuit specified by the expert problem solver module, or (3) a circuit retrieved by the replay manager from a log of a student's previous actions. The expert feedback modules do not need to know the source of the circuit test. This is possible because of the modularization chosen. Figure 3 shows how elements at various layers of the design are compressed into and embodied in the mystery box program modules.

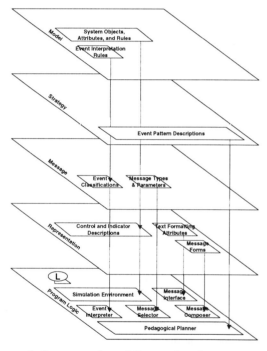

Figure 3: Compression of the mystery box structures

For some program modules there is a one-to-one mapping between them and a component at single other design layer. For others, decisions at multiple layers are embodied in a single program module. Modularization at the program logic layer comes from modularizations at the other design layers. This case study demonstrates how identifying appropriate design layers, modularizations, and alignments, facilitates creating product architectures that are adaptive, generative, and scalable.

A Design Framework for Interaction in 3D Real-time Learning Environments

Fanny Klett

Institute of Media Technology, Ilmenau Technical University, Germany
E: Fanny.Klett@RZ.TU-Ilmenau.de

Abstract

Currently, deficiencies in learning environments concern concepts, consistency, screen layout, interface design, navigation strategies, low-level of multimedia based interactions. In order to enlarge the communication capabilities in a learning environment, it is essential to focus on two fundamental aspects in the user interface: information presentation and information access. Pursuing this aim, this paper discusses first the design of hypermedia systems in its own unity of concept and design modeling for learner-beneficial support by the user interface. In relation to these explanations, it presents a learning system that benefits from the research about both the consideration of human visual and acoustic perception in the user interface for learning and the sufficient visualization of multidisciplinary contents for engineering educational activities, especially by 3D models, 3D animations and simulations, which facilitate the learner's immersion in a hidden world.

1. Introduction

The main potential of hypermedia systems that offer an extremely powerful way of accessing, organizing, presenting and interacting information challenges the traditional experience of shaping learning environments. Reflecting on their implementation for e-learning performance aimed at achieving the learner's optimal information acquisition in engineering education, this paper discusses an approach of learner-centered design along with layout considerations. Additionally, it describes a 3D hypermedia learning environment for real-time interactions - ED-MEDIA (Educational Media), which is preferentially designed for university students in media engineering courses.

The phenomenal growth of digital media over the last few years, coupled with the development of various multimedia applications enlarges the need for media knowledge. The learning concept is intended for the adaptation of mandatory and optional subject matter in lecture courses. It concerns the content-based and graphical implementation of knowledge subjects by taking into consideration learner-beneficial factors. With respect to newest findings in instructional, interface and visualization design, the primary aim of the highly interactive modular and open-ended learning environment is to motivate, point visually out the relevance of the knowledge to be studied, and emphasize self-controlled learning activities by a suitable range of 'modeling' [1] to control the cognitive overhead.

Therefore, the attention is being focused on the content structure, the content presentation, and the learning control, which have recently shifted the accent on the research field of knowledge construction, thinking and training. Dealing with this matter, the didactic and design concept of ED-MEDIA enables exploration of the subject domain for educational as well informative goals, enhances the motivation, and supports efficiently the learning and imaginative process of different learners.

2. Learning systems and interface design

Frequently, in hypermedia and presentation systems it is not easy to decide what is the nature of the preferred transformation of teaching and learning, how the vision could be given life or what media would show our progress toward that desired state. The correct goal-oriented selection of media formats for content presentation and the precise decision on integration of interaction opportunities are the cruces of the matter. This problem, which was originally a design problem, includes adequate decisions on content structure aspects of the learning environment and on specific human factors related to the cognitive information processing. Considering prior knowledge (external consistency) and experience (experience consistency), the way of placing the various media forms (text, sound, static and dynamic illustrations) and the interaction elements, which are simultaneously the communication elements for dealing with the application (internal consistency), presents an essential part of the navigation through the information space and the orientation in the learning environment. Therefore, the task of designing the user interface evolves into the task of creating the conditions for consistency, harmony, and unity. The user interface bears the responsibility for quick orientation and suitable gain in attention. An important aim is creating an intuitive interface, which should involve cognitive, structure and layout considerations.

2. Cognitive aspects of the learning design

Dealing with hypermedia learning environments, individual learner-relevant sequences emerge in consequence of the various paths for reaching an information node. Considering this fact, a deep cognitive processing along with recurrent revision of the subject matter could be achieved. Moreover, a context-related embedding of knowledge is maintained. Information nodes in various media formats (text, static illustrations, 2D and 3D dynamic illustrations, sound) support the learner in handling a variety of reality-close presentation forms in the user interface. They also allow the learner to create and extend the ability in competent situation-oriented acting.

Jonassen and Grabinger [2] summarize the learning processes that seem to be best aided by hypertext / hypermedia systems: information search, knowledge acquisition, and ability in problem solving. Duchastel [3] gives a more precise observation of cognitive processes during the information acquisition. He distinguishes between browsing (relatively free moving through the information space focusing on one's own interest and motivation), searching (goal-oriented pursuing of links by logical evaluation of their signal effect), integrating (relating new information to prior knowledge), and angulating (considering new information from different points of view and under varied conditions). The processes browsing and searching, integrating and angulating respectively concern distinctive information processing levels. Browsing and searching are strictly connected with the technical navigation concept. Integrating and angulating refer to the internal processing of distributed information.

Spiro and Jehng [4] look at the hypermedia features of information presentation and access as conceivable options for supporting advanced learning in complex heterogeneous domains. The web of nodes and the relations between these instances allow domains' crossing and domain analysis from different points of view. This is the way to recognize context-related meanings, irregularities and to build the ability to apply existent knowledge to new situations and subjects, which signifies the expert's cognitive flexibility in heterogeneous domains.

The modern theories of knowledge, which consider hypermedia systems as leading techniques for knowledge acquisition, construction and imparting served as an adequate source for a beneficial learning model targeted to the adaptation of mandatory and optional subject matter in lecture courses.

3. Structure aspects of the learning design

The flexible access to hypermedia information in various depth always holds a disorientation problem. By the assumption of a sufficient motivation, this was the reason for the consideration to exclude from the beginning the disorientation in hypertext during the conscious learning process or to reduce it appropriately. Therefore, the proposed model for adaptation to the subject matter is based on a general interlinked hypertext structure, which contains hierarchical and linear structures as subsets [5] to be achieved by a highly structured menu. Additionally, the conceptual model unites the mental models of active and passive learners who have different habits of working with learning environments.

The structured definition of documents carries built-in concepts of modularity and abstraction. Therefore, user interfaces require consideration of the human ability to memorize in the short-term and the long-term memory and to support suitable the knowledge transfer from the short-term into the long-term memory.

On the one hand, this problem refers to the granularity of documents. Patel and Kinshuk [6] also consider granularity from an interface design viewpoint. They decompose learning tasks into small components at varying levels of granularity with the perspective shift enabled through the user interface.

On the other hand, this problem signifies the cognitive overhead. Thus, it is connected with the hypermedia facilitating functions. Placing a wide range of orientation aids guides the learner through the information space and facilitates the use of the interface and the content exploration. Considering the implementation of functions aimed at the learner's sufficient and systematic interacting with the e-learning environment like history, thumb-tabs, footprinting, search, annotations, focusing [7], the learner has a consequential opportunity for motivation-based self-controlling of his own learning paths, dynamic self-regulating and content manipulating. Additionally, these functions support the learner in solving navigation and orientation problems within the same chapter or within other chapters and assist the context abstraction process that is very important for providing specific meaning and interpretation.

4. Layout aspects of the learning design

The layout considerations should ensure unity and harmony across the presentation of the learning system [8]. The established screen partition into three horizontal functional parts (identification area, learning area and control area) described by Strzebkowsky [9] and shown in Figure 1 served as a basis for further development presented in Figure 2.

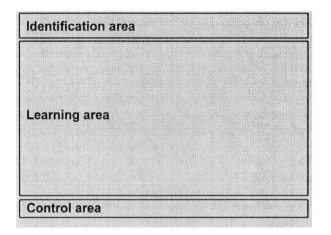

Figure 1. The established learning interface [9].

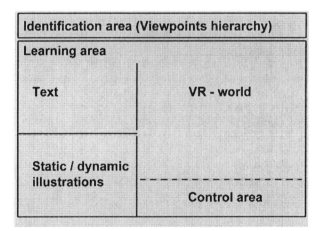

Figure 2. The developed 3D learning interface.

The most important aim in engineering education is the visualization of details, complex structures, processes, etc. With the aim of avoiding the learner's cognitive overhead and the layout overload, and to visually point out the significance of each medium, the offered subject matter is divided into a text and an illustration structures. The illustration structure includes 2D and 3D static illustrations, 2D and 3D dynamic illustrations, sound illustrations and a VR-world, which covers supportive VR-animations, too.

The identification area displays the viewpoints of the VR-world to be studied. They have to accurately reflect the content and the context of the studied domain. Their meaningful appearance helps to prevent the feeling of being lost in hyperspace.

As shown in Figure 2, the original learning and control areas run together to a unique area, which consists of four functional parts (text mode, illustration mode, VR-world and control area). The main area is occupied by the VR-world, which represents a 3D exploration menu. [10] The visually accentuated size of the VR-world illustrates its significant function as main navigation and addressing element in the learning environment.

Taking into account the design principle of the continuous line, the control elements are grouped and horizontally placed below the main area. Furthermore, this horizontal arrangement corresponds to the similar alignment predestined to the human field of view and supports a permanent connection to the subject matter and an optimal context-related orientation in the VR-world. In addition, the closeness of the control area to the VR-world concerns the design principle of proximity. The control elements, which simultaneously bear upon interaction functions, show additional information as text or illustrations only on request.

Selecting an element of the VR-world, you activate the available text and illustration structures. The description of the selected element emerges placed directly below the VR-window supporting related recognition and sufficient eye movement. Significant descriptions helpfully avoid disorientation problems in hyperspace. A reversible invisible making of irrelevant elements provides transparency to desired elements of the subject domain.

The location of the text and the illustration structures assumes both the cognitive propinquity of accompanying text medium, which supports the imagination process by studying the illustration, and presenting medium to each other and the importance of text and visualization expressed in the vertical arrangement.

The specific graphical concept illustrated in Figure 3 maps the presumptions of the cognitive, structure and layout considerations onto the learning interface. The resultant visual balance and visual flow help to guide the learner in his navigation and orientation tasks. The design of the included media should fit into this framework in order to achieve a corresponding design for learning activities, which enhances the individual motivation and serves the communication between author and learner most advantageously.

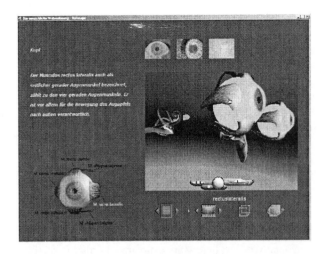

Figure 3. Screenshot of the 3D learning interface (module *Visual and Acoustic Perception*) [10].

5. Conclusion

By combining various presentation forms (text, sound, static and time-based illustrations), the complete e-learning application facilitates the learner in determined situations the access to information in various depth and in multiple dimensions (space and time).

The proposed educational hypermedia environment is designed to be modular and has different dimensions of use. Intended as a self-study environment and server application for learning via a local network, now a version for learning on the web is available. The 3D web-version enhances the usability by online access to the desired information in different file formats. Considering the disadvantages of existing web-based environments regarding delivery problems, required plug-ins, loading time, real-time sound support for 44 kHz samples, etc., it aims for keeping the benefits of the stand-alone environment. Generally, current web-based environments allow the integration of text documents and presentation formats similar to MS PowerPoint. The proposed 3D real-time learning application sufficiently supports dynamic illustrations as a VR-world and linked 3D-animations to facilitate the learners' imaginative and learning abilities. It allows the learners to freely move through the virtual world, to experiment with its elements, and to seek for help support, textual information, or linked animations.

The described project ED-MEDIA will be integrated in the thematic network THEIERE (Thematic Homogeneity in Electrical and Information Engineering thanks to pre-Requisites and ECTS), which is intended by the European Association for Education in Electrical and Information Engineering and funded by the European Union. About eighty European universities participate in this network, which allows students a cross European distance education.

References

[1] H.M. Niegemann, "Zum Einfluß von "modelling" einer computergestützten Lernumgebung", *Unterrichtswissenschaft 23*, 1994, pp. 75-87.

[2] D.H. Jonassen and R.S. Grabinger, "Problems and Issues in Designing Hypertext/Hypermedia for Learning". In D.H. Jonassen and H. Mandl (eds.) *Designing Hypermedia for Learning* [Proc. NATO Advanced Research Workshop on Designing Hypertext/Hypermedia for Learning], Heidelberg: Springer [CSS, Vol. 67], 1990, pp. 3-25.

[3] P.C Ducastel, "Examining Cognitive Processing in Hypermedia Usage", *Hypermedia, Vol.2, Nr.3*, 1990, pp. 221-233.

[4] R.J. Spiro and J.C. Jehng, Cognitive Flexibility and Hypertext: Theory and Technology for the Nonlinear and Multidimensional Traversal of Complex Subject Matter. In D. Nix and R.J. Spiro (eds.) *Cognition, Education, and Multimedia: Exploring Ideas in High Technology*, Hillsdale, N.J.: Lawrence Erlbaum Associates, 1990, pp. 163-206.

[5] F. Klett and H.-P. Schade, "Hypermedia – The Unity of Instruction and Design". In W. Frindte and T. Koehler (eds.) *Internet and Communication, Vol. 3*, Frankfurt: Peter Lang, (in press), 2001.

[6] A. Patel and Kinshuk, "Granular Interface Design: Decomposing Learning Tasks and Enhancing Tutoring Interaction". In M.J. Smith, G. Salvendy and R.J. Koubek (eds.) *Advances in Human Factors/Ergonomics – 21B- Design of Computing System: Social and Ergonomic Considerations*, Amsterdam: Elsevier Science B.V., 1997, pp. 161-164.

[7] F. Klett and D. Repschlaeger, "Hypermedia Prospects of Engineering Education – Learning while Interacting", *Proc. of the 11th Annual Conference on EAEEIE*, University of Ulm, 2000, pp. 142-147.

[8] F. Klett, "Hypermedia and the Communication Aims in Engineering Learning Environments", *Proc. of the UEF and IEEE Interdisciplinary Conference Electrical, Electronics & Computer Engineering Education in the Third Millennium*, Davos, 2000.

[9] R. Strzebkowski, „Realisierung von Interaktivität und multimedialen Präsentationstechniken". In L. Issing and P. Klimsa (eds.) *Information und Lernen mit Multimedia*, Weinheim: Psychologie-Verlags-Union, 1997, pp. 269-304.

[10] F. Klett, D. Dimitrov and S. Hoene, "Design of 3D Anatomical Models for Medical and Multidisciplinary Educational Activities", *Proc. of the World Conference on Telemedicine*, University of Toulouse, 2000, pp. 84-85.

[11] J. Nielsen, *Multimedia and Hypertext: The Internet and Beyond*, Cambridge: AP Professional, 1995.

[12] V. Bruce and P.R. Green, *Visual Perception: Physiology, Psychology and Ecology*, Hillsdale, N.J.: Lawrence Erlbaum Associates, 1993.

The RSVP Unibrowser: Bringing Humanness and Interactivity to e-Learning

Cupp, E.L., Danchak, M. M., Foster, K.J, Johnson, A.D., Kim, C.K., Sarlin, D.
Rensselaer Polytechnic Institute
danchm@rpi.edu

Abstract

RSVP (aka Rensselaer Satellite Video Program) initially delivered Master's level technical courses to the working professional via satellite. As technologies changed, so did RSVP. In particular, the World Wide Web forced a rethinking of the future of distance learning. Videostreaming is a natural replacement for satellite broadcast, but affords much more potential. A parallel development, called "GuideOnTheSide", added a human dimension to the asynchronous portion of the learning module. The introduction of the Unibrowser included a Videostream frame, a Content frame, a Communications frame and an Interactive frame. These function within the confines of web tools, such as WebCT. This paper will describe the evolution of these tools, their success and failure, and plans for future directions.

Background

Through a variety of delivery technologies, Rensselaer brings graduate courses, certificates, degree programs, and noncredit seminars and workshops to working professionals who pursue such learning while remaining fully employed at their work locations. Rensselaer's distance education program, RSVP, has been recognized nationally for its high-quality academic programs, successful partnerships with industry, and strong focus on customer service. In 1993, it was named "Best Distance Learning Program—Higher Education" by the US Distance Learning Association. In 1996, it received recognition from the same organization for its partnership with General Motors in the delivery of an MS in the Management of Technology. Currently, RSVP offers 14 graduate degrees and 17 certificate programs at a distance to more than 1000 students at 70 sites.

In 1987, the original delivery technology was satellite/video tape, hence the name Rensselaer Satellite Video Program (RSVP). As technology evolved, RSVP followed closely. The program was a very early adopter of videoconferencing for teaching and learning and was a founding member of CIVDL (Consortium for Interactive Video in Distance Learning). The motivation to move to videoconferencing was to provide as much instructor-learner interaction as the medium would allow. Hence pedagogy, not new technology, was the driving force behind the adoption. Because of cost and complexity, both of these technologies favored an established "site", usually located within a corporation. The corporation often provided additional services, such as proctoring, for their sponsored students. The customer in this case was the corporation and the student. Individuals not having access to a site were essentially excluded from participation.

As the Internet grew in popularity, RSVP hoped that it would allow us to better serve the individual student. While the convenience and accessibility of the web make it a very tempting medium, these advantages are offset by the lack of interactivity and "social presence" [1]. A conscious decision was made to NOT exclusively use asynchronous web delivery. We called this technique the "80/20" model, or blended learning [2]. We wanted a mix of both asynchronous (~80%) and synchronous (~20%) activities in our courses. Even the asynchronous portions should be interactive and personable. Typically, this is not possible in text-based web courses. However, the development of streaming media and Java changed all that!

Streaming Video

Our first use of streaming technology started in spring of 1998. Our distance-learning colleagues at Stanford University started using videostreaming as a replacement for the satellite broadcast. The same video equipment was used, but the camera shot showing the instructor was digitized and stored. Snapshots of course materials, such as instructor "blackboard" work, were also saved and later synchronized with the instructor videostream. It was then distributed on-demand over a very high-speed network from their streaming server. One major drawback was the difficulty of seeing the content material in the small windows available. Even one-quarter screen is substantially smaller than a video monitor!

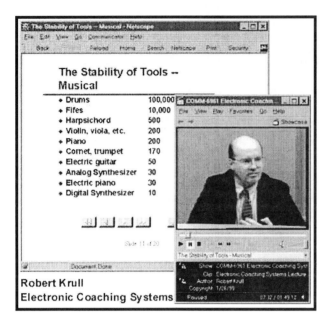

Figure 1. Original Videostreaming Configuration

RSVP created a variation on that theme. We devised a software program that allowed a real-time observer to mark the videostream being stored with tags at certain points in the presentation, such as when the instructor displayed a graphic. We insisted on good graphic design. Afterwards, the instructor's set of graphics were obtained, digitized and synchronized with the videostream using the Windows Media™ Indexer. A typical result is shown in Figure 1. Here the videostream of the instructor is in the standard Windows Media™ Player window while the slides are shown in a separate window. Thus, a student could access this stream "on-demand" and view the entire lecture, with slides, at their leisure. In the streaming window, you can see the controls for that stream and a drop-down menu that allows the student to jump ahead to another slide. The stream associated with that slide is then downloaded. This feature is standard with the Windows Media™ Player, but adds much to the experience. You can now fast forward over the "boring" parts!

A parallel project involved the "GuideOnTheSide" [3] concept of bringing a more human element to the on-line (asynchronous) modules. In this scenario, a small browser window popped up on the user desktop next to the content window they are studying. The purpose was motivation and guidance. At selected points in the lesson, a streaming video snippet of the instructor was used to set direction and motivate the student for this part of the lesson. When we wanted to give the student direction without distracting them from the content, a streaming audio snippet with a caricature of the instructor was used. Finally, a text window was used as a reminder when needed. All three flavors of the "Guide" are shown in the Figure.

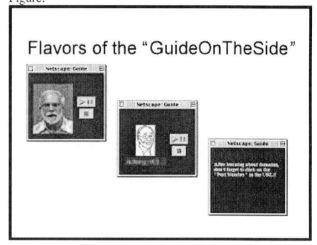

Figure 2. Variations of "The Guide on the Side"

This additional window complemented the WebCT™ window and gave us experience with augmenting WebCT™ with JavaScript functions.

The Birth of The Unibrowser

At this point, we were working with and trying to coordinate a number of browser windows. Then the idea of one browser – a Unibrowser – arose. Using frames within a single browser window solved a number of problems facing us at that time. As shown in the Figure 3, the Unibrowser has four distinct frames. The frame in the upper left is the "Videostreaming" Frame. It holds the videostream and the controls for that stream. The frame in the upper right displays the content associated with the stream and is called the "Content" frame. These two windows just represent a organizational tool, although there is value in not having to search for various windows. The lower left frame might be called the "Communications" or Navigation" frame because it provides the student with easy access to email connections to the course instructor or producer, as well as technical support

In the spring of 2000, we looked at how the Unibrowser was used by students in real courses. Typically, two one and one half hour lectures were posted per week and the streams represented most of the course material, aside from textbooks. While we did not have very sophisticated monitoring software, we did find that students tended to watch the streams in 15-minute chunks. In other words, they would connect and be on-line with the stream for a short period. We could not tell whether they bounced around within lecture or returned to the same lecture again later. Intuitively, we feel that they may be fast forwarding and concentrating their efforts where they think most appropriate.

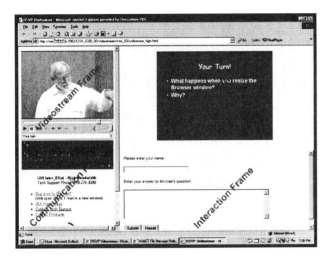

Figure 3. Current edition of the RSVP Unibrowser

The final frame, the "Interactive" frame, gives us a dedicated area in which to involve the student while the stream is in progress. Rather than passively "watching" the lecture, students can be asked to do something related to the stream. For instance, we have inserted applets that demonstrate a point made by the instructor. We asked the students to pause the stream and take some action with the applet. Afterwards, the stream is continued. Obviously, this works only if the course material lends itself to applets. Another use is for questioning. At the end of a videostreaming segment, we posed a question to the students that required some thought and opinion formation. A simple html form asked for their response. Students entered their responses and "submitted" them, receiving an instructor's answer in return. However, in this case the response was not recorded.

This version of the Unibrowser was used in a master's level course on GUI Building. Fortunately, the computer language used was Java and the material lent itself well to applets and interactive demonstrations. The course was designed quite differently from previous courses. We reserved synchronous interaction for asking questions, both of and by the students. Here we alternated audio conferencing with chat rooms on a weekly basis. Rather than videostreaming a long lecture, we prepared a videostream introduction for each module that lasted no more than 15 minutes. Since there was no local "class", these streams allowed the instructor to concentrate on the camera and talk directly to the student. Each module introduction was followed by an "on line" lesson that contained most of the content. Where possible, we incorporated applets in the lessons for interactivity and posed discussion questions related to the material.

Future Work

The Unibrowser continues to evolve. The questioning technique described above motivated us to develop a "QuikQuiz" program that goes beyond the canned version we used in the GUI course. While currently still in development, this client applet poses either discussion or multiple-choice questions at points in the videostream, using the interaction area. The student's response is sent back to a server, where it is saved for instructor review, and the instructor's answer is posted. Summary responses from other students can also be shared at that time, if so desired by the instructor. Although this feature may be added to any web page, we will be pilot testing it as part of the Unibrowser Interaction frame next fall.

The Unibrowser brings together a number of different functions that add humanness and interactivity to e-learning. Students are used to seeing the instructor and the verbal and non-verbal cues we receive helps establish a social presence. Interactivity is absolutely required for learning and we need to design it into all learning events, synchronous or asynchronous. The Unibrowser lets us do this. In this case, the whole (the Unibrowser) is greater than the sum of its parts (the individual frames).

References

[1] Gunawardena, C. N. & Zittle, F. J., " Social presence as a predictor of satisfaction within a computer-mediated conferencing environment", *The American Journal of Distance Education*, 11, (3), 8-26,1997.

[2] Lister, B.C., Danchak, M.M., Scalzo, K.A., Jennings, W.C., and Wilson, J.M., "The Rensselaer 80/20 Model of Interactive Distance Learning", *EDUCAUSE'99*, Long Beach, CA, October, 1999.

[3] Danchak, M.M., Jennings, William C., Johnson, Arlen D., and Scalzo, Kim A., "Teaching and Learning in a Technological World: The Rensselaer 80/20 Model for the Working Professional", *IEEE Frontiers in Education, 1999,*

Prototype of *Cyber Teaching Assistant*

Yoshiaki Shindo, Hiroshi Matsuda
Nippon Institute of Technology
shindo@nit.ac.jp

Abstract

*To improve the courses of computer-based education, we constructed the support facilities, which enable interactive communication between teacher and students. When student falls into the difficult situation, he can send the help request to the teacher's cockpit. When teacher detects the trouble situation, he can answer the question by voice communication or tries to resolve it by remote control operation. However, if many questions rush to the teacher's cockpit, he has to contact with student in order. Therefore, students have to wait for a long time. To improve this problem, we developed the prototype of **Cyber Teaching Assistant** (CTA). CTA is a Cyber Person software based on 3D computer graphics installed in the student's PC. CTA reads aloud the Japanese text by voice synthesizer with Facial Expression and plays specified body action according to the scenario saved in the network file server. Teacher can start any CTA with appropriate scenario by remote control operation. When teacher detects the student's question or help request and if student's difficult situation would be resolved by a typical or easy answer, he can start the CTA with appropriate scenario. As a result, CTA provides the reduction of student's waiting time.*

1. Introduction

We opened several courses of computer literacy and entry-level programming education [1][2][3]. To improve these courses and to aid the understanding, we constructed the computer seminar room. (Figure 1)

Student's Computers   Teacher's Cockpit
Figure 1. Computer seminar room

Some support facilities are equipped beside the student's computer desk those are as follows.
(1) Instruction screen located between two students.
(2) Small *keypad* including number keys and help key beside each student.

The other side, **Class Analyzer Computer** (*CAC*) is equipped in teacher's cockpit. *CAC* displays the **Seat-Map** of classroom and provides the support function for teacher, those are as follows.
(1) Teacher can send some visual subject to the student's instruction screen.
(2) Teacher can monitor any student's computer screen by clicking the seat on the Seat-Map.
(3) Teacher can detect the progress or difficult situation by watching the Seat-Map.
(4) Teacher can operate any student's computer by remote control operation.
(5) Teacher can talk with any student by microphone and headphone.

When student falls into the difficult situation, he can send the help request to the teacher's cockpit by pressing his **Keypad**. When teacher detects the trouble situation, he answers the question by voice communication or tries to resolve it by remote control operation.

However, if many questions rush to the teacher's cockpit, he has to contact with student in order. There is nothing to do for students but to wait for a long time. To improve this problem, we developed the prototype of **Cyber Teaching Assistant**. (*CTA*)

2. Cyber Teaching Assistant (*CTA*)

CTA is a **Cyber Person** based on 3D computer graphics who answers the student's question instead of the teacher when many questions rush to the teacher's cockpit.
(1) *CTA* is independent application software installed in the student's PC.
(2) *CTA* is a 3D shaped CG model with voice synthesizer.
(3) *CTA* reads aloud the Japanese text (Kana-Kanji strings) by voice synthesizer with facial expression.
(4) *CTA* plays specified body action to explain concretely, for example pointing to the keyboard picture.
(5) *CTA*'s scenario, which includes speaking text, facial expression and body action, is saved in the network file server. (Named **Scenario Database**)

Teacher can start any *CTA* with appropriate scenario by remote control operation. Because *CTA* is installed in each student's PC, teacher can start many *CTAs* with different scenarios in the classroom simultaneously. Figure 2 shows the example of *CTA*'s execution.

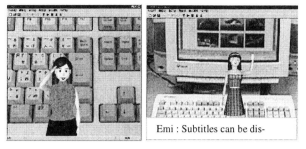

Figure 2. *Cyber Teaching Assistant*

When teacher detects the student's question or help request by watching *Seat-Map*, he can look at the student's computer screen. If student's difficult situation would be resolved by a typical or easy answer, teacher can start *CTA* with appropriate scenario. Otherwise, he may talk with this student by microphone and headphone. As a result, *CTA* provides the reduction of student's waiting time. Figure 3 shows the system outline of *CTA*.

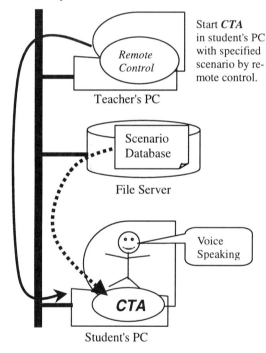

Figure 3. System Outline

As another way to solve these problems, VOD system was already reported [4]. However, it takes large cost to make and edit a digital video content. Moreover, digital video files occupy the large storage area in the network file server.
Our goal of researches is,
(1) *Scenario Database* can be written in text strings.
(2) It's easy to edit or modify the contents of *Scenario Database*.
(3) Storage size of *Scenario Database* must be small.

To accomplish these items, we developed the script language named *Cyber Person Scenario Language* (*CPSL*), which describes the scenario including the speaking text, facial expression and body action performance

3. CPSL (*Cyber Person Scenario Language*)

To describe the *Scenario Database* easily, we designed the <TAG> based script language (named *CPSL*) similar to HTML language for WEB page. The way to make a scenario is just to put an appropriate text string between a pair of <TAG> markers. Some <TAG> marker indicates the reading text and another one indicates the facial expression. Table 1 shows the examples of *CPSL* <TAG> function.

Table 1. Examples of *CPSL* <TAG> function

CPSL Tag	Function
<TALK>	Read text string with facial expression
<SPEECH>	Read text string without facial expression
<FACE>	Indicates to apply the facial expression.
<MOTION>	Indicate to play the body action.
<AU>	Indicate to apply AU/TTC sequence directly.
<PERSON>	Setup the Cyber Person
<CAMERA>	Setup or Move the Viewing Camera
<STAGE>	Setup the Background Photo

Figure 4 shows the example of *CPSL* coding (In fact, it is written in Japanese Kana-Kanji string.). This can be edited by using Text Editor (or Word Processor).

```
<MOTION PERSON="Nana">BOWING
    <SPEECH>Hello everyone. My name is Nana.
        <FACE>SMILE</FACE>
    I'm called Cyber Girl.</SPEECH>
</MOTION>
<SPEECH PERSON="Emi">My name is Emi.
    <FACE>SHY</FACE>
    <MOTION>RIGHT_TURN</MOTION>
    I'm Nana's elder sister.
    How do you do? I'm glad to meet you.
</SPEECH>
```

Figure 4. Example of CPSL coding

Our design goals of *CPSL* are as follows.
(1) The basis of *CPSL* is a *CTA*'s speech text.
(2) When *CTA* detects the *Registered-Word* within the Speech text, she reacts automatically this word. (Change the facial expression or play the body action performance.)
(3) <SPEECH>tag indicates to read speech text without automatic reaction.
(4) <FACE>tag indicates *Registered-Word* to change the facial expression without reading word.
(5) <MOTION>tag indicates *Registered-Word* to play the body action performance without reading word.

(6) <PERSON>tag creates the *Cyber Person*. It can create plural persons and provide the ability of selecting the active person.
(7) <STAGE>tag defines the background scene to display the facilities for explanation by 3D-CG models or Image photograph.)

To explain the grammar of *CPSL*, *CPSL* coding sample in English is showed as follows. (The real version supports Japanese text only.)

```
<SPEECH  PERSON="Nana">Hello everyone.
<MOTION>BOWING</MOTION></SPEECH>
<SPEECH>I get nervous now.
<FACE>SMILE</FACE></SPEECH>
```

CTA-Nana speaks "Hello everyone" making a bow, then speaks "I get nervous now" with smile face.

CPSL browser refers four kinds of Database to analyze the CPSL script strings. (See Figure 5.)
(1) Lyric Word Database.
(2) Facial Expression Database including AU sequence. (Describes at Section 4)
(3) Action Word Database.
(4) Body Motion Database including AU sequence or motion capture data.

Figure 5. *CPSL* system outline

Figure 6 shows the data flow, which generates the facial expression. *CPSL Browser* extracts the *Lyric word* by referring the *Lyric Word Database* and converts it to the *Facial Expression Name* (FEN). Then browser converts FEN to FACS-AU sequence (described at Section 4) by referring the *Facial Expression Database* to generate the facial expression.

Figure 7 shows the data flow, which generates the body action. *CPSL Browser* extracts the *Action word* by referring the *Action Word Database* and converts it to the *Body Performance Name* (BPN). Then browser converts BPN to BACS-AU sequence (described at Section 4) or *Motion Capture Data* by referring the *Body Motion Database* to play the specified performance.

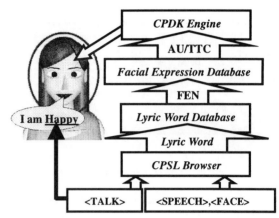

Figure 6.   Data Flow of Facial Expression

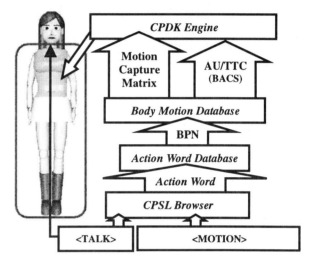

Figure 7.   Data Flow of   Body Action

4. Cyber Person Design Kit (CPDK)

To implement *Cyber Person* with facial expression and body action, we developed the *Cyber Person Design Kit* (CPDK), which is based on *OpenGL* platform.   It includes 3DCG-modeling tools, Script language of 3DCG shaped model, C++ class library for implementation and *Action Encoders*. In this paper, we describe the detail about *Action Encoders*.

4.1 Facial Expression Encoding

To encode the facial expression, *Action Unit* (AU) is defined, which is assigned number of moving parts of face muscle based on FACS code [5]. (Table2)

Table 2.   Samples of Action Unit (AU)

AU No.	Moving Parts
10	Lift up the both sides of eyebrows.
42	Lift down the right side of eyebrow.
50	Lift up the both sides of eyelid.
61	Lift up the left side of cheek.
430	Close the both eyes

A pair of AU and *Time Transition Code* (TTC) is a basic code of facial expression (Figure 8). The Sequence of this code and **FEN** are saved in *Facial expression Database*. Figure 9 shows the sample of facial expression.

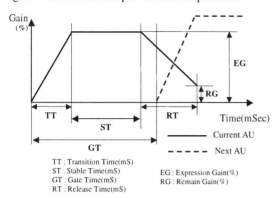

Figure 8. Time Transition Code (TTC)

(a) Smile (b) Angry (c) Sad

Figure 9. Sample of Facial Expression

4.2 Body Action Encoding

To encode the body action, another **AU,** a moving parts of body joint (named **BACS-AU**), are defined. **BACS-AU** also combines **TTC** in the same way of **FACS-AU**. Moreover, to encode the sophisticated performance, *Motion Capture Data* sequence is defined. It is a sequence of an affin transformation matrix assigned to the body joints; those are defined as a *Skeleton Link*. **BACS-AU** or *Motion Capture Data* sequence and **BPN** are saved in *Body Motion Database*. Figure 10 shows the encoding tool of **FEN** and **BPN**.

 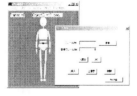

Figure10. **FEN** encoder and **BPN** encoder

5. Prototype Implementation

The specification of prototype version is; 54 codes of **FACS-AU** and 89 codes of **BACS-AU** and 41 entries of **Motion Capture** are defined. In addition, 266 **Lyric Words**, 23 entries of **FEN**, 232 **Action Words** and 130 entries of **BPN** are defined in each database.

We made use of *CTA* for some courses on trial. In computer literacy course, we made *CPSL* scenarios for

(1) How to use the keyboard and Mouse.
(2) How to operate the tool bar of Word Processor
(3) How to print the document

In computer programming course, we made *CPSL* scenarios to explain the compile errors and indicate the hint for correcting them. We are aware that most of students feel a good impression about *CTA*, but we did not acquire the effect of it objectively as yet. When we made a **FAQ** (Frequently Asked Questions) scenario as a conversation style, student's impression was better than our expectation. (Figure 11)

Figure 11. FAQ dialog scenario

6. Conclusion

We have described the summary of *Cyber Teaching Assistant* we have developed. Present version of *CTA* appears within the window frame with background, or on the desktop without background. Next, we are trying to develop the WWW Browser plug-in version for WEB presenter application. As a future research task, we are planning to enrich the contents of *Scenario Database* corresponding to the educational subject and investigate the effect of *Cyber Teaching Assistant*.

7. References

[1] Y. Shindo ": "Programming Education Based on Computer Graphics Animation", International Workshop on Advanced Learning Technologies (IEEE Computer Society Press), New Zealand ,pp.292-293,(2000).

[2] Y. Shindo, H. Matsuda & M. Mukuda: "Programming Education Based on the Computer Graphics Animation: Program Training Kit WinTK", 4th Global Chinese Conference on Computers in Education (GCCCE2000 Proceedings), Singapore, Volume 2, pp. 592-599,(2000).

[3] H. Matsuda, Y. Shindo & M. Mukuda: "Program Reading Practice using Programming Training kit WinTK".Advanced Research in Computers and Communications in Education (ICCE 99 Proceedings), Chiba (Japan), Vol.2, pp. 854-855,(1999).

[4] T.Okamoto, T.Matsui, H.Inoue, A.Cristea": "A Distance-Education Self-Learning Support System based on a VOD Server", International Workshop on Advanced Learning Technologies(IWALT2000 Proceedings, IEEE Computer Society Press), New Zealand ,pp.71-72,(2000).

[5] P.Ekman,W.V.Frisen: "Manual for the Facial Action Coding System",Consulting Psychologists Press,(1978).

Carpe Diem: Models and Methodologies for Designing Engaging and Interactive e-learning Discourse

Andrew Ravenscroft & Mark P. Matheson

Institute of Educational Technology
Open University, Walton Hall, Milton Keynes, MK7 6AA.
Email: a.ravenscroft@open.ac.uk

Abstract

In this article we argue that developments in e-learning dialogue should be predicated on pedagogically sound principles of discourse, and therefore, by implication, we need to develop methodologies which transpose—typically informal—models of educational dialogue into cognitive tools. A methodology of 'investigation by design' is described, which has been used to design computer-based dialogue games supporting conceptual change in science—based on the findings from empirical studies. An evaluation of two dialogue games for collaborative interaction—a facilitating game and a more informative elicit-inform game, has shown that they produce significant improvements in students' conceptual understanding, and they are differentially successful—depending on the nature of the conceptual difficulties experienced by the learners. The implications this study has for the role of collaborative dialogue in learning and designing interaction are discussed.

1. Why develop models of interaction and discourse?

Given the importance, and arguably the primacy, of dialogue in learning (e.g. [1,2]), how can we design cognitive tools that stimulate, support and mediate discourse processes that lead to conceptual development in collaborative e-learning contexts? Previous research has shown that a collaborative dialogue is often required to truly engage learners' conceptualisations in ways that support the refinement of knowledge and improvements in understanding through tutor-led lines of reasoning (e.g. [3, 4, 5]). These dialogues have clearly demonstrated the role and relevance of pragmatic level—or contextual—dialogue features. These include the relative and changing goals of the interlocutors, the asymmetrical roles they can play, the use of particular types of speech act—or dialogue 'move', and the socially implicit 'rules of the dialogue game' [6].

In contrast, current approaches to computer mediated communication (CMC) and dialogue design for educational contexts tend to be technology-led, and take virtually no account of the importance of such features [7]. Instead, these systems operate as mere conduits of dialogue, and fail to provide the structure and management that is often necessary to support and mediate effective educational discourse processes. Our experience at the UK Open University has shown that using generic CMC systems is problematic. They often lead to educational discourse that is superficial (e.g. incoherent and with no agreed closure), ambiguous (e.g. lack of shared meanings and little appreciation of different points of view) or simply unmanageable discourse (e.g. too many contributors and too much dialogue).

In addressing this problem, Ravenscroft & Pilkington [6] proposed a systematic, yet technology independent, approach to interaction design that focuses on pragmatic level dialogue features. This is particularly suitable for designing types of computer-based and computer-mediated collaborative dialogue for educational contexts. The scheme incorporates dialogue game theory [8, 9, 10] within a methodology of 'investigation by design' [6] to transpose—empirically derived—models of dialogue into collaborative discourse systems. The problem of transposing models of communicative interaction into usable educational technologies is a substantive issue in designing dialogical new media [11], because sound methodologies for achieving this are virtually non-existent.

2. A methodology: investigation by design

The 'investigation by design' (hereafter IBD) methodology takes a direct approach to the transposition problem. It incorporates a discourse analysis scheme called DISCOUNT [12]—that was developed by synthesising features of dialogue game theory, transactional analysis and rhetorical structure theory—to classify the relevant features of successful dialogue. Here, 'successful dialogue' refers to that which avoids some of the pitfalls of freer online or natural discourse—that were described above—or has accepted efficacy in achieving context related goals, such as "collaborative argumentation" [13, 5] "exploratory talk" [14] and "constructive conflict" [15]. Within the IBD scheme, the characteristics classified by DISCOUNT are then abstracted into a semi-natural set of dialogue features. These include participant goals (e.g. elaboration of knowledge, co-elaboration of knowledge, articulating and winning an argument), participant roles (e.g. explainer, inquirer, critiquer, proponent, opponent), dialogue tactics or moves (e.g. assertion, challenge, persuade, withdraw) along with the rules that guide and manage the interaction. These features are then designed as explicit components of a dialogue game. This dialogue model is then formally rendered to produce a system implementation.

Previous research has applied this methodology to produce an intelligent computer-based argumentation system called CoLLeGE (Computer based Lab for Language Games in Education) that has been reported in [16] and [6]. Other projects have used dialogue games to implement a computer-mediated argumentation system called DIALAB [17] and a computer modelling laboratory for investigating collaboration called CLARISSA [18].

The strength of this IBD approach rests in its pedagogical fidelity and flexibility. Other dialogue game approaches to interaction design (e.g. [19, 20, 21, 22]) usually take an existing dialogue model, such as the DC dialogue game [9]—for fair and reasonable debate, and test their applicability in educational discourse contexts. Instead, the IBD approach constructs models based on empirical evidence of successful dialogue in a particular context. It uses the features of descriptive (e.g. [8]) and prescriptive (e.g. [9, 10]) dialogue game theory—included in the DISCOUNT scheme—as 'building blocks' for specifying the models. And in specifying new dialogue games, the features of existing models are re-used, developed and synthesised. We argue that the rigorous and systematic nature of this approach explains the successful operation of the delivered dialogue games that has been found in empirical studies reported in Hartley & Ravenscroft [13], Ravenscroft [5] and the study described below.

3. Two dialogue games for collaborative discourse

Two models of collaborative inquiry dialogue to support conceptual change and development in science were developed using the IBD methodology. The models are specified as dialogue games involving a tutor-system and learner participant.

One of the games is a *facilitating* dialogue game (hereafter *f-dg*) for collaborative argumentation [5]. Within this scheme, the student is *questioned* and encouraged to express their understanding of a domain and to refine this in response to the tutor-system reasoning about the learner's explanations—examining their completeness, consistency and generality—and consequently *challenging*, *critiquing* and *probing* the student's explanatory model. A significant feature of this game is that the student cannot be informed of the correct answer. Instead, the qualitative 'logic' of the domain, or 'illogic' of the student's explanations, is reflected back to them in ways that encourage the identification of incompleteness or inconsistencies in their model, stimulating the student to refine their model for themselves. This game is typically conducted for up to forty five minutes, where the duration depends on how successful the student is in developing a complete, consistent and general model. It is described in detail in Ravenscroft [16]. This model has already been validated from an educational [5] and computational [6] perspective, and findings from these studies gave rise to a modified version of this game that was also evaluated in the reported study.

The empirical studies validating the *f-dg* demonstrated that a purely facilitating pedagogy, whilst technically attractive—as it avoids the necessity for a sophisticated and detailed domain model, would benefit from some additional tactics that were more 'didactic' in nature. Therefore, we modified the *f-dg* to develop a more informative *elicit-inform* game (hereafter *ei-dg*). In this case the student is *questioned* and encouraged to express their understanding of a domain—as in the *f-dg*. But, after reasoning about the learner's contributions, the tutor-system either *sanctions* their explanations, or points out that they were 'incorrect' and so *informs* them of a consistent, or 'correct' answer. So, this tutor-system follows the same strategy as in the *f-dg*, based on the same reasoning, but uses different tactics to address incompleteness and inconsistencies in the learner's explanatory model. This game is typically conducted for about twenty minutes, and is reported in detail in [23].

So, by evaluating and comparing the effectiveness of each model, we could systematically vary certain features—such as the type of tutoring tactic, whilst keeping other features constant—such as the tutor-system reasoning and dialogue strategy. This allows us to establish and examine the particular features that make collaborative dialogue successful in relevant contexts.

4. Evaluating the dialogue games

An empirical study evaluated these dialogue games in the context of a school physics curriculum. This study is fully reported in [23].

4.1. Context and research rationale

The study was integrated with a curriculum for fifteen to sixteen year olds, and concentrated on the topic of "forces and motion". This topic was chosen because it had been shown that a collaborative tutoring dialogue is necessary to support students in overcoming pervasive alternative conceptions, such as "a force implies motion", and assist the development of a more complete, consistent and general conceptual understanding [4, 24]. The thirty six students who participated in the study had all been taught the topic. The relative effectiveness of the *f-dg* and *ei-dg* was evaluated by comparing them with a group that received only the conventional teaching. A researcher performed the tutor-system role in both games. Twelve students were assigned to each condition, which had the form: pre-test - dialogue game - post-test - delayed post-test, with the delayed post-test being conducted about six weeks after each experimental session. In the conventional teaching condition, as a substitute to a dialogue game intervention, the students' received about forty minutes classroom teaching on an unrelated science topic between the pre and post test.

4.2. Results

The results of a quantitative analysis showed that the introduction of the dialogue games produced significant improvements in students knowledge of the topic compared with conventional teaching alone ($F(2,33)=7.97$, $p < 0.05$), with these improvements in the students' explanatory models being retained in delayed post-tests ($F(2,33)=3.32$, $p < 0.05$). Post hoc analysis showed there was a significant difference between *f-dg* and conventional teaching and a significant difference between *ei-dg* and conventional teaching at post-test and delayed post-test.

Chi square analyses showed that both dialogue games produced significant improvements in stimulating the students to develop a more complete understanding, by introducing friction into their explanatory models for example (*f-dg*: $x^2 = 6.1$, df=1, p<0.05; *ei-dg*: $x^2 = 10.3$, df =1, p<0.05). Although the *ei-dg* was marginally more effective in this respect, some slight regression—to an incomplete model—bought this condition in-line with the *f-dg* condition at post-test.

In addition, there were significant differences between conditions in addressing known alternative conceptions between pre to post-test ($F(2,33)=5.39$, $p < 0.05$) and pre to delayed post-test ($F(2,33)=3.32$, $p < 0.05$). Post hoc analysis showed significance for both dialogue games at post-test, and only significance for the *f-dg* at delayed post-test. So, although there was some slight regression to an inconsistent model at delayed post-test, there remained a significant improvement for the *f-dg* compared with the conventional teaching condition.

It was interesting that all students had difficulty in generalising their models by expressing them in terms of "net forces", because they did not understand the concept of "net". And for this reason, the dialogue games were shorter in duration than was anticipated, being conducted for up to thirty minutes.

4.3. Summary

These results clearly demonstrate the effectiveness of both dialogue games in stimulating improvements in the students understanding of the physics of motion. It was also noted that the dialogue games worked differentially in addressing the conceptual difficulties experienced by students. The *f-dg* was more effective in addressing alternative conceptions about the context, such as "force implies motion". In contrast, the *ei-dg* was slightly more effective in addressing incompleteness in the student's models, such as the exclusion of friction. The difficulties that students' experienced with the concept of "net" and consequently "net force" accords with previous findings reported by Hartley & Ravenscroft [24].

Ongoing qualitative analysis is further exploring the relative effectiveness of each dialogue game and the implications for interaction design.

5. Discussion

The outcomes of this research carry implications for the role of collaborative dialogue in learning and designing e-learning interactions.

Taking the results of the evaluation study collectively, a striking finding is that the addition of up to thirty minutes of a stylized collaborative inquiry dialogue about a subject produces significant improvements in students' knowledge and conceptual understanding compared with 'conventional' teaching alone. A more argumentative dialogue, employing *challenge*, *persuade* and *resolve* tactics to 'reflect back' the underlying logic, and the students 'illogic', was necessary to address pervasive alternative conceptions experienced by the students. Whereas *informing* was necessary to introduce key concepts that the students needed to integrate into their conceptualisations in order to develop a complete and—logically—consistent model. Current projects are exploiting these findings to develop the CoLLeGE system into an 'arguing tutor' and implement an interface supporting computer-mediated dialectical discussion about contraversial subjects [25].

The methodology reported in this paper is important because it is addressing the need for a 'science' of learning technology design, which incorporates an *implementation independent 'design level'*. This is an important advantage of current dialogue game approaches. Given that the pace of change of underpinning technology is unlikely to slow down, the need for relatively more stable and empirically founded interaction models is becoming increasingly important. This and other studies have shown that these models can be developed and tested systematically, irrespective of technological trends and changes.

In arguing for the above we are aiming for a much closer fit between empirical research, design, implementation and evaluation in educational technology research and development. One way to interpret this emphasis on empirically founded, and testable dialogue models is that we are actually treating *'design as theory'* [7]. That is, we are considering pedagogy, technology and context in the design of educational interactions, in ways that treat designs, like theories, as something that are developed, validated, evaluated and refined—rather than 'delivered'. The dialogue models are also prescriptive, so we can generate predications about the impact on learner knowledge and behaviour before and during the evaluation of implemented systems—rather than just 'trying the system out and seeing what happens'. Additionally, we are proposing an approach that accepts that a design does not automatically generalise across contexts, but instead, can be evaluated and systematically developed to address different situations. And given that there will be identifiable "family resemblance's" [26] between dialogue contexts, we argue that this is an economical way to achieve high quality educational interactions.

Acknowledgements

The authors are especially grateful to Dr. Richard Joiner at the University of Bath, for his advice and assistance with the statistical analysis.

References

[1] Vygotsky, L. (1962) *Thought and Language*. Cambridge, MA: Harvard University Press.
[2] Vygotsky, L. (1978) *Mind in Society*. Cambridge, MA: Harvard University Press.
[3] Pilkington, R. M. & Parker-Jones, C. (1996) Interacting with Computer-Based Simulation, *Computers & Education*, 27, 1, 1-14.
[4] Hartley, J.R. (1998) Qualitative reasoning and conceptual change: computer based support in understanding science. *Interactive Learning Environments*, 5, (1 & 2), 53-64.
[5] Ravenscroft A. (2000) Designing Argumentation for Conceptual Development. *Computers & Education*, 34 (2000), 241-255.
[6] Ravenscroft, A. & Pilkington, R.M. (2000) Investigation by Design: Developing Dialogue Models to Support Reasoning and Conceptual Change, *International Journal of Artificial Intelligence in Education*: Vol. 11, Part 1, pp. 273 - 298
[7] Ravenscroft, A. (2001). Designing e-learning interactions in 21C: Revisiting and re-thinking the role of theory, *European Journal of Education*, Vol. 36, No. 2 (in press).
[8] Levin, J.A. & Moore, J.A. (1977) Dialogue-Games: Metacommunication Structures for Natural Language Interaction. *Cognitive Science*, 1, (4), 395-420.

[9] MacKenzie, J. D. (1979) Question-Begging in non-cumulative systems, *Journal of Philosophical Logic*, vol. 8, 117-133.
[10] Walton, D. (1984) *Logical Dialogue-Games and Fallacies*. (published thesis), Lanham: University Press America.
[11] Cook, J. (1998) *Knowledge mentoring as a framework for designing computer-based agents supporting musical composition learning*. Unpublished PhD thesis, Computing Department, Open University.
[12] Pilkington, R.M. (1999) *Analysing Educational Discourse: The DISCOUNT Scheme*. Version 3, January 1999. CBL Technical Report No. 99/2.
[13] Hartley, J, R. & Ravenscroft, A (1999) Supporting Exploratory and Expressive Learning: A Complimentary Approach. *International Journal of Continuing Engineering Education and Lifelong Learning*. Vol. 9, Nos. 3/4, 275-291.
[14] Wegerif, R. (1996) Using computers to help coach exploratory talk across the curriculum. *Computers & Education*, 26 (1-3), 51-60.
[15] Kuhn, D., Shaw, V., & Felton, M. (1997). Effects of dyadic interaction on argumentative reasoning. *Cognition and Instruction, 15* (3), 287-315.
[16] Ravenscroft, A. (1997) *Learning as Knowledge Refinement: A Computer Based Approach*, Unpublished Ph.D. Thesis, Computer Based Learning Unit, University of Leeds, UK, 1997.
[17] Pilkington, R.M., Hartley, J.R., Hintze, D. & Moore, D. (1992) Learning to Argue and Arguing to Learn: An interface for computer-based dialogue games. *International Journal of Artificial Intelligence in Education*, 3 (3), 275-85.
[18] Burton, M., Brna, P. & Pilkington, R. (2000) Clarissa: a laboratory for the modelling of collaboration, *International Journal of Artificial Intelligence in Education*, 11, 79-105.
[19] Moore, D. J. (1993) *Dialogue Game Theory for Intelligent Tutoring Systems*. Unpublished Ph.D. thesis. Computer Based Learning Unit, University of Leeds, UK.
[20] Moore, D. J. (2000) A Framework for using multimedia within argumentation systems, *Journal of Educational Multimedia and Hypermedia*, 9/2, 83-98.
[21] Moore, D. J & Hobbs, D. J. (1996) Computational use of philosophical dialogue theories, *Informal Logic*, 18/2, 131-163.
[22] Maudet, N., & Evrard, F. (1998) A generic framework for dialogue game implementation, In Hulstijn, J. & Nijholt, A. (Eds) (1998), *Second Workshop on Formal Semantic and Pragmatics of Dialogue* (TWLT13) May 13-15, 1998, University of Twente.
[23] Matheson, M. & Ravenscroft, A. (2001) *Evaluating and investigating learning through collaborative argumentation: an empirical study*, Dialogue & Design Research Group Technical Report DDRG-01-01, The Open University, UK.
[24] Ravenscroft, A., & Hartley, J. R. (1999) Learning as Knowledge Refinement: Designing a Dialectical Pedagogy for Conceptual Change. In Lajoie, S. & Vivet, M. (eds.), *Frontiers in Artificial Intelligence and Applications Volume 50*, Amsterdam, IOS Press, 155-162.
[25] McAlister, S. (2001) *Argumentation and Design for Learning*, Dialogue & Design Research Group Technical Report DDRG-01-2, Institute of Educational Technology, Open University, UK.
[26] Wittgenstein, L. (1953) *Philosophical Investigations*. translated by G. E. M. Anscombe. Blackwell. Oxford. UK.

SAC: A Self-Paced and Adaptive Courseware System

Alvin T.S. Chan[*], Steven Y.C. Chan and Jiannong Cao
The Hong Kong Polytechnic University, Department of Computing
Hong Kong, SAR China
[*]cstschan@comp.polyu.edu.hk

Abstract

*This paper presents the design and implementation of a **S**elf-paced and **A**daptive **C**ourseware, in short SAC. The main focus of SAC is to formulate a model that encompasses the important requirements of supporting an adaptive learning courseware environment. The paper begins by discussing the issues and design requirements of developing an interactive and adaptive learning system that is able to individualize a student learning style, with the ultimate objective of maximizing his learning experience and effectiveness. The system has been designed and implemented on a three-tier web application architecture, which uses the AHAM approach to structure the domain, user and teaching model for use by the adaptation engine.*

1. Introduction

The widespread usage and popularity of the Internet and WWW have shifted information systems development from client-server based on a multi-tier, cross-platform and web-enabled application development environment. Such fundamental shift has also infiltrated to the development of courseware, in which WWW and the Internet provide unique opportunity to deliver distant education with wide accessibility at lower costs. In fact, the flexibility, ubiquity and cost of the technology have often been sited as strong motivations for its use. Despite the potentially significant advantages, web-based courseware packages tend to vary greatly both in quality and educational effectiveness.

Frequently such courseware, although initially novel to the learner, do not hold the learners' attention, fail to enhance his/her higher order learning skills and are not properly integrated within an educationally sound curriculum [1]. In most platforms, course materials are presented primarily in a sequential manner according to the author's perception of the organization of the materials, while the focus is solely on delivering multimedia contents to end users. Although this addressed the needs of some students, others may find it difficult to follow and that the flow structure may differ significantly from the student's preferred learning style. With the diverse backgrounds, knowledge and learning rate, it is reasonable to assume that each student may traverse the course materials according to his or her unique learning style. Unfortunately, most of the courseware available in the market do not readily consider this dimension of learning needs, in which the traditional style of delivering materials to a classroom of students is adopted. In particular, appropriate tools are not made available to students to help them make active choices of what course materials would be most conducive to their learning [4], while taking into consideration of individual needs. Occasionally, some of the courseware do provide limited preferences customization tools, dynamic hyper-linking to different references, but ignoring the most important element of a courseware – an adaptive linking, non-linear and self-paced capability. This paper presents the design and implementation of a **S**elf-paced and **A**daptive **C**ourseware, in short SAC. The main focus of SAC is to formulate a model that encompasses the important requirements of supporting an adaptive learning courseware environment. The paper begins by discussing the issues and design requirements of developing an interactive and adaptive learning system that is able to individualize a student learning style, with the ultimate objective of maximizing its learning experience and effectiveness. In section 3, we highlight some of the methods and techniques that are employed to develop adaptive hypermedia system (AHS). Section 4 presents the conceptual design and implementation of the SAC courseware. Finally, the paper ends with a conclusion in section 5.

2. Requirements for web-based educational courseware

A distinct feature of Web-based courseware is in its ability to be accessed in a wide area environment and across heterogeneous systems. In fact, this is the single most important driving factor for the increase awareness and opportunity presented by web-based courseware to provide *anywhere*, *anytime* and *anyplace* education platform. Unlike other teaching methods, courseware delivery through the WWW is highly heterogeneous, both in terms of the readers' knowledge and learning style, and also the information provided. Unless delivered in a closely monitored and observed environment, the academic standards of students are highly variable, such that no single learning style or program layout is suitable for all the students. In general sense, in fact, students are their own best guides for the learning process, in which online courseware should readily focus on learner-centered design. Such approach should emphasize on a constructive theory of learning wherein students learn through a process of building their own mental models[3], which is adaptive to their own learning progresses and academic backgrounds. Path-like mechanisms that support personal space organization dynamically evolve to adapt individual student needs for acquiring the ultimate learning objectives, according to their academic backgrounds, learning progresses, and preferences.

A dynamically adaptable educational system should provide personalized navigation guidance, knowledge construction assistance, and courseware analysis tools, which emphasizes on courseware structure and navigational behavior in an educational hypermedia environment. In formulating the personalized learning path, the system should take into consideration the courseware network to determine whether it is well structured, and automatically generate a hierarchical guidance map to help users navigate in a hypermedia environment [2]. The resultant hierarchical learning path allows opportunity for students to explore sequentially in an incremental manner, with adaptive capability to evolve the learning style as new knowledge is acquired.

An important feature of an adaptable educational hypermedia system is in its ability to monitor students' behaviors, learning styles and exploration of the learning resources and materials presented to them. An integrated mechanism is required to provide feedback as to the frequency of use of specific resources, and to identify trends in the student navigation of the course pages. In addition to supporting sequential browsing of the course contents, task and objective-oriented paths should be made available for students to acquire in-depth exposure to the specific subject matters.

The focus of this paper is to investigate the techniques and approaches to developing web-based multimedia courseware package, and to suggest the design and implementation of a highly adaptive, self-paced Internet courseware based on an improved Dexter-based Reference Model [7].

3. Approaches and Techniques for Adaptive Courseware

Adaptive techniques have been widely used by a number of researchers in an attempt to offer navigation guidance and orientation support for rich link structures in courseware systems. A number of techniques have been developed and reviewed in [6], most of which achieve adaptation of hypermedia documents based on user's state or level of expertise. Most techniques offer some variant of conditional text, resulting in variants of text fragments, whole pages, or even clusters of pages. Also, the link structure of a hyper document can be adaptive to offer personalized navigation guidance. Adaptive link structure can be positively guided by means of sorting available links according to user preference or relevance, or negatively guided by means of annotating, dimming or even hiding links which the user should not (yet) follow.

3.1. Structuring of Courseware

According to [5] in the implementation of the Modular Training System, an Internet-Based Courseware can be structured by organizing into three categories:
- *Course Nodes* (CN
- *Course Unit* (CU)
- *Course Material* (CM)

The network of *Course Nodes* defines a course structure and a pre-defined learning goal. While it defines the nodes dependencies and flow, the *Course Node* does not contain any information on how this learning goal can be achieved. Each node delivers a return value such that appropriate event and the next node can be selected. Each node may contain attributes, which indicate the objectives and pre-requisites of the node, as well as the available child nodes that are linked. The contents of each node can be dynamically linked to other child nodes to form a networked course structure according to the profile of individual students. Navigation through the course depends on learners' own path, which may be linked dynamically as a student traverses the course materials.

Course Unit defines the presentation units, exploration units and test units. They are self-contained, and contain references to other course materials. In this sense, *Course Units* form the containers of the course materials that deliver a return value indicating the learner's success when traversing the unit.

3.2. Adaptation Models

The Adaptive Hypermedia Application Model (AHAM) discussed in [7], encompasses most features supported by adaptive systems that exist today. The Adaptive Hypermedia Application Model (AHAM) is based on the *Dexter Model* [7], a widely used reference model for hypermedia. The model achieves personalized adaptation based on a *User Model*, which persists beyond the duration of a session. The system is comprised of three integrative models: *Domain Model*, *a User Model* and a *Teaching Model*.

The system keeps track of evolving aspects of the user, such as preferences and domain knowledge. This permanent and continuously updated record is captured in the *User Model*. It is used to guide the user towards learning goal and away from information the system considers to be inappropriate or irrelevant for the user. The AHS may do this by dynamically altering the hyperdocument's link structure, *Adaptive Link Annotation*, and/or by dynamically generating or changing the content of the information nodes, *Adaptive Linking*. Users can set certain preferences explicitly or initialize the user model through a registration form or "pre-test" so that the background knowledge can be captured for initial generation of the structure of the hypermedia documents. The *Domain Model* describes how the information is structured and linked together. The *Teaching Model* consists of pedagogical rules that define how the domain model and the user model are combined to provide ways to perform the actual adaptation.

4. Conceptual Design of Self-paced and Adaptive web-based Courseware (SAC)

This section describes the design and implementation of the Self-paced and Adaptive, web-based Courseware (SAC) that is designed to integrate the concepts of Domain, User, and Teaching models described in AHAM for adaptive hypermedia applications [7]. The course structure, materials and learning goals are captured in the abstraction of *Course Nodes*, *Course Units* and *Course Material* as described in the Modular Training System (MTS) [5]. The overall design the SAC is shown in Figure 1.

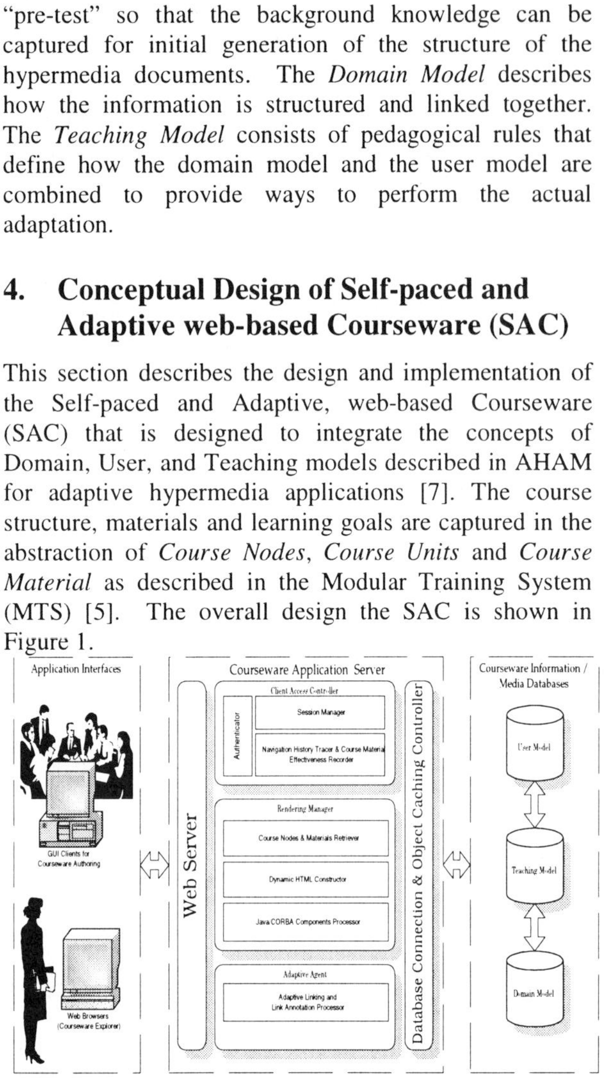

Figure 1: SAC Architecture

The courseware is implemented through a 3-Tier Application Model architecture. The first tier is implemented through the browser, which is concerned with presentation of the courseware. The middle tier is implemented and provided by a combination of the web server and an application server. The application server generates XML, DHTML and client-side JavaScript dynamically by using the model information stored in the 3rd tier backend database system.

The adaptive annotation feature of SAC is implemented through a dynamically generated XML document and a browser Java applet. Specifically, the courseware navigation is achieved by providing a tree-based menu, which is generated dynamically by the middle-tier application server in the form of a XML-formatted document. The tree-based menu reflects the current status of the overall learning progress and the adapted learning path for the student. All linked hyper-documents are annotated so that students can easily access the required course materials by surfing through the course nodes. According to student's navigation history and the assessments taken, the navigation menu of the courseware dynamically adapts to his or her learning experience and style. Suggested learning path, as described in the XML document, is parsed by an applet and display as a tree structure, such that student taking the course can view the learning map, with the ultimate objective of maximizing their learning experience.

4.1. Adaptive Mechanisms

The adaptive capability of SAC is provided by the Adaptive Agent. It is comprised of the following adaptive engines: pre-admission Adaptive Engine, Online Individual Adaptive Engine and the Domain Adaptive Engine.

The Admission Manager captures the educational background of individual student via pre-admission tests and SAC enrolment history. It is combined with the pre-defined course inter-node relationships stored in the Teaching Model to form students' academic background and course nodes pre-requisite analysis results, which is then stored into the User Model database.

By using the Student Preferences Customization function provided by SAC, individual student can input their own navigation preferences into the User Model. The Session Manager continuously keeps track of the Online Surfing Information and updates the Navigation History stored in the model.

The Online Individual Adaptive Engine retrieves individual navigation history, academic background and pre-requisites information for analysis. Subsequently, the Course Node Status information is formed which is used as the core information for rendering of the XML documents. The XML rendering and DHTML Generation process combine the Node Status information, user preferences and the Course Inter-Nodes Relationship to produce the Dynamic, Adaptive Link-Annotated web pages and returns to the SAC Browsers.

The batch mode Domain Adaptive Engine uses the historical Navigation Pattern from previous students and the default domain adaptive information to provide revision suggestions to the Teaching Model. The Domain Adaptive information will also be updated during the analysis process.

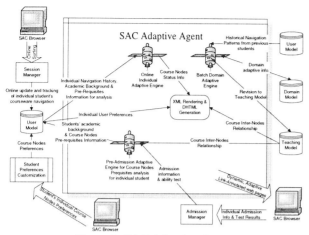

Figure 2: SAC Adaptive Agent

4.2. Dynamic Site Map

The dynamic adaptation path of the course guidance menu is generated by a Java applet residing within the client-side, which interacts actively with the server-side adaptation engine. Each SAC user is presented with their unique dynamic web content navigation guidance based on their learning style, pace and foreknowledge. Course nodes are presented as leaves and folders in the navigation menu. As the SAC learner surf through the course materials, the client-side Session Manager, implemented as a combination of Javascript and applet, will actively trace his navigation history and record the test results, if any, and finally update the User Model database via the application server. Next time when the user logon again, the generated tree menu will reflect the latest navigation history and test results by re-generating the required XML document as input to the tree applet, according to the Teaching & User Models stored in the backend database.

5. Conclusion

In this paper, we have highlighted the importance of courseware to incorporate the concept of adaptive learning by providing dynamic navigational guidance to students taking online courses. This is in contrast with most existing courseware systems, whose focus is on pure information delivery, which often leads to information disorientation and confusion for learners. As courseware systems are Internet-enabled to ride on the wide accessibility of Web infrastructure, the mix of students' backgrounds, knowledge and learning capabilities are often diverse and heterogeneous in nature. Online courses designed to be delivered across such open platform need to support the desire of students to self-pace their learning rate and style, and to help them to develop quality mental maps of the course delivery structure. This paper describes some of the approaches and development in the area of adaptive hypermedia guidance courseware. We introduce the design and development of a Self-paced and Adaptive Courseware (SAC) system in an attempt to develop an interactive and learning system that is able to individualize a student's learning style using adaptive link guidance approach. The system has been designed and implemented on a three-tier web application architecture, which uses the AHAM to structure the domain, user and teaching model for use by the adaptation engine. In the future, we are intending to develop a formal evaluation test on the effectiveness of the system in enhancing students' learning experiences using the adaptive framework of the SAC system. The results will assist us to better understand the nature and evolution of students' mental maps of the taught materials and to formulate ways in which the adaptation engine can be designed to develop guidance links to maximize their learning experience.

6. References

[1] Vincent P. Wade & Conor Power, Evaluating the Design and Delivery of WWW Based Educational Environments and Courseware, *ITiCSE '98* Dublin, Ireland, 1998 ACM 1-58113-000-7/98/0008

[2] Chuen-Tsai Sun, Chien Chou, and Bing-Kuen Lin, Structural and Navigational Analysis of Hypermedia Courseware, *IEEE Transactions on Education*, Vol. 41, No. 4, Nov 1998

[3] Owen Astrachan, Concrete Teaching: Hooks and Props as Instructional Technology, *ITiCSE '98* Dublin, Ireland, 1998 ACM 1-58113-000-7/98/0008

[4] Curtis A. Carver, Jr., Richard A. Howard, and William D. Lane, SM, Enhancing Student Learning Through Hypermedia Courseware and Incorporation of Student Learning Styles, *IEEE Transactions on Education*, Vol. 42, No. 1, Feb 1999

[5] Taofen Wang & Christoph Hornung, The Modular Training System (MTS) - A System Architecture for Internet-Based Learning and Training, *Fraunhofer-Institute for Computer Graphics (Fh-IGD)*, German, 0-8186-8150-0/97 1997 IEEE

[6] Licia Calvi, Improving the Usability of Hypertext Courseware through Adaptive Linking, *Hyprtes! 97*, Southampton UK @1997 ACM0-89791-866-5

[7] Paul De Bra, Geert-Jan Houbent and Hongjing Wu, *AHAM:* A Dexter-based Reference Model for Adaptive Hypermedia, *Hypertext 99* Darmstadt Germany Copyright ACM 1999 1-58113-064-3/99/2

Collaborative Analysis and Tutoring: The FACT Framework

Miguel Angel Mora
*Departamento de Ingeniería Informática.
Universidad Autónoma de Madrid
28049 Madrid, Spain.*
Miguel.Mora@ii.uam.es

Roberto Moriyón
*Departamento de Ingeniería Informática.
Universidad Autónoma de Madrid
28049 Madrid, Spain.*
Roberto.Moriyon@uam.es

Abstract

In this paper we introduce FACT, a Framework for the Analysis of the learning process and Collaborative Tutoring. The main feature of FACT is the use of representations of learning histories, analysis and related annotations, that enhance the guidance of the teaching process. FACT learning histories play an important role from the didactic point of view, since they allow tutors to review what students have done and to guide their work. Annotations to the histories also allow tutors to include links to simplified learning scenarios. In this way, students learn by working on simple examples related to their previous wrong decisions in their learning and practicing process.

1. Introduction

This paper is addressed to the study of ways to achieve guided collaboration tutoring systems that allow pupils to collaborate among themselves or with tutors. This kind of systems can increase both the speed students learn the tasks and concepts they are supposed to master, and the deepness they reach in their knowledge.

The development of telecommunications and multimedia technologies has incremented the ability to build such systems up to a level that was hard to think not long ago. However, traditional collaboration systems present some limitations about the role played by a tutor that in some cases can be crucial when using such a system in real life. The main limitation comes from the fact that a tutor who is in charge of several students can not be paying attention simultaneously to the work all of them are doing. A system that helps teachers to spend their time guiding those students that need more support at each moment, and to review afterwards what other students have been doing and start another guided collaborative tutoring session with some of them, giving them guidance based on their actions, can be of great help.

FACT, a framework for the development of guided collaborative tutoring applications that allows the analysis of learning sessions, will be introduced. The applications developed with FACT allow several people connected through a computer network like Internet, including a tutor, to participate simultaneously or asynchronously in learning sessions. The students can work isolated or by forming groups that share comments or analysis of different alternatives jointly.

In the group learning strategy, students can make proposals for the next actions to take by just putting them in effect and showing them to the other students afterwards. In this way, students get all the advantages of the classical group learning techniques, and the teacher has the advantage that s/he can revise their work, take into account the students actions and correct the most important mistakes that s/he finds out among them. The individual learning strategy has the advantage that it enforces students to act, but on the other hand they get less stimulus from the environment.

Independently of which of these two strategies is used, FACT gives bigger flexibility in the learning process in two main aspects: first, the students as a whole can get a more direct and deeper idea of what they have done in the wrong way or just what they can do better. In the second place, they can also get more easily an idea about how they can do their tasks in a better way. This second dimension is enhanced especially by their possibility to work on examples of simpler situations where similar problems arise, and practice with them, either by themselves or synchronously with the teacher.

Tutors can teach the students how to solve problems by making them follow synchronously their resolution. They can also guide the students directly while they work, by showing them synchronously alternatives to the decisions they made, or indirectly, by creating annotations that can include links to stored examples that illustrate similar situations or by creating alternatives to be studied by the students at a later time. In addition, they can review the analysis work of other students, reading their annotations. In order to facilitate the tutor's work, applications based on FACT can use agents that inform the tutor when a student or group of students needs help.

A Java application based on FACT, ChessEdu, [1], has been developed as a first step in the validation of the framework. ChessEdu allows students to practice and learn chess by playing individually or by groups, while a tutor follows several games and interacts with them. In order to use ChessEdu, a ChessEdu Session Server must be running in one machine. Each session represents a group of synchronous participants that work on a specific history or game. Several sessions can deal simultaneously with the same game, which takes place for example when a tutor is revising the way the students have played, or when someone is adding remarks or alternatives to previous decisions.

FACT is part of the EnCiTec project that includes among its much broader goals the development of infrastructure for the development of collaborative aids to guided computer assisted tutoring.

Learning histories that consist of video and audio have been used in asynchronous distance education, for example to reproduce the resolution of exercises by students or to replay a lecture given by a tutor [2]. The analysis of collaboration histories can be used to evaluate the social level structures and processes that take place in a CSCL environment [3]. Some tools, like CEVA [4], support synchronous and asynchronous collaborative analysis; in this case the analysis is based on a history that consists mainly of video data.

The rest of this paper is organized as follows: first, we will show some requirements to accomplish our main goals, and how the architecture of the FACT framework achieves these goals. Finally, we will expose some conclusions.

2. Requirements for Guided Collaborative Teaching

A framework for the development of tutoring applications based on guided collaborative teaching must give support for the common requirements that all these applications must satisfy. In this section we describe these requirements, grouped in two categories: user requirements are described in terms of the guided collaborative teaching process, while functional requirements concern the functionality that must be present in the tutoring applications.

We start by describing the user requirements associated with guided collaborative teaching. In order to allow tutors to guide synchronously students by reviewing with them their previous work, guided collaborative teaching applications must give support for the use and management of histories of collaboration. This also allows the teacher to analyze at any given moment the work of the students, adding comments or alternatives to be sent to them, who can play them as animations. This is the main feature that applications of this kind must provide.

Besides this, guided collaborative tutoring applications must satisfy the following user requirements:
• Students must be able to work by themselves and to collaborate among them. When several students are working together, they can share their comments, and analyze together different alternatives making proposals before any definite decisions. Comments from the students are attached to the context in which they are done, so that an asynchronous observer of their actions can see them when they are appropriate. Both comments and alternatives to be analyzed include multimedia components, like voice and animations of actions. Actually, any analysis of a working session can become an animation that can be included as part of the analysis of any moment during another working session.
• A tutor has privileges to guide the students he is teaching synchronously by enforcing them to know his proposals and comments and to accept his decisions with respect to the next decision to be taken. A tutor can also be an observer; in this case, he can add comments to the working history to be reviewed later, or he can just select some situations he wants to go back to at some other moment.
• Both the students and the tutor can switch seamlessly between being synchronized with the working session or working on the analysis of its history
• Tutors can get reports from agents that analyze the different working sessions, and call their attention about special situations that might need to be addressed directly between them and the students.

The functional requirements for guided collaborative tutoring applications are the following: first, the management of collaboration histories should include the support of collaborative Undo and Do. With this, the teacher can show a better solution to the students, going backward in the history if necessary, and producing a new branch in it. These actions form a tree whose nodes are actions taken in the resolution of a problem or in an alternative that is proposed to its development.

Other functional requirements of this kind of applications are:
- Dynamic collaboration sessions must allow groups of users to split and do their own analysis following different paths.
- The use of multimedia annotations, including synchronized voice and hypertext that can include references that in turn give access to states in the working process or in its analysis.

3. The Framework

FACT is a framework that achieves the previous requisites, the architecture of the system is formed of servers, generic clients, and an API that allows the integration of the user application with the rest of the framework; this architecture is summarized in Figure 1.

FACT is written in Java, and the applications communicate with the different servers in a transparent way by means of RMI, [5]. The framework can be used in the development of applications as well as applets, which can be started in a browser through the web.

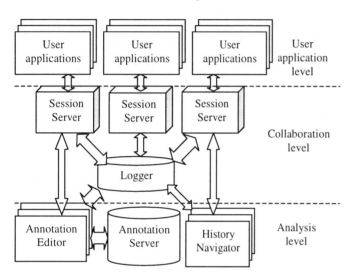

Figure 1. Framework Architecture

Each user application, including the annotation editor and the history navigator, communicates with a session server, forming groups of users that work synchronously. Moreover, session servers communicate with the central logger, and allow the asynchronous work between the different user groups.

The Annotation server and the annotation Editor allow the use of annotations, which can be associated to different nodes of the history stored in the logger. These nodes can be viewed with a history navigator, which also allows a user to navigate through the history.

The system can be divided in three different functional levels:
a) User application level: In this level the framework encapsulates the user applications with an undo/do abstraction of the application's behavior.
b) Collaboration level: This level provides the basic collaboration operations.
c) Analysis level: This level provides the analysis capability of the history of collaboration produced in the previous level, adding more information, such as annotations, reviewing or creating new branches in such history.

3.1. User Application Level

In order to create a collaborative application with FACT, the application must support undo/do operations and communicate changes to the session server, which acts like an Undo Manager. Briefly the requirements imposed to user applications are:
- Events must be communicated to the server of the session.
- A method must be provided in order to listen to and reproduce events from other participants. Both regular and redo events must be handled.
- A similar method has to be provided in order to listen to undo events.
- The events that are transmitted must include enough information to allow the performance of both undo and do actions.

In this way, user applications become shared applications, which will be used to produce new history nodes and to review nodes produced by other instances of the same application.

3.2. Collaboration level

This level provides the collaboration aspects required by the analysis level. FACT is not a generic collaborative framework, but a framework that provides analysis capabilities on top of a specialized collaboration layer, which can be used as a framework in order to provide guided collaborative tutoring. This collaboration layer has some special characteristics, like the use of a logger that store the events in a tree, rather than in a linear structure. In addition, each node of the tree can be referenced by a label, which allows the connection between the logger and the annotation module.

The logger is the central element in FACT, and it is responsible, together with the annotation server, of the asynchronous capabilities of the framework. It stores all the events produced by the different user applications through each session server. Special care must be taken in order to ensure that these events accumulate the necessary information to perform do or undo operations in any instance of the user application.

Session Servers are also parts in this level. Each Session Server supervises users access and maintains in synchronization its groups of users. This process of synchronization consists in the redo of the events produced in an instance of the user applications in the rest of instances connected to the same Session Server.

A Session Server connected to a logger points to a specific node in the tree-like history. This pointer is moved by the users when a new event is produced or when a user wants to review the history, producing undo or do events. When a new event is produced, the logger creates a new node in the history descending from the actual node in the tree, enlarging the actual branch or creating new branches if necessary.

A user can choose between working synchronously with other users, and working alone by choosing the appropriate Session Server or by launching a new one. This is done by a Session Manager, a special component that is included in user applications. Thanks to this Session Manager, the dynamic variations of the group that constitute each session are possible. The Session Manager is also in charge of the definition of access control rules. An example of such a rule is restricting the access to the rest of users, except for the tutor.

3.3. Analysis level

The analysis level is the fundamental part of FACT, this level allows the analysis of user interaction with the applications using generic tools, like the history navigator or the annotation browser, which can be used as stand alone applications or included in the user interface of the main application as components. This level provides also a generic API for application dependent agents to notify users when a special situation arises.

The history navigator borrows events from the logger, showing those events to the user, who can see a brief description of each event, and the annotations associated with it in the annotation browser. In the actual version of the framework users can choose between two version of the history navigator, one in which the user can see the whole tree that forms the history, and another one in which only a limited number of previous events can be seen, together with their alternatives, if they exist.

The navigator, like the rest of user applications, is associated to a session server and allows a group of users to navigate synchronously through the history.

The navigation process is very simple: when the user points to a specific node in the learning history the navigator sends to the session server Undo and Do events extracted from the logger, in order to switch to a different state of the application. Users can see the work done by other users because all share the same logger and the same initial state of the application.

Users can also make annotations, which can be seen by the rest of users in an asynchronous way. These annotations can be associated to a concrete point of the history in such a way that when a user passes through this point s/he could see the associated annotations.

The annotation server is the central repository of asynchronous communications between the collaborators. It stores multimedia objects, e.g. html pages, audio, or video. The annotation browsers/editors are synchronized, and the server notifies the clients when a new annotation has been added or deleted, or when the annotation that a user is viewing changes.

The Annotation Editor is like a web browser, which can also edit html pages/annotations. The html text can include audio, video, images, or links to different moments in the history. When the user follows a link, s/he is automatically conducted to the corresponding point of the history.

4. Conclusions

We have described the requirements that generic guided tutoring collaborative applications must satisfy, at both the user and functional level. We have also described FACT, a framework for the development of such applications. FACT has been used in the development of ChessEdu, a guided collaborative chess tutoring application.

FACT is based on the use of learning histories. From the didactic point of view this allows a tutor to point out to the students their mistakes, their consequences, and related simple situations they should look at in order to correct their mistakes, while the benefits obtained traditionally in group learning are kept. Moreover, by using this kind of applications, tutors are able to pay attention to a bigger number of students with essentially the same degree of effectiveness.

5. Acknowledgements

This work has been funded by the National Research Plan from Spain through the EnCiTec project, project number TEL-99-0181.

6. References

[1] M.A. Mora, R. Moriyón, "Guided collaborative chess tutoring through game history analysis", Proceedings SIIE'2000, Universidad de Castilla-La Mancha, Spain, 2000.

[2] P. Thomas, L. Carswell, J. Emms, M. Petre, B. Poniatowska, B. Price, "Distance Education over the Internet", Proceedings of the conference on Integrating technology into computer science education, ACM press, 1996, pp. 147-149.

[3] K. Nurmela, R. Lehtinen, T. Palonen, " Evaluating CSCL Log Files by Social Network Analysis", CSCL'99 Proceedings, 1999, pp. 434-442.

[4] A. Cockburn, T. Dale. (1997) "CEVA: A Tool for Collaborative Video Analysis", Group'97 Proceedings, ACM press, 1997, pp. 47-55.

[5] RMI API, Sun Microsystems, "Java 1.2 and JavaScript for C and C++ Programmers", John Wiley & Son, 1998.

Implementing Collaborative Learning Research in Web-Based Course Design and Management Systems

Kay J. Wijekumar
The Pennsylvania State University
kxw190@psu.edu

Abstract

The exponential increase in web-based distance learning environments for universities has spawned a parallel increase in web course design and management systems (WCDMS). These systems appear to serve the needs of designers, professors, and administrators by simplifying the creation, delivery, and maintenance of web-based distance learning courses. However, a review of the best selling packages and courses implemented using them shows that they fall short of meeting some of the basic findings from research on learning. This paper presents one area of research - collaborative learning that is frequently cited by the WCDMS as being supported by their products. This paper presents research on collaborative learning and methods for implementing them on the WCDMS and modifications to WCDMS are suggested.

1. Introduction

The growth of distance education (DE) has been pegged at 70% between 1997 and 1998 [4]. The numbers of educational institutions offering web-based DE has increased from 390 in 1997 to 798 this year [4]. A scan of the Chronicle of Higher Education shows no advertisements for WCDMS in 1995 to over 10 in the January 2001 issue. The need to compete in the web-based DE environments has placed educators in a frenzy to produce, market, and deliver web-based courses in a quick and efficient manner. WCDMS have therefore been developed and marketed to support this demand.

WCDMS have been created to provide a framework to develop, deliver, and manage web-based distance learning courses[1]. Course development functionality includes creation and maintenance of web-pages (notes, activities, syllabi) with multi-media (audio, video); methods of asynchronous and synchronous communication (bulletin boards, chat rooms); whiteboards, and assessment tools (multiple choice, true/false, essay). Course delivery and management functions integrated within these environments include class rosters, password security, orientation to products, automatic grading, logging of student's activities and time, and communication tools like email. They are described as "learning management software systems that synthesize the functionality of computer-mediated communications software (email, bulletin boards, newsgroups, etc.) and on-line methods of delivering course materials (e.g. the WWW) [2].

These course design and management systems offer "promise for improving the quality, flexibility, and effectiveness of higher education" [3]. They offer a broad toolset for the use of designers, professors, and administrators. Examples of these course design and management tools include Blackboard, E-College, JonesKnowledge, Prometheus, and WebCT. A review of the advertising and marketing literature for these products also show the focus on the toolset, ease of use, transferability, support, and commitment to learning.

Another view to the world of web-based course design and development products is based on the research in educational psychology, instructional technology, and learning. Reviewing the research on computer mediated communication, cognitive flexibility hypertext, and collaborative learning shows some remarkable findings that do not appear anywhere in the implementations of the WCDMS. In order to make sure that the technology developments in WCDMS are not driving the course development approach, this gap between research and application needs to be addressed immediately.

This paper presents collaborative learning as one of the concepts used frequently in WCDMS and identifies related research not addressed by the WCDMS. Some ideas are then presented on how the tools maybe modified or adapted to incorporate the research findings. Space limitations do not permit the research on active learning and cognitive flexibility hypertext to be presented here.

---

[1] These tools may be used for many other learning activities, however, this paper concentrates on web-based course development.

2. Collaborative Learning on the WWW

Collaboration is defined as "working together to accomplish shared goals" [12]. Collaborative learning in web-based DE is therefore an activity where learners work together in an educational environment designed to improve thinking skills, task skills, and human relation skills. To be effective, collaborative learning environments must establish and foster positive interdependence among members of the group, encourage active thinking, and verbalization. Finally, the learning environment should hold individuals accountable while maximizing the group productivity. These themes of collaborative learning come from social learning theories [1,27].

A review of advertising and training literature for the CDMS's suggests their tools foster collaborative learning. Collaborative activities suggested by the manufacturers of these products include bulletin board postings, chat room discussions, and email communications. For example, a course on Pollution Control may identify a topic and allow each student to research and post their facts and opinions to a bulletin board. A review of the postings by the group(s) is said to allow all students to learn from each other's postings.

A closer look at implementing collaborative learning using the WCDMS suggests a need for attention to the extensive research on collaborative learning [5-25]. These studies have shown collaborative learning is effective in discovery in science [14], encourages low-ability learners to perform better when paired with high-ability learners [9,20], is effective when specific activities/interactions are used in collaboration [18,24,25], and group dynamics are critical [11-16, 24,26]. The following questions are two of the many that need to be answered before collaborative learning in DE can become effective.

1. What are the effects of group dynamics in web-based courses? (Group composition as well as evolution of interactions)
2. What types of collaborative activities encourage learning? (Explanations, Accountability, Structure)

2.1 Group Dynamics

Research on collaborative learning has shown that the development of social, cognitive, and psychological adaptations to team activities is a critical component [26]. Two of the major issues in team dynamics are the composition of the group and the evolution of the groups.

Research on group composition has shown improvement in low ability students when grouped with high ability students [9]. The same study however showed that the high-ability students still outperformed the low-ability homogeneous groups. Many of the existing DE environments should concentrate on carefully grouping students to maximize the effectiveness of the groups.

Within the area of group composition the individual members' expectations and individual differences have also been shown to predict the effectiveness of groups. For example, low-achieving students have manifested high levels of passive behaviors in small group activities [14]. Students' communication expectations and apprehension have also been shown to affect the participation in collaborations [5,11].

In group evolution there are two major team development models that may be relevant. First, the 4 Phase Model [13], second, Drexel/Sibbet team Performance Model [6]. Both these models suggest that there are phases through which groups go through and the success of the groups depend on how they navigate these phases. For example the Drexel/Sibbet model suggests the following stages: Orientation, Trust-Building, Goal-Role Clarification, Commitment, Implementation, High Performance, and Renewal. Research on group problem solving [22] has also shown that members of groups spent as much time on problem solving as they do on getting to know each other. Additional research [15] comparing groups that worked in a division of labor, socially oriented, or goal oriented mode evolved into different patterns of communication and collaboration. Thus the composition of the groups is a critical component in the effectiveness of the groups.

Bringing together the research on group dynamics it can be suggested that for collaborative learning in web-based distance learning to be effective, the facilitators and designers must be aware of the composition of their groups and evolutionary nature of the members of their community and their interactions. It is probably wise to attend to most of the stages of group evolution models by conducting an orientation to the group's activities, allowing the members of the community to clarify their roles, goals, and preferences, and finally, achieve a re-generating balance.

2.2 Collaborative Activities

Collaborative learning has been researched extensively but there are mixed findings on its effectiveness. For example, research [9] showed positive results when high ability and low ability students were mixed in a heterogeneous group but the same did not hold for low ability homogeneous and high ability homogeneous groups.

A review of research with outcomes in favor of collaborative learning shows that the types of activities that are used in the collaboration are most important [16-21, 24-26]. These studies, related activities, and implementation using the CDMS are described next.

2.2.1 Explanation Giving

Explaining how to solve problems in groups improved learning in the person giving the explanation but the person requesting the explanation did not improve their performance [25]. Similar results have been reported in research on Computer Supported Intentional Learning Environments [19] and AI [24]. Implementing this type of activity requires the facilitator and/or group members to encourage explaining by group members. When bulletin boards are used to conduct a discussion it may be important to encourage members to explain their postings in greater detail.

Argumentation may be one of the methods that elicit more explanations and justifications by members [8, 24] and promote higher-order thinking. Research has shown than argumentation results in improved depth of knowledge [8]. The author has implemented argumentation using the WCDMS by creating a case-study to be researched by two groups. Each group presented their case-briefs to the facilitator for approval. The case-brief was then presented to the class on the bulletin board with a structured format including tags for each sentence [opinion, research citation, pros, and cons]. Each argument put forth in the case brief had to be refuted by the other group with references and within the structure laid out. This method was successful in eliciting meaningful interactions and explanation giving by members of the group.

However the bulletin boards were limited in the format for presentation. For example attachments had to be in html format, some graphics could not be viewed on the bulletin board. The addition of concept maps or some graphical synthesis of the debate is an additional feature that was not possible within the framework of the WCDMS. Finally, this implementation using bulletin boards was asynchronous by choice of the designer and facilitator. We felt that the quality of discussions would be improved when members had a chance to reflect on the postings. However, a case can be made for conducting an argumentation forum using chat rooms which are synchronous. Unfortunately, the text based chat rooms in the current WCDMS limits the quality of discussions.

2.2.2 Accountability in Collaboration

Accountability for each member of the group is critical in collaborative learning. Research has shown than low-achieving students participate less in collaborative activities [5]. The current WCDMS's provide bulletin boards but adding functionality for gated bulletin boards can be most effective in making sure each member does their own work. A common pattern observed in three courses that the author has facilitated on-line is that students paraphrase other's writings and wait for the deadline or till most of the discussions have been completed to participate. A gated discussion where all members must complete their posting before viewing other's postings can avoid this problem. A workaround the author implemented included assigning private bulletin boards for the instructor and each student. Students posted thoughts and the postings were moved to a public bulletin board after a deadline. This however was a time consuming and inefficient method.

Accountability can also be incorporated with the use of on-line journals that can be maintained privately by the students to communicate with their professors. These may be implemented in the existing WCDMS with private bulletin boards but those are limited in capabilities as described earlier in addition to the extraordinary amounts of time required to review the individual journals. The author implemented a course where learners were encouraged to summarize their journal postings each week with the creation of a living concept map. Each week the concept map was modified to include the learners' new knowledge and understandings. However, the concept maps had to be created outside the WCDMS and exported as GIF files to the professor. The addition of an on-line concept mapping tool to the WCDMS will encourage these types of knowledge and reflective exercises to aid in learning.

3. Conclusion

Even though the WCDMS available on the market today are making great strides to improve the creation and delivery of web-based DE environments they should be secondary to the research available. This paper presented research on collaborative learning that should influence the development of web-based DE environments.

Addressing the issues of group dynamics and activities for collaboration requires careful thought into planning as well as enhancements to the existing WCDMS.

Even though many of the studies cited here came from traditional learning environments they are the only source of information that currently exists in these areas. Similar reviews of research in the areas of cognitive flexibility hypertext and active learning are planned by the author.

References:

[1] Bandura, A. (1986). Social Foundations of Thought and Action: A Social Cognitive Theory. Englewood Cliffs, N.J. : Prentice-Hall.

[2] Britain & Liber, (2001). A Framework for Pedagogical Evaluation of Virtual Learning Environments. Available http://www.jtap.ac.uk/reports/htm/jtap-041.html, February 26, 2001.

[3] Dearing Report (1997). Available : http://www.leeds.ac.uk/educol/ncihe/, April 27, 2001.

[4] http://www.distance-Educator.com available February 26, 2001.

[5] Dobos, J.A. (1996). Collaborative Learning: Effects of Student Expectations and Communication Apprehension on Student Motivation. *Communications Education*. Vol. 45, Apr. 1996, 119-134.

[6] Drexel, A. & Sibbet, D. (1994). Drexel/Sibbet Team Performance Model.

[7] Engel, R.S. & Sathianathan, D. (1997). Using Teams to Achieve Course Objectives. *Center for Quality and Planning Publication* No. 33, Nov/Dec 1997.

[8] Gunawardene, Lowe & Anderson (1997) Online Debate/Interaction Analysis Model for CMC. Journal of Educational Computing.

[9] Hooper, S. & Hannafin, M.J. (1988). Cooperative CBI: The Effects of Heterogeneous vs. Homogeneous Grouping on the Learning of Progressively Complex Concepts. *Journal of Educational Computing Research*, Vol. 4(4), 413-424.

[10] Johnson, D.W. & Ahlgren A. (1976). Relationship Between Student Attitudes about Cooperation and Competition and Attitudes Toward Schooling. *Journal of Educational Psychology*, Vol. 68, No. 1, 92-102.

[11] Johnson, D.W., Maruyama, G., Johnson, R., Nelson,D. & Skon, L. (1981). Effects of cooperative, competitive, and individualistic goal structures on achievement: A meta-analysis. *Psychological Bulletin*, Vol. 89, 47-62.

[12] Johnson, D.W. & Johnson, F. (1991). Joining together: Group theory and group skills (4th ed.). Englewood Cliffs, NJ: Prentice-Hall.

[13] Litzinger, M.E. (1998). Team Development: What Is It and How Do We Get It?. Presentation: The Pennsylvania State University, January 1998.

[14] Mulryan, C.M. (1992). Student Passivity During Cooperative Small Groups in Mathematics. *Journal of Educational Research*, Vol.85, No.5, p. 261-265.

[15] Nyikos, M. & Hashimoto, R. (1997). Constructivist Theory Applied to Collaborative Learning in Teacher Education: In Search of ZPD. The Modern Language Journal, 81, iv. P.506-517.

[16] Ogata, H. & Yano, Y. (2000). Combining knowledge awareness and information filtering in an open-ended collaborative learning environment. *International Journal of Artificial Intelligence in Education*, Vol. 11, 33-46.

[17] Okada, T. & Simon, H.A (1997). Collaborative Discovery in a Scientific Domain. *Cognitive Science*, Vol. 21(2), 109-146.

[18] Ploetzner, R. & Fehse, E. (1998). Learning from explanations: Extending one's own knowledge during collaborative problem solving by attempting to understand explanationa received from others. *International Journal of Artificial Intelligence in Education*, Vol. 9, 193-218.

[19] Potashnik, M. (1998). Distance Education: Growth and Diversity. *Finance & Development* v35 n1 p. 42-45 Mar 1998

[20] Repman, J. (1993). Collaborative, Computer-Based Learning: Cognitive and Affective Outcomes. *Journal of Educational Computing Research*. Vol. 9(2), 149-163.

[21] Scardamalia, M. & Bereiter, C. (1996). Student communities for the advancement of knowedge. *Communications of the ACM* v39 n4 p. 36-37 Apr 1996

[22] Seeger, J.A. (1983). No innate phases in group problem solving. *Academy of Management Review*, 8, 683-689.

[23] Singhanayak, C. & Hooper, S. (1998). The Effects of Cooperative Learning and learner Control on Students' Achievement, Option Selections, and Attitudes. *Educational Technology Research and Development*, Vol. 46, No.2, 17-33.

[24] Soller, A.L. (2001 in press). Supporting social interaction in an intelligent collaborative learning system. *International Journal of Artificial Intelligence in Education*, Vol. 12.

[25] Townsend, M.A.R., Moore, D.W., Tuck, B.F. & Wilton, K.M. (1998). Self-concept and Anxiety in University Students Studying Social Science Statistics Within a Co-operative Learning Structure. *Educational Psychology*, Vol. 18, No.1, 41-53.

[26] Tuckman, B.W. (1965). Developmental Sequence in Small Groups. *Psychological Bulletin*, Vol. 63, No.6, 384-399.

[27] Vygotsky, L.S. (1978). *Mind in Society: The development of higher psychological processes*. Cambridge, MA: Harvard University Press.

[28] Webb, N.M., Troper, J.D., and Fall, R. (1995). Constructive Activity and Learning in Collaborative Small Groups. *Journal of Educational Psychology*, Vol. 87, No.3, 406-423.

[29] Webb, N.M. (1982). Group Composition, Group Interaction, and Achievement in Cooperative Small Groups. *Journal of Educational Psychology*, Vol. 74, No.4, 475-484.

[30] Yager, S., Johnson, D.W. & Johnson, R.T. (1985). Oral Discussion, Group-to-Individual Transfer, and Achievement in Cooperative Learning Groups. *Journal of Educational Psychology*, Vol. 77, No. 1, 60-66.

Structure of Training Cases in Web-Based Case-Oriented Training Systems

Torsten Illmann
Department of Multimedia Computing
University of Ulm
torsten.illmann@informatik.uni-ulm.de

Alexander Seitz
Department of Artificial Intelligence
University of Ulm
seitz@ki.informatik.uni-ulm.de

Alke Martens
Department of Computer Science
Institute of Applied Computer Science
University of Rostock, Germany
martens@informatik.uni-rostock.de

Michael Weber
Department of Multimedia Computing
University of Ulm
weber@informatik.uni-ulm.de

Abstract

In this paper we discuss the structure of training cases in web-based and case-oriented training systems. We combine the benefits of systems that support a structural representation of knowledge underlying a training case and those that allow a fine granularity of didactical aspects. Our results emerged from the development of a training system in medicine, "Docs 'n Drugs – The Virtual Polyclinic". The abstraction from the proposed training case structure offers a good basis for general case-oriented training systems, an easy exchange of tutoring data, and the web-based indexing of learning material.

1. Introduction

Even if computer supported learning has a tradition that can be traced back to the 50's [5], there is again a growing interest and popularity in the use of computers in educational settings today. Governments, universities and companies highly investigate in projects dealing with web-based training in areas like studying, life long learning, further training, vocational training, retraining and knowledge transfer within organizations. The World Wide Web combined with modern technologies provides advantages and new possibilities for establishing up-to-date learning strategies. Especially the field of web-based training, the integration of computer supported collaborative work (CSCW) and intelligent tutoring prevent users from feeling lost when learning with the system.

In order to obtain reusable learning material in web-based training systems, it should annotated according to meta data standards [2], defined [6] and integrated into training systems or media players through languages like SMIL [11]. As a result, search engines can be deployed to support learners, tutors and authors in easily finding needed electronic learning media.

We extend the notion of structured training material from an underlying systematic knowledge to descriptions of didactical aspects. The resulting structure can be applied to several fields like medicine, biology, law, repair services, or criminology. Only few work has been done to examine the field of domain-independent web-based case-oriented training concepts. Most published work uses domain-specific structures.

In this paper, we examine the general structure of cases in such systems. We introduce the project "Docs 'n Drugs – The Virtual Polyclinic", in which we are developing a web-based and case-oriented training system in medicine. Furthermore, we explain the three-layer structure of cases in our system. Next, we transfer the results to other subject areas and generalize them. Finally, we mention related work to this topic and point out the benefits for this approach in a summarizing conclusion.

2. Docs 'n Drugs – The Virtual Polyclinic

The aim of the project "Docs 'n Drugs – The Virtual Polyclinic" (DND) [4] is to develop a web-based case-oriented training system. Currently, the subject domain is restricted to the area of clinical medicine including knowledge and interaction patterns of medical treatment. Since many branches in medicine and computer science are participating in this joint project, system components and training contents in the form of training cases are developed simultaneously.

To complete and deepen the systematic knowledge provided by textbooks and lectures, medical education should also provide practically relevant case-oriented training in medical issues with emphasis on patient contact. However, a patient-centered education requires a large investment of time and human resources, which could hardly be satisfied. For that reason, a web-based and case-oriented training system is an important supplement for medical education.

Since early 2000, the training system is officially embedded in the medical curriculum at the University of Ulm and used by many medical students with different skills and knowledge.

The entire system (called tutoring system) consists of six main components that users are interacting with:

- Training system
- Authoring system
- Administration system
- Intelligent tutoring service
- Telecollaboration service
- Evaluation service

State-of-the-art. We have been developing all components as prototypes and evaluating the system for three students' terms. Currently about 80 cases were created with the system and about 300 users are registered. Since early 2001 the system is open to public through web access.

In the following chapters, we will have a closer look at the structure of training cases.

3. Case Structure

As far as the underlying models of training cases are concerned, we identified some important problems of existing implementations. With some systems [1], the incremental creation of general and case-oriented knowledge bases is not possible, upon which concrete training cases could be built. As a result of the lacking separation between case knowledge and its presentation, those systems cannot be easily transferred to alternative domains. Other systems [8] offer detailed expert knowledge models for general tutoring strategies but do not support linking special didactical information with model entities.

Regarding these problems, we divide training cases into three separate layers: the *medical knowledge*, the *case knowledge*, and the *tutoring knowledge*. Accordingly, we identify three different aspects of a training case: general, case-related, and didactical aspects, which correspond to the aspects of real-life human tutor-driven presentations of training cases.

3.1. Medical Knowledge

The medical knowledge defines the general knowledge of the medical application domain. It is case-independent and contains general classifications of main entities of medical cases, e.g. examinations, their results, diagnoses and therapies. Furthermore, relations between these entities can be established to rebuild the decision chain of a medical expert.

As the results of any (clinical, technical, lab, or anamnesis) examination we identified a general class of facts that may show some evidence for certain differential or final diagnoses. These include **r**isk factors, results of technical or lab **e**xaminations, **m**edicaments currently taken, and **s**ymptoms, briefly denoted by REMS. REMS comprise all types of facts that might play a role in the process of finding a diagnosis. They are associated to meta properties, which denote types of properties a REMS can have. Figure 2 illustrates the definition of REMS.

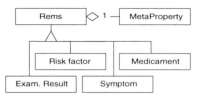

Figure 1. REMS: a generalization of symptoms

With this structure, we identify three important aspects:

1. All REMS are results of an examination. An examination is any medical examination: clinical, technical, lab or anamnesis.
2. Meta properties may be dynamically assigned to REMS. The medical domain knowledge is extremely complex and consequently it is impossible to define REMS completely in advance. This structure allows medical authors to change and to extend these definitions later on demand.
3. REMS can be associated to diagnoses by indication relations. To be more exactly, certain meta properties of a REMS can be related to a diagnosis. The relations are weighted by a set of positive and negative numbers as it can be found in [8], and describe how strong an abstract fact pleads for or against a diagnosis. In further extensions of our systems, these simple relations will be substituted by arbitrarily complex rules.

To integrate media items in the medical knowledge, we define a universal multimedia element type (document, image, audio, video, animation or movie) and allow possible associations to medical entities they describe. Thus, multimedia elements are potentially annotated with any knowledge item and thus may easily be retrieved by search engines later. The whole medical knowledge is based on the open world assumption and therefore incrementally extensible. Figure 2 shows an overview of the medical knowledge model.

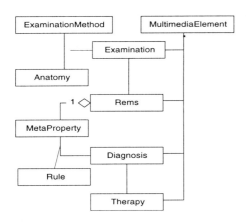

Figure 2. Medical knowledge layer (simplified)

3.2 Case Knowledge

The case knowledge is built on top of the medical knowledge. It identifies the part of the general medical knowledge that is covered by a special case. This includes all possible examinations, REMS, diagnosis, therapies, multimedia elements, and relations between those entities.

A case author has to integrate unimportant and important information into a training case. Unimportant information is required by the system to allow a learner to run into wrong paths when solving the case, as learning by doing mistakes is an important factor in case-oriented training.

Entries of the case knowledge may be annotated with case-dependent text or multimedia elements. The meta properties of REMS are instantiated to real properties with concrete values. All relations among entities (especially properties and diagnoses) may be also specified and weighted in the context of the case.

Referencing entries of the first layer implies that all cases use the same general knowledge (and terminology) and enables the system to support learners or authors during processing or creating cases [9].

3.3 Tutoring Knowledge

We employ two kinds of knowledge to support the tutoring process: didactical and pedagogical knowledge.

Didactical knowledge identifies pieces of shown information realized by multimedia and hypertext elements, and questions asked to the learner [7]. It describes how these entities can follow each other and what choices the learner has to navigate through the resulting entity network. Besides sequences and alternative choices also conditional branches and loops can be realized. We distinguish between guided, half-guided and unguided training cases, where the learner gets a different kind of freedom or responsibility to process cases. For example, guided cases, which only contain sequentially ordered pages, are ideal for beginners while unguided cases are suitable for experts.

Pedagogical knowledge allows the system to build up a model about the knowledge of the user when steering through a training case. For this purpose we allow authors to annotate learning entities with elements of the case knowledge. Based on those facts rules can be described by the author to say under which conditions didactical entities have to be presented.

4. Generalization

In this chapter, we generalize our case structure of the medical application domain to other subjects. When analyzing the three layers described above, only the first layer, the medical knowledge layer, is specific to the subject area. The second and third layers are general and may be transferred to other subjects without modification.

The medical knowledge layer has to be adapted to the specific entities of the destination subject. In general, we call this layer *domain knowledge*. Meta properties, rules, and multimedia elements are application-independent concepts. For the remaining entities, transformations to equivalent concepts in the destination subject can be easily found. Table 1 shows some examples of such a transformation.

Table 1. Examples of case concept transformations.

Medicine	Law	Chemistry
examination	question	lab examination
REMS	fact	result
diagnosis	law	evidence of material
therapy	judgment	measure
Criminology	**Repair Services**	**Generalization**
observation	inspection	inquiry/investigation
indication	irregularities	result/fact
crime	damage	problem abstraction/ identification
condemnation	repair	solution

The proposed general structure of training cases allows the development of tutoring systems for all domains where case-oriented learning occurs, and practicing the hypothesize-and-test strategy [10] for problem solving plays a main role.

Particularly, general pedagogical strategies can be implemented that use relations and rules between case entities. They allow the system to automatically correct and comment the choices of the learner when finding the right problem hypothesis. Furthermore, mechanisms can be realized that trace back argumentation steps to give hints to useful further investigations. The DND system implements both tutoring strategies.

6. Related Work

In research, there are many systems and projects that realize different aspects of computer supported learning. In Germany, for example, there are the project CASUS and ProMediWeb [1] – the latter is the web-based variant of the first one. In contrast to DND, CASUS and ProMediWeb are bound to the medical domain. They feature a fixed structure of a medical case, which aims at finding the correct diagnosis. There is no intelligent tutoring component and therefore learners may not be supported within the training process.

D3-Tutor [8] realizes aspects of intelligent tutoring, based on an expert knowledge model. It uses a case structure that focuses on finding correct diagnoses. In contrast to the other mentioned projects, D3-Tutor provides a framework that may be adapted to other learning areas. Exemplified knowledge bases for biology and car mechanics show the adaptability of their structure.

In the Adele system [3], probabilistic networks are used to represent domain knowledge for feedback and tip mechanisms. However, experiences with DND showed that authors prefer rather simple argumentation formalisms when formulating tutoring cases.

Finally, all these systems do not differentiate between and combine general, case-oriented and tutoring knowledge.

7. Conclusion

We generalized the DND approach to a general case structure and showed its applicability for other case-oriented training. Except from being application-independent, this approach entails further benefits:

- Only one of three layers has to be adapted when changing the subject area.
- The domain knowledge can be built up incrementally case by case and within the development of a single case.
- On each level, it is possible to exchange data with other training systems using that structure. An XML document type definition may simply be derived from the general structure.
- The general domain knowledge, including multimedia elements, may be used in meta data based search engines or in systematic knowledge databases.
- Authors and learners of different sub areas of the application domain have to agree upon and learn the same terminology.

7. Future Work

In the future, we extend the rule entity of the domain knowledge for more flexibility. We provide the intelligent tutoring service with more functionality like user modeling and tutor process adaptation, and improve the user interface for the training and authoring system. Furthermore, we want to build training cases in other subjects to practically prove this generalization.

8. References

[1] Baehring, T., Weichelt, U., Adler, M., Bruckmoser, S., and Fischer, M., "ProMediWeb: Problem based case training in medicine via the World Wide Web", *Proceedings of the ED-MEDIA World-Conference*, Freiburg, 1998.

[2] Dublin Core Metadata Element Set, Version 1.1: Reference Description http://dublincore.org/documents/1999/07/02/dces/, 1999.

[3] Ganeshan, R., Johnson, W.L., Shaw, E., Wood, B.P.: Tutoring Diagnostic Problem Solving. In Gauthier, G., Frasson, C., VanLehn, K. (Eds.): Intelligent Tutoring Systems, Springer, Berlin, 2000, pp. 33-42.

[4] Illmann, T., Martens, A., Seitz, A., and Weber, M., "A Pattern-Oriented Design of a Web-Based and Case-Oriented Multimedia Training System in Medicine", *Proceedings of IDPT Conference*, Dallas, 2000.

[5] Lelouche, R., "Intelligent tutoring systems from birth to now", *Künstliche Intelligenz*, 4/99, 1999, pp. 5-11.

[6] LOM. IEEE P1484.12, Draft Standard for Learning Object Metadata http://ltsc.ieee.org/wg12/.

[7] Martens, A. and Uhrmacher, A.M., "Modeling a Tutoring Process as a Dynamic Process - A Methodological Approach", *European Simulation Multiconference, Proceedings of ESM'99*, Warsaw 1.-4.6.1999, SCS, Ghent.

[8] Reinhardt, B., "Generating Case Oriented Intelligent Tutoring Systems", *AAAI Fall symposium, IST Authoring Systems*, Boston, 1997.

[9] Seitz, A., Martens, A., Bernauer, J., Scheuerer, C., and Thomsen, J., "An Architecture for Intelligent Support of Authoring and Tutoring in Multimedia Learning Environments", *Proceedings of the ED-MEDIA World Conference*, Seattle, 1999.

[10] Puppe, F., *Systematic Introduction to Expert Systems*, Springer, Berlin, 1993, p. 329.

[11] W3C, Working Draft specification of SMIL (Synchronized Multimedia Integration Language), http://www.w3.org/TR/WD-smil, February, 1998.

Stream-based Lecturing System and Its Instructional Design

Nian-Shing Chen
Department of Information Management
National Sun Yat-sen University
E-mail: nschen@cc.nsysu.edu.tw

Yueh-Chun Shih
Computer Center
National Kaohsiung Hospitality College
stone@cc.nsysu.edu.tw

Abstract

Web-based learning course lack the explanation and interpretation of teaching materials comes from teacher. In this study, we implement a stream-based lecturing system, named Media Master, which can fully integrate with web-based teaching materials and recur traditional lecturing situation in limited network bandwidth. We not only adopt this system for teacher lecturing but also for other instructional activities like collaboration/peer learning, term project report and topic discussion. We also conducted many courses by adopting this system for three years, and the result is sound.

1. Introduction

Nowadays, many colleges and universities also begin to put the distance education into practice through Internet, hoping to enhance the learning efficiency by conducting new educational methodologies from Internet media. The establishment of Web-based universities creates a whole set of Web-based instruction systems to simulate the real teaching-learning environment in the classroom via computer software and WWW based tools[1-2].

In general, e-learning adopts multimedia web-based teaching materials for long-distance learner, and does not adopt video-on-demand except for LAN or Intranet. Learner can study web-based teaching materials according to personal need and instructional schedule. Although web-based learning course are better than correspondence course, they also lack the explanation and interpretation of teaching materials comes from teacher. It seems that learning just change from books to web page.

2. Stream-based Lecturing System

In this study, we implement a stream-based lecturing system, named Media Master, which is a three-in-one system consist of the browser, stream-based lecturing recorder and player. The system can fully integrate with web-based teaching materials and recur traditional lecturing situation in limited network bandwidth.

Media Master consists of Stream-based lecturing recorder and player. There are three major functions as follows:

- Web browsing function
  This function is just like popular Web browser IE and Netscape.
- Stream-based lecturing recorder
  Instructor could browse web-based teaching materials and use graphical tools supported by Media Master to mark the focal point or add some text annotation. Then Media Master lecturing system could record video streams, audio streams and motion synchronously just like traditional classroom teaching (Figure 1).
- Stream-based lecturing player
  Learner have no need to download and could on-line browse the streams with web-based teaching materials, video, audio and motion created by Media Master recorder. The Media Master player will display web-based teaching materials, video streams, audio streams and motion synchronously (Figure 1).

Using Media-Master system, e-learning scenario will be as follows: First of all, instructor search, edit and design the web-based teaching materials as usual. Then instructor could begin to record and lecture on the web-based teaching materials by using Media-Master recorder. In the lecturing process, instructor could use graphical tools to mark the focal point or add some text annotation if instructor need. Media Master recorder could record video streams, audio streams and motion synchronously After recording, learner will use Media-Master player to display the on-line streams with web-based teaching materials, video, audio and motion synchronously. By using Media-Master system, we can guide and help the learner to catch on the focal point and improve learning effect.

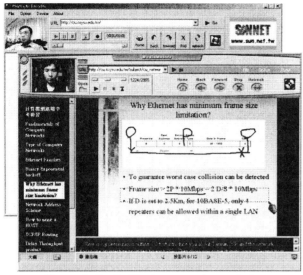

Figure 1. Media Master recorder and player

3. Instructional activities design

We not only adopt Media Master lecturing system to assist instructor lecturing but also adopt the same system to instructional activities. We have applied for collaboration/peer learning, term project report and topic discussion successfully (Figure 2).

Figure 2. Instructional design using Media Master

- Collaboration/peer learning

    Some topics or issues related course assigned by teacher dispatch to students who would divide into small groups. In addition to on-line forum, BBS, electronic White Board and e-mail, group members could apply Media Master lecturing system to discuss among group.

- Presentation of term project

    We apply Media Master lecturing system to make term project oral and paper report. Students transfer or write their reports as web pages. Then oral report can be done by applying Media Master recorder. Teacher and classmates can question and query about any group's work by using any communication tool or applying Media Master recorder. Reply to these questions can do in the same way. Project learning could be realized in web-based environment by applying Media Master lecturing system, and would be more convenient and agile than traditional way.

- Topic discussion

    Teacher or students can draw or post some issues or topics related courses to discuss by applying Media Master lecturing system. Students can reply by using any communication tool or applying Media Master recorder. Even teacher can take an electronic vote. This kind of instructional activities could draw more discussion and interest.

4. Conclusion

In this study, we implement a stream-based lecturing system, named Media Master, which can fully integrate with web-based teaching materials and recur traditional lecturing situation in limited network bandwidth. Media Master lecturing system can assist teachers easy to conduct and design web-based instruction, and provide them more choices and to enable them to produce fine teaching materials that are more inspirational and to design more attractive instructions and activities for students. We have conducted many courses by adopting this system for three years, and the result is sound.

Acknowledgement

The Study is supported by National Science Council of R.O.C.. (Project No. NSC89-2511-S-110-004)

Reference

[1]. Chen, N.S., Shih, Y.C., Dec, The Operational Model and Framework of Cyber School Learning Park, IWALT2000, New Zealand. , 2000, pp. 235-236.

[2]. Chen, N.S., Shih, Y.C., Instructional Framework and Its Design for the New Millennium, Advances in Infrastructure for Electronic Business, Science, and Education on the Internet, Italy, 2000

Learning through Ad-hoc Formative Paths

Vincenza Carchiolo, Alessandro Longheu, Michele Malgeri
Dipartimento di Ingegneria Informatica e delle Telecomunicazioni
Università di Catania – Italy
{ Vincenza.Carchiolo, Alessandro.Longheu, Michele.Malgeri } @iit.unict.it

Abstract

During the last decad , the interest for distance learning tools has grown thanks to availability of bandwidth and powerful computers. Thus the distance learning has moved from particular environments (as industry) to a larger community. The main challange of the last generation e-learning tools is to provide courses tailored to the different students backgrounds; this pushes the reasearch to create an adaptive environment able to just-in-time craft the best path for each student. This paper deals with this problem also presenting a web-based prototype of an e-learning tool to provide users with all paths moving from knowledge of a student to desired knowledge.

1. Introduction

The last decade has lead to significant changes in the role covered by computers, which may be now considered as enhanced terminals through which several distributed application can be accessed thanks to web technologies. Among these applications, the E-learning is one which will have great impact, allowing anyone connected to Internet to access lessons on the desired subject.

E-learning was initially developed inside specific environments, e.g. industries, where homogeneous contents were developed to homogeneous people communities. In this paper we want to consider students courses, where significant differences may characterize users, as the learning rate, personal capabilities and knowledge; this means that the same course must be tailored to each student.

Generally, it is possible to define a set of properties for a course to be followed inside an e-learning context. First, a course must encourage students cooperation as well as communication between students and the teacher, thus reducing the problem of students isolation which can characterize an e-learning course. Moreover, the use of computer-based interactions allows the development of more active learning techniques, in contrast with traditional lessons, where the flow of information is mainly directed from the teacher to students, with few possibilities to tailor course contents for each user based on his needs.

Based on these considerations, the system proposed in this paper aims to [1]:
- encourage active learning techniques;
- have an immediate feedback of each student activites;
- learn how to use students available time;
- adjust formative path to students capabilities.

The system must allow course modelling without fixed outlines (or with a limited number of constraints), from the teacher and the student point of view. The teacher adjusts the course to the class, adapting it to the average capability and knowledge of the students and modifying the way the lesson is proposed (examples, training, theory, etc.); the students point of view allows to take into account individual needs, accordingly to time availability and learning media. These two requirements are usually conflicting in traditional courses.

In order to allow maximum flexibility, the model chosen for courses is graph-based, where each node represents a subject matter and each arch measures the difficulty to face a successive one, in terms of knowledge and time needed. The target is to create, and so to suggest to the student, all the possible paths starting from his possessed knowledge, supplied explicitly or extracted from a personal profile, and arriving to the desired topic of interest; weighted routes allow eliminating those unsuitable because they are too long or too difficult for that student. The system can also adapt dinamically to student's needs, for instance suggesting if the student does not own an initially declared knowledge, the system could detect these missing information from exercises results and could suggests a new path to the student; or the system may change the path if the student's available time varies.

The prototype of the E-learining tool discussed in this paper has, moreover, the purpose to allow the teacher to reuse material already existing in other formats such as slides, electronic documents, images, or videos of the lessons. To satisfy this goal, a course is built putting together all the different pieces required in a web-based environment, where links among different elements are not directly embedded in the lesson itself but created thanks to the system. In this way, the editor of the course avoids the troubles connected to links maintenance, having indeed a greater degree of reconfigurability of formative paths.

Finally, to be independent from any hardware and software architecture, the E-learning system is totally based upon open source software [2], so each student only needs a browser (without any ad-hoc plug-in) to use the tool being developed.

In the following section the structure of E-learning tool and the tool guidelines are introduced. Section 3 shows the structure of the prototype that has been implemented, while section 4 presents conclusions and future work.

2. System Overview

The system is accessed through a web based interface, and allows a student to choose either to attend one of the available courses or simply search for a specific topic, creating in both cases his own personalized formative path. Potential paths are suggested, organized by the system which considers any acquired knowledge of the student, supplying information about the difficulty and the length of each specific route; in this way, the student will be able to adequately make the best choice on the basis of his own study style, level and time. At the end of each instructional unit, the student can perform the related exercises and tests to improve and verify his acquired knowledge. The results of the test will be then automatically evaluated, giving an overview of the student to the teacher, in order to estimate him and the followed path, providing so a feedback to the student itself.

After a careful analysis of the problem, it has been decided to organize the system with a single database containing several courses, in which topics can be correlated with each other, therefore, overcoming any interdisciplinary barriers. We may view the database as a *"one-argument lesson"* container, whilst a course is simply a path connecting the lessons.

This structure allows us a high degree of configurability and reusability. The most suitable structure to represent these relations is the graph: each subject matter is represented by a node, connected with the others through oriented and weighted arches. Such weights for each arch represent the difficulty to access a topic coming from a previous one. This logical relation of 'precedence and succession' is fixed and weighted by the teacher considering the real correlation between the subject matters, always retaining the possibility to modify them, even thanks to an accurate analysis of the student work and profile.

A formative route, finalized to the study of a specific matter, is mapped onto the graph structure with a path joining two nodes, the target and the a topic, meaning the beginning of a course having competence relatively to the matter.

To choose a path, the student selects it from a whole provided by the system, that offers all the possible connections between the nodes into account. In Fig.1, two path are shown as examples: the student can follow the path through boxes, which is shorter but harder, or select the path through circles, being a longer and easier path. Note that start and end topics are the same for both paths, which may also share other topics.

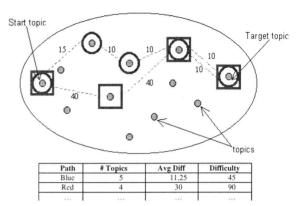

Path	# Topics	Avg Diff	Difficulty
Blue	5	11,25	45
Red	4	30	90
...	...	...	...

Routing table to reach the Target Topic

Fig. 1

This whole is further refined due to the eventually supplied history of the student with the already known topics; each route is ascribed with information about the relative difficulty for the specific student and with a numerical index weighting the "effort to learn" the target topic through it. From this basic idea, building a complete course means merging together a set of paths, each finalized to learn one objective of the chosen course.

Really, a set of questions about how a formative path is built and chosen should be addressed. Being this work in a first stage, we just describe such questions, which will be considered in detail in future works.

The first question is whether the difficulty of moving form a topic to the next one should be independent from each student's point of view. Clearly, it depends on capabilities and knowledge of a student, even if it may also depends on the specific subject the topics pair is about.

Another question is which factors the difficulty should be based on. Simply put, and considering each single arc, it may be related to the quantity of information making up a starting topic (prerequisites) and/or its correspoding target topic (goals); or it may also related to the estimated time needed to learn the set of information of a topic; it may also depends on a sort of "detail level" characterizing that lesson, e.g. an overview on a given subject should be easier than a complete description.

After having considered each transition (arc), a whole path should be defined using a criteria as $GOALS_i \cap REQS_{i+1} \neq \emptyset$, i.e. the knowledge needed to start a new lesson ($REQS_{i+1}$) should be in part provided by previous topic ($GOALS_i$), given that missing information should be already known by the user and stored into his profile. Another criteria to compose single lessons may be that the average detail level or average needed time over all lessons is constant or within a given range. All these criterias should be used both to make up paths proposed by the teacher, as well as to build new personal paths based on each student's needs.

For each topic, tests and exercises may be available to the student. A test consists of a series of multiple choice questions which are online checked and corrected, showing then the percentage of correctness. In case of unsatisfactory results the student is suggested to revise the topic just tested. An interesting feature of the tests is that they are dynamically generated, offering questions always in a random order and with different answers picked up from the database provided by the teacher. The exercises, differently from the tests, are not immediately evaluated, usually lacking an absolute answer, but often requiring the submission of a file representing the response. For both tests and exercises, it can be expected to download enclosed files to the question (with pieces of code, or text to analyse, etc.), easily accessible with a proper link.

The system has been planned to give a wide student-teacher interaction; explanations for any problem will be given via email by the teacher or directly checking a 'frequently asked questions' database, which grows with the system. Facilities to support student cooperation are mailing lists and chat rooms, including a profile database containing information about the topics each student is interested in, in order to inform whoever could be interested for a new room or thread when it is created.

3. The Prototype

3.1. Using the tool

To realize the system prototype [3], we choose to embrace the idea of Open Source [2], assuring that just a simple browser is needed, as well as gaining simple expandability, wide support and high performance. The prototype core is based on the Apache Web Server, completed with Php [4] for server-side scripting and MySql [5] for database management.

Following the general idea of the project, in this prototype, each course is realized as a set of topics, connected to each other with a "precedence and succession" relation (Fig.2), thereby, obtaining a map of subject matters as complex as needed.

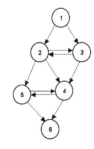

Fig. 2 A graph representing a course with six topics connected.

The topology best representing this structure is still a connected graph with oriented arcs; this allows realization of every kind of relation needed between topics of a course, considering, for sake of simplicity, unitary weight for each arch. Thanks to this structure it is possible to propose to the student all the paths available to attend a course, with an exhaustive route analysis of the graph, as well as partial paths directed to study a single topic in depth. The prototype has been realized with a web interface, through which, after a signup and a login procedure, the student can attend an entire course or just search for a specific topic. At the end of each topic it's possible to perform exercises and tests, which are immediately checked.

The page appearing to the student when reading a course topic (fig.3) is divided into two areas. The one on the left side is devoted to the student personal information and to the list of the topics. The wider area on the right is dedicated to display topics; on the bottom some buttons allow to perform a search of course topics, view correlated topics to the one displayed, choose a path or perform exercises.The student can anyway select a topic by choosing it from the index on the left.

The system has been designed to have three kinds of feedback: to the student, to the teacher, to the system itself. Immediate test results represent the first and more direct one for to the student; the others are realized due to log files analysis. Topics read and exercises done are tracked into these logs, retrieving from them the feedback for the teacher and the system, such as information about paths followed and results obtained by each student. From this analysis the teacher can eventually recalibrate exercises or modify the lessons design and content, while the system should online update student's paths.

To broaden maximally the configurability, any test or exercise is composed by several simpler elements; a question is endowed at least with a comment box where an answer in open form can be provided, optionally with up to four closed form answers. If needed, a link can be shown from which to download files containing data for the question in addition to the function of uploading a file as an answer itself. Any question can be related to different topics, allowing, during the test, reviews on past argument in the followed path and dynamic optimization of the actual route. In case of closed form answers, the student will receive a mark at the end of the test, but it is important to notice that the execution of the test does not stop the student from following the formative path in case of an insufficient mark, although the system suggests to review again the topic. This mark is expressed in percentage: the system calculates the maximum score obtainable as sum of the top quoted answers in each question and it compares this with the student score.

4. Conclusion

A great advantage for all students when learning would be to have a personal tutor adjusting the formative path for him. In this way each student can choose what and how to learn. By now, this dream is approached to become a reality thanks to the dissemination of Internet access and the band available that allows to exchange information with a high degree of structuring in interactive way.

The chance to supply courses based on the computer instrument is not a new argument, above all inside of specific contexts (business, course, etc.) is already used.

The proposal of this paper represents an attempt to create a general motor of E-learning facing the more common needs of a generic student (which, therefore, some characteristics are known). This goal imposes a structure that is able to adapt itself to the student needs in order to obtain the best result with minimum effort. Thus, a course must be available wherever and to leave the capability of the possessed computer apart.

Last but not least an E-learning tool must also allow teacher to set up the lessons and courses.

The idea on which the proposed prototype is based on is that the teacher prepares a container of lessons. Each lessons is characterized by required knowledge and by the goals. The teacher delegates to the tool the definition of the distances to suggest to the student in order to learn the wished argument. The student, based on his own preferences will choose one of the distances available.

The prototype allows to define the distances taking into account only the number of one-lesson of which is composed, and it needs also the processing of the teacher in order to establish the difficulty degree of every lesson.

The ideal E-learning tool is able to understand which is the best path of any student from information supplied by the same student (thanks to the exercises proposed both in the preliminary phase and during the interaction with the student).

In development course stage, the functions that allow to have a detailed feedback from the activity of the student are not implemented, in future release to such scope technical statistics eventually integrated with intelligence systems will be uses. It is developing also the interface to create courses that must allow to whichever teacher to insert lessons into the system and to define suggested learning path. Finally, a further and deeper analysis should be performed on all considerations mentioned in sec. 2 concerning how a path is built, and how its difficulty is defined.

Reference

[1] W. Chickering and S.C. Ehrmann, "Seven Principles for Good Practice in Undergraduate Education.", March 1987 AAHE Bulletin
[2] http://www.opensource.org/osd.html
[3] V. Carchiolo, A. Longheu, M. Malgeri, "Dynamic web pages for tuning formativa path", DIIT Report, Catania, July 2000
[4] http://www.php.net
[5] http://www.mysql.org
[6] Cavers, I et al., "Atlas: Serving Dynamic Courses Dynamically". Proceedings of WCCCE'99, Fourth Western Canadian Conference on Computing Education, May 1999.

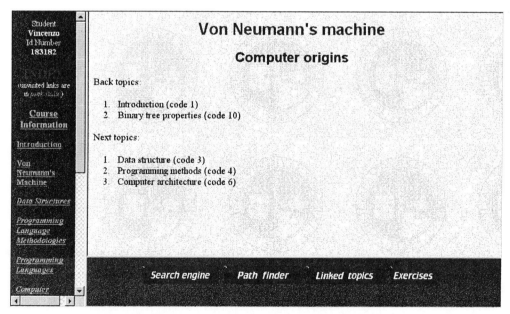

Fig. 3 Student view during the reading of a topic.

Theme 4: Teaching/Learning Strategies (Including Concept Mapping)

The Internet And The "Learning by Doing" Strategy In The Educational Processes: A Case of Study

Nicoletta Sala
University of Italian Switzerland
CH 6850 Mendrisio - Switzerland
Email: nsala@arch.unisi.ch

Abstract

The Internet represents a wonderful opportunity to modify the teaching methods, and to redefine our notions of education. The aim of this paper is to present our teaching project, named Web Learning Environment (WLE), which studies the use of the "Net" and its resources to develop a learning environment. We have involved a classroom (sixteen students, aged 16-17) of a technical institute in Italy (specialisation in Information Technologies). This work describes some educational and cognitive aspects in the realization of a cooperative hypertext (e.g., the educational and social targets, the modification of the role of the teacher inside the student's learning process). In this approach the role of the students is active inside their learning process.

1. Introduction

The starting point of our research has been to consider that the computer technology has become a fundamental part of education across the curriculum and will likely be more so in the future. Instead of simply being a source of information, computers are becoming multimedia workstations for students. Innovative use of technology combines telecommunications (e-mail, online resources, Internet navigators), multimedia authoring (home page editors), user-friendly convenient applications software and are turns students into producers as well as consumers of content. Using the new media and the Internet in the educational process we observe that the role of the classroom teacher is evolving from that of a giver of information to that of a facilitator of student learning [5]. The aim of our project, named Web Learning Environment (WLE), was to stimulate the active use of the Internet (and its technologies) in the educational process. For this reason the project has been divided in six different phases: the knowledge of the Internet (directly in the Web), the research of some educational information in the Internet, the access to the FTP (e.g. to download some interesting software tools to introduce in the educational process), the didactics with the Internet (using educational hypermedia online), the creation of a cooperative hypertext, and the communication using the "Net" [5]. In the WLE project the students were encouraged to work cooperatively in pairs while completing a WWW – based learning activity. In this paper we present the fifth phase of the WLE project dedicated to the development of a cooperative hypertext.

2. Hypertexts: an overview

Hypertext is described as non-sequential written text that allows branches and multiple paths to be selected by the reader. First idea of hypertext was presented by Vannervar Bush as early as 1945. His idea was to create a machine (MEMEX) that would link or associate material for reference purposes [1]. The evolution of the software and hardware technologies have permitted to the students to use the hypertext inside their educational process. We can give the following characteristics which permits to the hypertext to become a learning tool, in fact, hypertext mimics the way the brains work, it supports the connection of ideas and it permits to choose a personal learning path. Hypertext is also considered by many researchers to facilitate the human learning [3]. Using associative links and taking advantage of the structure of the information, learners are encouraged to explore and find the information they need, then progress to the learning activities.

3. Cooperative hypertexts in educational process

The students can realise cooperative hypertexts on the educational subjects. Using this approach we can combine two different learning strategies: the constructivist approach and the "Learning by Doing". The fifth stage of our WLE project has been dedicated to the creation of a cooperative hypertext on Bonaventura Cavalieri (1598 - 1647), an Italian mathematician, for an International Conference of Mathematics. We have established some didactic goals, for example to drill the students in a cooperative project, to train the correct use of some software tools, and some hardware devices directly in a real project context (in agreement with a "Learning by Doing" approach), to learn about the mathematics in the seventeenth century, to understand the Cavalieri's indivisible theory, and the connection between the mathematics and the physics in the seventeenth century.

The first step of our project was to propose to the students an entry multiple-choice test to control their basic knowledge on the mathematical subjects involved in the hypertext (e.g., "Do you know the Cavalieri's indivisible theory?"). The statistic analyses of the entry test have demonstrated that the students did not know this subject; in fact the percentage of correct answers has been 10% (13 correct answers on 128 answers). For this reason they were a reliable sample to study this "Learning by Doing" approach. During the hypertext development, we have explained to the students to put a particular care to study the hypertext organization and to make the user interface "friendly". These aspects are important to avoid the Conklin's problem "lost in hyperspace" [2]. For this reason, the students have organized a hierarchical hypertext, they have chosen the icons, they have also designed the visual interface (which involves the choice of colours, bottom shape). Sixteen students, organized in eight work groups, have developed the hypertext, in two languages: Italian and English, using the Hypertext Mark Up Language. It is available at the Internet address: http://www.verbania.alpcom.it/scuole/cavalieri/cav0e.htm.

This project is an example of "Learning by Doing" environment because the students have learnt the history of mathematics during the hypertext development, and they have trained themselves to the use of software tools and hardware devices. To evaluate our students' knowledge we have proposed a multiple choice test (some questions are the same of the entry test); 85% of the students have achieved the educational goals that we have established (which corresponds to the 109 correct answers on 128 questions).

4. Conclusions

Our project demonstrates that the Internet and the new technologies offer significant benefits in educational process, and it shows that there are clear educational advantages to be derived from cooperative student activities. Traditional education is based upon a paradigm normally called the "knowledge reproduction model". This model depends on verbal lecture, drill and practice sessions, structured classroom activities. In this paradigm the student is viewed as a passive learner "waiting to be filled" with knowledge. Knowledge is not static, there are multiple sources of knowledge, and students should not be passive learners. Students should be active in the learning process [4]. It is important to remember however that the teacher must be a facilitator in this cooperative learning environment, and not to fall into the trap of letting students totally "construct" their knowledge on their own. The teacher is still a very important resource and the computer is also a resource that will be used to help students in the acquisition of knowledge. In addition, the collaborative learning environments support "collaborative construction of knowledge through social negotiation, not competition among learners for recognition". Research frequently shows that there are clear educational advantages to be derived from collaborative student activities [6]. When students work in groups and small teams, the interaction s and activities frequently involve higher order and reflective thinking. Collaboration helps individuals to progress through their zone of proximal development through the communication and the joint activity in which they are engaged [7]. The use of new technologies to support engaged learning goes hand in hand with the philosophy of constructivism. There are a number of implications for teachers in a constructivist approach with respect to cooperative learning and technology. First of all, the teacher is a guide, not the lead, as students construct their own knowledge. In addition, due to various backgrounds, not all students will understand everything in the same way.

Our approach is in agreement with Seymour Papert that says "Better learning will not come from finding better ways for the teacher to instruct but from giving the learner better opportunities to construct" [4, p.3]. We have also noted that the students exhibited strong motivation in this hypertext development, and this is encouraging for the future application of our approach in the educational process.

References

[1] V. Bush, "As we may think", *The Atlantic Monthly*, 1945.
[2] J. Conklin, "Hypertext: An Introduction and Survey", *Computers*, 2(9), 1987, pp. 17-41.
[3] D. H. Jonassen, R. S. Grabinger, "Problem and issues in designing hypertext/hypermedia for learning". In D. H. Jonassen & H. Mandl (Eds.). *Designing hypermedia for learning*. Springer-Verlag, Berlin, 1999, pp. 3-25.
[4] S. Papert, "Introduction". *Constructionist learning*, Boston, MA: MIT Media Laboratory, 1990, pp. 1-8.
[5] N. Sala, "Web Learning Environment (WLE): An Example of the Use of The Internet in Educational Process", *Learning Technology*, Volume 2 Issue 4, 2000, Available at: http://lttf.ieee.org/learn_tech/
[6] R. E. Slavin, "Research on cooperative learning and achievement: What we know what we need to know", *Contemporary Educational Psychology*, 21, 1996, pp. 43 – 69. Boston: Allyn & Bacon, 1995.
[7] L. S. Vygotsky, *Mind in society*. Cambridge MA: Harvard University Press, 1978.

Effective Learning Strategies for the On-line Learning Environment: Including the Lost Learner

Mandi Axmann
Instructional Designer
Email: *axmannm@techpta.ac.za*

ABSTRACT

Much has been said about different learning styles, but it not always clear how learning strategies should be adapted to accommodate different learning styles in the on-line learning environment The question arises as to how does one ensure that there are no lost learners, especially working with extremely heterogeneous groups in the South African context, where learners may be from eleven different cultural and language groups, urban or deep rural settings, varying computer abilities and often from disadvantaged educational backgrounds. It indeed proves to be a very challenging task, and including first year learners from the subject Journalism Practice at the Technikon Pretoria, with the co-operation of the lecturer, Ms Wiida Fourie, I attempted to firstly design activities (instead of using an on-line survey) which in a fun way assessed these learner characteristics, and secondly using this information by giving students a choice of activities and assessments instead of limiting them to one or two methods. This study was aimed at encouraging learner autonomy, increasing learner confidence and trying to include the lost or "invisible" learner. This paper will report on the findings and conclusions from this study.

Introduction

The education literature suggests that students who are actively engaged in the learning process will be more likely to achieve success (Dewar 1995; Hartman 1995, Leadership Project 1995). It is therefore the continuing challenge of any instructional program to ensure that learners are involved, and one way to achieve this is to be familiar with the characteristics and learning styles of the learners.

There are a number of very well formatted on-line surveys available to determine learning styles, namely the Felder-Solomon Index of Learning Styles Questionnaire, the Modality Preference Inventory (Middlesex Community College) and many others, but they require one important ability, namely a good grasp of language and computer ability, which many of the new entrants to the Technikon Pretoria does not necessarily have, even though they will be included in the online training programs. Although the official modes of instruction are English and Afrikaans, the learners may be from eleven different cultural and language groups, urban or deep rural settings, varying computer abilities and often from disadvantaged educational backgrounds.

Instead of utilizing existing on-line surveys, I attempted to design easy-to-do activities based on the theories of learning styles to determine learner characteristics and learning styles. It was important not to stop there, but to incorporate this knowledge in the courseware by using different activities based on the results of these activities. Learners were then asked to compile an electronic portfolio of the specific activities which they have chosen, and this in turn was compared to the initial results to see whether or not the activities chosen indeed correlated with their learning styles. Learners also had to complete a learner evaluation during and after the course. Although not scientifically tested over a longer period of time and with different courses to establish validity and reliability, it may prove to be a more accurate way of determining learning styles when working with learners of this specific profile.

Looking at Learning Styles

The underlying theories related to learning styles which was considered for this study was namely the following. First Kolb showed that learning styles could be seen on a continuum. Hartman (1995) took Kolb's learning styles and gave examples of how one might teach to each them Again, each of us uses some of these styles when learning, but we tend to prefer a small number of methods to the rest. The MBTI has also been a very useful tool in contributing to our understanding of the role of individual differences in the learning process. Paulsen (1995) has organized the world of instructional strategies into four main types (based on the number and type of interaction there is between students, teachers and among students). Below you will find an outline of just some of the strategies he discusses in The Online Report on Pedagogical Techniques for Computer-Mediated Communication (1995).

Research

Including first year learners from the subject Journalism Practice at the Technikon Pretoria, with the co-operation of the lecturer, Ms Wiida Fourie, I attempted to firstly design activities (instead of using an on-line survey) which in a fun way assessed these learner characteristics, and secondly using this information by giving students a choice of activities and assessments instead of limiting them to one or two methods. This study was aimed at encouraging learner autonomy, increasing learner confidence and trying to include the lost or "invisible" learner.

Examples from the designed activities

Following are some extracts of the activity survey that the learners had to complete:

An audio script was included about the Missisipi River Boat. Some Web-based examples were also included, and the Solomon-Felder Index was used as a verifier.

Findings

In general, students who prefer sensing learning patterns prefer the concrete, the practical, and the immediate (Schroeder, 1996). These students often lack confidence in their intellectual abilities and are uncomfortable with abstract ideas. They have difficulty with complex concepts and low tolerance for ambiguity. Furthermore, they are often less independent in thought and judgement and more dependent on the ideas of those in authority. They are also more dependent on immediate gratification and exhibit more difficulty with basic academic skills, such as reading and writing. The path to educational excellence for sensing learners is usually a practice-to-theory route, - not the more traditional theory-to-practice approach.

Using the instructional strategies of Paulson as guidelines, the activities and assessments were designed to suit the different learners, by allowing the learners choice. These choices, however, had to be structured and planned carefully by the lecturer and the instructional designer. With slight variation, the learners seemed to prefer activities which related to their preferred learning style and characteristics, and the learners reported the learning experience as positive and gratifying. The learners felt that they were able to execute choices which made them feel more comfortable with the learning process. Some learners however, preferred a more structured approach and felt uncertain at times at to which activity was the more "suited" one, although all activities were designed to reach the same outcome, just in different ways.

It would be recommended that this study be repeated by different learner groups of the same heterogeneous background to ensure further scientific value.

Bibliography

Agogino, Alice M., and Sherry Hsi. 1995. Learning style based innovations to improve retention of female engineering students in the Synthesis Coalition. In *ASEE/IEEE Frontiers in Education '95: Proceedings*. Purdue University. http://fairway.ecn.purdue.edu/asee/fie95/4a2/4a21/4a21.htm

Belenky, Mary Field, Blythe McVicker Clinchy, Nancy Rule Goldberger, and Jill Mattuck Tarule. 1986. *Women's Ways of Knowing: The Development of Self, Voice and Mind*. New York: Basic Books.

Birkey, Richard C., and Joseph J. Rodman. 1995. *Adult Learning Styles and Preference for Technology Programs*. http://www2.nu.edu/nuri/llconf/conf1995/birkey.html:National University Research Institute.

Bodi, Sonia. 1988. Critical thinking and bibliographic instruction: the Relationship. *Journal of Academic Librarianship* 14, no. 3: 150-153.

Cantor, Jeffrey A. 1992. *Delivering Instruction to Adult Learners*. Toronto: Wall & Emerson. (pp. 35-43.)

Cranton, Patricia. 1992. *Working with Adult Learners*. Toronto: Wall & Emerson. (pp. 13-15 and 40-63.)

Dewar, Tammy. 1996. *Adult Learning Online*. http://www.cybercorp.net/~tammy/lo/oned2.html

Hartman, Virginia F. 1995. Teaching and learning style preferences: Transitions through technology. *VCCA Journal* 9, no. 2 Summer: 18-20. http://www.so.cc.va.us/vcca/hart1.htm

Schroeder, Charles C. 1996. *New Students--New Learning Styles*. http://www.virtualschool.edu/mon/Academia/KierseyLearningStyles.html

Analyzing Middle School Students' Use of the ARTEMIS Digital Library

June Abbas
*School of Library and
Information Science
University of North Texas
P.O. Box 311068
Denton, TX 76203-1068
jma0017@unt.edu
Voice: 940-565-2186
Fax: 940-565-3101*

Cathleen Norris
*School of Technology and
Cognition
University of North Texas
P.O. Box
Denton, TX 76203-1068
norris@tac.coe.unt.edu
Voice: 940-565-3790
x4189
Fax: 940-565-2185*

Elliot Soloway
*Dept of EECS
University of Michigan
1101 Beal Ave.
Ann Arbor, MI 48109
soloway@umich.edu
Voice: 734-763-6988
Fax: 603-415-1588*

Abstract

Research into student use of scaffolding features and representation issues encountered while using the ARTEMIS Digital Library is discussed. Research agenda, methods, preliminary findings, and future research directions are presented. Preliminary findings indicate that students exhibit some commonalities in their patterns of system and scaffolds use.

1. Introduction

Students using online resources often cannot balance learning to use the system with content understanding. The task then is to create tools that enable inquiry-based learning which are informed studies of children's use of digital environments and resources, as well as usability factors. Initial research into the scaffolding features of the web-based Artemis Digital Library research engine (http://webartemis.com:8080/artemis/index.adp), and our research questions and methods for addressing those questions is presented. This paper presents no conclusions, but rather focuses on questions that will be explored over the next twelve months.

2. Description of the Artemis Digital Library

The ARTEMIS Digital Library (ADL) web-based research engine provides access to the University of Michigan's Middle Years Digital Library (MYDL) collection. The collection now has approximately 4,800 age-appropriate web resources.

The Artemis research engine includes a variety of scaffolding features that make it more than a search engine. These scaffolds enable students to create a personal workspace that they return to each time they log on, thereby eliminating the need to start over each time; age appropriate representations of the resources to enable better information seeking; and a collaborative space for sharing ideas, comments, and resources. The three multipurpose scaffolds being examined in this research project include: **1.** Persistent workspace (space to save driving questions (DQ), comments, bookmarks, past searches, and past results), **2.** Website abstracts (age and topic specific descriptions of the resources), and **3.** Collaborative space (area to share Cool Sites, and to view and comment on other's Cool Sites and DQ's).

Approximately 65,000 log sessions from 2,000 student groups, comprised of two to three students each, have been collected over a seven month period. Outlined below are the methods used to begin analysis of the data.

3. Description of the Research

Log data gathered and analyzed in an earlier pilot study indicated some commonalities of use and representation obstacles while using the ADL. This research expands the sample set and further explores the results of the earlier study. The guiding research questions for this study are:

1. What strategies are the children engaging in to find information to answer their Driving Questions? Which scaffolds are being used and do they enable successful searching?
2. How does children's language relate to the language used to represent the documents in the system? Can children's language be used to represent documents within the collection? Will using student language within representations affect retrieval?

The two research questions and the plan to analyze the data collected to explore the questions is detailed briefly below.

4. Research Question 1

Children have information needs and information seeking strategies that differ from those of adults. Learning more about how they engage with systems can illuminate the cognitive processes in which they engage while seeking information. Modeling how children engage with systems may help us understand paths, processes, and obstacles they may encounter. Modeling children's engagement may provide insight into the need for additional scaffolds. Preliminary examination of the engagement patterns present in the log file data reveals that children exhibit some common activity patterns. These have been categorized into four categories as illustrated in Table 1.

Table 1: Activity Patterns, Scaffolds, and Use

Activity Patterns	Activities in Sequence	Scaffold Use
Exploration (3 activities)	View Shared DQ's View Shared Cool Sites Conduct First Initial Search	Each scaffold used between 1-2 times in beginning stages. Not used during intermediate sessions, but used 1-2 times at end of sessions.
Workspace Setup (4 activities)	Create DQ folders Create Past Searches folder Create Past Results folder Post DQ	Each scaffold used and repeated 1 to 3 times at the beginning of sessions. Not used during intermediate sessions, but may be used 1-2 times at end of sessions.
Beginning Search (4 activities)	Conduct Search View Abstract(s) View Website(s) Revise Search	Each repeated 3-4 times in an iterative sequence through majority of sessions.
Extended Search (4 activities)	Open Past Search folder(s) View Results from Past Searches View Abstract(s) View Website(s)	Each of these advanced scaffolds are used 1 to 2 times as the student learns more about the system features and the iterative nature of the search process.

After initial exploration and familiarization with the system, children begin using more of the scaffolds. As they become more proficient in their use, they begin using some of the more advanced scaffolds, such as the Past Search and Past Results folders. As we explore the data further, new activity patterns may be revealed.

One further potential means to interpret efficiency of the scaffolds may be to link teacher evaluation of student outcomes to the scaffolds used and the search history patterns. This analysis may also suggest further design and training issues.

5. Research Question 2

Question two examines one of the most problematic aspects of information seeking and retrieval, choosing words to search for the needed documents. Artemis provides user side scaffolds of representation, the DQ folders and spaces where users can save past searches and results. With the stored search words and driving questions it is possible to feed back into the system user side descriptors for documents, thereby increasing functionality and successful retrieval. This valuable resource can be used to enhance representations of the documents within the collection. The methods for examining this issue are detailed below.

1. Student search terms and the terms within the student groups' driving questions will be correlated. Student search terms will also be compared to the system side keywords and abstracts used in ADL.
2. Term frequency rates will be calculated and a list of frequently used terms will be compiled. The term frequency will be determined within the context of the subject topic for which the search is being conducted.
3. The list of compiled student terms will be used augment the indexing of the ADL resources. A sample of student searches will be re-conducted in order to evaluate the effect on retrieval.

Table 2 illustrates actual student search terms and system terms used in the document abstracts.

Table 2: Student terms versus system terms

Student's Driving Questions	Search terms used	System terms used
Is it true that Saturn is dense enough to float in a glass of water?	Saturn	saturn, planets "saturn", planets "moons" "saturn", density "volume", astronomy "stars & galaxies" "planets" "universe
What would happen if you put a furby in space?	space travel	astronomy "space exploration" "space shuttle" "orbits", astronomy "astronauts"
Do electro magnetic fields cause tumors in plants	electro magnetic fields, plants	biology, electromagnetism
Why do I have my mom's nose and my dad's hair?	genetics	genes, genetics "deoxyribonucleic acid -- dna"

6. Concluding remarks

The Artemis Digital Library provides us with a unique opportunity to study how sixth grade children make use of a scaffolded research environment. The data in the transaction logs are beginning to yield insights into scaffold use. Further analysis of the data over the next 12 months will enable us to develop better research engines to support student learning.

DIAL: serendipitous* DIAlectic Learning

Germana M. da Nóbrega   Stefano A. Cerri   Jean Sallantin

LIRMM – Laboratoire d'Informatique, de Robotique et de Microélectronique de Montpellier
161, Rue ADA 34392 Cedex 5 - Montpellier - France
{nobrega,cerri,sallantin}@lirmm.fr

Abstract

Killer applications, such as Excel and other ones, may be retrospectively considered as potential excellent Learning Environments in different specific domains and also in meta-cognitive skills even if there was no learning intention neither in the designer nor in the user. We briefly present a patented methodology for knowledge acquisition and construction widely and successfully used for five years, interpreted as a Learning Environment for its users engaged in domain-dependent interactions.

1. Introduction

Advanced Learning Technologies (ALTs) today are those that support Learning by stimulating it through Dialogues with Systems. These may range from the simplest, locally loaded programs to Web applications where Artificial and Human Agents collaborate to animate the Dialogues. A view of Human Learning as a potential, indirect side effect of Dialogues was the agreement resulting from discussions by eminent scholars, as it is summarized in [1]. This conjecture will be assumed here as an axiom.

A wide range of architectures may be conceived for Systems adequate to support Learning: from Tutoring Systems to Learning Environments [2]. We are addressing the last approach (LEs), where, among other properties, at least the following two may be considered applicable:

LE's popular property #1: Most LEs embody some "true" knowledge that is supposed to be acquired by the learner by interacting with the system. Socratic tutoring methods attempt to emulate the autonomous discovery process for the causes of inconsistencies by the learner as a consequence of challenging him/her with dialectic arguments. In spite of these pedagogical suggestions, few LEs are founded on these principles. Rather, most LEs developers wish "the truth in a domain" to be acquired by learners exposed to "the truth". Historical examples include the PLATO genetic simulations in the 70ties, simulations such as Thinglab or Cabri in the 80ties. Recent trends [3], on the contrary, privilege methods and tools facilitating the acquisition of meta-cognitive skills, the so called soft skills, with respect to domain dependent "true" knowledge and skills: the issue being that as a consequence of the rapid technological progress and the corresponding obsolescence of useful concepts and skills, one should learn to learn.

LE's popular property #2: Usually, developers of ALTs have a learning purpose in their mind before they design a System. Human learning is the explicit goal justifying and motivating the development of most LEs. In the reality, one may argue about the percentage of knowledge acquired by humans when exposed to settings purposely dedicated to teaching (e.g.: schools, training centers, …), with respect to the proportion acquired as a side effect of normal life activities.

Concerning 2, we adopt the approach opposite to the one of the majority of LEs developers. Concerning 1, we challenge the view that the "true" knowledge should be in the machine: knowledge to be learned is not in the LE but in the humans interacting (arguing) with DIAL.

2. DIAL: an overview

The DIAL system is an application of a methodology for knowledge construction, called *Phi Calculus* [4;5], which has been experimented since 1994, in collaboration with the Lawyer's Company Fidal-KPMG grouping 1200 lawyers in France. Lawyer's daily activity consists of understanding, proving and comparing contracts. The issue for innovation, for them, is that laws, norms, and events change continuously, so contracts have to be modified as well accordingly. In order to assist the lawyers in their activity, the Company has identified classes of contracts, and for each class has decided to offer lawyers a contract "template". It is the construction of these templates that is assisted by the methodology *Phi Calculus* and the tool it has generated, called fid@ct. This tool assists the iterative

---

* From http://lcs.www.media.mit.edu/people/foner/Yenta/glossary.html: a definition of serendipitous in an Agent's glossary: Serendipitous matches.

and interactive process of designing a template for each class of legal contracts: a team of two novice lawyers proposes and experiments a template, and a senior lawyer points out how to revise it. The cycle is repeated until a template is judged coherent by the senior. The Company has patented the methodology and the computer tool [6].

From a widespread perspective, *Phi-Calculus* may be seen as a methodology that organizes knowledge evolution in a situation of conception within an artificial agent supervised by a human agent. *Phi-Calculus* features are currently implemented in DIAL, in such a way to automate the assistance to the user by combining both machine learning techniques and the mechanism of constraint propagation.

The user should organize his/her universe of discourse in a Hierarchy of Terms. Then he/she, in relation to particular situations from his/her working domain, may build Examples as follows: for each example, *(i)* he/she builds a Formula as a conjunction of terms from the hierarchy, signing up, for each term of the Formula, whether it is observed as Present or Absent in the situation; then *(ii)* he/she classifies the Formula as a Positive/Negative Example of one of the Terms of the Hierarchy.

Once asked, DIAL may obtain Constraints out of Examples: within DIAL, Constraints are logical rules relating the Terms that appear in the Formulas of the given Examples. The user may choose what Constraints to keep.

Also, the validation of Constraints is provided: the user may propose to DIAL some situations different from those that the system already knows as Examples. By means of its Constraint Propagation mechanism, DIAL is capable to anticipate to the user what Terms he/she should use or not, as he/she formulates situations. Also, the system signs up Constraint Violations whenever the user does not follow its advertisements. This is an important issue: it may provoke the user to realize the necessity of revising Constraints, if, for instance, he/she disagrees with the behavior exhibited by DIAL due to its knowledge (Constraints).

3. Conclusions

We argue that learning may occur during interactions with DIAL by perspective users of any kind, in a fashion similar to the one testified by the activities of the Lawyers. At the moment, the only "real" users have been 20 teams of three Lawyers, that succeeded to build approximately 20 "contract templates", each consisting of 800 to 1800 terms constrained by 500 to 1000 constraints. Their "learning" in the "contract domain" is proved by their activity, converging to successful "descriptions" of template contracts, by their own self evaluation statements and by the commitment of about 400 Fidal-KPMG's lawyers willing to use the template contracts in their daily professional life. Since DIAL is "domain independent" we may argue that similar effects may occur in any domain.

The DIAL system, therefore, implements a dialectical approach where the LE represents a challenging, rational mirror [7] of the partial knowledge constructed and agreed by the partners of the conversations. Looking retrospectively, DIAL is an attempt to follow the guidelines of Soloway [8] that advocates for programming in the large "a research methodology that supports theory building, as opposed to methodologies that support theory testing".

Acknowledgements

Work partially supported by the European IST projects MKBEEM and SMARTEC on e-commerce and INCO-COPERNICUS LARFLAST on e-learning. The author G. M. da Nóbrega is a PhD student supported by CAPES (Brazil).

References

[1] S.A. Cerri. Models and systems for collaborative dialogues in distance learning. In M. F. Verdejo and S. A. Cerri, editors, Collaborative Dialogue Technologies in Distance Learning, volume 133 of ASI Series F: Computers and Systems Sciences, pages 119-125. Springer-Verlag, Berlin Heidelberg, 1994.

[2] J. R. Hartley. Effective pedagogies for managing collaborative learning in on-line learning environments. Educational Technology & Society, 2(2): 12-19, 1999.

[3] M. Baker, E. de Vries, K. Lund, and M. Quignard. Interactions épistémiques médiatisées par ordinateur pour l'apprentissage des sciences: bilan de recherches. In *Environnements Interactifs pour l'apprentissage avec Ordinateur* – EIAO (*to appear*), 2001.

[4] G. M. da Nóbrega, E. Castro, P. Malbos, J. Sallantin, and S. A. Cerri. A framework for supervised conceptualizing. In V. R. Benjamins, A. Gómez Pérez, N. Guarino, and M. Uschold, editors, *ECAI-00 Workshop on Applications of Ontologies and Problem-Solving Methods*, Berlin, Germany, 2000.

[5] S. A. Cerri, J. Sallantin, E. Castro, and D. Maraschi. Steps towards C+C: a Language for Interactions. In D. Dochey and S. A. Cerri, editors, *Artificial Intelligence: Methodology, Systems, and Applications*, volume 1904 of Lecture Notes in Artificial Intelligence, pages 34-48. Springer, 2000.

[6] Procédé et système de conception interactive d'une base de connaissance libre d'ambiguités ainsi que l'outil informatique pour la mise en œuvre du procédé et du système. FIDAL/CNRS, Patent deposed on November, 21 2000. No. 0014999.

[7] G. M. da Nóbrega, S. A. Cerri, and J. Sallantin. The rational mirror: Learning how to explicitly organize messy concepts. *Int. Journal of AI in Education* (submitted).

[8] E. Soloway. What to do next: Meeting the challenge of programming-in-the-large. In E. Soloway and S. Iyengar, editors, Empirical Studies of Programmers, pages 263-268. Ablex, 1986.

Learning Management in Integrated Learning Environments

Ildar Kn. Galeev, Sergey A. Sosnovsky, Vadim I. Chepegin
Kazan State Technological University
monap@kstu.ru, chepegin@knet.ru

Abstract

This paper discusses problems of Integrated Leaning Environments (ILE) creation. Architecture for ILE is suggested. An approach to learning management organization in such environments is described. The paper considers in details the situations, when the student does not achieve enough success in educational problem solving. In these cases the system sends him/her back to a hypertext textbook to learn theoretical material. Suggested approach is invariant for the broad class of domains. Instrumental tools MONAP-II support this approach.

1. Introduction

Learning management in ILE is a complex multidimensional challenge, which is in the center of attention for many researchers [1, 2]. It is necessary to note, that at present there is no common point of view on ILE architecture. What components have to be included in these environments and how do they have to be integrated [1]. For example, the paper [2] suggests integrating of the hypertext and the expert systems. Thus, first of all to form ILE architecture we have to formulate the purpose of its creation and define requirements, which this environment has to conform to. The purpose of the project, considered in this paper, is development and implementation of instrumental tools MONAP-II, which provide automation of ILE design for maximally broad class of domains. The requirements of maximal invariance for domain impose restrictions on the structure of ILE components. It is suggested that the structure of environment includes hypertext textbook, objective test and ITS as a kernel. Consequently, instrumental tools MONAP-II have to contain corresponding authoring tools for each of components, mentioned above.

2. Learning management

The instrumental tools MONAP-II are the development and extension of more early versions of the instrumental tools for ITS design of MONAP series: MONAP-MICRO, MONAP-PLUS and MONAP'99. Developed mathematical models, algorithms, architecture, software and practical experience of using the instrumental tools of MONAP series are described in details in a number of works of authors of this paper (for example, in [3,4]). Earlier versions of the instrumental tools of MONAP series provided automation of ITS design process and did not contain authoring tools for hypertext textbook and objective test creation. MONAP-II includes such authoring tools. As the base learning trajectory, supported by the instrumental tools MONAP-II, we suggest the following sequence of learning process stages.

Stage 1. Student learns theoretical material in the specific domain with the help of hypertext textbook. Learning control is determined by the structure of hypertext textbook developed by the teacher. Student himself/herself chooses the specific way for theoretical material learning. In this case ILE does not provide adaptation. This imperfection can be partly compensated by suggesting of several hypertext textbooks for different categories of students.

Stage 2. During this stage the system controls student's conceptual knowledge concerning learned theoretical material. Knowledge is controlled by the tests, developed by the teacher. Learning management is defined completely by the test structure. ILE itself does not provide the student with any adaptation. Similar to first stage this imperfection can be partly compensated by suggesting of several objective tests with different levels of difficulty for different categories of students.

Stage 3. This is the main stage of proposed learning trajectory, which is supported by the MONAP-II tools. On this stage student under the control of ITS solves the practical educational problems – acquires the practical skills.

On the each step of this stage ITS identifies the student's knowledge according to his results of problem solving and presents to the student the new educational problem with optimal difficulty value for the next educational step [4]. Thus, on this stage the ITS realizes an adaptive management of learning process. The didactic basis of this stage is the algorithmic approach to the

learning management. The idea of algorithmization of learning process consists in solving of following basic problems:
- development of algorithms of specific educational problems solving and training the students to apply this algorithms;
- development of algorithms of learning process itself, i.e. the algorithms, used by the teacher (human teacher or computer-based tutoring system) during the learning process.

In general case development of algorithmic instruction is feebly-formalized multi-objective problem, solved by the teacher. Variety of the types of operations, which are used by the student solving the problems in specific domain and correspond the algorithmic instruction, is symbolized by

$$Y = [y_1, y_2, ..., y_j, ..., y_J].$$

As the main component of student model the following vector is used:

$$P(k) = [P_1(k), P_2(k), ..., P_j(k), ..., P_J(k)]$$

where $P_j(k)$ is the probability of correct using the operation y_j at the k-th educational step.

For determination of $P_j(k)$ the system uses the Bayesian approach described in details in the paper [4]. At the each educational step, beginning from the $(k+\Delta k)$-th one, the system makes checking for abnormal termination. As the abnormal termination we mean sending the student from the third stage (educational problem solving) back to the first stage (learning the theoretical material with the help of hypertext textbook). The abnormal termination is realized if only one operation satisfies the system of inequalities:

$$\begin{cases} |P_j(k+\Delta k) - P_j(k)| \leq \Delta P_1; \\ |P_j(k_1+1) - P_j(k_1)| \leq \Delta P_2, \end{cases} \quad (*)$$

where Δk is the pre-abnormal (critical) number of educational steps, which are accompanied by the situation, when the student's knowledge can be not-increasing;

ΔP_1 determines the first interval of the vector $P(k)$ elements values variation;

k_1 varies from k to ($k+\Delta k-1$);

ΔP_2 determines the second interval of the vector $P(k)$ elements values variation;

Abnormal termination can be conditioned either by the weak student's knowledge or by the ineffective learning actions (comments on errors) or by both of them. Violating of second inequality in the system (*) prevents abnormal termination in the case, when student's level of knowledge was increasing but an unexpected breakdown took place. The value of this breakdown should be more then ΔP_2 ("threshold of stress") and the first inequality from the system (*), which is the necessary condition for abnormal termination, should be realized. Determining values of considered parameters the teacher expresses his/her own idea concerning the way of learning process realization. ITS will assume various personal characteristics (from "patient" and "tolerant" to "severe" and "strict") depending on parameter values, set by the teacher.

3. Conclusion

Suggested architecture of ILE is characterized by the balanced trade-off between the requirements of adaptation of the system to the specific student and the requirements of "openness" of the system. Realized in ILE learning management is characterized by the different levels of adaptation (to the specific student) depending on the learning stage. Maximal level of adaptation is provided on the stage of educational problem solving. On the each step of this stage ITS provides the student with educational problem, characterized by the optimal difficulty level. If there is no sufficient success in the problem solving, the system will send the student back to learn the theoretical material with the help of hypertext textbook. Considered approach to ILE design has been implemented at the designing of ILEs for Russian and German languages grammar.

4. References

[1] P. Brusilovsky, Intelligent learning environments for programming: The case for integration and adaptation. *In: J. Greer (ed.) Proceedings of AI-ED'95, 7th World Conference on Artificial Intelligence in Education,* Washington, DC, August, 16-19, 1995, pp. 1-8.

[2] Kinshuk & A. Patel, A conceptual framework for Internet based intelligent tutoring systems. *Knowledge transfer (volume II)* (ed. A.Behrooz), pAce, London, 1997, pp. 117-124

[3] I. Galeev, Automation of the ETS Design // *Educational Technology,* September-October, 1999. - V. XXXIX, No. 5. - pp. 11-15.

[4] I.Kh. Galeev, V.I. and Chepegin, S.A. Sosnovsky, MONAP: Models, Methods and Applications, *Proceedings of the International Conference on Knowledge Based Computer Systems (KBCS 2000),* Mumbai, India, 2000, pp. 217-228.

How Undergraduate Students' Learning Strategy and Culture Effects Algorithm Animation Use and Interpretation

Teresa Hubscher-Younger
Auburn University
Department of Computer Science and
Software Engineering
teresa@eng.auburn.edu

N. Hari Narayanan
Auburn University
Department of Computer Science and
Software Engineering
narayan@eng.auburn.edu

Abstract

Algorithm animation systems have not met the initial promise they seemed to hold for teaching algorithms to undergraduate computer-science students. A qualitative study of algorithm learning strategies of students and a usability study of a hypermedia algorithm animation system led to the conclusion that current algorithm animations may actually do more to hinder than help undergraduates develop a solid understanding of the algorithm, because of their learning strategies. Students relied on a single representation of an algorithm, misinterpreted the limitations and specifics of that representation, avoided using outside media sources and worked primarily within groups to understand algorithms. Thus, algorithm animation may have limited usefulness for this population. Therefore, to discourage students from relying on a single representation and to encourage them be more self-directed and engaged learners, we have students create, share and collectively interpret their own algorithm representations.

1. Introduction

Abstract procedural concepts, such as algorithms, are notoriously difficult for students to learn and difficult to teach. One cannot observe these algorithms directly, so many educators believe that using animation to visualize them would help students learn the concepts.

Currently, hundreds of algorithm animations and algorithm animation systems are available on the WWW. The goal of many of these systems is educational – to teach students new algorithms. The literature reports many educational benefits for algorithm animation, such as improving student motivation and developing student analytical skills [1].

However, an empirical study of learning an algorithm using animations by Stasko, Badre and Lewis [2] did not find a statistically significant difference in learning among students who used a textual description of an algorithm and those who used a textual description and animation.

Later studies (reviewed in [3]) also have not found reliable and robust results showing learning benefits for undergraduate students viewing algorithm animation.

The evaluation of algorithm animation systems have focused on either students' comments after using a system or whether students learn more from using the system than from more traditional methods of learning about the algorithm, such as listening to a lecture or reading a textbook [4]. What was not focused on was the type of problems undergraduate students typically have with learning algorithms and how they understand the representations presented to them by the algorithm animations.

Algorithm animation systems seem to have two implicit assumptions:

- Students work primarily with media, such as videos, textbooks and computer animations, in isolation (not in collaborative activity with peers) to understand algorithmic concepts.
- The best representations to give the students to understand these concepts should be based on those representations and metaphors employed by experts (i.e. teachers) in understanding algorithms and communicating with other experts.

A study of the learning strategies students typically use for studying algorithms and a study of how students use an algorithm animation system show that these assumptions are flawed.

2. Current learning strategies

To discover how students approach the task of learning algorithms a qualitative study was done with an introductory algorithm analysis class. The students were observed learning Quicksort, a sophisticated sorting algorithm, and were then interviewed about the strategies and tools they normally use to learn algorithms.

Six groups of two to three students were observed and videotaped, and 16 students have been interviewed, either in groups or individually.

For the observation part of the study, the students, in small groups, were shown a videotaped lecture on QuickSort, which resembled the lectures they normally attended, i.e. the videotaped lecture was presented by their instructor, who used only a white board and markers, his usual method of presenting the material. Then the group was given questions to answer about that algorithm. The participants were told to bring any materials they normally use to study algorithms. The students discussed the lecture and worked together to answer the questions.

The study found problems with both of the assumptions of algorithm animation systems. Most of the problems that students had in learning the algorithm had to do with their reliance on and misinterpretation of a representation of the algorithm presented in the video (a recursion tree diagram, a type of diagram commonly used by experts to explain recursive algorithms). Also, students avoided using other sources to learn the algorithm. They relied primarily on the lecture and their peers in understanding the algorithm, not other sources such as textbooks that were available to them.

2.1. Misinterpreted representation reliance

The main learning strategy employed by students was to imitate examples presented to them by the professor on the videotape. The students seemed to believe that if they worked examples, drawing diagrams similar to the ones they were given during the lecture, they would be able to answer the questions. They did this, even though not all the information they needed was in that representation of the algorithm.

Tree diagrams, such as the recursion tree diagram presented in the videotaped lecture, are often employed by experts to communicate algorithmic problems. However, this diagram interfered with the students' understanding of the algorithm. The problem of incomplete understanding of the algorithm occurred because they did not understand the *limitations* of this representation. They believed it captured everything they needed to know about the algorithm, rather than just showing a partial view of the algorithm's execution.

The partitioning strategy of the algorithm (how the elements in the input array are divided for further sorting) was not well explained by the representation they were given. The students did not realize the importance of this strategy and that they needed to learn more about it.

The test had several questions referring to this strategy and instead of seeking the answers from other sources they tried to *derive a plausible explanation* from a representation that did not contain any information that would be helpful with those questions. Most students did not seek out information from the textbook to supplement their understanding from the lecture, and if they did, most still did not understand that the original array was changed by the partitioning strategy. The students often said in the interviews that they only use the textbook when studying right before the test to help them remember material in their notes.

When students were observed using the textbook, it was clear that they were frustrated and were unable to integrate their knowledge from the lecture, their representation of the algorithm, and the description provided in the lecture with the descriptive material in the textbook. This might be due to the different styles of presentation in the text (which was dense) and the lecture.

Students were confused about concepts other than the partitioning strategy as well, including the order of recursive calls and the pivot picking strategy. These misconceptions might also have arisen from the students' primary reliance on *one* representation.

The students' attitude toward learning algorithms seemed to be that learning is something done in order to pass a test. They reported that they did not believe that the information learned in their classes would be applied outside of the classroom.

They seemed to also believe in an implicit contract between themselves and the teacher. In this fictional contract, the teacher tells them what exactly they need to learn, what tasks they need to do, such as homework, and what examples they need to be familiar with, and then the students are expected to know the material presented to them. They did not seem to engage in self-directed learning to further understand the algorithm.

The students' reliance on one representation is dangerous, because they did not realize that the diagram they were given was a *representation* of some aspects of the algorithm and that many other representations of the algorithm would also have explained these as well as other aspects. Furthermore, they did not realize that the representation simplified the algorithm, emphasizing some aspects (recursive calls) and deemphasizing (in fact hiding) other aspects (such as pivot picking and partitioning the input).

2.2. Learning from your peers

Students were grouped together randomly for the study. However, the students in the study were quite different from each other. The students' sex, race, work experience, previous schooling, nationality and age varied considerably. The study grouped students together who had never met and who did not normally study together.

Despite this, only one group decided to have the members of the group work independently on the test questions. The students usually chose to work together and learn from each other, despite their differences. Most students reported liking working with each other in the session, and most reported regularly studying or working with friends to learn the algorithms and complete homework assignments.

"Groupthink" was a problem for many groups. Often the students would convince each other that explanations based on a faulty understanding of the algorithm were correct. They convinced each other that they did not need to use the textbook to answer the questions, and in one instance even decided not to accept correct answers, when one member actually read the textbook during the session and proposed correct answers.

The belief of many algorithm animation system builders that individual students will engage in self-directed learning and use such systems outside of class to further understand the algorithms is questionable. For the students in this study at least, unless an information resource is integrated into the lecture, classroom activities or group study activities, it is unlikely to be used.

3. Evaluation of hypermedia algorithm animation system use

A usability test of a hypermedia algorithm animation system HalVis [4] led to some unexpected conclusions. The usability test was expected to find problems with basic controls, such as navigation and animation controls. What was not expected was that the small study would find so many critical problems with how students interpret and understand the graphical representations of data structures and algorithm operations in the animation.

The study was done using eight students from another introductory algorithm analysis class. The students were observed using the system to learn the same algorithm, Quicksort. They were instructed to use a "think-aloud" protocol while using the system, so that the observer could understand what usability problems might be occurring when the student behaved in an unexpected manner.

The study found that all the students using the system misinterpreted at least one of the representations of the algorithm presented to them.

On the first screen, the students found a short, simple animation introducing the main concept of Quicksort: A group of cartoon men separate according to how they compare in height to the "pivot" man, and then the subgroups recursively do the same process (Figure 1).

Figure 1: Image from animation introducing the main concept of Quicksort

Five of the eight participants had difficulty following the animation on this page. There seemed to be confusion about what was actually happening with the men on the screen and which dialog element belonged to which character.

The next screen with an animation of the algorithm had blocks with numbers representing different numbers of the array (Figure 2). The blocks moved around as they changed their position in the array. The blocks would turn different colors depending on their status.

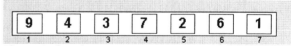

Figure 2: Starting array for animation of Quicksort

After watching the animation a few times, all but one of the participants reported that they felt they understood the animation. However, no participant could explain the color-coding scheme for the animation, when asked to do it. "I think the green numbers are less than the pivot, and the red are for numbers greater than the pivot. Wait it looks like they alternate color. I don't understand this at all," one participant said. The color-coding was critical to understanding the partitioning strategy of the algorithm.

The third animation screen was meant to show how the algorithm works on a large data set. Thin bars represented data elements, and the height of a bar corresponded to the value of an element. Squares were drawn around the bars to represent which elements were included in a recursive call, and the color of the bars changed depending on their status (Figure 3).

Figure 3: Image from animation of Quicksort sorting a large data set

Four participants did not understand the elements on the screen initially. One participant believed he was looking at a graph rather than data that would be sorted. He said, "I have no idea what this graph is supposed to be showing me. I wish it would tell me."

All of the participants, but one, reported that they would like to be able to interrupt the animation and would like to slow it down, even though they had set it for "slowest speed". Most participants blamed the speed for their inability to understand the animation. It seemed like the students were trying to study each movement and step, rather than get an overall impression of how the algorithm works on large data sets, given different pivoting strategies and different types of data sets.

Presenting students with animations of an algorithm is supposed to help them be able to mentally visualize the algorithm. However, successfully interpreting the graphical representations they are shown to help them develop this skill can often be much more difficult than anticipated by algorithm animation designers.

4. Current Work

The learning strategies and culture of the students observed in these studies imply that algorithm animations can be confusing, misinterpreted and under-utilized by the typical undergraduate computer science student.

Therefore, we are proposing that students should engage in an active process of algorithm representation *creation, sharing* and *collective interpretation*. Students are more likely to accept representations as being incomplete and partial when created by their peers, rather than by an authority figure. Thus they may be better able to understand that different aspects of the algorithm need to be understood, and that different representations deemphasize, as well as highlight, different aspects. By sharing their representations, they will be better able to check their understanding with others. Collective interpretation will hopefully help correct misinterpretation problems and help an instructor see common misunderstandings.

We are building a system called CAROUSEL (Collaborative Algorithm Representations Of Undergraduates for Self-Enhanced Learning) to help students do this. It was used in a small pilot study with 12 students in a beginning data structures course. The system collected and displayed student-created representations and collected the ratings students gave for certain characteristics of these representations. Students in the pilot study chose to work with a wide variety of media, including text, graphics, sound and animation, based on their personal preferences.

Students were tested before they created representations and after to measure their knowledge of the algorithm. Initial results from the pilot study suggest that this activity does help learning. For two of the three algorithms that were used in the pilot study, there was a significant positive correlation between creating and sharing a representation and test scores ($r=.635$, $p=.07$; $r=.663$, $p=.05$), compared to students who did not engage in these activities.

5. Conclusions

If educational technologies aimed at helping undergraduate students learn fail to address their natural learning strategies and culture, these will not be used or, worse, will confuse the student. Evaluating current student learning strategies, observing students using the technology, and tracking their understanding of the content presented to them are critical steps in developing advanced learning technologies. To ensure that a learning technology will actually be used, understood and helpful, one needs to understand the needs, culture and strategies of the learners.

Acknowledgment. This research is supported by the National Science Foundation under contract CDA-9616513.

6. References

[1] Gloor, P. *Elements of Hypermedia Design: Techniques for Navigation and Visualization in Cyberspace,* Birkhauser, Boston, 1997.

[2] J.T. Stasko, A. Badre, and C. Lewis, "Do algorithm animations assist learning? An empirical study and analysis", *Proc. INTERCHI '93 Conference on Human Factors in Computing Systems,* Amsterdam, Netherlands, 1993, pp. 61-66.

[3] C. D. Hundhausen, "A meta-study of software visualization effectiveness", Department of Computer Science, University of Oregon, Eugene, OR. Web document available at http://.lilt.ics.hawaii.edu/%7Ehundhaus/writings/MetaStudy.pdf.

[4] Hansen, S. R. *A framework for animation-embedded hypermedia visualization of algorithms,* Doctoral Dissertation, Department of Computer Science & Engineering, Auburn University, Auburn, AL, 1999.

Theme 5: Specific Applications

Domain-Expert Repository Management for Adaptive Hypermedia Learning System

Norazah Yusof
University of Technology Malaysia, Skudai, Malaysia
norazah@fsksm.utm.my

Paridah Samsuri
University of Technology Malaysia, Skudai, Malaysia
paridah@fsksm.utm.my

Abstract

This paper discusses about the design and development of a Domain-Expert Repository Management system (DERMs) that supports activities for maintaining the Domain and the Expert module of an adaptive hypermedia learning system.

DERMs involves two main activities: the domain knowledge manager which assists the instructors maintain their teaching materials; and the expert knowledge manager which assists the instructors maintain the expert knowledge for three different categories of learners namely poor, average and good.

1. Introduction

A research is being conducted at the University of Technology Malaysia (UTM) to develop an adaptive web-based learning system for teaching and learning computer programming language. The system comprises of three main components: the domain-expert model, the learner model and the adaptive engine. The domain-expert model contains all teaching materials and the teaching strategies for the different categories of learners. The learner model stores information of each individual learner. The adaptive engine acts as the interface between the user and the system, and between the student model and the domain-expert model.

One of the challenges in developing SPAtH is developing the domain-expert model. Materials must be prepared according to the system's specification and configuration. This task becomes more complicated if the instructor is collaborating with other instructors to collectively develop the subject content. To overcome this problem, a tool called Domain-Expert Repository Management system (DERMs) is proposed.

This paper discusses about the design and development of an intelligent Domain-Expert Repository Management system that supports activities of maintaining the Domain and the Expert module of the system. The main objective of DERMs is to help the instructors manage their teaching materials, as well as to manage the teaching strategies of the different category of learners.

2. The structure of Domain-Expert Model

The domain-expert model consists of two main parts: the domain knowledge and the expert knowledge. The domain knowledge stores the teaching materials consisting of notes, practices examples, check-point questions, help/hints and solutions to each questions.

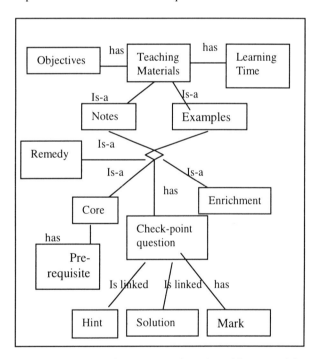

Figure 1. Semantic representation of teaching materials

Every notes, examples and check-point questions have their own specific objectives and estimated learning or answering time frame. They can be classified as core, enrichment, or remedy materials. Figure 1 shows the semantic network representation of the teaching materials and their relationship. Information of all teaching materials are stored in two different formats: document file storage and relational database. All teaching materials are prepared in active server pages (.asp) document files, as well as hypertext markup language (.html) files. Each file is considered as a node. Nodes are stored in three separate folders, which are Notes, Examples and Checkpoint folders.

The expert knowledge contains the teaching strategies that match the learner's knowledge acquisition status. It is stored in a relational database in three sets of expert tables to entertain the poor, average and good learners. At the moment, it lists out the unique node identifier that represents the sequence of nodes for each category of learners. In future, it may contain rules that describe the learning strategies for each category.

3. The Domain-Expert Repository Management System

The domain-expert repository management system (DERMs) lets the instructors maintain their teaching materials and set their teaching strategies. It supports remote client access, and instructors do not need to know the structure of the database or the location of the document storage. DERMs consists of two main components: the domain repository manager and the expert repository manager.

The domain repository manager provides the facility to add, edit and delete teaching materials by making selections and filling the form provided in html document. To edit or delete any note, the instructor may click on the link that indicates the specific note. The content of the note will be presented in a Front Page window. The instructor may edit and save the file in this mode. Due to hyperlinking of nodes, the system has to cater for all the broken links that may resulted from these operations.

To add new note, the instructor may click on the Add New Note button, and Add New Note interface will be displayed. To avoid inconsistency, every new note will automatically be given a unique note identifier (IndexNote) specified by the system. Before adding the content, the instructor is required to select the appropriate chapter and objective and provide other necessary information. The instructor may type a new note using the Front Page, or browse the existing note and then save it as a new file. Once the instructor clicks on the Add New Note button, DERMs will initiate a validation qualifier for error checking or incomplete form filling.

Next step, it will pass through a code converter. The code converter translates the information collected from the Add New Note form into a uniform code specified by the system and stores it into the appropriate table in the database. At the same time, the content of the new note will be saved as an asp or html document and uploaded into the appropriate document folder in the domain storage.

The instructor may list all notes, examples and checkpoint questions in the scrolled text boxes. He/she can then define the learning flow of the specified category of learner by selecting the required material and clicking on the right arrow button. The selected materials will be listed in the expert list box.

The instructor may also delete or reorganize the sequence of the list. Once the instructor finished defining the list, the expert repository manager will convert the expert information into specified code and record it into the appropriate expert knowledge table.

4. Conclusion

Maintaining the teaching materials and defining the expert knowledge of a domain in SPAtH is a difficult and time-consuming task for instructors who are familiar with the software development. It is almost impossible for instructors who do not know anything about software development.

DERMs is developed to make the job easier for all instructors regardless of their knowledge in software development. DERMs also enable collaborative work between instructors.

Although, development of DERM is still in progress, we foresee it as a promising tool to help solve the domain-expert maintenance problem.

5. References

[1] Lenon J. A., "Hypermedia Systems and World Wide Web and Beyond Applications", Springler, Germany, 1997.

[2] Brusilovsky P., Schwarz E., Weber G., "A Tool for Developing Hypermedia-Based ITS on WWW" ITS'96 Architectures and Methods for Designing Cost-Effective and Reusable ITSs, Montreal, June 10, 1996.

[3] Hashim S.Z, et al., "Design and Implementation of an Adaptive Web-based Hypermedia Learning System: A Prototype", International Conference on Organisational Development and Learship in Education, Kuala Lumpur Malaysia, Jan 30-Feb 2, 2001

[4] Yusof N., et al., "SPAtH: An Adaptive Web-based Learning System", International Conference of Science and Technology in Artificial Intelligent and Soft Computing, Banff, Canada, June 21-25, 2000.

Application of Web-Based Learning in Sculpture Curves and Surfaces

Wen-Tsai Sung,
songchen@ms10.hinet.net

Shih-Ching Ou
s9541010@cc.ncu.edu.tw

Automation & CAD Lab., Department of Electrical Engineering, National Central University, Taiwan

Abstract

VR technologies provide a unique method for enhancing user visualization of complex three-dimensional graphics and environments. The study attempts to apply virtual reality (VR) technologies to computer-aided design (CAD) curriculum by integration of network, cad and VR on a web-based learning environment. Through VR technologies, it is expected that the traditional two-dimensional computer graphics (CG) course can be expanded into a three-dimensional real-time simulation CG course. In this paper, the use of these modern education technologies can be used to improve the effectiveness of learning system dynamics in CG courses. We present these educational issues can assist and improve with the development of a web-based learning system, we call WebDeGrator (Web-based Interactive Design Graphics) .The design of a WebDeGrator learning system and related with theme of educational and VR technologies are presented and discussed. Future developments of the proposed web-based learning framework are also discussed.

1. Introduction

With the availability of inexpensive and powerful computers and advances in technology, the World Wide Web has become one of the major technologies in distance education. [1] The use of computers in undergraduate education is now widespread, both in the form of computer-based learning (CBL) or in the use of generic software tools in support of students' research, writing and presentation activities.[2] Courses are widely offered on the World Wide Web. Many distance learning systems, including EduCities [14], WebCT [15], TopClass [16], and Virtual-U [17], allow students to conveniently learn via the Internet. [2]Through the web-based learning system, students can perform various learning activities in a virtual classroom [25], In order to develop web-learning spaces that maximize educational benefits; the design of the entire course needs to be taken into account so that individual and social learning opportunities complement each other. [3] The learning environment needs to integrate web learning with activities that students can undertake on their own. To achieve this, we are developing an interactive web-based learning environment based on a constructivist approach.. Constructivists believe that learning is constructing knowledge from one's experiences rather than directly receiving information from the outside world (e.g. Collins & Green, 1992[4]; Resnick, 1987[5]; Brown, Collins & Duguid, 1989[6]; Collins, Brown & Newman, 1989[7]). Constructivism refers to a learning approach that places emphasis on the importance of experiential exploratory learning. It has evolved from the writings of Piaget and Bruner who together focused on the relevance of direct meaningful knowledge construction through one's experience of the world. Constructivism dove tails with VR learning environments in a fundamental and important way. The learning that occurs in VR is qualitatively experiential and direct. The knowledge acquired in a virtual environment (VE) is meaningfully constructed by the student via their exploration of the environment.[8] Recently the social negotiation of meaning has become more relevant through the development of networked VR..

Computer Graphics is an engineering course that teaches the fundamentals of graphical communication and how to use a specific computer aided design (CAD) software package. The course is often the first engineering course that students take, and many base their decision about their future as an engineering student on their experience in this course. A positive experience in this course commits students to the engineering program and motivates them toward completion of their degree.[9]our web-based learning system(WebDeGrator) has been successful to the point where students are able to explore and compare these algorithms of the sculpture curves and surfaces. The tutoring web page accommodates the different learning styles and abilities of students in the class by providing a wide range of instructional material (Carver et al., 1996, Ellis, 1996).[10] What was only partially accomplished in the lab with the constant presence of an instructor and lab assistant is now accomplished with a single lab assistant who spends the bulk of his time grading. The most recent stage of development in the research has been the addition of full multimedia lecture presentations. The entire course is now offered as an on-line course with interactive tutorial and review material that gives students a positive experience in engineering while requiring less faculty time. A number of examples in the area of engineering design can be found in the work done by Haugsjaa & Woolf, (1996)[11] and Hill et al. (1998)[12]. Work done by Suni & Ross (1997)[13] shows the effect of learning styles on student learning associated with the use of hypermedia materials developed to teach materials science. The successful development and use of web-based learning material is not only to integrate VRML with web–based learning system but also realizes 3D graphics and relevant applications in VR environment.

Figure 1 The hierarchy of these curves and surfaces

2.WebDeGrator (Web-based Interactive Design Graphics)

Curve and surface design is an interdisciplinary issue involving a theoretical background based on mathematics, computational algorithms, and engineering applications [13]. **WebDeGrator** provides students with an interactive environment so that they can grasp important concepts and algorithms. While a commercialized software product can be used for this purpose, we believe that a simple system to isolate the complexity and complicated operations from the students would facilitate the pedagogical process. Technically, WebDeGrator is a system designed to help both teachers and students to conduct the academic training interactively without the restrictions caused by different operating systems.

This work addresses all three types of parametric curves and surfaces i.e. Bézier, B-spline and NURBS. Our discussion starts with the hierarchy of these curves and surfaces:

3. The educational learning theories and technologies

Educational learning theories and technologies refer to a particular approach to achieving educational goals in which the focus is on the systematic development of teaching and learning procedures. Which are primarily based on behavioral psychology. In recent years, the fields of cognitive psychology, social psychology, psychometrics. Perception psychology, computing, and management have contributed to the development of education technologies. There are a number of learning theories and technologies that can be used to enhance the effectiveness of learning: for example, flexible learning,,

Many researchers and educational practitioners believe that VR technology has provided new insight to support education. For instance, VR's capability to facilitate activities of constructivism learning is one of its key advantages. In contrast, others have focused on alternative forms of learning -- such as visually oriented learners-- as potential to provide support for different types of learners. Still, others see the accessibility for learners, and educators to collaborate in a virtual class that transcends geographical boundaries as a major breakthrough.

4. Implementation and Illustrative Example:

The following diagrams provide illustrative examples when implementing the **WebDeGrator (Web-based Interactive Design Graphics)** system.

1 NURBS curve design and learning

Figure 2(a) Illustration of NURBS curve (Randomly Assign the Control Points and System Automatic Fitting)

(b) Real-time Local Modification (Real-time regular parameters by filling in left blanks)

2. 3D solid modeling simulation and real-time modified

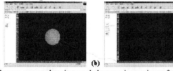

Figure 3 (a) Select a control point and degree in various forms.
(b) Illustration of a cylinder solid

3. Some examples combine 3D solid modeling graphics with database queried system

Figure4(a) Combine 3D VRML graphics and Database with a real-time 3D query interface.**(b)** A variety of queried merchandises.

5. Course evaluation:

Preliminary investigations into this learning system found those students using VR systems scored higher (pre-test) on practical examinations than those using more traditional learning systems (post-test). Another interesting finding was that students with access to VR revisited the site to refresh their skills and in computer graphics course testing retained a higher level of cognitive knowledge than those students did with no access to VR. Interestingly, students tended to view the VR learning system more in terms of a game that had instructional value as opposed to homework that was necessary for passing an examination required for graduation. The students found the learning environment VR created enjoyable and an extremely rewarding experience.

6. Conclusion and Future Work

Leaning theories tells us that experiential learning takes place when participants are engaged in tasks that embrace action, and reflection on action, and that students are able to achieve high levels of learning by being actively involved in the execution of tasks required for performing specific complex operations. The **WebDeGrator (Web-based Interactive Design Graphics)** system was designed to provide users and teachers with a convenient, easy and useful workspace in learning CAD. It provides a web-based learning environment including web-browser and VR-browser. The capabilities for user to walkthrough the 2D and 3D environment when simulating the designed curves and surfaces can increase the effect of learning. The VR learning system approach would appear to the author to be the next wave of education reform in high-tech workplace learning.

7. References:

[1] David G.Dewhurst, Hamish A.Macleod. *Independent student learning aided by computers: an acceptable alternative to lectures?* Computer & education journal.vol 35,,pp223-241,2000

[2] Gwo-Dong Chen, Chen-Chung Liu, Kuo-Liang Ou, and Ming-Song Lin, *Web learning portfolios: a tool for supporting performance awareness*, Innovations in Education and Training International (IETI), Vol. 38(1), 2000 in press. (SSCI)

[3] Sadhana Puntambekar, *An integrated approach to individual and collaborative learning in a web-based learning environment*. In Proceedings of the Computer Support for Collaborative Learning (CSCL) 1999 Conference, Dec. PP12-15, 1999

[4] Collins, E. & Green, J. L.. *Learning in classroom settings: making or breaking a culture*. In H. H. Marshall (Ed.), Redefining student learning: roots of educational change. Norwood, N. J: Ablex,. 1992

[5] Resnick, L. B. *Education and learning to think*. Washington, DC: National Academy Press.1987

[6] Brown, J. S., Collins, A., & Duguid,P. *Situated cognition and the culture of learning*. Educational Researcher, 18 (1), 32-42.1989

[7] Collins, A., Brown, J. S., Newman, S. E. *Cognitive Appreneticeship: Teaching the crafts of reading, writing and mathematics*. In L. Resnick (Eds.), Knowning, Learning and Instruction, essays in Honor of Robert Glaser. Erlbaum, Mahwah, NJ.1989

[8] Paul Cronin, *report on the applications of virtual reality technology to education*, doctoral dissertation,1997

[9] Stephen W. Crown, *Web-Based Learning: Enhancing the Teaching of Engineering Graphics*, interactive multimedia electronic journal,1999, http://imej.wfu.edu/articles/1999/2/02/printver.asp

[10] Ellis, A.. *Learning styles and hypermedia courseware usage: Is there a connection?* 1996 ED-MEDIA Conference on Educational Multimedia and Hypermedia. Boston, Mass. June 17-22, 1996..

[11] Haugsjaa, E. & Woolf, B. *3D visualization tools in a design for manufacturing*. 1996 ED-MEDIA Conference on Educational Multimedia and Hypermedia. Boston, Mass. June 17-22, 1996. 288-293.

[12] Hill, M., Baily, J., & Reed, A.. *Hypermedia systems for improving knowledge, understanding and skills in engineering degree courses*. Computers and Education, 31(1), 69-88. 1998

[13] Suni, I. & Ross, S. *Iterative design and usability assessment of a materials science hypermedia document*. Journal of Educational Multimedia and Hypermedia, 6(2), 187-199.1997

[14] EduCities http://www.educities.edu.tw/

[15] WebCT: *World Wide Web Course Tools*, URL is http://www.webct.com/.

[16] TopClass, http://www.wbtsystems.com/ by WBT Systems.

[17] Virtual-U, http://www.vlei.com/ developed by Simon Fraser University

Making Design Accessible

CN Lawrence & RG. Baird

Learning Technology Co-ordinators, LTSN Subject Centre Art, Design and Communication
cn.lawrence@ulst.ac.uk   rg.baird@ulst.ac.uk

Abstract

The paper describes the development of a multimedia programme, aimed at Art and Design Students, which is concerned with promoting awareness of how designers use drawing in their creative practice.

The programme was created in response to an identified need to support first year undergraduates applying their drawing skills within an educational environment. The paper illustrates how the team from the University of Ulster developed their section in terms of concept, structure and production methodologies.

Introduction

Drawing in design is crucial to the creative process. It enables the designer to understand form, surface, relationships, order, layout and the structuring of space, information and the three dimensional environment.

However the changing nature of art and design education in the UK has meant that the traditional Atelier type structure of teaching which encouraged a continuum of drawing as a core aspect of understanding, development and practice has given way to a much more modularised task based regime. This regime which still needs and employs drawing cannot maintain the support for it in the same holistic way. The consequence of this is evident in a poorer quality of visual expression and a lack of student confidence in using drawing in the solving of design problems.

In an effort to address this a consortium of Art and Design Institutions led by the London Institute and including Falmouth College of Art, Ravensbourne College of Design and Communication, and the University of Ulster collaborated to create a multimedia programme about drawing suitable as a general introduction to and reference resource for first year undergraduates in Art and Design. The programme was not intended to replace traditional teaching in drawing but support the student through a number of drawing perspectives which would describe a range of contexts, issues and applications and build to an awareness of drawing in a wider sense.

The programme recognises that one cannot teach drawing through a computer but that the computer is an effective tool for illustrating the rich multi faceted nature of drawing in a tangible array of exercises, exemplars and issue bases. It was decided that an appropriate way to reflect the eclectic nature of the drawing experience was to approach the programme from various perspectives to ensure that the breadth of approach would be of interest to first year undergraduates right across the spectrum.

Approach

The team from Ulster chose to address the needs of undergraduate designers and in particular illustrate how methods of drawing are based on a fundamental core that is reflected in specialist applications. This part of the programme was entitled Drawing as Method.

The section is not directly concerned with "how to" activity but focuses upon those forces which encourage one to draw, the intellectual process which sees, evaluates and translates observation into representation and the practical expressive skills which are developed as a corollary of the drawing process.

Structure

To work effectively as a learning tool it was important for the section to have a structure which enabled the student to understand the process in depth and breadth. A structure was created which allowed the student to choose how he/she navigated the section whilst having the option, if they choose, of following a recommended pathway. It was anticipated that the programme would be used both as an informing device and as a reference tool.

It consists of two elements; a core and a discipline section. To maximise the user potential the concept was to present a core element addressing the fundamental elements of the practice i.e. observation, evaluation, interpretation, formulation of the image and then reinforce these elements within a number of contexts across various design disciplines. The core is in the form of a presentation by an established ceramics designer using the drawing activity as a focus on fundamental aspects. A range of additional interviews complements this with discipline specialists who illustrate the use of drawing in the context of graphics, animation, product and environment. This approach reflects a well established pedagogy in Art and Design as research has shown that students respond well to practitioners who are prepared to discuss their work objectively.

The core consisted of an introduction, the process in action and a review of each section. The "core" video was shot as a complete drawing process; the designer described the selection of the subject, the reasons for using particular materials, the techniques employed in the drawing process and his thinking process during the activity. This was further extended by his approach to restructuring the drawings and using elements selectively to create his design forms. Finally the drawing was reviewed in the context of the three-dimensional ceramic object created. Each discipline was illustrated in sections entitled; why draw, approach, application, reference and style.

The short introduction is video based. The core presentation consists of five video clips and image support to complement the principle points. Each discipline section consists of video introduction, images for each of the five sections, audio support and text complement to major points. Each of these sections is consolidated by a short review to reinforce the principles.

Users can select how they view and can compare designers' work across the disciplines reinforcing the concept of the application of core elements. Together the two sections explore how analysis and synthesis operate within the process of drawing, and how these are conditioned by experiential factors that benefit the creative process.

Production system

The nature of the project meant that each institution had autonomy in the design and structuring of their particular section. The London Institute as lead partner was responsible for the composite development, software authoring and final production of the programme.

As visual people, contributing to a visually rich programme and working at a distance we found it more appropriate to create visual composites as a basis for discussing ideas and potential layouts. To save time these were created on the presentation programme Powerpoint. This enabled inclusion of video, image, audio, text and basic animation and although limited in some respects provided an excellent mock up of most of the screen layouts that led towards our final format. These were very useful in presentation form and as screen grabs were we needed to discuss individuals layouts.

The final work was sent in file form suitable for inclusion in Director. Video and audio material was created on DV and edited in Premiere and Edit DV and sent as compressed, flattened files suitable for cross platform application. Visuals were created through scanning and digital recording, modified in PhotoShop and coverted to jpeg. Text was in Word.

Conclusion

Much work is being done in the area of Art and Design as the potential for application of the complex image increases in educational material. Use of computing has increased rapidly as technology has made image manipulation significantly easier. This programme reflects this growth in both it's structure and content presenting the student with the diversity of drawing and it's use whilst providing a resource for study support.

Glasgow University and The London Institute are currently testing a beta version programme across a number of locations teaching Art and Design. Each of the contributing institutions is also testing the composite product through their own establishments.

References

1. J.S.Brown et al (1989), "Situated Cognition and the Culture of Learning", Educational Researcher **18**, 32-42
2. Kaye, A (Ed) Computer Mediated Communication and Distance Learning. In Mindweave: Communication, Computers and Distance Education. Pergammon Press Oxford.
3. M.Tarkka (1998) "Cuilture , Technology and Radically Hybrid Design" ARTTU! Nos5-6, 7-8
4. Laurillard, D (2001) Rethinking University Teaching. A Framework for the use of educational technology. Routledge London
5. L. Friedlander (1995)," Spaces of Experience on Designing Multimedia Applications", Contextual Media and Interpretation , Barret and Redmond 163-174

English Assistant: A Support Strategy for On-Line Second Language Learning

George R S Weir
Department of Computer Science
University of Strathclyde
Glasgow G1 1XH
UK
E-mail: gw@cs.strath.ac.uk

Giorgos Lepouras
Department of Informatics
University of Athens
Athens 157 71
Greece
E-mail: g.lepouras@di.uoa.gr

Abstract

This paper presents an approach to the provision of on-line assistance for learners of English as a second language. The basis for this design is the application of directed support in the form of 'dynamic annotation' to English information. Critically, we aim to provide aid where it is likely to be needed most. Our work falls into two parts. Firstly, we have a mechanism that facilitates the dynamic annotation of English content web pages. Secondly, we are investigating the application of criteria to drive the decision-making that underlies such support.

1. Introduction

This work seeks to provide a setting in which learners of English as a second language may use Web pages with English content as a basis for language learning. Clearly, an open Web context has considerable potential for self-guided learning. The major overhead is the need to provide learners with support for unfamiliar aspects of the language that they may encounter. If we can secure an environment in which learners are left to browse at will, yet receive appropriate language guidance as annotations to the incoming web content, this should go some way toward the desired facility, without the concurrent presence of dedicated teaching staff.

2. Language problems

Language learners face many obstacles to mastery of a second language On the one hand, learners meet words or constructs that are entirely new and unfamiliar. On the other hand, they meet contexts that appear familiar but are not what they seem.
Learners of a similar age and experience inevitably face similar obstacles to comprehension. Not least of their difficulties is the prospect of cross language confusion. Boulton notes four varieties of potential confusion [1]: malapropism, definitional differences, associational differences, and contextual misunderstanding. Additionally, a number of general factors are likely to cause difficulties for non-native users of English. These include 'special use' expressions, (e.g., jargon, slang, or idioms) which are likely to prove unfamiliar or elusive for second-language users.

Naturally, interpretation is coloured by our background and experience. In the context of second-language interpretation such colouring can become interference and lead to error. For example, in a survey of twenty Greek subjects at a Word for Windows seminar for novice users, some people were confused by the term 'font'. Several thought this equivalent to the Greek 'fonto' that means 'background' [2].

Our Web-based learner support seeks to address such scenarios through a combination of dynamic annotation and predefined knowledge-base facility. In addition, we have formulated several criteria as a basis for application of such support.

3. Annotation facility

Elsewhere we describe how second language assistance may be given in the context of computer applications [2]. English on-line help information can be supplemented with native language support - a technique that deploys targeted use of second language information, within the context of normal application use, i.e. adjacent to the English original. Beyond predefined local language supplements within Help files, some progress in dynamic second language support has been made for Windows-based applications. A dynamically accessed database of command terminology has been implemented for use in a PC environment. This allows supplementary Greek terms to be displayed adjacent to their English originals in the context met by the user (as illustrated in Figure 1).

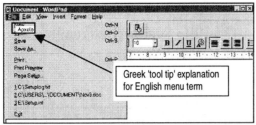

Figure 1 : Dynamic Greek support for menus

Our work touches on two languages that differ significantly from English: Greek and Chinese. Both require font sets that complicate their use in conjunction with English language displays. The Greek case is simpler than Chinese, since fonts are available that combine Greek with the Roman characters used in English. Such technical requirements for establishing native language assistance are prefaced by significant decisions on how to deploy such support. Full translation (e.g., to Chinese) may be given as accompaniment to English Web pages, but this entails significant translation overheads and requires advance selection and preparation of content. Targeted dynamic native language support seems preferable, especially if implemented in an 'open' Web context.

We have developed two approaches to Web-based native language support. The simpler variety uses concurrent frames to supplement English information with local language support. This affords flexibility in the local language support that may be implemented (see Figure 2).

Figure 2 : Frame-based Chinese support

This strategy is limited by the static nature of the second language support. All supplementary support texts must be prepared off-line and added (as frame-based components to Web documents or as associated pop-up definitions) to the original English information. To enhance the application of this support we have implemented a 'look up' system that dynamically checks word contents of the main English text and provides access to local language definitions and explanations.

4. Dynamic support

Our initial approach to dynamic support employed an on-line dictionary of informatics developed at the University of Athens to provide a language reference for English and Greek. Several technical means have been explored toward a dynamic link between this facility and Web page content. An early approach used a Java applet to 'read' terms from the loaded English Web page and seek corresponding Greek explanations from the dictionary. An alternative employs server-side scripts that receive a Uniform Resource Locator from the user, retrieve the document behind the scenes, and perform annotation before sending the result to the user's Web browser. Recently, we have moved to use a proxy server between the learner and the requested Web pages. Here, the user's Web browser is configured to use the proxy server when accessing external Web pages. The proxy requests an external Web page on the user's behalf but does not despatch the received data directly to the user's browser. Instead, the proxy server invokes an external program to scan Web page content for specified key words or phrases.

Based upon the successful matches, an annotated version of the Web page is created 'on the fly' and sent to the local user. In addition, the proxy server may be set to cache the annotated Web pages or the original documents.

The proposed approach to native language annotation promises significant benefits for individual learners. Customisation may be provided across two dimensions: level of language comprehension and modes of annotation. In the first case, the annotation system determines how much and what variety of second language support to add to the source Web pages. This level of support should be determined as a function of the user's English language ability, and based upon weightings attached to items within the annotation database. In the second case, a user may control the manner in which support is attached to the annotated Web pages. According to user wishes, this may use frame-based supplements in the local language, pop-up second language definitions and explanations, or added Web links to more detailed native language support, including dictionary and thesaurus.

5. Conclusion

By combining dynamic annotation techniques with metrics for English language complexity, we see scope to provide a virtual 'English Assistant'. The proposed system extends existing native language support facilities and looks to the plausible application of support criteria as a basis for focused on-line second language support.

6. References

[1] Boulton, M., *The Anatomy of Language: Saying what we mean*, Routledge Kegan Paul, London, 1973.

[2] Weir G.R.S., Lepouras, G. and Sakellaridis U., 'Second-Language Help for Windows Applications', *HCI'96 Conference on People and Computers XI*, Cambridge University Press, 1996.

Automating Repeated Exposure to Target Vocabulary for Second Language Learners

David Wible[*], Chin-Hwa Kuo[**], Feng-yi Chien[*], Nai Lung Taso[**],
Graduate Institute of Western Languages and Literature, Tamkang University,
dwible@mail.tku.edu.tw
Computers and Networking (CAN) Laboratory, Department of Computer Science and Information Engineering, Tamkang University, chkuo@mail.tku.edu.tw

Abstract

A web-based tool is described, called the Supplementary Reading Provider (SRP), devised to find related readings according to learners' lexical level which offer repeated exposure to target vocabulary. The SRP exploits text retrieval techniques based upon the hypothesis that there is a parallel between text similarity measurement on the one hand and the pedagogical task of providing supplementary readings which offer repeated exposure to new vocabulary on the other. Two criteria are compared for finding supplementary passages from a corpus of reading: one matches keywords of the original passage and those of the corpus passages, the other matches target vocabulary words without regard to keywords. The integration of SRP into an web-based language learning platform called IWiLL is described as well.

1. Background
1.1 Vocabulary Acquisition and Input

Research on reading and on vocabulary acquisition suggest the value of providing supplementary readings which offer learners repeated exposure to target vocabulary in context. Nation (1990) points out "...the effort given to the learning of new words will be wasted if this is not followed up by later meeting with the words." Thus, the "...increase in vocabulary size must be accompanied by many opportunities to put this vocabulary to use." (p.119) It is just such opportunities which are lacking in much high school EFL curricula under the pressures of time and the influence of vocabulary teaching traditions based upon rote memorization.

It is clear that the similarity of the supplementary readings and the repetition of target words in context form a meaningful reading for learners on the assumption that "...only contexts will fully demonstrate the semantic, syntactic, and collocational features of a word the learner has to process in order to establish the numerous links and associations with other words necessary for easy accessibility and retrieval" (Groot 2000. p.65).

2. Setting

The tool, called Supplementary Reading Provider (SRP), is designed to be one component in a larger integrated web-based language learning environment, IWiLL. IWiLL consists of several highly integrated components that support language learners, teachers, and researchers. The most mature component is designed to support English composition. A more recent module is the reading component. VoD (video on demand) tools designed by the IWiLL team are also integrated into the environment, allowing teachers and learners to do selective searches for specific sorts of linguistic input for the learners. SRP is intended as one tool within this platform. Currently the SRP is a stand-alone tool that provides teachers and student search capabilities for supplementary readings online or local reading search results. This is an initial stage in our development of the SRP. The SRP will have a web-based version whereby users need only a common commercial browser such as Internet Explorer or Netscape, and the tool will be accessible from the web. More specifically, we will incorporate it into a suite of tools available on our web-based language learning environment (IWiLL).

3. Method

To describe the method used in SRP, it will be helpful to imagine a particular setting. We assume a learner who is reading a lesson online in a web-based learning platform (e.g., IWiLL) and the passage has an accompanying list of target vocabulary items that appear in the passage. We will refer to the vocabulary token in the textbook passage as the target vocabulary item. The system will search a corpus of texts which we will simply refer to as the corpus. The aim is for the SRP to take a target vocabulary item as input and provide as output a set of texts from the corpus that contain tokens of the target vocabulary which resemble the original semantically and of course match it in part of speech.

We compare two algorithms for this search: one which relies on matching the keywords of the original textbook text and those of the texts in the corpus and another which matches the vocabulary list from the textbook text to the texts in the corpus rather than keywords.

For the purposes of testing the precision of this search technique, we use a sense tagged Brown corpus bundled in WordNet 1.6. One of the texts in the Brown corpus is selected as the "textbook text" and the remainder of the corpus serves as the corpus. The search technique makes no reference to the semantic tags, but we use those tags post hoc to evaluate the accuracy of the search results. If the semantic tag (the synset index) of the textbook vocabulary token matches the semantic tag of a matching token retrieved by the algorithm, we consider this a hit, that is, an accurate retrieval that would give the learner repeated exposure to the same target vocabulary word with the same sense encountered in the textbook.

In order to provide "similar" readings with repeated vocabulary items from high school coursebook materials, we use text retrieval technique to calculate the similarity between coursebook materials and supplemental readings. We use VSM (Vector space model) to build the feature vector of each document. Every document in teaching materials has two kinds of feature vectors: one is composed of a vocabulary list and the other is composed of keywords extracted automatically by the system. However, the supplemental readings only encode the second feature vector. Thus, we use a frequency list file to represent the feature vector. It is exemplified in the following list:

> *learning, 2*
> *english, 8*
> *easy, 1*
> *student, 1*
> *asks, 1*
> *questions, 1*

The first element is the keyword or vocabulary items and the second is the frequency of the word in the document. The feature vector composed by keywords is automatically extracted by the system. After extracting every single word in the document, the system will delete stop words from feature vector.

After the document preprocessing, we use a cosine measure to calculate the similarity between the textbook passage and each of the supplemental readings, first according to the vector calculated by keywords of the textbook passage and second according to the vector calculated by target vocabulary items in the textbook passage . The cosine formulation is listed below.

$$sim(D_1, D_2) = \frac{\sum d_{1k} \times d_{2k}}{\sqrt{\sum d_{1k}^2 \times \sum d_{2k}^2}}$$

where D_1, D_2 are the two documents, d_{ik} is the kth feature vector value of the D_i. We run these calculations on the Brown Corpus that has been semantically tagged according to WordNet senses. Results of the similarity calculations are compared to the Wordnet sense tags to test the comparative accuracy of using target vocabulary vs keywords as the evaluation standard for determining similarity.

4. Conclusion

Text similarity measurement and text retrieval techniques offer a range of ways to automate the task of providing learners with passages that give repeated exposure to target vocabulary sufficiently similar to the original usage to be helpful to the learner. Two specific approaches are tested. These techniques (and potentially a range of others) can improve upon the low precision of mere string matching KWIC searches for vocabulary yet circumvent the need for relying on heavily annotated, semantically tagged corpora to achieve this precision. Future research is needed to tease out a number of variables for improving this precision and applying the results to second language vocabulary acquisition.

5. References

Barnbrook, Geoff. (1996) *Language and Computers: A Practical Introduction to the Computer Analysis of Language.* Edinburgh: Edinburgh University Press.

Fellbaum, Christiane (1998) *Word Net: An Electronic Lexical Database.* Massachusetts: MIT Press.

Goethals, Michael. (1997) "How useful is word frequency information for the EFL teacher (and/or learner)?" Paper presented at EUROCALL Conference Dublin.

Groot, Peter J.M. (2000) "Computer Assisted Second Language Vocabulary Acquisition." Language Learning & Technology Vol4, No.1, 60-81. On-line. Available from Explore @ http://llt.msu.edu/vo14numl/groot/default.html

Krashen, S. (1995) *The Input Hypothesis: Issues and Implications.* Londons: Longman

Nation, J.S.P. (1990) *Teaching and Learning Vocabulary.* Massachusetts: Heinle &Heinle

Summers, Della. (1995) "Computer lexicography: the importance of representative ness in relation to frequency." In Thomas, Jenny & Short Mick, *Using Corpora for Language Research.*

Using Speech Analysis Techniques For Language Learning

Valery A. Petrushin
Center for Strategic Technology Research, Accenture
3773 Willow Rd., Northbrook, IL 60062
petr@cstar.accenture.com

Abstract
This paper presents an overview of speech analysis, visualization, and student response evaluation techniques that can be used for learning a foreign language by an adult learner. The general framework for student response evaluation is described. It is based on collecting experimental data about experts' and novices' performance and applying machine learning and knowledge management techniques for deriving evaluation rules. Application of the proposed approach to language learning tasks is discussed.

1. Introduction
Learning a foreign language is a difficult and time-consuming task. The best results are achieved in one-to-one interactions with a teacher who is a native speaker. Unfortunately, this approach is not affordable for most learners. Advances in speech technology resulted in proliferation of speech-enabled commercial language learning products that suppose to improve the quality and speed of language learning for a reasonable price. These products use commercially available speech toolkits, which were created for building voice-enabled applications for native speakers rather than for teaching foreigners. On the other hand, there are commercial products that can perform sophisticated low-level speech analyses that have been traditionally used for pathological voice evaluation. Many of these analyses can be successfully applied to foreign language learning. The main drawback of the current computerized language learning products is very weak feedback. Some systems can tell the user if his or her response is correct or wrong, or represent the quality of response on a scale "bad-satisfactory-good". The learner's frustration skyrockets when he gets several "wrong" or "bad" grades without any hints how to improve his performance. The learner needs a system that helps to visualize his performance, compare it to the teacher's performance, and provide feedback on how to improve it.

A digitized speech signal is represented as a sequence of 8-bit or 16-bit integer numbers sampled at a frequency that is twice larger than the maximal frequency of the signal. The graphical representation of speech signal as a function of time is called a waveform. This representation might be very confusing for learner, and I don't recommend showing it. The following more informative characteristics can be extracted or associated with speech signal: energy, pitch, formants, speaking rate, spectrum, cepstrum and transcription [1].

2. General framework
Learning to speak a foreign language involves the development of new motor skills, i.e. new movements of one's speech organs. To expedite the process, precise diagnostics of wrong movements and detailed description on how to fix them are necessary. The general framework for evaluating learner's performance for a particular task includes the following steps:

- *Create a descriptive model for the task.* The model describes gestures of the tongue, lips and jaw that are necessary to perform the task correctly.
- *Select acoustic features and create a quantitative model of the task.* For example, for learning vowels two formants F1 and F2 were selected as the features, and a two-dimensional Gaussian model was built for each vowel based on TIMIT database.
- *Collect experimental data from native speakers and learners.* For the vowel learning task performance data were collected and manually classified as correct or wrong. The recommendations on how to improve performance were created for each case of wrong performance.
- *Use machine learning and knowledge management techniques for creating a diagnostic system.* The diagnostic system contains a set of rules that tells how to compare the teacher and learner's data and gives recommendations how to fix learner's wrong performance. For the vowel learning tasks I used a decision three classifier that was based on the experimental data.
- *Use visualization techniques to present data to the learner.* In case of vowel learning, a F1-F2 chart was used for displaying teacher and learner's data.

3. Language learning tasks

3.1. Learning sounds

Sounds or phonemes are the basic elements of speech. There are two broad categories of sounds: vowels and consonants. Vowels are further divided into pure vowels and combined vowels – diphthongs and triphthongs. It is well known that vowels can be distinguished by their first (F1) and second formants (F2). Hence the best visualization for practicing a pure vowel is to calculate F1 and F2 for sequential speech frames of length 10 or 20 *ms* for both the teacher's and the student's utterances and display them on a F1-F2 chart. To estimate the student's performance the probability that the student's data belong to the model is calculated and compared to a threshold. If the probability is less than the threshold than the system uses rules to do diagnosis and give recommendations. Some rules can be derived from historical data using data mining techniques such as decision trees; the others are created by an expert. Knowledge of the student's native language can help significantly in creating precise diagnosis and useful recommendation because the typical mistakes are: 1) substituting the vowel by the closest vowel of student's native language, and 2) substituting a non-stressed vowel by the neutral vowel (schwa).

A F1-F2 chart for a diphthong contains initial vowel's data, transitional data and final vowel's data. Besides the F1-F2 chart, it is very important to track the dynamics of a combined vowel. To achieve this, I suggest comparing teacher's and student's F1 and F2 profiles. The diagnostic rules take into account the probabilities of the student's data to belong to initial and final vowels, and closeness of formant profiles.

Consonants are distinguished by their spectral and voiced/unvoiced features. The voiced/unvoiced indicator, spectrum and spectrogram can be used to compare teacher and student's sounds.

3.2. Syllabic intonation and stress

In polytonal languages, such as Chinese, Tai, and Vietnamese, the meaning of a word depends on syllabic intonation or tones. There are four tones in Mandarin Chinese, five tones in Tai, and eight tones in Cantonese. Learning tones is very hard problem for most of learners. But it could be much more easier if the system visualizes student's pitch profile

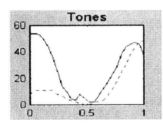

and compares it to the teacher's profile. The figure above presents a student's try to pronounce Mandarin word /wa/ with third (falling-rising) tone. Here the solid line represents a teacher's and dashed line represents a student's profiles. Both profiles are aligned, scaled, and normalized. The distance between profiles is calculated and compared to the threshold. The diagnostic rules take into account the duration and closeness of profiles. This approach can be extended to give an opportunity to the student to learn relationship among tones in multi-syllable words [2]. Most languages use energy stress, but some of them, such as Japanese, use tonal stress, i.e. a speaker raises his or her pitch to stress a word. Learning tonal stress versus energy stress and vice versa is a hard problem. Visualizing and evaluating the teacher's and learner's pitch and energy profiles can greatly improve the student performance. Another neat visualization of stress based on pitch, energy, and spectrum is proposed in [3].

3.3 Learning words, phrases and sentences

Learning to pronounce multi-syllable words, phrases and sentences is another hard problem. This problem has two sides: 1) how well recognizable is a learner's pronunciation, and 2) how "natural" is the learner's speech. The first side deals with quality of sounds. But the second side deals with prosodic features (intonation, stress, and pauses). Some language educators claim that the prosodic part of speech is more important for oral communication. In spite of some native speakers have problems in pronouncing some sounds they are perceived as native speakers based on their prosody and vocabulary. To address prosodic problems, visualization of teacher and learner's pitch and energy profiles is recommended. Diagnostic rules can be created for particular words, phrases and sentences. However, creating a comprehensive diagnostic system is a hard problem, which waits to be solved. But giving a learner a visual feedback itself can improve his or her performance. Using a speech recognition system allows estimating how recognizable is the learner's speech. Some speech recognition systems can output several possible transcriptions (N-best matches) ranked in accordance with their likelihood. The diagnostic rules can use this output in combination with prosody features for tailoring more precise recommendations.

References

[1] L. Rabiner, and B.-H. Juang *Fundamentals of speech recognition,* Prentice Hall, Englewood Hills, NJ, 1993.

[2] Zongji Wu, "From traditional Chinese phomology to modern speech processing. Realization of tone and intonation in standard Chinese.", In *Proc. 6th International Conference on Spoken Language Processing (ICSLP 2000),* vol. 1, pp. B1-B12.

[3] N. Minematsu, and S. Nakagawa, "Instantaneous estimation of prosodic pronunciation habits for Japanese students to learn English pronunciation", In *Proc. 6th International Conference on Spoken Language Processing (ICSLP 2000),* vol. 3, pp. 191-194.

Web Passive Voice Tutor: an Intelligent Computer Assisted Language Learning System over the WWW

Maria Virvou
Department of Informatics,
University of Piraeus,
80 Karaoli & Dimitriou St.
Piraeus 18534, Greece
mvirvou@unipi.gr

Victoria Tsiriga
Department of Informatics,
University of Piraeus,
80 Karaoli & Dimitriou St.
Piraeus 18534, Greece
vtsir@unipi.gr

Abstract

In this paper we describe Web Passive Voice Tutor (Web PVT), an adaptive web-based Intelligent Computer Assisted Language Learning (ICALL) program that is aimed at teaching non-native speakers the passive voice of the English language. The design of the system has been largely based on the results of an empirical study that was conducted at schools with the collaboration of human teachers. Web PVT incorporates techniques from Intelligent Tutoring Systems (ITS) and Adaptive Hypermedia (AH) technologies to provide students with individualised instruction and feedback. The system uses a combination of stereotypes and the overlay technique for the initialisation of the student model, which is then refined by observing the student while working with the system. The resulting student model is used for the annotation of the links to topics presented to the student. In addition, it is also used in the process of error diagnosis and the adaptation of feedback and advice provided to the student.

1. Introduction

Currently, a lot of research energy is put in the development of web-based educational software. Benefits of web-based education are independence of teaching and learning with respect to time and space [11]. Another important advantage of web-based educational software is platform independence of the application and easy access to it. These advantages ensure that the audience of web-based applications may be very large. However, most existing web-based educational applications lack the sophistication, interactivity and adaptivity of applications, such as ITSs [17].

On the other hand, ITSs are very good at individualising educational support to students, but they are criticised that they are often research products that are not accessible by many students. ITSs may benefit from Web technology to gain more accessibility, and web-based education may gain intelligence and adaptivity from ITSs. Therefore, the combination of these two technologies may produce very powerful and flexible educational programs. However, as Brusilovsky points out [3], web-based ITSs still constitute a rather small steam inside the ITS area. This is even more the case for Intelligent Computer Language Learning (ICALL) systems delivered via the WWW.

The existence of still few web-based ICALL systems does not undermine the educational impact that such systems may have if delivered via the web. Indeed, the domain of language learning is probably among the few domains that may be of interest to students internationally, irrespective of their mother tongue. Other ITS domains may have more restricted audience due to the dependence of the courses on the language they are written in. For example, a web-based course for the French language may be used by students of any mother tongue who learn French, whereas a web-based course for history written in Greek may only be used by history students who know Greek. This is an additional reason that advocates in favour of further research and development in the area of web-based ICALL.

In this paper we describe Web Passive Voice Tutor (Web PVT), an adaptive web-based ICALL that is aimed at teaching non-native speakers the passive voice of the English language. Web PVT has been based on an earlier standalone ICALL system, the Passive Voice Tutor [14, 15]. Prior to designing the web-based version of the system, an empirical study was conducted at schools with the collaboration of school-teachers.

Web PVT provides adaptive navigation support, intelligent analysis of students' solutions and individualised feedback and advice. The adaptivity of the system is implemented by presenting students with different, dynamically constructed HTML pages. The decision about the content of each page is based on the model of each individual student.

Brusilovsky in [2], outlines three stages in the adaptation process: collecting data about the user, processing the data to build or update the user model, and applying the user model to provide the adaptation. In the case of Web PVT, the system uses a combination of stereotypes and the overlay technique for the initialisation of the student model. The system uses stereotypes along several dimensions to provide initial values to attributes of the student model, which are then refined by observing the student while working with the system. The resulting student model is used by the tutoring module to provide adaptive navigation support while the student is presented with theoretical issues concerning the passive voice. Then the student model is also used for providing individualised problem solving support to students. This includes automatic error diagnosis to the students' answer to exercises and appropriate advice.

2. Requirements analysis and Design

As Mark and Greer point out in [9], the accuracy of ITS components such as domain knowledge should be ensured before a system is completed and assumed thereafter. One good way of ensuring this accuracy is to involve human teachers and students in the early stages of the development process [16].

The design of Web PVT has been based, to a large extent, on the results of an empirical study that involved school-teachers and students. These results were particularly useful for the requirements analysis of Web PVT. They showed what particular pieces of the domain knowledge seem difficult for students to comprehend fully, what misconceptions they may have while learning and what the explanations to those misconceptions may be [15].

3. System Architecture

The deployment of Web PVT was based on the "HTML-CGI" architecture. The user interacts with simple HTML pages and entry forms using her/his standard Web browser. Information entered by the user is sent to the Web server which forwards it to the CGI program which then replies with new HTML pages [1]. We have chosen this server centred approach, since the installation and maintenance of the Web-based ITS on the server side allows the designers and developers of the system to modify and update the software without the need of redistributing it to students.

Web PVT follows the main line of an ITS architecture, adjusted to the WWW technology (Figure 1). The system consists of four major components, namely the domain knowledge, the student modeller, the tutoring and the user interface [6, 13, 18].

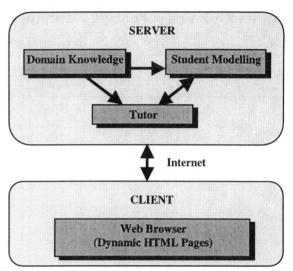

Figure1. Architecture of Web PVT.

The domain knowledge of Web PVT is represented using a semantic network depicting the relations between the concepts of the domain. Each node in the domain knowledge represents a certain category of concept, which may be further divided into smaller sub-concepts. There are three kinds of link between nodes: part-of, is-a, and prerequisite. A part-of relation points from a more general to a more specific concept, which is one of its parts. For example, the "verb tense conversion" concept, is a part of the mundane concepts. An is-a relation, points from an instance of a concept to the concept. For example, there is an is-a relation between the several verb tense forms and the "verb tenses" concept. A precondition relation points from a concept to another, which is its prerequisite. For example, in order to master the simple past tense, one should know how to form irregular verbs.

Web PVT uses a combination of stereotypes and the overlay technique for student modelling [8, 10]. The system uses stereotypes along several dimensions to initialise the student model. These dimensions concern the knowledge level of the student, an estimation of her/his carefulness while solving exercises and her/his prior knowledge of other languages. The student is initially assigned to one of the four distinct stereotypes, namely novice, beginner, intermediate and expert, according to her/his performance on a preliminary test. The system also comprises a long term student model [12], which records information about which concepts the student has mastered and to what extent. In addition, it records the kinds of error the student has made during past interactions as well as the most suitable explanation of each category of error. The information from the long term student model forms an individual model of the student, which together with the active stereotype are used in order to provide adaptive navigation support and perform intelligent analysis of the student's solutions to exercises.

The tutoring component is responsible for making several pedagogical decisions, based on the individual strengths and weaknesses of a student. The tutoring component of Web PVT consults the student model in order to provide adaptive navigation support and to individualise the feedback. Its full functionality is described in the next section.

The user interface of Web PVT consists of a set of dynamically constructed HTML pages and forms. The decision about the content and the form of each page is made by the tutoring component.

4. Functionality

Web PVT incorporates techniques from ITS and AH to provide students with individualised instruction and feedback. In particular, the system uses AH techniques, such as adaptive link annotation, to support the navigation of the student through the course material. Furthermore, Web PVT incorporates facilities for intelligent analysis of students' solutions to exercises. In case an error occurs, it tailors the feedback and advice to the individual student. In this way, advice is individualised to the particular needs of each student.

In the following subsections we analyse the approach taken by Web PVT, to address the above issues.

4.1. Adaptive Navigation Support

Due to extra navigational freedom they provide, hyperdocuments impose greater cognitive loads on users than linear documents, such as books on paper or on-line [4]. Therefore, although a good design of the navigation space may help, it is also necessary to provide more sophisticated mechanisms that modify the navigation alternatives by some sort of adaptation procedure [5].

Web PVT uses adaptive annotation of links to guide the student in the navigation process. The idea of adaptive link annotation is to augment the links with some form of comments which can tell the user more about the current state of the nodes behind the annotated links [2].

In particular, Web PVT uses different font types to provide adaptive navigation support. Whenever a link appears in the table of contents or other pages, the font type of the link is annotated so as to reflect the state of the concept behind the link, with respect to the student's current knowledge and misconceptions. The links that are in bold letters point to concepts that are ready to be learned and are recommended by the system. Links that are italicised correspond to concepts that have already been visited and are known to the student. Regular links point to concept pages that have been visited but the student cannot use them correctly while solving exercises. Finally, dimmed links correspond to concepts that are not ready to be learned, since some prerequisite concepts are not known to the student.

4.2. Intelligent Analysis of Solutions

In the WWW context, intelligent analysis of solutions is needed to perform error analysis, since the system has to deal with students' final answers to exercises. There are no significant delays, due to the fact that there is only one bi-directional interaction between the client and the server.

In Web PVT, the exercises are presented to students using HTML forms. The student is given a sentence in one voice (active or passive) and is asked to rewrite it using another voice. After solving an exercise, the student submits her/his answer to the server side, which is going to perform analysis of the solution.

ICALL systems have commonly based their error analysis (at least partly) on the mother tongue of the student. A web-based system, however, must adopt a more general scheme in order to accommodate international access [7]. Web PVT, apart from maintaining mal-rules associated with errors due to language transfer, also comprises knowledge about errors that are independent of the mother tongue of a student. Mal-rules related to language transfer are used in case the mother tongue of a student is known (e.g. Greek), while general mal rules are used even in cases we have no information about the native language of the student.

While performing analysis of the student's answer, the system ignores trivial typographic errors such as the absence of a fullstop at the end of the sentence, absence of any space between words, redundant spaces or commas, etc. If the student has made an error other than trivial typographic errors, then the system performs error diagnosis taking into account information that has been collected about the specific user in previous interactions as well as common mistakes that have been identified by the empirical study.

In some cases a mistake of the user may be attributed to more than one categories of error. For example, if a student has typed the word "teacher" instead of "teachers" in the converted sentence, this could either be an accidental slip or a singular/plural mistake. In cases like this the system takes into account the individual features of the user, that have been recorded in previous interactions, in order to formulate the kind of advice to give to her/him. For example, if the particular student has not previously made singular/plural mistakes but has made carelessness mistakes then the system favours carelessness as the most probable cause of the mistake. Based on the results of the error diagnosis, the system is able to provide advice and feedback tailored to the individual student.

5. Conclusions and Future Work

In this paper we have described Web PVT, an adaptive web-based ICALL that incorporates techniques from ITS and AH to provide students with individualised instruction and feedback. In particular, the system uses the link annotation technique so as to provide adaptive navigation support. Furthermore, Web PVT is capable of performing error diagnosis and ambiguity resolution based on the long term student model. The design of the diagnostic component has been based on an empirical study that has been held in real classrooms and has shown what the most common students' mistakes are.

The results of the empirical study have been used for the design of a first version of the system. However, one great advantage of web-based educational technology is that it produces systems that may be accesible by many students. This may serve as a means for collecting more data through the individual student models concerning the most common errors of students belonging to different categories (e.g. students having Greek as their mother tongue). Therefore, there are going to be further refinements based on iterations of empirical studies.

In addition, within the future plans of this research, is the improvement of the student modelling component of Web PVT, so that it may cope with temporal aspects of the student learning process.

6. References

[1] S. R. Alpert, M. K. Singley, and P. G. Fairweather, "Deploying Intelligent Tutors on the Web: An Architecture and an Example", *Journal of Artificial Intelligence in Education*, 10, 1999, pp. 183-197.

[2] P. Brusilovsky, "Methods and Techniques of Adaptive Hypermedia", *User Modeling and User Adapted Interaction*, 6(2/3), Kluwer, The Netherlands, 1996, pp. 87-129.

[3] P. Brusilovsky, "Adaptive and Intelligent Technologies for Web-based Education", *Künstliche Intelligenz*, 4, 1999, pp. 19-25.

[4] L. Calvi, and P. De Bra, "Proficiency-Adapted Information Browsing and Filtering in Hypermedia Educational Systems", *User Modeling and User Adapted Interaction*, 7(4), Kluwer, The Netherlands, 1997, pp. 257-277.

[5] R. Caro, E. Pulido, and P. Rodriguez, "Dynamic Generation of Adaptive Internet-based Courses", *Journal of Network and Computer Applications*, 22, Academic Press, 1999, pp. 249-257.

[6] J. R. Hartley, and D. H. Sleeman, "Towards intelligent teaching systems", *International Journal of Man-Machine Studies*, 5, Academic Press, 1973, pp. 215-236.

[7] T. Heift, and D. Nicholson, "Theoretical and Practical Considerations for Web-based Intelligent Language Tutoring Systems", In Gauthier G., Frasson, C., and VanLehn K. (eds.), *Lecture Notes in Computer Science: Intelligent Tutoring Systems*, Springer, Vienna New York, 2000, pp. 354-363.

[8] H. Hohl, H. Böcker, and R. Gunzenhäuser, "Hypadapter: An Adaptive Hypertext System for Exploratory Learning and Programming", *User Modeling and User Adapted Interaction*, 6(2/3), Kluwer, The Netherlands, 1996, pp. 131-155.

[9] M. A. Mark, and J. E. Greer, "Evaluation Methodologies for Intelligent Tutoring Systems", *Journal of Artificial Intelligence in Education*, 4, 1993, pp. 129-153.

[10] M. Murphy, and M. McTear, "Learner Modelling for Intelligent CALL". In Jameson, A., Paris, C. and Tasso, C. (eds.): *Proceedings of the Sixth International Conference on User Modeling*, Springer, Vienna New York, 1997, pp. 301-312.

[11] C. Peylo, W. Teiken, C. Rollinger, and H. Gust, "An Ontology as Domain Model in a Web-based Educational System for Prolog", In Etheredge J., and Manaris B. (eds) *Proceedings of the 13th International Florida Artificial Intelligence Research Society Conference*, AAAI Press, Menlo Park CA, 2000, pp. 55-59.

[12] E. Rich, "Users are Individuals: Individualizing User Models", *International Journal of Man-Machine Studies*, 18, Academic Press, 1983, pp. 199-214.

[13] J. Self, "The Defining Characteristics of Intelligent Tutoring Systems Research: ITSs Care, Precisely", *International Journal of Artificial Intelligence in Education*, 10, 1999, pp. 350-364.

[14] M. Virvou, and D. Maras, "An intelligent multimedia tutor for English as a second language". In Collis, B. & Oliver, R. (Eds.) *Proceedings of ED-MEDIA 99, World Conference on Educational Multimedia, Hypermedia & Telecommunications*, Vol. 2, AACE, Charlottesville, 1999, pp. 928-932.

[15] M. Virvou, D. Maras, and V. Tsiriga, "Student Modelling in an Intelligent Tutoring System for the Passive Voice of English Language", *Educational Technology and Society*, 3(4), 2000, pp. 139-150.

[16] M. Virvou, and V. Tsiriga, "Involving Effectively Teachers and Students in the Life Cycle of an Intelligent Tutoring System", *Educational Technology and Society*, 3(3), 2000, pp. 511-521.

[17] G. Weber, and M. Specht, "User Modeling and Adaptive Navigation Support in WWW-based Tutoring Systems", In *A. Jameson., C. Paris, and C. Tasso (eds.) Proceedings of the 6th International Conference on User Modeling*, Springer, Vienna New York, 1997, pp. 289-300.

[18] Wenger, E., *Artificial Intelligence and Tutoring Systems*, Morgan Kaufman, Los Altos CA, 1987.

The Development of CALL Environment on the WWW for Teaching Academic English

Jin Chen, Hisayoshi Inoue, Toshio Okamoto, Safia Belkada, and Alexandra Cristea
Artificial Intelligence & Knowledge Computing Laboratory
Department of Information Systems Science, Graduate School of Information Systems
The University of Electro-Communications,
1-5-1, Choufugaoka, Choufu, Tokyo, 182-8585, JAPAN
chenj@ai.is.uec.ac.jp

Abstract

This paper describes a course-support hypermedia language-learning environment, called "AcademicEnglish" based on communicative language teaching approach, for EFL/ESL academicians to improve their communicative competence in academic contexts, and shows the design and implementation of the courseware in this environment. To support our hypotheses in building such an environment, we conducted a questionnaire survey of the difficulties in academic language learning. Moreover, we have explored the research findings of SLA from the educators' and learners' points of view. The focus is to present our approach in communicative-oriented courseware development. Based on our hypotheses, we built an illustrative prototype system, which intends to allow learning goals communication, learning activities negotiation and collaboration between learners.

1. Introduction

The effectiveness of CALL systems has been proven by many previous studies (e.g., Yang, 1997). However, the main weakness of current CALL systems is that most of the video-based multimedia materials treat language as a set of linguistic structures. Thus, language is taught as words and sentences, not as discourses and interactions in context (Duranti &Goodwin, 1992). A communicative focus on the appropriate and meaningful use of language in a variety of contexts is still in its struggle in a computer environment, although hypermedia tools have offered a context for the successful acquisition of communicative competence with the aid of CALL. Another problem in present CALL systems is that there is little courseware for non-native English speakers to learn English for their academic purposes. Even if some systems provide examples, words, etc., from the academic life, there is no major difference between the support for social language learning and academic language learning in their courseware implementations. Context and cognitive complexity will be the two main factors, which affect academic language comprehension. So the current CALL systems cannot meet the academicians' requirements and needs. Therefore, our research goal is to respond to these needs, by building a hypermedia language-learning environment for teaching academic English via the World Wide Web.

2. Hypotheses

The learning environment and courseware that will be presented in this work is based on the following five hypotheses derived from our preliminary tests (see, Chen, 2001) and from various other research studies (e.g., Hoven, 1991): 1) interaction and negotiation are important features of communication, and therefore of second language learning; 2) students learn academic language and content more effectively with the explicit instruction of learning strategies; 3) hypermedia environment with well-defined courseware can play a mediating role between academicians and their context; 4) learning should be communicative, active, and dynamic.

3. AcademicEnglish: an illustrative system

Based on our hypothesis, we have developed a hypermedia language-learning environment called AcademicEnglish, which contains multimedia material targeted at undergraduate artificial intelligence (AI) and knowledge computing students. The environment allows students with some background knowledge structure of AI and education technologies (ET) in their own native language to learn to express and use this knowledge in English in academic context. The subject materials for the environment are topic-based conversations, which came from introductory AI courses, which are delivered via hypermedia language learning environment in accordance with the improvement of students' listening comprehension and furthermore with their academic language competence.

3.1 System structure

The architecture of AcademicEnglish consists of three layers (see figure 1): *visualization layer, funcitons layer*, and *data store layer*. The visualization layer presents the actual learning material and interacts with the user or a group of users through the Internet. The functions layer serves for the processing logic and allows the users to access the teaching resources in the data store layer via the search mechanism and to check answer and get feedback via the answer-check mechanism as well. The last layer is the data store layer (see figure1): like the *content data* (for more details, see Chen, 2001), the *student profiles*, in which students' personal data, learning behaviour, learning and communicative strategies, and actual performance will be stored, etc. Furthermore, during the learning, we provide three types of interactions: interaction between the user and the system, interaction between the user and other users in the same group, and interaction between user and the human teacher.

Student(s)

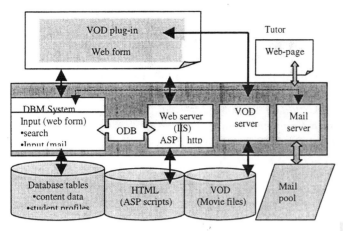

Figure1: System Structure

3.2 Courseware structure

When this environment is used, in the beginning, registration details, such as name, email address, and level of English learning of the user is required. After the registration, the user has two choices: to search the course via the VOD search mechanism (*autonomous study*), and to access the course via a placement-test (*automatic study*). When the user chooses one of them, the activity with VOD corresponding to the user's choice appears on a newly generated page. Each activity session consists of three modules: pre-viewing module, while-viewing module, and follow-up module. Besides, the user can test his/her performance in one session via Achievement test. After each session, the environment will update the relevant attributes of the student profile (e.g., current progress, learning behavior, communicative strategies, etc). Then it generates a plan of remedial actions and presents it to the student in the form of suggestions for further sessions and learning, and the user him-/herself can decide whether to continue the study or end it.

3.2.1 Activity design. The academic language function acquiring, language learning strategies and communicative strategies are naturally embedded in content activities. We defined the activities with three language levels: 1) intermediate, 2) upper-intermediate and 3) advanced, in addition, there is a continuum in the activities in and among each level, which are graded according to the cognitive and performance demands made upon the learner, moving from *comprehension* based activities to controlled *production* activities and finally ones which require the learner to engage in real *communicative interaction*, which are hierarchically organized into three modules (see figure2). Student performance in each activity contributes to the student profile. In the present courseware, we used communicative teaching strategies, like eliciting, personalization, information gap, problem solving, and game playing, etc., to promote opportunities to use language meaningfully. These types of activity are focused on extracting, recording and sharing information in various class groupings. This is exemplified as follows, where we see a unit, which is built around an information gap task. The activity contains the following steps.

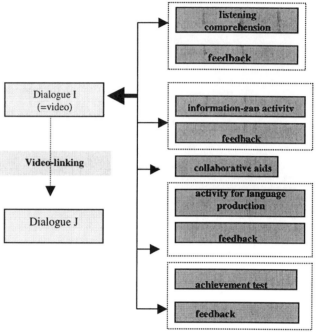

Figure2: Elements within one video

4. Conclusions

In this paper, we proposed a hypermedia language-learning environment based on the communicative approach. Our work is in the analysis and design phase and the actual system is in progress. What we need to do in the next steps is to implement the courseware, including answer-checking mechanism. We also need to continue to develop the collaborative learning functions. Finally, we plan to do real-life experiments to evaluate the effectiveness of our academic English teaching environment.

References

Chen, J. & Okamoto, T. (2001), The Development of Hypermedia Language Learning Environment for Teaching Academic English, ED-MEDIA2001, Finland.

Duranti,A. & Goodwin,C.(1992), Rethinking context: Language as an interactive Phenomenon. Cambridge, UK: Cambridge University Press.

Hoven,V. 1991, Towards a Cognitive Taxonomy of Listening Comprehension Tasks. SGAV Review, vol. 9, No. 2, 1-14.

Yang, J.C., and Akahori, K. (1997), Development of Computer Assisted Language Learning System for Japanese Writing Using Natural Language Processing Techniques: A Study on Passive Voice. In the proceedings of the 8th world Conference on Artificial Intelligence in Education (AI-Ed97).

OWLs in Flight: Online Writing Labs for Distance Learning

Melissa Ann Spore
University of Saskatchewan
Melissa.Spore@Usask.ca

Abstract

The online writing lab (or OWL) has emerged in the past decade as a major use of technology for the student; however, rarely does it serve the distance learner.

This paper considers (1) the needs of distance learners for academic support systems, (2) the failure of these services to be extended to technology-dependent students, and (3) the advantage of technology-mediated writing tutorials over face-to-face tutorials.

This topic demonstrates one instance in which technology-mediated instruction is advantageous, not compensatory, for distance students. It feeds a major concern (writing competence) in many schools as seen in remedial programs and Writing Across the Curriculum initiatives.

1. Introduction

Online writing labs (OWLs) provide composition tutoring that is directly related to course work and academic writing. OWL web sites offer handouts, style and grammar guides, and links to reference works. The Purdue University OWL has almost 4000 such pages; however, tutoring to individuals who cannot, or choose not to, access face-to-face services is the heart of the OWL. In the past 20 years, composition pedagogy and electronic communication have advanced in complimentary ways. Yet, most OWLs serve on-campus students and neglect distance education. This paper reports on the OWL phenomenon, considers the failure to expand OWLs to technology-mediated situations, and seeks to encourage OWLs for distance learning.

2. Need

Distance education students tend to fall into two groups: sophisticated adults continuing their learning and the educationally disadvantaged, who are marginalized by geography, part-time study, family and employment duties, and poor experiences in formal education. Such learners enter distance programs to change those experiences, yet cannot access the on-campus support systems (academic advising, study skills, adequate library services). Karuth [1] and Bates [2] argue that technology-based programs must integrate support services. The writing lab is a primary support for students, introducing the tacit conventions of academic, professional, and business communication.

3. Writing and Computers

Most writers build judgement and improve manuscripts through comments from readers and editors. Computers offer a convenient mechanism for this feedback and revision. While electronic communication no more ensures academic literacy than blackboards guarantee good lectures, computers encourage activities that were previously cumbersome: peer reading, co-authoring, tutor embedded comments.

Although the visual and interactive possibilities of technology are often emphasized, text continues to assert its power. Andrew Feenberg, a pioneer of internet education, has reflected on this phenomenon:

> Could it be that our earliest experiences with computer conferencing were not merely constrained by the primitive equipment then available, but also revealed the essence of electronically mediated education...For after all these years, the exciting online pedagogical experiences still involve human interaction and for the most part these continue to be text based. [3]

In the 1980s, composition theory rejected the atomized knowledge of behaviourism (drillpads and programmed materials) to focus on the writing process. More recently, constructivist pedagogy emphasizing audience and communication has become popular. OWLs apply these pedagogies, through electronic networks, to concentrate on the interactive tutorial.

4. Distance Education

OWLs develop from strong on-campus writing centres, rather than distance technology initiatives. The UK Open University provides "English Effectiveness" Web pages, but no interactivity or messaging. Canada's Athabasca University offers online writing courses for credit only. Deakin University in Australia supplies students with an essay writing video. In the U.S., some schools (such as Texas Tech) built OWLs for non-traditional and distance students, but dropped the off-campus service.

I have found only two services aimed at distance students and only one article discussing distance delivery.

The sole documented effort, at the University of Alaska, uses *NetMeeting*, for which students must book lab time at a regional centre—far from ideal for independent learners in remote locales. Unfortunately, "student interest and retention have languished…[and] the on-site computer support staff is often learning *NetMeeting* with the student" [4].

The reasons for this neglect are murky, but I can suggest possibilities: (1) both distance education and writing labs are marginalized and shy away from alliances with those similarly positioned; (2) both writing centre and distance students are somewhat suspect in the academy, thought of as unprepared or uncommitted and therefore undeserving of special effort; (3) the great success in technology-based learning has been professional continuing education, which can require minimum competencies; (4) academic support for distance learners is meager—without equal bridging programs, academic advising, and study skills, the OWL must seem like a luxury; (5) the designers of technology-enhanced learning may be too removed from the day-to-day problems of education, the trenches where writing problems dwell.

5. Online Tutoring

Online writing tutorials are among the easiest academic support services to mount. They demand little infrastructure, minimal administration, and good tutors (usually sessionals and graduate students). To understand this, let's look at what goes on in an OWL tutorial.

Students submit a work to which most tutors respond with an email message in three parts: front note, endnote, and intertextural commentary. The commentary—specific comments *within* the text—tends to be frank (often leavened with humour), which dispels ambiguity and contrasts with the delicacy of the front note. Students seem to accept honest, helpful criticism when it is removed from a fleshy authority. The combination of focused particularity and candor are the strengths of online tutoring. Tutors end up concentrating on the text rather than the students and they tend to use to informal language ("sorta", "little bit") and conditional modals ("You may want to…"). David Coogan has described his experience:

> I slowly realized that e-mail was allowing me to say things I might not have said to students face-to-face. E-mail enabled me to perform close readings of student work…What seems like a disadvantage (not seeing the student) can at times be an advantage. [5]

Metacognition is forced upon students in electronic tutorials. Weak writers are unaware of the process and structure of their own work. Email messages force them to articulate their concerns, analyse their text, and be aware of their choices. Experienced writers frequently imagine text changes and make mental comparisons between the original and revision. Inexperienced writers learn by manipulating the text on the page to compare versions. The isolation of reading and typing a response seems to encourage reflection.

As we all know, email is replete with poor phrasing, awkward grammar, and typographical errors; it resembles early drafts more than a finished text. By asking for, and commenting within, electronic drafts tutors can enter the writing process rather than comment on the (literal) margins. Papers can become communication, rather than the display of knowledge and the conventions that we usually see in the academy. If we want students to revise, perhaps every composition need not be judged by a single standard.

6. Conclusions

An OWL demands resources and commitment without many assurances. It merges the difficulties of distance learning—missing physical cues and spontaneity, onerous preparation and self-discipline, organizational and personal adjustments—with the imperfections of writing centres—learner confusion, inadequate time; a culture that offers only lip service to writing with acumen.

But the minimalist technology used in OWLs makes for an accessible, communicative application of an essential goal of education. Thus, writing instruction benefits from grounded pedagogy, relevant student practice, and accessible technology.

This article is based on a technical report, *Wired Writing: Issues in Electronic Writing Labs,* which can be found at www.extension.usask.ca/Staff/Spore/final%20Wired.pdf.

7. References

[1] Krauth, B. and J. Carbajal, *Guide to Developing Online Student Services*, Western Cooperative for Educational Telecommunications, 1999, http://www.wiche.edu/Telecom/resources/publications/guide/guide.htm (access 24/02/01).

[2] Bates, A.W., *Managing Technological Change*, Jossey-Bass, San Francisco, 2000, p. 74.

[3] Feenberg, A., *My Adventures in Distance Learning*, 1999. http://www-rohan.sdsu.edu/faculty/feenberg/TELE3.HTM (access 12/01/01).

[4] Thurber, J., "Synchronous Internet Tutoring: Bridging the Gap in Distance Education," *Taking Flight with OWLs: Examining Electronic Writing Center Work*, Eds. Inman, J. A. and D. N. Sewell, Lawrence Erlbaum Associates, New Jersey, 2000, p. 156–157.

[5] Coogan, D., "E-Mail Tutoring, A New Way To Do New Work," *Computers and Composition* 12.2, August 1995, http://corax.cwrl.utexas.edu/cac/Archives/v12/122 html/feature.l (access 19/01/01).

Bridging Gaps in Computerised Assessment of Texts

David Callear
Dept. of Info. Systems
University of Portsmouth
United Kingdom
David.Callear@port.ac.uk

Jennifer Jerrams-Smith
Dept. of Info. Systems
University of Portsmouth
United Kingdom
Jenny.Jerrams-Smith@port.ac.uk

Victor Soh
Dept. of Info. Systems
University of Portsmouth
United Kingdom
Victor.Soh@port.ac.uk

Abstract

A survey of major systems for the automated assessment of free text answers is presented. This includes the Project Essay Grade (PEG), Intelligent Essay Assessor (IEA) which employs Latent Semantic Analysis (LSA), and Electronic Essay Rater (e-rater). All these systems have the same weakness in that they are unable to perform any assessment of text content. The word order is not taken into account. In an effort to bridge the gaps in knowledge about this research problem, an introduction to a novel Automated Text Marker (ATM) prototype is given in this paper.

1. Introduction

A number of computerised free text assessment systems have been developed in the USA. Some based on the Latent Semantic Analysis (LSA) technique [6] have been independently tested in France [3] and there is at least one system being studied in the UK. None of the systems satisfactorily assesses text content.

This paper consists of a survey of the various major systems and an introduction to the authors' *novel* Automated Text Marker (ATM) prototype, primarily designed to assess text content.

2. Project Essay Grade (PEG)

Research in computerised assessment of students' essays dates back to the mid nineteen sixties, when Ellis Batten Page of Duke University in the USA developed the Project Essay Grade (PEG) [8].

Text content and word order are not taken into account, although PEG produced high correlations of around 80% between the computer-predicted and the human-assessed essay grades. This approach is based on the superficial surface features of an essay (counts of commas, semicolons, average word count, etc.) as indicators of its quality.

Whether PEG is suitable for assessing creative writing skills is debatable, although it is an indication that the assessment of essay style has been successfully automated. It is, however, unsuitable for assessing factual disciplines, in which text content is very important.

3. Intelligent Essay Assessor (IEA)

The Intelligent Essay Assessor (IEA) [5,6] was developed in the late 90s. It utilised the Latent Semantic Analysis (LSA) technique [6] which was originally meant for indexing documents and text retrieval in the late 80s. LSA was developed by Thomas K. Landauer of the University of Colorado, Boulder and Peter W. Foltz of the New Mexico State University in the USA.

LSA is not suitable to assess short answer questions and factual disciplines. The grammar and word order are not taken into account. LSA does not distinguish sentences such as, *"Country A bombed country B"* and *"Country B bombed country A"*, from each other.

4. Electronic essay rater (*e-rater*)

Jill Burstein of the Educational Testing Service (ETS) in the USA and others developed the Electronic Essay Rater (e-rater) [2,4] in 1998. The ETS is an organisation which conducts a number of world-wide standardised tests for the purpose of admissions to universities.

E-rater uses the Microsoft Natural Language Processing (MSNLP) tool [7] for parsing all sentences in the essays. In syntactic structure analysis, features identified include the numbers of complement clauses, subordinate clauses, infinitive clauses, relative clauses, subjunctive modal auxiliary verbs and others.

A grade prediction accuracy is determined by comparing human and e-rater grades across 15 test questions. The empirical results range from 87% to 94%. The system is similar to PEG, but the final linear regression model incorporates syntactic, rhetorical and some topical features. E-rater does not assess text content beyond spotting weighted keywords. The empirical results obtained are not from essays on factual disciplines.

5. Automated Text Marker (ATM)

The Automated Text Marker (ATM) prototype is designed for the purpose of automating the assessment of text content in detail. This adds up to a final summative score which reflects the detailed assessment of text content incorporating word order.

An infection is the invasion and multiplication of microorganisms on body tissue that produce signs and symptoms as well as an immunologic response.

Figure 1: Example Sentence

The two main components of ATM are the syntax and semantics analysers. ATM is written in Prolog. A model answer or an expertly written examiner answer to a closed-ended topic is automatically broken down into its basic concepts and dependencies, and the same is done with each student answer, then the two are compared.

DEPENDENCY GROUP 1
group(1,([infection] → [is] → [invasion])).
group(1,([infection] → [is] → [multiplication])).
group(1,([invasion] → [of] → [microorganism])).
group(1,([multiplication] → [of] → [microorganism])).

DEPENDENCY GROUP 2
group(2,([group(1)] → [in] → [body tissue])).

DEPENDENCY GROUP 3
group(3,([group(2)] → [produce] → [sign])).
group(3,([group(2)] → [produce] → [symptom])).
group(3,([group(2)] → [produce] → [immune response])).

Figure 2: CD Form of Example Sentence

Syntax analysers of varying complexity can be written in Prolog relatively simply and efficiently. The grammar can be augmented to include a wide-coverage, context-free and formalised grammatical description such as the Generalised Phrase Structure Grammar (GPSG) formalism [1].

A simple example of a sentence to be analysed is shown in Figure 1. For clarity, its conceptual dependency groups (CD) are shown in Figure 2 in output form, not the Prolog internal representation within the program.

Each fragment of concept is either totally independent (a dependency group by itself) or falls under a major dependency group, and is automatically given a numerical tag (number). Each numbered dependency group represents the context within which fragments of concept must be reclustered and segregated. These major dependency groups can be further related to each other so that successively larger dependency groups are generated and numbered automatically.

6. Conclusion

The difficulties in automating the assessment of text content incorporating word order are addressed in this paper. A solution is provided by ATM. Text passages are automatically broken down into their smallest viable units of concepts, and compared to an examiner's model answer.

7. References

[1] Bennett, P., *A Course in Generalised Phrase Structure Grammar*, UCL Press, London, 1995.

[2] J. Burstein, K. Kukich, S. Wolff, Chi Lu, M. Chodorow, L. Braden-Harder and Mary Dee Harris, "Automated Scoring Using a Hybrid Feature Identification Technique", *Proceedings of the 36th Annual Meeting of the Association of Computational Linguistics*, Montreal, Canada, 1998.

[3] P. Dessus, B. Lemaire and A. Vernier, "Free Text Assessment in a Virtual Campus", *Proceedings of the 3rd International Conference on Human System Learning*, Europia, Paris, 2000, pp. 61-75.

[4] Educational Testing Service (ETS), *E-rater*, website: http://www.ets.org/research/erater.html/, Princeton NJ.

[5] P. W. Foltz, D. Laham, T. K. Landauer, "Automated Essay Scoring : Applications to Educational Technology", *Proceedings of ED-MEDIA '99 Conference*, AACE, Charlottesville, 1999.

[6] T. K. Landauer, P. W. Foltz, D. Laham, "Introduction to Latent Semantic Analysis", *Discourse Processes*, 1998, vol. 25, pp. 259-284.

[7] Microsoft Corporation, *MSNLP*, website: http://research.microsoft.com/nlp/, USA.

[8] E. B. Page, "New Computer Grading of Student Prose : Using Modern Concepts and Software", *Journal of Experimental Education*, vol. 62, no. 2, pp. 127-142, 1994.

From Concrete Experiences to Abstract Formalisms: Learning with Interactive Simulations that Combine Physical and Computational Media

Marcelo Milrad
Framkom, Sweden & University of Bergen, Norway
marcelo.milrad@framkom.se

Abstract

In this paper I present my efforts in exploring the integration of physical and computational media for the design of interactive simulations to support learning about complex domains. My effort involves the design of interactive learning environments to integrate systems supporting alternative ways of interaction with simulations with an emphasis upon support for shared interaction to mediate social aspects of learning, knowledge construction, reflection and design. Some examples from the projects I am conducting in Sweden are presented.

1. Introduction

Advances in educational technology have led to interest in providing meaningful support for learning about complex domains [12]. Learning environments to support learning about complex, dynamic systems include SimCity™ for helping children learn about factors influencing the growth of urban area and Beefeater™ for helping adults learn about factors influencing the growth of a particular business, and a host of other such environments which are now frequently used to support various types of learning activities. Furthermore, the notion of interactive simulations is quickly gaining importance as a mean to explore, comprehend and communicate complex ideas [13]. Economics, ecology, epidemiology, and project management, all typically involve complex, dynamic systems and they are just some examples out of an infinite universe of questions that can be explored with simulations.

Regardless of the type or format of the simulation, the overriding purpose for simulating systems remains: to provide a learning environment that supports the learner to develop mental models about the interrelationships of variables; to test the efficacy of these models in explaining or predicting events in a system; and to discover relationships among variables and/or confronting misconceptions. However, the extent to which it is helpful to attempt to use interactive simulations to model reality in too many aspects is less evident [5]. While a number of features of the real world which are thought to be relevant to the learning process can be replicated to a certain extent by computer programs, others cannot, and indeed it may well be that maintaining a distinction between the real and the virtual is an important aspect of the transfer of learning from computer-based environments to the wider world. Frequently, the design of these simulation-based learning environments focuses exclusively on computers and the virtual environments they provide, excluding the physical environment. Moreover, few contemporary researchers or practitioners question the importance of interaction in computer-supported learning [7]. With the emergence of new technologies, and the continued refinements of existing technologies, design potential has expanded dramatically. What kind of interactions should be cultivated, for which types of learning tasks? How should differences in learning tasks influence the design of interaction strategies?

In this paper I present my efforts in exploring the integration of physical and computational media for the design of interactive simulations to support learning about complex domains. This effort involves the design of interactive learning environments to integrate systems supporting alternative ways of interaction with simulations - with an emphasis upon support for shared interaction to mediate social aspects of learning, knowledge construction, reflection and design. These interaction paradigms integrate the use of computationally-augmented physical objects [1] - to support and encourage face-to-face interaction among learners - with virtual objects - to provide computational support for the model underlying the simulation. Many models of learning and collaboration need to emphasize the creation of shared interaction, social structures, and cultural embeddings for meaningful learning [4]. In the next sections I illustrate my approach for designing interactive simulations to support learning about complex domains.

2. Design issues and conceptual framework

Land & Hannafin [8] point out that researchers and designers need to identify frameworks for analyzing, designing, and implementing interactive learning environments that embody and align particular foundations, assumptions, and practices. There is a need for learning activities that stimulate an interest for understanding *complex domains*, challenge current understandings and facilitate experience sharing between learners. Spector et al. [12] claim that instructional scientists and designers have not fully understood the socially situated learning perspective and its implications for human learning in and about complex domains. Complex domains are characterized as those with many interrelated components, with non-linear and fuzzy relationships that may change over time, with delayed effects, and with uncertainties due to dependency on human perceptions about some aspects of the system. It is well established that humans have difficulty in understanding complex systems [6].

According to Spector et al. [12] we lack a well-articulated design framework with sufficient detail to take us from a socially-situated, problem-based, collaborative learning perspective to the design of a particular learning environment for a particular subject domain. In my current research I am investigating how best to design and support learning in and about complex domains. More specifically, I am proposing a general approach which might best be characterized as socially-situated, problem-oriented learning in authentic and collaborative settings. This design framework is based on a experiential, problem-based and decision-based learning perspective. Figure 1 below illustrates the conceptual ideas of integrating all these approaches and their implication for design.

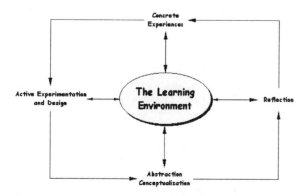

Figure 1. Design Framework

I suggest that the design of interactive simulations for complex learning should be guided by:

- *Authentic activities:* presenting authentic tasks that conceptualize rather than abstract information and providing real-world, case-based contexts, rather than pre-determined instructional sequences. Learning activities must be anchored in real uses, or it is likely that the result will be knowledge that remains inert;
- *Construction:* learners should be constructing artifacts and sharing them with their community;
- *Collaboration:* to support collaborative construction of knowledge through social negotiation, as opposed to competition among learners for recognition;
- *Reflection:* fostering reflective practice;
- *Situating the context:* enables context and content dependent knowledge construction; and,
- *Multi-modal interaction:* providing multiple representations of reality, representing the natural complexity of the real world.

The different components of this design framework are rotted in situated cognition [2] which emphasizes the importance of situating thinking with complex contexts. Learners are expected to generate problems to be solved and then, learn, develop and apply relevant knowledge and skills through progressive problem generation, framing and solving. The different learning activities which have been designed upon this framework require learners to identify research questions and variables, set hypotheses, build and construct experiments, test results, analyze observations and then refine hypotheses and casual variables accordingly. In the next section we present an interactive learning environment which has been designed based upon the ideas described in this framework.

3. Describing the learning environment

We have implemented an interactive learning environment (ILE) to introduce school children and adults to the computational and cognitive aspects of modeling, designing, and analyzing their own models of complex, dynamic systems. This rich technological environment, which involve some data modeling, LEGO construction and computer simulations, provides an experimental arena for learning in and about complex systems. Learners are engaged in an iterative process of model creation and scientific investigation as they explore important scientific principles and processes of complex systems. In order to promote meaningful learning in this kind of environment, we believe learners may begin with concrete operations, physically manipulating objects in order to solve specific problems. As these operations are mastered, they can then progress to more abstract representations and solve increasingly complex problems. Using this underlying framework, we describe below a learning environment for adults learning about issues related to marine biology. Figure 2 gives a few glimpses of the ILE by demonstrating some uses of the environment in the domain of water quality. Learners are first introduced to the problem of acid rain and water pollution in a lake, and thereafter to Model Builder [11], a data modeling tool. The tool supports relationship modeling and casual reasoning processes in ways that are conceptually and procedurally accessible for novice learners. Using this tool, learners identify different factors and relationships affecting water quality, and create qualitative models that represent their understanding of the problem.

After the modeling process, learners proceed by designing and constructing a device, using the LEGO

Using this material, groups of learners work together face-to-face to collaboratively model and design a solution to the specific problem. The students' effort has resulted in a device that is connected to different sensors (temperature and pH) to take input from its environment, to process data, and to control signals and devices involved in different processes.

In the next learning activity, learners are introduced to a system dynamics simulation, that we have developed using *Powersim*, a modeling and simulation tool. We provide learners with access to underlying models, including the opportunity for learners to explore the consequences of changing different parameters for key model components, and to reconstruct alternative models to explain observed behavior in this complex domain. Using this simulation, learners investigate and explore the impact of the acid rain in the fish population of the lake over a period of five years. Learners are asked to develop different policies to control the fish population by manipulating the different parameters and observing and analyzing the behavior of the system

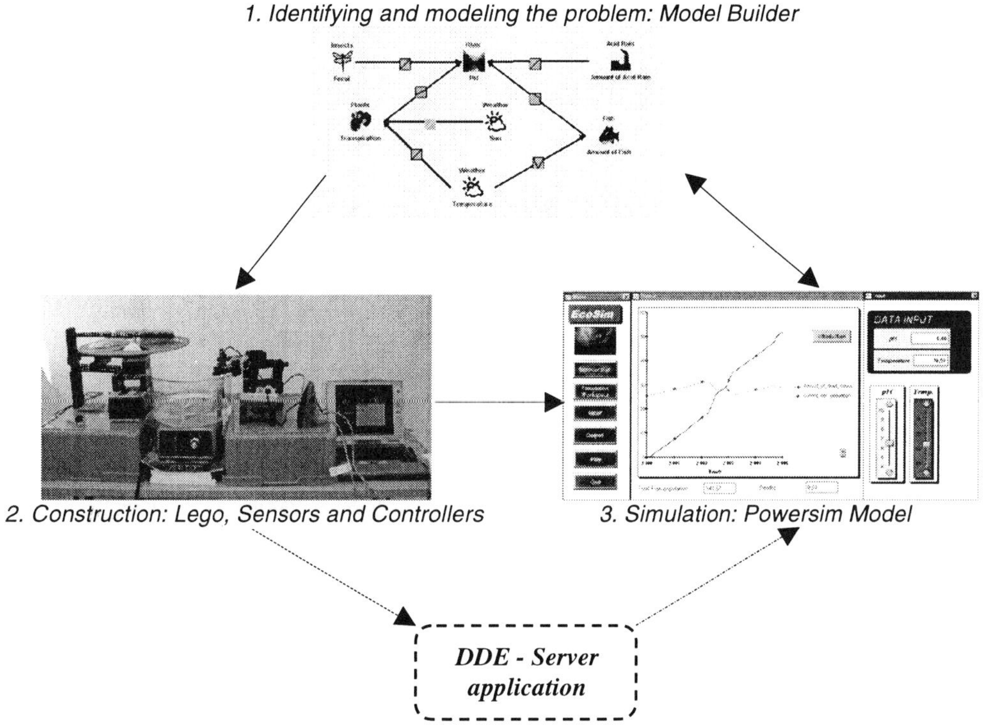

1. Identifying and modeling the problem: Model Builder

2. Construction: Lego, Sensors and Controllers

3. Simulation: Powersim Model

DDE - Server application

Figure 2. The different components of the learning environment

Robotics system, to monitor the temperature and the pH value of the lake. The LEGO robotics system is an autonomous microcomputer that can be programmed using the *ROBOLAB* programming language.

The different components of this ILE are all problem-based but address different aspects of problem solving activities and behavior. These aspects are related to problems directly associated with a concrete and specific environment, to problems associated with hypothesis formulation in a concrete setting, and then to problems associated with abstraction, generalization, and deep understanding of underlying structural causes for observed model and actual behavior. The preliminary findings of our user and classroom testing show some evidences [9] that our approach do indeed support learners' active construction of knowledge, and that learners can create meaningful models that affect and improve the understanding of the complex domain they are learning about.

In this type of situated context, learning occurs naturally as a consequence of the learner recognizing knowledge's practical utility as well as the need to use it in an attempt to interpret, analyze, and solve real-world problems [3]. Modoro [10] claims that exploratory learning environments should allow direct manipulation of concrete-abstract objects and the exploration of multiple representations of a phenomenon. Multiple representations can facilitate the process of creating meaning from representations if we assume that meaning is created essentially when learners relate to different representations.

We believe as stated by Land & Hannafin [8], that by designing and building physical experiments, varying parameters and hypothesizing outcomes, learners test assumptions and revise thinking based on resultant observations. Through the manipulation of the rich-technological environment, new experiences are enabled that offer counter evidence, and models of understanding change accordingly.

We are pursuing a new effort, in implementing a new piece of the learning environment, in addition to those presented before. This new component consists of a web based simulation using a system dynamics model to facilitate understanding of particularly difficult problems related to water quality (second order delays, non-linearities in the system, etc.). Furthermore, we are exploring the possibility of transferring real data from the sensors in the LEGO Robotics system to the Powersim model (see *figure 2*) in order to use this data during the simulation process.

4 References

[1] Blauvelt, G.; Wrensch, T.; & Eisenberg, M. (1999). Integrating Craft Materials and Computation. *Proceedings of Creativity & Cognition 99*, Loughborough UK. ACM Press.

[2] Brown, J., Collins, A. & Duguid, P. (1989). Situated cognition and the culture of learning. *Educational Researcher*, 18(1), 32-42.

[3] Barab, S & Duffy, T. (2000) From Practice Fields to Communities of Practice. In Jonassen, D. & Land, S. *Theoretical Foundations of Learning Environments*. Lawrence Erlbaum Associates, Publishers.

[4] Dillenbourg, P. (1999). *Collaborative Learning: Cognitive and Computational Approaches*. Pergamon.

[5] Dowling, C. (1997). Simulations: New "Worlds" for Learning? *Journal of Educational Multimedia and Hypermedia* 6 (3/4), 321-337.

[6] Dörner, D. (1996) (Translated by Rita and Robert Kimber). *The logic of failure: Why things go wrong and what we can do to make them right*. New York: Holt.

[7] Gavora, M; & Hannafin, M. (1995). Perspectives on the design of human-computer interactions: issues and implications. *Instructional Science* 22: 445-477.

[8] Land, S & Hannafin, M. (2000). Student-Centered Learning Environments. In Jonassen, D. & Land, S. (2000). *Theoretical Foundations of Learning Environments*. Lawrence Erlbaum Associates, Publishers.

[9] Milrad, M. (2001). Evaluating the Impact of Learning with Simulations that Combine Physical and Computational Media. In preparation, to be submitted to *Computers in Human Behavior*.

[10] Modoro, V. (1993). A Model to Design Computer Exploratory Software for Science and Mathematics. In *Simulation-Based Experiential Learning*. Springer-Verlag.

[11] Quintana, C., Fretz, E., Krajcik, J & Soloway, E. (2000). Evaluation Criteria for Scaffolding in Learner-Centered Tools, *Proceedings of CHI'2000*, 189-190.

[12] Spector, J. M., Davidsen, P. I. & Wasson, B. (1999). Designing Collaborative Distance Learning Environments for Complex Domains. *ED-MEDIA 99 Conference Proceedings*.

[13] Turkle, S. (1997). Growing Up in the Culture of Simulation. In P. Denning, and R. Metcalfe, (Eds), *Beyond Calculation: The Next Fifty Years of Computing*. Springer-Verlag. New York, Inc.

The Design and Evaluation of Simulations for the Development of Complex Decision-making Skills

Roger Hartley
University of Leeds, UK, LS2 9JT
j.r.hartley@cbl.leeds.ac.uk

Glen Varley
University of Leeds, UK, LS2 9JT

Abstract

The management of large-scale incidents, such as demonstrations that can affect Public Order involves careful planning that takes into account possible contingencies, and the effective control of police resources at strategic and tactical levels. The training and development of such skills presents problems since exercises on the ground are necessarily small scale, and tabletop map-based simulations are difficult to manage.

In association with the (London) Metropolitan Police a computer based simulation (CACTUS) was designed for improving strategic and tactical decision-making by Senior Police Officers. It incorporates a digitised map with active (iconised) police, crowd and hostile agents able to navigate the map and autonomously interact in ways which simulate aggression and disorder if the police resources and their instructions are not directed with some skill. Adaptive training scenarios are designed in CACTUS by the trainer/facilitator, and the training has phases of planning, event management and debriefing.

An evaluation study (over an eighteen month period) collected performance data and audio and video records of the interactions between trainer, trainees and CACTUS. These data gave useful insights into the decision-making processes and how the CACTUS simulation, through its design features, became a dynamic mediational tool in developing such skills.

1. Introduction.

Dealing with emergency and large-scale incidents such as demonstrations and marches requires careful and sensitive management. Such incidents can be volatile and, if they get out of control, can prove costly in damage to property, in personal injury, and in the erosion of public confidence in the police and associated emergency services.

But developing complex decision-making skills in Senior Police Officers who will manage such incidents presents considerable problems, not only in organisation, but in how such training should proceed. Pseudo-simulation exercises 'on the ground' are necessarily small scale and directed at operational tactics. Case-based discussions of strategies are useful but lack authentic feedback, and tabletop map-based simulations are extremely difficult to manage [1]. Instructions have to be given by trainees, then recorded manually, resources moved on the tabletop in a judged time interval, the consequences decided by the simulation manager and feedback given via information reports. Record keeping is difficult and such methods are variable in what they achieve.

Computer based simulations seem to hold several advantages; for example, in display features where types of groups and resources could be iconised and navigate a visual map, in interactivity between system, trainer and trainees, and in communication management and in record keeping. But how should such a simulator be designed to develop decision-making skills? And how should the system accommodate a variety of interventionist roles and commentaries by the trainer to ensure maximum benefits are gained from the training exercise [2]?

2. Design of the simulator.

To address these issues, a collaborative project was undertaken with the (London) Metropolitan Police Authority (Public Order Training) who deal continually with large scale demonstrations and emergency situations. The training was directed at Senior Police Officers managing a large-scale demonstration (march) at the strategic/tactical level from a control room [3]. The simulation exercises (typically involving several hundred marchers) should include pre-event planning, event management and a debriefing discussion. The system should be adaptive and enable the trainer/facilitator to easily prepare a variety of simulation scenarios.

In the computer simulation system (CACTUS) a digitised map of Central London is overlaid with the street map familiar to the Police. On the map a designated route for the march can be marked, 'sensitive' buildings highlighted and icons placed to locate extraneous incidents, *eg.* road accidents. Police resources (*eg.* crowd marshalling units, information gathering serials, special units) can be placed, each with their type of display icons, and are able to navigate the map in pseudo real-time, and act in response to the destinations and instructions given by the trainees. The Police actions include 'address crowd', 'cordon road', 'advance', 'withdraw', 'arrest', 'patrol', and commands for marshalling the crowd and calling up special units. Such instructions are given orally (by radio) and by information (text) templates (entered by the trainer/facilitator), and are recorded by the system in a communication log.

On the map (which has a 'zoom' facility) the crowd units (considered as groups of about twenty to fifty demonstrators) are placed at the start of the event. These groups include the 'leaders' (who navigate along the route-line), and the 'followers'—groups who make up the bulk of the demonstrators and tend to follow the groups in front of them. In addition, there can be 'hostile' groups who gravitate to the march and aim to disrupt proceedings. Each of these types of group is shown as an icon, distinctive in colour and shape, and to each is attached autonomous behaviours. For example, the 'leaders' were peaceable but the followers could 'slow down', 'stop' and 'sit down', 'shout abuse', 'march angrily', 'attempt to vandalise property', 'throw missiles', and 'attack police'. There were thirteen such behaviours arranged in a probabilistic network so their actions could escalate and de-escalate realistically (achieved after much discussion and trialling). At regular time intervals (usually 15 seconds) each crowd group assesses its behaviour. Those agents know their current state, but can 'sense' the behaviour state of the groups round them, the tactic of the police marshalling their section of the crowd, the presence of any nearby hostile groups and sensitive buildings. From these data (and probabilistic behavioural rules) each group selected its behaviour for the next time period.

The trainer's display screen contained all the dynamic map data noted above and the communication log, but had available tools to create and move icons, inject information and interrupt and move the state of the simulation under time-line control. These facilities were to be used in the discussion of tactics and to make the exercise probe the trainees' decisions. The trainee sees the dynamic display of the march and communication log, and should take responsive actions using his resources and issuing appropriate instructions to police the event efficiently. But trainees cannot manipulate the time-line, control, and CACTUS—through its rule set—decides on the consequences of the decisions giving feedback messages from the Police Units.

A view of the display is shown at the end of the paper.

3. Evaluation of the CACTUS simulation.

After some months of testing and trialling, CACTUS was ready for an initial evaluation within the training schemes for Senior Police Officers.

In the pre-event planning the trainee was given detailed information of the scenario (prepared by the trainer) before the session started, and then had to enter and place on the map display the chosen police resources (in number and type) for managing the march. These decisions had to be justified and 'what-if' contingencies considered. Following this discussion the simulation program was activated.

The principal interest of this paper is the evaluation of the event management performances. For this phase the trainer sat at his computer display out of sight of the trainee. All actions (including display changes) and the communication log were preserved by CACTUS so that the displays could be reconstituted under time-line control for the debriefing. The trainee was encouraged to talk as he made decisions; the training sessions were audio recorded with a video record being taken of the computer screen displays. Additionally the researcher/observer made notes on matters of interest.

The decision-making of the trainee was judged on four criteria: *(i)* appropriateness of a tactic in relation to an incident; *(ii)* level and type of resources in relation to the success of the tactic; *(iii)* timeliness—was the action prompt in its delivery? and *(iv)* prioritisation—was this well judged?

Additionally, five progressive themes were assessed across the whole incident:
 (i) the march—were the tactics employed to guarantee its safety appropriate and effective?
 (ii) the movement and management of resources—were the units active and well used, and was there a proper anticipation in their movement and placings?
 (iii) the management of hostile groups and threats;
 (iv) the proper enquiry and utilisation of information/intelligence;
 (v) the handling of extraneous events.

The trainees were experienced Senior Officers so it was unlikely their decisions would allow matters to get out of hand. This proved to be the case, though there was variability in the tactics employed by trainees, and how they justified these decisions in discussions with the trainer. It is this interplay and reflection supported by CACTUS which leads to the development of decision-making skills.

These data led to several conclusions. [A complete account of the analyses is provided in Varley [4].]
 (i) There was variation in the methods of 'protecting' the demonstrators from hostile groups. One strategy is to closely monitor such groups and employ resources to keep them well away from the march. In this case the march will be relatively lightly policed. In contrast the demonstrators themselves can be more heavily marshalled, but this gives more opportunities to the hostile groups. Both can be satisfactory tactics, and which decision is made depends upon trainees' judgements on the likely temper of the march, and the number and dispersion of the hostile groups.
 (ii) The movement and management of resources was critical not only for the current situation but the future progression of the march For example, some trainees used resources to corral hostiles, who were then released when the crowd had passed on. However major crowd trouble sometimes caused the march to be re-routed (an extreme measure), which brought the demonstrators and hostiles into proximity. In brief, peripheral police units have to be continually monitored and used.
 (iii) Hostile groups were dealt with in a variety of ways. They may be observed and tracked, they may be marshalled and kept from the crowd, or tucked in at the end of the crowd, or they may be corralled on the street. All have their advantages and disadvantages. Corralling can be effective but groups cannot be corralled indefinitely; they have to be arrested or released. Corralling therefore can tie up resources over a relatively long period.
 (iv) Seeking and utilising information is critical to decision-making, and can also involve informing other services, *eg.* the British Transport Police, or giving reassurance to crowd marshals if hostile groups are folded in to the end of the march.
 (v) Handling extraneous events was well done. Trainees were not deflected from their main purposes, preferring to leave such incidents to the local services.

The protocols of these discussions were analysed [4] and the type of interventions made by the facilitator classified. These data showed not only that the trainees had tactical knowledge but a developing awareness of how that knowledge was being used and improved, and monitoring/evaluating skills in terms of the overall progression of the task.

The evaluation reports of the trainees rated the CACTUS simulation exercises highly, both the system itself and decision-making the tasks required. Those who returned for further training with CACTUS were more fluent in using the system and showed improvements in performance. Equally important, the facilitators themselves became more perceptive in using the system facilities and have developed a wider range of challenging scenarios.

4. System developments.

As the training sessions developed the instructors wanted more interactive control of the system to expose or exploit suspected weaknesses in trainees' tactics. Accordingly a tool was designed so that the trainers/facilitators could define a region (not visible to trainees) on their display and bias the crowd behaviours, *eg.* towards fighting and aggressive behaviour.

A second notable change was the importance both trainer and trainee attached to reflective discussion, and trainers tended to introduce such debriefingss at crucial points during the simulation itself. Hence the trainer was given the facility to 'pull down' a transparent window over the map on which drawing, highlighting and labelling could be used to emphasize points in these discussions. A further facility allowed the trainer to mark points on the time-line, and to attach comments for further discussion in the debrief.

5. A concluding comment.

The CACTUS project can be summarised under an Activity Theory Framework where the overall scenario task carries the motivation and interest, and where the simulator acts as a common mediational tool. Through its operations and vocabulary, the trainee and trainer can mentally assemble the operations, triggered by the incidents of interest, into plans from which the actions/instructions are delivered. It is the design of the tool which enables actions to stimulate thought, and which allows tactics to be developed that seek to deliver the strategic intentions of the trainee. The experience of the trainer and the records of other trainees allow a range of competing actions and their consequences to be discussed and illustrated through time-line re-runs of the simulator. Thus the decision-making skills were developed through what became a collaborative and evolving learning experience for both trainer and trainee.

6. References.

[1] Dowell, J. "Coordination in emergency operations and tabletop training exercises", *Travail Humain*, 1995; 58, 1, 85-102.

[2] Lipshitz, R. "Converging themes in the study of decision-making in realistic settings", in Klein, G. A., Orasance, J., Calderwood, R, and Zsambock (eds), *Decision Making in Action: Models and Methods*. Ablex, New Jersey, 1993.

[3] Pearce, T. and Fortune, J. "Command and control in Policing: A systems analysis of the Gold, Silver Bronze structure", *Journal of Contingencies and Crisis Management*, 1995. Blackwell Publishers, Oxford, UK.

[4] Varley, G. A. "Formative evaluation of a simulation system for Public Order Training", 1999. Unpublished PhD Thesis, Computer Based Learning, University of Leeds, UK, LS2 9JT.

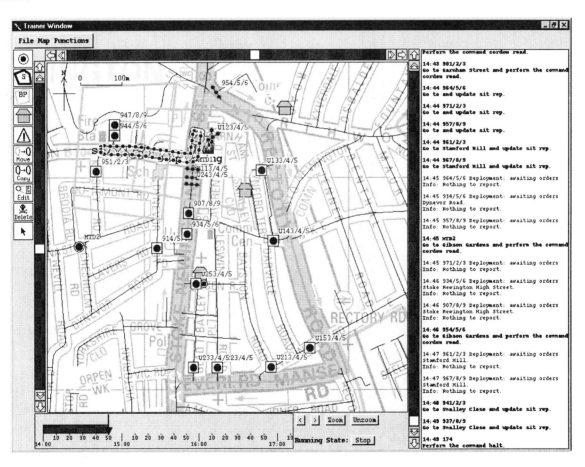

Evaluation of a Learning System – and Learning to Evaluate

Helen Keegan
University of Salford
H.Keegan@salford.ac.uk

Abstract

This paper describes the process of the formative evaluation of an acoustics computer-aided learning (CAL) package. The CAL package, entitled "Waves: Analysis and Synthesis", is a multimedia-rich learning support package for students on sound-related undergraduate courses. The evaluation took place over a two-month period in the School of Acoustics and Electronic Engineering at the University of Salford. In addition to being an evaluation of the CAL itself, this was also an exercise in learning to evaluate. The methodological mix is discussed along with the evaluation process as a whole.

1. The evaluation

This was effectively an exercise in learning to evaluate - accordingly there was no fixed methodology or set of tools from the onset. Advice was sought from 'expert' evaluators, along with a thorough survey of the evaluation literature from books and journals. Consequently this was an opportunity for the evaluator to employ different instruments and approaches in order to see what worked best (in this case). The sequence of the evaluation and the instruments used were mostly dictated by time and resource constraints.

The approach taken in the evaluation of the CAL was largely illuminative [1], using a combination of different evaluation techniques and exploratory data analysis to compare, triangulate and interpret findings. Quantitative methods were employed to obtain easily comparable data of fixed measures; for example, a post-task questionnaire (PTQ) was designed to elicit data on aspects of design, navigation and the user interface. Qualitative methods were employed in order to detect and explore unanticipated issues. In what was essentially an ethnographic approach, the learners were observed using the CAL. This proved to be an extremely rich source of data in terms of usability. Interviews and focus groups enabled us to follow-up the findings in more depth. Pre and post knowledge testing was used in an attempt to measure any learning gains as a result of the intervention.

In reality practicalities and time constraints largely dictated the combination of tools in each case. Five groups of learners participated in the evaluation of the CAL, which took place over a two-month period.

GROUP 1 - USABILITY: The approach to evaluation with group one was informal. The three volunteers were final year students who were asked to take the software home and use it in their own time, noting how long they spent using the package and making notes of any problems encountered as they worked through the CAL. All three students made comprehensive notes, which mostly related to usability. They also completed the PTQ before being interviewed on a one-to-one basis, which enabled us to discuss the issues that they had raised in more depth. The findings contributed to usability, design, and the user interface. Problems had been identified which were 'ironed-out' before the next stage of the evaluation in order to avoid re-evaluating the same usability issues.

GROUP 2 - USABILITY/PEDAGOGY: The evaluation approach was far more structured and intensive with group 2, which comprised of four final year students who had been away for a year on industrial placements. This stage of the evaluation was split into two sessions. During the first session the subjects completed Student Profile and Task Experience questionnaires before using the CAL, along with a pre-knowledge test. The same test was administered post-task in order to elicit two comparable sets of data to assess the pedagogical effects of the CAL. The post-task questionnaire, which focused on Content, Usability, Design and Navigation, consisted of Likert-scales and open-ended questions. Supplemental observation was employed in order to study the human-computer interaction. Any difficulties or unexpected problems were then recorded in the session logs, which were used to identify key issues surrounding the usability of the CAL. The logs were also used as a prompt in the follow-up interviews. The group interview allowed the subjects to discuss their experience of using the CAL, which proved to be a rich source of data.

GROUP 3 - USABILITY/PEDAGOGY: The approach taken and the instruments employed mirrored group 2, except that there was no group interview due to practical constraints. In this case the group comprised of four first year students, which enabled us to compare the pre/post knowledge tests of groups 2 and 3. By choosing 2 groups with a divergent range of abilities and domain knowledge we could observe the pedagogical effects of the package upon different groups of learners.

GROUP 4 – USABILITY: Group four comprised of eleven second-year students taking a Digital Audio module who used the CAL in place of a Sound Synthesis lecture. With this group there was no questionnaire or knowledge testing. We were concentrating on their open responses to the CAL, in light of the modifications that had been made to the software. The group were asked to write a 'letter to next years students' explaining the nature of the CDROM [2]. This proved to be an excellent way of evaluating their overall impressions of the CAL. However, this didn't tell us anything that we didn't know already, although it would have been valuable at earlier stages of the evaluation. The session ended with a focus group interview, where the group were asked specific questions related to issues that had been addressed through further development of the CAL in light of the findings from groups 1, 2 and 3.

GROUP 5 – USABILITY: Group five comprised of fourteen second-year students who were taking a Midi and Music Technology module as part of a Professional Sound and Video Technology HND. The CAL replaced a two-hour lecture on Sound Synthesis. This was the only stage in the evaluation where the evaluator was not known or present. This was intentional as the overall responses from the other groups had been consistently positive. It was decided that by implementing an 'anonymous' evaluation in which the lecturer (rather than the evaluator) explained and directed the study, we may be able to detect whether the findings were influenced by a Hawthorne effect. The learners were given post-task questionnaires. However, there were no significant differences or findings in the feedback (Open-ended questions) or the overall results (Likert scales). This does not necessarily mean that the prior evaluations were unbiased – perhaps the students also wanted to 'please' the lecturer.

2. Conclusion

This paper has described the process of a formative evaluation of an acoustics CAL package. The iterative nature of the evaluation resulted in a cycle of testing and amendments to both the CAL package and the evaluation tools. The evaluation progressed from being initially illuminative and open-ended to being more integrative in nature [3]. The evaluation took shape over the 5 sessions, as usability problems were detected and modifications were made before evaluating the CAL with the following group. The responses were easily anticipated by stage four of the evaluation, which indicated that the study had been fairly comprehensive. This process is shown schematically in Figure 1.

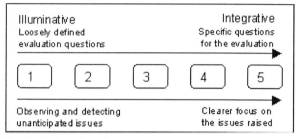

Figure 1. Evaluation process.

This framework facilitated a holistic perspective on the data, identifying which aspects of the CAL needed to be redesigned, and enabling the evaluator to ascertain the relative strengths and weaknesses of different evaluation techniques. The open-ended measures were by far the most valuable instruments employed in the evaluation. Learning gains proved to be extremely difficult to measure.

The initial stages in this phase of the evaluation were small-scale, highly domain dependent, and relatively low in authenticity [4]. Future work will involve the integrative evaluation of the CAL intervention as part of a longitudinal study.

3. References

[1] Parlett, M. & Hamilton, D. "Evaluation as illumination: a new approach to the study of innovatory programmes" In *Beyond the numbers game*, Basingstoke: Macmillan, 1977.

[2] Cowan, J. and George, J. *A Handbook of Techniques for Formative Evaluation*, Kogan Page, London, 1999.

[3] Draper, S.W., Henderson, F.P., Brown, M.I., & McAteer, E. (1996) "Integrative evaluation: an emerging role for classroom studies of CAL" *Computers and Education.*

[4] Oliver, M. "A Framework for evaluating the use of educational technology." BP ELT report no.1, University of North London. http://www.unl.ac.uk/tltc/elt/ 1997.

Further details

A comprehensive discussion of the CAL itself and the evaluation as a whole is beyond the scope of this paper. For the full version, please contact the author.

The Effects of Simulation Participation on the Perception of Threatening Cultural Dynamics in a Collaborative Virtual Learning Environment

Elaine M. Raybourn
Advanced Concepts Group
Sandia National Laboratories*
emraybo@sandia.gov

Abstract

Findings indicated that on the subject of power, participants in the computer-mediated, collaborative, intercultural DomeCityMOO simulation perceived its learning environment to be less threatening than did those who participated in the face-to-face Ecotonos simulation. Additionally, participants in the DomeCityMOO simulation perceived its environment as better for learning about identity and power differences than participants in the face-to-face Ecotonos simulation.

1. Introduction

Face-to-face interaction is frequently taken as the benchmark for ideal interaction in a computer-mediated environment [1]. However, instead of just striving to make computer-mediated environments more like face-to-face communication by adding video, audio, etc., we should investigate what makes the computer-mediated environment *unique*—the effects of (pseudo) anonymity on education and collaborative learning.

2. Background Research

Ecotonos is a face-to-face intercultural communication simulation designed to allow participants to learn to recognize power imbalances in groups. The simulation debriefing addresses participants' emotions and focuses on the skills necessary for collaboratively solving problems in multicultural groups. Raybourn demonstrated that *Ecotonos* participants perceived the setting as a threatening one in which to discuss topics related to identity and power, even though the simulation was designed to allow the 'safe' exploration those topics [2].

Raybourn then designed a multi-user social process simulation (*DomeCity*MOO) in a MOO (multi-user dimension object oriented) that also allowed participants to experience imbalances in identity and power [2]. Participants created cultural identities that reflected the power imbalances in society, and noted how their power and cultural identity were negotiated though their communication with others.

3. Methods and Procedures

The researcher tested three hypotheses, described below, by mean differences of independent sample t-tests calculated on the post-test scores of questionnaires completed by participants in *DomeCity*MOO and *Ecotonos* simulations [2].

4. Collaborative Virtual Environment Simulation Study Design and Respondents

Seventeen individuals responded to listserv solicitations for participation in one hour and 30 minute sessions in which they would log into the *DomeCity*MOO simulation. The median age was 27, the mean 29. Eighty-six percent were male, 14 percent female. None of the respondents had previously participated in the *DomeCity*MOO. Before the sessions, they completed pretests (5-point Likert-type scale). One or two days after participating, respondents complete and returned by email post- tests identical to the pretest.

5. Face-to-Face Study Design and Respondents

Ecotonos (face-to-face) participants were administered the same pre-tests and post-tests used in the *DomeCity*MOO study. The simulation was then conducted over two class periods. One day elapsed between the two class periods. At the second class period of the week, a debriefing of the simulation was conducted and the post-test then administered.

Participants in the *Ecotonos* simulation consisted of 77 University of New Mexico students. The median age was 22, the mean 23. Sixty-five percent were female, 35 percent were male. Ninety-two percent of the respondents had previously experienced educational simulations at least once, but none had previously participated in the *Ecotonos* simulation.

5. Results

Hypothesis 1 stated that *respondents who participate in a computer-based, multi-user simulation will perceive its environment as less threatening, concerning the subject of identity, than respondents who participate in a face-to-face simulation will perceive its environment, following participation in their respective simulations.* Post-test means, and t-values of the post-test independent t-tests for the two simulations are found in Tables 5.1 and 5.2, respectively. These findings do not support Hypothesis 1.

Table 5.1 Means and standard deviations for post-test scores for the *DomeCityMOO* and *Ecotonos* groups.

Source	N	Post-test Mean	s.d.
Identity			
DomeCity*MOO*	17	3.65	1.11
Ecotonos	77	3.22	1.07
Power			
DomeCity*MOO*	17	3.82	.73
Ecotonos	77	3.38	1.07
MOO Sim Better			
DomeCity*MOO*	17	3.53	.87
Ecotonos	77	2.83	1.07

However, a matched-sample t-test calculated on the pre- and post-test scores revealed that *DomeCity*MOO respondents did not perceive identity to be a threatening topic to discuss in a virtual setting at the time of the pretest, and they believed the *DomeCity*MOO simulation met its goal of creating an environment in which they could explore identity (t= -2.7, [df=16], p<.01).

Hypothesis 2 stated that *respondents who participate in a computer-based, multi-user simulation will perceive its environment as less threatening, concerning the subject of power, than respondents who participate in a face-to-face simulation will perceive its environment following participation in their respective simulations.* An independent sample t-test was calculated on the post-test scores of the *DomeCity*MOO and *Ecotonos* group (see Table 5.2). These findings support Hypothesis 2.

Hypothesis 3 stated that *respondents who participate in a computer-based, multi-user simulation which provides an opportunity for anonymity will perceive its environment as better for learning about prejudice-reduction and power differences than respondents who participate in a face-to-face simulation will perceive its environment, following participation in their respective simulations.* An independent sample t-test was calculated for the mean post-test scores of the *DomeCity*MOO and *Ecotonos* group (see Table 5.2). These findings support Hypothesis 3.

Table 5.2 Independent sample t-test values on post-test mean difference for DCM and Ecotonos (equal variances not assumed). *Significance at the 5 percent level.

Post-test Source	df	Mean Difference	t	p (2-tailed)
Identity	23	.43	1.44	.16
Power	32	.45	2.10	.04*
MOO Sim Better	28	.70	2.86	.00*

6. Conclusions

The results indicate that, following simulation participation, on the subject of power, participants *DomeCity*MOO perceived its environment to be less threatening as compared to respondents who participated in the *Ecotonos*. Moreover, analysis of post-test questions (not presented here for reasons of space) showed that the *DomeCity*MOO participants found it to have met its goal of creating an environment in which they could 'safely' explore power and identity.

Collaborative problem-solving and learning simulations designed for virtual environments may make it easier to learn the *essence* of identity awareness and intercultural relations skills by filtering out confounding variables such as biological markers (race, sex) and embodied face-to-face group affiliations, which bear on equitable, multi-cultural collaborative learning. To date, intercultural realtime simulations are only designed for the face-to-face environment. The *DomeCity*MOO is the first computer-mediated intercultural, multi-user, realtime simulation designed specifically to address issues of power and identity. The design principles of the *DomeCity*MOO challenge the popular belief that aspects of tacit culture, and intercultural awareness can only be taught face-to-face. See http://www.cs.unm.edu/~raybourn.

*Sandia is a multiprogram laboratory operated by Sandia Corporation, a Lockheed Martin Company, for the United States Department of Energy under Contract DE-AC04-94AL85000.

10. References

[1] J. Hollan, and Y. Stornetta, Beyond Being There. *Human Factors in Computing Systems CHI '92*, ACM Press, May 3-7, 119-25.

[2] E. M. Raybourn, An Intercultural Computer-based Simulation Supporting Participant Exploration of Identity and Power in a Text-based Networked Virtual Reality: DomeCityMOO. PhD dissertation, University of New Mexico, Albuquerque, New Mexico, 1998.

Computerised Problem-Based Scenarios in Practice – A Decade of DIAGNOSIS

Terry Stewart, Ray Kemp and Paul Batrum
Massey University
Palmerston North
New Zealand
T.Stewart@massey.ac.nz

Abstract

This paper discusses aspects of Problem-Based Scenarios (PBS) for teaching based on experiences with a computer package called DIAGNOSIS. This package has been used for teaching the diagnosis of plant pests and disease problems for a decade. The software is described, and its evolution in response to changing demands and software platforms is also covered.

1. Introduction

Problem-solving is an essential component of any worthwhile Instructional package. [3]. It lies at the core of Problem-Based Scenarios (PBS), a teaching style where students are given situations that they must assess, and act on. The use of PBS is a powerful approach to learning, one that can be facilitated by computer technology.

This approach is reflected in the crop protection-teaching tool DIAGNOSIS [1,6,7], used at Massey University and other institutions for a decade. This paper describes the value in the teaching approach it uses and how this teaching tool has evolved to accommodate expanded learning paradigms and software standards over the last 10 years. Although the software currently deals with crop problems, its approach is relevant to any subject domain.

2. How does DIAGNOSIS work?

Contrary to what the name might suggest, DIAGNOSIS is not an expert system. These systems embody the knowledge and rules of an expert. As a teaching tool, tutors may get students to use an expert system to assist in the diagnosis of a disease. The students feed the expert system the relevant observations, and then look over its shoulder as it offers a diagnosis. The system may explain why it came to a particular conclusion and in that way students learn something about the diagnostic process.

DIAGNOSIS is the reverse of an expert system. It assumes students are the experts and provides them with simulated observations of a plant problem. The students need to come up with the diagnosis themselves. Therefore the students, not the machine, carry out the interpretation and gathering of information. The software simply provides a PBS; a sort of virtual reality they can work in. It uses the metaphor of the computer "adventure game". In such games players moved from room to room, inspecting objects and clues and drawing conclusions, which could then enable them to progress further. A lot of thinking and deduction was involved. These same mental tasks also occur in trying to solve an unknown plant problem.

The core of a PBS is the teacher-constructed scenario. The scenario presents students with a problem

In the case of DIAGNOSIS, it may be wilting plants in a potato field, poorly-growing trees in an apple orchard or yellow leaves on wheat plants. The students are dropped into this problem, usually accompanied by a grower, who can answer questions on management and the history of the problem. The student must then interrogate the scenario by way of menu options, making observations of the plant and the environment, checking equipment and asking the grower questions. Students can collect objects such as plant parts, weeds and soil and take them to the lab for "forensics". Here they can gather further observations, look up weather information, try to extract the causal organisms and/or conduct a number of tests. Once students feel they have enough information to make a diagnosis, they type this into the computer, along with a justification and recommendation. Students can then be told the answer and given immediate feedback, or (more commonly) the student input will be extracted later by their teacher, printed out and graded. In the case of the latter, feedback is appended their input text. Feedback takes the form of the diagnosis, a correct interpretation of the significant observations, the common pathway an experienced diagnostician would have taken and a model recommendation to the grower. Some feedback is customised depending on whether or not students carried out particular tasks.

Many of the tasks, such as laboratory tests, have a price tag associated with them in DIAGNOSIS. Students are given a budget before the exercise and asked not to exceed it. This gives some incentive for the students to limit themselves to tasks and tests appropriate to the likely problem, rather than trying all the menu options. It mirrors the real world, where such procedures do indeed cost real money.

Scenarios then, are the "data" used by the DIAGNOSIS program. The software provides a "builder" for teachers, where scenarios can be constructed or altered to suit. A scenario "player" program then presents the problems to students.

3. The Importance of Good Scenarios

Problem-Based Scenarios should reflect reality and all its ambiguities, and include situations where the solution is not clear-cut. They should encourage some real detective work on behalf of the learner. They are best written with the aid of personnel actively dealing in the day-to-day decision-making pertaining to the domain. For example in DIAGNOSIS, collaboration with a field plant pathologist is important if the teacher is not actively involved in extension work.

Scenarios should incorporate whatever degree of difficulty the teacher feels appropriate for the course. Teachers can make interpretations for students or leave interpretation up to them. In DIAGNOSIS for example, if students send away samples for nutrient analysis, the scenario may return the results saying "the nutrient levels appear OK" or alternatively provide a list of values for each nutrient. In the former case, the teacher (or whoever built the scenario) has done the interpretation for the student while the latter requires them to understand the significance of the figures.

4. The Educational Value of DIAGNOSIS

DIAGNOSIS presents students with a problem to solve, captures their diagnosis, justification and solution and provides feedback. Students generally react very positively to the exercises [6]. They are forced to integrate and synthesise material learned elsewhere in other courses (soil science, plant physiology, entomology etc.) thereby looking at a problem in a holistic way. In this way they learn the "art" of pest and disease diagnosis [1]. The exercises were designed to be firmly embedded in a series of lectures on the diagnostic process. In this way the software is designed to complement, not replace, hands-on laboratory work.

The exercises seem to work best when the students undertake them in pairs or small groups. This way they can bounce ideas off each other, working and communicating as a team.

Another useful exercise is to get students to actually construct scenarios themselves, and then get other students to play and critique them. Just as having to teach material forces a new teacher to learn it well, constructing scenarios makes students think about how a plant problem may relate to management and the environment it exists in.

5. The Genesis and Evolution of DIAGNOSIS

5.1 Version 1

The original DIAGNOSIS was written in 1989 in Pascal. It was designed to be used under MS-DOS. The software could show images and used a simple verb-noun combination for interaction [5]. The relationship between this and early text-based adventure games was obvious. The builder program was simply a compiler, which parsed a formatted scenario text file and converted this to a data file for use in the student player.

5.2 Version 2

In the early 1990's MS-DOS was waning in favour of Microsoft Windows 3.1 so in 1992, a collaboration was formed with the CAL Program in the Co-operative Research Centre for Tropical Pest Management (now the Centre for Pest Information, Technology and Transfer), The University of Queensland, Australia, to produce a Windows 3.1 version (version 2.0). This version, Called DIAGNOSIS FOR CROP PROTECTION, used pre-set menus rather than verb-noun combinations (Fig 2). It allowed for a range of crop types, included pest diagnostic procedures, expanded on available laboratory tasks, accommodated pest problems and allowed video to be incorporated into scenarios. The builder was interactive and a lot more user-friendly. Version 2.1 incorporated the ability to save the student session while in progress, to add audio and export scenarios to a text file.

Figure 2. Some menu options in DIAGNOSIS V.2

A Macintosh version was also produced in 1992 using a generic plant (Fig 3) and laboratory which students could explore by dragging tools onto objects with the mouse. This version was discontinued due to the expense of supporting two different computer platforms.

Figure 3. Macintosh version of DIAGNOSIS Start-up screen

Versions 1 and 2 of DIAGNOSIS did not provide reference material to assist in the diagnosis. Students were expected to use reference material from elsewhere (books, their lecture notes, the web etc.). The software could present the PBS, but provide no help with them during the exercises. Feedback from users indicated that this would be a useful feature. Also, the software was looking old. The interface had been designed in 1993 for Windows 3.1, before the conventions (e.g. right-mouse clicking, standard file load/save dialogue boxes) forced by Windows 95 came into being.

The latest version, 3.0, is a radical re-write utilizing modern Windows technology and 98/NT/2000 conventions. The product is now called DIAGNOSIS FOR CROP PROBLEMS. It considerably enhances the DIAGNOSIS experience by allowing customisable menus and interfaces (Figs 4 and 5). It also allows tutors to embed, or hyperlink to, reference information either on the hard disk or the Internet. This feature enables teachers to provide not only the problem, but also information on the significance of what the student is considering at the time. In this way, its usefulness as a PBS engine is considerably enhanced.

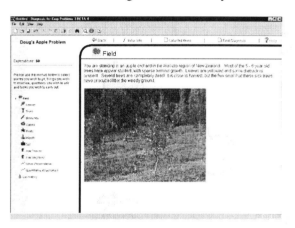

Figure 4. Version 3 Player Introducing a Scenario

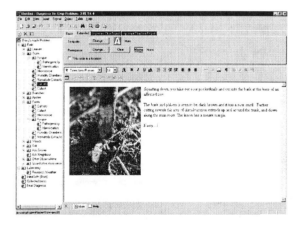

Figure 5. Builder Screen Shot from Version 3.

With the previous version of DIAGNOSIS, students were not aware of the significance of observations until the de-briefing at the end. Here, there is the option to provide feedback as they explore the scenario.

6. DIAGNOSIS for all Domains. A Generic PBS Program

DIAGNOSIS is not the only program designed to present PBS. Other programs, such as INDE [2] appear in the literature but examples of such programs are rare. There are a number of issues and difficulties with maintaining and supporting teaching software within an academic framework. The production and support of educational software for a specific narrow discipline is becoming increasingly difficult as operating systems become more complex, institutions become more commercially minded and programmers become more expensive. This has increased pressure for more generic software, which has a larger user base to support it. Problem-Based Scenarios can be constructed with general-purpose author ware such as Macromedia Authorware™, Director™ or HAVAS Interactive Hyperstudio™. The problem with this kind of powerful generic multimedia software is that the teacher needs to be, or enlist the aid of, someone skilled in the software itself. They are not built specifically for delivering Problem-Based Scenarios.

At the time of writing the authors are working on a new version of DIAGNOSIS, one that is very generic, can accommodate all domains yet has the PBS metaphor at its core. Also, research is also being conducted into software, which will allow tutors to easily storyboard scenarios, before including them in DIAGNOSIS. A kind of PBS "wizard'.

7. Final Comment

Problem-Based Scenario teaching programs such as DIAGNOSIS can enhance student learning. However, like all technology, it is how it is used that matters. The value of an PBS program as an educational experience is not the builder and the player software, useful though these are. In fact a "Diagnosis" exercise could be done using sealed envelopes, containing photos and descriptions of facets of the scenario. When a students wants to "examine the leaves" they could break into the envelope containing observations from this activity. Lo-Tech for sure, but it would provide a similar learning experience.

The real value of PBS is in the quality of the scenarios and how the exercises are used within a course. This quality and appropriateness of use comes from dedicated teachers and domain practitioners themselves. The experience, wisdom and knowledge of a domain practitioner, whether it be medicine, law, business or agriculture can be embodied in a scenario, and used by a skilful teacher to challenge students.

Technology can help, but it is an aid to not a substitute for, careful curriculum planning, inspired examples and good teaching paradigms. DIAGNOSIS as with any PBS progam is a useful tool, but it will always require good teachers to realise its full potential.

8. References

[1] R.G. Grogan, "The Science and Art of Pest and Disease Diagnosis", *Annual Review of Phytopathology 19*, Annual Reviews, U.S.A, 1981, pp. 333-351

[2] J.D. Knight, "DIAGNOSIS for Crop Protection – a computer-based teaching tool", *Integrated Pest Management Reviews*, Kluwer Academic Publishers, The Netherlands, 1995, 75-77

[3] M. D. Merrill, "Does Your Instruction Rate 5 Stars?" *Proceedings of IWALT 2000: International Workshop on Advanced Learning Technologies.* IEEE Computer Society, U.S.A., 2000, pp. 8-11

[4] C. Riesbeck, "INDIE: An Authoring Tool for Goal-Based Scenarios", [Online] Available http://www.ils.nwu.edu/~riesbeck/indie/, 1998. (January 30th, 2001)

[5] T.M. Stewart, "Diagnosis. A Microcomputer-based teaching-aid", *Plant Disease 76 (6)*, APS Press, U.S.A., 1992, pp. 644-647

[6] T..M. Stewart, B..P. Blackshaw, S. Duncan, M.L. Dale, M.P. Zalucki, and G.A. Norton, "Diagnosis: a novel, multimedia, computer-based approach to training crop protection practitioners" *Crop Protection 14: (3)*, Elsevier Science Ltd, U.K., 1995, pp. 241-246

[7] T.M. Stewart, "Diagnosis for Crop Protection" [Online] Available *http://www.diagnosis.co.nz* , 2001 (January 30th 2001)

Asynchronous Distributed Problem-based Learning

Frederick B. King
rking@uconnvm.uconn.edu

Hayley J. Mayall
hayley.mayall@uconn.edu

Neag School of Education
University of Connecticut

Abstract

This paper explores the formative and summative results of a problem-based, asynchronously conducted, graduate-level, educational psychology course on learning theory. EPSY was a 13-week, three credit, graduate level, educational psychology course conducted asynchronously at a distance. Multiple tools, methods, and data collection points were used in assessing this course. Data collected was both quantitative and qualitative. As part of the evaluation of this course qualitative data was collected from the participants at two separate intervals. Two open-ended surveys were filled out, one at the midpoint of the course, and the second at course completion.

This paper explores the formative and summative results of a problem-based, asynchronously conducted, graduate-level, educational psychology course on learning theory. The course (EPSY) was conducted online using personal computers. McConnell [10] (2000) states that "any form of cooperative learning communication that occurs over a network of computers" can be termed CSCL (p.33). For purposes of this paper, and according to McConnell's [10] own definition (p. 8), the terms cooperative and collaborative are essentially the same. Therefore, for this research, CSCL is used to denote Computer Supported Collaborative Learning [8] (Koschmann, 1996).

EPSY's primary purpose was to provide students with a working understanding of the major theories of learning. EPSY was based on social cognitive and sociocultural theories of learning and designed to support learners to be active, constructive, and intentional in authentic problem-base/case-based environments. Constructive, here, is defined as *meaning making* [4] (Bruner, 1990), rooted in the context of the situation [3] (Brown, Collins, & Duguid, 1989), whereby individuals construct their knowledge of, and give meaning to, the external world [6] (Duffy and Jonassen, 1992) as a product "shaped by traditions and by a culture's toolkit of ways of thought" [4] (Bruner, 1996). "Consistent with this view of knowledge, learning must be situated in a rich context, reflective of real-world contexts for this constructive process to occur and transfer to environments beyond the school or training classroom." [1] (Bednar, A. K., Cunningham, D., Duffy, T. M. & Perry, J. D., 1992, p. 22). One of the best ways to provide an authentic and rich context in an academic situation is through the use of problem-based learning (PBL). PBL has proven effective in simulating real-world contexts. In a problem-based environment learners work in teams reviewing, critiquing, testing each other's ideas, and engaging in collaborative knowledge building [11] (Scardamalia and Bereiter, 1994). Outcomes of a PBL collaborative environment have proven more effective in promoting retention, transfer, and reasoning strategies than the traditional method of instruction [9] (Koschmann, Kelson, Feltovich, & Barrows, 1996; McConnell, 2000). The problems/cases used in the educational psychology course were drawn from numerous, recent educational psychology texts, were actual classroom case studies, and were chosen to match the grade levels and experience of the EPSY learners.

EPSY was a 13-week, three credit, graduate level, educational psychology course conducted asynchronously at a distance. Fourteen learners, not including the instructor, took the course. All fourteen completed the course. All of these learners were in-service educators whose experience as K-12 teachers and administrators ranged from 6 to over 15 years. Most had never taken a problem-based course. The course met face-to-face for the first and last classes only, the rest of the courses were conducted over the Internet. The first class meeting was face-to-face and was used to explain the technology used and to establish student project teams. Four teams were formed, two having three members, and two having four members. Prior technological expertise of the students ranged from novice to intermediate. All were required, however, to be competent at using a computer to do basic personal productivity (word process, etc.), email, and access the World Wide Web.

Collaboration was a key element of EPSY and was conducted at two levels. Whole class collaboration occurred throughout the course, additionally, project team collaboration was necessary for some case studies and the final project. In both cases, asynchronous, distributed collaboration was conducted using WebBoard®. This communications tool allowed the instructor to set up discussion areas for the entire class as well as for each project team. WebBoard® automatically threaded the discussions using date/time stamping of all discussion contributions. Case studies were divided into two categories; those answered by the individual and those answered by project teams. Three cases were addressed by every student and "answered" on an individual basis. These cases were used as an introduction to PBL and self-directed learning. The instructor interacted directly with each individual, via email, on these case studies. Additionally, three cases were worked in small group/team format. Teams, consisting of three or four students, would present their solutions to the class and the class would respond to these postings, also in team format. All of the cases used in EPSY were specific to educational situations. Both instructor and the students facilitated threaded WebBoard® discussions through the use of questions, personal experiences, relevant literature, and observations and conclusions.

EPSY was conducted over the Internet using WebCT® as the technological backbone of the course and WebBoard® as the primary collaborative communications tool. For example: case studies were posted on WebCT but were discussed and "answered" using WebBoard®. Learners had access to multiple modes of communication. Email provided private conversations with the instructor or with other students. WebBoard® provided threaded discussion areas for each major topic and case study covered in the course and was also configured to provide private, project team only, discussion areas where the students could formulate their responses to the cases under study. All WebBoard® discussion areas were also chat capable if learners wished to communicate synchronously. The one drawback to using chat on WebBoard® was that chat discussions were not archived. Therefore, WebCT® was employed for the synchronously held (chat) weekly office hours with the instructor. These discussions were archived, edited, and distributed, via email, to all students the day following the office hours.

One of the desired outcomes of this dPBL was to encourage the establishment of a community of learners [10] (McConnell, 2000). To foster the establishment of a community of learners, Brown [2] (1997) believed that three activities must be performed: "(a) research, (b) in order to share information, (c) in order to perform a consequential task—are overseen and coordinated by self-conscious reflection by all members of the community" (p. 404). EPSY met these criteria through the use of case-based scenarios, projects and synchronous/asynchronous discussions. As the course progressed it was obvious that participants were actively engaged in both intra and inter team collaboration. Based on the discussions in the last part of the class it was apparent that a community of learners had been established within course parameters. However, no follow-up has been conducted to determine if the community of learners was promulgated outside the confines of the course.

Multiple tools, methods, and data collection points were used in assessing this course. Data collected was both quantitative and qualitative. As part of the evaluation of this course qualitative data was collected from the participants at two separate intervals. Two open-ended surveys were filled out, one at the midpoint of the course, and the second at course completion. Participants were asked to comment on various aspects of the course content and the delivery method. Several major themes emerged from this qualitative data, which appear to be consistent with the literature base regarding both distance education and problem-based learning. One theme that became apparent was that the students enjoyed the flexibility that came with a distance or distributed style course.

"The part I like best is being able to answer the questions at my convenience."

"I like being able to 'attend' when I have time. It fits well into my schedule."

"I like the freedom of choosing when to do the work for the course as far as when to go online instead of driving to class."

"I enjoy being able to complete this course from my house at any time and being able to access it from work also. I think it is exciting taking it over the internet and being part of a new way of learning."

A large component of the structure of this course was the use of collaborative learning. In an attempt to determine if this had been successful the participants were asked whether or not they thought the course supported collaboration. Their comments verify that the dPBL used in this course was effective

"The final group project and the collaborative reflective journals do support collaborative learning."

"For a course that is mainly done over the web I think that it supports collaborative learning the best that it can through the questions posted which students respond to and through chats and probably through the group project."

"Yes, it does support collaborative learning. We have been able to work together and 'use' one another's brains to pick and learn from."

Additionally, the participants were asked to comment on whether they found the use of cases as a basis learning the material helpful.

"The case studies I found to be a good measure of my learning. They gave me the opportunity to apply theories in a manner that was useful."

"I found the case studies to very helpful in applying the theories learned. They were also relevant to education and learning."

"The case studies were helpful. They made the content more meaningful because you could see the concepts in action."

"The analyses of the case studies was a way of self-testing. They let me evaluate whether I understood the information."

Overall, the EPSY learners provided a lot of useful information regarding this asynchronous, dPBL designed course. Their views about the course structure and pedagogy, and their attitudes regarding the effectiveness of this style of instruction are being analyzed. Participants in the course also completed both a pre- and post-course survey related to their study skills, self-efficacy and self-regulated learning. Preliminary results from these analyses indicate that the higher achieving students had better study skills, higher self-efficacy and better self-regulated learning behaviors.

In short, asynchronous dPBL appears to have been effective in this web based, educational psychology course.

References

[1] Bednar, A.K., Cunningham, D., Duffy, T.M, & Perry, J. D. (1992). Theory into practice: How do we link? In T. M. Duffy & D. J. Jonassen (Eds.), Constructivism and the technology of instruction: A conversation. Hillsdale, NJ: Lawrence Erlbaum Associates, Publishers.

[2] Brown, A. L. (1997). Transforming schools into communities of thinking and learning about serious matters. American Psychologist, 52, (4), 399-413.

[3] Brown, J.S., Collins, A., & Duguid, P. (1989) Situated cognition and the culture of learning. Educational Researcher, 32-42.

[4] Bruner, J. (1990). Acts of meaning. Cambridge, MA: Harvard University Press.

[5] Bruner, J. (1996). The culture of education. Cambridge, MA; Harvard University Press.

[6] Duffy, T. M., & Jonassen, D. H. (1992). Constructivism and the technology of instruction: A conversation. Hillsdale, NJ: Lawrence Erlbaum Associates, Publishers.

[7] Jonassen, D. H., Peck, K. L., & Wilson, B. G. (1999). Learning with technology: A constructivist perspective. Upper Saddle River, NJ: Merrill.

[8] Koschmann, T. (1996) Paradigm shifts and instructional technology: An introduction. In T. Koschmann (Ed.), CSCL: Theory and practice of an emerging paradigm. (pp.1-23). Mahwah, NJ: Lawrence Erlbaum Associates, Publishers.

[9] Koschmann, T., Kelson, A. C., Feltovich, P. J., & Barrows, H. S. (1996). Computer-supported problem-based learning: A principled approach to the use of computer in collaborative learning. In T. Koschmann (Ed.), CSCL: Theory and practice of an emerging paradigm. (pp.1-23). Mahwah, NJ: Lawrence Erlbaum Associates, Publishers.

[10] McConnell, D. (2000). Implementing computer supported cooperative learning (2nd ed.). London: Kogan Page Limited.

[11] Scardamalia, M., & Bereiter, C. (1994). Computer support for knowledge-building communities. The Journal of the Learning Sciences, 3, (3), 265-283.

Software Support for Creative Problem Solving

Aybüke Aurum
Sch of Inf. Syst,Tech and Mang
The University of New South Wales,
Sydney, Australia
Email:aybuke@unsw.edu.au

Jean Cross
Dept. of Safety Science
The University of New South Wales,
Sydney, Australia
Email: j.cross@unsw.edu.au

Meliha Handzic
Sch of Inf. Syst,Tech and Mang
The University of New South Wales,
Sydney, Australia
Email:m.handzic@unsw.edu.au

Christine Van Toorn
Schl of Inf. Syst, Tech. and Mang
The University of New South Wales,
Sydney Australia
Email: vantoorn@unsw.edu.au

Abstract

This paper describes a development effort and an empirical test of a software prototype for stimulating creative problem solving. The software was designed on the basis of a solo brainstorming technique that provided users with external stimuli and exposed them to a large number of ideas over a short period of time. The prototype was empirically tested in the context of issue identification for a natural disaster. The results of the test indicate that users were able to identify on average 40% of all critical issues individually, and over 95% of issues in nominal groups of 4-6 people. These results suggest that collaborative creativity may facilitate optimal performance and learning.

1. Introduction

There is a widespread recognition in the knowledge management literature of the importance of creativity and innovation for organisational success in the changing environment [9]. However, while change is inevitable, our adaptive response may be a choice between development or decay. The evolutionary principle proposes that those who are innovative survive, while those who are not become extinct. It is therefore not surprising that surveys show that creativity and innovation are among the top priorities for senior executives in industry [6].

Industry need for innovation requires an appropriate response from education and training. Great ideas need creative thinking. Some theorists believe that creativity is reserved for the gifted. Others (and us) see creativity as a skill that can be learned [11] both by individuals and groups. Therefore, the purpose of this study is to address the issue, by developing and testing a software prototype for stimulating creative problem solving in the context of issue identification for natural disasters.

2. Literature review

A number of decision making models recognise issue identification as a fundamental and explicit activity in the decision making process. Past studies identified three aspects of issue identification.

Firstly issue identification is used to initiate the problem solving process [14]. Identified issues give insights into the problem so that the decision maker can make a connection between the issues and the suggested solution [10, 12].

Secondly, the subsequent steps of decision making will depend on the nature of the issues identified previously. Identified issues create momentum and direction for the subsequent phases of decision making. These may potentially influence the later phases of decision making and, over time, may become a source of direction for thinking and behavior within the organization [10]. Errors in the issue identification process create potential problems and may increase the cost of a solution. Finally, the ability of key decision makers to identify issues accurately may affect the efficiency and productivity of other members of the organization [13].

It has been suggested in the literature that the solo brainstorming techique (SBS) may stimulate issue identification [2, 3]. SBS is an individual brainstorming technique in which the participant interacts with a set of documents and identifies issues from these. An SBS session involves 'reading' and 'editing' by following a protocol that encourages lateral thinking. The SBS protocol brings a formal setting to the process of issue identification. The main activities of the SBS session include reading documents, typing a summary of the documents, making lateral comments and nominating issues from them to be followed up [1, 2]. The SBS protocol also incorporates Osborn's [15] four basic brainstorming rules *i.e.* no criticism is allowed while interacting with documents, freewheeling is welcome, quantity is wanted, and combination and improvement are sought. In view of the above, the main objectives of the current study are (i) to develop a software prototype based on SBS, and (ii) to test it in the context of issue identification for a natural disaster.

3. Research methodology

3.1. Prototype design

A prototype software tool was designed to support creative problem solving. During the design process it was important that substantial cognitive resources not to be taken away from the task by the demands of the user-interface. The aim was to produce an interface that would have minimal impact on cognitive load, one which could be learned easily by a novice user and yet was comprehensive enough to satisfy the experienced user [4].

The technique employed in our study had three phases: (i) users documented the issues they considered relevant to the given problem; (ii) users interacted with the documents one at a time noting issues as they arose and generating ideas and (iii) users documented issues that occurred to them after interacting with the documents. The major aim of the first phase was to identify the issues that users were already aware of before interacting with the abstracts. The second phase was designed to capture users' reasoning, opinions, knowledge and lateral thoughts which went into their reaction to the abstracts. In this way, users could more confidently approach the task of identifying issues and generating ideas. The objective of the third phase was to enable users to express their ideas, comments and knowledge, combined with the data that they interacted with. At the end of phase 2 users were presented with a copy of their comments from the interaction so that they could go through the comments and identify the relevant issues. Since there was an initial set of issues produced by the users (in phase one), there could be a comparison made between the issues generated by interaction with those previously provided.

The tool was applied to a bushfire disaster situation. We considered the features that may (or should) have been included in a planning phase that occurred before such a natural disaster. In particular, we were interested in the features that should be included in planning for any future bushfires in NSW, Australia.

3.3. Measurement

Users' accomplishment in the task of issue identification was evaluated both individually and collectively based on experts' assessments. The specific aspects evaluated included: (i) the effectiveness of the tool in terms of the number of issue categories identified by individual users; (ii) the productivity of nominal groups, *i.e.* individuals who brainstormed alone and had their non-overlapping issue categories combined; (iii) the minimum number of people required in a group in order to be able to cover the maximum number of issue categories.

3.4. Subjects and procedure

A total of eleven users participated in the study on a voluntary basis. Subjects were mainly post-graduate students from the University of New South Wales, who were also working in industry.

Subjects received written instructions explaining their role in interacting with the data and the protocol they would need to follow whilst editing. On arrival, they also received an oral explanation and a small demonstration prior to commencing the experiment.

4. Results

Eleven users identified a total of 160 issues relevant in planning for a disaster or emergency situation. Most of their statements covered more than one issue. The method employed in the analysis of our experiments used expert opinion. The experts' task was to classify issues in a meaningful way so as to define a set of issue categories. A statement was counted as an issue if it was judged relevant and clear enough to be understood. Only three statements were ambiguous and thus were not counted by experts in their classification scheme, primarily because it was difficult to determine the intent within the given context. There was only one case of redundancy, where the same words were written before and after the interaction.

4.1. Individual performance

In order to understand the effects of SBS on the task of issue identification, we studied at an individual level the issues identified at the 'start' of the interaction and those identified 'after'. We found that in both cases the issues tended to fall into clusters.

With respect to issues written before interaction, there were two clusters. The first cluster of issues tended to contain the most common issues, examples of these included: equipment, fire-fighters, communication, insurance, and fire control. These issues were unlikely to be creative, however, they were of high quality because they were the most obvious issues. On the other hand, the second cluster of issues tended to be unique, because the contents reflected the users' own background and experience. Examples of these clusters included: bushfire education, control measures, maintenance of equipment, training of fire fighters, and hospital preparedness.

We found that most users wrote down the issues as brief phrases rather than long sentences. Issues consisted of a label of a few words which also had a certain degree of ambiguity. For instance, user-1 identified one issue as 'fire control during fire', user-2 wrote 'backburning' and user-5 wrote 'prevention (backburning)'. Although these phrases seem to be addressing the same issues, they could also be interpreted differently by their owners.

Issues written after the interaction could also be grouped into two categories. In the first group, issues were gathered from the abstracts. Examples of these include: the fitness of fireman, computer models for fire behavior, heat resistant garments, legal issues, and psychological effects of bushfire. In the second group, statements reflected the users' own interpretations of issues after the interaction. We observed this strongly in user-2's statements, for example, one such statement was the short one word phrase 'onlookers'. However, this issue did not come from the abstracts and made us think about it more carefully. Another statement from user-2 was 'Considerable resources to counsel fire-fighters seems to be wasted'. This user made such a statement because there were a few abstracts that talked about the psychological effects of bushfires on fire-fighters and the arrangements that had been made for counselling. Some users mentioned this subject as an issue that should be considered in planning. On the other hand, user-2 brought another perspective to this issue. This was an example of *lateral thinking*.

As mentioned above, users received a copy of their work when they were writing down issues after the interaction; however, most of them did not consult their documents. When examining their documents, we found that there were many issues that they identified from the abstracts that were not written in their issue lists. This prompted us to explore the users' documents in more detail [1, 2].

Users' individual achievements calculated based upon the experts' classification scheme which were showed that individual users identified on average more than 40% of the issues found by the full group (ranging from a minimum of 19% to a maximum of 63%).

4.2. Collective performance

To display a statistical comparison between results from individual users working alone, and individual users working collectively, nominal groups were formed from users who had brainstormed individually. From the experts' classification it was possible to compute several combinations of the users' outcomes.

To explore how individual users would behave collectively in the task of issue identification, we looked at the behavior for each combination in these groups of users. All the issues were combined and redundant issues were eliminated and counted only once. This process was applied to all the classifications, the target number in the resulting list was the total number of sub-issues generated by the expert.

Results indicate that optimum number of people in a group is 4-6. Groups of this size can cover between 95-100% of the issues at best (R-square = 0.882, $p < 0.05$), and 63-80% at worst (R-square = 0.997, $p < 0.05$).

5. Discussion

5.1. Main findings

The experimental results clearly indicate that a SBS-based software tool had a powerful effect on the task of issue identification and idea generation. The effect was evident at both individual and nominal group levels. In particular, users were found to identify around 40% of critical issues individually, and over 95% of issues in nominal groups of 4-6 people.

The results suggest that the SBS-based software may be a useful learning tool. It is likely to be most valuable in situations where the problem is unstructured, goals indistinct, and the outcome of an action cannot always be clearly identified. It uses a technique that can be applied to a variety of scenarios and can help people process documents whilst identifying issues. The documents act like a 'trigger' to stimulate ideas from users.

One of the most interesting findings from our study is the identification of the number of people who needed to participate in nominal groups in order to achieve optimal performance. We found that the optimal group size is 4-6.

Overall, the results of this study are consistent with the findings of other researchers, and support the notion of the usefulness of developing ideas in an individual session and then having ideas discussed in a group session [3, 5, 7, 8] Our emphasis here is not a comparison of the real group work with the nominal group work, but more the size of the knowledge base that can be formed by individuals at the end of the interaction session. Furthermore, if these individuals are exposed to group discussions after the interaction, they can also expand this knowledge pool by adding new comments and ideas. Note also that many of the common problems seen in traditional brainstorming sessions, *e.g.* production blocking and evaluation apprehension [7] can be avoided by using the tool before the group meeting. This can enable group members to express their ideas without forgetting, and with more confidence, they can also develop their ideas or thoughts with no interruption or social pressures. In addition, the 'housekeeping process' (*i.e.* removing redundant ideas), which is often seen in electronic brainstorming sessions, does not apply to our tool. This is because the participant filters out the issues (that occurred to them during the interaction) from the document they generated earlier.

5.2. Limitations and future research

The application of laboratory conditions is a limitation of this study, specifically because we identified the need for the environment to be conducive to the decision making process. We believe that in field work, the users' individual achievements would improve even further. We also speculate that the performance of users in an SBS session can be affected by their state of mind or previous experience. In some cases, the user may already be 'aware' of the issues and may have the desire to look for some particular aspects rather than identifying issues in general. There may also be other situations where the user may only want to see what the literature is about but not be interested in identifying issues. In such cases, the user will act differently in the course of interaction with the abstracts. Future research may address some of these issues.

6. Conclusion

This study aimed at development and empirical test of a software prototype for stimulating creative problem solving. The software was designed on the basis of a solo brainstorming technique. The essence of tool was to provide users with external stimuli and expose them to a large number of ideas over a short period of time. The tool was tested using the domain of natural disasters management. The results of the test indicated that the tool was quite helpful. Users were able to identify on average 40% of all critical issues individually, and over 95% of issues in nominal groups of 4-6 people. Future research is recommended to address current study limitations.

7. References

[1] Aurum, A., 'Validation of Semantic Techniques used in Solo Brainstorming Documents', *Information Modelling and Knowledge Bases X*, Jaakkolam H., Kangassalo, H., Kawaguchi, E., (eds), 1999, pp. 67-79.

[2] Aurum, A., *Solo Brainstorming: Behavioral Analysis of Decision-Makers*. PhD thesis. University of New South Wales, Australia, 1997.

[3] Aurum, A. and Martin, E.: Managing both Individual and Collective participation in Software Requirements Elicitation Process. *14th International Symposium on Computer and Information Sciences* (ISCIS'99), 1999, pp. 124-131.

[4] Aurum, A., Hiller, J., Warfield, S., 'User-Computer Interface Design For Support of Solo Brainstorming'. *6th International Conference on Human-Computer Interaction. HCI'95*, 44, 1995.

[5] Bouchard, T. J., 'Personality, Problem-Solving Procedure and Performance in Small Groups'. *Journal of Applied Psychology*, 53 (Part 1), 1969.

[6] BW, *Business Wire*, 14 December 1998.

[7] Diehl, M. and Stroebe, W., 'Productivity Loss in Brainstorming Groups: Toward the Solution of a Riddle'. *Journal of Personality and Social Psychology*, 53(3), 1987, pp. 497-509.

[8] Dillon, P.C., Graham, W.K., Aidells, A.L., 'Brainstorming on a 'Hot' Problem: Effects on Training and Practice on Individual and Group Performance'. *Journal of Applied Psychology*, 72(6), 1972, pp. 487-490.

[9] Drucker, P. F., *Innovation and Entrepreneurship: Practices and Principles*, Harper & Row, NY, 1985

[10] Dutton, J.E., Fahey, L. and Narayanan, V. K., 'Toward Understanding Strategic Issue Diagnosis'. *Strategic Management Journal*, 4, 1983, pp. 307-323.

[11] Ford, C.M., 'Theory of Individual Creative Action in Multiple Social Domains'. *Academy of Management Review*, 21 (4), 1996, pp. 1112-1142.

[12] Mintzberg, H., Raisinghani, D. and Theoret, A., 'The Structure of Unstructured Decision Process'. *Administration Science, Quarterly*, 21(2), June 1976, pp. 246-275.

[13] Moreland, R.L., Levine, J. M., 'Problem Identification by Groups'. *Groups Process and Productivity*. Worchel, S., Wood, W., Simpson, J. A. (eds), Sage, London, UK, 1992.

[14] Nutt, P., 'Types of Organizational Decision Processes'. *Administrative Science Quarterly*, 1984, 29, pp. 414-450.

[15] Osborn, A.F., *Applied Imagination: Principles and Procedures of Creative Thinking*. Charles Scribner's Sons, New York, 1957.

Effect of using Computer Graphics Animation in Programming Education

Hiroshi Matsuda, Yoshiaki Shindo
Nippon Institute of Technology
hiroshi@nit.ac.jp

Abstract

We developed a new lecture style for elementary programming education named *Program Reading Practice* by using the *Computer Graphics Animation* programs. (Named *WinTK*). In this practice, students read the source codes of various kinds of Computer Graphics Animation programs and challenged to modify or extend these programs instructed by teacher (*Program Re-Write Practice*). We have done this practice for six years and investigated various kinds of results in order to acquire the effect objectively. In this paper, we describe about the investigation way and the several results of our practice.

1. Introduction

We opened the course of elementary programming education using the C language. To improve these courses and to aid the understanding of the essence of computer programming, we developed a new lecture style named "*Program Reading Practice*" [1][2][3].
In this practice, students read the source code of training programs based on Computer Graphics Animation (Named *WinTK*) and challenge to modify or extend these programs instructed by Teacher (*Program Re-Write Practice*). We have done this practice for six years and investigated various kinds of result in order to acquire the effect objectively. In this paper, we explain about *Program Reading Practice* briefly at first. After that, we describe about the investigation way and the several results of our practice as follows.
(1) Comparative questionnaire about understanding level of learning subjects before and after this practice.
(2) Comparative investigation by using paper examination of learning subjects before and after this practice.
(3) Follow up survey of low-scored students in previous paper examination.
(4) Questionnaires of student's impression.

2. *Program Reading Practice* with *WinTK*

WinTK provides four kinds of training kit programs for *Program Reading Practice*. *WinTK-1* is the *Rabbit animation* (including 14 source codes) those are most basic training programs. *WinTK-2* is *Interactive Paint Tool* (including 10 source codes) to learn how to implement the event handlers of keyboard and mouse. *WinTK-3* is *Computer Aquarium* animation (including 7 source codes) based on the photo-realistic animation. *WinTK-4* is *Dragon Adventure* animation (including 21 source codes) using *3D Computer Graphics* animation based on OpenGL framework. Figure 1 shows the sample of *WinTK-4*. *Program Reading Practice* gives a new lecture style instead of the traditional way [4]. Students can master the construction method of computer programs in many cases by reading and re-writing these source codes.

3. The Final Work Programs

During the last course, as their *final work*, students have to write their own programs by using the *WinTK* framework; those get a marked evaluation. The teacher suggests several themes and students have to choose one of them. Figure 2 shows an outstanding *Final Work* program. In order to examine the students' acquisition level, we investigated the program contents.

Figure 1 Sample of WinTK   Figure 2 Final Work Program

4. Results of *Program Reading Practice*

We conducted a various kinds of investigation to acquire the effect of *Program Reading Practice* objectively.

4.1 Ratio of Understanding Level

We investigated about the understanding level of learning subjects by comparative questionnaire before and after this practice. Figure 3 shows the ratio of students who understand the *Control Structure* (It means a *Repeat Block* by using *while* or *for* statement and *Conditional Block* by using *if* and *else* statement).

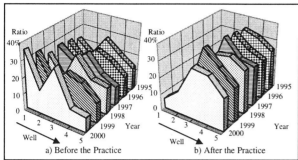

Figure 3 Ratio of Students who understand the **Control Structure**

After **Program Reading Practice**, the greater part of students ranked form 3 to 5. Students, who ranked zero before our **Practice**, decrease to less than 5%. We have conducted this investigation for 6 years. It is very meaningful that each year evaluation indicates the similar result.

4.2 Results of Paper Examination

In 2000, we lectured the programming education using character based console mode framework at first. After that, we conducted the paper examination. Dotted line in figure 4 shows the result. Then we opened the course of Program **Reading Practice** using **WinTK**. After that, we conducted the paper examination again. Solid line in figure 4 shows the result.

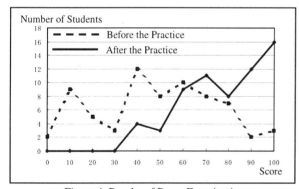

Figure 4 Results of Paper Examination

The attention points in this result are as follows.
(1) Before our **Practice**, the average point was **55.1**. However, after the **Practic**e, it increased 74.3.
(2) Students who got a perfect point increased 8 times.
(3) After our Practice, nobody got less than 30 point.

4.3 Follow up survey of low-scored students

We calculated the rate of increase about points of paper examination between before practice and after practice. The obvious facts of it are as follows.
(1) The average rate of increase was **1.66**.
(2) 86% of students increased their point.
(3) One of students increased more than **5** times.

Moreover, We traced students individually who get less than 50 point in paper examination before our practice. Figure 5 shows that more than 90% of students increased and 50% of students got more than 2 times.

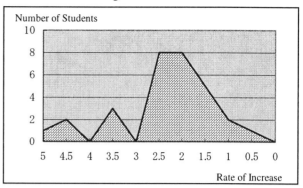

Figure 5 Rate of Increase traced individually

5. Conclusion

In another questionnaire, 85% of students want to study by not using character based console mode framework but by using **Graphics-Based** window **framework**. Although many students thought that **Program Reading Practice** is very hard, they tried to develop the excellent **Final Work** programs exceeded teacher's expectancy.

Let us add one more result for caution' sake, our analysis during 6 years indicated the similar results. Now we are sure that, the usage of **Computer Graphics Animation** in Programming Education facilitates the student's learning and understanding.

6. References

[1] Y. Shindo: "Programming Education Based on Computer Graphics Animation", International Workshop on Advanced Learning Technologies (IEEE Computer Society Press), New Zealand, pp.292-293, (2000).

[2] Y. Shindo, H. Matsuda & M. Mukuda: "Programming Education Based on the Computer Graphics Animation: Program Training Kit WinTK", 4th Global Chinese Conference on Computers in Education (GCCCE2000 Proceedings), Singapore, Volume 2, pp. 592-599, (2000).

[3] H. Matsuda, Y. Shindo & M. Mukuda: "Program Reading Practice using Programming Training kit WinTK". Advanced Research in Computers and Communications in Education (ICCE 99 Proceedings), Chiba (Japan), Vol.2, pp. 854-855, (1999).

[4] A.Kashihara, Atsuhiro Terai and Jun'ichi Toyoda: "Making Fill-in-Blank Program Problems for Learning Algorithm." Advanced Research in Computers and Communications in Education, ICCE'99, Vol.1 pp.776-783, (1999).

X-Compiler: Yet Another Integrated Novice Programming Environment

Georgios Evangelidis, Vassilios Dagdilelis[*], Maria Satratzemi, Vassilios Efopoulos
Department of Applied Informatics, [(*)]*Department of Educational and Social Policy*
University of Macedonia, Thessaloniki, Greece
{gevan, dagdil, maya, efop}@uom.gr

Abstract

This paper presents a simple programming language, called X, and an educational programming environment, called X-Compiler, designed to introduce students to programming. X-Compiler can be used to edit, compile, debug and run programs written in X, a subset of Pascal. X-Compiler could be didactically interesting because of the following features: (a) users can watch the intermediate steps of the execution of a program: source code compilation, correspondence of source and pseudo-assembly code during execution, registers content, and intermediate values of user and temporary system variables; also, they can edit the produced pseudo-assembly code and re-execute it, (b) there are many detailed and explanatory messages that can guide novice programmers when debugging their programs and, in general, help them write better programs.

1. Introduction

For many years now, the most common methodology [4] for teaching principles of programming languages is based on the use of a general purpose programming language, like C or Pascal, and a commercial programming environment. But, this approach does not appear to be didactically effective, since novice programmers are expected to become familiar at the same time with many concepts related both to the structure and operation of information systems and the programming techniques. Du Boulay [2], for example, mentions the following general factors that hinder the learning process:
1. The way students understand and control the "mental" machine that executes their programs and its relationship with the actual machine (the hardware).
2. The rules of the programming language (the syntax and semantics of the language affect and/or extend the behavior of the mental machine).
3. The need to first comprehend the language control structures.
4. The need to master a problem solving technique (students have to design, implement, test, and debug a program using a predefined set of tools).

Below, we further elaborate on the factors mentioned above:
- Program execution is a kind of mechanism that accomplishes a certain task. It may be very hard for students to distinguish between the program and the mechanism it describes.
- A computer (hardware) and a programming environment (software) comprise together a mechanism that is used to create other mechanisms, that is, programs. Students should be aware of the way such a computing system operates, i.e., what exactly happens in the internal of a computing system, or they may develop their own theories of operation that are usually insufficient and erroneous.
- In most programming environments, it is usually hard for students to understand the information presented on the computer screen. This information may refer to a previous interaction between the student and the programming environment, to the input the student has just entered, or to the output of the execution of some code.
- Students have to master procedures regarding the editing, loading and saving of programs.
- The internals of the "mental" machine are usually hidden. Students cannot "see" what is happening during the compilation and/or execution of their code.
- The code produced by students has to adhere to strict syntactic and semantic rules or it cannot be "understood" by the system. This is a source of frustration for students because they usually tend to lend human characteristics to the system.
- The use of English words as keywords is another potential source of anomaly because it lends the system a kind of intelligence. Also, these words have different meanings when used in programs and when in natural language.
- Most programming environments do not provide adequate and user-friendly error handling and reporting [7].

Considering the above-mentioned problems, we believe that it is essential to develop alternative didactic methodologies for introducing students to programming. This can be aided by the use of specially designed programming environments. In this paper we introduce a programming language we have designed and we call X, and its corresponding programming environment. In Section 2, we present the basic design principles we used when designing our programming environment (X-Compiler). The specifications of the languages (X and pseudo-assembly) are given in Section 3. In Section 4 we describe the programming environment. Finally, Section 5 gives an insight on the didactic use of X-Compiler. We conclude with a summary.

2. Design Principles

Programming environments for novice programmers should be effective tools for achieving certain didactic goals. Below, we list a number of principles [12] we considered essential while developing the programming environment for X.

Minimalism. The programming language should be as simple as possible. We avoided the use of types and we don't require variables to be declared before use. Also, the programming environment does not present unnecessary information.

Simplicity. Novice programmers are asked to program a mental machine they barely know and understand. This is a machine whose nature is determined, or better, implied by the programming language. It is essential that the mental machine is as simple as possible, i.e., it should consist of a small number of components that interact in a well-defined and clear manner [3, 9, 13].

Stepped execution and control through visual feedback. Instant feedback can help novice programmers implement and debug their programs. A graphical debugger is useful even for correct programs: it can help novice programmers understand the way their programs work. A programming environment should help novice programmers test, debug, and execute their programs [8]. It is essential that the programming environment include a low-level debugger and a code execution tracer together with data visualization [7, 10, 11, 14].

3. Languages X and pseudo-assembly

We have designed a Pascal-like language, called X. The language supports the **assignment**, **if ... then**, **while ... do**, **read**, **write** and **compound** statements. Identifiers and numbers are integers and all, possibly nested, arithmetic expressions evaluate to integers. X supports only three relational operators: **>**, **=**, and **<>**. A comment is text enclosed in curly brackets ({}). In Table 1 that follows you can find the full specification of X.

Table 1. Specification of X

program	BEGIN {statement;}* END.
statement	id := expr | READ id | WRITE expr | IF rel_expr THEN statement | WHILE rel_expr DO statement | BEGIN {statement;}* END
id	any string consisting of letters, digits and underscore and starts with a letter
expr	id | number | expr op expr | (expr)
op	+ | - | * | /
number	any long integer between
rel_expr	expr rel_op expr
rel_op	> | = | <>
comments	anything enclosed in curly brackets

The assembly language used is a pseudo-assembly that runs on a virtual machine with two registers and includes the basic LOAD, STORE, COMPARE, JUMP, ADD, etc., instructions needed to implement the source language. In Table 2 below you can find a detailed description of the instructions of the pseudo-assembly.

Table 2. The pseudo-assembly used in X-Compiler

instruction	Explanation
BLOCK v	declare an integer variable or a memory position with name v
LOAD(r, v)	store contents of memory location v in register r (r can be 0 or 1)
LOADN(r, n)	store number n in register r
STORE(r, v)	store contents of register r to memory location v
ADD_R(r1, r2, r3) SUB_R(r1, r2, r3) MUL_R(r1, r2, r3) DIV_R(r1, r2, r3)	add/subtract/multiply/divide contents of registers r1 and r2 and store the result in register r3
COMPARE (r1, r2, r3)	compare the contents of registers r1 and r2 and store the result in register r3; the result is –1 if r1 < r2, 1 if r1 > r2, and 0 if r1 = r2
INC(r)	increment the contents of register r
DEC(r)	decrement the contents of register r
NEG(r)	negate the contents of register r
my_label:	declare a label with the name my_label
JUMP_ZERO(r, lb)	jump to lb if contents of r = 0
JUMP_NEG(r, lb)	jump to lb if contents of r < 0
JUMP_POS(r, lb)	jump to lb if contents of r > 0
JUMP(lb)	unconditionally jump to lb
READ(r)	store user input to register r
WRITE(r)	print contents of register r to the output window

4. X-Compiler programming environment

The X-Compiler programming environment has been implemented on the Microsoft Windows platform using Macromedia Director 7 and the compiler construction tools LEX and YACC [1].

X-Compiler allows users to edit, debug, and execute their programs. It consists of five windows (1-source-code, 2-assembly-code, 3-system-registers&tempvars, 4-user-vars, 5-output), and has two modes of operation (novice and advanced) (see Figure 1). In the novice mode only windows 1 and 5 are active, whereas in the advanced mode all windows are active. Of course, users can activate or deactivate any window any time. The provided menu-bar and window-specific toolbars allow the intuitive use of the programming environment (open, save, and edit source or assembly code, compile, execute, or step-execute either type of code, arrange windows, get help on the operation of the programming environment or the X-language). Here, we should mention that double-clicking any keyword, operator, or delimiter on the source code window provides help on the specific feature of the X-language.

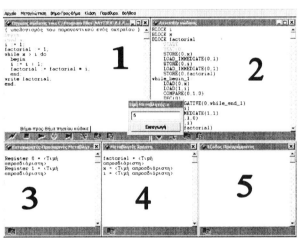

Figure 1. The X-Compiler programming environment

During compilation, syntactic errors in the source code trigger a pop-up window that contains two drop-down lists, one for the detected errors and one for the warnings issued by the compiler. Users can choose the list element they desire to get an explanation of the type of the error or warning. At the same time the appropriate line of the source code is highlighted.

Once users succeed in compiling their code they can either execute it or step-execute it so that they are able to examine what actually happens during execution. For each source code statement the corresponding assembly code statement(s) are highlighted and at the same time the appropriate system registers, temporary variables, and user variables get updated if necessary. Input statements are handled by using a pop-up window that allows users to enter the desired value for their integer variables (see in the center of Figure 1). The output window displays the output generated by the WRITE statements of the user programs.

An interesting feature of the assembly code window is the ability to edit/alter the compiler produced assembly code that corresponds to a given source code fragment and execute it. Since the compiler produces non-optimized assembly code this feature can allow teachers to guide their students in manually optimizing their assembly code. Alternatively, users can write their assembly programs from scratch.

5. Didactic features of the X- Compiler

The didactic objective of X-Compiler is to offer students a lightweight programming environment with simple high level and pseudo-assembly languages and clarify the phases of compilation and program execution that usually constitute a "black box" in professional programming environments [3, 4].

X-Compiler offers interesting didactic features. Users get detailed feedback on the errors encountered during compilation, and are always aware of everything that happens to the internals of the mental machine during program execution (by seeing the correspondence between source and assembly code, the intermediate values of the machine registers, the system generated temporary variables, their own variables, and the contents of the output window). Moreover, users can alter the produced assembly code and then execute it.

We provide teachers and students with the appropriate manuals that contain a series of educational activities on the use of X-Compiler. We have designed the included activities based on the findings of the research community and our teaching experience on the difficulties encountered by students that are novice programmers.

For example, the following case of "cognitive transfer" [5, 6] could be a potential source of difficulties for novice programmers trying to solve problems in a traditional programming environment. Some students may believe that the following code computes the area of a parallelogram:

    area := base * height;
    read(base);
    read(height);
    write(area);

They will be surprised to realize that **area** is not computed correctly. In the X-Compiler programming environment they can see why the above program is not correct by observing the intermediate values of their variables.

Now, consider the code fragment below that swaps the values of variables A and B.

TEMP:=A
A:=B;
B:=TEMP;

The teacher can observe that one can get the same effect without using the extra variable TEMP, as shown in the following code:

A:=A+B;
B:=A-B;
A:=A-B;

Students can examine the intermediate values of the variables and understand why this solution is correct. The teacher could then show that this solution is slower (because it uses more assembly instructions than the previous solution) and also it does not always work correctly (when we have integer addition underflow or overflow).

Those two examples demonstrate the didactic capabilities of our programming environment. Students can not only examine whether their programs produce the correct output, but also discover easily and fast the syntactic and semantic errors they make.

6. Summary

X-Compiler is already being used by the Greek Ministry of Education in a number of secondary education schools and in the entry level university courses on programming we teach (especially its assembly language features). Currently, we are in the process of implementing some additions to the software concerning, (a) a small extension of X to include strings the procedures, and (b) the creation of a "smart" advisor on the logical errors made by students (see first example with the calculation of the area of a parallelogram).

7. References

[1] A. V. Aho, R. Sethi, and J. D. Ullman, "Compilers: principles, techniques, tools", Addison-Wesley, 1988.

[2] B. Du Boulay, "Some Difficulties Of Learning To Program", *Studying The Novice Programmer*, E. Soloway and J. Sprohrer (Eds.), Lawrence Erlbaum Associates, 1989, pp. 283-300.

[3] B. Du Boulay, T. O'Shea, and J. Monk, "The Black Box Inside the Glass Box: Presenting Computing Concepts to Novices", *Studying The Novice Programmer*, E. Soloway and J. Sprohrer (Eds.), Lawrence Erlbaum Associates, 1989, pp. 431-446.

[4] P. Brusilovsky et al, "Mini-languages: a way to learn programming principle", *Education and Information Technologies*, 2, 1997, pp. 65-83.

[5] V. Dagdilelis, "Conceptions des eleves a propos des notions fontamentales de la programmation informatique en classe de Troisieme", Memoire D.E.A., Universite Joseph FOURIER, Grenoble, France, 1986.

[6] V. Dagdilelis, "La validation en programmation: a propos de conceptions des etudiants", actes V Ecole d'ete de Didactique des Mathematiques et de l'Informatique, Plestin-les-Greves, France, 1989.

[7] S. N. Freund and E. S. Roberts, "THETIS: An ANSI C programming environment designed for introductory use", *ACM SIGSCE '96*, Philadelphia, PA, USA, pp. 300-304, 1996.

[8] C. DiGiano, R. Baecker, and A. Marcus, "Software visualization for Debugging", *Communications of the ACM*, Vol. 40, No. 4, pp. 44-54, 1997.

[9] P. Mendelsohn, T.R.G. Green, P. Brna, "Programming Languages in Education: The Search for an Easy Start", *Psychology of Programming*, J. Hoc, T. Green, R. Samurcay, and D. Gilmore (Eds.), Academic Press, 175-200, 1990.

[10] S. Mukherjia and J. Stasko, "Applying Algorithm Animation Techniques for Program Tracing, Debugging, and Understanding", *IEEE 0279-5257/93*, pp. 456-465, 1993.

[11] S. Mukherjia and J. Stasko, "Toward Visual Debugging: Integrating Algorithm Animation Capabilities within a Source-Level Debugger", *ACM Transactions on Computer-Human Interaction*, Vol. 1, No. 3, pp. 215-244, 1994.

[12] J. F. Pane and B. A. Myers, "Usability Issues in the Design of Novice Programming Systems", *Technical Report CMU-CS-96-132*, School of Computer Science, Carnegie Mellon University, 1996.

[13] M. Ruckert and R. Halpern, "Educational C", *ACM SIGSCE Bulletin*, pp. 6-9, 1993.

[14] R. S. Sangwan, J. F. Korsh, and P. S. LaFollette, "A System for Program Visualization in the Classroom", *ACM SIGSCE '98*, Atlanta, GA, USA, pp. 272-276, 1998.

How students learn to program: Observations of practical tasks completed

Pete Thomas and Carina Paine
Computing Department
Open University
Milton Keynes, MK7 6AA
p.g.thomas@open.ac.uk, c.b.paine@open.ac.uk

Abstract

Students on a distance education course in introductory object oriented programming to engage in a number of practical, computer-based activities. Each activity consists of a significant number of small tasks packaged together into sessions. This paper analyses students' attempts at the tasks in terms of the number of attempted and the time taken to complete them. The observed student behaviour is then compared with the behaviour recommended by the course designers. The results reveal both expected and unexpected behaviours and provide some useful feedback on the design of practical activities. The amount of data collected is too large for manual analysis making an automatic analysis tool essential. Therefore, the paper also describes the tool and shows how it has been used to identify student behaviours.

1. Introduction

The AESOP project [4, 9] has developed an electronic observatory for researchers to observe how students tackle the practical activities that form part of their undergraduate studies in computing [9]. Students are taught object-oriented programming using the LearningWorks software system [7] that has been adapted for Open University use. The adapted system uses a book metaphor in which the software presents itself to the student as sequence of LearningBooks each of which contains a set of exercises (practicals) that ask the student to attempt a number of computer-based tasks.

The electronic observatory records, in a text file, the actions that a student takes when interacting with a LearningBook. A recording consists of a series of time-stamped events that can be replayed at a later date using another piece of software, the Replayer. Detailed descriptions of these components can be found in [1, 3].

The *Tasks Completed Tool* is used to identify the times that students spend carrying out practical activities and which activities they perform. To evaluate the effectiveness of the tool and to devise suitable measures to be used in a large-scale investigation, a small-scale trial was conducted in 1999 in which the tool was used to extract data from a set of recordings provided by 30 students working on 8 LearningBooks. Analyses of this data have identified some interesting student behaviours that suggest that all students make comprehensive attempts at the practicals but there are limitations on how much time students are prepared to spend on this work.

2. The Tasks Completed Tool

A LearningBook (LB) comprises a number of *sessions*, are subdivided into *practicals*. The practicals ask students to perform 'small grained' programming activities known as *tasks* (for example, entering a fragment of Smalltalk code for the system to evaluate).

The *Tasks Completed Tool* analyses recordings of students' practical activities. The recordings, whilst small in terms of file size, contain a great deal of information that makes them unsuitable for extensive human evaluation. As we are aiming for a large-scale trial that would involve several thousand recordings, an automatic means of analysing them was required. Therefore, the *Tasks Completed Tool* compares the tasks that students have completed successfully (tasks completed) with the tasks they were asked to complete (tasks set).

In order to specify the tasks contained in an LB, we adopted the notion of an *ideal recording* – a complete attempt at the activities in an LB – that would be used as a benchmark against which student attempts would be measured. The output from the tool gives the number of tasks completed successfully.

3. Research questions

We noticed, from an initial scan of a few recordings that students did not seem to complete all the tasks set. Also, in LBs with multiple sessions, students tended to try the first session and then either gave up or attempted all sessions in the LB. Therefore, we wished to know whether or not students completed all the tasks set and, if not, was there any pattern in their behaviour.

Table 1 shows the average percentage of tasks completed for all LBs analysed (we looked at only those

LBs in which students performed programming activities).

Table 1: Tasks completed in each LearningBook

LB	Time allocated (hours)	No. of Sessions	No. of Tasks	No. of Students	Average % of Tasks Completed
9	4	4	83	6	69.28
10	4	5	43	7	90.37
12	1	1	57	8	78.95
13	4	3	66	13	90.33
14	4	3	72	14	89.48
15	2	3	58	15	92.64
16	3	3	33	16	82.01
20	4	1	49	8	64.29

Table 1 shows that, on average, students complete over 60% of the tasks set in all LearningBooks. Also, while some students (under 10%) complete all the tasks set, in general the majority of students do not do everything in a LearningBook (80% completed, on average). This raises two questions.

First, does the number of tasks set in an LB affect the amount of tasks a student completes? That is, is the setting of more tasks off-putting to students? Perhaps the apparent volume of work puts students off. We hypothesised that the more tasks there are to do in an LB, the number of tasks completed would be lower.

Second, is the time spent studying an LB related to the number of tasks set? It would seem obvious that the length of time spent studying would be related to the amount of work set, but we wondered whether the perceived amount of study required affected the students' willingness to complete all the tasks.

Throughout the investigations we have examined both the number of tasks set as well as the number of tasks completed (as reported by the *Tasks Completed Tool*). The tool is designed to identify all successfully completed tasks that match either the expected outcome or an outcome that is close to the expected one. Thus, it underestimates the number of tasks attempted. Thus, the number of tasks completed as reported by the tool provides a lower bound on tasks attempted. However, since it is not possible to attempt more than all tasks set, this provides an upper bound on tasks attempted.

4. Results

Figure 1 compares of the number of tasks set with the number of tasks completed in each of 8 LBs. The upper dotted line indicates the maximum number of tasks that could be completed (the number of tasks set). A regression line for the data is also shown.

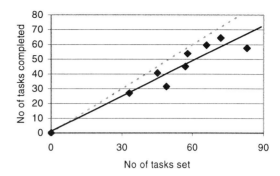

Figure 1. Tasks completed compared to tasks set.

Spearman's *Rho correlation test* applied to the data of Figure 1 yields a value for er_s of 0.9047 (the critical value for er_s for a two-tailed test when N=8 is 0.738 with 95% confidence). This confirms a correlation between the number of tasks completed and the number of tasks set. The regression line has a slope of 0.796 which suggests that, across all LBs, students complete around 80% of tasks set, as observed earlier.

Figure 2 compares the percentage of tasks completed with the time spent in an LB.

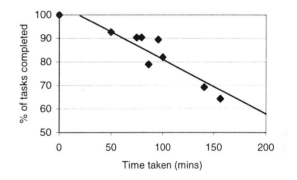

Figure 2. Percentage of tasks completed

This time Spearman's Rho correlation test shows a strong negative correlation ($er_s = -0.9048$). This suggests that the more time a student spends, the smaller the proportion of tasks set are completed.

We looked at the percentage of tasks completed in each *session* of an LB to see whether there was a relationship between the number of tasks set and the number of tasks completed. Figure 3 shows that there was a significant relationship.

The Spearman Rho correlation test gives $er_s = 0.9726$, confirming a strong relationship. That is, the number of tasks completed is proportional to the number set.

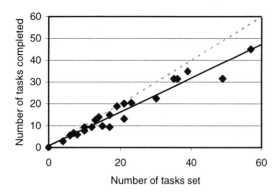

Figure 3. Tasks completed in a session

A plot of the number of tasks completed against the time spent studying a session is shown in Figure 4.

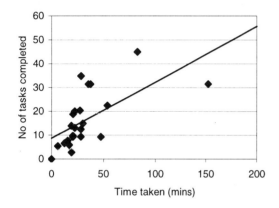

Figure 4. Time spent and tasks completed in sessions

Spearman's Rho test produces a value of $\varrho_s = 0.6897$, confirming a significant relationship. We conclude that, on average, the time taken to study a session is proportional to the number of tasks set which is a relationship that we did not find for LBs.

The analyses discussed above assume that all tasks are of the same complexity either in terms of the time taken to complete or degree of difficulty. However tasks can vary in levels of difficulty, and this could be significant when comparing different LBs. Therefore, we attempted to obtain a view of difficulty by analysing the tasks completed in each LB.

The first step was to define a metric for task difficulty. Initially, we devised three metrics for *task difficulty* as follows:

A: the average time (minutes) spent per task set;
B: the average time (minutes) spent per task completed;
C: the recommended time (minutes) per task set.

Metrics (A) and (B) are based on student behaviour whereas metric (C) is based upon the course team's perception of the tasks. Values for the three metrics for each of the 8 LearningBooks investigated are shown in Table 2.

Table 2. Values of the three metrics for task difficulty

LB	Metric A	Metric B	Metric C
9	1.75	2.53	2.89
10	1.84	1.93	5.58
12	1.51	1.91	1.51
13	1.13	1.25	3.64
14	1.33	1.48	3.33
15	0.86	0.93	2.07
16	3.03	3.70	5.45
20	3.19	4.96	3.85

Since the tool underestimates the number of tasks completed, metric A provides a minimum value and metric B a maximum value. Together, metrics A and B provide a range within which the ratio of tasks attempted in the time spent lies. The data in Table 2 shows that the course team generally over-estimates task difficulty (the middle column in each category).

More revealing is a comparison between the task difficulty and the percentage of tasks not completed in a LearningBook shown in Figure 5.

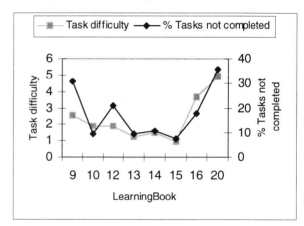

Figure 5. Task difficulty and tasks completed in LBs

Spearman's *Rho correlation test* yields a value for ϱ_s of 0.762, and confirms that there is a correlation between the number of tasks not completed and the task difficulty.

It was expected that a student would complete a session in a single continuous interaction with the computer. A question naturally arises as to the extent to which students followed this pattern given the variation in the number of tasks per session.

If we define a *sitting* as a period of time that a student interacts continuously with the computer without a significant break in activity we can see whether the number of sittings varies with the number of sessions in each LB. The identification of sittings is not straight forward because a recording is simply a list of time-stamped events. We cannot be certain whether the time elapsed between two events represents a break in study or an activity, such as reading, that is related to the practical activity being undertaken. Therefore, a gap between two successive events greater than a given threshold was taken to be a break in study and divides one sitting from the next. In earlier work [2], we concluded that a gap of 10 minutes or more is an indication that a student has broken off from their studies.

On average, for LBs with multiple sessions, students either follow the instructions precisely and study each session in a single sitting, or take one or two more sittings than recommended.

Figure 6 plots the time per sitting with three different indications of study breaks. We observe that there is a limit to the amount of time students are prepared to spend in a single sitting. For the 10 minute gap, the limit is around 25 minutes.

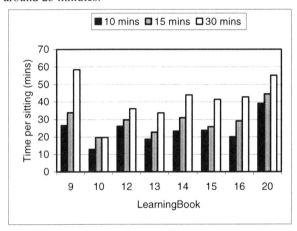

Figure 6. The time per sitting in each LearningBook

This suggests that sessions should be studied in a single sitting designed to take not more than 30 minutes.

5. Future work

During the 1999 presentation, data was collected from a relatively small number of students: 30 students took part in the experiment but each one did not provide a complete set of recordings for the 8 LBs investigated. Therefore, the experiment has been repeated in 2000 with around 200 students taking part and initial investigations support the results reported here.

We wish to investigate in detail is how students tackle their assignments. The LBs we have investigated so far have asked the students to engage with a series of well structured but small-scale tasks. In their assignments, students have more freedom of choice and it will be interesting to see how the analyses we have performed so far compare with student behaviour in a less constrained environment. Analyses from this year's study will provide more detail about tasks *attempted* by students.

Investigations of any relationships between the amount of practical work a student carries out in a LearningBook and their resulting TMA and exam scores will also be conducted.

6. Conclusion

We have found that the most significant relationships were found when we examined LB sessions. Our analyses suggest the following student behaviour:
1. students generally make a good attempt (greater than 80% on average) at all practical activities but seldom complete them all;
2. students usually spend significantly less time on their practical work than expected by the course team;
3. the more time students spend on practical work the more likely they are to stop before completing a significant proportion of the work;
4. the students' perception of the difficulty of tasks is often different from that of the course team;
5. students normally tackle each session of a multi-session LB in a single sitting as recommended (although many take more sittings), but students tackle single-session LBs in multiple sittings (3 or 4);
6. in general, a sitting usually extends for not more than about 30 to 40 minutes.

The results suggest that the *Tasks Completed Tool* gives a reasonable estimate of the actual number of tasks attempted. However, the tool underestimates the number of tasks attempted so we shall be improving its accuracy and using it to perform similar analyses on a more substantial and representative collection of recordings.

7. References

[1] MacGregor, M, Thomas, P.G., and Woodman, M., "*Recording and Analysing User Actions in a Smalltalk Programming Environment*", TOOLS (USA), 1999.
[2] Paine, C. and Thomas, P.G., "How students learn to program: observations of study time behaviour", Research report: 2000/02, Computing Department, Open University, 2000
[3] Thomas, P.G., Martin, M., and Macgregor, M., "Observing Students as they Learn", ITiCSE 98, Dublin, 1998.
[4] Thomas, P.G., "An Electronic Student Observatory", Proceedings Frontiers in Education 98, p1120, IEEE. Phoenix: USA, 1998.
[5] Smith, J.B., Smith, D.K. and Kupstas, E. "Automated Protocol Analysis", *Human-Computer Interaction*, Vol. 8, pp 101 – 145, 1993.
[6] Kivi, M.R., Grönfors, T. and Koponen, A. "MOTHER: System for continuous capturing of display stream", *Behaviour and Information Technology*, Vol.17, No. 3, pp 152-154, 1998.
[7] LearningWorks 0.7 © Adele Goldberg and Neometron, Inc and ParcPlace-Digitalk, Inc
[8] M206, Computing: An Object-oriented Approach, The Open University, 2000.
[9] AESOP: http://www.open.ac.uk/aesop

C-VIBE: A Virtual Interactive Business Environment addressing Change Management Learning

Albert A. Angehrn, Thierry Nabeth
{albert.angehrn, thierry.nabeth}@insead.fr
INSEAD - Centre for Advanced Learning Technologies (CALT)
Bd. de Constance, F-77300 Fontainebleau France
http://www.insead.edu/CALT/

Abstract

C-VIBE is an advanced learning system taking advantage of simulation, multimedia, virtual reality, agents/avatars-based, and multi-user, distributed communication technologies to deliver a realistic learning experience addressing the dynamics of change and innovation processes in organizations. This paper illustrates and discusses the pedagogical effectiveness of the core layers of C-VIBE, the design of its VR and multi-user components, as well as its application in the domain of management learning.

1. Introduction

Managing efficiently and effectively change and innovation processes in organizational contexts has become a key challenge for managers in organizations world-wide. Accordingly, as discussed in [1], research has focused on how to help managers to develop skills such as being able to learn [2], able to innovate [3], able to design and drive organizational simplicity [4], able to manage ambiguity [5], and manage and thrive on change [5, 6, 7].

We describe here the key features of an advanced learning system aimed at helping managers to understand the challenges of organizational change and facing the natural resistance to innovation and change latent in organizations. The system, called 'Change VIBE' (C-VIBE), supports experiential learning through a multi-user multimedia simulation taking place in a Virtual Interactive Business Environment (VIBE) in which the users have to accomplish a change management mission within an organization. In this environment, users are able to initiate actions to achieve their goal which consists in convincing, over a given time period, the (simulated) top management team of a (simulated) company to adopt a major innovation. Throughout this experience, users can initiate different organizational actions (change management initiatives) and see dynamically the impact of these actions, thus learning (by doing) in a highly realistic way how to manage effectively change in an organizational context.

This document describes and discusses the three layers of C-VIBE (see Figure 1). The focus of section 2 is on the kernel of the learning experience, the so-called EIS Simulation [8, 9], which has been extensively tested in top schools and universities world-wide to train managers in the theory and the practice of managing change and organizational transformation. We also provide there details on the pedagogical effectiveness [1] of the first layer of C-VIBE and the underlying concept of Virtual Interactive Business Environments (VIBEs) [10].

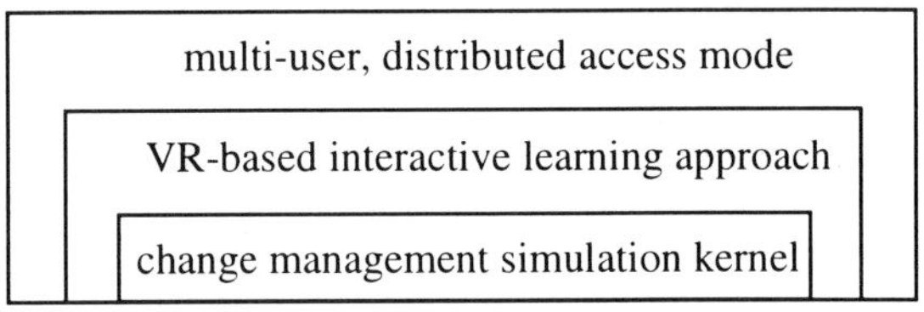

| multi-user, distributed access mode |
| VR-based interactive learning approach |
| change management simulation kernel |

Figure 1: The 3 C-VIBE Layers

In section 3 we describe the other two layers of C-VIBE, i.e. the virtual reality (VR/VRML) and avatar-based environment we are designing to provide a highly interactive learning approach and to increase the realism of the experience, and the distributed multi-user dimension aiming at increasing and extending the way in which individuals and teams can access C-VIBE. With these two additional layers extending the simulation kernel, C-VIBE aims at providing a learning experience taking full advantage of the combination of VR/VRML and Internet/distributed multimedia networks technology.

2. The kernel of C-VIBE: The EIS Simulation and the underlying VIBE concept

The kernel of C-VIBE consists of the inference engine of a simulation of organizational dynamics called EIS Simulation [8, 9], which adopts an experiential learning approach to force realism into the theoretical and conceptual change management debate and to improve the effectiveness of the development of change management and implementation skills and competencies.

The EIS Simulation provides typically the basis for half-day or full-day workshops in which managers and students are allowed, like in a flight simulator, to have a first-hand experience as a member of a change agent team intervening in a company, to test the effectiveness of their implementation strategies and initiatives, and then to debrief this experience in groups.

The challenge for the change agent teams playing this simulation is to gain the commitment of the top managers of a company for a major innovation: the implementation of a corporate computer-based information, communication and reporting system called EIS. The team operates hence within a virtual company in which they have to spend 6 simulated months interacting with virtual managers with very realistic profiles, behaviour and different ways of resisting the innovation. These virtual managers gradually develop a positive attitude towards the adoption of the targeted innovation as a function of the initiatives undertaken interactively by the users (see Figure 2 for screens displaying the organizational chart of the virtual company, information about initiatives/tactics the change agent teams may use in any way they choose (or not choose) at any time during the simulation, as well as a screen through which the users can take new decisions such as trying to arrange a face-to-face meeting with one of the virtual managers).

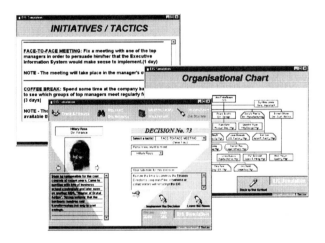

Figure 2: Sample screens from the EIS Simulation

Workshops or sessions based on the EIS Simulation start with a brief introduction taking place just before breaking into teams and starting the simulation. The introduction typically highlights:
- why being able to manage change is necessary in today's business environment and the apparent levels of corporate effectiveness in doing so,
- how individuals usually adopt change or innovation at differing rates and are sensitive to different approaches,
- how each virtual manager in the simulation has been modelled with backgrounds, personalities and the ability to react to initiatives and tactics used by the change agent team as well as to other events - such as the influence of other managers in the virtual organization,
- what information and communication based initiatives and tactics are available to the participants,
- how they will receive immediate feedback about the effects of their actions,
- the necessity for developing a strategy for how their change team will approach the implementation before "getting stuck into it", and
- a reminder that they could hear a build-up of frustration from some of the team members as they implement their strategy as they will encounter the same hostile behaviour and resistance patterns as in the "real-world".

After the introduction, it is up to the teams to first develop a strategy to address the change management mission, and then to start operating within the virtual company to gain the virtual managers' commitment over a period of 2 to 4 hours (simulating a 6 months intervention in the company).

At the end of the time period set for creating and implementing their change strategy (and of a typically very intensive intellectual and emotional experience), each team is requested to reflect on the key Do's and Don'ts they have learned in relation to managing change, and then to present and discuss their experience. In order to take the learning experience from single-loop learning into the more effective, double-loop learning process for each participant, a debriefing session is used to link the learning generated from the simulation to the participants' personal experience of change and resistance. Questioning techniques enable the participants to discuss and reflect on their own experiences as change strategists, change agents or change recipients including their own resistance to change techniques. They are also challenged to think about how they will now handle situations upon return to their own organizations.

As reported in more detail in a recent qualitative study of the effectiveness of the EIS Simulation [8], learning points about managing change raised during the debriefing sessions which relate to the direct use of the simulation nearly always encompass:
- Surprise at how realistically the simulation captures people and their reaction in a normal business situation
- Need to identify and communicate with key stakeholders and key influencers
- Necessity of information gathering
- Power of informal networks
- Awareness that receptiveness and resistance to change is different for each person

- Maintaining communication flows in all directions
- Persistence in maintaining momentum of change
- Need to understand the dynamics of teams
- Having a strategy and then being flexible and able to adapt strategy based on the feedback being received
- Time management
- Effectiveness of a combined top-down, bottom-up strategy

A closer view based on the framework introduced in [1] provides better insights into the pedagogical value of the simulation. In particular, the framework helps identifying which pedagogical objectives are best addressed. As a large number of users (and trainers) of EIS Simulation workshops report in the structured questionnaires administered over the last few years, the realism of the approach helps managers becoming more "streetwise" in the context of organizational change and innovation initiatives, i.e. it provides a risk-free environment to experiment with different ways of succeeding in complex organizational contexts. "Action"-oriented skills (typically acquired through extensive experience) are particularly emphasized, also given that the simulation is one of the first executive learning tools designed based on the Business Navigator Method [10]. This method provides guidelines for the design of learning systems enabling managers to be projected into virtual interactive business environments (VIBEs) which are "highly interactive and realistic... in which he/she will experience the difficulties of thinking, moving, understanding and acting in the diverse, socially complex, information and knowledge intensive, competitive and co-operative reality of today's businesses" [10].

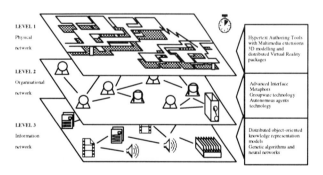

Figure 3: Architecture of Virtual Interactive Business Environments (VIBEs) based on the Business Navigator Method

In order to design these effective management education tools, the Business Navigator Method is built upon the principles of:

- the processes that people go through to learn - the Experiential Learning Model and their preferred style for learning - the Learning Styles Inventory (LSI) and their approach to problem solving/opportunity identification;
- adaptation of learning theory for more effective adult learning - that is ensuring that reflection is a key part of the learning cycle and building in motivation, involvement, curiosity and interest and even fun, novelty and mystery;
- benefits of group dynamics and teamwork;
- providing the learners with more flexibility and control over their own learning experience;
- providing a risk-free learning environment;
- building on the Case Method (a non-experiential learning approach) in which learners are exposed, individually and through facilitation, to a business situation and theoretical frameworks and given an opportunity to exercise analytical skills;
- modelling the complexity of the current business environment and the effects and inter-relatedness of decision-making within organizations.

The Business Navigator Method also takes full advantage of current and emerging information, knowledge and communication technologies such as multimedia, virtual reality, avatar-based navigation, intelligent agents and bots, and artificial intelligence.

3. The two other layers of C-VIBE: VR and distributed, multi-user dimensions

In C-VIBE, the inference engine of the EIS Simulation is extended with a VRML interface consisting primarily of (1) a Virtual Board Room, and (2) a Change Agent Avatar.

Users (working individually or in co-located teams) start the learning experience by connecting to the Virtual Board Room, where an introduction similar to the one described in the previous section is delivered to them via a set of videos displayed in the Virtual Board Room in which the users are represented simultaneously by avatars of their choice. As an additional option, a trainer-controlled avatar provides the opportunity for a Q&A session taking place in the Virtual Board Room. During this introductory phase, users also meet the Change Agent Avatar (a bot) which will provide them assistance throughout the simulation.

The Virtual Board Room hence provides users with a 3D environment in which they can access (directly or indirectly, through the Change Agent Avatar) information of different kinds and initiate organizational actions. The functionality embedded in the Virtual Board Room corresponds to the one provided by the current 2D interface of the EIS Simulation and includes 3D visualization and control panels through which users can monitor dynamically the progress of their organizational intervention.

It is within the Virtual Board Room that users can proceed with the next steps of the learning experience, i.e. with the formulation of a strategy (resulting from a real-time discussion taking place in the chat space integrated in the Virtual Board Room) and with the actual intervention in the simulated company, which takes place over a time period of 2 to 4 hours as described in the previous section. During this intervention, users are able to make decisions on which actions to undertake and then 'send' their Change Agent Avatar to implement their decisions (e.g. a face-to-face meeting with one of the managers, a workshop, or a memo to selected managers) in the organization. After each action, the Change Agent Avatar returns to the Virtual Board Room and reports on the outcome of the initiative (impact on the organization) as well as on other events taking place dynamically in the organization. The information represented in the Virtual Board Room changes dynamically providing a continuously updated view on key organizational parameters.

After completing the simulation (which typically results in a 'relative failure' given the strong resistance to change the users encounter), the users proceed to the debriefing of their change management experience, as described in the previous section. This debriefing session also takes place in the Virtual Board Room, and concludes the overall learning experience. It is in this last phase that the distributed, multi-user dimension of C-VIBE plays a crucial role as it allows distributed teams of users and experts/trainers to connect simultaneously to the Virtual Board Room (instead of being co-located) and discuss together the learning experience.

The key assumption behind the two layers of C-VIBE illustrated in this section is that the VR and the distributed, multi-user dimensions will enhance significantly the realism of the experience, contributing to the value of the learning process and additionally extending the users' competencies in the domains of:
- representing static and dynamic information,
- navigating efficiently,
- interacting with others (as well as with bots), and
- reaching consensus on decisions in distributed VR environments.

4. Conclusions and Next Steps

We have described the key features of C-VIBE, an advanced learning system based on three layers taking advantage of simulation, multimedia, virtual reality, agents/avatars-based and multi-user, distributed communication technologies to deliver a realistic learning experience addressing the dynamics of organizational change and innovation processes. Furthermore we have discussed the pedagogical effectiveness of its core layer (the simulation component) and motivated its extension with two further layers.

After the technical completion and fine-tuning of the VR and multi-user, distributed components of C-VIBE, we will proceed with extensive user testing using management teams to validate the value added by these two additional layers. Measurements will include the extent to which these components increase the efficiency and effectiveness of the learning process in the domain of managing organizational change (helping people to become better change agents) as well as in the domain of managing virtual teamwork (helping people to become more effective in co-operating with others interacting through virtual environments).

Further steps we are planning include the extension of C-VIBE with additional VR components allowing even richer interaction within the virtual company during the simulation. Such extensions will enable users to experience virtual dialogues with bots playing the role of virtual managers as well as the possibility to engage in avatar-based role plays to further extend the learning scope of C-VIBE and evaluate its effectiveness as a learning approach for other, broader domains of management development.

5. References

[1] Angehrn, A.A. and Jill Atherton, "A Conceptual Framework for Assessing Development Programmes for Change Agents," *INSEAD/CALT Working Paper*, 2001.
[2] Argyris, C. "Teaching Smart People How To Learn," *Harvard Business Review*, May-June, 1991, pp. 99-109.
[3] O'Reilly III, C.A. and M.L. Tushman, "Using Culture for Strategic Advantage: Promoting Innovation Through Social Control," *Managing Strategic Innovation and Change: A Collection of Readings*. Oxford University Press, New York, 1997.
[4] Jensen, B. "Make it simple! How SIMPLICITY could become your ultimate strategy," *Strategy and Leadership*, March/April, 1997, pp. 35-39.
[5] McCasker, M.B. *The Executive Challenge: Managing Change and Ambiguity*. 1982, Pitman, Boston.
[6] Hiltrop, J.M. "Preparing people for the future," *Strategic Change*. 7 July, 1998, pp. 213-221.
[7] Ulrich, D. "A New Mandate for Human Resources," *Harvard Business Review*, January-February, 1998, pp. 125-134.
[8] Manzoni, J.-F. and A.A. Angehrn, "Understanding Organizational Dynamics of IT-Enabled Change: A Multimedia Simulation Approach," *Journal of Management Information Systems*, 14, 3, Winter 1997-1998, pp. 109-140.
[9] Angehrn, A.A. and T. Nabeth, "Leveraging Emerging Technologies in Management Education: Research and Experiences," *European Management Journal*, 15, 3, 1997, pp. 275-285.
[10] Angehrn, A.A., Y. Doz and Jill Atherton, "Business Navigator: The Next Generation of Management Development Tools," *Focus*, 1, 1995, pp. 24-31; or 93/37, *INSEAD/CALT Working Paper 1*.

How to Design Web-based Counseling Systems

Marja Kuittinen, Susanna Pöntinen, Erkki Sutinen
Dept. Of Computer science
University of Joensuu
P.O. Box 111
FIN-80101 Joensuu, Finland
{marja.kuittinen,susanna.pontinen,erkki.sutinen}@cs.joensuu.fi

Abstract

The purpose of a web-based counseling and evaluation system is to serve students and other interest groups of virtual university. To develop a competent system requires careful design. We describe the requirements of our system and specifically, we discuss the criteria we have created to evaluate existing counseling systems. Criteria work as guidelines when designing a counseling system. The evaluation process helps us to construct technically more solid solutions for the needs of the Finnish virtual university counseling and evaluation system.

1. Introduction

In Finland, the virtual university consortium consists of all the Finnish universities. Later, the goal of the project is to connect more research institutions, polytechnics and enterprises along [7]. Virtual university can be defined as the infrastructure providing students with a learning platform and related support services to complete a degree partially or totally online [1]. It also offers to faculty members resources to effectively teach and research online. In addition, virtual university connects different institutions together. Enhancement of the virtual university requires that counseling is well organized and users get and find information, services, and tools they need.

However, there are no criteria for the design or even evaluation of web-based counseling systems. Evaluation of existing systems gives perspectives to construct better systems and also to develop existing ones [8]. For these reasons, we developed evaluation guidelines for the design of a web-based counseling system.

2. Definition of Counseling and Our System

In the Finnish virtual university project, counseling refers to two things; firstly, student guidance to pass her/his studies in the virtual university, and secondly, instructional guidance in virtual courses. In addition, counseling has to take into account the whole personnel of the university and its external interest groups.

The purpose of our research and development project is to implement a web-based counseling system for the Finnish virtual university. Primarily it serves students to get information, help and services in a way which best supports her/his to pass the studies in the virtual university. In addition, it is a resource for the staff which is responsible for student information, counseling and guidance.

3. Evaluation Guidelines

The following evaluation guidelines are based on our definition of counseling and systems requirements. Moreover we have investigate general evaluation criteria of computer aided instruction, web-based instruction and virtual learning environments (e.g., [2], [3], [4], [5], [6]). We have also explored existing counseling systems. Our criteria give detailed technical information about the counseling system and aid to analyze how technical features support counseling practices.

Evaluation of technical aspects

In our evaluation criteria we focus on four technical aspects: functions, content, type of pages, and user-interface (see Figure 1).

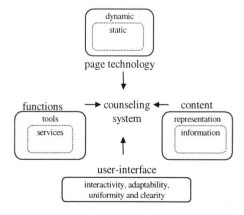

Figure 1: Technical aspects of evaluation guidelines.

Functions (e1), content (e2) and page technology (e3) consist of two connected aspects. Services could be considered as a subset of tools, static page technology as a subset of dynamic technology and information as a subset of representation of content. User-interface (e4) includes three equal and separate aspects to evaluate. Computer scientist are experts in designing, implementing and evaluating tools, representation of information, page technology and user-interface. Counseling professionals are experts in designing and evaluating services and information.

In the design and evaluation process we emphasize the importance of functions. By functions, we refer to various services and tools of a counseling system. By utilizing a *service* a user gets information according her inquiry (including navigation) or she can save data. However, she/he needs a *tool* to be able to manipulate data, e.g., list of study attainment versus maintain personal curriculum.

Effectiveness of Counseling
With the following questions we analyze, in terms of technical aspects, the system as a whole and how the system supports good counseling practices. In our criteria good counseling practices are based on our definitions of counseling, system requirements and quality checklist of student counseling [9] which is compiled by The National Union of Finnish Students (SYL).

Table 1: Effectiveness of counseling.

Counseling perspective	technical aspect
To whom system is made for and for what purpose?	e2
What types of information is available?	e2
What kind of possibilities are in page maintaining?	e1,e3
How does the system support different ways to seek information?	e1, e4
How does the system support different kind of users?	e1, e2, e3,e4
How the system support different kind of data manipulation?	e1,e3,e4
Level of services? (versatility, adequacy, functionality)	e1
Level of tools? (versatility, adequacy, functionality)	e1
Level of intelligent tutoring?	e1, e3, e4
Level of interactivity?	e1, e4
Level of communication possibilities among interest groups?	e1, e4
Level of possibilities to give feedback?	e1, e4
Level of possibilities to handle feedback and get reports?	e1
Level of entirety of the system?	e1,e2,e4

When evaluating effectiveness of the counseling system it is important to know the goals of the system. Analysis always includes subjective aspects and the result of the evaluation process depends on evaluator's opinions of good counseling practices.

4. Conclusions

We are constructing web-based counseling system which is based on educational theory, research on educational technology and computer science. Evaluation of existing counseling systems gave ideas to our design process. Good counseling practices (see Table 1) are still difficult to implement. A functional counseling system requires sufficient resources that meet the needs of individual guidance. In our criteria we answer to this challenge by evaluating e.g., personal tools, information seeking opportunities, and communication possibilities among interest groups. This means that the future counseling systems have to *activate* a virtual university student and we have to emphasize this aspects in our evaluation criteria. In general, the existing systems are mainly passive repositories of guidance material or cover just a few features of a functional counseling systems.

References

[1] A. Kumiko, and D. Poggroszewski, "Virtual University Reference Model: A Guide to Delivering Education and Support Services to the Distance Learner", *Online Journal of Distance Learning I*, 3, 1998, http://www.westga.edu/-distance/-aoki13.html (5.2.2001).

[2] D. Squires, and J. Preece, "Usability and Learning: Evaluating the Potential of Educational Software", *Computers and Education 27*, 1, 1996, pp. 15-22.

[3] G.A. Hutchings, W. Hall, J. Briggs, N.V. Hammond, M.R. Kibby, C. McKnight, and D. Riley, "Authoring and Evaluation of Hypermedia for Education", *Computers and Education 18*, 1-3, 1992, pp. 171-177.

[4] J. Nielsen, *Usability Engineering,* Boston (Mass.) Academic Press, 1993.

[5] M. Kopponen, *CAI in CS,* University of Joensuu. Computer Science, Dissertations 1. Yliopistopaino, Joensuu, 1997.

[6] Open Learning and Information Network, *A Web Tool for Comparative Analysis of Online Educational Delivery,* http://www.olin.nf.ca/landonline (5.2.2001).

[7] Opetusministeriö, *Koulutus ja tutkimus vuosina 1999—2004 kehittämissuunnitelma,* 2000, http://www.minedu.fi/-julkaisut/KESU2004/KESU.html (5.2.2001).

[8] T.C. Reeves and P.M. Reeves, "Effective Dimensions of Interactive learning on the World Wide Web", B.H. Khan, *Web-based Instruction,* Engelwood Cliffs, New Jersey, 1997, pp. 59-66.

[9] The National Union of Finnish Students (SYL), *Quality Checklist of Student Counselling,* http://sylsrv.syl.-helsinki.fi/english/laatulistauusin_en.htm (5.2.2001).

Application of Multipoint Desktop Video Conferencing System (MDVC) for Enhancement of Trainee-Supervisor Discourse in Teacher Education

Myint Swe Khine, Leslie Sharpe, Hu Chun, Lachlan Crawford
S. Gopinathan, Moo Swee Ngoh, Angela Wong
National Institute of Education
Nanyang Technological University, Singapore
mskhine@nie.edu.sg

Abstract

Videoconferencing technology has been in use for teaching, training and communication purposes over the past several years. Recent advances in digital streaming technology, increasing bandwidth and software engineering allow the use of video conferencing in the most efficient and cost effective ways. In the teacher training environment, the use of video conferencing has potential to improve the discourse between the trainee and supervisor and thus increase the overall effectiveness of the training process. This paper describes a research project which utilizes the multipoint desktop video conferencing at the National Institute of Education in Singapore to investigate the feasibility and pedagogic values of the technology in teacher training programs. Some preliminary findings are reported in this paper.

1. Introduction

Educators are aware of the fact that application of technology in teaching is not just introducing hardware and technical system into the curriculum. It is the use of technology which epitomizes constructivist approach that matters. Jonassen, Peck, & Wilson (1999) urged that the use of technology is to help the students articulate and reflect on the previous information and facts and personally construct the knowledge together. The most important fact is that in order to obtain a maximum benefit, technology should be used by learners to engage in active, constructive, intentional, authentic, and cooperative learning.

This paper reports on the recent findings of the "Teaching Practice Discourse and Computer Communication Technology Project" which explores the feasibility of combing videoconferencing and digital technology to enhance the trainee-supervisor discourse in the practicum component of the teacher training program.

2. Background to the project

The Teaching Practice Discourse and Computer Communication Technology Project of the National Institute of Education (NIE), Singapore evolved from the previous research attempt that investigated the discourses between trainee teachers and their university supervisors and school-based co-operating teachers (Sharpe *et al.*, 1994). Discourse analysis were made using a modified version of the Zeichner discourse inventory and it was concluded that there was a preponderance of low-level "factual" and "prudential" discourse and a relatively low level of higher "justificatory" and "critical" discourse. The project has two major aims. One of the aims concerns technical and developmental aspects which involve the practical feasibility of deployment and installation of a reliable MDVC system that fully exploits the state-of-the art technology and ICT infrastructure. The other deals with the pedagogic question on how MDVC can add value to the Practicum by enhancing the professional preparation of trainee teachers (Sharpe *et al.*, 2000). MDVC makes use of the broadband service of SingTel Magix, a local commercial Asymmetrical Digital Subscriber Line (ADSL) to deliver Asynchronous Transfer Mode (ATM) through the existing telephone wires.

3. Active and Meaningful Learning

Since the project started in early 1999, a number of trials and experiments were completed. During these experiments different cohorts of trainee teachers were engaged in video conferencing sessions with their respective supervisors. Each session of video conferencing accommodated between four to five trainees from different schools and the supervisor. Video conferencing sessions were carefully planned so that trainee teachers were actively involved in the discussions with their colleagues and their supervisor. Trainee teachers were given some tasks prior to the video conferencing to prepare themselves for the sessions. For example, trainee teachers were assigned to record different parts of their practice teaching on a digital video camera. Each video clip lasted for three minutes and episodes covered three important teaching competencies. These video clips were later sent to the project site from their respective schools by File Transfer Protocol (FTP).

In addition, trainee teachers also took digital still photographs of their activities, curriculum resources, classroom arrangements and environment and other relevant visuals. During the conference session, these pictures can be placed on the "whiteboard" facility of CU-seeMe software to further aid discussion. These activities allow the trainee teachers to participate actively and in a meaningful way.

During the video conferencing sessions, trainee teachers saw their colleagues in the same school cluster (usually five trainee teachers) and their supervisor on the CU-seeMe screen. While they were discussing a particular topic, relevant video clips were accessed from the project server. The video clip was displayed in the Window media player and appeared superimposed on the CU-seeMe the screen. The video clips could be stopped, rewound and replayed during the discussion.

During the first semester of academic year 2000/2001, twenty two trainee students were involved in weekly MDVC sessions. After every session, trainee teachers were asked to provide feedback on various aspects of the conferences. Preliminary results show that more than 90% of the students found it easy to record their practice teachings on video tape. They also reported that it was easy to access the video clips from the project web-site. The value of the video clips is also highlighted by the trainee teachers. The participants found that the process was fun, enlightening and make them more self-conscious. Some reported that they were nervous and concerned that these might give wrong impression on their practice teaching. Despite some initial technical problems, most of them agreed that it was good for personal and professional reflection. They felt that discussions took place in a collaborative learning environment, and thus provided an opportunity to learn and help each other. They also felt that when video clips were used to discuss a point during the conference it became more informative, reflective and corrective. Some of them were reluctant to give negative opinions on others and felt embarrassment, but eventually they built up confidence and saw the value in sharing the experience among peers.

Some trainees expressed a view that sharing the experiences, discussing with other trainees on pertinent issues and seeing how others conduct their classes enabled them to pick up some new skills. Multipoint videoconferencing made it possible to watch the video clips and discus the point of interest in real-time online with all participants in the group. This provided an avenue to obtain ideas from other trainee teachers with the moderation from the NIE Supervisor. In addition to gaining new knowledge and skills about teaching, they were also acquiring information and communication technology proficiencies by handing various computer related equipment. More detailed analysis of the feedback is in progress and the results will provide better understanding of the benefits of these video conferencing sessions.

4. Future Direction

The multiple desktop video conferencing (MDVC) allows participants to interact in real time and synchronous mode. The preliminary information gathered from the trials indicates that if the conferencing sessions are properly structured students do engage in active, constructive, intentional, authentic, and cooperative learning. The students also acquire a new set of information and communication technology skills and prepare them to face the networked society of the 21st century.

Video conferencing technology is advancing very rapidly. New products, hardware and software are becoming available on the market. With the improved algorithms, compressing and streaming technology, affordable and reliable web-based video conferencing is now becoming reality. It is envisaged that once the technology improves to greater reliability and cost effectiveness, the use of MDVC will become a regular feature of the teacher training process in this context.

5. References

[1] D.H. Josassen, K.L. Peck, and B.G. Wilson, *Learning with technology: A constructivist perspective,* Merrill, New Jersey, 1999.

[2] L. Sharpe, S.N. Moo, L. Crawford, and S. Gopinathan, *Teacher supervision patterns of discourse,* National Institute of Education, Singapore, 1994.

[3] L. Sharpe, H. Chun, L. Crawford, S. Gopinathan, S.N. Moo, S.N, and A. Wong, *Enhancing multipoint desktop video conferencing (MDVC) with lesson video clips.* Paper presented at annual conference of Educational Research Association. Singapore, 2000.

Theme 6: Authoring and Development of Systems

Authoring and Delivering Adaptive Web-Based Textbooks Using WEAR

Maria Moundridou
*Department of Informatics,
University of Piraeus, Greece*
mariam@unipi.gr

Maria Virvou
*Department of Informatics,
University of Piraeus, Greece*
mvirvou@unipi.gr

Abstract

In this paper, we describe the adaptive courseware authoring capabilities of WEAR, an ITS authoring tool for Algebra-related domains. The system allows the authoring of electronic textbooks and delivers them over the WWW. Then learners are offered navigation support adapted to their individual needs and knowledge. The domain knowledge and the information kept in learner models are used in WEAR to provide adaptive navigation to students and also to support the instructional designers in the authoring process. An instructor model incorporated in WEAR's architecture contains additional information that is exploited by the system to achieve the goal of authoring support.

1. Introduction

What best describes Adaptive Hypermedia Systems (AHS) is their ability to adapt the interaction with each individual user. Adaptation is based on the model they maintain of each user's goals, preferences and knowledge. Education is a popular application area for AHS since previous knowledge on the subject being taught, rate of progress, learning goals and other characteristics may vary a lot among learners. This is even more the case with Web-based education, which aims at reaching a much more heterogeneous group of learners. This challenging goal urged in the recent years a number of research groups to engage in research on adaptive Web-based educational systems. As a result, quite a lot of Web-based adaptive hypermedia systems exist, making use of various methods and techniques.

Brusilovsky in [2] provides a detailed review of adaptive hypermedia methods, techniques and systems and distinguishes between two main adaptive hypermedia technologies: (i) *adaptive presentation*, the case where adaptation is performed at content level and (ii) *adaptive navigation support*, which is performed at link level. Both these technologies have been evaluated and the results offer strong evidence that their use in an educational AHS can have a positive effect on students' learning and comprehension of the domain (e.g. [4], [9], [1]).

However, the development of a Web-based course is a quite time consuming task that cannot be carried out by the instructor or the author of the teaching material but rather requires the involvement of programmers and other experts. A way to overcome these problems is to develop a Web-based course using an authoring tool. Among the available tools of this kind, there are quite a few that support not only course creation and delivery but also other functions and utilities such as bulletin boards, course management, chatting etc. ([7], [14]). Unfortunately, the courses created with these powerful, commercial tools do not adapt in any way to the individual learner and are rather static. On the other side, there are tools originating from research efforts that support the development and delivery of adaptive Web-based courses. Among them, we cite Interbook [5], AHA [6] and MetaLinks [10].

In this paper, we will describe the adaptive courseware authoring capabilities of a tool we are developing and is called WEAR (WEb based authoring tool for Algebra Related domains). WEAR is a Web-based Intelligent Tutoring System authoring tool, which is mainly concerned with problem construction and solving in Algebra-related domains [12]. However, the adaptive textbook authoring facility it offers could also be utilised for the creation of courseware in other domains too, even if they are not Algebra-related.

In particular, WEAR allows the authoring of electronic textbooks by instructors and delivers them over the WWW to learners. These textbooks offer navigation support to students, adapted to their individual needs and knowledge. To achieve these aims, WEAR is based on three models: the domain model (representing knowledge about the domain of the subject matter), the learner model (representing knowledge about the individual learner) and the instructor model (representing knowledge about the instructor). Using information derived from these models the system can provide individualised support to instructors concerning the authoring of the course. The remaining of this paper is organised as follows: Section 2 describes the models of the system; in section 3 we present how adaptive navigation is performed and in section 4 we describe the course authoring process and the support provided to designers with regard to adaptive navigation.

2. Models of WEAR

The *domain model* containing knowledge about the subject matter is structured as a network of hierarchically organised topics (textbook sections). Links between nodes of that network represent relationships between topics. At the moment, two types of relationship are used: *is_prerequisite_of*, to describe a topic a learner should know before accessing the more advanced one, and *is_related_to*, to describe that these two topics are in some way related to each other. Each topic has an associated *difficulty level* ranging from 1 (very easy) to 5 (very difficult). Finally, problems and/or tests examining the knowledge that must be acquired by studying a particular topic are associated with it; these associations are also part of the domain knowledge.

For each topic contained in the domain model, the individual *learner's model* stores two attribute-value pairs. These are: (i) *read* (true or false), indicates if this topic has been visited by the student, (ii) *knowledge weight* (ranging from 0 to 1), is an estimation of the student's knowledge level on this topic; it is calculated taking into account both the student's performance in solving the problems associated with this topic (if such problems or tests exist) and also the value of the *read* attribute.

The part of the learner model that is used for adaptive navigation is a combination of a stereotype and an overlay student model, similarly with other systems, such as [8]. The stereotype model (formed either directly by the instructor or after a preliminary test posed to the student) classifies initially the student according to his/her knowledge of the domain. As a result of this, each student is assigned to a stereotype (novice, beginner, intermediate or expert). The stereotype model also defines initial values for the overlay student model described above, taking into account each topic's *difficulty level*. If for example the stereotype model indicates that a student is "intermediate", then the initial value of the attribute *knowledge weight* will be 1 for all topics with difficulty level 3 or lower. The underlying assumption in this is that a student considered "intermediate" probably knows every topic which is not rated as difficult or very difficult.

Each time a student visits a topic, solves a problem or does a test, his/her user model is modified to reflect his/her current knowledge state in the domain being taught.

WEAR -unlike other tools- incorporates an instructor modelling component in its architecture [13]. With regard to the adaptive textbook authoring, the *instructor model* mainly holds information obtained explicitly by the instructor. Such information is the instructor's long-term preference concerning the difficulty of the course. The instructor is also asked to specify how the students' level of knowledge will be calculated. For example, an instructor may state that s/he wishes the reading of a topic by a student to be given a weight of 20% and the rest 80% of the knowledge level to be obtained from student's scores in problems and tests.

3. Adaptive navigation support

Brusilovsky in [2] describes several methods of adaptive navigation support, such as adaptive link sorting, annotation, or hiding, map adaptation and direct guidance. WEAR at the moment provides adaptive navigation support through adaptive link annotation. In particular, using the information stored in each learner's model and the domain knowledge, WEAR generates a table of contents consisting of links to every topic of the domain. These links are annotated in order to inform students about the educational appropriateness of the topic behind them. In that way, five different states of links can be distinguished in a Table of Contents (TOC) generated by WEAR, as shown in Table 1. To annotate different states of links WEAR uses different icons. For example, a checkmark next to a link means that this topic's state is "Visited and Known", whereas a "no-entrance" icon implies that this topic is "Not Ready" to be visited yet.

When a student visits a topic, s/he is offered the choice of seeing a list of related links. These links are also annotated in the same way as the links in the Table of Contents.

Table 1. Topic states

State	Rules
Visited but Not Known	If this topic is visited but its knowledge weight is lower than a threshold
Visited and Known	If this topic is visited and its knowledge weight is higher than a threshold
Not Ready	If this topic is either not visited or Visited but Not Known AND there is at least one prerequisite topic that is either not visited or Visited but Not Known
Ready and Highly Recommended	If this topic is Visited but Not Known and all prerequisite topics are Visited and Known
Ready and Recommended	If this topic is not visited and all prerequisite topics are Visited and Known

4. Authoring for adaptive navigation

Authoring an ITS with WEAR involves creating and structuring the teaching material, constructing problems and tests and managing the student records. In the subsequent sections we will deal with the former (the textbook construction) since this is mostly related to adaptive navigation support.

4.1. Authoring procedure

Although most of the existing authoring tools for adaptive educational textbooks approach the adaptivity issue in quite similar ways, they differ a lot in the authoring process they impose to their users (authors). For example, in Interbook [5] the author should provide a specially structured, annotated MS-Word file. In AHA [6] the author should write annotated HTML files. MetaLinks [10] on the other hand, provides a GUI interface for authoring all aspects of the electronic textbook.

In WEAR, we address authoring in a way that in its first steps resembles the simple one adopted by commercial tools like WebCT [7]. In particular, the authoring procedure is the following: The author should prepare HTML files for the topics that would be contained in the electronic textbook. This is very trivial since there are tools both for creating HTML files and for generating HTML from other document formats. The next step is to use WEAR's facilities for uploading these files to the WEAR server. For each uploaded file the author must also specify a title, a difficulty level and the position that it should have in the topics hierarchy. Finally, the author must edit the is_prerequisite_of and is_related_to relationships between topics. To perform this, the author is presented with the hierarchy of topics and s/he should write in two columns next to each topic the section numbers of its prerequisite and related topics.

The author may also create multiple choice tests or problems and associate them with the appropriate topics. S/he may also create a preliminary test to be set to students in order to classify them in a stereotype and initialise their user model. In that case, the author should state which ranges in scores obtained from the preliminary test correspond to which stereotype. Alternatively, the author could manually define for each student of the virtual class the stereotype s/he belongs.

4.2. Authoring support

Some problems linked to the authoring of adaptive textbooks by instructors through the use of authoring tools are the following:
(i) Instructors may face several difficulties during the design process (e.g. they may not be sure about the structure their course should have), (ii) they may provide inconsistent information to the tool that may lead to the generation of adaptive textbooks with problematic behaviour (e.g. they may define prerequisite relationships in such a way that some sections of the textbook would be unreachable or never recommended), (iii) in order to be domain independent and generic, most authoring tools embody predefined pedagogical rules which cannot even be configured; instructors are obliged to accept them even if these rules contradict the way they perceive instruction.

To overcome some of these problems, several approaches were proposed. For example, Wu et al. in [15] describe support tools that help authors create usable and consistent adaptive hypermedia applications. Brusilovsky in [3] introduces a concept-based course maintenance system which can check the consistency and quality of a course at any moment of its life and assist the course developer in some routine operations. A more sophisticated approach is presented in [11] by Nkambou et al.: In order to provide designers with support that focuses on the expertise for building courses, they propose to use an expert-based assistant integrated with the authoring environment. The expert system reasons on a constraint base that contains constraints on curriculum and course design that come from different instructional design theories. In that way, the expert system validates curriculums and courses produced with the authoring tool and advises the instructional designer accordingly.

In WEAR, instructors are assisted in authoring a course along various dimensions. First of all, when building the course they are provided with tools that verify the course's consistency and report possible problems or errors (such as the case when the prerequisite relationships impose that a topic indirectly requires the knowledge of itself). To offer more intelligent and individualised help to instructors WEAR relies on the information provided by an instructor modelling component that it embodies. The information of this model and the learner model is used by WEAR to support instructors in the authoring process in the following ways:

Instructors are offered the choice to see what other instructors have done. The information that is presented to the user in that case, is the structure of a similar course (in terms of the domain to which it belongs and in terms of the difficulty level assigned to it by its author) created by another instructor. In particular, the instructor may see an enriched TOC presenting not only the topic hierarchy but also the prerequisite and is_related relationships between topics. In that way, instructors who may be novice as course designers could be assisted by more experienced peers who have previously used WEAR.

While students are working on the course, the system collects evidence to build reports and offer advice that may be of interest to the instructor. If most students are not doing well and the instructor's goal (as recorded in his/her user model) is to offer an easy course, then s/he is notified of the inconsistency. Furthermore, WEAR also performs more thorough checks: for instance, if the majority of students fail to comprehend a specific topic (indicated by low scores in the corresponding tests), then the instructor is informed and given some suggestions concerning this situation (e.g. the underlying reason for the students' failure may be the misplacement of the specific topic in the curriculum, or otherwise it may be that the test was too difficult). By receiving feedback concerning the efficiency of the course they constructed, instructors can redesign it; multiple iterations of this process may lead to the construction of optimal courseware for their class.

5. Conclusions

In this paper we described WEAR, an authoring tool for adaptive Web-based courses. The system based on a domain model authored by the instructional designer and on individual learner models, provides to students navigation support adapted to their own knowledge and needs. WEAR also deals with modelling not only students but also the other class of its users: the instructors. By combining the information of all of the above-mentioned models, WEAR besides providing adaptive navigation to students, it supports instructors in authoring consistent and efficient courses.

6. References

[1] C. Boyle, and O. Encarnacion, "Metadoc: An adaptive hypertext reading system", *User Modeling and User-Adapted Interaction*, 4, Kluwer, The Netherlands, 1994, pp. 1-19.

[2] P. Brusilovsky, "Methods and techniques of adaptive hypermedia" *User Modeling and User-Adapted Interaction*, 6 (2-3), Kluwer, The Netherlands, 1996, pp. 87-129.

[3] P. Brusilovsky, "Course Sequencing for Static Courses? Applying ITS Techniques in Large-Scale Web-Based Education", in G. Gauthier, C. Frasson and K. VanLehn (eds.): *Intelligent Tutoring Systems, Proceedings of the 5th International Conference on Intelligent Tutoring Systems*, Lecture Notes in Computer Science, 1839, Springer, Berlin, 2000, pp. 625-634.

[4] P. Brusilovsky, and J. Eklund, "A Study of User Model Based Link Annotation in Educational Hypermedia" *Journal of Universal Computer Science*, 4 (4), Springer Science Online, 1998, pp. 429-448.

[5] P. Brusilovsky, J. Eklund, and E. Schwarz, "Web-based education for all: A tool for developing adaptive courseware", *Computer Networks and ISDN Systems*, 30 (1-7), 1998, pp. 291-300.

[6] P. De Bra, and L. Calvi, "AHA: a Generic Adaptive Hypermedia System", in P. Brusilovsky and P. De Bra (eds.): *Proceedings of 2nd Adaptive Hypertext and Hypermedia Workshop* at the 9th ACM International Hypertext Conference - Hypertext'98, Computing Science Report No. 98-12, Eindhoven University of Technology, Eindhoven, 1998, pp. 5-11.

[7] M.W. Goldberg, S. Salari, and P. Swoboda, "World Wide Web - Course Tool: An environment for building www-based courses" *Computer Networks and ISDN Systems*, 28, 1996, pp. 1219-1231.

[8] H. Hohl, H. Böcker, and R. Gunzenhäuser, "Hypadapter: An Adaptive Hypertext System for Exploratory Learning and Programming", *User Modeling and User Adapted Interaction*, 6 (2-3), Kluwer, The Netherlands, 1996, pp. 131-155.

[9] T. Murray, J. Piemonte, S. Khan, T. Shen, and C. Condit, "Evaluating the Need for Intelligence in an Adaptive Hypermedia System", in G. Gauthier, C. Frasson and K. VanLehn (eds.): *Intelligent Tutoring Systems, Proceedings of the 5th International Conference on Intelligent Tutoring Systems*, Lecture Notes in Computer Science, 1839, Springer, Berlin, 2000, pp. 373-382.

[10] T. Murray, T. Shen, J. Piemonte, C. Condit, and J. Thibedeau, "Adaptivity in the MetaLinks Hyper-Book Authoring Framework", in C. Peylo (ed): *Proceedings of the International Workshop on Adaptive and Intelligent Web-based Educational Systems* (held in Conjunction with ITS 2000), Technical Report of the Institute for Semantic Information Processing, Osnabrück, 2000, pp. 61-72.

[11] R. Nkambou, C. Frasson, and G. Gauthier, "A new approach to ITS-curriculum and course authoring: the authoring environment", *Computers & Education*, 31, Elsevier Science, 1998, pp. 105-130.

[12] M. Virvou and M. Moundridou, "A Web-Based Authoring Tool for Algebra-Related Intelligent Tutoring Systems" *Educational Technology & Society*, 3 (2), 2000, pp. 61-70.

[13] M. Virvou and M. Moundridou, "Modelling the instructor in a Web-based authoring tool for Algebra-related ITSs", in G. Gauthier, C. Frasson and K. VanLehn (eds.): *Intelligent Tutoring Systems, Proceedings of the 5th International Conference on Intelligent Tutoring Systems*, Lecture Notes in Computer Science, 1839, Springer, Berlin, 2000, pp. 635-644.

[14] WBT Systems, TopClass, http://www.wbtsystems.com

[15] H. Wu, G.J. Houben, and P. De Bra, "Authoring Support for Adaptive Hypermedia Applications", in B. Collis, and R. Oliver (eds.): *Proceedings of ED-MEDIA 99, World Conference on Educational Multimedia, Hypermedia & Telecommunications*, AACE, Charlottesville VA, 1999, pp. 364-369.

Using the REDEEM ITS Authoring Environment in Naval Training

Shaaron Ainsworth, Ben Williams and David Wood

*ESRC Centre for Research in Development, Instruction & Training, School of Psychology,
University Park, University of Nottingham, Nottingham, NG7 2RD, UK.
{sea,bcw,djw}@psychology.nottingham.ac.uk*

Abstract

REDEEM is an ITS authoring tool that allows instructors to create simple ITS from existing CBT. In this paper we consider the application of REDEEM to Naval training. Two Subject Matter Experts created learning environments for Naval Cadets and Reservists. They did so operating within a very different context to our prior experience of authoring for the school classroom. In this paper, we consider if the adaptive pedagogical features that REDEEM provides are appropriate to the needs of adult trainers.

1. Introduction

In the last few years, there has been an increasing attempt to transfer advanced educational technologies from the laboratory to practical application in schools and colleges (see [1,2]). In so doing, systems are required to be effective, robust, usable, and appropriate to the pedagogical and curricula needs of their intended milieu. At the forefront of such developments lie Intelligent Tutoring System (ITS) authoring tools. By definition, they were conceived as a solution to the challenge of cost-effective, quick and large-scale development of ITSs. By allowing end-users with appropriate subject and pedagogical experience to create ITSs, it is also hoped to improve effectiveness and ensure better fit to organizational requirements. Consequently, practical deployment of ITS authoring tools must account for the needs of authors in addition to those of the learners.

REDEEM is an ITS authoring tool that was initially developed in the context of the school classroom. The ITS tools take extant Computer Based Training (CBT) and allow teachers and subject matter experts (SMEs) to overlay their instructional expertise. The REDEEM shell uses this knowledge, together with its own default teaching knowledge, to deliver the courseware adaptively. REDEEM ITSs are limited by the domain content of the CBT and have a small number of tutorial actions. Thus, REDEEM ITSs are less adaptive than a normal ITS.

However, a teacher with no computing experience can use REDEEM to create such an ITS from CBT in substantially less time than that reported for other ITS authoring tools ([3]), at around two hours per hour of instruction. REDEEM has been judged to be among the most usable of ITS authoring tools. We argue that for the vast majority of teaching situations where ITSs are not available, minimally adaptive ITSs designed by experienced teachers will be more effective than non-adaptive CBT. We are exploring the validity of this claim by conducting empirical studies in a variety of situations.

In this paper, we discuss the experience of using REDEEM in a context that tests many of our assumptions concerning its design. It reports on how REDEEM has been used by Naval Trainers to create ITSs for use in Naval Colleges and for Naval Reservists. We focus on whether the underlying philosophy of REDEEM's design is appropriate to this context.

2. Authoring with REDEEM

REDEEM involves a number of authoring stages, but each is designed to be relatively straightforward. Essentially authors create a domain model by describing characteristics of the existing material to allow the ITS Shell to sequence and structure it. They supplement the material with additional questions and provide hints to their solution (see Figure 1). Authors then individualize a course to meet the requirements of their specific learners. They classify their students into categories (based upon any dimensions that they like) and then assign different content to these categories. For example, an author could stretch their high performing students by assigning more difficult content or make sure learners with reading problems have limited text to read. Different teaching strategies are created by manipulating dimensional sliders of eight components of instruction (*e.g.* amount of student control, position and amount of question, amount of help, number of attempts at questions). They also identify questions of appropriate styles and difficulty to strategies. These strategies are then assigned to different student categories. Thus, there are essentially two stages

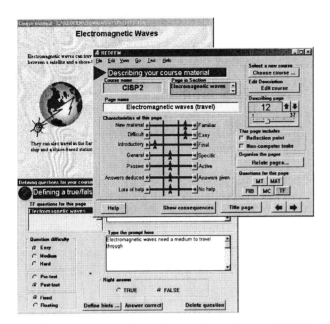
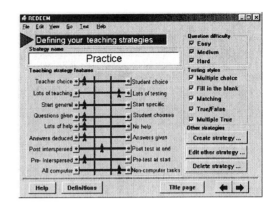

Fig 1. Authoring a Page of CISP and Developing a Teaching Strategy

to REDEEM authoring; in the first stage the domain material is enriched but it remains essentially non-adaptive CBT; in the second stage the CBT is individualized to the perceived needs of learners by macro-adapting the teaching strategies and the content to student categories. This can then dynamically adjust during a student's interaction with the system.

3. Authoring in Real World Contexts

Previous studies with primary and secondary school teachers and teacher-trainers have shown that teachers use the enriching and adaptive features that REDEEM provides to turn CBT into learning environments that are customized to their class. The teachers created a number of student categories and assigned both different content and strategies to these student categories [3,4]. The REDEEM ITSs that resulted from this process were differentiated by perceived student ability and by learning goals such as revision or initial learning.

However, adult training in a Naval context provides a very different set of opportunities and constraints to the school classroom. Positively, the Navy has substantial amounts of legacy CBT. Such software contains much good material, but some of it may have dated and can be very limited in its interactivity. The CBT has been developed over a number of years and hence employs a number of styles. REDEEM imposes a common house style, which minimizes time to learn by students. The legacy CBT rarely allows multiple modes of operation. Such software will have either been designed to support whole class teaching or individual practice in an onshore classroom. But, it cannot adapt to new situations. For example, there is a growing recognition of the need for just-in-time training delivered on ship. Thus, if REDEEM proves appropriate to the Navy's needs, it should allow such software to be reused in a variety of new contexts with minimal redevelopment costs.

Conversely, there are set of constraints imposed by Naval Training that interact with REDEEM's intended operation. One such constraint is an organizational context that precisely specifies the content of CBT during a formal Quality Assurance procedure. Thus, the authors now working with REDEEM are not normally allowed the flexibility to sequence and structure material that the tools provide. Secondly, Naval trainers tend to be SMEs rather than trained teachers. The way that REDEEM decomposes pedagogical processes may not be appropriate to this user group. Thirdly, at least some of the authoring stages with REDEEM are based upon teachers having an intimate knowledge of their students. This is less likely in Naval Training which has many short courses with different trainers being the norm. Finally, there is the fundamental issue of whether Naval trainers want to shift from conventional CBT to the knowledge-based adaptive teaching that systems such as REDEEM provide.

4. REDEEM and Navy Training

The current study involved two authors working independently on the same material; SME1 teaches at a Naval College and SME2 teaches at a Naval Reservist Centre. The course material that was REDEEMed was a seven-chapter course on Communication and Information

Systems Principles (CISP). This consists of ToolBook courses written by the Royal Navy School of Educational and Training Technology. It teaches by presenting declarative material with an occasional question. The courses have limited interactivity and contain high-quality video, animation etc. They are organized in a linear fashion and learner control is very limited. Their content was based upon an original 'chalk and talk' course (C&T), and this analysis of authoring took place as part of a learning outcomes study comparing the effectiveness of C&T, v CISP, v REDEEMed CISP.

To discover how the use of REDEEM is influenced by the constraints described above, we analyzed the REDEEMed courses to discover if Naval trainers use the adaptive features that REDEEM provides. Table 1 summarizes these findings. The most striking first impression is the way that both authors essentially replicate the linear sequence of the underlying CBT. This is in contrast to previous studies where schoolteachers imposed their own views on appropriate course structures for groups of learners [4].

The reasons for this are two-fold. The first is the inherent linear sequence in the original courseware. It was designed with a strong prerequisite structure which is reflected in the resultant REDEEMed courses. Given this rigidity, it made authoring with REDEEM a time-consuming way to return to where the CBT started. This is a strong limitation on the REDEEM philosophy, which suggests that if REDEEM is to be used with such linear courseware in future, we need to create a quicker way for authors to achieve this. Additionally, REDEEM may be most appropriate for material that does not already have a linear narrative. The second reason may be authors' familiarity with CBT rather than ITSs. They were keen to ensure that the course for the most part was delivered according to a specification document and were used to a storyboarding model for learning environment design.

However, there were many ways that the authors did change the courseware. Even with linearly structured content, not all learners need to see the same material. SME2 who trains reservists identified two types of topic in CISP - Naval specific issues and general information about information systems. He used REDEEM to ensure that learners with good general knowledge of the underlying physics and information technology could concentrate solely upon the Naval specific content.

Both authors added significant numbers of additional questions and these were also used for different purposes. For example, SME1 might add a question as a pre-test (i.e. before the material was covered) when he felt that learners would have covered the content in an earlier course or when he wanted them to realize that material about to be covered was complicated and that they needed to pay attention! SME2 added questions to remind learners that they already have the necessary knowledge to understand upcoming complicated materials. Along with the addition of reflection points, hints and feedback, it can be seen that the courseware that resulted from this process is far more interactive than the original CBT and that this interactivity is focused around key pedagogical issues. The authors who have significant experience in teaching on these courses took account of cognitive and motivational characteristics of the learners, unlikely to be known to a centralized software creation team. This can also be seen in the teaching models they created. There is an implicit teaching strategy in the CBT: no student control, questions immediately follow material, no help on questions, and a mix of questions that require multiple attempts or just one attempt. Furthermore, this is true for

Table 1: Comparing REDEEMed Courseware to the CBT

Similarities between REDEEM and CBT	Dissimilarities between REDEEM and CBT
Domain Model Underlying sequences very similar to original CBT Limited use of non-computer tasks No differentiation by content by SME1 **Teaching Model** No differentiation by strategy for SME1	**Domain Model** Flat structure replaced with a hierarchical one that provides summaries of progress and introductions to topics Reflection points to indicate where students should take notes. Significantly more questions added (around a factor of three). Questions were of the same type (e.g. multi-choice, true-false) but were used for different purposes. Contingent hints to some questions provided. **Teaching Model** Both SMEs created their own teaching strategies, which differed to the (implicit) one in the CBT. SME2 authored multiple strategies for different learner needs. Allowed trainers to monitor trainees' performance more successfully (e.g. answers to questions, times etc).

all learners in all situations. The authors in this study were able to impose their own teaching strategies. They tended to agree about no student control, but used questions either before material or as a refresher later as well as immediately post material. They provided help both on error and request and were able to enforce consistency about whether answers to questions should be deduced or given. SME2 also created more than one strategy. He developed a standard strategy for most occasions and an advanced strategy for learners who were revising, which included more student control and more challenging questioning.

5. Conclusion

In considering whether REDEEM is appropriate to the constraints imposed by adult training in the Navy we have found a generally positive response that may have wider implications for other authoring tool use. Most importantly, we asked if Naval Trainers wanted the adaptability that REDEEM provides and we obtained a mixed response. In contrast to the school classroom studies, we have found less individualization to the needs of specific learners and a limited use of the more adaptive features of REDEEM. This may in part be due to the specific CBT that was created with a strong, underlying narrative. It is also likely that this reflects wider organizational goals. All trainees taking CISP at the Naval College are expected to achieve the same outcome (the faultless operation of Information & Communication systems). With such a goal there may be less need or less opportunity for the individualization that REDEEM and other ITSs provide. However, Naval Reservists have differing experiences and objectives. In this situation, REDEEM's adaptive and individualization features were more widely used and seemed more appropriate. This serves as a timely reminder to us not to treat a large organization as if it has monolithic training needs.

Both authors were more concerned to achieve flexibility to adapt to different functions (revision, CAI, just-in-time teaching) rather than flexibility to meet individual students needs and used REDEEM to achieve this in a way that the CBT could not achieve. In addition, their knowledge of previous students taking the course allowed them to author for different proto-typical students, even if they did not have knowledge of the specific learners taking the course.

The selective use of REDEEM's functionality apparent in this study did not undermine the utility of the approach. Both trainers stated that they thought the ITSs they had produced would provide more effective teaching than the underlying CBT, despite their limited course re-sequencing. In this case, it may have been preferable to provide trainers with a partially authored solution, which they could then individualize to their students' needs. In this way REDEEM's decomposition of the teaching process could be retained but different experts could take responsibility for appropriate aspects of ITS creation.

In addition, REDEEM played some positive roles that we had not initially envisaged. The authors used it to create far more interactivity than the original courseware, hopefully enhancing its effectiveness. It also was useful as a rapid prototyping environment. Authors were able to to try out many of their pedagogical decisions, altering a feature and then running the ITS shell to experience the consequences of that decision. Authoring also provided a good way of identifying missing or erroneous information. This suggests that REDEEM could be used earlier in the development of a course.

The REDEEMed courses created by Naval trainers in this study are a compromise between full-blown ITSs and non-adaptive CBT. They maintain the requisite structure of the underlying CBT, but benefit from the local expertise of experienced trainers who enhanced the CBT and then customized it to their different situations. As such REDEEM is one solution to producing successful partnerships between ITSs and creative teachers.

6. Acknowledgements

This research was sponsored by the Office of Naval Research under grant no N000 14-99-1-0777at the ESRC Centre for Research in Development, Instruction and Training. The authors would like to thank CPO Pete Skyrme at HMS Collingwood, CRS JJ O'Neil at HMS Sherwood, and all the team at RNSETT without whose time, support and enthusiastic participation this project would not have been possible.

7. References

[1] K. R. Koedinger, J.R., Anderson, W.H. Hadley & M. A. Mark. Intelligent tutoring goes to school in the big city. *International Journal of Artificial Intelligence in Education*, 1997, pp 30-43.

[2] J. Ward Schofield *Computers and Classroom Culture*. Cambridge University Press: Cambridge, NY, 1995

[3] T. Murray, Authoring intelligent tutoring systems: An analysis of the state of the art. *International Journal of Artificial Intelligence in Education,* 1997, pp 98-129.

[4] S.E Ainsworth, S.K Grimshaw & .J. D. Underwood Teachers as designers: Using REDEEM to create ITSs for the classroom. *Computers & Education*, 1999, pp, 171-188

[5] S.E Ainsworth, J.D. Underwood & S.K Grimshaw Using an ITS Authoring Tool to Explore Educators' Use of Instructional Strategies. In G. Gauthier, C. Frasson & K. VanLehn (Eds.) *Intelligent Tutoring Systems*, 2000, pp 182-191

Adaptive Presentation of Evolving Information Using XML

Mária Bieliková

Slovak University of Technology, Dept. of Computer Science and Engineering
Ilkovičova 3, 812 19 Bratislava, Slovakia
bielik@elf.stuba.sk

Abstract

In this paper we describe the proposed framework for representation and presentation of evolving information. As an example we use administrative information about an educational module. Evolution is expressed by a set of scenarios and is used for adaptive presentation of information represented as hypermedia. Representation of meta-information is based on the XML (eXtensible Markup Language). We experimented with the proposed approach and implemented a software prototype, which enables the time view and the user view of the educational module information. All presented data are stored in the database. Each element has associated constraints at the level of presentation, intended group of readers and presentation time.

1. Introduction

In this paper we deal with a problem of adaptive presentation of evolving information represented as hypermedia. Our attention is devoted to presentation of information about an educational module, its objectives, reader, lab requirements, schedule of lectures, conditions to obtain a grade, etc. This kind of information is necessary for any kind of education delivery. Usually it is presented on the paper (on whiteboards), or electronically in the form of non-structured documents, or as a hypertext document on the Web (represented in the HTML).

Methods and techniques of adaptive hypermedia are largely designed and evaluated within educational environments. They often focus on the educational materials (e.g., adaptive hypermedia books). Some of them also incorporate administrative information [1], [7], but use the same presentation techniques as for educational materials.

The content of administrative module information is usually solid. The problem is that each overlooked word can cause a student later problems to fulfil requirements to obtain a grade. Although reading all available information can be useful in some cases, we have experience that for the most students it is helpful to see details when they are current (i.e., needed).

Selected example of administrative module information presentation enables discussion related to the specific requirements for such application. Proposed framework can be used for other application domains, which produce evolving documents and this evolution is important for the effective presentation.

Moreover, we exploit the fact that presented information is used repeatedly. This feature is specific for educational environments, where modules run several times (e.g., in subsequent years). Though the context of a particular module changes each time, some presentation patterns can be found and reused.

2. Context-Based Presentation

An important feature of effective delivery of administrative module information is the adaptivity based on the current presentation context. The adaptive presentation is one of the generally recognized ways of adaptation in adaptive hypermedia [6]. Irrelevant information and links that overload the user's memory and screen are hidden, labeled as not recommended, or removed. Often additional information related to recommendations for a reader is presented by different color [8].

Adaptive application uses information about the user (a *user model*) to deliver the content, which is the most interesting for the user (e.g., his knowledge about a particular domain, preferences, previous visits). Adaptive application should define also a *domain model*, describing how the information content of the application is structured ([5]). In adaptive applications with evolving information also a *presentation context* is important. For example, desirable administrative information depends more on time of reading rather than on a complex user model, which represents the user's knowledge related to the information content. Often a static *user profile* provides sufficient representation framework. Adaptation based on the current presentation context would improve delivering information to a reader (e.g., a student).

The presentation context is defined as a collection of data, which depend on the current state of a presentation. The context model does not depend on the user model. However, both models relate concepts from the domain model in order to define an *adaptation model*. For the purposes of administrative module information delivery, the presentation context can be reduced to considering time of information reading. However, the context model can represent also data important for adaptive presentation in general. Example of such data is browser or language used. Recognizing the language is interesting in the case of a multi-lingual educational module, which is often the case in non-English speaking countries. In this case, we need several variants of the presented document and have to deal with the problem of hypermedia document version management (see for example [2]).

Keeping modifications introduced to the pages is also important feature of a good educational module information presentation system. In traditional learning environments, the modification of the module material is not frequent and usually not interesting for a student (student receives the package with learning material, and modifications are usually incorporated into a new version of learning material used in the next module run). Modification is frequent and important in presentation of organizational aspects of the module. Visibility of a single modification is related to both user model and presentation context.

3. Example of Adaptation

For adaptive presentation the granularity of the information content is important. In the case of administrative information about an educational module, a fragment is defined as a text paragraph (or sequence of several subsequent paragraphs) together with embedded media (e.g., pictures or video sequences). Another important entity is a hyperlink, which should be manipulated separately. Each of identified fragments (a sequence of text paragraphs, or a link) is described by several attributes, which state fragment's visibility according the user and context models.

In our example we consider a simple user model based only on a specification of several groups of users. The user model does not evolve, i.e. the system is not personalized. Issues related to the user modeling aspects according to adaptive presentation or adaptive navigation are widely discussed in the literature (e.g., [5], [7], [8]) and can be used for extension of the proposed approach to the adaptation.

The presentation context is expressed by the current time. The information content is structured to fragments and fragments are grouped according an evolution scenario. The evolution scenario defines steps of presentation according the progress in the course of a module run. It represents a kind of pattern for information delivering during some period of time. The scenario can be defined in a week term for traditional course delivery. In the case of distance education, a user's achieved state is more relevant criterion for moving to the next step in the scenario.

Figure 1 illustrates part of the appearance of presented information for the Knowledge-Based Systems module (a master course at the Slovak University of Technology). Each page starts with a generated header, which contains text defined in the first element of the fragments' hierarchy and the current date. Header is automatically included in each page. The page content is generated according to defined domain model, which is the XML-based document. Each page contains a frame on the right hand side, which serves as a content navigation within a current page. Different data formatting can be expressed by a style sheet. Alternative style sheets can be applied at any time, changing the format to suite intended audience (e.g., format of the university web pages) and the capabilities of the used publishing media.

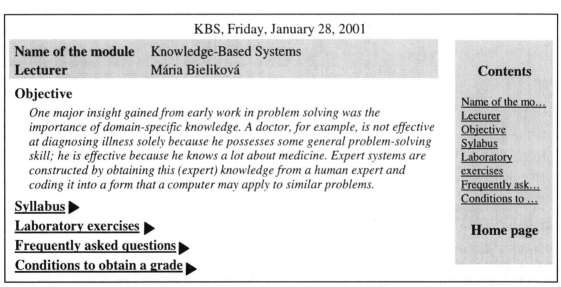

Figure 1. Example of module information presentation.

Presentation is adapted both on the content and navigation according the user profile and the context model. In order to reduce the amount of information displayed at once, the system can create for each fragment a link navigating to information content represented by a particular fragment, or split the content into several parts presented independently and reached by the hypertext links.

4. Data Representation

The module information is represented by the XML (eXtensible Markup Language). The XML is a meta-language, which describes a data format suitable for storing structured and semistructured text. The XML-based document is logically structured into named units and sub-units called elements.

5.1. Content

Each element represents a fragment in a domain model. It has defined adaptation conditions in the form of required values of defined attributes. Elements can form an arbitrary hierarchy and can be named by the user (who is responsible for the module development). Presentation of information copies this hierarchy. In order to present 'pages' with reasonable length, some elements with long text and/or many offsprings can be displayed by links.

Granularity of a document (the degree to which element's content is organized into offspring elements) is determined by a developer. The presentation method only requires declaring conditions (as predefined attributes of elements) and the content of elements. Some elements however, do not have content in the form of text – they just name the hierarchy. This kind of elements is useful in case of need to split the presentation into several pages.

Each educational module has its own Document Type Definition (DTD), which defines the elements allowed in a particular document of the module type. Several modules can use predefined common DTD.

Bellow is an example of part of DTD declaration:

```
<!DOCTYPE Module [
  <!ELEMENT Module (Header, Body, Foot)>
  <!ELEMENT Header (Module-name,
                    Module-Abbreviation?,
                    Study-Type?)>
  <!ELEMENT Module-name (#PCDATA)>
  ...
  <!ELEMENT Body (Info?, Objectives,
                  Syllabus, Labs, Faq?)>
  ...
  <!ELEMENT Foot (Modification-date,
                  Administrator-name?)>
  ... ]>
```

A major decision in producing adaptive hypermedia systems is the decision about structuring information in such a way that it will be possible to adapt presentation. In our approach, the XML element according to the module data type definition is the smallest information chunk that can be adapted. However, the structure of the module information and the structure of the content to be adapted and presented can be of different granularity. Developer can define tags, which reflect the structure of presented information (e.g., paragraph tag <p>), but are not related to the structure of the module information.

Information related to a new module can be developed by the use of predefined templates or from the scratch.

5.2. Meta-data

Actual data related to the particular module are represented by the XML-based document according to its DTD. Each element contains several conditions, which define the model of presentation according to a given context. Attributes and their values within each element represent conditions. We have experimented with attributes related to the type of a user, granularity of presentation, time and splitting of the presented information to several pages. Set of available attribute-value pairs can be enhanced. In this case, their interpretation should also be defined.

Bellow is an example of meta-data related to the `syllabus` element:

```
<syllabus description="Syllabus"
 from="01/01/2001" to=""
 user_operator=">=" user="guest"
 mode_operator="=" mode="medium"
 visible="yes" link="yes" include="no">
  <B>Background:</B>
  A brief history of expert systems;
  expert system concept;
  ...
</syllabus>
```

All meta-data defined in the `syllabus` element are evaluated according to the current state. If date of presentation is after the 1st of January 2001, this element is presented to any user (*user >= guest*) who have selected the medium granularity (*mode = medium*). When the `syllabus` element is presented, the text "Syllabus" is displayed. This text is marked as link (*link = yes*), which results in hiding the actual text of syllabus into the separate page (reachable through this link, see Figure 1). When the `link` attribute is set to "no", all text defined for this element is displayed.

6. Adaptive Presentation

We use the conditional technique for the adaptation [6]. With this technique, all information about a module is linked to the condition at the level of a user or context model. When presenting information in the particular context and for the particular user, the system presents only information with the condition evaluated to be true. A simple example is hiding detailed information about the lab exercises in the first week of a semester. The purpose of hiding this information is not to disable view of the information, but prevent information overload at the beginning of the semester. Solution to this requirement is to disable some information in short or medium views in the first week, but enable it in the full view. Later, when information is current, it is presented also in short or medium views. When the presentation of the module information is required, attributes of each element are evaluated and compared against the current context and adapted module information is displayed.

We developed a software prototype for adaptive presentation of administrative module information [3]. The software prototype uses the following data about the current state: group of the users, granularity of presentation and context represented by the current time.

We also developed a tool for supporting the design of the adaptive hyperdocument, including definition of fragments, conditions of their adaptation and the content. A developer can create and modify a hierarchy of elements and define meta-data and the content of each fragment.

Design and implementation of a prototype was oriented mainly toward prototyping the proposed framework for adaptive presentation of the module information. However, the effective using of such system in the Web environment requires effective storing and manipulation of the data about several modules, effective storing and manipulation of the data about users and some level of security (in particular at the level of authentication for modification of the XML documents related to the particular module). Due to the aims of the prototype development, the problem of security was not solved. Data manipulation efficiency is tackled by the design and implementation of a relational database, which serves for storing the whole content of data elements together with the data about users and developers. We used the database server MySQL.

7. Conclusions

In this paper we discussed possibilities of improving current state in presentation of evolving information. As an example we used administrative information related to an educational module.

A domain model in our system is represented as a hierarchy of fragments. It is sufficient for this specific application area. However, in the case, where more complex relations between fragments are required, advanced representation be used ([1], or [7]).

We use the conditional content adaptation technique for adaptation, similarly like the AHA system [4]. Unlike the AHA system, where conditionals are encoded in the structured HTML comments, we represent conditions as the attribute-value pairs in the XML. Conditions are expressed declaratively as values of attributes. They are interpreted in our prototype by the PHP3 scripts.

An evolution scenario can be defined manually or semiautomatically by developing a software agent watching on the first manual adaptation of presented information. We are working now on a development of such a software agent.

Acknowledgements

The work reported here was partially supported by Slovak Science Grant Agency, grant No. G1/7611/20.

References

[1] L. Aroyo, and D. Dicheva. Domain and User Knowledge in a Web-based Courseware Engineering Course. *Knowledge-Based Software Engineering*, T. Hruška, M. Hashimoto (Eds.), IOS Press Amsterdam, 2000, pp. 293-300.

[2] M. Bieliková, and P. Návrat. Modelling Versioned Hypertext Documents. *SCM-8 Symposium*, B. Magnusson (Ed.), Springer Berlin, 1998, pp. 188-197.

[3] M. Bieliková. Adaptive Presentation of Educational Module Information using XML. *Knowledge-Based Software Engineering*, T. Hruška, M. Hashimoto (Eds.), IOS Press Amsterdam, 2000, pp. 301-306.

[4] P. De Bra, and L. Calvi. AHA: A Generic Adaptive Hypermedia System. *Adaptive Hypertext and Hypermedia*, Pittsburgh USA, 1998, pp.5-11.

[5] P. De Bra, G.J. Houben, and H. Wu. AHAM: A Dexter-based Reference Model for Adaptive Hypermedia. *ACM Hypertext'99*, Darmstadt, Germany, 1999, pp. 147-156.

[6] P. Brusilovsky. Methods and Techniques of Adaptive Hypermedia. *User Modeling and User Adapted Interaction*. Vol. 6, No. 2-3, 1996, pp.87-129.

[7] N. Henze, and W. Nejdl. Extensible Adaptive Hypermedia Courseware: Integrating Different Courses and Web Material. *Adaptive Hypermedia and Adaptive Web-Based Systems*. P. Brusilovsky et. al. (Eds.), Springer, 2000, pp.109-120.

[8] G. Weber, and M. Specht. User modeling and Adaptive Navigation Support in WWW-Based Tutoring Systems. *User Modeling*, A. Jameson, C. Paris, C. Tasso (Eds.), Springer, 1997, pp. 289-300.

A Collaborative Courseware Generating System Based on WebDAV, XML, and JSP

Changtao Qu
Inst. of Computer Engineering
University of Hannover
Appelstr. 4, 30167, Hannover
Germany
qu@kbs.uni-hannover.de

Johann Gamper
Faculty of Computer Science
Free University of Bozen
Mustergasse 4, 39100, Bozen,
Italy
jgamper@unibz.it

Wolfgang Nejdl
Inst. of Computer Engineering
University of Hannover
Appelstr. 4, 30167, Hannover
Germany
nejdl@kbs.uni-hannover.de

Abstract

Today, with the use of global cooperation in education, more and more courses are given directly on the Web in a collaborative manner among geographically dispersed universities. Consequently, Web-based courseware generation is becoming increasingly complicated as this cooperation also introduces additional challenges for the courseware authoring and publishing process. In this paper, we present a courseware generating system which focuses on facilitating the courseware generating process taking advantage of recent Internet protocols and industry standards. We adopt a collaboration-friendly Internet protocol, WebDAV, to support collaborative courseware authoring, XML to represent meta-data of course contents, and JSP to realize dynamic courseware presentation. With its simple syntax, XML can on the one hand simplify the courseware authoring process in company with WebDAV, on the other hand, as a neutral meta-language, it can also separate course contents from courseware presentation in company with JSP.

1. Introduction

Since the summer semester 1999, the CS1 course "Introduction to Java Programming" (Info-1 for short) has been given on the Web in three German universities and one university in Italy. In this paper we will describe the latest incarnation of our courseware system used for the joint course Hannover/Bozen, which has been developed within the Virtual Campus Project and the cooperation with Bozen University. While such sort of Web-based distance education is on the one hand reducing the teaching costs, increasing teaching opportunities, as well as improving the collaboration among the partner universities, it raises additional challenges when we design the Web-based courseware in a collaborative manner among geographically dispersed lecturers.

First, we have to find an efficient mechanism in order to support collaborative courseware authoring.

The authoring of our joint Info-1 courseware is performed by different lecturers situated at Hannover and Bozen. To reflect the latest technology developments, we have to continually adjust and revise the courseware over the whole semester. In this process, an efficient mechanism is urgently needed to facilitate, as well as secure the collaborative courseware authoring process, e.g., facilitating opinion exchange among lecturers, preventing "overwriting" on each other's work, etc. According to our experience, the use of existing simple tools designed for supporting collaborative work, e.g., FTP, e-mail notification, etc., is difficult and cumbersome, while version management systems, which are providing required functionalities, require new client side programs and new commands, which distract the lecturers from the main task of actually writing the courseware, especially when the authoring process occurs over the whole semester. By adopting the collaboration-friendly Internet protocol WebDAV (Web-based Distributed Authoring and Versioning)[1] we have found a way to facilitate courseware authoring.

Second, we have to find an efficient mechanism in order to separate course contents from courseware presentation.

Because the course contents is continually changing during the teaching process, we need a courseware publishing engine which is cleanly separated from course contents and is able to immediately reflect any modifications of course contents. By adopting XML (eXtensible Markup Language) as the standard data interface, as well as JSP (JavaServer Pages) as the core of the publishing engine, we can easily separate course contents from dynamic courseware presentation.

Finally, we have to choose appropriate technologies in order to shorten the development process and test-cycle.

Courseware generating should be a "building" process rather than a repetitive "programming" process. By adopting some standard and "reusable" technologies, particularly XML Data-Binding Specification: JSR 031 [3] and JSP tag libraries, we have achieved above design goal.

2. Collaborative courseware authoring based on WebDAV and XML

Our courseware generating system consists of a courseware authoring module and a courseware publishing engine. The courseware authoring module comprises a courseware repository which stores course script files, and an XML file which represents the courseware structure. The latter, which is restricted by the pre-defined DTD, serves also as the standard data interface between authoring module and publishing engine. Any courseware validated by the DTD can be directly rendered by the publishing engine without the need of any reconfiguration.

The courseware authors can access the courseware repository and the courseware structure file via WebDAV protocol using a WebDAV-enabled authoring tool, e.g., Microsoft FrontPage 2000, Office 2000, or Adobe GoLive 5.0, etc.

Taking advantage of WebDAV, the lecturers can "in-place" (directly on the remote server) implement most activities needed for collaborative courseware authoring, e.g., editing course script files stored in the courseware repository, manipulating the repository's namespace, utilizing locking mechanism to prevent "overwriting", or manipulating properties of a specific course script file in order to exchange ideas and opinions among lecturers [2]. Since all WebDAV-enabled tools are aware of WebDAV's methods (Propfind, Lock, etc), they can "in-place" handle all "format-compliant" course script files without the need of explicit download and upload.

3. Courseware publishing based on JSP

The courseware publishing process has to go through three steps. First, the XML-based courseware structure has to be translated into corresponding Java objects (JavaBeans) utilizing JSR 031 [3]. Second, based on the generated JavaBeans, JSP tag libraries, which encapsulate commonly used JSP presentation templates, need to be developed. Finally, JSP tag libraries are further used to construct JSPs which are then directly accessed by Web browsers. In Fig.1. we illustrate the courseware publishing process.

Actually, the first and second step are the key to produce dynamic and reusable courseware presentation. JSR 031 is a Java Specification Request proposed by Sun Microsystems Company for defining an XML data-binding facility for the Java platform. Our purpose for applying JSR 031 is mainly to automate the Java XML data-binding process and to reduce programming work. In fact, together with pre-defined DTD, the JavaBeans generated according to JSR 031 can essentially ensure the reusability of JSP tag libraries and JSPs.

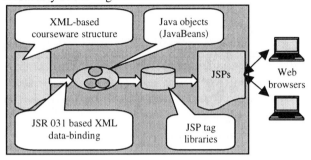

Fig. 1. Process of courseware publishing

Furthermore, as we can see in Fig. 1., JSPs applied in Info-1 are not directly constructed utilizing JavaBeans, but JSP tag libraries. Although JavaBeans plus Java in-line code can be directly applied in JSPs for realizing dynamic courseware presentation, we have to use relatively more Java code in JSPs to control presentation logic because we want to generate multi-views of Info-1 based on the same course contents. By utilizing JSP tag libraries, we can reduce the necessity to embed large amounts of Java code in JSPs and make the courseware publishing engine more robust and reusable

4. System implementation

The whole system has been built on Windows 2000, but also runs on Solaris. The standard Web server is Apache Web Server 1.3.14. The basic capabilities of WebDAV are implemented by Apache mod_dav 1.0.2 [4]. The JSP support is provided by Apache Tomcat 3.2.1, which is installed as an add-on to the Apache Web Server. In addition, the Apache XML parser Xerces 1.2.3 is installed to handle XML syntax.

5. References

[1] Goland, Y. Y., E. J. Whitehead, A. Faizi, S. Carter, and D. Jensen, "HTTP Extensions for Distributed Authoring-WEBDAV," RFC 2518, Feb. 1999.

[2] Qu, Ch., T. Engel, and Ch. Meinel, "Implementation of a WebDAV-based Collaborative Distance Learning Environment", in *Proc. of ACM SIGUCCS Fall 2000*, Virginia, USA, Oct. 2000.

[3] Reinhold, M., "An XML Data-Binding facility for the Java Platform", Available at: http://java.sun.com/xml/docs/bind.pdf

[4] Stein, G., "mod_dav: a DAV module for Apache", Available at: http://www.webdav.org/mod_dav/

Use of RDF for Content Re-purposing on the ARKive Project

Andy Dingley
HP Laboratories,
Bristol

andy_dingley@hpl.hp.com

Paul Shabajee
Graduate School of Education & Institute
for Learning and Research Technology,
University of Bristol
paul.shabajee@bristol.ac.uk

Abstract

This paper reports on prototype systems to provide an infrastructure for the dynamic and flexible re-purposing, of multimedia resources held in a large database. The database, called ARKive, holds film, stills, audio and text about globally endangered and native UK animal and plant species as well as their habitats. It aims to offer a wide range of users customised access to both the core multimedia data, and full integration of core data with external educational resources.

Aspects covered in the paper include; designing for re-purposing with respect to specific audiences, storage and querying using RDF, XSL, SMIL and related technologies. The advantages of the approaches taken are discussed and key issues are highlighted.

1. The ARKive Project

ARKive [1] is a Wildscreen Trust [2] initiative to build a Web-based digital archive of all UK, and the world's endangered, animals and plant species. The project brings together thousands of films, videos, sounds and photographs of threatened and recently extinct species.

HP Laboratories [3] are supporting ARKive by funding a research team to develop the technical infrastructure, including the content re-purposing architecture.

The effective and efficient educational use of these materials is fundamental to the aims of the project. This paper describes the on-going work to re-develop the existing ARKive website, so as to offer content tailored to the dynamic needs of multiple educational user groups.

1.1. ARKive's Requirements

The publishing aspect of ARKive needs to query the large content database (6000+ species) and generate an appropriate set of content for each use. This content is then presented accordingly, as HTML, streaming video etc. Additional metadata may be attached to this presentation, describing its content, target audience etc.

There are additional re-purposing requirements on this publishing, such that:

- Valuable items of content may be re-used in different contexts; i.e. a habitat image may be accessible from the descriptions of many different species, the habitat or a geographical location. These items may be either narrative descriptions, or media assets.

- Content may be targeted to an audience (public, schoolchildren, pre-school, scientific researchers)

- The target age group and language skills may influence the targeting of content, independent of the content's complexity.

- Descriptive metadata is stored with the content, and attached on publication. Appropriate metadata may be created dynamically, i.e. a PICS rating indicating any gory content.

- The delivery format of content (e.g. streaming video) may be targeted to the browser device's capabilities.

The current ARKive site is based heavily on the "Species Page", a pre-packaged assembly of content that describes a single plant or animal. The text content may be varied for varying audience age groups, but the basic set of content is fixed. There is no access to this content on another axis (e.g. by searching for habitats) and content items are not re-used between pages.

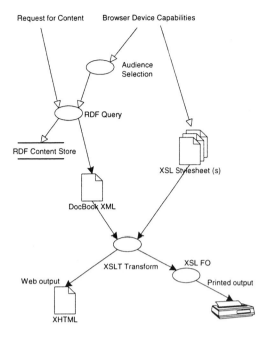

2. A Design for Re-purposing

To encourage re-purposing of content, ARKive's content, metadata and rich-media asset stores are to be completely re-engineered. The old design was traditional, based on SQL and a rigid data model of the "biodiversity" knowledge model, with media assets stored in the filesystem.

The research purpose of re-purposing here is to extend the audiences and the usage models of this content. It includes re-purposing to different device or browser types, but this is just a small and previously well-studied component.

Figure 1   Publishing data flow

The redesign takes a much more generic approach to storing content. Media assets will still be stored separately, because of their large size and a need for accessibility by time-code. The remainder of the content is to be treated as a flattened landscape of "content units", each carrying substantial metadata to identify and describe them. This metadata allows the units to be joined into an arbitrary, although broadly tree-shaped, graph. Some units represent leaf nodes; links to media assets, or blocks of descriptive text. Others are more structural; the set of description sections for a species page, or a list of species inhabiting a habitat.

Content units are represented and interlinked by use of RDF (Resource Description Framework) [5]. Their structure will be described by one of the related schema tools; RDF Schema or DAML+OIL [7].

2.1. Publishing

Most of ARKive's publishing is either to the Web (as XHTML), or to local devices such as printers, but still using web-based technology. All can be performed by a single publishing engine, based on RDF and XSLT.

The content to be delivered can then be selected according to the format requested, and the target audience. Typical formats might be a species page, a video of a behavioural ethogram, a habitat description, or an arbitrary collection, e.g. "Big Cats of Africa". The target audience is characterised by learning context, content-target age, "reading age" and preferred language. Identification is by standard web techniques, such as an explicit URL or navigation to a children's section, supplemented by use of cookies.

Using the request and the audience target, the data store is queried and returns an RDF document. This might represent, "A Description of the Golden Toad for Schoolchildren aged 7-11". It is audience and subject specific, but still device independent. Embedded within it are references to all the relevant media assets, including their multiple formats and image sizes.

An appropriate XSLT stylesheet can then be selected, according to the browser device, and used to transform the document into XHTML. The stylesheet, is parameterised somewhat to hint at device screen size and ability to play streaming video, so that appropriate image sizes of the media assets may be selected for a reasonable size on screen. Streaming media is in both Real and Windows Media formats, and these too may be selected according to the device capabilities.

Current browser devices are categorised simply as "HTML 4 / CSS capable" and "HTML 3.2 without CSS", although this could be developed in the future to include WAP/WML devices or SMIL.

Where a browser is capable of client-side XSLT transforms (currently only Internet Explorer 5), the stylesheet and the document are supplied directly to it, for a local transform. As 70% of the audience are now using this browser, this represents a substantial saving in server processing load. Overall, the processing cost of XSLT publishing is no more than that of traditional page generation.

High quality printed output may be available locally, by following the XSLT transform with an XSL FO (XSL Formatting Objects) processor like Cocoon [9].

2.2. PICS Ratings

Some content may be excessively gory or sexually explicit for some age groups. This may be indicated using a PICS rating scheme. PICS filtering is used within the site, to suppress this content at source. A default PICS rule is implied by each targeted age group. Squeamish users may also choose to further restrict this.

2.3. Implementation with RDF

It is planned that the prototype will hold all content as RDF, according to a representation described in RDF Schema. Large media assets are held as references into a media asset store, rather than directly in RDF. Within these content units, the content will be held according to an XML Schema; DocBook or XHTML for text formatting, and MPEG-7 for structural description of video. To aid audience targeting, content units are also marked-up according to the LOM (Learning Objects Metadata, [10]) educational schema.

ARKive describes the biodiversity information of its species according to a simple 9 topic model (Locomotion, Reproduction, Habitat etc.), with varying sub-topics according to one of 10 quite broad schema classes (Mammals, Vascular Plants, etc.). . These techniques are still quite new, and so the trade-off of broad but stable classes was chosen, rather than a schema with a more refined structure. This would have required many more classes and a need to revise them frequently as more species were added.

The current RDF data store is a prototype, built above SQL Server 7.0. Much research remains to be done in this area, particularly in efficiently supporting large data sets. This implementation is far from complete.

2.4. Dynamic Queries

Once ARKive's content is held in a large, flat, content store with adequate metadata, this then becomes a rich territory for providing query facilities beyond the simple pre-structured species page. Instead of displaying several pages, all with "plumage" descriptions, it should be possible to query for "The appearance of all birds with blue wings". ARKive's metadata will be rich enough to support this. The query for "blue plumage" may be mapped to an RDF query for all species which have plumage, for the plumage colour being blue, and to return only the appearance description of these species. This can then return an RDF document with the matching species and appearance nodes, and their attached text. From this point on, publishing will be identical to that for the species page. Because the publishing mechanism has no dependency on the content's structure (beyond the XML schema), there is no need to rework it for content with different structure.

2.5. Audience Targeting vs. Device Targeting

The ARKive publishing process makes a distinction between audience and purpose-based filtering before the content document is produced, and device-based filtering that happens afterwards. This distinction appears valid, particularly where some transformations, especially those that are device-dependent, may be effected at the client

3. Video

Central to ARKive's species holding is the "ethogram", a behavioural record of a species. Part of this will be approximately 10 minutes of video footage (for species where such exists). Like the text, the video is also intended to be audience-targeted. This can change the content included, and will allow a subtitled or narrated version of the video to be offered, improving accessibility, especially for younger users.

3.1. SMIL

A future direction for ARKive is in the use of SMIL (Synchronised Multimedia Integration Language) [11], an XML language for authoring multimedia. Like HTML, SMIL is supplied to the client's player device, but rather than rendering a static page from text and images, SMIL renders a presentation from video, audio, text or other media types.

The intention is for SMIL to fulfil a similar role within ARKive's video as DocBook or HTML does for the static pages. In the current process, an audience-targeted video with subtitles or narration requires expensive post-production. This may be done instead, and very much more cheaply, by including a subtitle element or an audio narration track in the SMIL document and still referencing the same video footage.

SMIL also opens up the possibility of custom-assembled video as a response to dynamic queries, serving up a composite presentation of those clips matching the search.

4. Metadata Publishing

There is a requirement to publish metadata, attached to the ARKive content. Although this is based on the same content store, the *publishing of* metadata is quite distinct to ARKive's main task, *dynamic publishing of content by the use of metadata.*

Metadata publishing addresses two audiences; the commonplace public web search engines (e.g. Google) and more focussed metadata clients, capable of a deeper understanding.

At present (April 2001) there are few effective "harvesters" of such metadata. In one view of the future [12], clients will use high quality metadata to give, "a Semantic Web in which machine reasoning will be ubiquitous and devastatingly powerful". In the meantime, the only interestingly capable metadata consumers are in the educational field.

4.1. Schema Translation

Publishing of ARKive's metadata to the LOM schema will be simple, as that is its native format. Translating to a lowest common denominator (e.g. Dublin Core) is also simple, as this may simply be hard-coded.

Where other schemas must be supported (typically other educational schemas), research is in progress to develop automatic translation mechanisms, following the lead of [8]. This work is based on describing both schemas in a common format, then manually describing an ontological mapping between their elements. This mapping is then used to automatically generate an XSLT stylesheet, which may then perform the translation mechanically.

Obvious areas of future study are in the description language for this mapping, its capabilities for transformation, and in reducing the human workload to author it.

5. Integration with External Education Resources

The system will allow the dynamic creation of teaching and learning resources integrated with materials external to the core species and habitat data by combining the external resources with the core data (e.g. images, text, video clips) via the querying and publishing system described above. These materials can then be targeted to particular user groups and can potentially offer end users the ability to create their own resources drawing on the ARKive data

6. On-line 'Museums'

The core technology developed within the ARKive project is a powerful solution to this project's requirements, but is also capable of being applied to any on-line museum or information repository. The use of a semantically agnostic data store, rather than a hard-coded database structure means that the same code-base can be applied to any site with similar requirements, not just natural history.

7. Conclusions

The use of an RDF content store instead of a rigid SQL data model aids the production of a system that supports re-purposing, rather than just one single rigid view of the data. This re-purposing addresses new audiences and usage models, rather than the device re-purposing that is already commonplace.

Infrastructure tools for handling RDF are as yet immature. Production-grade large-volume content stores and schema translation tools are obvious areas in need of development.

XML and XSL based publishing is already a mature and stable technology, with much to offer the publishers of complex and varied content. Techniques for device-based targeting can address problems of supplying content to a population of widely varied capabilities.

Metadata consumers are still rare. Those that do try to harvest metadata (web search engines) have a naïve view of it and are more confused by sophisticated schemas than they are aided by them. Our expectation is that this will improve, especially if the Semantic Web concept becomes popular.

8. References

[1] ARKive project, http://www.arkive.org.uk

[2] Wildscreen Trust, http://www.wildscreen.org.uk

[3] HP Laboratories, http://www.hpl.hp.com/arkive/

[4] BBC Wild, http://www.bbcwild.com

[5] Resource Description Framework (RDF), http://www.w3.org/RDF/

[6] Institute for Learning and Research Technology, http://www.ilrt.bris.ac.uk

[7] DAML+OIL, http://www.daml.org/

[8] Combining RDF and XML Schemas to Enhance Interoperability Between Metadata Application Profiles, Hunter J. & Lagoze C., WWW10, Hong Kong, May 2001, http://archive.dstc.edu.au/RDU/staff/jane-hunter/www10/paper.html

[9] Cocoon, http://xml.apache.org/cocoon/

[10] Learning Objects Metadata (LOM), http://grouper.ieee.org/p1484/ltscdocs/wgc/LOMdoc2_5a.doc/

[11] Synchronized Multimedia Integration Language (SMIL), http://www.w3.org/AudioVideo/

[12] T. Berners-Lee, "Semantic Web Road map", September 1998, http://www.w3.org/DesignIssues/Semantic.html

From CD-ROM to Web-Served:
reverse engineering of an interactive Multimedia Course

Daniele Maraschi
LIRMM, Montpellier, France
maraschi@lirmm.fr

Stefano A. Cerri
LIRMM, Montpellier, France
cerri@lirmm.fr

Gianna Martinengo
Didael SpA, Milano, Italy
martinengo@didael.it

Abstract

The Web offers advantages for Learning Technologies, but the interactivity necessary for the learner requires technical features that are currently hardly supported. In the paper, we briefly describe how a traditional Multimedia Course, usually marketed as off-line CD-ROMs, has been reverse engineered to a server-based, multi-client Web application in order to have an identical behaviour at the learner's site. The new Course consists of XML/XSL documents programmed by server-driven Java functionalities managing the conversations with the remote learners. Abstraction and generalisation to other Courses and/or interactive Web applications or services, become now feasible, easy and re-usable, because all middleware technologies are Open Source.

1. Introduction

Advanced Learning Technologies (ALTs) today are those that support Learning by stimulating it through Dialogues with Systems[1]. These may range from the simplest, locally loaded program, to Web applications where Artificial and Human Agents collaborate to animate the Dialogues.

A wide range of architectures may be conceived for Systems adequate to support Learning: from Tutoring Systems to Learning Environments; each of these may be based on a client-centred or server-centred approach. In this paper we consider an existing available Course, designed, developed and marketed since years on CD-ROMs; we present why it is convenient to transform it into a dynamically generated, interactive, multimedia, server-centred Web application and we summarise how we did the porting to the Web.

2. A conversational view of the Web in ALT

The Web as an interactive, multimedia Communication medium is currently hindered by a major limit: its behaviour is mainly page-centred as HTTP is a stateless protocol. In interactive Web applications, such as those that support Human Learning, this limit has to be overcome.

In recent papers [e.g.: 2] a serious approach to advocate a computation-centric view of servers (in opposition to the usual page-centric) has shown that continuations may model multiple conversations between users and a Web server preserving the state of the conversation by means of an extremely simple mechanism of continuations like the one available in the Scheme programming language. We agree with this view, so deeply that we have proposed a perfectly complementary model - called STROBE - for modelling Agent to Agent Conversations [3-5]. However, Lisp or Scheme are not widely used applicative languages, even if we believe they are best adequate for Web applications [6].

Looking at the literature, the porting to the Web of available Intelligent Tutoring Systems (ITS) has been reported in cases where a LISP-HTTP server has been used, compatible with the Lisp code of the ITSs [7-8]. Surveys about Web-Based Education do not present other examples of Web-Generated Courses at the moment [9]. On the contrary, a rapidly growing number of examples of Web-Generated Interactive Web Sites are currently reported [10].

3. How we did the porting to the Web

In order to publish XML documents on the Web without any kind of plugin, we have chosen a servlet-based Java framework, Cocoon [10]. In such a way, the two steps of a) parsing the XML content and b) interpreting the XSL layout in order to obtain HTML pages, are executed on the Web server each time the page is requested by the user.

The use of Java servlets allows to manage user sessions in a simple yet efficient fashion. This management requires the persistence of at least one data structure, modelling the user state during the whole period of interaction between the user and the Web application.

At the moment the Cocoon framework allows to program both XML and XSL documents using the Java programming language. Starting from three CD-ROMs containing a Toolbook application, generating a proprietary Multimedia Course to learn the Italian Language, we have classified all the possible types of pages (the menus, the presentations including multimedia components and the exercises) and, for each of them, we have built the corresponding XML structure along with a series of XML/XSL documents apt to generate the output layout of the XML content. The new source code amounts to about 750Mbytes.

We encountered two types of obstacles, each corresponding to a class of problems: conceptual and engineering problems. The conceptual ones concern the reverse design of an XML structure reflecting «potentially» the results of a Toolbook application.

We could have adopted here: a) the approach of redesigning the XML tree structure with our own mental model of how, later, the XSL and Java code would have had to be developed in order to have the final HTML pages, or b) the approach of maintaining the «semantic» structure underlying the Toolbook application, so that the final Web product would have exactly the same appearance as the original CD-ROM based Course. We have preferred the second option for market reasons. Since the CD-ROM Course is a success, clients are used to it, so that in this first experience we preferred to limit as much as possible the modifications on the interface.

The engineering problems concerned the acquisition and mastery of the relevant languages and tools and their integration. This was not really beyond our scope, as we are currently developing an integrated Environment for both Agent architectures and Web Applications [11]. Among the interesting engineering challenges, we implemented a Java-based protection functionality that allows specific XML files to be decrypted-encrypted on the fly at runtime.

4. Foreseen developments and conclusions

The product will be delivered in the next weeks.[1]

The evident advantages of the porting in terms of reuse of methods and tools in the design and development phases are clear from the separation, in the reported architecture, of the concerns regarding the document structure (XML), the page layout (XSL) and the user interaction (Java). Once a Course is published on the Web, it is easily conceivable that other features can be integrated, such as those offered by the ATHENA© Platform [12]. The evident advantages of the porting to a server-based Web application for the Course delivery phase, consist of the world-wide offer to learners that, on turn, may well feed back the results of user's evaluations (managers, teachers as well as students) to the designers and developers for generating subsequent, more powerful versions.

The next endeavour is to integrate tools so that remote human teachers, experts, designers may modify dynamically the « teacher Agent » according to evolving specifications, in such a way that a minimal technological expertise is needed even for updating the Course [13]. After 30 years a content independent Authoring System becomes a feasible goal.

We foresee, however, that such an achievement will require really integrated Agent Technologies [11].

As a conclusion, we argue that these innovative operations require, at the moment, highly skilled human resources. The difficulty is not in terms of person-months (six p.m. in our case) but in terms of « culture » ; technical awareness and familiarity with a variety of concepts and tools that Web processing requires/offers (see for instance www.sourceforge.net).

---

[1] We hope to be able to demonstrate it at the Conference, including its performances, expected to range between 100 and 1000 interactions per minute.

These concepts and tools are not in the standard background of our graduates in Computing, at least insofar they are not part of standard curricula but have to be acquired by self learning.

5. Acknowledgements

We are grateful to Didael SpA for having stimulated, supported and facilitated our endeavour. The work described in this paper was part of the EU INCO-COPERNICUS project LARFLAST (LeARning Foreign LAnguage Scientific Terminology).

6. References

[1] S. A. Cerri, "Models and Systems for Collaborative Dialogues in Distance Learning", *Collaborative Dialogue Technologies in Distance Learning*, vol. 133, *ASI Series F: Computers and Systems Sciences*, M.F. Verdejo and S. A. Cerri, Eds. Springer-Verlag, 1994, pp. 119-125.

[2] C. Queinnec, "The Influence of Browsers on Evaluators or, Continuations to Program Web Servers", *International Conference on Functional Programming*, ACM. Montréal, Canada, Nov. 2000.

[3] S. A. Cerri, "Computational Mathetics Toolkit: architecture's for dialogues", *Intelligent Tutoring Systems*, vol. 1086 in LNCS, C. Frasson, G. Gauthier and A. Lesgold, Eds. Springer-Verlag, 1996, pp. 343-352.

[4] S. A. Cerri and D. Maraschi, "Dialogues among Distributed Agents", *Workshop on Advances in Languages for User Modelling, 6th International Conference on User Modelling,* S. A. Cerri and V. Loia, Eds., Chia Laguna, Sardinia, Italy, 1997.

[5] S. A. Cerri, "Shifting the focus from control to communication: the STReams OBjects Environments model of communicating agents", *Collaboration between human and artificial societies*, vol. 1624 in LNAI, J. Padget, Ed., Springer-Verlag, 1999, pp. 71-101.

[6] S. A. Cerri, "Dynamic typing and lazy evaluation as necessary requirements for Web languages", *European Lisp User Group Meeting*, Amsterdam, NL, 1999.

[7] P. Brusilovsky, E. Schwarz, G. Weber, "ELM-ART: an Intelligent Tutoring System on World Wide Web", *Intelligent Tutoring Systems,* vol. 1086 in LNCS, C. Frasson, G. Gauthier and A. Lesgold, Eds. Springer-Verlag, 1996, pp. 261-269.

[8] A. Mitrovich and K. Hausler, "Porting SQL-Tutor to the Web", *Workshop on Adaptive and Intelligents Web-based Education Systems, ITS'2000,* Montreal, Canada, 2000. pp. 37-44.

[9] S. R. Alpert, M. K. Singley, P. G. Fairweather, "Deploying Intelligent Tutors on the Web: an Architecture and an Example", *International Journal of AI in Education*, 10, 1999, pp. 183-197.

[10] Cocoon. http://xml.apache.org/cocoon/

[11] D. Maraschi, "Jaskemal: a Java/Scheme based integrated Framework to develop Web Application and Intelligent Multi-Agent Systems", *PhD Thesis*, LIRMM, University of Montpellier II, France, 2002. (in preparation)

[12] The Atena Platform. http://www.didael.it

[13] S.A. Cerri, S. Dikareva, D. Maraschi, S. Trausan-Matu, "Web Server Based Architectures for Language Learning: LARFLAST Agents generating CALL Dialogues", *Conference on Computer Assisted Language Learning*, Univ. of Exeter, UK, September 2001. (accepted)

A Matter of Life or Death: Re-engineering Competency-based Education through the Use of a Multimedia CD-ROM

Joanne Wilkinson
Austin and Repatriation Medical Centre
Melbourne, Australia
joanne.wilkinson@armc.org.au

Abstract

Organisations and their clients have limited tolerance for inept performance, particularly when it's a matter of life or death.

Service organisations rely on competencies of individual staff to produce and deliver core activities. In these organisations employees engage directly with clients in complex interactions as part of process. High level performance is required in challenging environments – and it doesn't occur by chance. Competencies are learned.

In situations where the risk to individuals and the organisation is high, education and credentialling of employees in key competencies is warranted. Educators are engaged to teach competencies that are relevant and known enhance an organisation's performance.

Competency-based education underpins effective organisational risk management in service delivery organisation, most notably perhaps in organisations that deliver human services. However, the cost of competency-based staff development and evaluation programs for large groups is often high.

This paper describes:
- *A scenario of a large multi-campus Medical Centre that requires competency-based education in basic life support for 4,500 students and staff.*
- *Process that entails the re-engineering of competency-based credentialling services through the design, integration and use of a multi-media CD-ROM.*
- *Outcomes of enhanced learning and cost savings coupled with reduced organisational risk.*

1. Introduction

Health service organisations are required to deliver quality services for humanitarian as well as business imperatives. It's often a matter of life or death – and never more so than in life support situations where an individual's life may be contingent upon the ability of clinical staff to respond appropriately to emergency situations. Development of quality performance, enabling staff to make the right response quickly in life support emergencies, is the subject of this paper.

Quality in health has assumed the form of evidence (read research) and competency-based services. This focus on evidence and competencies creates many prescriptions on both content and method in the delivery of continuing professional education. The problem is compounded when large numbers of services are required or when the ideal delivery mode is perceived to incorporate expensive sophisticated technology and has high establishment costs. It is challenging to be able to demonstrate cost effectiveness in health education using new technologies within a quality frame.

Re-engineering of process is an ongoing activity to keep pace with developments in the science and technologies. Forces that define as well as confound the re-engineering process often drive health delivery organisations. The transition from traditional face-to-face education to mixed mode delivery, incorporating the use of a Multimedia CD-ROM in education for clinicians, illustrates this complexity.

2. Context

The Austin and Repatriation Medical Centre is a multicampus and multilevel health service organisation situated in Melbourne, Australia. The organisation is fully accredited and has a reputation for quality care. Underpinning standards for care delivery is a policy that all nurses and various other staff are credentialled annually in the performance of Basic Life Support interventions. Traditionally the education and testing program has been conducted face-to-face and over many sessions for the various learner groups. The program was time consuming, repetitious for the educators and costly. Program evaluation also suggested that the program was dry and hence 'forgettable'. It is here that the challenge of re-engineering educational process in basic life support was taken on.

3. Reengineering Process

The ultimate intent was to deliver the program on line with competency testing occurring face-to-face. The preferred delivery method would correspond with principles of adult learning and provide opportunities to stimulate the interest of learner through interactive learning activity. Telematic approaches met the requirements for the knowledge building component of the program but this option was not achievable at the time the project was conceived. Clinical information systems weren't sufficiently rigorous to support higher end programs required for evaluation, many clinical staff possessed only beginning levels of computer literacy and educational staff lacked the skills to design and deliver sophisticated interactive programs on-line. An alternative that would approximate the desired outcomes was chosen for an interim period. The decision was taken to develop a transitional program entailing a number of stages, each of which would lead to the next. This approach would also accommodate concurrent and rapid evolution in the sciences and technologies.

Development of the transitional product entailed the preparation of the learning package written in HTML and produced as a CD-ROM enabling ease in on-line delivery at a later date. This objective shaped subsequent activities. A production team of consultants was assembled and introduced to the subject matter. The team of specialist nurse educators was taught to write for on-line delivery. The chosen strategy reduced the cost of production of the CD-ROM and fostered competencies in staff that would equip them for an envisaged future of telematic education.

The first part of process re-engineering entailed close scrutiny, review and upgrade of program content. This was achieved through a thorough review of the literature, preceding the development of a print-based self-directed learning package to delineate and trial the key concepts and related information required by learners. The knowledge component was to designed to be supported by elective face-to-face tutorial and practice sessions as well as mandatory testing of competencies. Testing was streamlined to ensure validity and consistency while enabling the load to be shared among a wider group of evaluators. A train-the-trainer model was adopted to enable shift workers with high patient ratios to meet competency requirements.

Various stakeholders were invited to participate in the process. This enabled the designers to achieve relatively high increases in the scope of the target audience and levels of satisfaction in response to small changes in learning materials and approach.

The issue of leaner centricity arose during the early phase of the project. Within learner group diversity arising out of differing levels of education and experience, inter-professional differences and retention/forgetting factors confounded the problem. We deliberated to great length as to how a single program might serve these individual learners well. There was a clear need to break away from an information-giving model as the organising paradigm toward a model that would support the development of the individual, interacting with concepts at their own pace and through their preferred directions i.e. learner centric. Mapping of concepts and design based on a 3-tier multimedia design model (Chu and Wilkinson 1999) resolved the dilemma by placing the learner in control of the navigation, learning activities and self assessments.

A prototype CD was produced, trialed and critiqued again by targeted learner and other stakeholder groups. Following minor revisions the final version was burned.

A parallel trial was conducted with some learners using the CD-ROM and others in the traditional program. While enthusiasm was high, numbers were as yet too small to validate observations. It was clear early that distribution and quality of hardware and use of information technology were uneven across the Medical Centre. Many staff were cautious, if not technophobic, and required support. Issues of access to hardware in busy clinical units for intervals of adequate duration were also uncovered.

The CD-ROM is currently being delivered on the Medical Centre intranet and is supported with face-to-face testing of competencies. Two research projects focusing on program evaluation are underway- one focusing on product design and the other a controlled comparative study investigating efficacy of the traditional delivery mode against the use of the CD-ROM.. Analysis of teacher and learner feedback is ongoing.

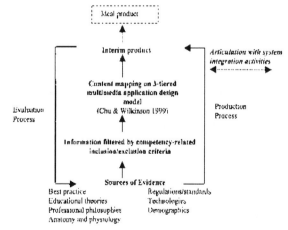

1. Process re-engineering: Product development and review

In summary the reengineering process has entailed two distinct but closely articulated groups of activities: development of the new educational tool and redesign of the context into which the tool was introduced. Both groups of activities entailed dimensions of content validity, cost containment, systems integration and quality improvement.

Figure 1 describes product development and review activities.

Content validity

The validity of content was largely assured through the use of evidence-based information. The literature is a rich source of information about health, education and technology that has been subjected to rigorous investigation. Careful selection of content was paramount. However, the art of ensuring content validity extends well beyond content selection, particularly in the development of competencies. Unlike skill-based programs that are prescriptive, competency-based education entails learning that includes the application of principles and thoughtful discrimination among alternatives. Competencies are multidimensional, comprised of knowledge, skill and psychosocial factors that are brought together in complex behavioural responses to environmental cues. Content validity directed toward mastery learning of competencies requires information inputs in forms that assist in recall and in identifying performance benchmarks.

Competency benchmarks exist for life support interventions. These benchmarks are founded on local, national or international standards. Some variation does exist, requiring decisions to be made the about the appropriate standard for the organisation. Choice is influenced by the nature of the target audience. We chose the national standard to ensure high inclusivity. Evidence was progressively uncovered to filter and validate the field of life support information and standards for best practice were set out in an information framework. Content was then analysed to tease out elements of major concepts and subjected to review by various expert panels.

Honing for precision occurred at the content-delivery medium interface. Participation and review by educators, who were also clinical specialists, ensured that there was no inadvertent change in focus during the writing up phase of the product development process.

Cost containment

The working group addressed cost-benefit tradeoffs and philosophical issues pertaining to target audience, size and composition. As this was an early venture into the field of online delivery, it was decided to maintain costs at a conservative level, to curtail adventurous experimentation and to exploit what we already did well. It was further decided

- to develop a new product as an iterative staff development exercise. Staff were expert clinicians and qualified educators but novices in writing for on-line delivery. Preparation of the educational team to write for on line delivery was a significant measure in reducing costs. A consultancy team was employed to prepare and shepherd the educational staff throughout the writing phase and to produce the prototype CD-ROM.
- that a community service rather than a profit motive would drive that product development. Low price was to be offset by cost savings rather than profit making. The widest possible learner audience was accommodated with cross-disciplinary and cross-institutional boundary audiences benefiting from the value-added approach.
- to produce a generic rather than a specialist learning package. The design of the product would facilitate easy and low cost modifications.

The centrality of these early directions became increasingly apparent as subsequent decisions were taken.

The strategy of transforming print and face-to-face delivery materials into a multimedia CD-ROM entailed front-end investment to reduce recurrent operational costs. This is anticipated to achieve an annual savings of an estimated $50,000 - $60,000 Au.

Quality improvement

The transmission of valid content, in a robust form that supports the learning process, is an ongoing quality improvement concern. In the design of this product, alternative approaches were mapped on the quality framework of evidence and competence. Knowledge and skill acquisition and performance evaluation were examined together and apart. Issues of demographics, geographic dispersal, service volume, content and costs were all considered. Decisions were taken regarding the best configuration of content and use of educational media

Implications for the adult learner were clear. The material needed to be engaging, interactive, enabling self-pacing and providing opportunities for periodic affirmation of progress. Appropriate feedback mechanisms were recognised as being crucial to learning effectiveness.

A strategic plan was developed to enable improvements to occur in a stepwise manner using mixed methods that would support the independent development and validation of education and evaluation components. In these progressive stages, the education material is being reviewed, extended and further validated. The expert panel comprises a number of clinicians from various disciplines internally as well as external community representatives.

Although performance assessment, in the early phase of development, was to occur in a simulated setting, it was decided to provide checks in the form of information self-tests. Initially these were presented simply as statements of questions and answers at the conclusion of sections of the learning program. It was anticipated that the static presentation would later be superceded with

Systems integration

Figure 2 illustrates the components of the loosely coupled system that has been achieved through process re-engineering.

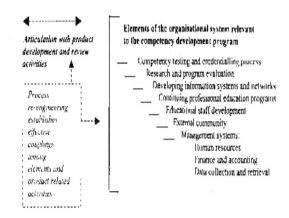

Figure 2. Process re-engineering: Systems integration

Features of the product development and review process are articulated with elements of the greater system through a process of loose coupling (Orton and Weick 1990). The method entailed design of a system characterised by perpetual responsivity among components of the basic life support program. This dynamic network of organisational features continue to support the educational strategy.

Systems integration has necessitated the development of linkages among program components, with organisational infrastructure with evolutionary change within the organisation and wider social trends. These linkages are apparent in various agenda, committee structures and programs and futures scenarios.

4. Conclusion

This project has demonstrated the feasibility of re-engineering competency-based education in health through the use of a purpose-designed multimedia CD-ROM. The product adds practical value to the educational process through effective engagement of the individual with the subject matter.

The early implementation phase has highlighted increased penetration of the clinician group numbers with a concurrent growth in mastery learning in basic life support. It has been demonstrated that front-end investment in computer assisted education can offset recurrent operational costs.

Analysis of critical incident data will, cumulatively, highlight the program effectiveness as quality of staff performance in the real world of emergency life support interventions.

References

[1] Chu Stephen & Wilkinson, Joanne 1999, 'Building learner-centric self-directed, multimedia learning applications' Proceedings of Health Informatics Conference 99, ISBN 0-9585370-4-6.

[2] Orton, J. D. & Weick, K. E. 1990, 'Loosely coupled systems: a reconceptualisation', Academy of Management Review, 15(2), 203-23.

[3] Australian Nursing Council Incorporated (ANCI), National Competency Standards for Registered Nurses (3rd edition) 2000.

[4] Wilkinson, Joanne (Director) 1999 'Basic Life Support' CD-ROM, HealthQuest International.

[5] Linnard-Palmer, L.R. (1996) 'The effect of a skills algorithm on nursing students' response rate, skill accuracy and reported attention management during simulated cardiopulmonary arrests', University of San Francisco ED. D.

[6] Fisher, M. L., Hume, R. and Emerick, R. (1998) 'Costing nursing education programs: it's as easy as 1-2-3', Journal for Nurses in Staff Development, 14(5):227-35, 255.

[7] Villaire, M., McManamen, L. and Hendricks, L. (1996) 'Telemedicine: tuning in on critical care's future?', Critical Care Nurse 16(3): 102-7.

From Educational Meta-Data Authoring to Educational Meta-data Management

V. Papaioannou, P. Karadimitriou, A. Papageorgiou, C. Karagiannidis and D. Sampson
Informatics and Telematics Institute (I.T.I.)
Centre for Research and Technology – Hellas (CE.R.T.H.)
1, Kyvernidou Street, Thessaloniki, GR-54639 Greece
Tel: +30-31-868324, 868785, 868580, internal 105
Fax: +30-31-868324, 868785, 868580, internal 213
E-mail: {vickyp, karadim, tp, karagian, sampson} @iti.gr
www.iti.gr

Abstract

Educational meta-data can significantly assist the effective and efficient retrieval of educational resources. Currently, there are a number of tools for educational meta-data authoring, facilitating the creation and modification of educational meta-data files. This paper outlines the requirements, and proposes an architecture for the development of a complete EMD management toolkit which go beyond meta-data authoring, supporting the educational community in the full management of educational meta-data.

1. Introduction

Information and telecommunication technologies facilitate mouse-click access to enormous knowledge repositories and educational resources, unconstrained from time, location, etc. The full exploitation of this mass body of knowledge in education can be, however, compromised, by the difficulty in describing, classifying and maintaining these knowledge sources in such a way that they can be retrieved in an *"educationally efficient and effective way"*.

Educational Meta-Data (EMD) are attracting increasing attention in this context, since they facilitate the description of educational resources, so that they can be easily retrieved [1], [2]. A number of international efforts have been initiated during the past few years, aiming to define EMD standards for the common description of educational resources. These standards include fields that are considered necessary for the description of educational resources – such as the type of the resource (i.e. whether it is an experiment, simulation, questionnaire, assessment, etc), the target learner age, difficulty level, estimated learning time, etc – as opposed to "general purpose" meta-data standards (e.g. the Dublin Core, purl.oclc.org/dc), or standards that have been developed for different fields of knowledge (e.g. geo-spatial meta-data standards, badger.state.wi.us/agencies/ wlib/sco/metatool/mtools.htm). The most well-known international EMD standardisation initiatives are the IEEE LTSC (ltsc.ieee.org), IMS (www.imsproject.org), AICC (www. aicc.org), ARIADNE (ariadne.unil.ch), and CEN / ISSS (www.cenorm.be/isss/Workshop/lt/) [3].

At the same time, a number of tools have been developed for the description of educational resources through EMD. These tools can be roughly classified into two major categories: *"XML authoring tools"*, and *"educational meta-data authoring tools"*.

The former category includes tools for the creation of files in XML, which is the most commonly used format for meta-data files in general (i.e. not only for educational purposes). They usually support a number of functionalities, including the mapping of XML documents to other meta-data formats (e.g. RDF), updating, validating, searching and manipulating XML documents, etc. These tools are not specifically developed for educational purposes, but they can be used for the creation of EMD files, if the user imports the specific DTDs required for respective EMD standards. This, however, requires a *substantial* meta-data technologies expertise. Examples of these tools include EditML (editml. homepage.com) and XMLSpy (www.xmlspy.com).

The latter category includes tools that have been specifically developed for educational purposes. That is, they usually facilitate (through a user-friendly interface) the creation of *educational* meta-data files that are based on a specific *educational* meta-data standard. However, these tools usually do not support the mapping, management and modification of meta-data files, validation on structure and data entries, etc. Examples of these tools include the EUN Resource Description tool (www.en.eun.org) and the Reggie Metadata Editor (metadata.net/dstc).

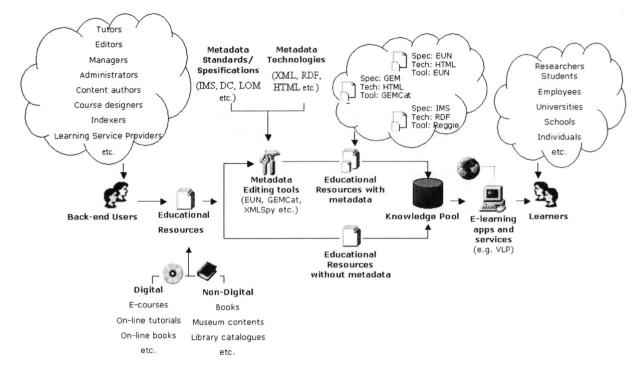

Figure 1 – Requirements for Educational Meta-Data Management

In summary, the educational community has not yet exploited the full potential of EMD, since:
1. still, a lot of educational resources are available in the Internet without a meta-description, i.e. without EMD files for their description
2. for those educational resources that are described through EMD files, their description can be based on *different standards*, and/or *different file formats*
3. even when EMD files are based on the same standards and file formats, they can still differ if they are developed through *different meta-data authoring tools*

This paper focuses on the requirements for educational meta-data *management*, and proposes an architecture for the development of a complete EDM *management toolkit*, which overcome the above limitations, supporting the educational community to fully exploit the potential of EMD.

2. Requirements for Educational Meta-Data Management

The main requirements for educational meta-data management can be derived from Figure 1, where the different people, technologies, tools, etc, involved in meta-data management are shown. As it is shown in this figure, meta-data management, towards a collective and harmonised EMD repository, requires: creation of new and modification of existing EMD files, validation of meta-data information, support of *all* the EMD standards / specifications, creation of *new* meta-data specifications for specific requirements (e.g. for including additional fields which are necessary for a specific application), support of all meta-data technologies, mapping of EMD files between different EMD specifications/standards, etc. These basic requirements are briefly elaborated below.

Creation of new educational meta-data files: this is the most basic function in EMD management. The user should have the option to define the new EMD file according to *any* of the existing EMD standards and specifications (e.g. create an IEEE LOM, or an IMS EMD file). Moreover, since this function is mainly targeted to educational resources *authors*, who are not necessarily experts in meta-data technologies, it should be supported through a user-friendly interface (e.g. through wizards), providing help concerning the information that needs to be inserted into each EMD field.

Modification of existing educational meta-data documents: this function concerns the update of data entries and modification of the document structure of existing EMD files, by inserting or removing values in EMD fields.

Support of all educational meta-data standards: EMD management should support the creation of EMD files in any existing or emerging EMD standard/specification. Moreover, EMD management should support the definition of new EMD sets, through adding / removing fields in the EMD standard, and saving the new EMD set.

This would ensure that all existing and emerging EMD standards are supported.

Mapping of meta-data standards: the EMD files can be created according to a number of EMD standards. Therefore, EMD management requires that the user is able to map EMD files that are based on a specific EMD standard (e.g. IEEE LOM) to any other EMD standard (e.g. IMS).

Validation of semantic educational meta-data: one of the main problems with EMD files is that they can include inaccurate information. Therefore, EMD management should facilitate the validation of the information included in EMD files, when this is possible. The user should be informed if the entries in the fields are unacceptable (e.g. when text is inserted in fields where a number is expected). In addition, EMD management requires that the validation of the structure of EMD files, concerning their conformance with to the selected EMD standard / specification.

Meta-data document management: EMD management would also take into account the needs of EMD repository managers to find, update, delete, sort and group any set of EMD files through multiple document selections, multiple editing in EMD files, and with the help of a graphical interface including drag & drop features.

3. Architecture for Educational Meta-Data Management

This section proposes an architecture that can can support the requirements for EMD management that have been described in the previous section. The architecture is depicted in Figure 2, and each different sub-component is briefly described below.

Repositories: they are used for storing and manipulating all the different types of EMD files (XML Schema, DTD, XML files). The "XML files repository" is used for storing the XML files, the "XML Schemas repository" for the XML Schemas, the "DTD repository" for the DTDs, "XML file databases" (one for every different EMD standard/specification) to group the XML files and allow manipulation of their elements by using queries, the "maps database" to store the maps used for mapping of documents between the available EMD standards/specifications, etc.

Associations: associations between the repositories, and the databases are achieved by using the Document Object Model (DOM). This is a platform- and language-neutral interface, which defines the way a document can be accessed and manipulated. With DOM implementation, a programmer can create a document, navigate its structure, and add, modify, or delete its nodes (elements). The DOM represents a tree structure of the XML document. A program called XML parser can be used to load an XML document into the memory of the computer. Similarly, the DTD and XML Schema documents can be loaded. When a document is loaded, its information can be retrieved and manipulated by accessing its DOM. In addition, by accessing the DOM, any format of files can be generated (XML files, DTD, XML Schema and DB XML files) and therefore one format can be converted to another through DOM.

Editors: the "XML Editor" provides the interface for creating new XML files or loading and modifying existing ones, and the generated XML files are stored in the "XML file" repository. Similarly the "XML Schema Editor" and the "DTD Editor" facilitate the creation and modification of new or existing XML Schemas and DTDs, respectively, and both the generated documents are stored in the respective repositories.

Validation: the "Validation" component of the architecture provides two different types of XML file validation: structure validation and data validation. Structure validation checks if the XML files conform to the element structure and hierarchy of the associated DTD or XML Schema files. While, data validation checks if XML files conform to the associated XML Schema (if any) in terms of data type. Validation is taking place every time an XML file, DTD or XML Schema is generated or modified, and the outputs of this component are validated XML files or DTD/XML Schemas stored in the respective repositories.

Mapping: the "Map generator" and the "mapping" components allow conversion of XML files from one educational standard to another. The "Map generator" component requires as input two XML files of different standards/specifications. The outcome of this operation is a map that allows conversion between these two standards /specifications. The generated map is stored into the "Map database". The inputs to the "Mapping" component are the XML file for conversion and the map. The outcome of this component is a new XML file of the required standard /specification and it is stored into the "XML file repository".

Wizard: the wizard is an optional interface layer between the user and the "XML Editor". This component provides the available features of the "XML Editor" but with some restrictions: it does not allow modification of the XML file structure and therefore the modification of the DTD and the XML Schema files are restricted. The aim of the wizard is to encrypt the complicated operations of EMD management, and to provide an interface where the user just completes values in a number of fields. It provides guidance for creating XML files that conform to the three main IMS, DC and LOM.

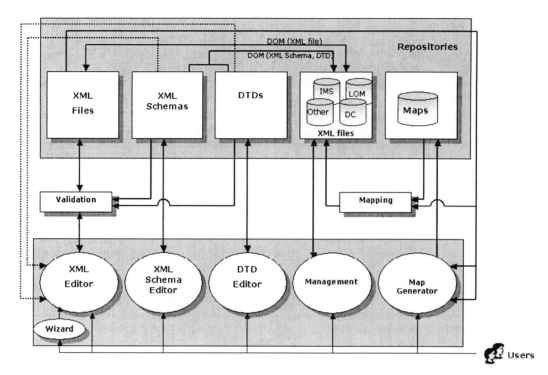

Figure 2 – Architecture for Educational Meta-Data Management

Management: this is a user interface component allowing manipulation of multiple XML files. The component and the "XML files database" are associated, allowing the retrieval, update and sorting of XML files by their element values. The "Management" interface provides the user with a graphical representation of all the XML files stored in the repositories. The user is enabled, with simple commands (sort by, update, select, edit, etc) or with graphical features (e.g. drag and drop), to manage the existing XML files.

4. Conclusions and Future Work

Educational meta-data management is very important for managing educational resources, so that educational e-content can be effectively and efficiently retrieved by on-line learners through e-Learning applications and services. This paper has outlined the basic requirements for EMD management (as opposed to EMD authoring), and has proposed an architecture that can support these requirements.

The ideas presented in this paper have been initiated during the development of an EMD editor (*the EM² tool* [4]). This tool facilitates the creation of new, and the modification of existing EMD files of any EMD standard.

Our on-going and future work in this area involves the extension of the EM² tool, towards the full implementation of the proposed EMD management architecture; as well as the localisation of EMD management tools, so that they can take into account the different linguistic, cultural, etc, requirements of different countries.

Acknowledgements

The work presented in this paper is partially funded by the European Commission Information Society Technologies (IST) Programme through the IST No 12503 KOD "Knowledge on Demand" Project.

References

[1] Greenberg, J. (2000). *Metadata and Organising Educational Resources on the Internet*. Journal of Internet Cataloging, (3)1.
[2] Bourda, Y., & Helier, M. (2000). *What Metadata and XML can do for Learning Objects*. WebNet Journal, (2)1.
[3] CEN/ISSS Learning Technologies Workshop (2000). *A Standardization Work Program for Learning and Training Technologies & Educational Multimedia Software – CWA*, electronically available at www.cenorm.be/isss/Workshop/lt.
[4] Sampson, D., Karagiannidis, C., Karadimitriou, P., Papageorgiou, A., (2001). *EM² – an Educational Meta-Data Management Tool*, 13th World Conference on Educational Multimedia, Hypermedia and Telecommunications (ED-MEDIA 2001), Tampere, Finland.

Design and Implementation of WBT System Components and Test Tools for WBT content standards

Kiyoshi NAKABAYASHI,
Yukihiro KUBOTA
NTT-X, Inc./ALIC
naka,kubota@nttx.co.jp

Hiroshi YOSHIDA
Hitachi Electronics Services Co.,Ltd.
yoshida@hitachi-densa.co.jp

Tatsuo SHINOHARA
Transmedia K.K.
sinoha@trans-media.co.jp

Abstract

This paper describes the design and implementation of library module components for WBT systems conforming to the WBT standards such as AICC CMI specification or ADL SCORM specification. These components provide high level APIs for WBT application program to handle WBT content data models defined in the specifications.

1. Introduction

Since a learning content requires high development costs, there is a high demand for the standards to achieve learning content interoperability and reusability. Such learning content standards are being developed by the organizations like AICC (Aviation Industry CBT Committee) or ADL (Advanced Distributed Learning) initiative as AICC CMI (Computer Managed Instruction) specification[2] or ADL SCORM (Shareable Content Object Reference Model) specification[3]. These standards are not necessarily easy to understand or implement even for high-skilled system vendors. Moreover, it is difficult to avoid the ambiguities or implementation dependent terms in the specification document itself. This means that the systems or contents developed carefully to conform to the standards, when connected to each other, are not guaranteed to work together.

Considering these status, ALIC (Advanced Learning Infrastructure Consortium), a Japanese initiative for e-Leaning promotion, planed to design and implement program library module components and test tools to support WBT system vendors and content vendors to employ WBT standard specifications for their products. This paper describes the design and implementation of library module components and test tools.

2. WBT courseware standards

The target specifications for the developed library module components and test tool are AICC CMI specification[2] and ADL SCORM specification[3].

Both specifications deals with the WBT content executed on the server/client type WBT systems. In the specifications, it is assumed that WBT system consists of server-side sub-system (CMI) and client-side sub-system (CBT). CMI has the capability to load the tree-structured courseware definition file and track learner's activity along with the courseware definition file. CBT is a multimedia content such as HTML file, Javascript program, Java applet, or plug-in program which is executed on the WWW browser. Each CBT corresponds to each page (leaf node) of the tree-structured courseware. CBT is launched by CMI, and communicates with CMI to exchange information on learner's performance.

To support this execution model, the specifications defines the following data formats and protocols:

(1) Course structure format

It defines the data models of tree-structured courseware including sections called block and pages called AU (Assignable Unit). The data model also includes learning objectives and logical statements for prerequisites and completion requirements. CMI specification provides the binding of the data model to CSV file while SCORM provides the binding to XML.

(2) CMI/CBT communication protocol

In the CMI specification, HTTP-based protocol called HACP (HTTP-based AICC CMI Protocol) is defined. It includes CBT launch protocol and CMI/CBT communication protocol. In the newer version of the CMI specification as well as SCORM, Javascript API for CBT contents is added which provides for contents high level API to hide the details of HACP implementation.

(3) CMI/CBT communication data model

It defines data models for learner information such as learner's name, ID, past score, etc. from CMI to CBT and spent time, answers to the exercise questions, current score, etc. from CBT to CMI. Implementation using Windows INI file format and CSV file format is also provided as well as the implementation of data model handled with Javascript API in the newer version and SCORM.

(4) LOM

In the SCORM, LOM specification is exploited to index and describe every content element in the content structure format such as sections and pages as well as SCO. LOM data model comes from IEEE LTSC specification[4] and binding to XML comes from IMS specification.

3. WBT library module components

Four library module components are designed to implement the WBT content standards described in the previous section. These components provide high level APIs (Application Program Interfaces) to handle WBT content data models, hiding implementation details of CSF files or HACP communication from application to reduce system/contents development cost and to improve interoperability.

3.1. CSF import/export module

This module provides API to handle CSF (Course/Contents Structure Format) data model defined in AICC CMI and ADL SCORM specification as well as LOM data model. The API provides similar functionality as of DOM API for handling XML data model[6]. With the API, data objects corresponding to blocks, AUs, and objectives can be created, deleted, and inserted relating to other element. It is also possible to traverse the tree-data structure to access any data elements. Properties of the data elements including prerequisites and completion requirements are also set, read and modified using the API. This module can be built in to the course structure format authoring tool to edit the CSF file, or to WBT server to read the CSF file and map the CSF to server's internal data structure. Physical CSV or XML file format is hidden from application.

3.2. CMI/CBT communication module

CMI/CBT communication module consists of API for handling CMI/CBT communication protocol and data model. The API provides the object representing the communication data model. WBT server application creates the instance of the object corresponding to each AUs in the CSF. CBT associated with the AU can be launched and communication data is sent and received by calling the instance methods. The module communicates with CBT using HACP although its implementation is completely hidden from application.

3.3. CBT-API module

CBT API module implements Javascript API which is defined in the specification for CMI/CBT communication on the CBT side. The module communicates with CMI using HACP although its implementation is completely hidden from CBT content.

3.4. CBT Interface module

Although CBT content can be rather easily developed with the CBT-API module, it is preferable to provide more application-oriented API for typical CBT contents such as true/false or multiple choice exercise. CBT interface module implements these high level API for typical exercise hiding detailed data model of CMI/CBT communication.

4. Conclusion

Using library components, test tool to check WBT content's standard conformity is also implemented. Interoperability experiment was conducted involving several WBT system and content vendors. The result proves that the specifications and the library module components as their implementation are effective to achieve the interoperability.

Acknowledgements

The authors thank members of TBT consortium who joined the interoperability experiment and ALIC to support this project. This work is granted by IPA (Information-technology Promotion Agency, Japan).

References

[1] K.Nakabayashi, "Internet-based Educational Environment and Learning Technology Standard", Proc. NTCL2000, Awaji-shima, Japan, Nov. 2000, pp. 1-8.
[2] "CMI-001: CMI Guidelines for Interoperability", 3.0.2, May. 2000, Aviation Industry CBT Committee.
[3] "SCORM: Shareable Content Object Reference Model", 1.1, Jan 2001, Advanced Distributed Learning Initiative.
[4] "Learning Object Metadata (LOM)", P1484.12/D4.0, Feb 2000, IEEE Learning Technology Standard Committee.
[5] "Learning Resource Metadata XML binding specification", 1.0, Aug 1999, IMS.
[6] "Document Object Model (DOM) Level1 Specification", 1.0, Oct 1998, W3C.

UNIVERSAL - Design and Implementation of a Highly Flexible E-Market-Place for Learning Resources

Stefan Brantner*, Thomas Enzi+, Susanne Guth+, Gustaf Neumann+, Bernd Simon+

*INFONOVA
E-mail: stefan.brantner@infonova.at
Seering 6, A-8141 Unterpremstätten, Austria

+Vienna University of Economics and BA
E-mail: universal@wi.wu-wien.ac.at
Augasse 2-6, A-1090 Vienna, Austria

This paper illustrates the design and implementation of a highly flexible, electronic market-place for learning resources called UNIVERSAL. Integrating learning resource related data in a (semi-)automated way demands a flexible data model. The paper elaborates on components of an educational market-place model such as learning resources, agents, rights and delivery systems. The central data formats of UNIVERSAL are based on RDF. We argue that the flexibility of RDF provides a high level of adaptability to future changes in the data model and a maximum level of openness. UNIVERSAL aims at contributing to the idea of a semantic web of universities, pursuing the vision of having data on the web defined and linked in such a way that it can be used by machines not just for display purposes, but for automation, integration and reuse of data across various applications.

1. INTRODUCTION

The European IST-Project "UNIVERSAL - Universal Exchange for Pan-European Higher Education" (http://www.ist-universal.org/) is an attempt to demonstrate an open exchange of learning resources (LRs) between higher education institutions across Europe [see 1]. The business-to-business oriented brokerage platform is designed to support offers, enquiries, booking and delivery of LRs.

UNIVERSAL will enable collaboration among leading Higher Education Institutions (HEIs), by providing exchange services. The aim of UNIVERSAL is to increase the quality and competitiveness of HEIs through a cooperative – and, at the same time, competitive – environment [see 2, 3].

The subjects of exchange on the UNIVERSAL platform are LRs. LRs are more than a link to a faculty member's homepage; they are defined as academic content devoted to a specific subject. An LR can be a part of a lesson, but can also span several lessons. It is not necessarily identical with a course, a unit which is restricted by the academic calendar of an HEI. The platform supports synchronous as well as asynchronous LRs. The following examples of LRs are taken from the preliminary version of the UNIVERSAL catalogue:

- A recorded session of the course "Corporate Strategy in Emerging Markets" dealing with an ethical case study on Henkel in the Asian-Pacific Area;
- A live session of the course "International Marketing" dealing with an case study on Levi Strauss;
- Slides, exercises, and support material from an introductory course in "Information Technology and Management Communication" dealing with the creation of effective documents.

2. THE UNIVERSAL SYSTEM ARCHITECTURE

The UNIVERSAL platform comprises various engines (see Figure 1) providing services to users and external information systems such as local learning management systems, web-based content databases and ERP solutions of HEIs. An engine is related to a particular application providing a set of services and an interface to other applications.

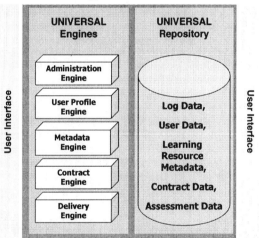

Figure 1: UNIVERSAL Engines

When it comes to the provision of LRs, the *metadata engine* facilitates the maintenance of LR related data. The engine interacts with LR providers via a web-based user interface, or uses a protocol to exchange data with local learning management systems. The LR metadata engine also facilitates access to the LR repository. Thus, on the one hand, the system has to provide flexible search services, allowing users to enter and combine queries as powerful search engine does. On the other hand, it enables users to browse a well-structured directory based on an international taxonomy.

The *contract engine* is involved in the process of placing offers and requests for LRs, matching offers with requests and booking of LRs. During these transactions the system deals with issues such as user authentication and transaction supervision; in the future maybe even billing will be handled.

The *delivery management engine* arranges for the delivery of LRs by granting access rights. Since UNIVERSAL does not offer content storage itself, delivery systems are connected to the brokerage system. A thin generic interface layer provides communication functionality across this connection, e.g. by providing authentication and authorization services, delivery negotiation and delivery supervision. UNVERSAL will offer a fully implemented delivery interface to the following restricted set of delivery systems, but is also open to others by providing a generic delivery interface:

- Apache web server (asynchronous packaged content)
- Hyperwave's e-learning suite (learning management system),
- RealNetwork's Realserver (asynchronous streaming content),
- Isabel (synchronous collaboration tool and video conferencing system)

The *user profile engine* is in charge of user administration issues such as user registration and cancellation. The user engine provides services to other engines (especially the LR metadata engine) in order to maintain user profiles. Enhanced usage logging allows UNIVERSAL to provide personalized access to its LR repository.

Finally, the *administration engine* maintains all logging data and tracks changes in the UNIVERSAL repository. In addition, the administration engine provides platform assessment services.

3. MODELING AN E-MARKET-PLACE OF LR

UNIVERSAL can be viewed as a medium which enables communication about artefacts between human and artificial agents [see 4]. The necessary prerequisite for communication is a common language and understanding [5]. Defining a common syntax and semantics is, therefore, a crucial activity in the design of an e-market-place. Learning resources and other artefacts must be described using structured metadata in order to enable an effective query of the repository [6]. The structured metadata provides a knowledge base that can be used to facilitate an open interface for various standards issued by organizations such as IEEE, IMS, Dublin Core, ADL, and AICC.

On the e-market-place LRs are traded by offers comprising the components shown in Figure 2. Each component represents a type of artefact which can be glued together with others. All artefacts are modelled within RDF [7].

Figure 2: Components of a LR Offer

3.1. Modelling Learning Resources and Related Information

The central artefacts of UNIVERSAL are *learning resources*. This component contains attributes describing the learning resource and providing hints on its usage. The general attributes are divided into two levels depending on their importance. Most of these attributes are equivalent to the attributes suggested by the IEEE LOM standard [10] and IMS [11]. However, some modifications to these suggestions, as well as more in-depth definitions, were necessary in order to serve the requirements of the platform. A learning resource in this context is an abstract definition which can be associated with more than one physical resource specified by the delivery system it requires. *Annotations* are used for storing reviewing data provided by registered users.

Taxonomies provide schemas for the organization of the UNIVERSAL LR directory [see 12]. An LR metadata instance can refer to multiple categories within a taxonomy. LRs can be classified multiply using different taxonomies. By treating classifications as components, taxonomy interrelations can be expressed more easily (e.g. linkage between taxonomies in different languages).

3.2. Modelling Agents

The metadata on 'agent' comprises information about human or software agents. Universal supports role types for Learners, Provider, Consumer, Evaluators, and Local Registration Authorities [for detailed discussion, see 4].

The metadata on a human agent is based on the widely used vCard [13] standard. The graph in Figure 3 illustrates an RDF instance of an human agent.

One important part of a platform framework that trades with digital goods is the Digital Rights Management. This provides services that secure digital content, and manage the use of the secured content in accordance with the rights and interests of all parties in digital commerce.

The UNIVERSAL platform will allow assignment, to every learning resource, of a set of *rights* to grant usage privileges to an agent. This set of rights is based on the developing industry standards of right(s) expression languages: The "Extensible Rights Markup Language" (XrML) [14] defined by Xerox and the "Open Digital Rights Language" ODRL [15] developed by Ianella.

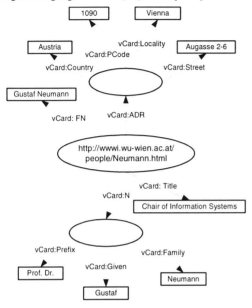

Figure 3: RDF Instance of Human Agent

The Rights Management System contains: Rights Expression Language, Rights Messaging Protocol and Policy Enforcement. The Rights Expression Language specifies user privileges regarding LRs; the Rights Messaging Protocol provides means to exchange rights expressions among agents. The Policy Enforcement ensures the application of the policies implied by the digital rights expressions.

3.3. Modelling Offers and Related Information

To create an *offer scheduling*; information can be added to specify the exact delivery dates and the *delivery system*. Delivery system metadata contains information on how to access the LR offered. The following elements are encapsulated within the delivery system description:
- name of the delivery system;
- URL of the delivery systems' UNIVERSAL interface layer;
- method of authentication and authentication details; and
- contact addresses of agents responsible for the delivery system.

With this data, the UNIVERSAL brokerage platform can access the delivery system and negotiate the terms of the actual delivery, preparing the way for actual consumption of the LR by a potential platform user. When it comes to LR delivery, the functionality of the delivery system interface guarantees fulfillment of the offer terms specified in the Rights Expression Language and in *scheduling*.

4. IMPLEMENTATION OF THE RDF APPROACH

Formal presentation of the artefacts is modeled by means of the Resource Description Framework (RDF) [7]. The foundation of RDF is a general model to represent the named properties and property values of resources. The Extensible Markup Language (XML) [8] is used to encode the RDF-modeled learning resource metadata on the UNIVERSAL platform. The integration of a wide variety of different learning asset delivery systems, most probably based on heterogeneous description schemes, demands a metadata approach which is both highly flexible and essentially distributed in nature. By using XML, metadata descriptions can easily be understood, reused or transformed by using standard tools (see Figure 4). RDF provides the framework to describe metadata assets in a flexible way, using description elements from a variety of different description schemas. The RDF approach further ensures that metadata can be accessed and searched in a distributed manner.

The UNIVERSAL brokerage service holds learning resource offer descriptions as XML:RDF instances in a metadata repository, which is built as file store as well as an RDF triple database. Consequently accessing UNIVERSAL metadata can be accomplished in a fast and standardized manner, and metadata search can be carried out very flexibly. The RDF triple approach is further used for entering metadata in user interface forms.

Figure 4 illustrates the "metadata lifecycle" for the process of learning resource metadata provision. Metadata on learning resources can be entered into the platform in two different ways. On the one hand, providers can directly enter data via the user interface. On the other, the platform allows direct import of metadata from an external XML or XML:RDF instance. In this case metadata is mapped to a triple representation and stored in the triple database. Alternatively, a foreign metadata instances can be stored as an XML:RDF instance which is compatible with the UNIVERSAL metadata format.

This feature facilitates the open exchange of learning resource descriptions from third party delivery systems. It allows learning management systems to search the repository via an RDF query protocol. The above process can be similarly mapped to the other types of metadata used in the platform. Data from the triple

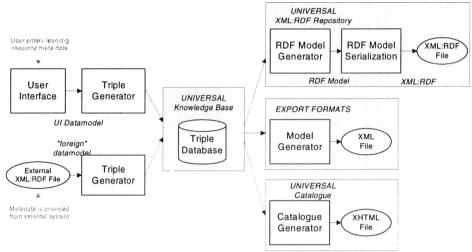

Figure 4: Metadata Lifecycle for the Process of Metadata Provision

database can be mapped onto an XML:RDF instance at any time, and vice versa.

5. CONCLUSION

This paper has discussed modeling components of an electronic market of learning resources. The XML:RDF approach was chosen in order to ensure openness and flexibility in modeling artefacts such as learning resources, agents, rights, taxonomy, annotations, scheduling data and delivery systems. Openness is required in order to enable efficient data interchange among various systems, such as those mentioned above or student modeling agents (e.g. the student assistant agent of the MADE system [9]). On the one hand, the XML:RDF files can be parsed easily by third party systems; on the other, RDF triples (i.e. sets of subject, predicate, object) provide an effective tool for searching.

UNIVERSAL is currently available in prototype form. However, new challenges are already arising, such as providing generally accepted taxonomies, multi-language support and establishing quality assurance mechanisms. At the time of writing, trials are taking place with professors outside the development team.

6. REFERENCES

[1] M. Hämäläinen, A. B. Whinston, and S. Vishik, "Electronic Markets for Learning: Education Brokerage on the Internet," Communications of the ACM, vol. 39, pp. 51-58, 1996.

[2] D. Tsichritzis, "Reengineering the University," Communications of the ACM, vol. 42, pp. 93-100, 1999.

[3] P. Meier and B. Simon, "Reengineering Undergraduate Teaching by Introducing Internet-based Learning Information Systems," in Proceedings of the 8th European Conference on Information Systems, vol. 1439-1444, Vienna: Springer, 2000.

[4] S. Guth, G. Neumann, and B. Simon, "UNIVERSAL - Design Spaces for Learning Media," in Proceedings of the 34th Hawaii International Conference on System Sciences, Maui (USA), Maui, USA: IEEE, 2001.

[5] K. Stanoevska-Slabeva and B. F. Schmid, "Requirements Analysis for Community Supporting Platforms Based on the Media Reference Model," International Journal of Electronic Markets, vol. 10, pp. 250-257, 2000.

[6] F. Arcelli and M. De Santo, "An Agent based Internet Infrastructure for Learning Commerce," in Proceedings of the 33rd Hawaii International Conference on System Sciences - 2000, O.A., Ed. Maui, 2000.

[7] D. Brickley and R. V. Guha, Resource Description Framework (RDF) Model and Syntax Specification: W3C Recommendation: REC-rdf-syntax-19990222, 1999.

[8] T. Bray, J. Paoli, and C. M. Sperger-McQueen, Extensible Markup Language (XML): W3C Recommendation: REC-xml-19980210, February 1998.

[9] S.-G. Han, J.-B. Park, J.-E. Jung, and G.-S. Jo, "Intelligent Gathering of Contents on Distance Education using Mobile Agents," in Proceedings of the International Conference on Electronic Commerce 2000, Seoul, Korea: 2000, pp. 267-273.

[10] "IEEE Learning Technology Standards Committee (LTSC) Learning Object Metadata - Draft Document v3.8" http://ltsc.ieee.org/doc/wg12/LOM3.8.html, 1999.

[11] T. Anderson, D. McArthur, S. Griffin, and T. Wason, IMS Meta-data Best Practice and Implementation Guide: Educause, 1999.

[12] B. Simon and G. Vrabic, "Learning Resource Catalogue Design of the UNIVERSAL Brokerage Platform," in Proceedings of ED-MEDIA 2001, C. Montgomerie and J. Viteli, Eds. Tampere, Finnland: AACE, 2001, forthcoming.

[13] F. Dawson and T. Howes, vCard MIME Directory Profile. The Internet Society: http://andrew2.andrew.cmu.edu/rfc/rfc2426.html, 1998.

[14] X. Wang, Extensible rigths Markup Language (XrML): http://www.xrml.org, 2000.

[15] R. Iannella, "Open Digital Right Languages (ODRL)," W3C Workshop on Digital Rights Management, 2001.

Theme 7: ITS and Agent Based Methods

Investigation of Learning Object Metadata and Application to a Search Engine for K-12 Schools in Japan

Mayumi Okamoto, Masanori Shinohara
NTT-East Corporation
okamoto@mm.bch.east.ntt.co.jp
shinohara.m@east.ntt.co.jp

Yasuhiro Okui, Shigeki Terashima
Nihon Unitec Corporation
Yokui@utj.co.jp
Tera@utj.co.jp

Mari Hashimoto
Center for Educational Computing
Hashimoto@cec.or.jp

Abstract

Standardizing Learning Object Metadata (LOM) has been investigated by IEEE[1], IMS[2] and other organizations. In this project, we customized LOM to make it consistent with Japanese K-12 school education under the guidelines of the Advanced Learning Infrastructure Consortium (ALIC)[3], which is currently proceeding with standardization related to e-learning in Japan. Expanding LOM structure and creating Japanese-specific vocabularies were investigated to make effective use of the educational contents used in the new search engine. Characteristic metadata based on the customized LOM was created for a large number of educational resources produced in national projects conducted in accordance with government policy and registered in the search engine. This search engine will be open to the public on the Internet in May 2001. The educational resources will also be available over the Internet. We project that LOM use will lead to more effective use of educational resources in K-12 schools.

1. Introduction

There is an enormous variety of educational resources. Since they are not classified, there is confusion resulting from the many different degrees of difficulty, use objectives and resource types. Educators and students alike face various problems when trying to use these learning resources at school. It takes a great deal of time to find resources which match the learner's skills and/or needs. In addition, there is the issue of copyright infringement when learners use these resources at school. LOM has been investigated and will be standardized as one of the primary solutions to these problems. However, it has not yet been put into actual use in search engine applications. In order to put LOM to practical use, the customization to make it applicable to Japan's educational system is necessary.

Educational reform is now proceeding [4] in Japan. New National Curriculum Standards (NCS)[5] for teaching and information technology will be introduced into schools. Furthermore, many educational resources have been produced in various national projects. Looking at this from the standpoint of educational reform, now is the chance to put LOM into practical use. Since LOM use is seen as timely and important, we determined LOM specifications and vocabularies taking into account Japanese K-12 educational conditions and trends.

In this paper, we have described how to determine LOM specifications and vocabularies, and have included some samples. Furthermore, we show how to realize effective retrieval by applying LOM to newly developed search engines.

2. Process development committee for LOM determination

A committee of 55 educational authorities — including K-12 schoolteachers, university professors and corporate software developers — was established to determine LOM specifications and vocabularies. Referring to the emerging LOM standards (IEEE LTSC 1484.12[1]and IMS Version1.1 [2]), LOM specifications and vocabularies were determined through discussions from the standpoints of both users and resource developers. The LOM reference model mentioned in this paper is that reported by IEEE Version 4.0[1].

After determining the major problems in the use of learning resources in Japanese K-12 schools, LOM was customized to resolve them. Based on this customized LOM, we designed interfaces and search engine functions to let teachers and students from primary to senior high schools find the resources they are looking for.

3. LOM specifications and vocabularies used for problem solution

3.1. Learning Resource Classification

Learning resources in the LOM 5.2 reference model are classified in five main categories — materials, presentation type, use location, use application and use objective — to let users to find desired resources quickly and accurately. For example, some students may want to find a picture for use in a chemistry experiment. Some teachers may want to propose an issue to their students or help them understand a trend using statistical data. Other teachers may want to share their lesson plans with other teachers only. The five vocabulary categories in "Learning Resource Types" developed to meet these needs are shown in Table 1. Some new vocabularies have been added. "Use Objective" is included in the comments on how the resource is to be used in the LOM 5.10 "Description,"(Table2) with the addition of some mandatory vocabularies.

Table 1. LOM5.2"Learning Resource Type" vocabulary

Element	Vocabulary
Material	Movie, Picture, Sound, Text, Software
Presentation Type	Graph, Index, Slide, Table, Diagram, Simulation, Animation
Scene of Use	Research, Experiment, Observation, Examination, Appreciation, Discussion, Exercise
Role of Use	Lesson Plan, Case Study, Statistical Data, Reference, Questionnaire, Quiz, Workbook

Table 2. LOM5.10"Description" vocabulary

Educational Objective	Subject Discovery, Study Guide, Knowledge Complement, Understanding, Brain Storming, Evaluation, Ability Development, Activity Introduction, Lesson Planning, Experimental Learning, Cooperative Learning

3.2. Investigation of learner's skill level

A large volume of educational resources — with various degrees of difficulty and ranging from children to adult use — currently exist on the Internet. As a result, it is difficult for learners, especially children, to find resources they can understand. This problem can be solved by creating LOM user specifications like those shown in the reference model. As user specifications, we used both the LOM 5.5's "Intended End User Role" and the LOM 5.7's "Typical Age Range." But since most educational resources for schools in Japan are arranged to correspond to grade rather than age, we included grade vocabularies in the "Typical Age Range."

Here, we defined the LOM 5.5 as the intended user corresponding to the degree of difficulty of the educational resources and the LOM 5.7 as the grade in which these resources are studied. This means that if the resource contains information to be studied in 5th grade but has a high degree of difficulty, the vocabulary for "teachers" in the LOM 5.5 and "grade 5" in the LOM 5.7 should be applied. If the resource has a low degree of difficulty, if it is easy for children to understand, the vocabulary for "learners" in LOM 5.5 is applied. Using these LOMs makes the degree of difficulty of the resources clear for learners.

3.3. Measures Prevent Copyright Infringements

Students often download educational resources to their personal computers, revise them or upload them on their school's web sites without the author's permission. This may constitute a copyright infringement, because many teachers mistakenly see their actions as being permitted for use at school. If the boundaries of use approval are clearly defined for each resource, these problems can be avoided. To classify the scope of use approval for all resources, the category "Use Restrictions" was created as LOM 6.4 in the "Rights" section of LOM 6.

The following four vocabularies were established according to how educational resources are used in school lessons, as shown in Table 3.

Table 3. LOM6.4" Use Restrictions" vocabulary

Permission of Use	Scene of Use of Contents at School
Watch	To Watch Contents Through Browser or Print
Reproduce	To Download Contents with PC
Revise	To Revise Contents after Downloading
Distribute	To Upload Contents on School Web Site or Distribute Them as Printed Matters

3.4. Classification for easy retrieval

3.4.1. Classification in accordance with National Curriculum Standards (NCS) guidelines. Almost all schools in Japan provide instruction according to NCS guidelines, which describe what teachers should teach their students. Teachers and students are confused on the subject of finding learning resources for their lessons because these resources have not been classified according to these guidelines. LOM 9 "Classification" can be used for quick and easy retrieval. What's more, NCS guidelines will be revised in 2002 and remain unchanged for the following 10 years. Therefore, classification based on the revised NCS guidelines is seen as the best way for Japanese learners to find proper resources easily.

By gathering important keywords from the NCS, we created new vocabularies covering curricula from elementary to senior high schools and registered these as the recommended vocabularies in the LOM 9.2's "Taxon Path" specifications. These vocabularies have hierarchical structures. Since the number of these vocabularies is very large, one sample is shown as reference in Fig. 1.

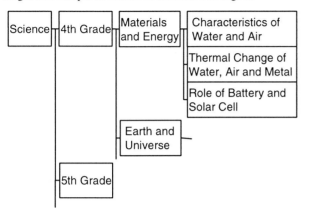

Figure 1. One of the vocabularies of "Science" according to the NCS

3.4.2. Classification by 5W1H. Another classification is describing features relevant to the questions, "Where", "When", "What", "Who" and "How." Students, especially primary school pupils, tend to retrieve educational resources from search target features relevant to 5W1H as well as from the curriculum [6]. As a result, this classification has been adopted to let them find these resources from search target features quickly and easily.

Vocabularies showing these features were extracted to cover words and phrases to be learned in primary school. This would appear to be the correct approach, since creating words covering all fields would be impossible. This classification also features a hierarchical structure, like that of the NCS. LOM specifications relevant to "When" and "Where" — temporal and geographical elements — were included in LOM 1.7's "Coverage," which covers time, geography and region. In the "When" vocabulary, name of eras such as the Edo era, which is peculiar to Japan, is used in the same way as centuries. In "Where," names of countries, prefectures in Japan and living places are included in the vocabulary.

LOM specifications related to "What," "Who" and "How" are included in LOM 9's "Classification" section. The "What" vocabulary is classified into living things, materials, earth and universe, environment, culture, industry and other categories. Each category is then further broken down into several sub-categories. The "How" vocabularies include "how to" (how to breed, how to live, how to make things, etc.), "mechanisms" (structure, function, causes, influences, etc.) and "things for the purpose of" (things for the purpose of eating, things for the purpose of living, things for the purpose of using things, things for the purpose of decorating, etc.).

Using these classifications makes retrieval easy. When a pupil wants the find information on the breeding habits of the insect a beetle, for instance, he or she can get it by selecting "insect" from "What" and "how to breed" from "How". Information on necessities of life in the Edo era can also be easily found by selecting "Edo era" from "When" and "things for the purpose of living" from "How."

4. LOM applications for search engines

4.1. Search engine characteristics

This search engine was developed for effective retrieval of educational resources using LOM and intended for students and teachers in elementary through senior high schools. LOM in the form of extensible markup language (XML) was automatically transferred to the search engine database as the search target. LOM use has made the search engine appropriate for a wide range of age groups by providing different retrieval interfaces suitable for different user grades. Retrieval by selecting LOM vocabularies as well as by normal keyword is also available. Titles, descriptions and other resource information also appears in the retrieval output display.

4.2. LOM application for effective search and retrieval

LOM use enables us to distinguish educational resources by resource type, school curriculum, degree of difficulty and use objective. This means that we can retrieve what we need by selecting these categories. In addition to LOM explained above, showing information on other related educational resources described in the LOM 7 "Related" section should be an effective way to expand learners' curiosity about another fields. We have also controlled users' rights of access by assigning different IDs and passwords. Using this access control, it is possible to the display the different retrieval interfaces suitable for each user grade. Moreover, resources necessary for teachers only, like lesson plans, can be displayed on the browser only when a teacher accesses it.

Search engine functions enabled by LOM applications are shown in Table 4.

Table 4. Search engine functions and LOM applications

Function		Applied LOM
Selection	To Select Educational Resources in Accordance with Material, Purpose of Use, Scene of Use, and others	LOM 5.2:Learning Resource Type LOM 5.10:Description
Selection	To Select Range of Permission of Using Resources	LOM 6.4:Permission of Use
Retrieval	To Retrieve According to the NCS	LOM 9:Classification
Retrieval	To Retrieve by Feature Relevant to 5W1H	LOM 9:Classification
Retrieval	To Retrieve by Keyword	All Information in the LOM and Text Written in Resource Itself
Retrieval	To Retrieve Suitable for Learner's Skill	LOM 5.5:Intended User LOM 5.7:Typical Age Range
Retrieval	To Expand Learner's Curiosity	LOM 7:Relation
Display of Output	To Restrict the Accessible Resources	LOM 5.2:Learning Resource Type
Display of Output	To Display Several Kinds of Information	LOM 2.2:Title LOM 2.5:Description Etc.
Display of Output	To Link Resource on the Internet	LOM 4.3:Location

4.3. Practical use LOM and search engines

LOM metadata was created for all educational resources (about 12,000) produced under the auspices of national projects in fiscal year 2000 and registered in the search engine database using a new LOM entry tool. By solving the problems described in the introduction of this paper, effective retrieval has been accomplished. This search engine will be open to the public in May, 2001, with many educational resources available over the Internet.

5. Conclusion

Based on the LOM reference model investigated by IEEE, we customized the specifications and vocabularies to fit Japanese K-12 school conditions. Effective retrieval has been successfully accomplished by LOM application to search engine targets. LOM was created for 12,000 educational resources and registered in the search engine. These resources will be available to students and teachers through Internet search engines. In the interests of LOM standardization for Japanese K-12 school educational resources, we will go on to conduct experiments at schools and evaluate LOM efficiency in depth.

6. Acknowledgments

This work was done as one of the projects supported by the IPA (Information-technology Promotion Agency) in fiscal year 2000. The authors would like to thank Professor Isao Kondo of Okayama University, Professor Kazuo Nagano of Sacred Heart University and other educators cooperating in the Konet collaborative research project [7].

7. References

[1] IEEE Information Technology - Learning Technology - Learning Objects Metadata (P1484.12) "Learning Object Metadata Working Draft (WD5) ", http://ltsc.ieee.org/doc/wg12/LOM_WD5.pdf.
[2] IMS Learning Resource Meta-data Information Model Version 1.1 - Final Specification, http://www.imsproject.org/metadata/mdinfov1p1.html.
[3] Advanced Learning Infrastructure Consortium (ALIC) http://www.alic.gr.jp/eng/index.htm.
[4] For example, http://www.cec.or.jp/e-cec/e-image/miti01.pdf
[5] National Curriculum Standards (in Japanese), http://www.mext.go.jp/b_menu/houdou/11/03/990302.htm.
[6] M. Shinohara, K. Tokuhata, M. Okamoto, F. Miyake, and K. Nagano, "Development and evaluation of the search engine for educationally useful resources having interface for children", Transactions of Japanese Society for Information and System in Education (to be published) (in Japanese).
[7] M. Shinohara, H. Saito, W. Adachi, S. Wada, S. Arai, N. Shinagawa and A. Nakazawa, "School Networking in Japan - Activity of Konet Plan", Advanced Research in Computers and Communication in Education, IOS Press, pp.1049-1052, 1999.

K-InCA: Using Artificial Agents to Help People Learn and Adopt New Behaviours

Albert Angehrn, Thierry Nabeth, Liana Razmerita, Claudia Roda
{albert.angehrn, thierry.nabeth, liana.razmerita, claudia.roda}@insead.fr
INSEAD - Centre for Advanced Learning Technologies
Bd. De Constance, F-77300 Fontainebleau France
http://www.insead.edu/CALT/

Abstract

This paper describes an artificial agent system, designed to help people in organizations to learn and to adopt knowledge-sharing practices. The agent, a personal assistant, continuously analyses the actions of the user in order to build and maintain a "behavioural profile" reflecting the level of adoption of the "desired" behaviours. Using this profile, and relying on a model borrowed from change management theories, the agent provides customised guidance, mentoring, motivation and stimuli, supporting the gradual transformation of the user's behaviours.

1. Introduction

In recent years, organisations have increasingly recognised the value of the "knowledge capital" distributed amongst their members. Managing this knowledge effectively represents a key factor for company's success in a competitive economy. In particular, support is needed for the phases of (1) knowledge acquisition, (2) knowledge manipulation, (3) knowledge sharing and (4) knowledge exploitation and transformation into value.

Setting-up effective Knowledge Management (KM) processes is not only a matter of time and resources. The process impacts on the structure of the organization and on the role of its members [3, 4]. One effective strategy to overcome this problem is to guide the members of the organization to the adoption of the desired KM behaviors.

The system described in this paper (Knowledge Intelligent Conversational Agents - K-InCA) relies on the idea of offering to each user an artificial agent that helps him to progressively adopt new behaviours. The agent continuously inquires the user in order to build and maintain a "behavioural profile" reflecting the level of adoption of the different "desired" behaviours. This agent provides guidance (such as suggesting some resources), mentoring (such as proposing some exercises to complete) and stimuli (such as questioning provocatively the user's beliefs [1]), leading to the progressive adoption of the behaviours.

2. Designing K-InCA: Key concepts

K-InCA agents rely on some key concepts and models borrowed from different domains such as Organizational Behaviour (change management), KM (knowledge sharing behaviours & attitudes) and Artificial Intelligence (user modelling, interface agents).

2.1. K-InCA as a change management agent.

The change management model embedded in the K-InCA system is based on the work of Everett Rogers on innovation diffusion [5]. Roger describes the innovation-decision process as "the process through which an individual [...] passes (1) from first knowledge of an innovation, (2) to forming an attitude toward the innovation, (3) to a decision to adopt or reject, (4) to implementation of the new idea, and (5) to confirmation of this decision" [ibid, p.161].

K-InCA agents support this model by guiding the learners through a set of adoption stages (awareness, interest, trial and adoption), and by providing them with information and stimuli customised to their profiles and levels of adoption. Rogers' model also distinguishes different categories of users according to their attitude towards innovation. K-InCA agents choose an "interaction style" adapted to the attitude towards innovation of the user, providing more information to the innovators, whereas the social component will have a more important role for less innovative categories of population.

2.2. K-InCA Behaviours and User Model

K-InCA agents can be thought of as customisable agents, supporting the adoption of behaviours within a given domain. In our first implementation, K-InCA agents are designed to support KM behaviours. Behaviours are stored in a hierarchical repository. Action descriptors are associated to the leaf behaviours of the hierarchical structure. K-InCA agents use these descriptors to recognise the behaviours that can be attributed to a user who has performed a given action.

K-InCA agents maintain a user model in order to adapt their actions to the specific level of adoption of the user and to his preferences. The components of the user model most relevant to this discussion are:

- **Change state** - the user's state with respect to the adoption of the desired behaviours (e.g. the user is unaware of the "I recognise expertise" behaviour).
- **KM agenda** - KM activities that have been proposed to the user by the agent.
- **Preferences** - user's interests and skills.

Other components of the user model include descriptions of the user's personality (e.g. resistance to change, learning preferences), his social network (friends, acquaintances, superiors, peers, etc.), and a history of his actions.

3. A K-InCA Scenario

The following scenario presents a typical exchange between a user and his K-InCA agent. Figure 1 is a screenshot of the interaction space at the end of the scenario.

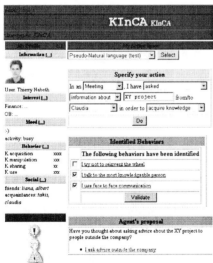

Figure 1 – K-InCA system screenshot

The left frame reflects the content of the user model[1]. The user reports an action he has performed and expects to receive a diagnosis on how well he is managing his knowledge, how he could improve his management process. Table 1 shows a sample of the user-KInCA agent interaction. Following the initial user input, from the *Specify your Action* panel, the agent attributes a set of behaviours to the user and displays them in *Identified Behaviours* panel and updates the user model

```
The User: In a 'Meeting', I have 'asked'
'information about' 'XY project' to/from
'Claudia' in order to 'acquire knowledge'.
The Agent: identifies the associated behaviours
(the users uncheck the wrong ones):
  e.g: * I talk to the most knowledgeable person
       * I use face-to-face communication
Claudia is added to user's acquaintances.
The User: Validates the identified behaviours
The Agent's proposal: Have you thought about
asking advice outside the company? etc.
```

Table 1 - Sample of user-KinCA agent dialogue

The user may continue by supplying a new action description, or by clicking on any of the behaviours names to browse the behaviours' repository, or to update the user model.

4. Conclusions

A first prototype of the system was realised using Zope/Python web application server system. A second version is being developed in java (using servlet technology). However, the K-InCA system is in an early stage to evaluate the soundness of the approach from an individual and organizational perspective (will it not be considered too intrusive for the user? what will be its substantive pedagogical value for the organization?). However, this work seems to indicate that it is possible to create learning tools offering deep support of the learning process, and covering the entire knowledge acquisition cycle.

5. References

[1] Angehrn, A.A., "Computers that criticize you: Stimulus-based Decision Support Systems," *Interfaces*, 23, 3, 1993, pp. 3-16.

[2] George Lawton, "Industry Trends: Knowledge Management: Ready for Prime Time?", *Computer*, Vol. 34, No. 2, February 2001.

[3] Jonathan D. Day and James C. Wendler, "Best practice and beyond: Knowledge strategies" *The McKinsey Quarterly*, 1998 Number 1.

[4] Jean-François Manzoni, Albert Angehrn, "Understanding organizational dynamics of IT-enabled change: a multimedia simulation approach" *Journal of management information systems*, vol. 14, no. 3, (winter 97-98) pp. 109-140

[5] Everett M. Rogers, "Diffusion of Innovation", 4th edition, Free Press, NY, 1995. First edition by Everett 1962, same title.

This research work is funded by Xerox Corporation.

Human Teacher in Intelligent Tutoring System: A Forgotten Entity!

Kinshuk, Massey University, New Zealand
Alexei Tretiakov, Massey University, New Zealand
Hong Hong, Massey University, New Zealand
Ashok Patel, De Montfort University, United Kingdom

Abstract

Intelligent Tutoring Systems (ITSs) have not yet proved very successful and one major reason seems to be that research on ITSs has largely failed to recognize the role of the teacher in the ITS design process. This paper discusses an undergoing project at Massey University, which is incorporating a 'Human Teacher Model' in an ITS prototype to teach Japanese.

The project identifies the teacher attributes and formulates them into a coherent teacher model. They are then applied in the prototype, which offers adaptivity to teacher at two levels: presentation based adaptivity and navigation base adaptivity. We believe this work will substantially improve the applicability of ITSs in real academic environment.

1. Introduction

Although computers are being used at all levels in curriculum, their use for automated tutoring has been quite limited, in spite of decades of research in Intelligent Tutoring Systems (ITS), Expert Systems and Hypertext as well as continuing attempts at more procedural and simple systems. The researchers in the field are now hoping that the new avenues of Multimedia, Hypermedia and Internet will somehow resolve the lack of real progress. It appears that inadequate attention has been paid to the area of application, and the user modeling aspect has received more than its fair share of attention at the expense of adequate consideration of the fundamental educational issues such as what constitutes knowledge, what are the methods of knowledge acquisition for these constituents, how suitable are computers for knowledge transfer in these constituents and how to test, using computers, the level of acquired knowledge [1].

The attention received by the user modelling aspect has failed to address the diversity and richness of the educational environment, even in terms of the typical user attributes - nursery to post-graduate schools, adult education, distance education and diverse subject disciplines are indicative of the size of the problem.

Kinshuk [2] pointed out that an intelligent tutoring system (ITS) is a 'joint cognitive system', where the human psychology of the student and teacher and the cyber-psychology of an ITS - which reflects the psychology of the ITS designers - all interact. The student-ITS interaction, where the ITS is non-trivial, extends far beyond the scope of the routine Human-Computer Interaction (HCI) studies as it is a convergence of the human psychologies of a student and a teacher and the cyber-psychology of an ITS - reflecting the psychology of the ITS designers including their perception of students, teachers and the learning process in an interactive but nevertheless automated ITS.

Therefore, we argue that ITSs have not yet proved very successful and one major reason seems to be that research on ITSs has largely failed to recognize the role of teacher (except perhaps providing few pedagogic rules) in the ITS design process [3]. In recent years, whereas the work on student modelling has benefited by the user modelling research in the field of HCI, the research on the role of a human teacher as a collaborator in the computer integrated learning environments is almost non existent.

This paper describes an ongoing project at Massey University funded by MURF Funding that is researching, developing, and evaluating the incorporation of a "Human Teacher Model" in the design of an ITS.

2. Motivation for the work

The over ambitious nature of the traditional ITS research also manifests itself in the almost complete exclusion of a human teacher's role in the process of computer assisted learning. In the absence of an ITS that can fully replace a human teacher, it is essential to view any tutoring system as a joint cognitive system [4]. The teacher plays various roles including those of setting the scene, providing context, selecting and scheduling other educational technologies, managing the curriculum and overseeing the learning progression. In the ensuing power relationship, the preferences of a tutor may prove to be more important than the learning style of a student. Identifying these preferences is a difficult task as teachers have different personality and different teaching styles born out of their traditional, progressive or vocational outlooks and possibly their own learning style [5].

For introducing the teaching styles in the design of ITS, and effective collaboration between the teacher and an ITS in the process of student education, it is necessary that the ITS can adjust to different teaching styles, at least in a broad sense. The current practice of a group of teachers using authoring tools to create an ITS is likely to produce an ITS that is biased towards a particular teaching style if the group thinks alike or a compromise that may fail to satisfy anybody. Unless the ITS has some framework is to provide adaptivity to the teachers by monitoring the behavior of how the teachers use the system for creating, adding and customizing the courses.

3. The prototype

The following section will present the detailed architecture of the project prototype, based on the architecture of the existing ITSs.

Based on the classical ITS architecture, we have added a Human Teacher Model within the ITS framework which interacts with other components to provide adaptation not only to suit student preferences and competence level (based on student model) but also to incorporate teacher's

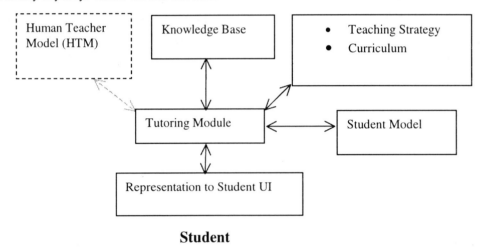

Figure 1. Architecture of ITS with Human Teacher Model

conception of a teacher through a teacher model and enables its configuration to suit the collaborating teacher, it may not find easy acceptance. Alternatively, if research shows that certain areas of a subject discipline are taught in a similar fashion by a very large body of teachers and there is an intrinsic affinity between a particular area of learning and the tutoring style adopted by an ITS then the teacher model can remain static and implicit [2].

Our objective has been to identify the teacher attributes required in the ITS design process, then to develop and incorporate a 'Human Teacher Model' in the design of an ITS. For demonstrating how the 'Human Teacher Model' works, we have created a framework, which allows teachers to create on-line courses based on a set of pre-fabricated teaching units by combining units in a WYSIWYG tool. At the price of limiting possible relationships between units, we strive to achieve easy to use interactive interfaces, which allow teachers with very minimal computer skills to create original courses, based on re-used content. The most important feature of the

attributes during the learning process (Figure1).

In order to test and evaluate, we created a basic test bed prototype by incorporating the Human Teacher Model into an ITS for teaching Japanese.

3.1 Material structure

As a testing prototype to be implemented, a kit for teaching Japanese language at beginners level has been built. Material can be organized by the teacher as a collection of units.

There are 3 types of units: section, subsection and normal.
- Sections are the highest-level units. Each section comprises a complete domain topic;
- Sub-sections relate to various concepts within a topic, may be multi-level in hierarchy and contain the related normal units;

- Normal units are contained in sub-sections and carry the course content with multiple outgoing links.

Section and subsection units have no content and no outgoing links. When section and subsection units are rendered, they can show titles and summaries of the units they contain. Normal units carry the course content. A normal unit has content (HTML) and multiple outgoing links. It can contain images, Java applets, sound and any other multimedia. The presentation type of course (e.g. using sound or not, in the Figure 2 teacher selected to use sound indicated by speaker) can be decided by teacher, or by the system based on monitoring of the preferences of teacher.

A unit can be included in the course or can be kept in reserve list. Section units can contain subsection units, and subsection units contain normal units. All units are added in sequential fashion; a unit can be added once all of its prerequisites have been included in the course.

A link has its destination (the ID of the target unit), title (text), teacher's annotation, and indication if the link is enabled.

3.2 User interface

The user interface has two panes (Figure 2): Course Structure pane (left), and Unit Contents pane (right). Course structure pane shows all enabled/added units as a tree of icons, and Unit Contents pane displays the currently selected unit's course content.

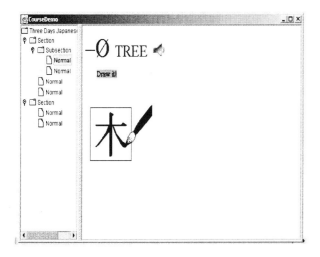

Figure 2. A screenshot of user interface

4. Adaptivity to the teacher in the prototype

4.1 Teacher operations

The main operations offered to teacher are:
- including/removing/replacing units
  - ♦ Include a reserve unit into the course (under an appropriate container);
  - ♦ Remove a unit from the course, putting it in reserve;
  - ♦ Replace a unit by the one in reserve (placing the current one in reserve).

All operations list above should preserve prerequisite structure.
- changing units:
  - ♦ Annotate unit;
  - ♦ Enable link;
  - ♦ Disable link;
  - ♦ Annotate link.
- course rendering: at any time, teacher should be able to render the course as an HTML site or a PDF document.

4.2 Adaptation levels

When a teacher attempts to create a new unit, adds various presentation components, or customizes previously available units, system monitors the interaction of teacher and stores the profile of teacher's preferences.

The adaptation occurs at two levels:
- first level, presentation based adaptivity where the system monitors and stores the preferences of teacher for certain types of presentation items and offer the similar items to the teacher at later time. As mentioned above, the system offers choices within the units for alternative presentations;
- second level, navigation based adaptivity where the system classifies the units, monitors the type of units teacher is adding preferentially, and then offer similar units at later time (leading to a stage where system can offer completion of a course half-way through). This adaptation takes place when system has a good understanding (in system's assumptions) of teacher's "teaching style".

5. Conclusion

The project we outline in this paper attempts to identify the teacher attributes required in the ITS design process, and formulates these attribute to develop a framework, which provides two levels of adaptivity to the teacher – presentation based adaptivity and navigation based adaptivity.

We believe that the project will help not only in opening up a new and very promising research area, but also in moving more attention from technical possibilities and the fascination on to the basic educational issues, which are essential to produce any substantial and usable tutoring system.

To determine the feasibility of the human teacher model, the prototype will soon be evaluated by some Japanese teachers. Also, at this point, the prototype contains a simplistic Human Teacher Model. Once the approach is validated, more sophistication will be added to the model for better and advanced adaptation.

6. References

[1] Cumming, G. (1993). A perspective on learning for intelligent educational systems. *Journal of Computer Assisted Learning*, 9 (4), 229-238.

[2] Kinshuk & Patel, A. (1996). Intelligent Tutoring Tools: Redesigning ITSs for Adequate Knowledge Transfer Emphasis. In C. Lucas (Ed.) *Proceedings of 1996 International Conference on Intelligent and Cognitive Systems*, Tehran: IPM, 221-226 (ISBN 964 90010 3 4).

[3] Kinshuk, Patel, A., Oppermann, R. & Russell, D. (2000). Human Teacher's Role in Emerging Internet-based Intelligent Tutoring. *Supporting the Learner through open, flexible and distance strategies: Issues for Pacific Rim Countries*, Wellington: The Distance Education Association of New Zealand, 209-217.

[4] Dalal, N. P. & Kasper, G. M. (1994). The design of joint cognitive systems: the effect of cognitive coupling on performance. *International Journal of Human-Computer Studies*, 40, 677-702.

[5] Entwistel, N. (1981). Teaching Styles, Learning and Curricular Choice. *Styles of Learning and Teaching*, John Wiley & Sons Ltd., 225-242 (ISBN 0 471 10013 7).

Multi-Media as a Cognitive Tool: Towards a Multi-Media ITS

Fawzi Albalooshi
Department of Computer Science
College of Science, University of Bahrain
P. O. Box 32038, Isa Town, Bahrain
fbalushi@batelco.com.bh

Eshaa M. Alkhalifa
Department of Computer Science
College of Science, University of Bahrain
P. O. Box 32038, Isa Town, Bahrain
ealkhalifa@sci.uob.bh

Abstract

Cognitive tools are instruments integrated into a learning environment to provide representational support to learners and allow them to focus on the subject matter [5]. Two of these; namely, animation and verbal representation are in a constant contest of effectiveness, without any noticeable differences, except in particular questions concerning conceptual versus procedural knowledge. Therefore, the effects of individual media may change according to both subject matter as well as individual preferences. Here, we offer students a Multi-Media representational module, which has been designed with the aim of incorporating it into a full scale Multi-Media Intelligent Tutoring Systems. Results show that allowing animation and verbal descriptions to interact within the same system may result in improvement of student levels up to 40% from their post classical- classroom session. This is indicative that perhaps, the two media may have ambiguous internal factors that affect learning in a way that disappears when each is analyzed in isolation of the other.

1. Introduction

Intelligent Tutoring Systems (ITS), offer a great deal of flexibility in control, making them highly adaptable to individual student progress. This makes them excellent candidates to play the role of "Cognitive Tools" [5]. These tools are capable of supporting learners by explicitly representing information. They allow learners to see the structure of the cognitive process by externalizing it and freeing memory for the more important learning task at hand. The simplest form of a tool is a pen and paper, where students can write notes to remind them of the numbers involved when performing addition. Therefore it should not be surprising that computer based educational systems impose themselves at the top of the list of Cognitive Tools.

However, the sudden growth of multi-media computer systems necessitated the need for a deeper understanding of the characteristics of each of the different media. Norman [8] indicates that each media has "affordances" and "constraints" that would be either beneficial or counteractive to educational goals. Complexity grows exponentially when the aim of the selection is to include these media into a shell representing an adaptable ITS system. The shell itself would be flexible to individual student needs. Therefore, it should not be surprising to see research start off in highly controlled specific cases.

For instance, Sharples and du Boulay [10] argue that learning medical concepts is normally acquired through induction. This is done by showing students several scenarios and allowing them to generalize their own models over the possible cases. The problem with this approach is that it leads students to over generalization because they are not always exposed to the extreme possibilities. However, when students are exposed to a controlled set of images through a computer-based tutor then highlighting the extreme cases becomes possible and the problem is alleviated.

Another experiment tested if individual differences have any effect on solving syllogistic reasoning problems. These are usually in the format, A is related to B with a premise, B and C are related with a second premise, then the subject's task is to say what, if anything, follows from the given information. These problems can be solved either through drawing a diagram, Euler's circles or through natural deduction using symbols. Monaghan and Stenning [7] categorized subjects according to their performance in the paper-folding test (PFT) as designed by French, Ekstrom & Price [3]. The test requires subject to visualize the array of holes that result from a simple process. A paper is folded a certain number of folds, a hole is made through the folds then the paper is unfolded. Students are asked to select the image of the unfolded paper that shows the resulting arrangement and results are discriminated along a median split as high versus low visualization abilities. Students of both groups where then split into two groups and taught how to solve syllogisms either through Euler's circles, which is graphical, or through natural deduction, which is serial.

Following this, students were given a test with 8 syllogisms each, selected to cover a range of difficulties. They were instructed to solve them in one of the two ways according to the way they were taught. Results showed that those who scored high on the PFT test made fewer errors when taught to solve them using Euler's circles than their serialist counterparts, who scored low on the PFT test when given the same teaching method. Oddly enough, this influence only seemed to take place in the final stage or in the translation of the results from the graphical modality into sentential form. The most important result is that most subjects would either perform better when taught verbally or when taught through diagrams according to their abilities or preferences. Additionally, those with visualization abilities seem to need a stage of "translation" from one modality to the other. This leads to one main conclusion. The ideal way of describing a process to students must allow for possible individual differences.

2. Verbal/Pictorial vs. Animated

Now that both methods of representation seem necessary, a question arises as to whether one can subsume the other and present itself as the "ideal" method of teaching the behavior of processes. In short, **is animation the "ideal" way?** Well, evidently from research, there seems to be a serious difficulty in getting clear-cut results to say that animation is more effective than verbal/pictorial representation or vice versa.

Pane, Corbett, and John [9] ran a detailed study to assess the effects of dynamics representation in a computer-based system that teaches developmental biology. They compared animation to a textual description that is enriched with carefully selected still images. They found no difference in student performance when declarative questions are given. Another study [6] showed that with respect to teaching algorithms, "active laboratory" sessions seemed to result is better student performances. During these sessions students created their own algorithms and saw them animated. They performed better in "procedural questions" as compared to students who were exposed to animations of previously selected examples.

Two other experiments showed that animations might aid students in procedural knowledge by allowing them to "predict" the next step in an algorithm's behavior. However, similar results were found when students were asked to predict algorithm behavior from static diagrams [2]. Then what role does animation play?

> *"When the perceptual system cannot directly perceive change over time, it will seek out implicit evidence of change."[4]*

Perhaps these findings are not as surprising as they may seem at first sight, if we presume dynamic mental representations. Freyd [4] showed through several experiments the existence of a memory distortion that represents a shift forward to the next expected state when even one image is shown. One of her experiments involved two static images of a man jumping off a wall. A subject is shown one image first. The subject would then be shown another image and asked whether they are the same. For example, if in the first image, the man is in the air, then subjects would readily identify that the image of the man standing on the wall is not the same. On the other hand, they would take longer to identify the difference if the order of the images was reversed.

This implies that subjects "expected" the second image to follow the first one temporally. This order was maintained in the experiments described above. In fact, the first study on developmental biology [9] replaces an animation with a sequence of four images that show different screen shots of stages in the animation. These images, according to dynamic representation are no different from exposure to the animation itself because a dynamic cognitive representation would fill in the gaps.

If the two representations appear similar, then where lay the difference? The clearest difference is that when images are presented as a cognitive tool, the externalised representation carries less information. Cognition has to account for the "expected" stages to recreate the complete animation, which is cognitively taxing. A more interesting difference though, seems to lie in the predictive ability of animation in showing students the "direction" of thought when images can frequently be unordered. When a sequence is shown, and then an image, a student may be more readily prepared to "predict" as is the usual requirement in procedural type questions.

3. More Choices with Multi-Media?

So far, work has shown that there is a strong reason to believe the existence of individual differences. Therefore, an educational system that provides the two representations that are associated with these differences is unlikely to do worse, than those that include either one or the other. The idea is to cover for individual differences in preferring one representation to the other as well as to provide them both in parallel. Only through this, can any interaction between the two be assessed. If there is no interaction, then one would render the other redundant and the total impact will be no better than that obtained in the positive experiments described. If on the other hand, an interaction does exist and is a negative one, then each of the representations would negatively affect the other. Performance would worsen following a classroom lecture as "confusion" may result. However, if the interaction is a form of fortification then an improvement that exceeds expectations will result. This would imply the existence of a positive interaction between the two modalities.

4. Multi-Media Data Structures Tutoring System

The selected subject matter was Data Structures including the concepts of Stacks, Queues, Lists and Trees. The system is a precursor for a Multi-Media Intelligent Tutoring System, which is currently under development. These topics were presented to students in both media simultaneously; animation and verbal description. The screen was therefore divided into two windows; one containing a carefully written description of the concept and the other an animation that the student can start, stop, and partially control.

The module itself is represented as a Java Applet with the aim of placing the system on the Internet. It has several sections each concerned with one of the topics listed above and each in turn has several screens associated with it representing Terminology, Operations, Examples and Quiz. The Terminology page explains the basic terminology students need to learn for that data structure and is purely verbal. The Operations page shows and explains through text and animation the basic operations that can be performed. Examples include preset examples represented again through both animation and verbal representation. The Quiz page is a student self-assessment exercise.

Students are given the full navigational freedom to go to any page they wish and repeat the animations included as many times as they wish within the specified time allotted for the experiment. They also had the ability to control the speed of the animation by selecting a number from 1 to 6. This was included because students complained about the slow speed in the experiments run by Pane, Corbett, and John [9]. The loading time of the applet was somewhat slow but the running time was appropriate since the subject matter covered only the basic essentials of each topic. Students were urged to then think of new possible cases with the basics they were shown.

5. Evaluation of the Module

The multi-media module was tested through an experiment that compared its effects on student performance to standard classroom lectures. Additionally, the experiment tested its effects on students who already attended the classroom lecture. Predictions, made are that it will not result in a lower level of performance than the classroom lecture, while it will be able to result in a highly significant improvement in student performance from their post-lecture test levels.

Students were distributed into three groups based on a quiz they were given earlier in the course to ensure that all groups are comparable. Group 1 attended the lecture only while group 2 attended the lecture and the following day used the module. Group 3 on the other hand, did not attend the lecture and just used the module on the second day. Both groups 1 and 2 took a test at the end of the classroom lecture. All groups took the second test on the second day, which was highly similar to the first test with a difference in the order and wording of the questions.

5.1. Subjects

45 students from the University of Bahrain volunteered to participate in this experiment in exchange for class credit. They were distributed evenly into three groups of 15 students each.

5.2. Materials

Materials included in this experiment concentrated on Stacks as a data structure. They included one classical lecture given to groups 1 and 2. Additionally, use of the multi-media module for that particular data structure by students in groups 2 and 3. Then the tests included 7 questions, which tested comprehension of the various parts of the presentation as well as the ability to recreate or imagine new uses or applications of stacks.

5.3. Results

Group 2, showed a highly significant improvement in test results following using the system when compared to their post-classroom lecture levels. An ANOVA test showed $F= 9.19$ with $p< .005$. No significant differences were found between group 1 who attended the classroom lecture only and group 3 which used the system only. In this case, $F=.598$ with $p<.446$ which shows that they are extremely comparable.

When a comparison is made in individual questions, an interesting phenomenon seems to take place. Students who attended the lecture only, group 1, were only better at the question "Using an example, explain briefly the stack concept and its possible uses?" than students of group 3 who used the system only. This was with a significance of $p< .03$.

Additional information was found by comparing student performance in similar questions in the pretest before using the system and the post-test after using the system. An ANOVA test compared the grades of the same students in both situations on a per question basis. Most of the difference or improvement came from the questions "Using an example, explain the stack concept and its possible use?", "How could we implement a stack in a program?" and "List the data variables and operations associated with the stack?". The significance was $p< .000$ in all three cases.

6. Discussion

Results seem to indicate strong positive interaction between animation and verbal representation with results in such a strong improvement in student levels. This provides strong support to the predictions made here in that having the two modalities in parallel may have better results than having each on its own. The question at which classroom only students did better was interested in showing how representation may imply a "limiting" effect to imagination. When students are presented with examples that take some form, it becomes more difficult for them to break out of the boundaries of that example and find another. Students who used the system seemed to be directed towards how a stack functions rather than application areas and showed this in their responses.

This implies that the effects of animation require further analysis to identify if they include a "channeling" effect that restrains students from broadening their scope of imagination. This is beneficial with respect to some domains and a disadvantage with respect to others. For example, a graphical model of a GRE problem is shown [1] to prohibit common errors simply because it includes "directionality" represented by arrows. The diagrammatic method allows students solving a problem to follow only existing arrows from their start to end. In this case, students are "channeled" into thinking in the correct direction.

The current system, exhibits some of that effect in a beneficial fashion. It also shows the positive effects of including both animation and verbal descriptions in parallel with a highly significant degree of improvement. This is highly promising since this project aims at utilizing this module as the heart of an Intelligent Tutoring System.

7. Future Directions

The tested system was implemented as a Java Applet, which did not take long to show its limitations with respect to flexibility, functionality and speed when placed on the Internet. A follow up system was designed and is being implemented as Java Servlets. These are server side processes that are responsible for filling up the different parts of a frame-based page. These areas are divided into three main areas, one for the verbal description, the second for the animated Flash file and the third for interaction with students. In short, Servlets offer the ability to call any one of the verbal descriptions and the matching animation according to student progress ensuring adaptability. It would be interesting to find out if students would prefer a particular type of animation for a particular section and prefer verbal descriptions for another. Additional tests could be done to students similar to those for individual differences to test if any interaction occurs with the multi-modal representation.

Acknowledgments

The second author would like to thank god, first and foremost for insight, and for giving her a guiding star. Both authors wish to thank Prof. Waheeb AlNaser for his support. This research is supported by a grant from the Deanship of Scientific Research, University of Bahrain.

References

[1] E. M. Alkhalifa, Why Re-Invent the Wheel? Insights from Cognitive Research on Continued Education in Logic, *A Millennium Dawn in Training and Continuing Education (conference and exhibition)*, 2001.

[2] M. D. Byrne, R. Catrambone, J. T. Stasko, Evaluating animation as student aids in learning computer algorithms, *Computers and Education*, 1999, v. 33(4) pp. 253-278.

[3] J. W. French, R. B. Ekstrom, L. A. Price, *Kit of reference tests for cognitive factors*, Princeton, NJ: Educational Testing Services, 1963.

[4] J. Freyd, Dynamic Mental Representations, *Psychological Review*, 1987, v.94(4), pp.427-438.

[5] W. R. van Jooligan, Cognitive tools for discovery learning, *International Journal for Artificial Intelligence in Education*, 1999, v.10, pp. 385-397.

[6] A. W. Lawrence, A. N. Badre, J. T. Stasko, Empirically Evaluating the Use of Animation to Teach Algorithms, Georgia Institute of Technology, Technical Report GIT-GVU-94-07, Atlanta, 1994.

[7] P. Monaghan, K. Stenning, Effects of representation modality and thinking style on learning to solve reasoning problems, In *Proceedings of the 20th Annual Meeting of the Cognitive Science Society*, Madison Wisconsin, 1998.

[8] D. Norman, *The design of everyday things*, New York: Currency/ Doubleday, 1988.

[9] J. F. Pane, A. T. Corbett, B. E. John, Assessing Dynamics in Computer-Based Instruction, In *Proceedings of the 1996 ACM SIGCHI Conference on Human Factors in Computing Systems*, Vacouver, B.C. Canada, April 1996, pp.197-204.

[10] M. Sharples, B. du Boulay, Knowledge representation, teaching strategy and simplifying assumptions for a concept tutoring system. In *Proceedings of European Conference on Artificial Intelligence*, 1988, pp. 268-270.

A Case-based Agent Framework for Adaptive Learning

CHIEN-SING LEE, YASHWANT PRASAD SINGH
Faculty of Information Technology,
Multimedia University, 63100 Cyberjaya,
Selangor, Malaysia.
e-mail: {cslee, y.p. singh@mmu.edu.my}

Abstract

An adaptive learning system centres on the learner's needs and meeting those needs at his or her level or pace. Designing the nature and quality of adaptive interaction that will facilitate association of prior knowledge and current stimuli is thus crucial. Case-based reasoning is proposed as the mode of inference or learning in an agent-based context. An agent is chosen because of its ability to reason like a human and its ability to take on the human role of mentor. Considering that information is often chunked in granules in the form of concepts, the case-based agent's feedback should strive to encourage the formation of associative memory among these granules that will meet respective learning goals through various forms of media and representations of several types of knowledge. Concepts are situated in contexts and the granules for contexts are cases denoting a problem situation. Hence, quality interaction design will have to involve a cognitive interface that provides salient details, which will trigger associations with experiences and previous cases. In cases where case-based reasoning can only provide possible alternatives, induction serves as a bridge to communicate the underlying propositions or units of meaning in the working memory.

1. INTRODUCTION

Intelligent multimedia presentations that adapt to the user's specifications of goals or tasks, abilities and pace of learning will not only reduce the problem of disorientation but also the problem of cognitive overload. The application of artificial intelligence techniques will provide alternatives in terms of feedback to the user and navigational choice. This improves the quality of interaction and meaningful learning.

Meaningful learning further requires active dialogue and collaboration between the user and the computer system. However, the designer's perception of the learner's needs and suitable strategies for achieving the learning goals may be biased to the designer's experiences or personal preferences.

Hence, the designer may not have taken into account all factors affecting the learner. The learner can sometimes be trapped in the system model (the computer interface) that externalises the underlying design. However, with intelligent interfaces, the system now adapts to the learner and helps the learner to interact meaningfully by providing more control to the learner.

It is not possible to anticipate each and every path that the learner will undertake. Therefore, an intelligent application that adapts its content should be able to infer gaps in the user's approximation of the expert model and generate explanation facilities, which would aid the learner to navigate with better learning strategies. Considering that quality interaction lies in involving and engaging the learner in a dialogue that meets his or her learning needs [1], a design process that generates high quality interaction will undoubtedly encourage the learner to immerse himself or herself in learning. Since most of us learn not only from rules but also from experience, case-based reasoning is chosen as the control mechanism upon which inference will be made in cases where there is uncertainty or where rules do not provide any exact answer.

This paper thus considers the rationale for case-based agents, the granularity of information as exemplified by concepts, concept classification, the relationship between goals, tasks and concepts and the role of quality interaction in facilitating information pickup.

2. CASE-BASED AGENTS

The agent in artificial intelligence refers to a system capable of interacting independently and effectively with its environment via its own sensors and effectors (data input-output devices) in order to accomplish some goals or self-generated tasks. An agent is considered here as it exemplifies human abilities of reasoning and can provide an emotional association as well as provide feedback as a mentor or coach [2]. The agent is able to monitor the learner, provide feedback, examples and explanations. Furthermore, the agent takes on a human character or personality -- someone that the learner can relate to who is neither condescending nor too fast-paced as the agent adapts to the learner's ability and pace. Granularity in presenting the reasoning process follows the cognitive apprenticeship approach. Presentation of lesson or feedback graduates from modelling to coaching. Finally when the learner has reached an adequate level of problem-solving skills, the apprentice takes control and determines his or her own strategies based on prior learning [3].

Inferencing mechanisms underlying feedback to the learner are usually rule-based. However, the inferencing capability of knowledge-based expert systems mainly caters to well-defined domains. As such, knowledge-based systems may constrain the learner only to predetermined declarative knowledge. However, real world problems are ill-structured. The schematic representations called for will thus vary to a greater extent. Hence, when rules are not found as a means of determining the next step of action or to draw conclusions, the learner needs to refer to the most pertinent cases from memory for previously experienced solutions (cases) in human problem solving. These cases can be used to draw inference to fit the new situations resulting in new solutions.

Since events around us seldom change at a rapid rate and often recur, it is probable that similar solutions can be applied to almost similar kinds of problems. Therefore, what is important is to recognise the similarities in the problems faced in order to classify the situation correctly and find an available match in previous solutions or to regard the previous solution as a starting point for exploring possible solutions [4]. In cases where there may be more than one solution, the utility of the solutions need to be weighed and compared.

Prior solutions with any representation scheme (data) can also serve to provide pointers regarding possible reasons or factors that will lead to failure. Case-based reasoning is therefore more viable in situations where rules are not sufficient as cases are less likely to contradict each other compared to inferencing from rules, which may result in contradiction among rules or between hypotheses and evidence. This however, does not discount the use of rules in inferencing as rules can influence similarity judgements and summarise cases.

Since cases are dependent on inherent concepts, greater accuracy in the classification of concepts will lead to better conceivable classifications of situations and a better basis for further induction.

3. CONCEPT ACQUISITION

Concepts can serve multiple functions [5]. Of interest to this paper are the cognitive economy, linguistic, inferential and epistemological functions.

Cognitive economy is similar to the inheritance characteristics of object-oriented programming whereby the objects inherit attributes from their parent classes. If a new concept is classified as belonging to a certain category, then it takes on the attributes of that category. This aids association of concepts as hooks are created to existing concepts whether to prerequisite or to related concepts. Linguistic functions are another aid to memory associations as they provide synonyms to existing words. This provides an equivalence relationship that facilitates retrieval from the conceptual index.

Perceptual information provides the stimuli to inferencing non-perceptual information. Multiple forms of media facilitate inferencing when one medium supports the other in what Najjar calls dual coding [6]. For example, a graphic or a simulation of a process helps the learner to visualise textual presentation of concepts. One media thus provides affordance to the other whereby one media reminds the learner of the related links to the other media. This is representative of the underlying semantic web. However, the inferencing function can only lead to accurate associations of concepts if the epistemological function indicates clearly classification criteria that filters which entity is an instance of which category.

The problem arises when there are inadequate features to classify them according to existing categories in declarative knowledge [7]. Therefore, a probabilistic view of concepts is suggested whereby membership in a certain category is graded instead of a binary yes or no. The more number of similar features to a category will attach a higher membership value to the new stimuli. This concept is similar to the certainty value attached to hypotheses in rule-based systems. The other possible view, the exemplar view, is similar to case-based reasoning except that case-based reasoning deals with situations whereas the exemplar view deals with the micro level of situations i.e. concepts. In the exemplar view, new input is classified according to the number of similarities with known examples of that particular category.

An integration of both views will result in an initial categorisation effort according to the exemplar view failing which the probabilistic view comes in. In other words, where case-based reasoning does not supply a probable answer, further induction based on the solutions arrived at by case-based reasoning, becomes the alternative.

4. GOALS, TASKS AND CONCEPTS

Information processing in the human framework is basically an attempt to contextualise knowledge in terms of concepts and situations based on associations and rules. This belief situates cognitive processing in the context of activities or experiences. A combination of case-based reasoning and induction can serve to situate learning and therefore lead to situated actions that will achieve the intended goals.

The GOMS (Goals, Operations, Method and Selection) model attempts to describe the interaction between goals, operators, methods and selection rules but presupposes that the learner is sufficiently skilled to make the right choice.

Furthermore, the tasks involved are serial in nature and seldom consider differences among learners. External factors that affect the learner such as his or her environment and the context in which the task is situated are often neglected [8].

Adaptive learning requires a flexible framework that relates concepts, goals, representation and the presentation material i.e. the media. A connectionist approach is suggested [9]. Similar to a neural network process, domain knowledge is distributed among three layers. The first layer is the knowledge goal, the second layer the concepts and the third layer the educational material or the presentation media.

Each goal in the knowledge goal layer is linked to a subset of concepts in the second layer. The cluster of concepts may be the prerequisite, related or the outcome concepts. Each concept is then externalised in the form of examples, exercises etc. within the corresponding knowledge module (Figure 1).

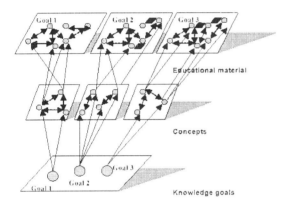

Figure 1. The connectionist-based structure of the domain knowledge of a course.

Papanikolaou, Magoulas and Grigoriadou (2000, p.190)

After determining the knowledge goal, concepts, and representations related to the goal are chosen. Each concept corresponds to a Concept Node called the Relationships Storage Network, which is trained to form connections among the concepts of a knowledge goal. Associations, which are found to be of less relevance or are no longer applicable or true, will not survive the evolutionary process of classifying, organising and assessing the current information structure.

The resulting patterns are generated in view of the strategies for planning the sequence and content of the lesson. A lesson comprises of a display of the outcome concepts, links to short presentations of prerequisite concepts and links to the related concepts in a glossary.

Assessment of mastery of the concept then follows. Evaluation is categorised into six categories: extremely insufficient, insufficient, rather insufficient, rather sufficient, almost sufficient and sufficient. For example, the strategy may take into account that the learner has successfully mastered the prerequisite concepts (assessment value of 0.1 - sufficient) and therefore can proceed to a higher level of difficulty by studying the outcome and related concepts. If however, the learner fares poorly in certain outcome concepts (assessment value of 1 - extremely insufficient), then the learner will have to retrace his steps, review and reassess the prerequisite and related concepts. Based on the results of the evaluation, presentation media to be displayed will take into consideration the various concepts involved e.g. main, prerequisite and related concepts, the difficulty level, the document format and the interactivity level.

Since the basis for any associative memory is accurate classifications of concepts, considerations of the various functions of concepts will form more accurate relational networks. As the knowledge goal changes, the agent differentiates between actions that will lead to the achievement of goals or that will mislead the learner. Based on the associations established among concepts, the agent provides advice that will enable the learner to undertake the optimal path. This is possible if prior to learning, the expert path has been determined and the learner's path is constantly monitored in contrast to the expert's.

Rules are not sufficient as a basis for inference as solutions are constrained to existing rules. A higher level of autonomy will utilise experience gained regarding the suitability of a particular action in a certain situation to guide the problem-solving process by exploring possibilities [10]. Case-based reasoning thus presents a possible means to situate and contextualise associations of concepts as well as to form the basis for further induction.

5. CONCLUSIONS

Quality interaction is optimal when learning adapts to the learner's pace, ability and existing knowledge in a context that the learner can relate to. Since cases are situated in contexts, case-based reasoning provides a possible means for capitalising on past experiences in search of optimal solutions to similar cases. Where more than one solution is found, the utility value is weighed.

Where solutions cannot be found, inference needs to be further fine-grained to the conceptual level. Inaccurate granules need to be weeded out. New concepts and associations need to be formed by considering the degree of probability of a concept belonging to a certain classification.

In addition, a greater degree of saliency will increase the likelihood of information pickup as the learner interacts with the system. In view of the importance of concept classification and association, the saliency of new perceptual stimuli that hooks onto concepts becomes pertinent. Since interaction influences the degree of saliency, the quality of interaction that provides for engaging learning must be carefully designed [11].

Adaptive instruction that focuses on the underlying architecture of information will facilitate transfer of learning to different and multiple contexts. Its value is especially appreciated in distance learning courses where face-to-face interaction and communication is sometimes minimal and the learner is expected to be independent. In cases where learners are not sufficiently skilled to monitor and plan their own learning, association of concepts and mapping of mental models become frustrating. Learning becomes disorientated and cognitive overload overwhelming.

Hence, the design of adaptive interaction needs to consider the size of granular information i.e. concepts to be presented, that will suit different learning styles and abilities. In addition, the potential of capitalising on case-based reasoning and induction should be further taken into account.

REFERENCES

[1] R. Sims, *Interactivity: A Forgotten Art?* [Online]. Available http://intro.base.org/docs/interact/, January 27, 1997.

[2] A. Baylor, "Beyond Butlers: Intelligent Agents as Mentors", *Journal of Educational Computing Research*, 22 (4), 2000, pp. 373-382.

[3] A. Patel, Kinshuk, and D. Russell, "Intelligent Tutoring Tools for Cognitive Skill Acquisition in Life Long Learning", *Educational Technology and Society*, 3 (1), 2000, pp. 32-40.

[4] I. Gilboa, and D. Schmeidler, "Case-based Knowledge and Induction," *IEEE Transactions on Systems, Man, and Cybernetics*, 30 (2), 2000, pp. 85-94.

[5] P. Davidsson, "A Framework for Organisation and Representation of Concept Knowledge in Autonomous Agents", *Scandinavian Conference on Artificial Intelligence - 93*, Oxford, IOS Press, 1993, pp. 183-192.

[6] L. J. Najjar, "Multimedia Information and Learning", *Journal of Educational Multimedia and Hypermedia*, 5 (2), 1996, pp. 129-150.

[7] P. Davidsson, "Concept Acquisition by Autonomous Agents: Cognitive Modelling vs. Engineering Approach", Lund University Cognitive Studies 12, ISSN 1101-8453, Lund University, Sweden, 1992.

[8] J. R. Olson and G. M. Olson, "The Growth of Cognitive Modelling in Human-Computer Interaction since GOMS", *Human Computer Interaction*, 5, 1990, pp. 221-265.

[9] K. A. Papanikolaou, G. D. Magoulas and M. Grigoriadou, "A Connectionist Approach for Supporting Personalised Learning in a Web-Based Learning Environment," In P. Brusilovsky, O. Stock, C. Strappavrava (eds.), *AH 2000*, LNCS 1892, Berlin, Springer-Verlag, 2000, pp. 189-201.

[10] F. Heylighen, "Autonomy and Cognition as the Maintenance and Processing of Distinctions", In F. Heylighen, E. Rossel, and F. Demeyere (eds.), *Self-steering and Cognition in Complex Systems: Toward a New Cybernetics*, New York, Gordon & Breach Science Publishers, 1990, pp. 89-106.

[11] C. S. Lee and Y. P. Singh, "Interaction Design: A Systemic Review", In B. Furht (ed.), *Proceedings of the International Conference on Internet and Multimedia Systems and Applications*, IASTED, Las Vegas, Nevada, Nov. 19-23, 2000, pp. 361-367.

A Web-Based ITS Controlled by a Hybrid Expert System

Jim Prentzas, Ioannis Hatzilygeroudis, C. Koutsojannis

University of Patras, School of Engineering
Dept of Computer Engin. & Informatics, 26500 Patras, Greece

Computer Technology Institute, P.O. Box 1122, 26110 Patras, Greece
prentzas@ceid.upatras.gr, ihatz@cti.gr, ckoutsog@ceid.upatras.gr

Abstract

In this paper, we present the architecture of a Web-based Intelligent Tutoring System (ITS) for teaching high school teachers how to use new technologies. It offers course units covering the needs of users with different knowledge levels and characteristics. It tailors the presentation of the educational material to the users' diverse needs by using AI techniques to specify each user's model as well as to make pedagogical decisions. This is achieved via an expert system that uses a hybrid knowledge representation formalism integrating symbolic rules with neurocomputing.

1. Introduction

Intelligent Tutoring Systems (ITSs) have become extremely popular during the last years and have been shown to be quite effective at increasing users' performance and motivation. The emergence of the WWW increased the usefulness of such systems [1].

In this paper, we present the architecture of a Web-based ITS for teaching the use of new technologies (e.g. Internet) to high school teachers. It is the output of a research project aiming to develop methods and tools dealing with the teletraining of persons with different knowledge backgrounds and skills.

Figure 1 depicts the basic architecture of the ITS. It consists of the following components: (i) the domain knowledge, containing the structure of the domain and the educational content, (ii) the user modeling component, which records information concerning the user, (iii) the pedagogical model, which encompasses knowledge regarding various pedagogical decisions and (iv) the user interface.

The ITS is based on an expert system aiming to control the teaching process. The expert system employs a hybrid knowledge representation formalism, called neurules [2]. In the following sections, we elaborate on the system's key aspects.

2. Domain Knowledge

The domain knowledge contains knowledge regarding the subject being taught as well as the actual teaching material. It consists of two parts: (a) the *knowledge concepts* and (b) the *course units*.

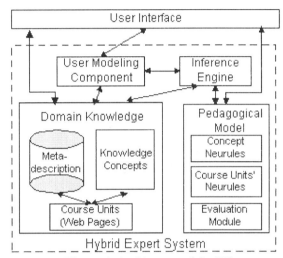

Figure 1. Architecture of the ITS

The knowledge concepts refer to basic pieces of knowledge concerning the domain. A concept can have links to other concepts. These links denote its prerequisite concepts. In this way, one or more *concept networks* are formed representing the pedagogical structure of the domain to be taught.

The course units constitute the teaching material presented to the system users as Web pages. Each course unit is associated with a number of knowledge concepts. The user is required to know the concept's prerequisite concepts in order to be able to grasp the knowledge contained in the corresponding course unit. A course unit may present theory, may be an example or an exercise. The system keeps variants of the same page (course unit) with different presentations.

To facilitate the selection and ordering of the course units, the domain knowledge includes a *meta-description* of the course units based on their general attributes. Main such attributes for a course unit are its level of difficulty, its pedagogical type (theory, example, exercise), its multimedia type, the required Internet connection, etc.

3. User Modeling Component

The user modeling component is used to record information concerning the user which is vital for the system's user-adapted operation. It contains models of the system's users and mechanisms for creating these models.

The user model consists of four types of items: (i) *personal data* (e.g. name, email), (ii) *interaction parameters*, (iii) *knowledge of the concepts* and (iv) *student characteristics*. The student characteristics and the knowledge of the concepts directly affect the teaching process, whereas the interaction parameters indirectly.

The interaction parameters form the basis of the user model and constitute information recorded from the interaction with the system. They represent things like, the type and number of the units accessed, the type and the amount of help asked, the answers to the exercises etc.

The student characteristics include items such as the multimedia type preferences (e.g. text, images, or animations) regarding the presented course units, the domain knowledge level, the learning ability level, the available Internet connection, etc. The student characteristics are mainly represented with the stereotype model meaning that the user is assigned to predefined classes (stereotypes). Based on the way they acquire their values, student characteristics are discerned into two groups, *directly obtainable* or *inferable*. The directly obtainable characteristics obtain their values directly from the user whereas the values of the inferable ones are inferred by the system based on the interaction parameters and the knowledge of the concepts. A neurule base containing *classification neurules* is used to derive the values of the inferable characteristics. The user models are updated during the teaching process.

The user's knowledge of the domain is represented as a combination of a stereotype and an overlay model. The stereotype denotes the domain knowledge level. The overlay model is based on the concepts associated with the course learning units. More specifically, each concept is associated with a value denoting the user knowledge level of this concept.

4. Pedagogical Model

The pedagogical model represents the teaching process. It provides the knowledge infrastructure in order to tailor the presentation of the teaching material according to the information contained in the user model. As shown in Figure 1, the pedagogical model consists of three main components: (i) *concept neurules*, (ii) *course units' neurules* and (iii) *evaluation module*.

The task of the concept neurules is to construct a user-adapted lesson plan by selecting and ordering the appropriate concepts. This is based on the user's knowledge of the concepts, the user's domain knowledge level, the concepts' level of difficulty and the links connecting the concepts. According to the plan constructed by the concept neurules, the course units' neurules select and order the course units that are suitable for presentation. For this purpose, the student characteristics of the user model as well as the meta-description of the course units are taken into account.

The evaluation module evaluates the user's performance and updates accordingly the user model. More specifically, it assigns knowledge values to the concepts based on the interaction parameters and updates the inferable student characteristics based on the classification neurules of the user modeling component. When the user gains an acceptable knowledge level of the concepts belonging to the initial lesson plan, a new plan is created.

5. Expert System

The expert system's knowledge representation formalism is based on neurules, a type of hybrid rules integrating symbolic rules with neurocomputing. The attractive feature of neurules is that they improve the performance of symbolic rules [2] and simultaneously retain their naturalness and modularity [3] in contrast to other hybrid approaches.

Neurules are constructed either from empirical data (training patterns) or symbolic rules. Each neurule is individually trained via the LMS (Least Mean Square) algorithm. In case of inseparability in the training set, special techniques are used [2],[3]. In this way, the neurules contained in the pedagogical model and the user modeling component are constructed. The inference mechanism is based on a backward chaining strategy.

6. References

[1] P. Brusilovsky, A. Kobsa and J. Vassileva (Eds.), *Adaptive Hypertext and Hypermedia*, Kluwer Academic Publishers, Dordrecht, Netherlands, 1998.

[2] I. Hatzilygeroudis and J. Prentzas, "Neurules: Improving the Performance of Symbolic Rules", *International Journal on Artificial Intelligence Tools*, World Scientific, 2000, 9(1), pp. 113-130.

[3] I. Hatzilygeroudis and J. Prentzas, "Producing Modular Hybrid Knowledge Bases for Expert Systems", *13th International FLAIRS Conference*, AAAI Press, 2000, pp. 181-185.

Acknowledgements

This work is supported by the Greek GSRT within the framework of the research project PENED99ED234.

E-learning from Expertize: a Computational Approach to a non-textual Culture of Learning

Giovannina Albano, Ferrante Formato
DIIMA University of Salerno
Albano,Formato@diima.unisa.it

Abstract

The importance of "culture of learning" has been pointed out by many authors and is an emerging feature of e-learning. Our basic assumption, as Bruner has pointed out, is that learning is fully embedded in the cultural roots of the community in which people live.

In this paper we propose a computational model of learning from expertise based on the theory of approximated reasoning developed by Gerla in [2] and on the studies on language made by Vygotskij and its school ([6], [5]). In our interpretation, expertise is an ability acquired mostly by experience. Our research is also supported by a software prototype that we are developing at University of Salerno, that we use for methodological purposes.

1. Introduction

Learning is mostly a socio-cognitive activity and therefore it is deeply connected with the culture of a community.

Culture and cultural context deeply influence the way learning takes place, both through the way people communicate and represent knowledge, through the contents that are taught/learnt. This manner of learning is the most suitable for learning complex and highly structured notions, such as bridge or concrete structure design, control systems and even theoretical subjects such as mathematics and physics itself. None on the less, by identifying learning with "institutional learning" one could mistakenly identify learning with "just" one type of culture of learning: i.e. the "textual culture".

It is a matter of fact that in early times, and still at presents, a craftsman gained his skill not much by studying on a book but rather by improving its abilities through experience. As an example, the Gothic Cathedrals in North-West part of France were built up by very skilled carpenters that where not used to codify their knowledge into books.

Still, today we often are confronted with tasks that can be learnt only by a simple system of rules and a lot of "expertise", such as driving a car or playing soccer.

In this paper we formulate the hypothesis that this kind of learning is deeply relying on some properties of natural language that, at present, have not been fully investigated. As stated by Vygotskij's school, the interplay between a word and its meaning its crucial, even though, in our case, the nature of this connection appears to be thoroughly peculiar. We observed that, unlike its institutional counterpart, learning by expertise is an adaptive process, and therefore it needs to be supported by a flexible language, where the meaning of words can be "tuned up" according to the particular context in which they are used. As a matter of fact, this features are embodied by natural language, that currently deal with vague notions.

The theory of approximated reasoning ([2] [7]) seems to be a suitable tools for an analytical investigation of this subject. As it is common in cognitive science, we have also formulated a computational model and we have developed a system called SciLog, an Integrated development Environment that has been built at University of Salerno using Mathematica and Visual Basic. We have chosen these two systems because they combined rapid prototyping with a rich and handy mathematical library.

2. A computational model for learning by expertise

Approximated reasoning has been developed in the framework of fuzzy logic for control theory ([7], [4]) and has been recently given an adequate inferential apparatus by Gerla ([2]).

We send the reader to [5] for the terminology of logic programming. Given a set X, a fuzzy set, or a fuzzy subset of X, is a map $s:X \rightarrow [0,1]$.

We now consider an example of fuzzy program. Let $X=[0,30]$, $Y=[100,600]$ and let the following fuzzy sets be given:

$Cold$: $X \to [0,1]$, $Mild$: $X \to [0,1]$, Hot: $X \to [0,1]$
$Slow$: $Y \to [0,1]$, $Moderate$: $Y \to [0,1]$, $Fast$: $X \to [0,1]$

We now introduce a new predicate name, as an example the predicate *"Good"* and we interpret $Good(x,y)$ as:
"if the input variable is x, then y is good value, according to the expert, for the control variable".

We now consider the following fuzzy program

$Cold(x) \land Slow(y) \to Good(x,y)$.
$Mild(x) \land Moderate(y) \to Good(x,y)$.   (*)
$Hot(x) \land Fast(y) \to Stable(x,y)$.

This program represents a control of a cooling system of a PC; the set X represents the values of temperature of the Cpu and the set Y is the domain of velocity of cooling fan. Let $\underline{x} \in X$, $\underline{y} \in Y$, $\lambda_1 = Cold(\underline{x})$, $\lambda_2 = Slow(\underline{y})$. Let $\otimes$ be a t-norm, i.e. a binary operation on [0,1] that is commutative, associative, monotonic and such that $x \otimes 1 = x$. For example, in this case, we choose the t-norm of the minimum. We now describe an example of approximated reasoning:

$Cold(\underline{x})$ $Slow(\underline{y})$ λ_1, λ_2
―――――――――――――――――――
$Cold(\underline{x}) \land Slow(\underline{y})$ $Cold(\underline{x}) \land Slow(\underline{y}) \to Good(\underline{x},\underline{y})$ $\lambda_1 \otimes \lambda_2$
―――――――――――――――――――
$Good(\underline{x},\underline{y})$ $\lambda_1 \otimes \lambda_2 \otimes 1$

I.e, in the case $\otimes$ is the t-norm of the minimum, if $\underline{x}$ is cold at least with a degree λ_1 and $\underline{y}$ is slow at least with degree λ_2, then the two values are good at least with degree **Min**$[\lambda_1, \lambda_2]$.
By repeating the reasoning for the other three rules, we can derive the predicate $Good(x,y)$. In fact, for any x,y in X and Y, respectively, we have :
$Good(x,y) = $ **Max**$[\text{Min}[Cold(x), Slow(y)],$
    [**Min**$[Mild(x), Moderate(y)]$,
    [$Hot(x), Fast(y)$]].
Once the predicate $Good$ is known, and we have seen that it is possible to do so in a computational manner, it is possible to derive the ideal function $\underline{f}$:$X \to Y$ that is supposed to regularized to cooling system of the PC and can be computed, when it is possible, with the methods developed by classical calculus.

3. A case study: Learning to control a System by Expertise

In this section we describe the use of Scilog to model how an agent, either human of artificial, succeeds in mastering a control system, learning how it works by the use of his expertise.
Consider the problem of the regulation of temperature in a PC described in the previous section. We assume that, by a series of experiments and measurements, the expertise acquired by the agent about domains X and Y is represented by the families of fuzzy set of Figure 1
Suppose that, by a further series of experiments, the agent learns that the "laws" that manages the control system are represented by the set of rules (*). In our case, the predicate *Good* stand for "stability" of the system according to the experience of the agent.
Then Scilog infers, i.e. computes the fuzzy predicate *Good* and this represents the expertise that the agent has learnt about the working of the control system.
We remark that in this process of learning knowledge has not been transmitted as something that is existing in an objective manner, but has being "built" in a constructive manner by the learner. We argue that, as stated by situationist theories, in a typical learning process learning is, de-constructed negotiated and re-constructed between a learner and a teacher, but this will be the subject of another work.

Bibliography

[1] Bruner J. *The Culture of Education*, Harvard University Press, 1966

[2] Gerla G. *Mathematical tools for fuzzy logic*, Kluwer Academics, 2000.

[3] Lloyd J. *Foundations of Logic Programming*, Springer Verlag, 1987.

[4] Kosko B. "Fuzzy Neural networks ", Mac Graw Hills

[5] Kuhn T. *The cognitive development* in "The cognitive,Ling. and Perc. Development" Borenstein, Lamb Eds. Engelwood Cliff, 1988.

[6] Vygotskij L. S. *Language and Thought* , (Transl. From Russian) Harvard University Press, 1966.

[7] Zadeh L. "Computing with Words" Fuzzy Sets and Systems, 9, 1996.

An Adapted Virtual Class Based on Intelligent Tutoring System and Agent Approaches

Arturo Hernández-Domínguez
Federal University of Alagoas - UFAL, Brazil
Information Technology Department - TCI
ahd@vcnet.com.br

Aleksandra do Socorro da Silva
University of Amazônia – UNAMA, Brazil
alekas@brhs.com.br

Abstract

This paper describes a virtual class, which is based on the principles of ITS (Intelligent Tutoring Systems). The virtual class allows to a group of learners to participate into training sessions of an adapted teleteaching system (adapted virtual class), this system takes into account progression rhythm different into a community of remote learners. The virtual class allows adapting the teaching system in a flexible, individual, and collective way. Agent-oriented modeling is being used to specify and implement the architecture of adaptable virtual class. The implementation of virtual class is being developed in distance education context using reactive agents, Internet and Java.

1. Introduction

The principle of the architecture proposed for an adapted training service is to allow the adaptation of knowledge transmission from a teaching function in a virtual class context (figure 1). The architecture proposed (figure 2) takes into account the principles of ITS (Intelligent Tutoring Systems) [6], [2]. Teaching function is distributed taking into account the participation of teachers, system (intelligent tutoring system in this paper), resources and learners in some cases (when a learner has the knowledge or experience necessary for playing the teacher's role).

In this paper, teaching function is represented by a set of group controllers and a class controller (figure 2). We propose for each group controller a specification based on agent-oriented approach.

2. Group Controller

An Adapted Group Training Service (AGTS) or group controller belongs to cooperative service layer (figure 2).

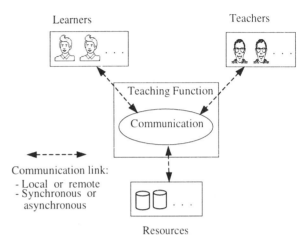

Figure 1. Teaching function in a virtual class context.

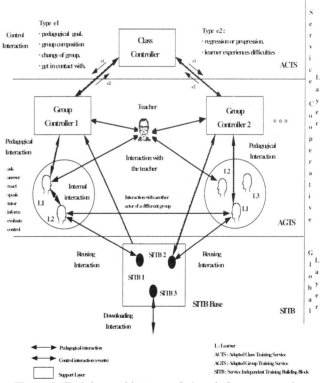

Figure 2. Training architecture of virtual class proposed.

The group controller has to manage: - the communication with the ACTS (Adapted Class Training Service), - the reusing interaction with the server of SITBs (Service Independent Training Building Block) and - the didactic control takes into account a didactic strategy. One group controller allows to consider the behavior of group by behavior zones (which were specified in [3] and two kind of profiles: learner's profile and profile of group). A group controller was specified (based on principles of Intelligent Tutoring Systems [6]) taking into account agent-oriented paradigm. This specification was developed applying the agent-oriented methodology of Wooldrige [7]. The intelligent tutoring system is represented by a set of agents. The agents and their interactions into tutor are represented in figure 3.

Learner profile is represented by personal information (name, e-mail, age, sex,...) and a set of scores such as global score, sum of right responses, sum of wrong responses, sum of empty responses and sum of exercise executed. Group profile is represented by knowledge level, number of learners, name of each learner, arithmetic mean of global score, arithmetic mean of right responses, arithmetic mean of wrong responses and arithmetic mean of empty responses.

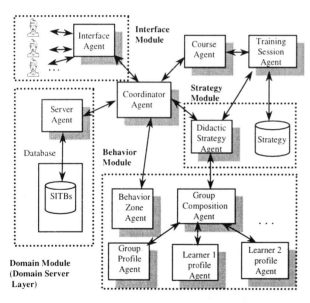

Figure 3: The agents and interactions in group controller based on intelligent tutoring system architecture.

Each agent must interact with others agents by sending and receiving of messages. Agent behavior is determined by receiving of events and a treatment correspondent. Each agent in group controller has three events (start, message receiving, and finish) based on [1].

3. Conclusion

The architecture of the virtual class proposed is a response to traditional virtual classes (using Internet) which are not adapted during training session. The virtual class proposed is considered adapted by using learner and group profiles, which are analyzed in the strategy module (by decision rule tactic) during tactic execution.

A server of reusable SITB was developed in the object-oriented programming domain. This server was implemented using Java and Internet. The same server will be used for each group controller. A group controller was developed, the agent-oriented paradigm (in particular reactive agent) was used for modeling and implementing it (in Java). Some pedagogical patterns may be implemented by tactics proposed in this paper, in particular the pedagogical pattern LDLL (Lab Discussion Lecture Lab) (http://www.lifia.info.unlp.edu.ar/ppp) was specified by a set of tactics proposed and it was taken into account as a strategy in the AGTS developed [5]. Others pedagogical patterns are being analyzed for implementing them as strategies into tutor developed..

4. References

[1] B. F. T. Azevedo, "MutAntIS - Uma Arquitetura Multi-Agente para a Autoria de Tutores Inteligentes", 115 p. Dissertação (Master) – CT/UFES, Brazil, 1999.

[2] M.F. Canut, and M. Eloi (1994). "Vers la généricité: spécification d'un Shell pour un environnement d'apprentissage basé sur l'entraînement". IX congrès Reconnaissance des Formes et Intelligence Artificielle, RFIA 94, Paris, France, 11-14 January.

[3] A. Hernández-Domínguez, "An adapted virtual class based on two approaches: CSCW and intelligent tutoring system", International Conference Educational Multimedia and Hypermedia ED-MEDIA'95, AACE, Graz, Austria, Juin 1995.

[4] A. Hernández-Domínguez, "Specification and implementation of an adaptable virtual class", ED-MEDIA/ED-TELECOM 97 World Conference on Educational Multimedia and Hypermedia and on Educational Telecommunications, Calgary-Alberta, Canada, June 97. pp. 467-472.

[5] A. Socorro da Silva. "TUTA - A tutor based on Agents", Dissertação de mestrado em informática (master). Campina Grande DSC/UFPb, Brazil, February 2000.

[6] E. Wenger, "Artificial Intelligence and Tutoring Systems: Computational and Cognitive Approaches to the Communication of Knowledge". Morgan Kaufmann Publishers, Inc. California, USA, January 1987.

[7] M. Wooldrige, N. R. Jennings, D.Kinny, "A Methodology for Agent-Oriented Analysis and Design", Proceedings of the third annual conference on on Autonomous Agents May 1-5, Seattle, USA. 1999.

Theme 8: CMC, Dialogue, Collaborative Learning

An Analysis of Critically Reflective Teacher Dialogue In Asynchronous Computer-Mediated Communication

Mark Hawkes, Ph.D.
Dakota State University
mark.hawkes@dsu.edu

Abstract

This study explored the professional development experiences of 28 practicing teachers in ten Chicago suburban schools involved in a two-year technology supported Problem-Based Learning curriculum development effort. Asynchronous computer-mediated communications (CMC) were featured as teacher communication tools of the project. The computer-mediated discourse produced by the teachers was compared with the discourse produced by teachers in face-to-face meetings. Research methods including discourse and archival data analysis were applied to determine the nature of the teacher discourse and its reflective content. The results show that while the computer-mediated teacher dialogue was less interactive, it was significantly more reflective ($t = 4.14$, $p = .001$) suggesting that the value of computer-mediated communication lies in its ability to facilitate professional collaboration between teachers and encourage critical reflection on educational policy and practice.

1. Introduction

Computer-mediated communication (CMC) presents teachers with new opportunities for communication. Though the use of CMC suggests more convenient access to professional colleagues, it does not ensure professional growth and learning experiences. The purpose of this study was to determine if and how critical reflection—as a meaningful professional development objective—arises from computer-mediated collaborative dialogue.

Professional development for teachers constitutes formal and informal processes of knowledge and skill building. Though it had taken a permanent place in school culture, professional development lacked, for the most part, the ability to improve learning and teaching. Such professional development activities have been characterized as nothing more than the delivery of an assortment of relatively abstract ideas providing little support to learning. The call to improve the quality of inservice teacher development ardently encouraged opportunities for teachers to collaborate with peers to make sense of the teaching and learning process [1].

2. Collaboration to Critical Reflection

Collaboration is generally described as a process of willing cooperation with peers and colleagues to reach educational objectives. An important element of collaboration is conversation. Giving utterance to an idea or summarizing a thought forces a cohesive explanation of interrelating ideas. The process of articulating ideas, beliefs, or reservations enhances retention, clarifies one's position on an issue, forces the learner to take a stand on an issue in the presence of peers, and commits the conversant to evaluate and assess that knowledge in light of new information [2].

Collaborative conversation encourages relational knowledge that links what teachers learn and understand about their practice to other conditions that impact student learning such as family influences and the educational setting. Perhaps most important, collaborative settings are the likely contexts in which critically reflective exchanges about learning and instruction can take place between teachers.

Schön depicts reflection as a social-professional activity in which teachers adapt knowledge to specific situations [3]. As the very essence of professional activity, reflection is a process that engages teachers in framing and reframing problems while designing and evaluating solutions. This kind of discourse leads to critical theory where teachers raise questions about equity and power on issues such as student assessment, ability tracking, and classroom management.

3. Computer-Mediated Reflective Dialogue

Computer-mediated communication (CMC) facilitates person-to-person or person-to-group contact by means of computer networks. CMC can be synchronous, where participants interact in real time, or asynchronous, where communication turnaround may be delayed for hours or days. Because of the speed with which the medium sends and retrieves messages to support participant interaction, the time and place independence of the medium, and the storage capacity of the technology, CMC is thought to be especially suited to the task of linking teachers together in experiences that may be both professionally and personally rewarding [4].

Despite CMC's ability to connect teachers, little is known about the technology's ability to facilitate teacher collaborative reflective processes. Studies that do address reflection are usually done in the highly controlled context of preservice teachers development [5]. Only a few studies address the reflective quality of computer-mediated discourse for practicing teachers. Of those studies, little description of the reflective processes or outcomes of collaborative teacher discourse is offered.

The excitement that computer networks generate around communication coupled with the role of critical reflection in improving teacher practice suggest that more inquiry is necessary to illuminate the potential of computer networks for hosting collaborative, collegial reflection. This study attempts to address this knowledge gap by determining at what level asynchronous computer networks are able to host critically reflective dialogue.

4. The Study Setting

The two-year project involved 28 teachers (3 male, 25 female; 18 yrs. average teaching tenure) from ten elementary schools in a Chicago metropolitan suburb. The teachers were generally novice technology users but worked in schools with substantial technological capacity.

The primary program goal involved building teacher capacity for developing Problem-Based Learning (PBL) curricula. Teacher teams completed and delivered their first PBL unit in the spring of the first project year. The focus of the second year of the initiative was to use new technology tools to expand instructional practices and skills in PBL curricular development. Project teachers provided input on the development of an electronic toolkit located in a district server file folder where electronic tools were retrieved by teachers to develop and refine their PBL units. The kit included database, communication, graphics, word processing, and program /multimedia authoring tools.

5. Methods

A multimethod, quasi-experimental approach to data collection and analysis is applied. Comparative content and statistical analysis of computer-mediated and face-to-face discourse comprise the bulk of inquiry into reflective quality of the two mediums.

To determine what levels of collaborative reflection are present when teachers interact under usual circumstances, face-to-face work meetings of school teams consisting of two to five teachers were audiotaped. Teachers were told that their meetings were being taped as a part of the data collection for project evaluation. Teachers were unaware that the level of reflective discourse was the variable of interest in this study. Researchers recorded six teacher meetings from February to May of the final project year. This schedule closely corresponded to the timeline that teachers had for developing their second integrated PBL unit, delivering it, and documenting the unit for archival purposes. The recording of the face-to-face meetings also ran concurrently with the collection of group computer-mediated communication.

While analysis identifies the flow, frequency, and volume of communication activity and the nature of the computer-mediated and face-to-face dialogue, it centers on the reflective attributes of the discourse. A seven-level reflection rubric based on Simmons' taxonomy for assessing reflective thinking is used to analyze the discourse [6]. The rubric is one in which low-level reflective responses are those which merely describe events and appear disconnected from the observer. More reflective responses richly describe events and attempt to explain them in light of theory or principle.

To prepare the face-to-face discourse for analysis it was "chunked" into comparable frames. The chunking of face-to-face discourse was guided by principles of distributional accountability suggesting that utterances are related to each other by their relationship to a common theme [7]. Chunking resulted in 222 distributionally accountable portions of face-to-face discourse ranging in size from 1 to 12 exchanges (see Table 1).

Table 1. Dialogue content comparison.

	Number of Words	Utterances /Messages	Chunks/ Messages	Words Per Chunk/Post
Face-to-Face Discourse	19,000	846	222	86
Computer-Mediated Discourse	19,303	179	179	108

After all identifying information was removed from electronic messages and transcripts, A team of three indepentent raters judged each of the chunked exchanges in the face-to-face (n=222) and computer-mediated (n=179) messages using the seven-level reflection rubric.

Data analysis proceeds by comparing the presence of reflective overtures, comments, and observations of the face-to-face dialogue against reflective qualities of the CMC dialogue. Interrater correlates are determined to note the consistency of ratings, and statistical analysis includes a t-test for dependent means. Time is also examined as a function of reflectiveness.

6. Results

Assessing Participant Interaction. The comparison of face-to-face and computer-mediated discourse began with the analysis of the conversation on selected dimensions of interaction. In conversation, interaction involves the selection and delivery of words that unite the speaker, listener(s), and content. Specific strategies such as conversational involvement tactics ("wh" clauses: who, what, where, etc.), conversational cooperation (ratio of answered to unanswered questions), and sequential accountability (coherence of utterances in discourse) can be employed to heighten the interaction of participants in the dialogue. Table 2 compares the discourse on the two mediums using the *chi*-square test of association between the observed proportions in each interaction variable.

Table 2. Chi-square values on discourse interaction variables.

Involvement Strategies	Parameter	CMC (n)	Face-to-Face (n)	X^2_1	P
"Wh" clauses	Words	220	326	22.9	.001
Indefinite pronouns	Words	435	474	2.4	.065
Amplifiers	Words	220	288	10.4	.001
Conversational Cooperation	Questions	25	95	10.0	.002
Sequential Accountability	Utterances	15	43	3.0	.064

df=1 for each variable

In all cases, the number of observations made of the selected interaction variables in face-to-face discourse exceeds the interaction of CMC. In three of the five variables: "wh" clauses, amplifiers, and conversational cooperation, results are significant at the $p < .05$ level. The results show that face-to-face discourse generally rates higher in interactivity than does CMC.

Assessing Reflective Content. Reflection is also an indicator of the medium's ability to promote interactivity between discourse participants. Talking, sharing, exploring, and analyzing are important interactions in sense making and, by themselves, constitute key components in the critical reflection process. Reflection is distinct from interaction, however, in that it requires a certain amount of self-disclosure about professional beliefs and practice.

A comparison of the ratings (independent *t*-test for equality of means) between the reflective levels of the two communication mediums shows that CMC has significantly higher ratings on the seven-level reflection scale than does face-to-face communication ($t = 4.14$, $p = .001$). A test of rater consistency produced an inter-rater reliability statistic of .87 on the face-to-face discourse achieved and an item alpha level of .80 on the CMC discourse for the three raters involved.

A breakdown of the ratings of reflectiveness in Figure 1 shows the percentages of ratings assigned to each of the seven reflective levels. The majority of messages (70% for the face-to-face, 63% for CMC) were rated at the first and second of the seven levels of reflection and generally show that neither CMC nor the face-to-face discourse is abundantly reflective. The fact that different distributions appear for each discourse mode with positive skews of very different strengths, suggests that critical reflection may have a more purposeful role in CMC facilitated discourse than in face-to-face discourse.

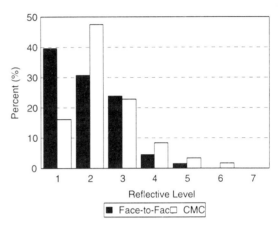

Percentage of Rater Observations at Each Reflective Level

Reflection Over Time. A final analysis explores the possibility of any increase in discourse reflectiveness over the duration of the project timeline. Face-to-face teacher meetings recorded at the beginning, middle, and end of the project period were compared with each other. A similar comparison was made of the CMC discourse. Each of the 179 CMC messages were chronologically ordered and then divided into thirds to roughly match the meeting intervals. Comparisons of the first third of the messages were made against the second and final thirds.

The face-to-face discourse shows gradually increasing reflectiveness over time from one observation period to the next. When the means scores for the observations at the three periods were tested for statistical significance, one-way analysis of variance shows no significant difference at the $p>.05$ level (see Table 3).

Table 3. Analysis of face-to-face discourse over time.

Source	SS	df	MS	F	Sig.
Between Groups	5.35	2	2.67	2.94	.055
Within Groups	199.11	219	.91		
Total	204.47	221			

CMC discourse shows a curious pattern over the three observation periods. Between the first and second periods reflective scores decreased, while the second and third periods show a substantial increase. When the differences between the reflectiveness at these three time periods were tested, a one-way analysis of variance indicates a significant difference (see Table 4).

Table 4. Comparison of CMC discourse over time.

Source	SS	df	MS	F	Sig.
Between Groups	36.68	2	18.34	20.7	.00
Within Groups	155.88	176	.89		
Total	192.56	178			

The increase in CMC reflective scores occurs only after a significant drop in the second of the three observations over time. Differences over time are possibly explained by the early initial use of CMC by participants who were experienced and skilled at using the medium. As less confident and skilled teachers forayed into the discourse, their contributions reduced the overall reflection scores produced. More comfort with CMC possibly increased the reflective scores at the time of the third observation.

7. Summary

Independent rater assessments show that computer-mediated discourse achieves a higher overall reflective level than reflections generated by teachers in face-to-face discourse. Although more reflective, CMC proved not to be as interactive as face-to-face discourse. When time is a factor in the analysis, there is no significant difference in three time-differentiated observations made of reflectivity in the face-to-face discourse. In CMC the difference is characterized by a dip in the reflectivity scores from the first to the second observation. Reflectiveness at the third and final observation then rises to levels significantly above scores at the previous two observations.

In CMC facilitated discourse, follow-up semi-structured interviews found that the convenience, quality, breadth, and volume of peer-provided information improved teachers' knowledge of educational theory, policy, and the educational community. The use of computer networks to share experiences and beliefs, creates a new interactive dynamic for teachers. This study suggests there is an evolution of teacher CMC use that begins with observation; advances to application of the medium for low-risk and low-importance uses; and progresses to more substantial, revealing, and reflective uses.

A probable key to moving CMC participants through this continuum of use, beginning at caution and eventually arriving at confidence, is the "climate" of interaction. Climate is defined here as the extent to which teachers in the project feel comfortable in the open exchange of ideas, opinions, and feelings over the medium. In this research, self-disclosure appeared to be the catalyst for many reflective contributions to on-line discourse, embedding teacher learning in real experiences. At times, though infrequently in this study, this discourse linked what teachers already know to new conditions of educational practice to produce relational knowledge.

As for critically reflective dialogue, the transcripts of a selected number of face-to-face meetings show that teacher work was highly task focused. However, face-to-face discourse failed in several dimensions of analysis to reach equivalent breadth and depth of teacher reflection on CMC. This finding suggests that when there is specified work to be done, such as identifying instructional strategies for a particular learning outcome, the task drives the agenda. Unless critical reflection is planned as a part of the process or becomes an objective of a face-to-face meeting itself, reflective activities will be sporadic at best. In the CMC discourse examined, task driven or not, the time independence of the medium gave participants a greater chance to weave reflective processes into curriculum development tasks by asking: Why are we doing this? What are the long-range consequences of this decision? What meaningful relationship does this activity have with key learning outcomes?

8. References

[1] D. Sparks, "A paradigm shift in staff development", Journal of Staff Development, 1994, (15)4, 26-29.

[2] T.D. Koschman, CSCL, Theory and Practice of an Emerging Paradigm, New York, Teachers College Press, (1997).

[3] Schön, D. A. (1987). Educating the reflective practitioner. San Francisco: Jossey-Bass.

[4] M. Honey, "Online communities: They can't happen without thought and hard work", Electronic Learning, 1995, 14(4), 12-13.

[5] R.F. Kenny, B.W. Andrews, M. Vignola, A. Schilz, and J. Covert, "Toward guidelines for the design of interactive multimedia instruction: Fostering the reflective decision-making of preservice teachers", Journal of Technology and Teacher Education, 1999, (7)1, 13-31.

[6] J.M. Simmons, G.M. Sparks, A. Starko, M. Pasch, A. Colton, and J. Grinberg, "Exploring the structure of reflective pedagogical thinking in novice and expert teachers: The birth of a developmental taxonomy", Paper presented at the annual conference of the American Educational Research Association, San Francisco, CA. 1989.

[7] D. Schiffrin, Discourse Markers, Cambridge: Cambridge University Press, 1987.

Facilitating Entry into Professional Life with CMC Support

David R. Walker
McMaster University
walkerd@mcmaster.ca

Abstract

A graduate course in music analysis was designed to assist graduate students enter professional life with the support of computer conferencing and the web. The students produced a peer-reviewed internet journal of their work using simple computer technologies, significantly enhancing learning. Career enhancement and interest led to extra time spent on course work, and excellent results.

1. Introduction

Graduating students encounter a gap between school and professional work. This educational design experiment [1] investigated facilitating this transition at the university graduate level, focussed on the social aspects of learning. Based on Lave and Wenger's [2] theory of Legitimate Peripheral Participation (LPP), the goal was to study learning in a local community and the efficacy of contact with the community of professional practice. Computing was used to facilitate social interaction and exchange of ideas, and as a medium for the display of work, but not as a teaching medium.

2. Situated Learning and LPP

Educational theorists advance the notion of situated learning [3][4], that what is learned is directly related to its context. Cognitive apprenticeship [5] is refined to include more of the social dimension, with a caveat that school work often merely teaches how to "do school" [6]. In LPP, learners construct their understanding of the world from the interaction of their own agency, their activities, and the larger world in which they practice these; by participating in communities of similarly directed individuals. Participation is legitimate as it is meaningful both to the individual and the group, although peripheral rather than "leading edge." Groups in which learners take part – "communities of practice" – interest current scholars [7],[8]. For this study the local community was the class: students, professor, and researcher as a participant-observer [9]. The larger community was practicing professionals.

3. Qualitative Methods in this Study

As Brown [10] has shown, qualitative methods of inquiry allow for detailed investigation into learning phenomena. Case study is the most common tool for exploring new designs [11], allowing for an in-depth study of the specifics of the class within a holistic context [12]. The greater detail of context provides more information about questions of "how" and "why" phenomena occur [13]. Here the flexibility of the method allowed the investigation of emergent issues identified as important by the participants.

As a participant-observer the researcher attended classes and took field notes from classroom and on-line discussions. Research instruments included bi-weekly surveys and a 90-minute semi-structured exit interview with each participant. Data were triangulated and member checked: field notes were compared with survey answers, and interim conclusions were presented to each participant at the start of each interview, with a request for corrections.

4. Use of Computing in the Course

The participants in the study were all eight students in a masters-level class in music analysis, a required course in a degree program in music criticism. Class met once per week for a regular 3-hour seminar, with lectures, discussion, debate, and analysis. Five progressive assignments led to a final article for a student-run peer-viewed journal. From their local class community, they contacted the professional community by "publishing" the on-line journal. Each student reviewed two peers' articles, and worked on one of the journal committees.

FirstClass conferencing software was used to communicate between classes, from asking simple questions and sharing insights to posting draft essays and critiques. The students were enthusiastic about using conferencing, especially at the start of the semester, when getting acquainted with each other and with the subject matter. As well as posting notes, they posted assignments on-line. Use of conferencing declined toward the end of the semester, when students were busy with individual articles and peer reviews. At this time they also learned to create HTML pages for the web journal, which was published during the final week of classes.

5. Findings

The students rated the course a great success because they learned a good deal and produced work they were proud to exhibit. All found the perspectives of the other students displayed on-line to be very valuable, as models both positive and negative. Discussions led up to the next class and helped them prepare. The professor noted much better in-class discussions than in past years. Quieter students appreciated their ability to take a greater part in discussion on-line, while all liked working at their own pace and schedule.

More important to the students were aspects of the course that enhanced their career prospects. Their best experiences were those most relevant to their careers, particularly learning the process of journal writing and editing, and learning computing skills. They addressed musical academia and prospective employers, rather than other graduate students.

The students cited their enjoyment of opportunities to synthesize their research interest with course work as the major incentive for spending extra time on this course. This and the expected enhancement to career led most students to spend a great deal of extra time on both course work and learning new computing skills. All students were excited about making a good impression on the professional community and reported taking extra care with these articles.

The professor also rated the course a success, citing the journal articles as the best set of essays received from a graduate class in over a decade of teaching. He felt that the review process led them to write more drafts than normal.

6. Evaluation of Technology Use

Modest computing resources aided learning significantly, adding to the sense of community in the class as well as to learning. With this means for communication between classes, issues remained alive during the week, and debate spilled over from the classroom to the on-line forum, and back into the next week's class.

The most important aspect of the class was the on-line journal. It provided a focus for work over the entire semester, and engaged the students. Because they felt that it was worthwhile for their future career as well as their present learning, and since it interesting to do, they did a great deal of extra work for the course and reported enjoying it most of all their courses.

Students and teacher considered the course design a success. Preparing their work for the professional community drew out the best from the students, and most exceeded the expectations of the professor. The journal received good reviews from the professional community, and the professor plans to use this same basic framework for this course again this year.

References

[1] A. Collins, "Toward a Design Science of Education." *New Directions in Educational Technology.* E. Scanlon & T. O'Shea (Eds.) Springer-Verlag, Berlin, 1992, pp. 15-22.

[2] J. Lave and E. Wenger, *Situated Learning: Legitimate Peripheral Participation.* Cambridge University Press, Cambridge, 1991.

[3] P. Honebein, T. Duffy, & B. Fishman. "Constructivism and the Design of Learning Environments: Context and Authentic Activities for Learning." *Designing Environments for Constructive Learning.* T. Duffy, J. Lowyck, & D. Jonassen (Eds.). Springer Verlag, Heidelberg, 1993.

[4] J. Brown, A. Collins & P. Duguid. "Situated Cognition and the Culture of Learning." *Educational Researcher*, Jan-Feb, 1989, 32-42.

[5] A. Collins, J. Brown & S. Newman. "Cognitive Apprenticeship. Teaching the Crafts of Reading, Writing, and Mathematics." *Knowing, Learning, and Instruction: Essays in Honor of Robert Glaser.* L. Resnick (Ed.). Lawrence Erlbaum Associates, Hillsdale, NJ, 1989.

[6] M. Scardamalia, & C. Bereiter. "Schools as knowledge building organizations." *Developmental health and the wealth of nations: social, biological, and educational dynamics.* D. Keating & C. Hertzman (Eds.). Guilford, New York, 1999.

[7] H. McLellan. (Ed.) *Situated Learning Perspectives.* Educational Technology Publications, Englewood Cliffs, NJ, 1996.

[8] J. Brown, & P. Duguid. *The Social Life of Information.* Harvard Business School, Boston, 2000.

[9] J. Spradley. *Participant Observation.* Holt, Rinehart and Winston, Toronto, 1980.

[10] A. Brown, "Design experiments: Theoretical and methodological challenges in creating complex interventions in classroom settings." *The Journal of the Learning Sciences,* 2(2), 141-178.

[11] S. Merriam. *Qualitative Research and Case Study Applications in Education.* Jossey-Bass, San Francisco, 1998.

[12] R. Stake. "Case Studies." *Handbook of Qualitative Research.* N. Denzin & Y. Lincoln (Eds.) Thousand Oaks, CA: Sage Publications, 1994.

[13] R. Yin. *Case Study Research: Design and Methods.* Second Edition. Sage Publications, Thousand Oaks, CA, 1994.

Facilitating Computer-mediated Discussion Classes: Exploring Some Teacher Intervention Strategies

S.A. Walker and R. M. Pilkington
Computer Based Learning Unit
University of Leeds
S.A.Walker@cbl.leeds.ac.uk

Abstract

This paper examines the strategies used by two teachers facilitating computer-mediated debate within an out-of-school class to help children aged 11-14 overcome a literacy deficit that inhibits the fulfilment of their educational potential. In order to assist the children to express themselves more fluently in writing, especially the writing of argumentative texts, the class uses synchronous test-based computer mediated communication to debate topics of interest. Computer generated logs of the debate have been analysed to determine the types of strategy used by the teachers both to facilitate the discussion and to manage the class within the computer-mediated context. The responses of the children are also examined in order to determine the effectiveness of different types of strategy. The paper concludes with discussion of the implications of this study for teaching in a CMC environment and for further research into the facilitation of CMC debate.

1. Introduction

The computer-mediated classroom offers a different type of teaching environment from the traditional face-to-face setting. In order that teachers should be able to use this new environment to greatest effect it is important to consider what types of teacher strategies are most useful in the computer-mediated context. This paper reports on the strategies used by two teachers using computer-mediated communication to develop discussion and argumentation skills in young people and evaluates the types of teacher intervention which are most effective in advancing the debate concluding with recommendations for classroom practice and for further research.

2. Research context

The study was conducted at CHALCS, an out of school project in an inner city district of Northern England. The project serves the needs of local young people, most of whom come from Asian or Afro-Caribbean backgrounds – many speaking a first language other than English. The project provides extra tuition in maths, science, computer skills and English/literacy. Every classroom in the project is equipped with computers running Windows 95/98 on a Novell network; the project aims to use these to motivate the students as well as to provide the new technology skills that the students will need for higher education and for employment.

A review of the English/literacy programme at CHALCS [1] found that it was not meeting the needs of the older students (aged 12+) and a new programme, making more use of the CHALCS computer facilities, was designed for this age group. This new programme, called DaRE (for Discussion and Reporting Electronically) uses the chat tool of WebCT, a virtual learning environment, for a class discussion. This is followed by the collaborative writing of a report, usually undertaken by a small group of students, which is then posted to the WebCT group presentations area for the class and the teacher to read.

2.2 Using CMC to teach argumentation for writing

In the early years of school children are taught to write narrative texts, either stories or reports. However, as they mature, they are expected to write argumentative texts. As Andrews [2] points out, the progression from narrative to argumentation is not clear but requires scaffolding. One way of doing this may be to conduct classroom debates and, indeed, it has been shown that discussion may improve argumentation skills [3]. Within CHALCS, debates are commonly used to help students develop argumentation. Class discussions conducted through computer mediated communication have the added advantage of taking place in writing. Furthermore, the use of computers has, in itself, been shown to improve writing (for example, [4] [5]).

A pilot study [6] [7] showed that, over time, in CMC debate the children appeared to write more fluently; to become more focussed in the discussion, became more responsive to each other and became more skilled at justifying their arguments. However, in that first study, there was minimal tutor participation and thus little opportunity to evaluate the types of strategy that were most effective in advancing the debate.

3. Rationale

CMC classes are not the same as face-to-face lessons. Firstly, the seating arrangements in the two types of class are likely to differ. For example, at CHALCS, in a face-to-face discussion, the tutor is likely to seat the pupils in a horseshoe arrangement (see Figure 1) which means that members of the group can make eye contact with each other. In a CMC, class, however, each participant is seated at his/her own computer (see Figure 2) and concentrating on the screen.

Figure 1: Face-to face discussion class

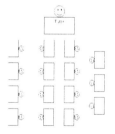

Figure 2: CMC discussion class

An important difference between face-to-face and CMC class is in the way that talking time is shared between teacher and pupils. Sullivan and Pratt [8] and Warschauer [9] both found that students participated more equally in CMC classes than in face-to-face classes. They also found the CMC classes were less teacher dominated than the face-to-face classes. This is borne out by a simple comparison of two CHALCS classes taught by the same teacher; one a face-to-face debate and the other a computer mediated discussion. Figure 3 shows the proportions of teacher and student talking time in a face-to-face debate, measured by continuous observation using unit blocks of 15 seconds. Figure 4, on the other hand, shows the proportions of teacher and student talking time in a CMC discussion as measured by a word count. The two figures show that in the face-to-face class, 80% of the talking was by the teacher with the most talkative student speaking for only 7% of the time. In the CMC class, however, only 31% of the talking was by the teacher.

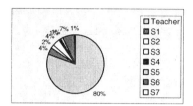

Figure 3: Proportions of teacher talking time and student talking time in a face-to face class

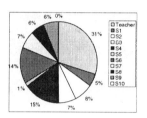

Figure 4: Proportions of teacher talking time and student talking time in a CMC class

The differences between face-to-face and CMC discussions mean that the teacher may need to use different intervention strategies when facilitating a CMC class from those s/he would use with a face-to-face group. A study of the strategies used by teachers in a CMC class should show what types of strategy teachers use in this environment and whether or not these moves are successful in advancing the discussion.

4. Method

4.1 Principles of facilitating discussion

There are many books which provide guidance on the facilitation of argumentation and discussion. Fisher [10] and Lipman [11], for example, suggest types of question and statement that teachers may use to help children develop thinking through argumentation. Fisher, in particular, lists Socratic interventions such as asking students to provide supporting information (e.g. explanations or examples), encouraging students to explore reasons and evidence, and looking at alternative points of view. It would therefore be expected that, when collecting interventions actually used by teachers in practice, a high proportion of these would be Socratic questions.

4.2 Collection and analysis of data

All the CMC discussion classes were logged by computer. Three sessions were taken for detailed analysis. One of these sessions was facilitated by the usual class teacher working alone. One session was taken by the researcher, acting as tutor, whilst the third session was taken by the class teacher with the researcher acting as a second tutor. The teacher's turns from each of these sessions were analysed, using the DISCOUNT coding system [12], with slight adaptations, in order to determine the type of dialogue move that the tutor was making in each turn. These moves were compared to responding student moves in order to evaluate which types of teacher intervention received the greatest number of appropriate responses from the students. Student responses which were not appropriate to the teacher move were not counted.

5. Results

Table 1 shows the total numbers of moves made by the teachers, according to move type. Metastatements are interventions which refer beyond the topic content of the discussion, for example, interventions to focus the task or to manage the class. A breakdown of metastatements is provided in Table 2. After metastatements, the most common types of move were 'probe' (asking questions to elicit more information from a student) and 'challenge' (questions to encourage a student to justify an opinion or argument). The teachers also used a high number of 'inform' moves (giving information), largely in reply to questions from pupils. Figure 5 shows the number of times a single student responded to the teacher's move relative to the number of times that more than one student responded. The numbers of student responses are expressed as a percentage of the numbers of teacher moves in order to provide a comparison of the efficacy of different move types. For example, the teachers made 26 'challenge' moves of which 57.7% received a response from a single student and 15.3% received responses from more than one student. Altogether, 73% of challenge moves were effective in stimulating student activity. 'Probe' moves achieved a student response level of 75.9% whilst the pupils responded to 85.7% of 'open' moves (not restricted to greetings but inviting students to open a new topic). 'Inform' moves, on the other hand, generated a student response level of 41.7%, metastatements generated a response level of 28.6% and 'encourage' moves generated a response level of 18.7%.

Table 1: Numbers of teacher intervention moves and student responses

Move Type	Total	Receiving a single response	Receiving multiple responses
Agree	1	0	0
Ask-clarify	3	1	0
Challenge	26	15	4
Counter	1	0	0
Critique	2	0	0
Encourage	16	1	2
Inform	24	9	1
Instruct	2	0	0
Justify	2	1	0
Metastatement	35	8	2
Open	7	2	4
Probe	29	18	4

Table 2: Types of metastatement

Apology/self-correction	3
Class control/management	11
Joke	2
Spelling correction (of pupil)	3
Task	16

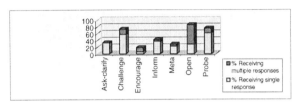

Figure 5: Single (bottom) and multiple (top) student responses to different types of teacher move

6. Discussion

The tables and figures show that the dialogue moves most commonly used by the teachers were metastatements, 'probe', 'challenge', 'inform' and 'encourage'. Of these, 'probe' and 'challenge' moves, where pupils are asked to create or defend a line of argument, can be said to fall within the categories of Socratic intervention recommended by Fisher [10]. 'Inform' moves tended to be in direct response to questions asked by students. 'Probe' and 'challenge' moves seem to be relatively successful in advancing the discussion as both received a high rate of student responses. 'Inform' moves, on the other hand, received a relatively low student response rate. The student questions that were answered with 'inform' could probably have be used to better effect had the teachers responded with a countering 'probe' or 'challenge' move instead of providing a simple answer to the question. Another possibility might be for teachers to follow the 'inform' with a 'justify' move, thus providing a model to the students of how to support a line of argument. However the teachers were very unlikely to use 'justify' moves; only two were used in total.

'Open' moves, where the teacher where the teacher invited the class to contribute to a new topic or extended the scope of the discussion, were used relatively infrequently but were very successful in stimulating student participation. In particular, 'open' moves tended to generate multiple student responses.

'Encourage' moves and metastatements both appeared to have a low rate of student response. However, this could be because little formal acknowledgement is expected after such moves or because students might tend to respond indirectly, with a change in behaviour, rather than with a verbal response. For example, when a teacher tells a student that s/he has made a good contribution, the pupil might not acknowledge the teacher's remark but could react by praising another student's contribution later in the discussion. This type of response is difficult to quantify, as there is no means of knowing how the student might have behaved had s/he not received the positive feedback from the teacher. However earlier findings suggested that, over time, students tended to increase their use of positive reinforcement and this may, in part, be due to the model provided by the tutor.

7. Conclusions and recommendations

The study analysed the intervention strategies of teachers leading a computer-mediated discussion class and compared these to the levels of student response. The analysis showed that the types of move which were most successful in generating student responses were 'open', where a topic was opened or extended to the class, 'challenge', where students were encouraged to justify a line of reasoning and 'probe' where students were asked to provide more information about a topic. Teachers also used a high number of 'inform' moves but these were less likely to encourage students to contribute to the discussion.

Teachers aiming to encourage more student contributions in CMC debate, therefore, should perhaps endeavor to use a strategy which combines 'open' 'challenge' and 'probe' moves. Use of 'inform' moves could be replaced with 'challenge' or 'probe' where possible or, alternatively, accompanied by 'justify' in order to model argumentative reasoning to the students.

It should be noted that the teachers used very few 'counter' moves presenting an alternative line of argument and also tended not to critique the students' reasoning. More research is needed in order to determine whether or not this was deliberate (i.e. if the teachers felt that this type of move would not be useful) and what effect these moves would have on the debate. More research is also needed in assessing the overall quality of the debates *as sound argument* and in determining the types and proportions of different teacher moves that lead to the production of sound argument, here the emphasis has been on encouraging development of the skills of argument and equal participation from students rather than the strength of the arguments put forward.

References

[1] Ravenscroft, A. & Hartley., R. *Evaluation of Chapeltown and Harehills Assisted Learning Computer School (CHALCS)*. Leeds, Computer Based Learning Unit, University of Leeds. 1998

[2] Andrews, R. *Teaching and Learning Argument*. London, Cassell. 1995

[3] Kuhn, D., Shaw, V. & Felton, M. "Effects of Dyadic Interaction on Argumentative Reasoning" in *Cognition and Instruction 15*(3) 1997 pp 287-315.

[4] Phinney, M. "Computer-Assisted Writing and Writing Apprehension in ESL Students." in P. Dunkel *Computer Assisted Language Learning and Testing* New York, Newbury House. 1990

[5] Pennington, M. "Writing the Natural Way: on Computer" in *Computer Assisted Language Learning 9*(2-3). 1996

[6] Walker, S. A. & Pilkington., R. M. "Networked Communication and the Collaborative Development of Written Expression at Key Stage Three". *Proceedings of the Second International Conference on Networked Learning* Lancaster, Lancaster University and Sheffield University. 2000

[7] Pilkington, R. M. &. Walker, S. A. "Overcoming 'Literacy Deficit': Using CMC to Develop Written Argument" in Andriessen, Baker and Suthers *Arguing to Learn: Confronting Cognitions in Computer-Supported Collaborative Learning Environments* (forthcoming)

[8] Sullivan, N. and E. Pratt. "A Comparative Study of Two ESL Writing Environments: A Computer-Assisted Classroom and a Traditional Oral Classroom" *System 29*(4). 1996

[9] Warschauer, M. "Comparing Face-to-Face and Electronic Discussion in the Second Language Classroom" *CALICO Journal 13*(2) 1996 pp7-26.

[10] Fisher, R. *Teaching Thinking. Philosophical Enquiry in the Classroom*. London, Cassell. 1998

[11] Lipman, M., Sharp A. M., Oscanyan, F. S. *Philosophy in the Classroom* (2nd Edition). Philadelphia: Temple University Press. 1980

[12] Pilkington, R. M. *Analysing Educational Discourse: the DISCOUNT Scheme*. Leeds, University of Leeds. 1999

How to Learn the Many Unwritten "Rules of the Game" of the Academic Discourse: A Hybrid Approach Based on Critiques and Cases to Support Scientific Writing

Sandra M. Aluísio*#, Iris Barcelos*, Jandir Sampaio* & Osvaldo N. Oliveira Jr.‡#

*Instituto de Ciências Matemáticas e de Computação, USP, CP 668, 13560-970 São Carlos, SP, Brazil

‡Instituto de Física de São Carlos, USP, CP 369, 13560-970 São Carlos, SP, Brazil

#Núcleo Interinstitucional de Lingüística Computacional, ICMC-USP, CP 668, 13560-970 São Carlos, SP, Brazil

{sandra,iris}@icmc.sc.usp.br, massa@grad.icmc.sc.usp.br, chu@ifsc.sc.usp.br

Abstract

We present the computational and composition theoretical bases for the design of a collaborative writing tool, based on the critiquing approach, to assist non-native novice researchers in the understanding and production of the structure of scientific papers. This Critiquing Tool is embedded in a suite named AMADEUS that caters for various needs of non-native English users to produce a first draft of a paper, relying on the reuse of contextualized linguistic material as input for the user. Our emphasis here is in the architecture and methodology to build the linguistic resources for the Tool. Though originally targeted at non-native authors, the Critiquing Tool may also be useful for novice native English writers and as a teaching resource for English for Academic Purposes practitioners.

1. Introduction

Scientific writing poses enormous difficulties for non-native English authors to publish in international journals. Help can be found in books, online textbooks and software tools – mainly for post-writing evaluation, but an important difficulty remains: the production of a first draft that can be later polished. These users may face difficulties not only at the lexical and syntactic levels, but most importantly at the textual level where major problems are the use of rhetorical structures from their mother language, and misuse of logical relations. One decade ago, we realized that the reuse of written material in the form of expressions or passages of authentic texts produced by competent writers of English could be implemented as resources in computational settings. We then conceived a suite of tools and resources named AMADEUS (AMiable Article DEvelopment for User Support) aimed at catering for distinct users' needs, providing users with linguistic input for writing correct sentences and passages employed for specific purposes in a scientific paper.

Among the tools in the AMADEUS suite, the Reference tool [1-4] was the first to be developed. It works as a lexical database consultation where the units, instead of words, are expressions and standard sentences annotated according to the schematic structure of paper sections: abstract, introduction, review of the literature, methodology, results, discussion and conclusion. It was aimed mainly at (1) researchers who are familiar with the academic writing but still need some feedback, (2) proficient or near-proficient English users who need to write under the constraint of time. Tests with this tool performed at the University of São Paulo, in São Carlos, showed that the tool was successful with students that were already familiar with the scientific discourse, but failed to help less experienced writers, mainly because such users had difficulties in localizing expressions, collocations and cohesive links appropriate to their needs.

For the latter type of user, we developed the Support Tool [5-8], a case-based system (see ref. [9]), which works in co-operation with the user through a three-step procedure: i) gathering of features, in which the user selects the features intended for his/her Introduction; ii) selection of the best-match case, following case recovery by the system through appropriate similarity metrics; iii) revision/adaptation on the selected case in order to insert the user's own text. The self-reviewing stage of the Support Tools is based on a Systemic Functional Linguistic (SFL) interpretation of textual revisions [10]. The implementation of this tool required a comprehensive corpus analysis because — despite its relatively well-defined schematic structure (see refs. [11-12]) — a scientific paper can be organized in different ways. Owing to the complexity of the linguistic analysis, this module covers only the Introductory sections of papers in Physics. The detailed schematic structure comprises 8 components, 30 rhetorical strategies and 45 different types of message (standard linguistic expressions).

A limitation observed with the Reference and Support tools was that students sometimes required feedback to their choices, which prompted us to develop the Critiquing Tool — the object of this paper. The Critiquing tool works collaboratively in cycles — the user presents a product (a dual structure of a paper, see below) to the system and the system gives feedback through critiques for improving the product. Using structural knowledge at the supraparagraph level, the tool indicates the most appropriate sequences of the schematic components, suited for a particular audience and purpose, in the various sections of a paper. The papers were also annotated with components from another structure (called here "specific criteria for papers submission" components) as each different type of paper, e.g. empirical, experience, system, theory, and methodology papers, possess different content in each component from the schematic structure, allowing a more fine-grained recovery suitable to the user needs. We have, therefore, adopted a dual structure to represent different types of papers of a particular scientific community. Here, the community chosen was the researchers from Computer-Human Interaction and the annotated papers came from the CHI'96 short papers session. If we had chosen the learning technology community and a particular instance of it, say ICALT conferences, we would be working with three different types of submissions: theory, empirical and survey papers.

The following sections present the computational approach we employed (critiquing systems) and the composition approach that inspired us, the architecture, and methodology to build the linguistic resources of the Critiquing Tool.

2. Theoretical background

A critiquing system comprises agents, generally one computer and one user, working in collaboration. Both contribute with their knowledge about the domain to which the problem to be solved belongs. The user's role is to generate a first product/solution, and then modify it following the advice of the computer system. The role of the system is to identify deficiencies in the product or in the user's action, and not necessarily solve the problem; indeed most critiquing systems only make suggestions of how to improve the product. With such suggestions/critiques, the user starts the next iterative stage of the process, either solving the problem or seeking additional explanations. One agent that is able to criticize is classified as a critic, and may be a machine or a human being [13-15]. Even though the roles of the critic and user are different, i.e. there is a distinction at the cognitive processes level, there is a coordinated, synchronous activity that is the result of a continued attempt to maintain a shared conception of the problem, differently from what occurs in a co-operation process.

From the composition research point of view, there has been increasing interest in the nature of "success" in writing and the diverse factors contributing to it [10]. This is, in fact, the third main stage in the research of English for Specific Purposes, since its early beginnings in the 1960s. In the first stage, focus was placed on the surface forms of the language (whether at sentence level, or in register analysis, or in discourse analysis), which was called product based. Then research followed another path, with the cognitive or process-based approach, in an attempt to look below the surface and consider not the language itself but the thinking processes that underlie language use [16]. The latter phase received several criticisms from researchers working within the genre-based framework of English for Academic Purposes, e.g. Swales [11], because it overemphasizes the individual and neglects the sociocultural context. The third main phase/paradigm for composition theory then attempts to fill this sociocultural vacuum. In this phase, the written product is considered a social act that can take place only within a specific context and audience.

This approach to composition theory has inspired the design of the Critiquing Tool presented here. However, we have attempted not to get too tied up to this perspective, so that we could still take advantage of the findings and efforts of the first and second phases. To paraphrase Hayes [17] in response to those who believe that cognitive studies are out: "Our research problems are difficult. We need all the available tools, both social and cognitive". Accordingly, the first stage has inspired the use of schematic structures, the second the main processes of planning, composing and revising, and the third the analysis of the product of writing as a social act addressed to a specific purpose and audience.

3. Architecture and methodology to build the linguistic resources of the Critiquing Tool

As Fig. 1 illustrates, the critiquing process starts as soon as the user presents the tool with a product. In order to criticize the product, the tool must identify the user's object(s). This can be done through two mutually exclusive processes: (1) **Identification of the objects** through the user's choices, or (2) **Acquisition of the**

objects, which is done by employing explicit information provided by the user. After getting the user's objects, in our system two types of analyzers are employed to evaluate the product: **Analytic and Differential.** The first analyzes the product using *guidelines* while the Differential method employs a case generated with the Case-based reasoning approach to be compared with the product. The Differential Analyzer follows the specifications from Aluísio & Oliveira [6], where the cases are now represented by two structures: the schematic and that of specific criteria (Fig. 2).

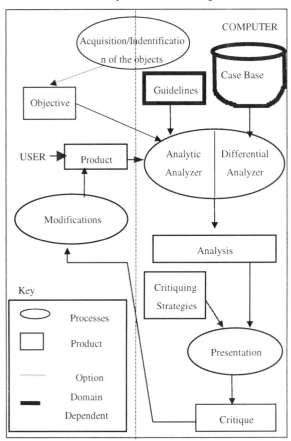

Figure 1. Architecture of the Critiquing Tool

case(system1,
    schematics(intro(related_work(s), domain,
    background, purpose, what_has_been_done,
    structure)
    ),
    specifics(system, system_behaviour,
    system_architecture, system_behaviour,
    what_has_been_implemented)
).

Figure 2. Schematic structure and specific criteria for a paper represented in Prolog.

The schematic features, for example, are taken from the schematic structure components of the CHI'96, which are: domain, contributions, contents, consequences, and structure of the paper, where "contributions" has the following subcomponents: background, related work(s); "contents" has the subcomponents: purpose, what's been done, methodology, and rationale, and "consequences" has the subcomponents: future work(s), lesson(s) learned, and innovation(s).

The Analytic process makes use of guidelines for revising the paper structure, which are rewritten in the form of rules. The Differential process works with a base of papers and verifies the components and their organization, such as the most likely order for the components in a given structure for a specific type of paper. The Differential analyzer is also responsible for showing the user the most similar case retrieved after the modifications in the structure take place. The product analyzers generate facts regarding the comparison of the product with the case recovered, and the guidelines. The component **Presentation** will show the user these facts based on **Critiquing Strategies,** which include five types of critiques (praise, direct criticism, indirect criticism, direct suggestions and instructions), similar to those used by referees. The user then generates a new version of the product and the cycle starts again. Furthermore, general suggestions for improving the paper structure are also provided. Fig. 3 shows the Critiquing Tool in use.

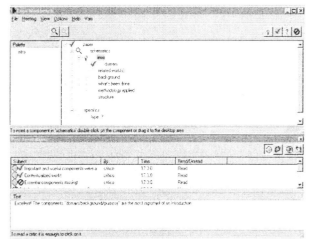

Figure 3. Screendump of the Critiquing Tool in use. The author had selected various topics for his/her intended Introduction taken from the palette. The tool provides feedback, with compliments on the choice of important components and context of the work, but critique on the absence of essential components

Most of the knowledge embedded in the tool can be reused in new domains, by following the steps below:

1. Perform corpus analysis of papers from a given scientific community to identify the components of the

dual structure of the papers. Annotate the corpus according to the components identified. The annotated corpus is to be used by the Differential Analyzer to recover similar cases. Display the cases with a presentation tool/browser (using XML, HTML, or SGML tags).

2. Rewrite guidelines from the Call for Papers or Instructions for Authors sections in the form of rules, which are to be employed by the Analytic Analyzer.

3. Employ the shell of the collaborative tool to obtain the interface and appropriate interaction mode.

4. Test the tool, now including new knowledge, in terms of the communication with the user.

4. Final remarks

In this paper, we described an approach for the development of writing tools based on collaboration, targeted at specific purposes and audiences. The advantages of a collaborative paradigm include (1) it places the learner as one of the agents (actor); (2) it allows the learner to choose when to access the system; (3) it encourages interaction, thus improving the learning process. The paradigm is ideally suited to coaching novice writers in achieving a cohesive schematic structure for their papers, since problems in this domain do not have well defined answers. For organizing material for a paper is not a prescriptive task, and the collaborative learning systems may help exploit the several ways through it such a task can be performed. It should also be mentioned that our emphasis has been on laying the basis for collaborative tools to be developed. We are aware of the need of extending the existing resources to a much wider variety of areas, not only to provide adequate linguistic input for a specific field but also to cover idiosyncrasies in the schemata of a scientific paper in different areas. All the tools in AMADEUS architecture, and the Critiquing Tool in particular, were developed to allow facile extension and customization, and we welcome the feedback of users and collaboration for extending the existing resources to new areas and/or different formats.

Acknowledgments

The authors acknowledge the financial support from CNPq and Fapesp (Brazil).

References

[1] Oliveira Jr., O.N., Aluísio Caldeira, S.M. & Fontana, N. Chusaurus: a writing tool resource for non-native users of english. *Ricardo Baeza-Yates and Udi Manber (eds.) Computer Science: Research and Application*, New York: Plenum Press, 63-72 (1992).

[2] (Caldeira) Aluisio, S.M., De Oliveira, M.C.F.; Fontana, N.; Nacamatsu,C.Y. & Oliveira Jr. O.N. Writing tools for non-native users of English. *Proceedings of the XVIII Conference LatinoAmericana de Informatica*, 224-231 (1992).

[3] De Oliveira, M.C.F., (Caldeira) Aluisio, S.M, Masiero, P.C. & Oliveira Jr., O.N. A discussion on human computer interfaces for writing support tools. *Proceedings of the XII International Conference of The Chilean Computer Science Society*, 223-233 (1992).

[4] Fontana, N., (Caldeira), Aluísio, S.M., De Oliveira, M.C.F. & Oliveira Jr., O.N. Computer assisted writing - applications to English as a foreign language. 145-161. *CALL (Computer Assisted Language Learning Journal)* **6**(2), 145-161 (1993).

[5] Aluisio, S.M. & Oliveira Jr,. O.N. A detailed schematic structure of research papers introductions: an application in support-writing tools. *Revista de la Sociedad Espanyola para el Procesamiento del Lenguaje Natural* 19, 141-147 (1996).

[6] Aluisio, S.M. & Oliveira Jr. O.N. A case-based approach for developing writing tools aimed at non-native English users. *Lectures Notes in Artificial Intelligence* 1010, 121-132 (1995).

[7] Aluísio, S.M. & Gantenbein, R.E. Towards the application of systemic functional linguistics in writing tools. *Proceedings of International Conference on Computers and their Applications*, 181-185 (1997).

[8] Aluísio, S.M. & Gantenbein, R.E. Educational tools for writing scientific papers. *VIII Simpósio Brasileiro de Informática na Educação*, 239-253 (1997).

[9] Kolodner, J.L. *Case-Based Reasoning*. Morgan Kaufmann (1993).

[10] Gosden , H. Success in Research Article Writing and Revision: A Social-Constructionist Perspective. *English for Specific Purposes*, 14(1), 37-57 (1995).

[11] Swales, J. *Genre Analysis - English in Academic and Research Settings*. Cambridge University Press (1990).

[12] Weissberg, R. & Buker, S. *Writing Up Research - Experimental Research Report Writing For Students Of English*. Prentice Hall Regents (1990).

[13] Fischer, G. et al. Critics: An Emerging Approach to Knowlwdge-Based Human-Computer Interaction. International J. of Man-Machine Studies 35, 695-721, 1991.

[14] Silverman, B.C. (1992). Survey of Expert Critiquing Systems: Practical and Theoretical Frontiers. CACM, Vol.35, N4, 107-127.

[15] Rankin, I. (1993). Natural Language Generation in Critiquing. The Knowledge Engineering Review, Vol 8:4, 329-347.

[16] Hayes, J.R. & Flower, L.S. Identifying the organization of writing processes. *L.W. Gregg & E. R. Steinberg (eds.), Cognitive Processes In Writing*, Hillsdale, N.J. Erlbaum. 3-30 (1980).

[17] Hayes, J.R. A new framework for understanding cognition and affect in writing. *C. Michael Levy and Sarah Ransdell (eds.), The Science of Writing – Theories, Methods, Individual Differences, and Applications*, Lawrence Erlbaum Associates, Inc., 1-27 (1996).

Helping the Tutor Facilitate Debate to Improve Literacy using CMC

P.A. Kuminek and R. M. Pilkington
Computer Based Learning Unit, University of Leeds
P.A.Kuminek@cbl.leeds.ac.uk

Abstract

Recent research has shown that Computer-Mediated Communication (CMC) in the form of synchronous chat can help students develop their debating skills. However, there are difficulties in learning to use chat to debate. The roles of the tutor in such debates was observed with the aim of designing chat tools that better support the tutor in facilitating students' learning. In particular a design for a system CHATTERBOX is proposed which is aimed at providing support (a) through the use of dynamic graphical displays that monitor student participation and (b) through the use of sentence openers that not only assist the composition of turns but help provide feedback on the balance of different kinds of contribution made by students and the tutor during chat.

1. Introduction

The attraction of Synchronous CMC chat as an *interpretative zone* that enables participants to share and debate multiple perspectives lies in its potential to offer shy or more reflective students an environment where they need not worry about interrupting the flow of a conversation that has "moved on" [1, 2]. CMC allows students more opportunities to speak than in traditional classrooms, increasing the number of occasions that they are able to practice debating skills. These CMC "advantages" stem from an apparent "disadvantage" - the violation of normal turn-taking caused by not being able to "see" or "hear" other participants in the process of "speaking". The result is the non-adjacency of related turns and many overlapping and parallel threads of conversation. Together with the phenomenon of *screen roll-off* this means that coherence is affected. It follows that the tutor is faced with a number of *unique* management problems arising from using the medium.

2. The rationale and aims of the course

This report is based on observations of an on-line literacy course for students aged 13-15 at the Chapeltown and Harehills Assisted Learning School (CHALCS), an after school learning centre situated in a deprived inner city area of Leeds, UK. The project serves the needs of local young people, most of whom come from Asian or Afro-Caribbean backgrounds – many speaking a first language other than English. The programme, DaRE (for Discussion and Reporting Electronically), makes use of a WebCT™ virtual learning environment (VLE) which incorporates a synchronous chat tool. The course aims to develop students' understanding of a topic and debating skill by using the chat tool to help them express themselves in writing. The class tutor acts as a facilitator of chat debate. Recent research on the effectiveness of the course "showed improvements in fluency, confidence, argumentation and awareness of audience" [3]. This study observed on-line tutors facilitating chat debate. Observations are based on 2 classes each with networked computers sufficient for the tutor and a maximum of 14 students.

3. Problems in facilitating debate

From the observation of the tutors, it was evident that facilitating debate is affected by non-adjacency of turns and screen roll-off. These phenomena make the tutor's role cognitively demanding. This is compounded when the tutor's concentration is interrupted by the need to intervene to deal with a behaviour problem. When this occurs, (he) she will have to either:
(a) do nothing and hope that they will pick up the gist of the argument later on;
(b) scroll the chat screen back to read intervening dialogue.
The latter strategy risks missing continuing conversation and is limited since the screen record can only be rolled back so far - early conversation eventually disappears altogether from the screen (though it is kept in a log file). Our aim is to assist the tutor with such problems through the redesign of chat tools.

4. Using colour to help monitor discussion

One of the main difficulties for the tutor in monitoring participation is locating previous messages and attributing them to individuals. Given the monochrome nature of the current WebCT™ chat window, one solution is to differentiate each student message by assigning a unique colour to their text. Preliminary results based on the use of Yahoo Messenger™ indicate that this solution offers some advantages. Our design differs from Yahoo Messenger™ and similar tools in enabling the tutor to control the assignment of colour. In our design tutors are able to set permanent colours to each participant so that they can identify and refer to each student based on the colour of observed text, not just in one chat but throughout the course. By observing colour-coded messages the tutor can judge the proportion of the conversation contributed by each student. In addition, such a device could be used to help tutors to quickly find individuals' previous messages.

5. Using meters to help monitor discussion

Another difficulty in effectively facilitating discussion is in identifying those who are inactive in the chat – *lurkers*. A tool that could dynamically provide the tutor with visual clues to student interaction patterns is proposed. The tool is based on Babble's "social proxy" meter [4]. Instead of representing *persistent* numbers of threaded synchronous and/or asynchronous conversations that participants can move between, it represents the degree of active participation in a single synchronous debate. In this debate (large circle in Figure 1), each participant is represented by their colour and image on their "marble", with the colour of each marble matching the tutor's assignment of colour for viewing messages in the conversation window. When a participant sends a message their marble moves towards the centre of the circle and then drifts outward with inactivity. This should help bring inactive students to the tutor's attention. The tutor can then encourage these students to participate.

6. Using "sentence-openers" to coach debate

Many students at CHALCS lack debating skills. A number of researchers have pointed towards using predefined lists of "*sentence-openers*", opening phrases that can be grouped into *move* types such as statements, checks, challengers, counters and conclusions [1]. The use of sentence-openers by students has been shown to improve student reflection [5]. As part of an initial investigation into their use, one of the tutors at CHALCS is being encouraged to use *sentence openers* to model and describe appropriate example turns. Tracking their use provides an easily automated way to assess the balance of preferred moves adopted by students to help monitor their changing skills.

7. The CHATTERBOX design

Combining the ideas outlined above, our proposed chat tool called CHATTERBOX is illustrated below. A prototype is in the early stages of implementation and is expected at the end of 2001.

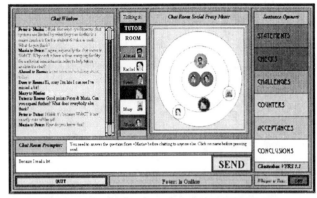

Figure 1. A design for CHATTERBOX

8. Conclusions

The job of the tutor in facilitating high quality debate is extremely cognitively demanding. A design for a system CHATTERBOX is proposed which is expected to assist the tutor via the use of dynamic graphical displays and the categorization of sentence openers to identify individual student contributions and monitor levels of their participation. This should, in turn, release the tutor to concentrate on maintaining class focus, modeling and scaffolding the acquisition of key debating skills.

References

[1] Veerman, A.L., Andriessen, J.E.B. & Kanselaar, G., "Learning through synchronous electronic discussion", *Computers & Education*. Vol. 34, 2000, pp. 269-290.

[2] Robertson, J., Good, J. & Pain, H., "BetterBlether: The Design and Evaluation of a Discussion Tool for education", *International Journal of Artificial Intelligence in Education*. Vol. 9, 1998, pp. 219-236.

[3] Walker, S.A. & Pilkington, R.M., "Networked Communication and the Collaborative Development of Written Expression at Key Stage Three". In *Proceedings Networked learning2000 Lancaster 17th-19th April* 2000, pp. 28-37.

[4] Erickson, T., Smith, D.N., & Kellogg, W., Laff, M., John T. Richards, Bradner, E., "Socially Translucent Systems: Social Proxies, Persistent Conversation, and the Design of ``Babble''. *CHI 1999*, pp. 72-79.

[5] Baker, M.J. & Lund, K., "Promoting reflective interactions in a computer-supported collaborative learning environment", *Journal of Computer Assisted Learning*, Vol. 13, 1997, pp. 175-193.

A Study of Social-Learning Networks of Students Studying an On-line Programme

Gurmak Singh
Subject Leader
g.singh@wlv.ac.uk

John O'Donoghue
Research Project Manager
j.odonoghue@wlv.ac.uk

Abstract

This paper reports the findings of a research project exploring the socio-learning environment of students studying an on-line programme. The findings point to the key areas of conflict between the developed on-line teaching and learning systems and the socio-learning environment of the learners. The main contention of the paper is that whilst current frameworks provide useful insights into the socio-learning environments of the learners, many important aspects have not been fully explored. The findings support and build on current studies which propose that the level of on-line interaction received by students is indicative of the success of the learning process. Furthermore, the differing levels of interaction aid learners to consolidate their knowledge by communication with academic support groups, tutors, family and friends.

The findings are reported using the social network theory specifically addressing four levels of relationships. The first level relationships consider the learning relationships between the academic tutors and the learners. The second level relationships explore the interaction between the learners and other parts of the institutions, such as learning centre staff and pastoral care counsellors. The third level relationships are between the different learners and the fourth level relationships are between learners and the outer community, such as friends and family members.

1. Introduction

Over the last decade there has been a significant increase in the use of information communication technologies in higher education institutions. The advantages of computer-mediated learning environments over traditional approaches have been well documented; incorporation of individual learning strategies (Paterson and Rosbottom, 1995) shift the degree of control to learners (Naidu, Barrett and Olsen, 1995) encouraging active learning flexible delivery and a learner-centred approach . On-line interactivity through computer technology has already received much attention from educators (Harasim et al., 1995; Mason & Kaye 1989; Berge and Collins, 1995). Compared with face-to-face interaction, online interaction is usually asynchronous and hence adds the benefit of being more reflective. While it lacks the uses of the face-to-face context, and the immediacy of feedback of synchronous communication, it does create a record of the interactions in a series of messages which can be re-read and even quoted in assignments (Mason, 1998). During this period there has been shift in focus from the individual towards a new emphasis on social contexts for learning in terms of collaboration and co-operation (Glaser, 1990; Kaye, 1992; Fowler 1999). There is a growing recognition that much of the learning may occur outside the formal classroom environment. Fowler and Mayes (1997) refers to this 'out-of-classroom' learning as situated learning. There are many views on situativity. Lave and Wenger (1991) emphasise a wider social context exploring relationships between the wider identifiable groups of people. For situated learning the increased interactions amongst learners, groups of learners, instructors and tutors is provided by electronic communications supported by learning technologies. Such environments have been shown to enhance learning outcomes in many different ways, including improvement in the quantity and quality of the learning experience (Grabinger, 1995). Furthermore, such technological environments remove the logistical problems encountered by traditional approaches and improve collaboration between learners. Whilst online technologies offer many perceived benefits they will not in themselves improve or cause changes in learning. What improves learning is well-designed instruction (Paterson and Rosbottom, 1995) and learners who are used to working independently, and who have personal motivation to learn. More recently, Fowler and Mayes (1999) emphasised social network theory to explain and model learning relationships. They recognised three types of relationships: explorative, where learning is about discovery; formative learning, the building of understanding through guided activity; and comparative, where the learner becomes an accepted member of the community.

The aim of the research project is to explore the nature of social relationships of learners using Fowler and Maye's (1999) framework. These insights are used to develop a theoretical model that identifies the nature of the relationships by identifying their dimensions. The research method involves investigation of a single - referred to as intrinsic case study by Stake (1995). The case is examined in detail, its contexts scrutinised, its activities investigated.

Five exploratory group interviews were carried out with sixteen respondents. The main aim was to develop an understanding of the meanings attached to learner experiences in an on-line environment. Respondents were asked to share their experiences, to describe whatever events seemed significant to them, and to provide their own definitions of their situations. The second interviews were semi-structured with twenty-one respondents. These were designed to explore the causality of relationships among the concepts identified in the first interviews. The key anchor themes were identified from the raw causal statements. The concepts mentioned frequently were grouped together into the framework identified by Fowler and Mayes (1999).

2. Details of the case

The Wolverhampton Open Learning Framework (WOLF) is a purpose built computer based learning environment developed by the University of Wolverhampton. Through close consultation between academic staff and developers, an integrated system has evolved which enables students to access course notes, related resources, support materials and collaborative tools quickly and easily.

The WOLF system uses streamed Internet based technology to bring together a wide range of powerful tools to create a multi-structured aid to learning. One copy of all learning material and associated resources is stored centrally and streamed to any user with access to the Internet, on demand. The key feature here, apart from being available when needed by the user, is the simplicity in updating and amending material, given only one copy is kept. Freeing lecturers from their previous paper chase ensures that all students on the course have the 'correct' learning and assessment materials. Lecturers are free to keep improving the learning environment and adding to the learning materials. It is based on the simple acceptance of "Anytime, Anywhere Study". The development of content involves no programming (HTML etc.), which means that lecturers can update their material themselves or submit it electronically to a course administrator.

3. Profile of Respondents

The respondents were predominantly mature students: 7 male and 14 females with an average age of 26. They came from a variety of disciplines and possessed a range of educational qualifications. All were studying part-time and most had been with their organisations for just over three years. All had received training of using the on-line system. The study was undertaken in autumn semester 1999 with part-time students in their final year Business Administration Award.

4. First Order Relationships: Tutor and Student

Whilst the relationship with tutors was considered to be essential for effective learning, respondents claim it varied depending on the teaching methods of the tutor, assessment task and the learning material available. Furthermore, they claimed that in the majority of cases the course material available on the system appeared to be designed to replace textbooks and lecture notes. The on-line system provided all the necessary material (module guides, assessment, lecture notes and references to key reading material) in an easily accessible form using this approach. Perhaps surprisingly, learners were comfortable with this approach as it was easier for them to translate their expectations of a traditional approach to the 'on-line' system. The tutor-learner communication mainly occurred when the students needed re-assurances and guidance on the requirements of the assessment. Communication in this context was one-way, static and 'on-demand'. Learners suggested that the tutors replied to their queries in-depth but the response sometimes took over three to four days and by that time the other avenues had been found. Learners suggested that whilst tutors were keen to help them with the assessment task they were not always so forthcoming trying to 'teach across the wire'. The advantages of working in your own time, at your pace, without the need for irrelevant material outweighed the need for learning experience. This 'outcome' based approach was evident in students and tutors.

5. Second order relationships: Student and Student

All students agreed that the majority of their time was spent on completing their assessment through group interaction, even when the assessment was an individual piece of work. The individual tasks offered advantages in that there was no reliance on other group members, however the learners complained it did create a *'feeling of isolation'* and *'working in the dark'*. To overcome this remoteness the learners chose to discuss their progress with their tutor-group. Initially, the interaction was difficult as the learners did not know each other, but once relationships were built they became vital in all aspects of their study. Initially all emails were copied to all group members but later in the course selective relationships were built. These group members became aware of each others' working patterns and knew when to expect replies to their communications. Learners were prepared to share material that they found, discuss the progress they had made but, not surprisingly, were reluctant to share their final work. Some learners suggested that there was pressure to circulate their work to group members and, in some cases, had done so to remain a legitimate member of the group. Whilst in other groups there was a feeling of shared responsibility to help colleagues; success was measured by group success even with individual assessments. Each member felt they had a role to fulfil within the group in order to justify their position in the group even though this meant taking on extra tasks.

This strong group membership resulted in an absence of clear assessment criteria. The learners came to a consensus and decided themselves the requirements of the assessment. Such group decision-making was seen as vital and the decision reached as the agreed 'absolute'.

6. Third order relationships: Student and Other University Units

The designated 'course buddy' allowed the learners to contact a *'human-being'* when problems arose or clarifications were required. The form of communication was informative in that students required guidance on 'where to get information' and 'who to contact'. The accessibility was the key issue; learners wanted a person on the other side of the phone line 9-to-5. The Course Manager, who was involved in other activities and frequently away from the office, could not provide the service required.

The learning centre was a key resource as most assessment tasks required some form of research. Learners suggested that as part-time working students they did not have large amounts of time to spend seeking information in the learning centre. So, although the learning centre had staff available all the time, it was important for the learners to identify a key person with whom they could build some form of rapport. This would result in a more personalised service where the learning centre staff would recognise them and be aware of their needs without having to repeat the process every time they made contact.

7. Fourth order relationships: Students and Wider Society

Whilst learners were able to share much of the material with their group members they did not think it appropriate to share their final [individual] assessments. Work colleagues who specialised in specific functions, such as IT, Marketing and Human Resources were used to feedback on completed and draft versions. The form of feedback generally resulted in the 'final check' before submission of the assessment. Most learners agreed that they did not receive any extra help from their work colleagues, but their involvement reassured them that they had not omitted any key issues. Many students also suggested they used members of their family to read through their work.

First Level	Second level	Third Level	Fourth Level
Communication *(formative)*	Social environment *(formative)*	Routing information *(formative)*	Specialist guidance *(comparative)*
Learning experience *(formative)*	Group Membership *(comparative)*	Instant access (	Use of IT
Feedback *(formative)*	Accountability *(comparative)*	Focused guidance *(formative)*	Verification *(formative)*
Variances in approaches	Confirmation *(formative)*		

8. Discussions

The students are perhaps more aware of the change in learning environments when using learning technologies than their tutors. This is not surprising as research suggests that learners select such approaches because the method fits their circumstances and they have the appropriate communication and motivation skills. The issue is three-fold. First, there is a need to raise awareness amongst tutors of where and when learning occurs, certainly not in the classroom and not solely through tutor-student interactions. The most disturbing finding was that students adopt alternative learning strategies when the designed infrastructure fails. Second, although social learning research has made significant steps forward, it is in vain unless it becomes an implicit part of the teaching and learning strategies of the on-line systems. Third, the learning technologies are particularly useful for part-time mature students who are self-motivated and self-disciplined but they need assurances, especially in the early stages, from the human tutor.

The findings suggest that learning may take place in a wider context than suggested by the Fowler and Mayes (1999) model. First order relationships appear to be the most explorative focused where the learners are establishing the boundaries and context of their learning. In this context the relationship between the learner and the tutor is passive in that the latter provides the information and the former collects as much information as possible. In the second order relationships there is more formative learning between the learners. Initially the relationships are fragmented and cautious but later these become more selective and comparative.

9. References

Berge, Z. and Collins, M. (eds.) (1995). Computer Mediated Communication. Vol I, II and III. Hampton Press Inc., New Jersey.

Fowler, C. and Mayes, J. (1997), Applying telepresence to education. BT Technology Journal, 14, 188-95.

Fowler, C. and Mayes, J. (1999) Learning relationships from theory to design. Association for learning Technology Journal, 7, 3, 6-17.

Glaser, R. (1990), The re-emergence of learning theory within instructional research, American Psychologist, 45, 1, 29-39.

Grabinger, R. and Dunlap, J. (1995) Rich environments for active learning: a definition. Association for learning technologies Journal, 3, 2, pp5-35.

Harasim, L., Hiltz, R., Teles, L. and Turoff, M. (1995). Learning Networks. MIT Press, Cambridge.

Kaye, A. (ed.) (1992), Collaborative Learning through Computer Conferencing: The Najaden Papers, Heidelberg, FRG: Springer Verlag.

Lave, J. and Wenger, E. (1991), Situated Learning: Legitimate peripheral Participation, Cambridge: Cambridge University Press.

Mason, R. and Kaye, A. R. (eds.)(1989). Mindweave: Communication, Computers and Distance Education. Pergamon, Oxford.

Mason, R. (1998). Globalising Education. Trends and Applications. Routledge, London.

Naidu, S., Barrett, J. and Olsen, P. (1995). Improving instructional effectiveness with computer-mediated communication. Association for Learning Technology Journal, 3,2, 63-76

Paterson, P. and Rosbottom, J. (1995) Learning style and learning strategies. Association for Learning Technology Journal, 3, 1, PP 5-12

Stake, R. (1995). (ed) The art of case study research. London, Sage

Wenger, E. (1998), Communities of Practice: learning, meaning and Identity, Cambridge, Cambridge University Press.

Proposal of a Collaborative Learning Standardization

Toshio Okamoto, Mizue Kayama and Alexandra Cristea
University of Electro- Communications Graduate School of Information Systems
{okamoto, kayama, alex}ai.is.uec.ac.jp

Abstract

This paper reports on considerations and steps towards standardization [2] of the collaborative learning environment. This standardization will extend and widen the field of applications possible within the collaborative learning paradigm, and will make possible the usage of the fruits of years of research and individual implementations of the concept of collaborative learning, from our own laboratory and from others.

1. Collaborative Learning Support

Distributed [4,5] collaborative learning [1,3,6,7,8] support is a research domain that tries to find out ways to support the collaboration of multiple learners on the network (CSCL - Computer Supported Collaborative Learning), in problem solving or other cooperative curriculum activities, according to the used LT (Learning Technology). Compared to CSCW (Computer Supported Cooperative Work), CSCL has as a goal not so much the working efficiency, but the learning achievement efficiency, and the promotion of deep understanding of the subject field by the learner, combined with the recognition or meta-recognition of the achievement of this ability by other persons. The regular CSCL groupware implementation provides usually two types of activity space: a private and a collaborative working space, where the learners can exchange information in a synchronous or asynchronous manner.

2. Primitive Activities and Resources

Primitive activities in collaborative learning are: Dialogue (with Interaction), Data/Idea sharing, Observing/ suggesting, Turn-taking, Coordinating/ Control, Planning/ Executing, Initiative/ Supervising. The resources required in collaborative learning are: Dialogue Channel, Shared Workplace (shared object space), Technologically mediated remote communication (audio & visual), Personal Workplace.

3. Collaborative Environment Structure

For collaborative learning, we can differentiate between *learner-to-learner dialogue* (communication) and *other activities* (problem solving, etc.). When a learner faces a problem that s/he cannot solve, s/he can, in collaborative learning, exchange meaningful information via interactions with his/her learning companion(s). This can lead to understanding other persons' perception ways and also help in finding eventual inconsistencies in ones own judgments. Present researches analyze such interactions, their catalyzators and effects. Moreover, as the learning efficiency has been shown to increase in such situations, many systems try to positively encourage them via computer implementations.

4. Essential Structural Elements

Learners can belong to one or more groups and can be involved in projects or parts of projects together, therefore sharing a particular space, and work privately for the rest. The shared working place (***collaborative workplace***) contains the *dialogue support objects* for dialogue and information exchange support, the *collaborative working objects* for activity support, and the *collaborative memory*, for reference and information accumulation. On the other hand, the ***private working place*** contains the working depository of the *private working objects*, and the *private memory* for consultation and accumulation of private activities related information. Moreover, the information referencing layer contains information oriented towards individual and collaborative learning goals, learning materials, various educational data, libraries, educational applications, all-purpose tools, market applications, etc.

There are 6 essential structural elements of the collaborative learning environment standardization:
1. *collaborative learning environment expression*
2. *collaborative workspace expression*
3. *collaborative learning resource(s) expression*
4. *collaborative workplace expression*
5. *learner group model in collaborative learning*
6. *collaborative memory structure expression*

Due to the lack of space, these items are not detailed here.

5. Info Retrieval in Collaborative Learning

The collaborative learning support system has to able, at the learner's request, to send and receive information on the essential elements of both collaborative and private workspace. E.g., a loading function is necessary, which fetches collaborative/private work objects, requested by the group or by individual learners, from various resource(s) (*load_into* relation). This relation is defined within the essential structural elements of the collaborative learning environment. Another relation ensures the sending and receiving of problem solving communication data within the collaborative working place, between the dialogue support objects and the collaborative working objects (*link_to* relation). Another relation ensures the inserting/ saving of objects, results and information from the private workplace of the private workspace as collaborative work objects of the collaborative workplace (*insert_in* relation). The relations between the collaborative work objects and the collaborative memory are "*store_to*", for storing work objects into the collaborative memory, and "*refer_to*" when referring objects already stored. We are gradually building the essential functions, which can be extended to serve any collaborative learning environment.

6. Interface

Standardization of the interface means defining the 5 interfaces below.
1. between learning resource(s) and collaborative work object(s)
2. between dialogue support object(s) and collaborative work object(s)
3. between the private workspace and collaborative work object(s) of the collaborative workplace
4. between collaborative work object(s) of the collaborative workplace and the collaborative memory
5. between the collaborative memory and group model

Due to the lack of space, these items are not detailed here.

7. Data Exchange

One of the essential structural elements of the collaborative learning environment is the *virtual agent*. The information exchange between the other essential structural elements is done via agent(s). The attribute(s) of the appropriate essential elements are stored in the collaborative memory as well as the learning log developed during the collaborative learning curriculum. Furthermore, depending on the request from group member(s) and collaborative work object(s), agents refer the information in the collaborative memory and integrate the exchanged information into a defined form. The concrete function of agents is to cope with the behavioral differences of the essential structural elements. Moreover, the information exchange protocol content varies, according to the transmission source and reception destination, and according to the behavior or functions of the bi-directional structure of the essential elements. However, the basis functions and structure of the agents in the collaborative learning environment are defined simply as the *exchange*, *deletion* and *addition* of essential structural elements.

8. Collaborative learning Agent

The standardization target delimited by the hypothesis is represented by the 5 items below.
1. collaborative learning environment agent(s) structure
2. collaborative learning environment agent type(s)
3. essential elements of 1.
4. attribute(s) of 3.
5. relation(s) of 3.

8. Conclusion

The standardization of the collaborative learning environment is a collective effort and an ongoing process. We have outlined here some of the basic considerations and steps we intent to take in the future towards this standardization.

Further required functions for the collaborative learning environment are as follows.
- *Coordination* (constrained and mediated by external environment)
- *Reification* (material evidence in the external environment)
- *Illustration* (external representation)
- *Storage* (in later use, for the purpose of reflection)

Examples of general tools for supporting collaboration are as follows.
- Concept Mapping tool
- Editors for argumentation network
- Work flow (planning tool)
- WYSIWIS (What you see is what I see)

At the university of Electro-Communications, Japan, we have integrated a few parallel projects that have related goals concentrated around distance-learning and life-long learning, also known under the name RAPSODY [4] and RAPSODY-EX [5]. The proposed standards are being gradually implemented and brought to life within these projects. Real-life feedback will be used to correct, improve and fine-tune the proposed standard.

Reference

[1] du Boulay, B., Can we Learn from ITSs?, International Journal of AI in Education, 2000, 11, 1040-1049.
[2] http://lttf.ieee.org
[3] Van Joolingen, W., ed., Collaborative discovery learning in the context of simulations, Proceedings of the workshop at ITS 2000, 11, 1030-1039.
[4] Okamoto, T., Cristea, A.I., and Kayama, M., ICCE 2000 Invited Talk, Towards Intelligent Media-Oriented Distance Learning and Education Environments, ICCE 2000 proceedings, http://icce2000.nthu.edu.tw/Proceedings.hi
[5] Okamoto, T., The Distance Ecological Model to support Self/Collaborative-learning in the Internet Environment, ISSEI 2000, Bergen, Norway, Section V; Workshop no.: 511 (appeared as CD-ROM).
[6] Pikington, R.M., ed., Anlysing educational dialogue interaction: towards models that support learning, Proceedings of the workshop at AIED 9910, 1040-1049.
[7] Soller, A. L., Supporting Social Interaction in an Intelligent Collaborative Learning System, International Journal of AI in Education, 2001, 12 (to appear).
[8] Wasson, B., Identifying Coordination Agents for Collaborative Telelearning, International Journal of AI in Education, 1998, 9, 275-299.

Cosar: Collaborative writing of argumentative texts

Jos G.M. Jaspers   Gijsbert Erkens   Gellof Kanselaar

Department of Educational Sciences
Utrecht University
Heidelberglaan 2
NL-3584CS Utrecht
The Netherlands

J.Jaspers@fss.uu.nl   G.Erkens@fss.uu.nl   G.Kanselaar@fss.uu.nl

Abstract

In the COSAR project (COmputer Supported ARgumentative writing) we study electronic collaborative text production regarding the relation between characteristics of interaction on the one hand and learning and problem solving on the other. A groupware program has been developed to support collaborative writing. This system consists of a basic environment that can be extended with a combination of tools. The experiments with the basic environment have been completed. The first data of the evaluation forms will be presented. The second phase of the project in which tools are added to the basic environment in different combinations is currently in progress. Students are generally enthusiastic about the system.

Based on the observations we have made during the experiments we will adapt the system in order to improve the usability and extend the range of subject areas.

1. Introduction

A recent Dutch educational law has transformed the program of the final three years of college preparatory high school. Among others, schools are required to provide support for students to do increasingly independent research, in order to prepare them better for university studies. Working and learning actively, constructively and collaboratively are seen as important elements of this curriculum. The computer-supported collaborative writing environment developed in the COSAR project fits well within this new program, as the active and interactive nature of the Information and Communication Technology (ICT) involved emphasizes both the constructivist and collaborative aspects. Computer- and telematics-based environments seem especially suited for collaborative learning by the variety of possibilities they possess: they may integrate multimedia information sources, data processing tools and systems of communication (time and place independent) in one single working environment [2,7]. Computer Supported Collaborative Learning systems (CSCL) are assumed to have the potential to enhance the effectiveness of peer learning interactions [6,1].

Central research topic of the COSAR project is how groupware programs can facilitate collaborative learning.

2. Description of the environment

In the COSAR project (COmputer Supported ARgumentative writing) we study electronic collaborative text production regarding the relation between characteristics of interaction on the one hand and learning and problem solving on the other (http://owkweb.fss.uu.nl/cosar). A groupware program (TC3: Text Composer, Computer Supported & Collaborative) has been developed that combines a shared word-processor, chat-boxes and private access to internal and external information resources to facilitate collaborative distance writing. The program is meant for pairs of students (16-18 year old) working together on argumentative essays based on provided information resources, within the context of the Dutch language curriculum. The assignment is to choose a position pro or contra a current topic (cloning or organ donation) and to write a convincing text addressed to the Department of Welfare, Public Health and Culture. The information resources provided are recent articles and commentaries from Dutch quality newspapers. The texts should count 600 – 1,000 words and are graded anonymously by the students' own teachers. Each partner works at his/her own computer and, wherever possible, partners are seated in different classrooms. The window of the basic program displays several private and shared windows.

The basic environment consists of four main windows (see Figure 1):

1.   INFORMATION (upper right): The assignment, relevant information sources and TC3 operating instructions can be accessed in a tabbed window. Sources are divided evenly over the partners.

2. NOTES (upper left): A notepad in which each student can make private, non-shared notes.
3. CHAT (lower left): The lower chat box shows the student's current contribution, the one above it shows the incoming message of his partner as it is being typed (WYSIWIS: What You See Is What I See). Pressing the enter key will clear the textbox and add the line to the scrollable window that shows the discussion history marking the contribution of each partner in a different color. This chat history can be scrolled to view the complete chat from the beginning of the assignment.
4. SHARED TEXT (lower right): A shared word-processor (also WYSIWIS) in which the common text can be composed by taking turns. Whenever this text changes the system ensures that the area in which the change takes place is visible to the other partner.

Two planning tools have been be added to the basic TC3 environment. We are currently experimenting in the schools to access the effects of each tool and of the combination of both tools:

5. DIAGRAMMER: A shared tool for generating, organizing and relating information-units in a concept map. With the diagrammer the students can make a graphical summary of the paper. (see figure 2) The boxes are color coded. Clicking one of the buttons on top of the screen will create an empty new box.
6. OUTLINER: A shared tool for linearizing content in a text outline structure, similar to a table of contents. It consists of a window in which the lines will automatically be numbered like a table of content.

On the bottom part of the screen is an area that contains several groups of action buttons:

Layout: In the default layout the screen is divided by a horizontal bar. This layout can be changed by moving this bar up or down to a fixed position. This enlarges the chat and common text or the private notes and the information window respectively.

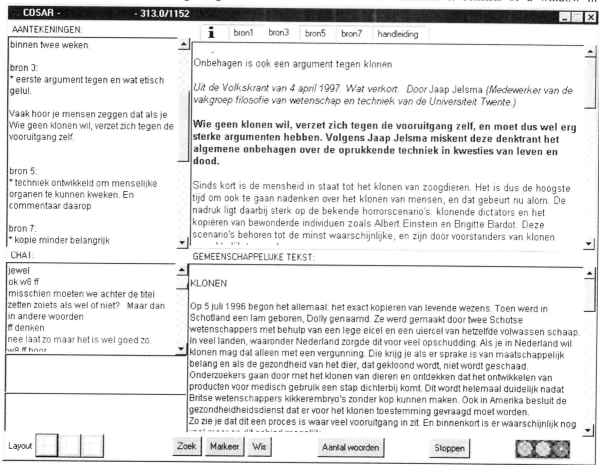

Figure 1 Screen dump of the COSAR environment

Marks: (search, mark, erase) Enables the student to mark the selected text from the information sources with a bright yellow color. The search button will show the student the next marked area in a cyclic fashion. The erase button will unmark the selected area.

Word count: displays a dialog box with the number of words in the shared text.

Traffic light: During the session only one student at a time is allowed to change the common text, diagram or outline. The other student can request his/her turn by clicking the traffic light. The traffic light of his/her partner will start to flash yellow, and by clicking it, the turn goes to the other partner.

3. Technical information

The COSAR-system consists of a server and a client program. Both programs are written in the scripting language TCL. [4] We use the standard internet protocol to communicate. The server can handle at least 90 simultaneous connections i.e. 45 pairs of students. In order to minimize the efforts on part of the system administrators in the participating schools the Cosar program is compact (<1 Mbytes) and virtually installation free.

Dutch schools depending on the local circumstances differ in the manner and the bandwidth they establish their connection to the internet. In order to guarantee a stable

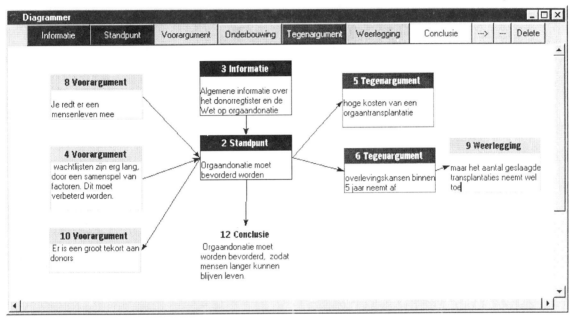

Figure 2 The Diagrammer

A complete record of all interaction is stored on the server. From this log the entire session can be reconstructed or even replayed. All actions in the different text areas such as text insertion, deletion, scrolling, marking etc. are logged. This logfile is processed by a separate program to create a more condensed version of the interaction. This condensed protocol is processed by MEPA (Multiple Episode Protocol Analysis) [5]. The purpose of MEPA is to offer a flexible environment for creating protocols from verbal and non-verbal observational data, and annotating, coding and analyzing these. This program is also capable to compute a number of statistics and summaries of the protocols.

network connection to the server and to create a comparable network performance we installed the same server in the local network of the schools.

4. Results

In the first study, pairs of students from two college preparatory highschools wrote one or two argumentative texts on the topics cloning and organ donation in the basic TC3 environment. After completing the assignment a total of 109 students completed an evaluation form on different aspects of the assignment and the environment. The evaluation by the students showed that, although criticizing technical flaws and drawbacks of the program (mainly in the first session), they were rather enthusiastic about this way of computer-supported collaborative learning. On a five point scale ranging from "very boring" to "much fun" the average score was 3.46. There was a significant (p<.05) difference in appreciation between the sexes. Males are more enthusiastic (mean 3.63) than females (mean 3.23).

Sexes also differed in their appreciation of the information window. On a three point scale females were slightly but significantly (p<.05) more positive (mean 2.77) than males (mean 2.53). Students were less positive about the ease of use of the buttons in the lower part of the screen. Females (mean 1.43) had significantly (p<.05) more problems than males (mean 1.84) on a three point scale.

When correlating the different items in the evaluation form with the quality of the resulting text the following items showed a significant (p<.05) correlation:
- Evaluation of the notes window (r=.261)
- Evaluation of shared text editor (r=.204)
- Evaluation of turn taking mechanism (r=.222)

This means that the satisfaction with the environment was related to the quality of the collaboratively produced text. Students enjoying the environment wrote better texts and seemed to use the environment more fully.

In a second study we are experimentally introducing the planning tools in order to determine their effect on the argumentation in the discussion and the resulting essay.

4. Future plans

The environment has specifically been tailored to meet the research issues relevant to the COSAR project. A number of restrictions have been incorporated in the system in order to standardize the environment for the purpose of our experiments. Presently the COSAR environment is limited to two students working synchronously. In an educational setting, this is limiting the utility of the system. Students are not necessarily available at the same time. Furthermore, the groups working on a research project will often be comprised of more than just two students. In a school context, the assignments are often spread out in time over several weeks. The teacher currently has no means to keep track of progress.

In a follow-up research program, we are adapting the program to remedy these shortcomings. The program will be extended to allow a group of students to collaboratively create a research report. We intend to support a wide range of subject areas e.g. a research project about a specific historical theme or event. Students can access and modify this report synchronously an a-synchronously. A messaging system will allow the students to communicate relevant messages to each of the participants and the teacher. The teacher can monitor progress and answer questions or comment on the current state of the research report. In addition to this detailed view, a summary of vital statistics of all the research reports of his/her pupils is provided. This enables the teacher to quickly identify possible problems.

We hope that this environment will facilitate collaborative learning for a range of subject areas in a broad learning community.

References

[1] Andriessen, J.E.B., Erkens, G., Overeem, E., & Jaspers, J. (1996). *Using complex information in argumentation for collaborative text production*. Paper presented on the First Conference on Using Complex Information Systems (UCIS'96), Poitiers, 3-6 Sept 1996.

[2] Bannon, L.J., *Issues in Computer Supported Collaborative Learning*. In C. O'Malley (Ed.), *Computer Supported Collaborative Learning* (pp. 267-283). NATO ASI Series, Vol. 128, Berlin Heidelberg: Springer-Verlag, 1995.

[3] Dillenbourg, P., *Introduction: what do you mean by "Collaborative Learning"?* In: P. Dillenbourg (ed.), *Collaborative Learning: Cognitive and computational aspects* (pp. 1-19). New York: Pergamon,1999.

[4] Ousterhout J., *Tcl and the Tk Toolkit*, Addison-Wesley, New York 1994.

[5] Erkens G., (Tabachneck-) Schijf,H, Jaspers, J.& van Berlo,J., *How does computer-supported collaboration influence argumentative writing?*, paper presented at EARLI SIG-Writing Conference 2000, Verona

[6] Katz, S., *Identifying the Support Needed in Computer-Supported Collaborative Learning Systems*. In J.L. Schnase & E.L. Cunnius (Eds.), Proceedings of CSCL'95; The First International Conference on Computer Support for Collaborative Learning (pp. 200-203). Bloomington, Indiana: Indiana University, 1995.

[7] Van der Linden, J.L., Erkens, G., Schmidt, H., & Renshaw, P. *Collaborative learning*. In: P.R.J Simons, J.L. van der Linden and T. Duffy (Eds.). New Learning. Kluwer Academic Publisher, (2000).

The Knowledge Management for Collaborative Learning Support in the INTERNET Learning Space

Mizue KAYAMA & Toshio OKAMOTO

University of Electro- Communications Graduate School of Information Systems
{kayama, okamoto}@ai.is.uec.ac.jp

Abstract

The purpose of this study is to support the learning activity in the Internet learning space. In this paper, we examine the knowledge management and the knowledge representation of the learning information for the collaborative learning support. RAPSODY-EX is a remote learning support environment organized as a learning infrastructure. RAPSODY-EX can effectively carry out the collaborative learning support in asynchronous/synchronous learning mode. Various information in the educational context is referred and reused as knowledge which oneself and others can practically utilize. We aim at the construction of an increasingly growing digital portfolio database. In addition, the architecture of the learning environment which includes such a database is researched.

1. Introduction

The development of the recent information communication technology is remarkable. As an effect of this, the education environment is being modified to a new environment which differs qualitatively from the previous one[14]. The new education environment contains not only computer but also communication infrastructures such as the information communication network represented by the Internet[4,5]. We call this learning environment the Internet learning space. Information is transmitted for the learner in this learning space from the external space. The information quantity that is available to the learner is enormous. However, there is a limit to the information quantity which the learner can process. The imbalance of this information processing quantity is a peculiar phenomenon in post modern ages. Secondary phenomena are also triggered by this problem. These phenomena become factors which inhibit the sound transmission of knowledge and the progress of learning[12,13]. In asynchronous learning, the transformer of knowledge and the transformee of knowledge communicate with a time lag. In such a situation, more positive support is required to realize an effective and efficient learning activity. We need to build a learning infrastructure with learning spaces with various functions.

2. The purpose of this study

We investigate the mechanism of transmission and management of knowledge for the development of the knowledge community in the learning space, within the educational context. In this paper, we examine the knowledge management and the knowledge representation of the learning information for the collaborative learning (CL) support. The purpose of this study is to support the learning activity in the Internet learning space. RAPSODY-EX (Remote and Adaptive Educational Environment : A Dynamic Communicative System for Collaborative Learning) is a remote learning support environment organized as a learning infrastructure[16]. RAPSODY-EX can effectively carry out CL support in asynchronous/ synchronous learning mode. Remote learning is a learning style where individual learning and CL are carried out on the multimedia communication network. In the remote learning environment, arrangement and integration of the learning information are attempted to support the decision making of learners and mediators. Various information in the educational context is referred and reused as knowledge which oneself and others can practically utilize. We aim at the construction of an increasingly growing digital portfolio database. In addition, the architecture of the learning environment including such a database is researched.

3. Remote education/learning system

Remote education/learning support systems are classified into 2 types. One type are systems using the advanced information network infrastructure to realize smooth communication for remote education/learning. The other type represents systems which positively support various activities of remote education/learning.

As an example of the former, we can indicate the Open university which provides a correspondence course in the United Kingdom[2] and the universities for remote education in North America (University of Phoenix[17] and Jones International University[18]). In the United Kingdom, there exists the "National Grid for Learning" project[11]. Within this project, an environment for remote education and the learning resources for remote education (within the lifelong education) are offered. In addition, ANDES which is a satellite communication

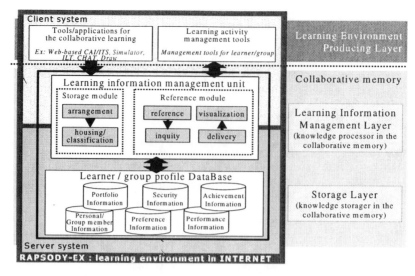

Figure 2: The mechanism schema of the RAPSODY-EX

remote education system using digital movies[21] and the synchronous remote education system which apply Web browser sharing technology[10] should also be mentioned.

As an example of the latter type, there are the "TeleTOP" that was researched and developed at the University of Twente in the Netherlands[3], and a remote education system which does not utilize the WWW technology, at the University of Trento in Italy[9]. There are other examples. The "Electronic Learning" project at the university of Southern California in North America[19] provides an intelligent tool which supports designing of the learning course and authoring of the teaching material for remote education. A remote education system with synchronous and asynchronous learning mode at IBMJapan[20], applies the Web operation recording technology.

4. CL with RAPSODY-EX

A learner group which guarantees the smooth transmission of knowledge can form a community (the knowledge community) by sharing and reusing common knowledge. Learning activities which occur within this group are as follows : the achievement of learning objectives as a group; the achievement of the learning objectives of each learner; the achievement of the learning objectives of the learner group which consists of multiple learners. RAPSODY-EX supports the transmission of knowledge in the learner group and the promotion of the learning activity. It is indispensable that RAPSODY-EX has the following functions :
1) the function which controls learning information for the individual learner and the group.
2) the function which manages learning information of the learner for mediation.

The learner and group information are produced from the learning space. This information will be stored in the collaborative memory. This information is defined as learning information. We also define the method of information management of such information and the structure of the collaborative memory.

5. The management of learning information

The simple mechanism of the management of learning information developed in this study is shown in Figure 2. The processing mechanism consists of two components. The first one is a module which offers the learning environment. The second one is a collaborative memory which controls various information and data produced from the learning environment. In the learning environment, 2 types of functions are offered. One is the monitoring function for the learning progress. The other is the tool/application for the CL. The former function controls the learning history/record of individual learners and the progress of the collaborative group learning. The latter tool/application becomes a space/workplace for collaborative synchronous /asynchronous learning. The learning information which emerged from such a learning environment is handed to the collaborative memory. The collaborative memory offers 2 types of functions. One is the knowledge processing function, and the other is the knowledge storage function. In the former, input learning information is shaped to the defined form. In the latter, for the formatted information, some attributes related to content are added. The complex information processing takes place in the collaborative memory.

6. The knowledge management in RAPSODY-EX

In this study, the processing described in the previous section is considered as a process of the knowledge management in the learning context. Knowledge

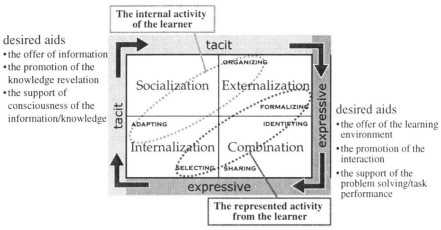

Figure 3 : The SECI model in the educational context

management is defined like follows[14]. The knowledge management is "*the systematic process of finding, selecting, organizing, distilling and presenting information in a way that improves an employee's comprehension in a specific area of interest.*"

Nonaka arranged the process of knowledge management as a SECI model[8]. Figure 3 shows the SECI model. The SECI model is expressed as a conversion cycle between tacit knowledge and expressive knowledge. Tacit knowledge has a non-linguistic representation form. Expressive knowledge is a result of putting tacit knowledge into linguistic form. Tacit knowledge is shared with others by converting it into expressive knowledge. In the SECI model, socialization (S) / externalization (E) / combination (C) / internalization (I) of knowledge is expressed.

The knowledge management in educational context is defined as follows : "*the systematic process of finding, selecting, organizing, distilling and presenting information in a way that improves a learner's comprehension and/or ability to fulfill his/her current learning objectives.*" In RAPSODY-EX, the C (combination of knowledge) phase is supported. RAPSODY-EX also promotes the knowledge conversion to the I (internalization of knowledge) phase from the C phase and the knowledge conversion to the C phase from the E (externalization of knowledge) phase. The learning information contains the expressive knowledge of the learner. This expressive knowledge is a result of expressing tacit knowledge of the learner using the language. This knowledge is converted from the learner's tacit knowledge.

In this situation, what we have to consider is as follows :
- What are the knowledge resources in the learning group ?
- What is the gain for the learning group ?
- How are the knowledge resources controlled to guarantee the maximum gain for the learning group?

7. The collaborative memory

In the collaborative memory, information generation / arrangement / housing / reference / visualization are the management processes of expressive knowledge in the learning space. RAPSODY-EX is a learning environment which possesses a knowledge management mechanism. In this environment, 1) the review of the learning process, 2) the summarization of the problem solving process and 3) the reference of other learners' problem solving method are realized in the learning space. Learning information is expressed by an unified format. Then, that information is accumulated in the collaborative memory. This information becomes the reference object of the learner. The generation and the management of the information on the learning performance and the portfolio of the learner and group are main objects of the knowledge management. In this study, learning information is obtained from the application tools for the CL. It is necessary to control the learning record, the reference log of the others' learning information and the log of problem solving and learning progress. To realize this control not only techniques based on symbolic knowledge processing approach, but also techniques based on sub-symbolic knowledge processing approach are used.

The collaborative memory consists of two parts. One is the information storage unit. The other one is the management unit of the stored information. The information storage unit mainly processes 4 kinds of information.

1) Learning information,
2) Information on the learner,
3) Information on the setting of the learning environment and
4) Information on the learning result.

The information management unit deals with the reference / arrangement / integration of learning information. The Individual learner profile information is composed of information following the IEEE Profile information guidelines[6]. The group information is expressed by the expansion of the individual learner profile information. The conversion from the learning log data to learning information is necessary to develop this profile database. The information which should apply in learning information is as follows :

- Information and/or data on its learning context and/or learning situation
- information about the sender and the sendee of the information
- significance and/or outline in the educational context
- Information on the relation structure of the learning information
- the reference pointer to individual learner and group who proposed or produced the information
- the relation with other material

By adding these information, the learning information is arranged into an unique form. If a learner requires some information related to his/her current learning, RAPSODY-EX shows the (estimated) desired information to the learner.

8. Conclusion

The purpose of this study is to support the learning activity in the Internet learning space. We examine the knowledge management and the knowledge representation of the learning information for the CL support. RAPSODY-EX is a remote learning support environment organized as a learning infrastructure.

In this paper, the management of learning information in RAPSODY-EX is described. RAPSODY-EX is an integrated learning environment with a remote education infrastructure and supporting tools/applications for the CL. Also, in this paper, learning technology for the learning support is proposed. The knowledge management mechanism in the educational context is showed. The detailed knowledge management technique will be realized in the future, and it will be integrated with the current learning support environment.

9. References

[1] ADLNet, "*Shareble Courseware Object Reference Model:SCORM*", Ver.1.0, http://www.adlnet/org/ , 2000.

[2] A. R. Kaye, "Computer Supported Collaborative Learning in a Multi-Media Distance Education Environment", in Claire O'Malley (Ed.) *Computer Supported Collaborative Learning*, pp.125-143, Springer-Verlag, 1994.

[3] B. Collis, "Design, Development and Implementation of a WWW-Based Course-Support System, *Proceedings of ICCE99*, pp.11-18, 1999.

[4] G. Cumming, T. Okamoto and L. Gomes, "*Advanced Research in Computers in Education*", IOS press , 1998.

[5] J. Elliott,: "What have we Learned from Action Research in School-based Evaluation", *Educational Action Research* ,Vol.1, No.1, pp175-186 , 1993.

[6] IEEE, "*Draft Standard for Learning Technology -Public and Private Information (PAPI) for Learner*", IEEE P1484.2/D6, http://ltsc.ieee.org/, 2000.

[7] IMS, "*Learning Resourcde Metadata : Information Model, Best Practice and Implementation Guide*", IMS Ver1.0, http://www.imsproject.org/ , 1998.

[8] I. Nonoka, "*The Knowledge-Creating Company*", Oxford University Press, 1995.

[9] L. Colazzo, and A. Molinari, "Using Hypertext Projection to Increase Teaching Effectiveness", *International Journal of Educational Multimedia and Hypermedia*, AACE, 1996.

[10] M. Kobayashi, M. Shinozaki, T. Sakairi, M. Touma, S. Daijavad, and C. Wolf, "Collaborative Customer Services Using Synchronous Web Browser Sharing", *Proceedings of CSCW 98*, pp.99-108, 1998.

[11] The United Kingdom Government, "*Connecting the Learning Society*", The United Kingdom Government's Consultation paper, 1997.

[12] S. McNeil et al., "*Technology and Teacher Educa-tion Annual*", AACE ,1998.

[13] T. Chan et al., "*Global Education ON the Net*", Springer-Verlag , 1997.

[14] T. Davenport, "*Working Knowledge*", Harvard Business School Press, 1997.

[15] T. Kuhn, "*The structure of scientific revolutions*", University of Chicago Press ,1962.

[16] T. Okamoto, A.I.Cristea and M. Kayama, "Towards Intelligent Media-Oriented Distance Learning and Education Environments", *Proceedings of ICCE2000*, 2000.

[17] University of Phoenix Home Page: http://www. uophx.edu/

[18] Jones International University Home Page http://www.jonesinternational.edu/

[19] W. L. Johnson, and E. Shaw, "Using Agents to Overcome Deficiencies in Web-Based Courseware", *Proceedings of the workshop "Intelligent Educational Systems on the World Wide Web"* at the 8th World Conference of the AIED Society, IV-2, 1997.

[20] Y. Aoki and A. Nakajima, "User-Side Web Page Customization," *Proceedings of 8th International Conference on HCI*, Vol.1, pp.580-584, 1999.

[21] Y. Shimizu , "Toward a New Distance Education System", *Proceedings of ICCE99*, pp.69-75, 1999.

Theme 9: Virtual Learning Environments

Virtual Classroom

Artur Krukowski
University of Westminster, London, UK.
Applied DSP and VLSI Research Group,
Email: krukowa@westminster.ac.uk

Izzet Kale
University of Westminster, London, UK.
Applied DSP and VLSI Research Group,
Email: kalei@westmisnter.ac.uk

Abstract

This paper presents the results of the "Virtual Classroom" project sponsored by the University of Westminster, Educational Initiative Centre (EIC) for delivering lectures on-line. The ultimate aim was to provide a portable hardware and easy to use software toolset as well as easy to follow guidelines on how to propel the lectures from the conventional dull chalk and talk environment to the realm of all interactive computer assisted web based electronic classroom. The additional goal was to minimise the number of staff required to give the lectures and allow them to use their valuable time for other academic duties.

The proposed hardware/software setup is a fully automatic solution controllable from a single PC. It allowed full student interaction and use of multimedia in the lecture. Although the best quality was achieved by use of multicasting, the students connecting via modem can also take interactive part in the lectures.

1. Introduction

The work presented in this paper has concentrated on establishing the infrastructure required to deliver lectures remotely. This means two-way audio and video communication between the lecturer and his/her students. The current setup allows students to take part in the lecture from different physical locations and enables lecturers to undertake joint lectures for people located at different geographical locations. The setup is also suitable for use in a two-room single lecturer scenario to eliminate the problem of very large student numbers and small lecture theatres currently faced on some courses. It can be applicable for doing lectures from locations inaccessible for student for security or other reasons. Such setup is an attractive solution for Virtual Universities. There are a number of them created around the world [1]-[3]. However, most of them do not provide on-line interactive multimedia teaching, so important for good perception of the teaching material.

In the proposed system all the multimedia communication is done via multicast server (connected to MBONE [4]) allowing an uninterrupted two-way 4Mbps transfer of multimedia streams for clients connected to a similar server. This way the quality of reception is the highest. It is also possible to connect using standard TCP/IP network or via modem when connecting from home or while travelling. The system has been designed in the way to allow the full control of the AMX [5] enabled multimedia hardware from the single controller PC during the presentation by the lecturer himself. This includes such tasks like video and audio source selection and control of cameras, projector and amplifier. The hardware control is done using windows compatible PC Touch [6] software via familiar Graphical User Interface.

2. Hardware Setup

The demonstration system was designed to allow two equally equipped lecture rooms to share a joint lecture. They can both serve both as servers and as clients. The system diagram is shown in Figure 1.

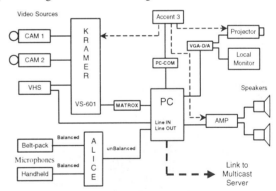

Figure 1. Schematic diagram of the hardware setup.

The main elements of the system are:

- High performance PC serving as a controller of the system. It uses a Matrox Millennium G200TV video capture card to receive audio and video signals from camera(s) and microphones in the lecture room. The

PC is connected to the multicast server (connected to MBONE) via 10/100Mbps 3Com509 network card. *Approximate cost £1000.*

- Handheld (for students) and belt-pack (for lecturer) wireless UHF microphones connected via Alice Match pack converter to the audio input of the PC. *Approximate cost £280/microphone.*

- Accent-3 Multi-port Integrated Remote Controller allowing communication with its AMX slave devices via IR ports (one-way RS-232), I/O channels providing automatic-sensed IR power control and RS-232/422/485 ports operating at up to 115,200 baud. It was used to remotely control the input video selector, InFocus projector and Denon PMA-535R power audio amplifier. The AMX PC-COM AXLink to PC Communication Port Interface was used to convert commands from PC's RS-232 COM port to AXlink control operations, emulating up to four AXlink panels. Combined with PC Touch software turns any Windows PC into a color AMX Touch Panel. *Approximate cost £1500.*

- The Kramer VS-601N Vertical Interval Switcher providing truly effortless switching between six Video/Audio-stereo inputs and two sets of parallel Video/Audio-stereo outputs. Flicker-free switching between synchronized sources is done during the Vertical Interval. *Approximate cost £350.*

Figure 2. Photograph of the hardware setup.

The content of the lecturer's computer screen was sent to the InFocus XGA video projector in the same lecture room. A the same time he was able to choose which video signals would be sent to other lecture rooms. The hardware was made compact and easily configurable allowing portability when needed (Figure 2). Although stationary cameras were used in the demonstrator, the system and software were designed to support remotely controllable cameras.

3. Multicasting and MBONE

Today, the majority of Internet applications rely on point-to-point transmission. The utilization of point-to-multipoint transmission has traditionally been limited to local area network applications. Over the past years new applications appeared on the Internet that rely on multicast transmission. Multicast IP conserves bandwidth by forcing the network to do packet replication only when necessary. It allows viewing the same real-time data over Internet without overloading it, as it is enough to send one copy of the data for many people to receive it.

The Internet Multicast Backbone (MBONE) [4] is an interconnected set of subnetworks and routers that support the delivery of IP multicast traffic. The MBONE is a virtual network that is layered on top of sections of the physical Internet. It is composed of islands of multicast routing capability connected to other islands by virtual point-to-point links called "tunnels". The tunnels allow multicast traffic to pass through the non-multicast-capable parts of the Internet. Since the MBONE and the Internet have different topologies, multicast routers execute a separate routing protocol to decide how to forward multicast packets. The MBONE carries time-critical "real-time" audio and video multicast communication.

The most important reason for using MBONE in our design was being able to allocate guaranteed 4Mbit/sec channels for video communication, hence a perfect picture and sound at any time for computers connected to the similar multicast server. It is important to note here that all other communication is done via the same link, hence downgrade of the maximum network speed for standard data communication. The second disadvantage is that if other data needs to be transmitted together with video, for example lecture presentation, containing video of its own, the video quality may be affected. Therefore a careful selection of the video and audio coding is important for achieving high quality lecture broadcast.

4. Software Setup

One of the main objectives of the project was to allow the reception of the lecture at minimum hardware expense from the viewer. Therefore all the communication has been done via public domain and Freeware software. Voice only connections were also accepted. The core element was NetMeeting [7], available free from Microsoft, bundled together with Internet Explorer few the past few years. NetMeeting includes a number of features very suitable for remote lecture scenarios:

- Support for the H.323 audio and video conferencing standard, and the T.120 data conferencing standard. It can place calls to H.323 multipoint conferencing units (MCUs) and participate in multipoint audio/video conferences. Very often students lack the ability to transmit their own video. However, they can still receive video calls in the NetMeeting video window.
- NetMeeting allows working easily with other meeting participants by sharing programs. Only one computer needs to have the program, and all participants can work on the document simultaneously. This gives a great chance for the teacher to involve students in solving problems during the remote lecture. In addition, people can send and receive files to work on.
- Using the Whiteboard, the lecturer can explain concepts by diagramming information, using a sketch, or displaying graphics. He can also copy areas of his desktop or windows and paste them to the Whiteboard.
- Remote Desktop Sharing allows the lecturer to access a computer at one location from a computer at another location. This way he can setup a computer in the laboratory and by controlling it from the lecture room he can do remote demonstrations. In the demonstration system the remote desktop sharing has been used to minimise the network traffic by running the lecture presentation at the same time on the remote computer. On the Digital Signal Processing courses the lectures are on multimedia processing and presentation often contain sound and video samples. Therefore limiting the data transfer to video-conferencing only allowed to sustain high quality of video conferencing throughout the lecture.

The example screen shot from the controller PC is shown in Figure 3. It was taken during the preparations for the pilot lecture. It can be noticed that there are two members of the conference, both transmitting videos. There is a window with lecture presentation in Microsoft Power Point on the right of the screen. At the same time the contents of the screen was projected to the screen for students. The lecturer had control of the presentation in both lecture rooms with one click of the wireless Radio Frequency (RF) mouse. Using wireless RF keyboard allows the lecturer to avoid sitting at the desk during the lecture, thus getting better attention from the audience.

An important part of the software package running on the controller PC was PC Touch. As it was mentioned before in section 2, it was used to provide the control of the multimedia hardware in the lecture room from the computer screen via Graphical User Interface. It can be seen at the top of the window in Figure 3. The similar software needs to run on other controller PCs that taking part in the conference. In this particular example PC Touch has three buttons on the left for selecting video sources (two cameras and video player), two horizontal buttons for audio volume control and two more buttons for opening additional windows for video projector settings and remote camera control.

5. Example Lecture

The pilot lecture was done by Dr Izzet Kale and Dr Artur Krukowski on the 21st of January 2000 on the topic of Balanced Model Truncation Techniques for the DSP Masters Industrial Short Course module at the University of Westminster. More lectures took place at later times. The purpose of the first one was to prove the concept and test different teaching scenarios (Figure 4).

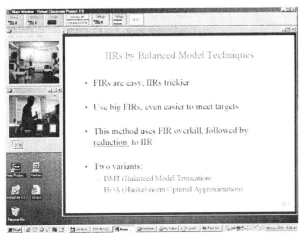

Figure 3. The screen shot from the controller PC.

Figure 4. On-line lecture for DSP Masters Industrial Short Course students.

Therefore the lecture was divided into two parts. During the first part the main lecturer was in the same lecture room with the students and the supporting lecturer was pretending to be a remote student in another room. During the second part of the lecture the main lecturer was away from the room where the students were, and was teaching remotely from his office.

In the first part the "remote student" took active part in the lecture both asking questions verbally and making use of the Whiteboard to present his comments graphically. This also simulated the scenario of the lecture given jointly by two lecturers located far away from each other.

The second part simulated the scenario when the lecturer is not able to be present in the lecture room (when away for the conference, doing industrial work or even working at home). The students can still have full communication with the lecturer and can benefit fully from the lecture.

The students took active part on both parts of the lecture, both asking questions and drawing their ideas on the whiteboard or in the computer, independent on the location of the lecturer. After the lecture they were asked to comment on the idea of such remote lecturing. They all agreed that although the teaching material is received very well, the presence of the lecture in the same lecture room with students is preferable and helps to concentrate and focus on the subject. They also agreed that such technology should be used only when there are objective reasons for the lecturer not being in the class and to help students who are unable to attend the lecture.

From the presenter's point of view the lack of physical teacher's presence in the classroom caused difficulty in maintaining discipline. Students were more likely to be late for the class and doing other things than watching the lecture. Although the teacher was able to monitor events in the classroom, he was limited to voice reprimands only.

6. Summary

During the project the lecture system has been designed to allow the lecturer to give remote lectures having different scenarios. The lecturer can do the remote demonstration either alone or with a help of a technician or assistant. He can do the joint lecture with another lecturer at a distant location. He can make a remote demonstration in the laboratory, while being in the lecture room with students. The system can be used to allow remote students or those unable to attend the lecture to take active part in it. It also allows decreasing the number of teaching staff required to give the lecture, letting others to concentrate on research and industrial work, an important part of the University life.

In all the cases the system can be controlled from the single controller PC. This has been achieved by using programmable AMX-enabled hardware and a software written in-house running under MS Windows environment to control this hardware via Graphical User Interface.

The system has been designed to be scalable. The simplest working solution can cost as low as £2500/system for the specialised AMX hardware. The cost of setting up and programming are considerable (around £500/day), which can average down when a number of systems are to be set-up. The system can be easily expanded adding a variety of video and audio sources, audio/video mixing table, remotely controllable cameras and projectors etc.

The technical concepts have been tested and proved to work properly. The student response was that although the technology behind the lectures is new and exciting, the presence of the lecturer in the same room with students is and will be very important for the good perception of the teaching material. The new technology should be only used for situations when either the lecturer or the student is not able to attend for objective reasons. It could be also used to allow publishing of the lecture on-line for wider audience. Apart from teaching the system can be used for on-line discussion forums, point-to-point multimedia communication with customers and co-workers.

Plans for the future are to integrate our remote lecturing system with Internet based teaching to support the idea of the Virtual University, the new cross-university initiative in the United Kingdom, to promote on-line teaching and assessment at the University level. Some approaches have already been implemented, but there is still a need for a uniform and tight system of lecturing, examination, assessment and marking to make the idea of the Virtual University a reality.

7. References

[1] Virtual University at the Michigan State University:
http://www.vu.msu.edu/

[2] Clyde Virtual University:
http://cvu.strath.ac.uk

[3] Neil Pollock & James Cornford, "Theory and Practice of the Virtual University", *report on UK universities use of new technologies*:
http://www.ariadne.ac.uk/issue24/virtual-universities/

[4] Frequently Asked Questions on the Multicast Backbone:
http://www.cs.columbia.edu/~hgs/internet/mbone-faq.html

[5] AMX Corporation WEB site:
http://www.amx.com/

[6] PC Touch:
http://www.panja.com, http://www.amx.com

[7] NetMeeting:
http://www.microsoft.com/windows/netmeeting/

A Virtual Learning Environment for Short Age Children

Pascual González, Francisco Montero,
Víctor López, Antonio Fernández-Caballero
Regional Development Institute
University of Castilla-La Mancha (Spain)
pgonzalez@info-ab.uclm.es

Juan Montañés, Trinidad Sánchez
Psychology Department
University of Castilla-La Mancha (Spain)
jmontanes@psic-ab.uclm.es

Abstract

In this paper we introduce a project that validates educational capabilities of a game. This game, called Prismaker, incorporates two versions: a physical version and a virtual version. Thus, we want to find out the real potentialities of games in learning processes and to evaluate a single game from two points of view: the physical and the virtual version that is being developed.

1. Introduction

New information technologies offer new tools that can fully modify the way we nowadays think of teaching. The creation of software for children is usually linked to the binomy computer games and visual interfaces. Their playing issues and their attractive interfaces make these kinds of tools especially interesting for educational environments. There is a lack of scientific knowledge about the factors that have an effect on motivation, enjoyment and satisfaction [3]. New works have appeared lately suggesting that computer games increase motivation for children learning at school. Some works [2] state that playing in children is equivalent to working in adults.

According to Sedighian et al [4] computer games supply meaningful learning, goals, success, challenge, cognitive artifact, and association through pleasure, attraction, and sensory stimuli. These factors are even enhanced when using 3D environments and attractive visual interfaces. Computer games can play a significant role in forming children attitudes toward computers and offer the possibility of learning a wide variety of contents, both in and out of school.

2. Virtual-Prismaker project

A capital goal in this project is to evaluate the educational potentiality of Prismaker[tm] game from two points of view: the physical game, and the virtual game. This project is being carried out in conjunction with the corporation that created the physical game and it tries fundamentally to analyze the game's utility inside educational environments. In order to achieve this goal a multidisciplinary team has been created, formed of computer engineers, psychologists, and pedagogues. Psychologists and pedagogues are designing educational activities. At the same time, children and teachers from different primary schools are collaborating to validate how good this game may be for short age children. This validation will begin next academic year. Up to date, we have introduced this project to primary school teachers, who are collaborating in the final design of all activities.

Prismaker[tm] is a construction game that provides a reduced number of kinds of pieces to build with (figure 1). Besides, you can use logos to assign a meaning to these pieces. Activities where playing is an important factor improve the learning capacity of children in each of their evolving states [5]. This improvement is achieved as children activity is performed in an imaginary situation. Personality parameters influenced are affectivity, mobility, intelligence, creativity and sociability.

Activities (figure 1) have been incorporated using a set of forms where any Prismaker[tm] material used to carry out the game is explained. Each form shows how to play a specific game, which capabilities are enhanced by this activity, what are the general goals, what are the specific goals, and the curriculum parts where it could be used. Any available form is structured in different sections to help understanding the activity. Besides this, a questionnaire has been designed to allow teachers to evaluate each pupil's activity as well as the activity itself.

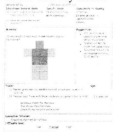

Figure 1. Prismaker[tm] and activity form

3. A new virtual playground

In the current virtual version we provide children with a working environment as similar as possible to the way things exist in the real world. The designing process methodology is meant to be participative and multidisciplinary, where children, teachers, pedagogues and software and graphics designers work together to achieve the final goal [1].

Virtual-Prismaker is a 3D environment based game (figure 2) that simulates the possibilities offered by the physical version of Prismaker^{tm} game. This environment uses the different kinds of pieces of Prismaker^{tm} system to develop many different games. All games inside our 3D virtual environment are carried out in a playing room. Of course, there are different rooms. These rooms have different difficulty levels, in the way that the level the child finds in the activities in a given room agrees with his age. Children can even customize the look for some of the objects in the room: the walls, the table, etc.

Virtual-Prismaker has been designed for people without computer knowledge. To make these people easier to use Virtual-Prismaker, we are engaged in removing computer concepts, such as "load/save", "menus", and so on. We have designed a user interface based on metaphor use that offers the concepts the way a child is familiar with.

Figure 2. Virtual-Prismaker playground and book

Game instructions are collected in a virtual book. Any available game is represented by its icon and a piece of text. When you select an icon, its associated text is automatically read and that game's instructions pages are shown (figure 2). The book pages are composed of text, sounds, images and videos. The text in the pages may be automatically read by means of a voice synthesizer. In some specially hard to play games, or in games for very short age children, we have included videos to visually explain how to play.

One of the most difficult games is the construction game. In this case, we have replaced standard I/O operations (load/save) with some shelves where users can store their work. To add new blocks to the game, children take them from a box and drag them to the playground, avoiding this way the use of toolbars or menus.

One of our goals is to allow competition and collaboration by playing through Internet.

To make teacher's work easier, our system allows him to design new games based upon the basic activities we provide, so they can adapt activities' contents to the didactic unit they are teaching. Apart from all this, the system is able to track, not just the final result of each activity, but the sequence of tasks the user has performed to achieve that result. Thus, it allows the teacher to gather information about the reasoning process and he can discover children "reasoning patterns". Besides, a user is able to learn without the physical presence of a teacher. Our system evaluates any performed activity and is able to find out if that activity was successful.

4. Conclusions

We introduce an educational project where we want to analyze the advantages of playing a game from two different points of view: a physical and a virtual game. An interesting goal is to find out the way to take advantage of all the chances supplied by information technology in learning processes. We show some of them: tracking the learning process, introducing virtual multimedia help, and, of course, enhancing the motivation children find in playing a computer game.

We think that Virtual-Prismaker incorporates many educative features, and that it is flexible enough to let teachers use it in their necessary curriculum parts. Virtual-Prismaker interface is attractive enough to increase children's motivation in their learning process. Our first prototype has been evaluated by a group of teachers that found it useful, and we hope to get the same results later on this year when the virtual game will be tested in school.

5. Acknowledgement

This work is supported in part by the CICYT–FEDER 1FD97–1017 grant and by the 2000020264 JCCM and European Social Fund.

6. References

[1] A. Druin, *The Design of Children's Technology*, Morgan Kaufmann Publishers, Inc., 1999.

[2] C. Garvey, *Play*, Cambridge, M.A, Harvard University Press.

[3] D.A. Norman, *Things that make us smart: defining human attributes in the age of the machine*, Addison-Wesley, 1993.

[4] K. Sedighian, A.S. Sedighian, "Can Educational Computer Games Help Educators Learn About the Psychology of Learning Mathematics in Children?", *18th Annual Meeting Int. Group Psychology of Math. Education*, Florida, 1996.

[5] L.S. Vigotszki, *Thought and Language*, MIT Press, 1986.

Using a Virtual Learning Environment in Higher Education to Support Independent and Collaborative Learning

C. L. Bennett
Department of Psychology
University of York
E-mail: C.Bennett@psych.york.ac.uk

R. M. Pilkington
Computer Based Learning Unit
University of Leeds
E-mail: R.M.Pilkington@cbl.leeds.ac.uk

Abstract

A case study in the use of a Virtual Learning Environment (VLE) to support small-group teaching in Psychology is reported and discussed in the wider context of the ASTER[1] project, which aims to establish and disseminate good practice in the use of communications and information technologies for small-group teaching. The aim was to create a cultural shift in students' learning away from rote learning of lecture notes - "knowledge telling", towards a deeper engagement and ownership of the content - "knowledge transformation". To this end, a VLE was used for group communication and to provide a gateway to linked resources. Group-work was used to encourage students to work collaboratively. Some embraced the approach enthusiastically whilst others expressed anxiety. The success of the approach is discussed in relation to student attitudes and performance. Some recommendations for shifting the learning culture toward independent learning are suggested.

1. Introduction

In the modern world, education is no longer viewed as a 'once for all' option in which we complete a number of years of study that fit us for life. Instead we are asked to embrace the concept of life-long learning [1]. The consequences of such a cultural shift are that "what we learn about" is no longer as important as "learning to learn". To effect a shift towards independent learning we need to change education. Students, even in Higher Education, are comfortable with existing lecture-based methods that operate according to a "knowledge transfer" model of teaching and learning in which they rely on copying and rote learning large volumes of content material to be reproduced in essays and examination [2].
Vygotskian theory [3] and its extensions e.g. activity theory suggest that language is a cognitive tool mediating action. When text is used to represent joint understanding of progress towards a goal it becomes a mediating artifact [4]. Writing as a mediating tool in knowledge building requires a deep engagement with content that yields not only richer understanding but encourages a sense of ownership, placing the author within a discourse community [5, 6] that is actively engaged in knowledge construction.

This is in keeping with what Berieter and Scardamalia [7] call the "knowledge transformation" approach to writing, with the writer involved in a creative and generative process – manipulating, constructing and reconstructing ideas. We need to encourage this deep engagement with content if students are to develop independent knowledge building skills.

2. WWW and knowledge transformation

The use of language as a cognitive tool has always been linked to the technologies used in its reproduction. The internet provides a reservoir for group information enabling access to group resources outside class hours but also creates new genres of writing which are changing discourse communities. Computer-mediated communication (CMC) blurs the distinction between writing and talking by using text as a mediating artifact.

Ekeblad [8] analyses the use of listserve discussion and describes a knowledge transforming layer of activity called "multilogue" in which the common objective is to increase understanding of relevant issues on the discussion topic. Anyone who is a member of the community can contribute to this process but "lurkers", those who read but do not send messages, do not contribute towards the *knowledge building* that is the objective of this system.

 Traditional face-to-face teaching can lack the sense of joint ownership and equal participation of the ideal "multilogue". Typically, the teacher dominates, often being responsible for over 80% of the conversation. CMC seminar participation is often much more equal with the teacher typically responsible for less than 50% of the discussion and with many more participating students [9]. CMC studies also report relatively more on-task talk and more relevant and thoughtful contributions [10]. For these

---

[1] ASTER (http://cti-psy.york.ac.uk/aster/) is a consortium project funded by the Higher Education Funding Council for England under Phase Three of the Teaching and Learning Technology Programme.

reasons and because the course content would also discuss some of these issues, we wanted to embed the redesigned course within a VLE.

3. Structure of the course

"Language and Cognition" is an undergraduate final year course with 22 students in the year 2000. It is compulsory for Cognitive Science students (6 students) and optional for Psychology (15) and Management Studies students (1). The course ran for 11 weeks with 22 hours of face-to-face teaching in 2-hour blocks each week. The aim is for students to develop a critical awareness of how the study of language relates to the study of thinking and reasoning and the changing roles of communication media in the communication of knowledge at a distance. Course assessment is by written essay (33%) and two written examination essays (67%).

The course considers the relationship between new technologies, language and cognition and so has always had a large Communication and Information Technologies (C&IT) element. From 1996 onwards, WWW materials and an e-mail list have supported the course. In 1999 the course was moved to the WebCT™ VLE to provide a "one-stop-shop" for course materials, news, reading resource lists and communications technologies including e-mail, asynchronous bulletin board and synchronous chat discussion rooms. However, the structure of the course remained largely unchanged: the first hour was a tutor-led lecture and the second was a computer-lab class in which students were guided in the use of C&IT tools.

In 2000, dissatisfied with some students over-reliance on course notes, and lack of evidence of interaction with course materials, the tutor took the decision to change the teaching method to a more discursive approach. The aim was to encourage students to engage more critically and independently with the learning material. Following familiarization with WebCT™ in week 1, students were asked to prepare for each session by reading the week's course notes and two research papers, and to come prepared to discuss the topic question. A face-to-face or electronic discussion was held in the first hour, replacing the traditional lecture. Where possible, the discussion topics were integrated with the C&IT exercises that followed. This often included students posting reflections on the key issues to the bulletin board.

From week 5, students were asked to form collaborative groups of 3 to 6 to build a web-based resource. Each student chose one of the five assignment topics to study in more depth so that each group of students collected material sufficient to plan the answer to three questions, one for the essay assignment and two for the examination. Students were guided in the use of online library and journal resources and WWW search engines, with the aim of supporting them in generating and selecting suitable reference material. Over the next five weeks they were further guided in building web-pages that integrated their own notes with hyperlinks to their collected resources. Each individual student's pages were integrated with those of their group and uploaded into the WebCT™'s group presentation space, where all students could view them. The idea was to encourage the development of a discourse community whose members would discuss critically and constructively the issues of the course and with whom they would share resources collaboratively to support each other's learning.

4. Results

The course was evaluated using the following sources of information:
- Student feedback from questionnaires and a videotaped focus group held in week 10;
- Student collaboration, from bulletin board use and peer and tutor reviews of student group-work;
- Summative performance, from essay and examination marks.

4.1. Student feedback

The standard module questionnaire designed by the department of Psychology was handed out to all students in weeks 9 and 10. There were no questions specifically directed at the C&IT aspects of the course and the wording of questions assumed that the course was traditional in style asking students to comment on the quality of "the lecturer" and "lectures". There were 14 returned forms representing 63% of students. The following comments were received:

Positive comments:
- Liked building web-pages; enjoyable, and good for employment (9 students);
- Liked having course notes on the web (5);
- Liked references, good and easy to access (3);
- Content of course interesting and informative (3);
- Structure of course transparent and clear (3);
- Seminar material easily available (1);
- Technical support good (1).

Negative comments:
- Disliked the fact that the web-resource built was not assessed towards grade (4 students);
- Disliked the lack of face-to-face lecture (3);
- Disliked the structure of the face-to-face sessions (3);
- Disliked online chat as a discussion medium (2);
- Disliked building web-resource (1);
- C&IT content too much (1);
- C&IT content too little (1).

When asked to rate the overall effectiveness of the course on a 5 point scale (excellent, v. good, good, satisfactory and poor) the mean score was 3.4 (between good and very good) whilst for the "overall effectiveness of the lecturer" the mean was 3.1 (closer to good). Since students didn't receive *lectures* as such, the lower score for the quality of the lecturer may reflect the fact that some students felt cheated of "lectures". Averaging over 12 sub-questions on the quality of the materials and their presentation by the lecturer achieved a mean of 3.8.

Five students agreed to take part in a focus group in week 10. Students described feeling unsure of the structure of the course and what was expected of them and said they were anxious that the lack of formal lectures meant they were missing out on content material. They admitted that as the course progressed they felt less worried but one still commented:

"It was interesting and dynamic [but] in some cases a more extensive lecture clarifying the points would have been more helpful."

Although one student felt that the C&IT content had been too much and remained unsure of its value, others commented positively that it helped to make the course content concrete and that it would be useful for their CV. One commented:

"I would have liked to have seen even more [C&IT practical work] and maybe for it to count more. I don't think that the coursework should be an essay, its not appropriate."

Others agreed and one added that they would have put a lot more effort into the groupwork if it had been assessed. However, the group also raised many of the difficulties inherent in assessing collaborative work.

4.2. Student collaboration

Table 1. Comparison of messages read and posted to the bulletin board for 1999 and 2000.

	no. students	mean no. read	mean no. of posts	total posted/ total read
2000	22	46.5	3.3	0.07
1999	26	34.6	1.5	0.04

From table 1 it can be seen that students made greater use of the bulletin board in 2000 than in 1999 with more messages read and posted and a higher rate of posting relative to reading suggesting a more active and balanced participation by members of the group. This probably reflects the tutor's efforts to encourage use of the bulletin board in a more focused way by setting topic questions and asking students to post responses to topic issues. In the following extracts students demonstrate listening and responding skills as they address each other's points critically, constructively and positively:

SR on Mon Jan 31, 2000 09:51
Yes I agree that animals such as dogs have their own codes. However perhaps what is unique to humans (and possibly their direct ancestors) is conscious cognitive mediation and intention.
It is likely that lower animals simply react in hardwired displays to given stimuli, for instance a specific call to signal the approach of a snake. If this is termed language ,is aggression also 'language', whereby an animal reacts to defeat by communicating submission signals. This is an example of an agreed upon (probably innately) code used in interaction.
Another key aspect of language is its recursive nature, which makes language generative and therefore limitless. Anyone any examples of this in codes used in lower animals?

SH on Mon Feb 07, 2000 16:31
I think your last points of language as generative and limitless are crucial.
This is surely what separates human language from birdsong or "responses of mice to aggression" which fail to convince me if they're language rather than communications in the first place....

This suggests some success in creating a learning culture more like the ideal *discourse community*.

In week 10 students and the tutor used the bulletin board to give feedback on each other's groupwork. Students put considerable effort into their group-pages. They also gave empathetic, constructive and critical feedback as the following extract shows:

IL on Thu Mar 30, 2000 12:08
This site looks very useful - it's good to have the lecture & reading material explained in someone else's words because this often makes it easier to understand. There aren't many links to other resources but it's got lots of information in itself, and it covers several areas of the course. The pages are nice to look at as well (especially the one with bananas on it. Yum yum.)

4.3. Summative performance

From table 2 it can be seen that there was no significant difference between examination results for the course running in 1999 and 2000.

Average essay performance was marginally higher in 2000, with relatively higher numbers of upper second class marks relative to lower second class marks but this was not sustained in the examination. The tutor subjectively felt that the change in the teaching method had resulted in less rote learning of lecture notes evident in the content of essays, but this did not in itself guarantee a better mark. In this respect student anxieties would seem justified. Student perceptions were that they worked harder to achieve this result and that a large part of that work - the groupwork - went unrecognized.

Table 2. Essay and exam 1 results by degree classification for years 1999 and 2000.

	1	2.i	2.ii	3	Fail	N/A[2]	Mean (sd)
Essay '00	3	14	4	0	0	1	64.2 (5.1)
Essay '99	3	12	8	1	0	2	63.6 (6.9)
Exam 1 '00	1	10	9	1	0	1	60.8 (6.5)
Exam 1 '99	4	12	9	0	0	1	64.2 (9.8)

5. Conclusions and recommendations

The results of the evaluation suggest that the tutor succeeded in raising the level of active participation and engagement with the materials through the use of electronic discussion and groupwork and that students were able to respond critically but constructively both in debate and in peer-reviewing each other's groupwork.

The ASTER Project has conducted over 30 detailed case studies on the use of C&IT to support small-group teaching, with a specific interest in collaborative learning. In such situations, students iteratively interact with each other or a tutor and engage in dialogue for analysis, reflection or critical thinking. Encouraging a deeper approach to learning is a common motivation for the use of C&IT seen in many of these examples [11].

Assessment, however, presents a very real problem. C&IT and debating skills are often talked of as highly desirable but students are still assessed on a final written examination and teaching must remain sensitive to the needs of students, with class time spent on tasks that prepare them for this. These problems of assessment are also raised in other ASTER case studies where the tutor seeks to achieve a similar shift in learning through an emphasis on collaboration and groupwork [11]. Students and tutors alike feel constrained by current assessment methods and it is clear we need new approaches if we are to fairly reward the skills acquired in groupwork.

The change in teaching method showed some indications of shifting the culture of learning toward the community of discourse which the tutor desired. The examples illustrated show that computer-mediated debate can encourage active, constructive and independent learning. However achieving this was not without problems and lessons learnt include the need to:
- change gradually and to familiarize students with the new environment and the objectives of change;
- reassure students that necessary course material and tutor explanations are still at hand.

It is reasonable to ask what the real incentive to innovate is, given the lack of obvious impact on performance. To answer this requires a detailed look at the quality of student participation online and recognition of the soft-skills gained through collaborative groupwork.

6. References

[1] McCombs, B.L., "Motivation in lifelong learning", Educational Psychologist, 26 (2), 1991, 117-127.
[2] Laurillard, D., "Rethinking University Teaching: A Framework for the Effective use of Educational Technology", Routledge, London and New York, 1993.
[3] Vygotsky, L.S., "Mind in Society: The Development of Higher Psychological Processes", Harvard University Press, Cambridge, Mass., 1978.
[4] Engeström, Y., "Learning by Expanding: An Activity-Theoretical Approach to Developmental Research", Orienta-Konultit Oy, Helsinki, 1987.
[5] Swales, J.M., "Genre Analysis: English in Academic and Research Settings", Cambridge University Press, Cambridge, 1990.
[6] Wenger, E., "Communities of Practice: Learning, Meaning, and Identity", Cambridge University Press, Cambridge, 1998.
[7] Bereiter, C. and Scardamalia, M., "The Psychology of Written Composition", Lawrence Erlbaum and Associates, Hillsdale, New Jersey, 1987.
[8] Ekeblad, E., "Contact, community and multilogue - electronic communication in the practice of scholarship", paper presented at the Fourth Congress of the International Society for Cultural Research and Activity Theory, Denmark, June 7-11 1998. http://hem.fyristorg.com/evaek/writings/iscrat98/cocomu.html
[9] Sullivan, N. and Pratt E., "A Comparative Study of Two ESL Writing Environments: A Computer-Assisted Classroom and a Traditional Oral Classroom", System 29(4), 1996.
[10] Chun, D., "Using computer networking to facilitate the acquisition of interactive competence", System 22, 1994, 17-31.
[11] ASTER, "Investigating the use of electronic resources in small-group learning and teaching", November 2000.
http://cti-psy.york.ac.uk/aster/resources/publications/publications.html

---
[2] Not submitting work or not attending the examination due to illness or other causes.

Applying Interactive Mechanism to Virtual Experiment Environment on WWW with Experiment Action Language

Rita Kuo, Maiga Chang and Jia-Sheng Heh
*Dept. of Information Computer and Engineering,
Chung-Yuan Christian Univ., Chung-Li, 320, Taiwan*
rita@mcsl.ice.cycu.edu.tw, maiga@ms2.hinet.net, jsheh@ice.cycu.edu.tw

Abstract

To build an interactive mechanism for distance learning environment such as a scientific experiment, the architecture of learning environment and assisted agents should be taken into consideration first. Another issue for constructing interactive mechanism is protocols between the environment and agents. An Experiment Action Language is designed for this purpose. Besides, the way for agent inferring rules is also presented in this paper.

1. Experiment Environment in WWW

A disadvantage of distance learning on Web, the lack of interaction between user and computer, makes the teaching procedure inefficient. Hence, a learning environment, like Virtual Experiment Environment (V.E.E.), with interactive mechanism supported for scientific experiment on WWW is needed.[6][7]

Some distance learning systems now propose agents for managing system files [10], notifying grades in on-line quiz system [8], diagnosing problems in problem solving system [1][2][5], a recorder of the system [3][4], and so on. This paper will support an interactive mechanism for agents to let agents own the ability to guide and assist students by getting information from V.E.E.. Besides, the protocol used to communicate with both V.E.E. and agents will also be discussed.

2. Architecture of Interactive V.E.E

To build a proper environment for student experiment on web, there are four components in V.E.E. as shown in Figure 1. Accompany with an Interactive V.E.E., an agent called Experiment Agent is used to assist learner during experimenting. Figure 2 illustrates that there are three sorts of interactions among V.E.E., Experiment Agent, and User.

In order to have the assistant ability, the Experiment Agent should be able to infer. There are five components make the architecture of experiment agents completely presented in Figure 3. To make an Experiment Agent intelligent, Knowledge Base is a necessary component and makes the Inference Engine get facts more properly. The results reasoned by Inference Engine will be transmitted to the Knowledge Translator, and then will be translated to appropriate Experiment Action Language before sent out to V.E.E. through using Communication Device.

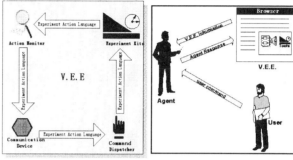

Figure 1.
Architecture of I.V.E.E.

Figure 2.
Interactions

Similarly, the Experiment Action Language transferred from V.E.E. will also be received by the Communication Device and then be sent to Knowledge Translator. Information Filter deletes the garbage information translated by the translator and store useful parts into Knowledge Base.

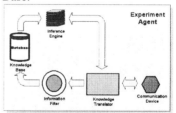

Figure 3. The Architecture of Experiment Agent

3. Interactive Protocol Develop

Experiment Action Language is a language for agent to control V.E.E. and record each step proceeded from learners. To let most of learners realize what command they call easier, scripting language is selected to be the language type.[9] The Experiment Action Language format lists as below:

```
command [object] [-option] [parameter]
```

Inference Engine will divide the knowledge translated and pass from the Knowledge Translator into three conditions, including correct, inexactitude and error listed in Table 1.

Condition \ Command	Command	Object	option	parameter
Correct	Correct	Correct	Correct	Correct
Inexactitude	Correct	Correct	Error	-
	Correct	Correct	-	Error
Error	Correct	Error	-	-
	Error	-	-	-

Table 1. The relationship between determined conditions and EAL (symbol, '-', means "don't care.")

To implement the interactive mechanism of V.E.E., Visual Lab is chosen for the experiment environment in this paper.[11] Taking a real action in Visual Lab for example, a learner moves the vertical ruler to the new position (x1, y1) could be written as `move Ver_Ruler -u x1 y1` in EAL. This command asks agent to "move" the object "`Ver_Ruler`" (Vertical Ruler) to the new position "`x1, y1,`" and the option "`-u`" let agent know that (x1, y1) is the position of the upper-left corner of Vertical Ruler.

By using the interactive mechanism analyzed and designed above, an Experiment Agent can either cheer students or provide suggestion when they doing correct or inexactitude experimental procedure as Figure 6 and Figure 7 shown.

Figure 6. correct step in V.E.E.

Figure 7. inexactly step in V.E.E.

4. Conclusion

This paper proposes an interactive Virtual Experiment Environment and the architecture of Experiment Agent. Besides, an intercommunication protocol, Experiment Action Language, is also designed to realize the communication issue between V.E.E. and agents. However, since the interactions will also happen between agents, Knowledge Query Manipulate Language (KQML), is necessarily taken into consideration future. Beside the formation of rules and the intercommunicate protocol, the relations between Inference Engine and Information Filter might need further analysis for the automatic learning issue.

References

[1] Janie Chang, Maiga Chang and Jia-Sheng Heh (1999), "Applying the Evaluation Model of Problem Solving to Agent-based Instructional System ," *Proceedings of the 7th International Conference on Computers in Education*, Chiba, Japan, 1999, pp. 141-148

[2] Janie Chang, Maiga Chang, Jeng-Lun Lin and Jia-Sheng Heh (2000), "Implements a Diagnostic Intelligent Agent for Problem Solving in Instructional Systems," *Proceedings of the IEEE Int. Workshop on Advanced Learning Technologies*, Palmerston North, New Zealand, pp.29-30, Dec. 4-6, 2000

[3] Chi-Wei Huang, Shiao-Ting Sun, Chia-Chin Chang, Kuo-Chang Jan and Kun-Yuan Yang (1998), "Education Software and Hardware", *Proceedings of the 2nd Global Chinese Conference on Computer in Education*, Hong Kong, 1998, pp.50-57

[4] Chi-Wei Huang, Chang-Kai Hsu, Maiga Chang and Jia-Sheng Heh (1998), "Designing an Open Architecture of Agent-based Virtual Experiment Environment on WWW," *World Conference on Educational Multimedia, Hypermedia & Telecommunications (ED-Media), 1999*, Seattle, WA, USA, pp.270-275

[5] Jia-Sheng Heh (1999), "Evaluation Model of Problem Solving," *Mathematical and Computer Modelling*, Vol.30, 1999, pp.197-211

[6] Yu-Wei Jeng, Maiga Chang, Ivory Chung and Jia-Sheng Heh (1996), "Designing Objects for Virtual Experiments," *OOTA'96*, Taiwan, 1996, pp.331-339

[7] Li-Ping Kuo, Da-Xian Dong, Chang-Kai Hsu and Jia-Sheng Heh (2000), "Design an Enhanced Virtual Experiment Environment Using Science Process Skills on WWW," *World Conference on Educational Multimedia, Hypermedia & Telecommunications (ED-Media), 2000*, Montreal, Canada, Jun. 25- Jul. 1, 2000, pp.1785

[8] A. Okada, H. Tarumi and Y. Kambayashi (2000), "Real-time quiz functions for dynamic group guidance in distance learning systems," *Proceedings of the First International Conference on Web Information Systems Engineering, 2000*, Vol.2, pp.188-195

[9] Robert W. Sebesta (1999), *Concepts of Programming Languages*, Addison-Wensley, 1999

[10] A. Shimano and H. Kuramae(2000), "Design and construction of educational computer system using self-maintenance system for files and user identification agent," *Proceedings of the 9th IEEE International Workshop on Robot and Human Interactive Communication, 2000. RO-MAN 2000*, pp.23-28

[11] Chao-Chiu Wang, Maiga Chang, Chang-Kai Hsu and Jia-Sheng Heh (2000), "Visual Lab - A Multimedia Virtual Experiment Environment on WWW," *World Conference on Educational Multimedia, Hypermedia & Telecommunications (ED-Media), 2000*, Montreal, Canada, Jun. 25- Jul. 1, 2000, pp.1509-1510

A Criterion-Referenced Approach to Assessing Perioperative Skills in a VR Environment

Wayne O. Witzke
University of Kentucky
wwitzke@immersa.uky.edu

Donald B. Witzke
University of Kentucky
dbwitz1@pop.uky.edu

James D. Hoskins
University of Kentucky
jdhosk00@pop.uky.edu

Michael J. Mastrangelo
University of Kentucky
mmast@pop.uky.edu

Uyen B. Chu
University of Kentucky
ubchu2@pop.uky.edu

Ivan M. George
University of Kentucky
imgeor1@pop.uky.edu

Adrian E. Park
University of Kentucky
apark@pop.uky.edu

Abstract

The need for technical competence brought about by the advent of minimally invasive surgery (MIS) has caused educators to reevaluate the methods used to train surgeons and to judge their competence. Additionally, the extensive training required to become competent is an attendant problem. To help solve these problems, we have begun to develop a method for training and evaluating perioperative MIS surgical competency using an immersive virtual reality trainer for surgical preparation. This paper presents and summarizes the results of our pilot-test development of a criterion-referenced approach to evaluating performance. Because of the high degree of internal consistency of responses from master surgeons using our system, we have confidence that this approach is viable.

1. Introduction

The methods used to train and evaluate residents' surgical skills have changed very little since formal medical education was implemented; an apprenticeship model continues to be used today. However, surgical education is under critical review due to a confluence of forces reshaping the practice of medicine. These forces include funding changes in surgical education, the need to reduce training costs, the need to reduce medical errors, the high costs of using attending surgeons' time, the high costs of using the operating room (OR) to train, the influx of high technology into surgery, and a shift from open surgery to minimally invasive surgery (MIS) for many procedures [1]. Like all surgery, MIS requires highly trained, skilled surgeons. Unlike open surgery (OS), where incisions can be any length needed to receive a surgeon's hands and surgical instruments, MIS surgeons use long, small diameter instruments (usually 2mm to 5mm in diameter). These instruments are used in trocars or ports inserted through "keyhole" incisions (usually 2mm to 10mm long) [1]. MIS techniques require a considerably higher degree of psychomotor skills--skills that can only be mastered by many hours of practice [2,3,4]. Specifically, surgeons in training must learn to manipulate MIS surgical instruments through ports inserted into (for general surgery) a body cavity, manipulate these instruments while having limited motion, and maneuver instruments while watching a two-dimensional display and receiving limited tactile feedback.

MIS requires the surgeon to place into the patient's body ports through which the MIS instruments will be inserted. Well-located ports maximize the ease and efficiency with which the procedure will be performed. In contrast, poor port placements may increase patient risk, the possibility of complications, and the potential for needing to convert to open surgery. To date methods used to teach perioperative skills have been limited to what residents observe or practice under supervision. However, these methods may place the patient at increased risk and extend the time needed for the procedure. To overcome the shortcomings of traditional training we suggest the use of an immersive virtual reality (VR) system to train and evaluate physicians in perioperative skill development.

Examples of VR systems to train surgeons in various MIS surgical skills have been well documented in the last four Medicine Meets Virtual Reality proceedings [5,6,7,8]. However, only a few researchers have suggested methods for measuring performance to evaluate the degree to which the skills have been mastered. The number of researchers mentioning the reliability and validity of such evaluations is smaller yet. To determine the psychometric properties of the assessment method, up-to-date measurement theory should be employed. For example, current thinking holds that estimating reliability and validity is a process, rather than a product [9,10]. Therefore, we must move beyond calling the "test" reliable and valid to establishing what these two attributes are whenever we use data to make decisions about people using scores derived from the method. In short then, the use made of a set of scores is the focus of concern, not the scores or measures themselves. To this end, we are testing a VR system, along with its measurement capabilities.

Our scoring system uses a criterion-referenced model of measurement, since we are interested in absolute levels of performance rather than a norm-referenced model. We will use expert's responses to the simulation as the criterion to which we will compare trainee performance.

We are using experienced MIS surgeons to set performance standards on our VR simulation. We are dealing with a finite universe of "correct" performances and have generated the standards we are willing to use to evaluate this performance. The perioperative skills we are training and evaluating in this simulation are those required to prepare the patient and OR for a laparoscopic cholecystectomy. These skills include the placement of surgeons, placement of monitors, patient position, table tilt, placement of a Veress needle, placement of instrument ports, and other perioperative decision-making.

We have collected pilot-test data for two expert surgeons. We will show that we can use these data to define a criterion-reference model for trainee evaluation. Extracting usable data posed a challenge, since several of the variables, including placement of the surgeons, monitors, Veress needles and ports and the tilt of the operating table, have no fixed reference for correct location. Once these data are processed, we will be able to use the resulting criteria as a reference in the simulation so that trainees may be evaluated and provided feedback in real time.

2. Materials and Methods

The Trainer for Surgical Preparation [11] was written to work on an Octane MXE (Silicon Graphics Incorporated) Dual R10000 250 MHz Processor computer with 256 MB of memory, connected to an ImmersaDesk (Fakespace Systems) using a SpacePad tracking system (Ascension Technology Corporation), a 6 DOF WorkWand input device (Fakespace Systems), and CrystalEyes eyewear (StereoGraphics Corporation). The Trainer for Surgical Preparation is written in C++ using the CAVE Library version 2.6 (VRCO) and the OpenInventor Library version 2.5 (TGS, Inc.), and compiled using the MIPSpro C++ Compiler version 7.30.

User input into applications on our ImmersaDesk VR system is done through the WorkWand, which has three buttons as well as a joystick that provides independent x and y coordinate input, and through the SpacePad tracking system, which tracks sensors attached to both the wand and a pair of "driver's" CrystalEyes goggles that the user wears while operating the system. The SpacePad tracking system provides both the position and orientation of the sensors connected to the wand and the "driver's" goggles.

The CAVELibrary handles most of the implementation of immersive features expected of a VR system, including the simulation of head motion parallax and stereoscopic display, as well as a number of more fundamental program implementation issues, such as window management and input device management. The OpenInventor Library handles geometry and rendering.

The Trainer for Surgical Preparation was designed to be a stepwise procession through the decisions that must be made prior to surgery, in this order: perioperative considerations, patient positioning, accessory device selection, surgeon and monitor placement, Veress needle placement, and laparoscopic port placement. Before the beginning of the procession, and between each step in the procession, the user is presented with an introduction to the section. After each step in the procession, feedback will be given to the user based on their performance in the preceding step as compared with some set of criteria, and if the user made mistakes the program will be put into a state such that the mistakes do not corrupt the later steps in the procession. Once all steps have been completed, the user will be presented with an overall performance evaluation. The program tracks every discreet decision made during the simulation, and once a decision is made, it cannot, as the program stands now, be changed. Once the user is finished with the entire stepwise procession, those decisions are saved to a standard Unix text file. The information collected includes: the name of the user; the date and time of the decision; the exact position of the menu selection made in the menu hierarchy, if such a position was relevant; the name of the decision being made; and a position or orientation associated with the decision, if applicable.

The interface for the program was based on the realization that we needed two mechanisms for presenting information to the user. First, we needed a generic interface component that could be used to present the user with instructions, reference points, and a familiar decision-making interface element. Second, we needed a general scene component that would provide visual data to the user for interpretation and possible manipulation. The generic interface component displays title text for the current step in the program, cue text to help guide the user in the task that they are to accomplish in the current step, instructional or feedback text, a progress wheel that shows how much of the stepwise procession has been completed, a wand icon that provides a simple description of how the wand is used in the current step, a menu system for basic decisions and input into the program, and a pointer to interact with the menu system. The general scene component displays everything that the interface does not display, including the operating room, operating table, patient, ports, and other simulation geometries. In general, the user interacts with the interface component, except in situations when the user makes a choice through the menu that would allow manipulation of the scene component.

Several of the steps in the procession presented a problem with determining a way of evaluating the user. Surgeon, monitor, Veress needle, and port placements all required that the user place an object in the program, and there was no way of relating the placement of the object, captured as position (in feet) and orientation (in degrees) provided by the SpacePad tracking system, to any real world measurement of correctness. To resolve this problem, the program was designed so that it could be used without an evaluation of user performance, although user decisions were still tracked and logged. Using the program in this configuration we input master responses into the program and use the resulting data to create a set of evaluation criteria that can later be written into the program to evaluate user performance. Since different master surgeons might make different decisions in the program, and since we did not want to preemptively eliminate any valid set of input, we chose to use this criterion-reference approach with all decision points in the program that we were interested in evaluating.

Access to geometric primitives in the scenes used in this program was not part of the program design or part of supported library functionality. This shortcoming complicated determination of port placement coordinates, since the actual intersection between the port and the patient model could not easily be determined. To compensate for this without major program redesign, we assumed that the port or Veress needle would be placed so that its shaft was intersecting the surface of the patient, and that, therefore, the middle of the shaft of the port or Veress needle would be a reasonable approximation of the actual intersection of the port or Veress needle with the surface of the patient model. This approximation could affect the data, in the worst case, by offsetting the actual point of intersection into the abdomen by a distance of as much as half the length of the shaft of the port or needle (at most 1.5 inches).

To collect these data, time was scheduled with surgeons to use the Trainer for Surgical Preparation. Ideally, each surgeon would contribute data on several different occasions. Each surgeon was instructed to make decisions they believed were correct. After each run through the program, data were collected and compiled with earlier data for analysis. A possible complication to this process was that surgeons were presented with no clear patient history. Programmatically, no assumption was made as to what responses the surgeons would provide.

In analyzing these data, each variable was assumed to be independent. We will continue to assume this until we have collected enough data to make an empirical determination.

3. Results

During the analysis of the master data, we expected patterns to emerge from the decisions. Specifically, we expected a general consensus on perioperative considerations and accessory device selection, general consensus and internal consistency in patient and table positioning, patterns for placement of surgeons and monitors following one or more of the established practices, and patterns in Veress needle and port placement following established recommendations in port placement. The data we obtained indicate that this was indeed the case. The placements of surgeons and monitors were internally consistent, with a standard deviation of 2.4 inches for one surgeon and 13.8 inches for the other surgeon. However, comparison between the two surgeons proved impossible, because the data clearly showed that each surgeon had used a different set of procedural guidelines for determining the correct configuration for the operating room.

Similarly, the port placements made distinct clusters of data points in space, consistent with expected patterns.

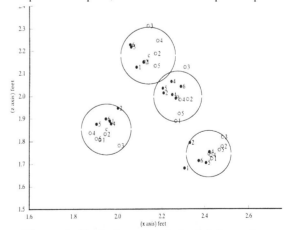

Figure 1. Port placements by trial number for two surgeons from an overhead view

The standard deviation from the mean for each cluster was calculated for each surgeon and across surgeons and clusters. An overhead view of the results across surgeons, clusters, and trials is projected orthographically onto the XZ plane in Figure 1, along with circles representing the standard deviations of each cluster from their cluster mean. Although not shown, similar graphs representing the XY and YZ planes were constructed with similar results. The standard deviations for the clusters from left to right are: 1.3 inches, 1.5 inches, 1.3 inches, and 1.2 inches. The overall standard deviation for one surgeon was 1.3 inches, and for the other surgeon .98 inches.

4. Discussion

Based on the analysis of our pilot-test data, we are confident that we can construct a set of criteria for determining whether or not residents meet minimal training standards for perioperative decision-making skills. While we have arbitrarily set one standard deviation to delimit a range of acceptable performances based on an uncued performance by two master surgeons, we can adjust this standard as we collect data from additional experts and determine that adjustment is required. In spite of the complications we encountered in this study, responses were consistent, especially where surgeon, monitor, Veress needle, and port placements were concerned, indicating that skills are meaningful and measurable in our VR Trainer for Surgical Preparation. We also believe that many of the complications encountered in this pilot study can be overcome in the next version of this program.

Since we have determined that using a criterion-referenced approach to train surgeons in perioperative skills is feasible, we will begin more aggressively to pursue establishing more robust sets of criteria. Once an acceptable criterion set is established, we can begin to apply the criteria to evaluating MIS surgical competence before residents begin to work in the OR. We anticipate applying rigorous psychometric analyses as part of our continuing efforts at refining the evaluation system.

We also believe that our criterion-referenced approach could be used in other VR and immersive applications. We plan to use this approach in new applications as they are developed.

5. References

[1] P.J. Quilici. *Online Laparoscopic Technical Manual*. Last Update Feb 12, 2001 Burbank (CA): TransMed Network; 1997 [cited 2000, Feb 26]; Available from: URL: http://www.transmed.net/lapnet/lapmenu.htm

[2] The Southern Surgeons Club, M.J. Moore and C.L. Bennett. The learning curve for laparoscopic cholecystectomy. *Am J Surg*. 1995:170:55-59.

[3] D.I. Watson, R.J. Baigne, and G.G. Jamieson. A learning curve for laparoscopic fundoplication: Definable, avoidable, or a waste of time? *Ann Surg*. 1996:234:198-203.

[4] J.C. Rosser, L.E. Rosser, and R.S. Savaigi. Skill acquisition and assessment for laparoscopic surgery. *Arch Surg*. 1997:132:200-204.

[5] N. Taffinder, C Sutton, R.J. Fishwick, J.C. McManus, and A. Darzi. "Validation of VR to Teach and Assess Psychomotor Skills in Laparoscopic Surgery: Results from Randomized Controlled Studies Using the MIST VR Laparoscopic Simulator," in J.D. Westwood, H.M. Hoffman, D. Stredney, and S.J. Weghorst (eds.). *Medicine Meets VR: Art, Science, Technology: Healthcare (R)evoution*, IOS Press, Amsterdam, 1998, pp. 124-130.

[6] A. McCarthy, P. Harley, and R. Smallwood. "Virtual Arthroscopy Training: Do the 'Virtual Skills' Developed Match the Real Skills Required," in J.D. Westwood, H.M. Hoffman, R.A. Robb, and D. Stredney (eds.). *Medicine Meets VR: The Convergence of Physical & Informational Technologies: Options for a New Era in Healthcare*, IOS Press, Amsterdam, 1999, pp. 221-227.

[7] P. Oppenheimer, S. Weghorst, L. Williams, A. Ali, J. Cain, M. MacFarlane, M. Sinanan. "Laparoscopic Surgical Simulator and Port Placement Study," in J.D. Westwood, H.M. Hoffman, G.T. Mogel, R.A. Robb, and D. Stredney (eds.). *Medicine Meets VR 2000: Envisioning Healing: Interactive Technology and the Patient Practitioner Dialogue*, IOS Press, Amsterdam, 2000, pp. 233-235.

[8] D. Berg, J. Berkley, S. Weghorst, G. Raugi, G. Turkiyyah, M. Ganter, F. Qintanilla, and P. Oppeneimer. "Issues in Validation of a Dermatologic Surgery Simulator," in J.D. Westwood, H.M. Hoffman, G.T. Mogel, D. Stredney and R.A. Robb (eds.). *Medicine Meets VR 2001: Outer Space, Inner Space, Virtual Space,* IOS Press, Amsterdam, 2001, pp. 60-65.

[9] S. Messick. Validity, in R.L. Linn (ed.). *Educational Measurement: Third Edition*, Collier Macmillan Publishers, London, 1989, pp. 13-104.

[10] L.S. Feldt, R.L. Brennan. Reliability, in R.L. Linn (ed.). *Educational Measurement: Third Edition*, Collier Macmillan Publishers, London, 1989, pp. 105-146.

[11] D.B. Witzke, J.D. Hoskins, M.J. Mastrangelo Jr, W.O. Witzke, U.B. Chu, S. Pande, and A.E. Park. "Immersive VR Used as a Platform for Perioperative Training for Surgical Residents," in J.D. Westwood, H.M. Hoffman, G.T. Mogel, D. Stredney and R.A. Robb (eds.). *Medicine Meets VR 2001: Outer Space, Inner Space, Virtual Space,* IOS Press, Amsterdam, 2001, pp. 577-583.

Theme 10: Support and Tools Systems

Student Adaptivity in TILE: A Client-Server Approach

Kinshuk, Binglan Han, Hong Hong, and Ashok Patel*
Massey University, New Zealand
*De Montfort University, United Kingdom
Kinshuk@massey.ac.nz

Abstract

The paper presents a client-server approach to student model that is implemented using agent technology. The student model is divided into client and server parts to facilitate adaptivity in various configurations such as online, offline and mobile scenarios. The implementation of this approach to student model is discussed in the TILE project, which is researching, evaluating and developing an integrated system for education at a distance, with powerful adaptivity for the management, authoring, delivery and monitoring of such material.

1. Introduction

Web-based education is currently an important research and development area and it has opened new ways of learning for many people.

But it is important that a web based educational system should have *adaptivity*. According to Brusilovsky [1] adaptivity is important for web based educational applications since distant students usually work on their own. In addition, being adaptive is important for web based courseware because it has to be used by a much wider variety of students, without assistance of real teachers [2]. We are applying adaptivity concepts in the TILE project, which is researching, evaluating and developing an integrated system for education at a distance.

2. Adaptation in web-based educational systems

The concept of 'adaptation' is an important issue in the research for learning systems [2][10]. It has been shown that the application of adaptation can provide better learning environments in such systems but many issues need to be resolved before an effective and efficient adaptation in learning systems is possible [9]. There have been many attempts in last decade to include user models and adaptation features within systems with a view to improve the correspondence between user, task and system characteristics and increase the user's efficiency.

Two kinds of systems have been developed for supporting the user in his/her tasks:
- Systems that allow the user to change certain system parameters and adapt their behavior accordingly are called *adaptable*.
- Systems that adapt to the users automatically based on the system's assumptions about the user needs are called *adaptive*.

The dynamism in the system requires consideration of following criteria [6]:
- Adaptation with respect to current domain competence level of the learner;
- Suitability with respect to domain content; and
- Adaptation with respect to the context in which the information is being presented.

The fulfillment of these criteria requires the development of a student model, which captures the interactions of students with the system to extract information about their competence level for various domain concepts and tasks represented in the system.

According to the potential of web-based systems and the criteria of the dynamism in the web-based systems, we are developing a suitable way to capture interactions over the Internet and to provide a continuous interaction pattern for a given student, even in off-line mode. In web-based systems, the interactions between client and server normally take place using Hypertext Transfer Protocol (HTTP). HTTP is a stateless protocol, which makes it difficult to track the students progress and hence to analyze the mental processes of the student [5]. However, using a judicious mixture of SQL and HTML protocols, we can be more precise about browsing behaviour. Since part of the student model resides on the server and part on the user's machine, off-line adaptation with intermittent update of server side student model as and when possible has also become a possibility.

3. Student model in TILE

We are developing a student model and an inference module that determines the state of the student's knowledge. It differentiates between miss-conceptions and missing conceptions [6] and determines the best possible action on the part of the system and guides the student's path [11].

The TILE system models individuals and classes of users (extracted by a cumulative analysis of the individual student models) to maintain *individual student model* and *group student model*.

3.1. Individual student model

Individual student model allows adaptive behavior of the system by providing information about the learner. It is simple in construction but is quite detailed to facilitate rapid updates and better context sensitive guidance. It contains four main components:

- Global preferences of the learner (behavioral component): the global preferences are applied to the whole system.
- Specific content presentation related preferences of the learner (behavioral component): it is applied to specific contents.
- Domain competence related information about the learner (domain based component)
- Student's working history with annotated system feedback (only in problem solving scenarios, for the problems which have sequential processes)

The individual student model is used by the system primarily:

- to provide adaptive navigation guidance - based on prioritized successors and learner model
- to select coarser/finer granularity of domain content
- to provide context based excursions to other learning units
- in making analogies with previously learnt material
- in making direct references to previous learnt material
- to provide dynamic messaging and feedback, for example:
    Navigation related system messages
    Content related system messages
    Dynamic progression recommendations based on learner's domain competence and current context

3.2. Group student model

The system summarizes students' common behavior and preferences to create and maintain the group student model. Students using the system are categorizes into different groups by matching their behavior and domain competence with that reflected by a certain group student model. The structure of the groups is multi-dimensional; one student can belong to one or more such groups (e.g. in a serialistic learning style group and in an active learning group). The mutuality of the group student model and the individual student model highly improves the effectiveness and accuracy of TILE system's adaptivity. It also offers more reasonable default setup and help for newcomer individual students.

The update of group student model takes place at every startup of the program.

4. Client-server architecture of student model in TILE

The learning process in the TILE environment benefits from the client-server approach which has also been extended to the adaptation process. Consequently, the student model is divided into host and client bases. The host base contains the group student model, and partial individual student model, whereas client base contains only individual student model. This approach allows the system to be much more flexible, particularly in the web-based environment, where connectivity between host and client is not always guaranteed, and the quality of the connection often suffers from traffic congestion. The client side student model facilitates adaptivity even in offline mode, whereas host side student model allows adaptivity based on new information available from the domain experts and better adaptation procedures resulting from group student model. Figure 1 shows the schemata of the adaptation mechanism in the TILE environment.

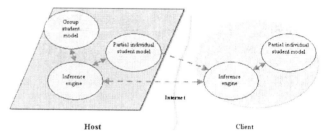

Figure 1. Adaptation mechanism in TILE environment

To improve the quality of communication between client and server, mobile agents are used in the implementation of the TILE adaptation mechanism. A mobile agent interacts with client side inference engine to pick up data, which relies on individual student model at client side. Then mobile agent moves to the host (or server) side. At host side, mobile agent performs all the processes needed, such as updating the partial individual student model based on summary of client side student model (based on the data brought by mobile agent), interacting with group student model to determine if that student model also needs to be updated. After mobile agent finishes all actions at host side, it gathers all the information it needs, and returns to the client side. Then it updates the client side individual student model. This mobile agent approach works even in the intermittent connectivity between client and host because mobile agent can be dispatched when the connection is available and then the agent works autonomously without requiring continuous connection.

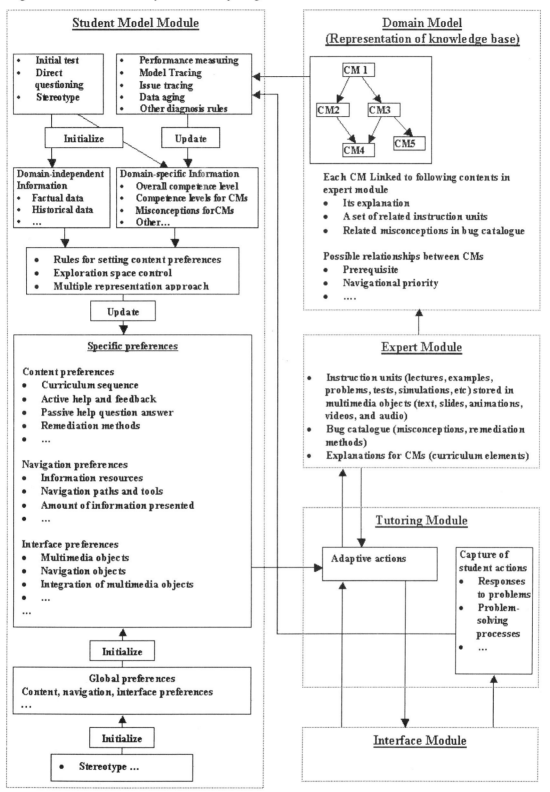

Figure 2. Summary structure of student model in TILE

5. Structure of student model in TILE

Figure 2 shows the summary of student model structure in TILE. The highlights of the model are the 'Exploration Space Control' [4] providing adaptation for exploration space by limiting information resources, number of searching paths and tools, and amount of presented information; and 'Multiple Representation Approach' [8] for adaptation of domain content presentation.

6. Conclusion

Adaptivity in web-based educational systems has been proved very important, particularly in the distance learning scenarios where direct contact between human expert and learner is not ensured. A survey of various techniques, used in such systems to provide adaptivity, is presented before discussing one such implementation in TILE project. Adaptivity features are being integrated in the TILE project at Massey University. We believe that the future of education, not only within continuing education, will more and more be reliant on web-based, on-line educational material. The TILE project has a major thrust at providing an integrated system for the management, authoring, delivery and monitoring of such material. It is based on a sound research foundation and is being developed in a manner that provides simple-to-use interfaces in order to reach out to the educators in developing archives of indexed material that may be accessed adaptively by the students in the same way as they would interact with a tutor or mentor.

Reference

[1] Brusilovsky, P., "Adaptive and Intelligent Technologies for Web-based Education", In: C. Rollinger and C. Peylo (eds.) *Künstliche Intelligenz, Special Issue on Intelligent Systems and Teleteaching*, 4, 1999, pp.19-25.

[2] Brusilovsky, P., Schwarz, E., & Weber, G. "A tool for developing Adaptive Electronic Textbooks on WWW". *WebNet 96*.

[3] Hoschka, P. *"Computers as Assistants: A New Generation of Support Systems"*, Mahwah, NJ: Lawrence Erlbaum Associates, 1996, pp.336-340.

[4] Kashihara A., Kinshuk, Oppermann R., Rashev R. & Simm H. "A Cognitive Load Reduction Approach to Exploratory Learning and Its Application to an Interactive Simulation-based Learning System", *Journal of Educational Multimedia and Hypermedia*, 9 (3), 2000, pp.253-276.

[5] Kinshuk & Patel, A. "A Conceptual Framework for Internet based Intelligent Tutoring Systems". In A. Behrooz (Ed.) *Knowledge Transfer*, Vol. II, London: pAce, 1997, pp.117-124.

[6] Kinshuk & Patel, A. "Knowledge Representation and Processing in Web Based Intelligent Tutoring". In S. P. Lajoie & M. Vivet (Eds.) *Artificial Intelligence in Education*, Amsterdam: IOS Press, 1999, pp.720-722.

[7] Kinshuk, Oppermann, R., Patel, A. & Kashihara, A. "Multiple Representation Approach in Multimedia based Intelligent Educational Systems. In S. P. Lajoie & M. Vivet (Eds.) *Artificial Intelligence in Education*, Amsterdam: IOS Press, 1999, pp. 259-266.

[8] Kinshuk, Oppermann,R., Patel, A., & Kashihara, A. "Multiple representation approach in multimedia based intelligent educational systems". *Artificial Intelligence in Education* (Eds. S.P.Lajole & M. Vivet), IOS Press, Amsterdam, 2000, pp. 259-266.

[9] Milne, S., Shiu, E. & Cook, J. "Development of a model of user attributes and its implementation within an adaptive tutoring system". *User Modeling and User-Adapted Interaction*, 6, 1996, pp.303-335.

[10] Nikov, A. & Pohl, W. "Combining User and User Modelling for User-Adaptivity Systems". In H.-J. Bullinger & J. Ziegler (Eds.) *Human Computer Interaction - Ergonomics and User Interfaces*, New Jersey: Lawrence Erlbaum, 1999.

[11] Patel, A. & Kinshuk, "Granular Interface Design: Decomposing Learning Tasks and Enhancing Tutoring Interaction". In M. J. Smith, G. Salvendy & R. J. Koubek (Eds.) *Advances in Human Factors/Ergonomics - 21B - Design of Computing Systems: Social and Ergonomic Considerations*, Amsterdam: Elsevier Science B. V., 1997, pp.161-164.

MachineShop: Steps Toward Exploring Novel I/O Devices for Computational Craftwork

Glenn Blauvelt and Mike Eisenberg
Department of Computer Science and Institute of Cognitive Science
University of Colorado, Boulder
zathras@cs.colorado.edu; duck@cs.colorado.edu

Abstract

The notion of "computational crafting" focuses on the numerous ways in which computational media may be used to expand the expressive range of traditional educational crafts. One important dimension of this approach involves a close re-examination of an issue often taken for granted in educational technology—namely, the design and use of I/O devices. The next decade is likely to produce a fascinating array of novel I/O devices and technologies; these in turn offer substantial promise of augmenting the power of computational tools for children's craftwork.

This paper describes initial work toward developing an educational crafting application for the design of mechanical toys and automata. Our application, MachineShop, is intended to allow students to create mechanical parts (e.g., cams, gears, and shafts) that may be customized and simulated on the computer screen, and finally "printed out" on a laser cutter for realization in materials such as wood and foam core. We describe the current (early) state of the application and discuss its implications for the design and use of novel or unorthodox I/O devices in educational technology.

1. Introduction

At the risk of a slight degree of caricature, there is something of a "mainstream view" of educational computing in which the notion of the "computer" has remained qualitatively unchanged for at least two decades. In this view, the heart of the computer is the CPU/memory combination portrayed in standard textbooks of computer architecture; and input/output (I/O) devices are the means by which communication with this calculating engine is effected. Undoubtedly, the details of this picture may change—processors get (much) faster, memory (much) larger, screens more expansive, applications more complex, and so forth—but the basic model of the computer remains constant.

While this mainstream portrait of the "educational computer" has provided a foundation for a generation of magnificently productive work, it nevertheless conceals a variety of superficially plausible but potentially limiting assumptions. One such implicit assumption—reflected uncomfortably in the standard vocabulary of the computer industry—is that I/O devices are "peripheral" to the true core of computer science. A corollary of this view is that there need only be a few standard, general-purpose means of communicating user input to the device (keyboards and a variety of pointing/selection devices), and a few means of communicating the device's response back to the user (screens and printers).

This is not by any stretch the only problematic assumption of the traditional portrait: others include the notion (now rapidly fading) that a computer is inevitably a desktop device; or that a larger and more powerful computer is necessarily preferable to a smaller, simpler device; or that computer languages, because they are designed to exploit the large desktop devices in which they run, are therefore vast and sprawling exercises in complex notation. Over the past several years, our explorations in the area of "computational craftwork" have led us to re-examine all of these assumptions (among others). For the purposes of this paper, however, we will focus on the issue of I/O devices. In particular, we argue here that viewing the computer as one element of a larger system of craft technologies leads quite naturally to an increased attention to the space of materials and objects that can be output from (or even, conceivably, input to) computational media.

By way of example, we describe our initial work in creating an application called MachineShop; this is a software system whose purpose is to assist students in the creation of physical mechanical devices and automata. MachineShop is conceived as an application through which objects can be realized in a variety of materials, including cardboard, wood, and foam core. It thus represents a sample excursion into the vast (and largely unexplored) territory in which computational media and educational crafting may eventually intersect.

The remainder of this paper is organized as follows: in the second section, we give a brief outline of the notion of "computational crafting" in education, and sketch some of our representative earlier work in this area. The third section presents an example of cam construction using our current version of MachineShop, illustrating the conceptual path from on-screen design to "printing out" in wood. In the fourth and final section, we use MachineShop as a springboard for a much broader discussion of the potential role of novel I/O devices and technologies in educational computing.

2. Blending computational media and educational crafts

The basic idea behind "computational crafting" is that it seeks ways in which computers may be used to enhance the expressive range of traditional (and in some cases, nontraditional) educational crafts. While activities such as geometric papercrafting, making string figures, and mechanical design have a venerable history in mathematics and science education, they are usually regarded as "enrichment" activities, not to be accorded too much time or attention; this unfortunate status has been exacerbated by the reputation that craft activities have of being "low-tech" and thus somehow anachronistic in an age of burgeoning educational technology.

It is our view that there are tremendous educational benefits to be reaped, both by advocates of craftwork and advocates of technology, by blending techniques from both cultures. Crafters gain new tools, notations, and materials to work with; technologists gain fascinating new domains to explore, and fascinating research issues that accompany those domains.

In practice, computational crafting takes several major forms. One line of work, pioneered by Resnick and his colleagues in the development of the "programmable Lego brick"[1], focuses on embedding computational capabilities within craft objects. In our group, we have pursued this line of thought by creating prototypes of programmable craft objects (hinges, tacks, and ceramic tiles): in each case, small computers are used to provide customized behavior to the object in question. For example, the "programmable hinge" may be sent commands to open and close in user-controlled patterns. [2] Another line of work, closer in spirit to MachineShop, is the design of software applications to enrich the use of (otherwise traditional or noncomputational) materials. Examples of our group's earlier work in this area are HyperGami, a program for the design of paper polyhedral models [3], and HyperSpider, a program for the design of mathematical string sculptures [4]. The basic notion behind such applications is that the user creates models, on the computer screen, of objects that will subsequently be the targets of physical design. In the case of HyperGami, for instance, the typical scenario is that a user creates a three-dimensional polyhedron on the screen; the program "unfolds" that shape into a two-dimensional pattern; the user then decorates and prints out this pattern on a color printer; and finally, the pattern may then be cut out and folded into a physical object.

MachineShop is representative of this second style of work in that it is conceived as a design application for (generally noncomputational) mechanical automata. This particular domain, however, brings to the fore a number of interesting problems for the application designer—problems not encountered in the creation of HyperGami and HyperSpider. One obvious element is that the objects produced by MachineShop are dynamic; thus, our design application must provide at least a rough simulation capability so that students can see how their creations will move. Another issue—not quite as glaringly apparent—is that the objects produced by MachineShop are not themselves "whole" creations (as in the case of the earlier two programs), but are rather imagined as parts to be used in larger creations. When one creates an automaton, the mechanical elements are crucial; but they are "support structure" for what is often a much more elaborate endeavor in representing (e.g.) a moving animal or human figure. (Generally, the sorts of creations that we have in mind here are currently exemplified by the brilliant automata on display at London's Cabaret Mechanical Theatre. [5])

One thus thinks of MachineShop as a tool for the creation of "kits" of moving pieces that can be customized and re-used for producing specific types of motions in various automaton-design projects. Importantly, the pieces created may (for various reasons) be realized in materials such as paper, cardboard, foam core, plastic or wood, depending on the nature of the particular design project involved. MachineShop therefore must produce specifications of objects that can be "printed out" in a variety of appropriate media. We return to this topic in the following two sections.

3. MachineShop: a brief sample scenario

In this section we present a short scenario illustrating the creation of a customized cam using the current MachineShop prototype. (At present, the system is still very much a work-in-progress, and only the cam module is sufficiently developed for purposes of demonstration.) Figure 1 shows the MachineShop screen interface for the construction of a snail cam.

The cam design window of Figure 1 is divided into three sections. At upper left is a graph representing the path that a cam follower would take as the cam turned; in this case, the follower would move up steadily, followed by a short vertical drop at the notches in the cam. At lower left are various pull-down menus to specify cam parameters, and buttons for saving and loading files. At right is a line drawing showing the cam profile; this corresponds to the parameters specified and the graph at upper left. Finally, there is a button for animating the profile so that the user can view the cam's motion; this causes the profile to rotate (soon to be accompanied by an animated cam follower), permitting the user to see how the physical part will move once constructed.

One of the themes of MachineShop's design is that the user should, wherever feasible, be able to create parts to produce some desired pattern of motion. To this end, the rectangles on the graph representation of Figure 1 can be moved interactively on the grid, allowing the user to change the dwell and lift of any lobe while seeing these changes as part of the follower's path. As the user moves these points the cam profile is updated to reflect the cam's

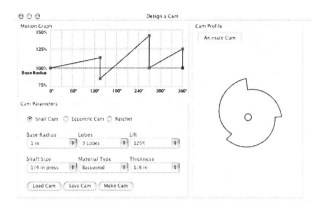

Figure 1. The MachineShop interface for creating a snail cam. In this scenario, we are creating a customized 3-lobed cam as described in the text.

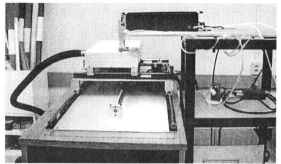

Figure 2. The cam being cut from basswood on a CO_2 laser cutter.

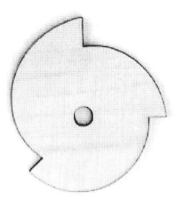

Figure 3. The finished cam.

new shape. For the types of snail cams shown, the drop from the high point of one lobe to the low point of the next is vertical. This is captured in the graph by constraining points that share a common horizontal position to move horizontally in unison. The thicker horizontal grid line indicates the base radius of the cam with each division above and below it representing a 25% change in that radius, from 75% to 150%. Each vertical division represents 30 degrees of rotation.

The various pull-down menus toward the left allow the user to customize the cam in a variety of ways: she can change the base radius for cam size, the number of lobes, the offset of the shaft hole (for use with eccentric cams, the second cam type currently provided by the software) and the size of the shaft hole. The control buttons just above these menus provide for creating and retrieving files of cam profiles and parameters; the Save and Load buttons perform the obvious functions, while the Fabricate button creates a file that can be employed by a computer-controlled machine tool (in this case, a carbon dioxide laser cutter) to fabricate the cam.

Figures 2 and 3 show a continuation of this scenario in which the customized cam designed in Figure 1 has been translated to a laser-cutter fabrication file, and "printed out" in wood. (Figure 3 shows the final object itself.) Once this step is complete, the cam could be incorporated into larger mechanical-design projects. It should be noted that if alternative materials are desired, the laser cutter shown in Figure 2 is also capable of producing the object in materials such as cardboard and foam core.

4. Re-thinking the notion of "peripherals" in educational computing

We believe that it is worth dwelling for a moment on the implications of the scenario in the previous section. Output is not merely an afterthought to computation, as the term "peripheral" might suggest. The fact that new tools such as laser cutters are now becoming increasingly affordable and accessible implies that they could well be a growing presence in classrooms over the next decade or so. Indeed, we would like to believe that the burgeoning presence of color printers (in both homes and classrooms) is a harbinger of a similar pattern for other sorts of output devices in the near future. Whether this in fact occurs will in part depend on the presence of appropriate educational software to exploit the new devices; there is something of a cycle of cause and effect here between applications and devices, in which the presence of useful applications (such as MachineShop) encourages the spread of certain devices (such as laser cutters) which in turn encourages the development of still more expressive applications.

Traditionally, children's exposure to both tools and materials for building the kinds of objects described here have been limited to shop and art classes. These are inadequate for the mastery of either, and children tend to encounter great frustration when trying to make all but the simplest of objects. The fact that a device such as a laser cutter is capable of producing output in (e.g.) wood suggests that many craft design activities could now be within the purview of elementary school classrooms. Exploiting a device such as the one shown in Figure 2, we can envision systems for the design of mathematical and scientific craft objects such as (just to name a few): customized wooden sliding tile puzzles or tangram sets;

balancing wooden toys that perch on the edge of a shelf; "jumping jack" puppet figures made of wood and string; jigsaw puzzles in both two and three dimensions; articulated shadow puppets; and so forth. All these examples require nothing more than a relatively simple one-dimensional laser cutter (in which the beam can only cut material in one direction); such devices produce only "flat" forms as output. A slightly more ambitious prospect would be to employ two-dimensional laser cutters as output devices; in this case one could envision classroom applications to produce (e.g.) wooden burr puzzles; etched designs on the surfaces of wooden polyhedra or spheres; customized connectors for three-dimensional wooden "erector set-like" construction kits; and so forth.

The upshot of this speculation is that the types of educational applications that software designers can create may be vastly expanded by the advent of novel, accessible output devices. That is, output devices potentially hold a central—not a peripheral—place in our imaginations as designers. There is nothing outrageously futuristic about this speculation, except perhaps in that word "accessible"; laser cutters are currently prohibitively expensive for the majority of classrooms. Still, as suggested by the analogy with color printers (which were also prohibitively expensive for classrooms ten years ago), there is no a priori reason to believe that laser cutters may not become as plentiful (and safe) as other classroom technology. Moreover, such a development could potentially put to rest another common worry of educators: namely, that the presence of computational media necessarily leads to an increasingly "virtualized" existence for children, and a corresponding diminution of experience and pleasure with real-world materials and physical objects. On the contrary, we believe that applications such as HyperGami and MachineShop can enrich children's experience with the physical world.

Indeed, there is still another point to be made here on the subject of children's experiences with physical objects. In many cases, when educators discuss the use of physical materials such as construction kits, tile puzzles, balancing toys, and so forth, the assumption is that these are commercial items to be purchased; that is to say, they are *not* craft items, but are instead products to be sold to classrooms and children. Our belief, in contrast, is that the advent of appropriate output devices and software can expand the landscape of what we conceive of as "children's crafts". Objects that at one time might only have been available commercially (such as balancing toys or jigsaw puzzles) could soon be the province of artistic and creative children. Far from impoverishing the space of children's work with physical materials, then, technology can potentially endow children with expressive control over realms that are currently beyond their typical experience.

Of course, we needn't curtail our speculation at the use of laser cutters as output devices. Current research in a variety of commercial three-dimensional printing technologies (cf. the highly readable discussion in Gershenfeld [6]) suggests that still more expressive output devices may be on the not-too-distant horizon. By using these devices, still other craft activities and materials could easily be within the purview of children's crafts: using plastic materials, one could make snap-together bricks, or customized spinning tops, or realistic animal or human figures, or chemical models, or myriads of other objects. Going just a bit further, we believe that it would be especially useful for educators themselves to speculate, proactively, on the sorts of I/O devices that would hold special utility for children's crafts. One object on our own "wish list" would be a device to output yarn with specified color patterns: in a sense, such a device would be to yarn what a color printer is to blank paper. With a device of this sort, we could output complex color patterns for yarn to employ in sculptures such as those designed by users of HyperSpider. Another possibility would be to think of new input, as well, as output, devices. One might imagine, for instance, "readers" for at least certain classes of three-dimensional objects to be used as templates for craft projects (e.g., a solid figure might be read in; converted to a discretized polyhedral representation; and then output to a HyperGami-like folding net).

All this is of course a long way from our early work with MachineShop; and there is still much more short-term development that needs to be done with our current system. Nonetheless, even our very earliest experiences with novel output devices suggest that the realm of children's craftwork has an exciting and healthy future.

Acknowledgments

We are indebted to the ideas and encouragement of Gerhard Fischer, Mitchel Resnick, and Carol Strohecker, among many others. Ted Chen designed the HyperSpider program; Tom Wrensch and Ann Eisenberg have collaborated on the work described here. This work has been supported by a generous gift from the Mitsubishi Electric Research Laboratories in Cambridge (MERL) and by NSF grants CDA-9616444 and REC-961396.

References

[1] Resnick, M.; Martin, F.; Sargent, R.; and Silverman, B. Programmable Bricks: Toys to Think With. *IBM Systems Journal*, 35:3, pp. 443-452. 1998.

[2] Wrensch, T. and Eisenberg, M. The Programmable Hinge: Toward Computationally Enhanced Crafts. *Proceedings of UIST 98*, San Francisco, November, pp. 89-96. 1998.

[3] Eisenberg, M. and Nishioka, A. Orihedra: Mathematical Sculptures in Paper. *International Journal of Computers for Mathematical Learning*. 1(3): 225-261. 1997.

[4] Chen, T. HyperSpider: Integrating Computation with the Design of Educational Crafts. M.S. Thesis, University of Colorado, Boulder. 1999.

[5] Onn, A. and Alexander, G. Cabaret Mechanical Movement. London: Cabaret Mechanical Theatre. 1998.

[6] Gershenfeld, N. *When Things Begin to Think*. New York: Henry Holt. 1999.

Supporting Knowledge Communities with Online Distance Learning System Platform

Correa, Juarez Sagebin; jsagebin@crt.net.br
Fink, Daniel; dfink@crt.net.br
Moraes, Candida Pasquali; cpmoraes@crt.net.br
Sonntag, Alexandre A.; asonntag@crt.net.br
CRT Brasil Telecom

Abstract

The Online Distance Learning System from CRT Brasil Telecom (Brazilian Telephone Company), has showed that technology and education found a commum point to perform good educational methodologies. Contents and ODL (Online Distance Learning) programs reached a very good point of performance and moreover, internal culture is much more prepared to use Online Learning now than ever. The next target now is: how to steer employees to get good results on a "sea" of knowledge. CRT Brasil Telecom Knowledge Communities are assembled to give a natural orientation specifically prepared for both professional and applied educational needs.

I. INTRODUCTION

CRT, a recently privatized Telecommunication Company in Brazil, is now very well positioned at the global scenario of the competitive Corporations. Considering the increasingly need for education methodologies [6], CRT has developed a Distance Learning Strategy in order to continuously educate professionals from different areas of the Company and prepare them for the future. Starting with the launch of a Distance Education Program and the establishment of a development team, now CRT has a full ODL (online distance learning) structure running over its Intranet. The next developments points are: how to create good learning paths for each career (or workgroup) and how to assure that knowledge has been applied for results improvement on the job.

II. OBJECTIVES

The CRT Online Distance Learning structure, named SEND, was created in order to contribute on the search for an educational model for a Telecommunication Company, specifically adapted to aspects of its internal culture. This search started in December, 1999 and opened a full set of new possibilities for employees development. These possibilities could not be planned on, considering traditional paradigms of education. That is why it was necessary to embody aspects like continuing education and non presential training for the ODL development team and changes on internal culture for CRT employees[7].

In short, CRT is an year old company using training over its Intranet on a model that seems like to be definitive in terms of improvements on distance learning events.

These events reflect specific needs for technical, commercial (business) and administrative areas of CRT. That is why programs were created with different approaches in order to adequate the training to the employee. Contents, for example, are connected to students in the form of Learning Objects, according to Industry patterns.

The promotion of interaction among members of the same project must allow ideas sharing with experts and managers. So, business processes can be periodically reviewed and upgraded according the perception of each element of the team.

The key for this objective is the strengthening of the knowledge community through an oriented educational plan. This plan represents a set of Learning Objects like e-learning contents, video, books, articles, synchronous events [1] and even presential training.

III. CONSTRAINTS

Knowledge communities are composed by groups of professionals that share the same information and act on similar areas. As an example, it is pointed out the Consultants Team, which takes care of CRT businesses costumers. This group handles a lot of different technologies regarding telecommunication systems and also has to be well developed on human communication, interpersonal and sales skills.

As a multidisciplinary group, training for this community is far away from exposing a deep set of distance learning contents and waiting for the conclusion of all events. As an example showing them a big library without orientation, employees usually do not have the aptitude of conduct their own research and development.

So, the CRT Distance Learning System as being the system responsible for their education and development, has the mission of connecting suitable contents to specific knowledge communities.

IV. IMPLEMENTATION

A. Integrating Knowledge Communities

Some aspects must be considered in order to create and integrate a Knowledge Community. CRT Training and Development area have developed human resource instruments like education tracking and expert identification.[3] This information is assembled in data bases where it is possible to identify similar education skills and job areas.

Training Consultants contribute with the analysis for development needs, strategic plans and targets for each business department. With these information, it is possible to define communities with their respective development plan and the nature of information needed to support a good job performance and the teams communication flow.

Distance Learning programs developed under SEND can now be connected specifically to each knowledge community. It represents the content element of the diagram shown on Fig.1.

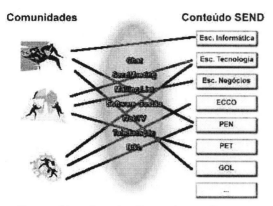

Communities   Learning Connections   Contents
Fig. 1- Knowledge Community concept

Connection elements are SEND services, that represent the interaction ways, giving access and further information for all online students.

B. Connection Elements
The implementation of interaction tools represents the students channel that enables him to keep in touch with other students and with Online Tutors during synchronous or asynchronous events. These elements must be positioned around the employee and must be accessible at any place [5].

Teleducation		Online Library
Chat		CD ROM
Web TV	SEND	Home Access
Mailing List	Student	Virtual Classroom
SENDMeeting		Forum

Fig. 2- Systems available to an ODL Student

C. Developed Programs and Communities

C.1 – Continued Commercial Education Program
Commercial area professionals from CRT were divided in two communities in order to have a better design of the development programs to the following activities: costumer care and Commercial Outlets.
Characteristics of the costumer care team, pointed that there is a high employees turnover level and some non-employees from partners companies. Premises for development skills of these students are the commitment with the Company by learning about telecommunications business, products & services and costumer care and needs. This community is developed with the production of instructional materials based on Video on Demand in the format of Learning Objects (small pieces of information for continuous education).
With the use of Learning Objects it was possible to assemble an education path for each student in order to organize an individual professional career.
As a motivational aspect, students feel more comfortable being certified with their professional improvements.

C.2 – ADSL Team Community
Implementation of new Telecommunications services motivated the creation of this community specific for Network Installers. New professional skills should be developed for these teams, since these employees were not skilled in dealing with Computers Architecture and Data Communication. During the installation process of an ADSL (for costumers) they must give all professional information about installation levels in order to reach goals, downloads of documents, contacts with equipments providers, access to Products and Services informations and costumer care proceedings. Also, the exchange of news and troubleshooting among professionals are shared with the use of mailing lists and forums accessible in a web site specially designed and developed for them.

C.3 – Tutor Services
Virtual Communities in CRT have tutorship, which is responsible for that community. This tutorship shows a new model of learning-teaching (that came with the concept of virtual communities) : now people bring their needs to the community, and, other peers can help them in these needs. It is a process where all people are involved and where everybody can teach and, at the same time, learn something[2]. The old teaching concept had to change: formal tutors are worried about how people work in a virtual community, that means, that students became "actors" in their learning process, instead of the method of having instructors giving the contents courses to the students and answering questions.
The virtual community success point is: people have to interact and feel that they are part of a community and responsible for its nourishing and growing as well as for the contents' application in the organization.

V. RESULTS AND CONCLUSIONS
SEND already trained about 3000 students, on 12 different development titles and 4 distance learning programs. On Knowledge Communities methodology we have 380 students in 2 groups each having its own learning connections using 11 different interaction services. For the near future, content development will not be the main success point for a learning organization, but the way students access all contents.
Technology development in distance learning [4,5] is also a technique used to conduct and steer professionals in order to support performance and efficiency on the job. SEND might be the most efficient one for a Telecom Company Environment. The new concept results' showed that people learn more, get more involved and participate more on their own learning process, besides that, this process helps the company to manage its knowledge and also to get more results on its business.

VI. REFERENCES
[1] Tiffin, John and Rajasingham, Lalita; "In search of the Virtual Class"; Routledge, 1995.
[2] Holmes, Neville W.; "The Myth of the Educational Computer"; IEEE Computer Maganize, 1999
[3] Hall, Brandon; "Web-Based Training Cookbook"; Wiley Computer Publishing, 1996.
[4] J.R. Bourne, A. F. Mayadas, J.° Campbell; "Asynchronous Learning Networks: An Information-Technology-Based Infrastructure for Engineering Education"; Proceedings of the IEEE, vl. 88, no.1, January 2000, pg 63-69
[5] J.M. Wilson, W. C. Jennings, "Studio Courses: How Information Technology Is Changing the Way We teach, On Campus and Off"; Proceedings of the IEEE, vl. 88, no.1, January 2000, pg 72-79
[6] Ulrich, Dave; "Recursos Humanos Estratégicos"; Editora Futura, 1999, pg 347
[7] Meister, Jeanne C., "Educação Corporativa"; Makron Books

Support Tools for Graphs in Computer Science Education

Victor N. Kasyanov
*A.P. Ershov Institute of Informatics Systems,
Lavrentiev pr. 6, Novosibirsk, 630090, Russia*
Email: *kvn@iis.nsk.su*

Abstract

The main thesis of this paper is that intricate nature of software systems and computer education systems can, and in our opinion should, be represented by graph models. Graphs are used in computer science and computer education almost everywhere, since graph is a very natural way of explaining complex situations on an intuitive level.

In the paper, methods and systems for visual processing of graphs and graph models are presented. Our series of books intended to be a general guide on graph algorithms is outlined. A dictionary on graph theory and its electronic version are described.

1. Introduction

We are entirely convinced the future is 'visual', and the graph models are the best formalism for visual representation of information of complex and intricate nature. Visualization of a conceptual structure is a key component of support tools for complex applications in science and engineering. Moreover, the information that is interesting us in computer science and computer education is of structural and relational rather than quantitative nature.

In the paper, our projects on the development of methods and tools to support graphs in computer science and computer education are presented.

The work is partially supported by the Russian Foundation for Basic Research under grant RFBR 00-07-90296 and the Ministry of Education of Russia.

2. Methods and systems for visual processing

In some application areas, organization of information is too complex to be modeled by a classical graph. To represent a hierarchical kind of diagramming objects, more powerful graph formalisms have been introduced, e.g. higraphs, compound digraphs and clustered graphs.

One of the recent nonclassical graph formalisms is hierarchical graphs and graph models. They can be used in many areas where strong structuring and visualization of information is needed [1]. A hierarchical graph $H=(G,T)$ consists of an underlying graph G and an inclusion tree T. The underlying graph G can be any undirected graph, digraph or hypergraph. The inclusion tree T represents a recursive partitioning of G into fragments. A hierarchical graph model is defined as a set of labelled hierarchical graphs together with an equivalence relation between them.

In the paper, the HIGRES and VEGRAS systems that support visualizing, editing and processing of hierarchical graphs and graph models are presented. The systems are implemented in C++ and work under Microsoft Windows 95/98/NT.

HIGRES is a visualization tool and an editor for attributed hierarchical graphs and a platform for execution and animation of graph algorithms. The system is available at <http: // pco.iis.nsk.su/higres>.

Every fragment in HIGRES is represented by a rectangle and can be closed or open. When a fragment is open, its content is visible; when it is closed, it is drawn as an empty rectangle with labels inside. A separate window can be opened for each fragment. Other visualization features are the following: various shapes and styles for vertices, both polyline and smooth curved edges, various styles for edge lines and arrows, a flexible technique to visualize object labels, color selection for all graph components, possibility to scale graph image to arbitrary size, edge text movable along the edge line, external vertex text movable around the vertex, font selection for label text, two graphical output formats, a number of options that control graph visualization. HIGRES has a user interface with almost unlimited number of undo levels, optimized screen update, automatic elimination of object overlapping, automatic vertex size adjusting, grid with several parameters, a number of options that configure user interface, online help available for each menu, dialog box and editor mode. HIGRES supports construction and execution of a wide range of algorithms for visual processing of hierarchical graph models. In particular, it provides a special C++ API that can be used to create new graph algorithm as well as a run-time, repeated and backward animation of graph algorithms.

VEGRAS is a universal and simple-to-use editor of attributed graphs, including hierarchical, oriented to support of construction of qualitative graph illustrations. It supports exchange of the graph illustrations with other Windows applications, including the HIGRES system.

3. A guide on graph algorithms

As a rule a specific problem can be solved on the base of known efficient graph processing algorithms and methods, but to find (and understand) the needed algorithms among a great number of scientific papers can be an unsolvable problem for an end-user.

We believe that a general guide on graph algorithms would be useful for programmers and decided to prepare a series of books intended to be not only a reference manual of graph algorithms but also an introduction to the part of the graph theory and its applications to computer science and programming. We thought that it is reasonable to group graph algorithms into certain classes which deal with the same type of graphs.

In contrast to Donald Knuth who used the assembly language of the so-called MIX computer in his fundamental books "The art of computer programming", we decided to use in our book series a high-level and language-independent representation of graph algorithms.

In our view, high-level and language-independent approach to the algorithm presentation is preferential, as it allows us to describe algorithms in a form that admits direct analysis of their correctness and complexity, as well as a simple translation of algorithms to high-level programming languages without disturbing correctness and complexity. Besides, this approach allows the readers to understand an algorithm at the informative level, to evaluate its applicability to a specific problem, and to make all its modifications needed for correct application of the algorithm.

At present, the series consists of three books: "Graph theory: algorithms for processing trees" (1994), "Graph theory: algorithms for processing acyclic graphs" (1998) and "Reducible graphs and graph models in programming" (1999).

The first book contains a high-level and language-independent description of the methods and algorithms on trees, the most important type of graphs in computer science. It has been translated into English [2].

The book consists of three parts. In the first part, we present the main notions, properties of trees, and some basic algorithms on trees, such as depth-first and breadth-first traversals of trees, the algorithms of coding and generation of trees, etc. The second part deals with applications of trees to problems connected with structuring of programs, unification problem, term rewriting systems, syntax analysis, etc. The third part is devoted to the problems of data storage and retrieval.

The second book is devoted to the directed acyclic graphs (or DAGs) which simulate posets and, like trees, form an important class of graphs that is widely used in computer science and programming. In this book, some basic techniques and algorithms connected to different applications of DAGs to computer science are considered. Then, the elements of the theory of posets, lattices and semilattices are given. Finally, algorithms for the semantic analysis and the object code generation are presented.

The third book consists of two parts. The first part is dedicated to the class of algorithms for reducible (or regularizaible) graphs that expand DAGs and are the most common graph models of the structured programs. This class of graphs plays a very important role in software systems, e.g. many compiler optimizations are simpler, more efficient or applicable when the control flow graph is reducible. In the second part of the book, some graph models widely used in computer science (such as program schemata, Petri nets, intermediate program representations, etc) are considered.

4. Dictionary on graph theory in computer science and programming

The problem of terminology is one of the main problems in application of graph methods to programming and computer science.

The dictionary [3] recently published is the first attempt to solve this problem. The most common terms on graph theory and its applications to computer science and programming are collected in it. The preliminary version of the dictionary was published in 1995-96 in the Novosibirsk State University. The Web-dictionary based on this version is named GRAPP and is available at <http:// pco.iis.nsk.su/grapp>.

5. Conclusion

The author is grateful to all colleagues from IIS who took part in the projects considered in the paper, first of all, to Prof. Vladimir Evstigneev, and also to Ivan Lisitsyn, Tatyana Merdischeva, Ekaterina Merdischeva and Vasilij Kazantzev.

References

[1] V.N. Kasyanov, I.A. Lisitsyn. "Hierarchical graph models and visual processing", *Proceedings of Conference on Software: Theory and Practice. 16th IFIP World Computer Congress 2000*, Beijing, 2000, pp. 179-182.
[2] V.N. Kasyanov, V.A. Evstigneev. *Graph theory for programmers. Algorithms for processing trees*, Kluwer Academic Publishers, 2000, 432 p.
[3] V.A. Evstigneev, V.N. Kasyanov. *Explanatory Dictionary on Graph Theory in Computer Science and Programming*, Nauka Publ., Novosibirsk, 1999, 288 p. (In Russian).

A Template-Based Concept Mapping Tool for Computer-Aided Learning

Ana Arruarte, Jon A. Elorriaga and Urko Rueda
University of the Basque Country (UPV/EHU)
649 P.K., E-20080 Donostia, Basque Country
E-mail: jiparlaa@si.ehu.es, jipelarj@si.ehu.es, 0100urko@euskalnet.net

Abstract

In this paper, the authors study the use of concept maps as support resources useful for different tasks in the computer-aided learning area. These applications include an auxiliary cognitive tool for the student, the design of new knowledge presentation and evaluation activities as well as the gathering of the domain knowledge. In addition, a generic concept mapping tool designed for these aims is presented.

1. Introduction

"Since Novak [7] placed concept mapping on the educational agenda, many studies have shown that learners benefit from the use of both concept mapping as a method and concept map as a teaching aid" [6]

Although originally *concept mapping* was defined as a method to present knowledge and information, it has become an increasingly popular advanced teaching and learning tool. Concept mapping is the process of constructing concept maps. A *concept map* (CM) is a graphical way of representing and organizing knowledge (Figure 1). It is comprised of nodes and links, arranged in some order to reflect the information domain being represented. Nodes represent concepts, and links represent the relationship between concepts; both concepts and links are labelled and may be categorized.

Concept mapping can be used either as a planning and teaching tool for teachers or as a learning tool for students. Several purposes are: to design or produce curriculum, to organize knowledge, to generate or communicate ideas, to identify relationships among concepts, to aid learning by explicitly integrating new and old knowledge, to observe changes in students' understanding of concepts over the time, to assess or diagnose students' misunderstanding.

Over the last years, a number of products have emerged to support computed-based concept mapping. Computer-based concept mapping software enables much easier production and modification of concept maps [5].

Inspiration (http://www.inspiration.com) is one of the most popular computer programs for creating concepts maps. Inspiration offers an intuitive interface and integrates dynamic diagramming and outlining environments to help in the knowledge organization. *Axon* is another program for building concept maps (http://web.singnet.com.sg/~axon2000/index.htm). It includes a powerful graphic user interface and offers the possibility of relating actions to the graphical elements *(zoom possibilities, to run a multimedia file, to open another conceptual map, etc)*. In (http://www.ozemail.com.au/~caveman/Creative/Softwar e/swindex.htm) a library of software resources for creativity and idea generation is collected.

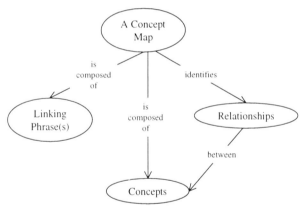

Figure 1. Example of a concept map (from http://www.inspiration.com)

2. Research context

The goal we pursue in our research is to develop and integrate computer educational tools for automatic instruction. Automatic instruction applies computer technology to education; it establishes a cooperative process between two agents, the automatic tutor and the learner, with the related goals of teaching and learning respectively. In order to obtain cooperation between the automatic tutor and the learner, different components must be developed:

- Knowledge representation component, or knowledge about the content and learning techniques intended to be transferred to the learner.
- Student module, or information about the student's skills and knowledge acquired during the teaching-learning process.
- Pedagogical component, which selects information and decides how to present the information to the student.
- Communication interface, which takes charge of supervising tutor-student interactions.

Trying to facilitate the building of such automatic tutors we have already developed the IRIS shell [2]. IRIS facilitates teachers' attempts to build Intelligent Tutoring Systems (ITS) for those domains in which they are experts. It has the following functionalities:

- to assist with data acquisition related to semantics of the teaching domain, the nature of the desired tutoring system, and the final users of the tutoring system;
- to determine a tutor architecture derived from the specified requirements;
- to provide the final tutoring system by selecting the necessary set of rules and objects in terms of the previously specified properties of the tutor, domain and final users;
- to incorporate all the necessary pedagogic resources (presentation forms -texts, graphics, sounds, ...).

The information required for building the domain knowledge is organized in four groups: meta-information for characterizing the domain in a general way, basic learning units, instructional objectives and requisite relationships. For modelling the learner, IRIS requires a curriculum, a learner profile, and learner goals. Finally, the necessary pedagogical information includes the specification of the instructional method that the target tutor will use, the motivation resources and the support tools. Support tools [1] can be thought of as cognitive tools where the student can realise the desired learning actions during the whole teaching-learning process. Some of these tools include: word processing programs, drawing tools, etc.

IRIS generates tutors for procedural and declarative domains but it lacks the necessary components to carry out the diagnosis of procedural skills. On the other hand, DETECTive [4] is a generic and customisable diagnostic system valid for evaluating conceptual and procedural skills, as well as guiding and assessing the users. To generate a diagnostic system for a particular domain the instructor must define the functional domain knowledge, in particular its functional objects, procedures and exercises with their most significant solving models. DETECTive can be used both as an autonomous tool or integrated in a system. Currently, we are developing the first prototype of IRIS-D. IRIS-D results from enhancing the IRIS system with DETECTive in order to obtain an authoring tool capable of building teaching-learning systems that can diagnose procedural skills.

3. Uses of Concept mapping tools in computer-based learning

Despite of *Concept Mapping Tools* (CMTs) have proved as useful resources in learning process as well as in knowledge representation, their possibilities have not been exploited in learning/teaching computer-based environments. In this section we will point out some tasks in which CMs would help both in the learning process and in the preparation of the teaching material.

1. First of all, a CMT is a useful <u>complementary tool to support the learning process</u>. Therefore, any kind of learning or teaching environment can incorporate a CMT providing the student with a tool for drawing the CM of the learnt subjects. If the learner is involved in this task he will fix better the subject's concepts relating them properly and attaching a graphic view of the domain. On the other hand, the students can use these CMs as a mean to review learnt concepts. In addition, the learning system can involve various students in building CMs in a collaborative way. Finally, the CMs of the learners can be used for evaluating their knowledge. Although the automatic evaluation of CMs is a difficult issue [5], it is possible to infer some conclusions and, of course, the teacher can revise them manually.

2. Computer-Based learning systems can include CMT in order to widen the set of <u>learning activities</u>. An automatic tutor can show the CM of the current subject in order to summarise it. In addition, if the system inserts the learnt concept in a wider CM, the learner can put it in the appropriate context and create new links between previously acquired knowledge and the new concept. Another important point is the use of CMs in exercises. Most of the ITSs lack a wide variety of evaluation activities. Due to the easiness of automatic treatment, tests and fill-gaps are the most used resources to evaluate the student's knowledge. CMT can be used to design different evaluation activities. As mentioned above, it is difficult to evaluate a freely generated CM. Thus, the instructional designer must fix some of the features of the CM and the expected solution(s) in order to automate the evaluation. Such exercises can be especially useful resources to evaluate structural and taxonomical knowledge of the learner. Next some exercises are listed:
   - Draw the structure or composition of a complex concept, for example the components of a computer or the parts of the human hearth.

- Draw taxonomies, for example the taxonomy of the animal kingdom, or the means of transport.
- Sorting exercises, for example to sort the steps to perform a procedure.
- Linking exercise, for example link devices with their functions.
- ...

The instructional designer must choose the nodes and relations to use in the CM together with the solution or solutions. She can determine optimal and suboptimal solutions and also typical errors. The tutor would use this information to evaluate the student's answer. If the student's solution doesn't fit any recorded solution, the system can perform some evaluation taking into account: number of nodes and relations, missing or extra nodes and relations, the importance of the errors, etc [5]. In addition, the instructional designer can conceive exercises to involve the learner in the analysis of a wrong CM or in the completion of a CM.

3. The CMs can be used within an authoring tool in order <u>to represent the domain knowledge</u> in a graphical way. A graphical representation of the domain is a more natural representation for the teacher because she can check the completeness of the subject matter and also mark the parts (nodes) of the domain that have to be developed. Our experience with IRIS, teach us that a classical form-based interface is not very appropriate for teachers to gather the domain knowledge.

4. Finally, the CM of the subject matter can be the <u>backbone of a discovery-learning environment</u>. The system shows the CM of the subject matter and the student navigates through it by clicking in the nodes and links.

4. Design of a generic Concept Mapping Tool

Taking into account the various uses of CMTs in the computer-based learning area and, due to the work developed by our team in ITS shells, we think that we can profit from this kind of tools. The shell can improve its way of gathering the domain knowledge including a graphical view of it. The ITSs built with the IRIS-D shell can incorporate the CMT as an auxiliary cognitive tool (*support tools*) and include new learning activities (*presentation and evaluation activities*) based on CMs. Within the IRIS shell, the CMT is going to be used in different ways, so we decided the construction of a generic concept mapping tool that should be specialised for each task. The specialization of the tool is implemented by the **template** concept, the tool have a different template for each use. A template specifies the kind of nodes, relations and labels that are going to be used in a category of CMs.

The generic CMT has been designed using the object oriented paradigm. Next some of the most important objects are shown:

```
Concept_Map (is-a CMT_component)
  Name: <string>
  Subject: <string>
  Purpose: <string>
  Template: <CM_template>
  File: <file_name>
Template (is-a CMT_component)
  Name: <string>
  Purpose: <string
  Node_types <list of Node_type>
  Link_types: <list of link_type>
  Label-types: <list of label_type>
Node_type (is-a Template_component)
  Purpose: <string>
  Label_type: <Label_type>
  Colour: <Colour>
  Form: <Form>
  Size: <integer>
Node (is-a Node_type, CM_component)
  Label: <string>
  Position: <Node_position>
  Links: <list of Links>
  Note: <file_name>
```

Table 1. Some of the generic CMT's objects

Our CMT has two levels of working. In the first one, the instructional designer works with the template. The generic CMT provides facilities to define new templates and reuse or modify them from the tool's library of templates. In the second level, the user can draw the CMs based on the selected template.

The interface of the tool is intuitive and friendly. It has been designed with the aim of sharing the same treatment in both levels. Thus, in the first level the user chooses the template from the library and the tool shows it as a CM allowing the user to make modifications on it. The interface is composed of a working area in which the concept map is shown and modified. The operations are included in the menu bar, in tool bars and in contextual menus. Thus, the user can choose his preferred way of working.

Templates for the tasks mentioned in the previous section have been defined. For example in figure 2 the CMT is specialised for the domain knowledge gathering in an ITS Shell. Concretely, the tool shows the concept map of the mathematical derivative domain.

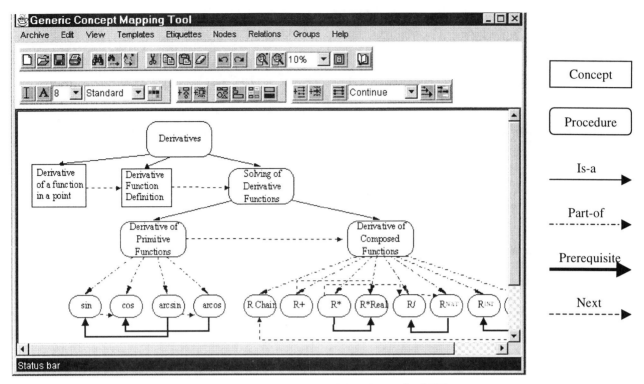

Figure 2. Interface of the Generic CMT

5.- Conclusions

In this paper the authors have explored the possibilities of concept mapping in the computer-based learning area. The aim has been to widen the learning resources within a shell for building Intelligent Tutoring Systems. Concept maps are useful for visual learners who can memorize information contained in a picture [3], therefore a tutor should include it as an auxiliary tool. On the other hand, concept maps can be the basis to design other kind of exercises different from typical tests and filling-gaps. In addition, the teacher can benefit from concept mapping when transferring the subject matter to the shell.

We are currently developing a Java prototype of the template-based tool that supports the above mentioned applications. In the future we plan to integrate this tool in the IRIS-D shell.

6. References

[1] A. Arruarte, I. Fernández and J. Greer, "The CLAI Model. A Cognitive Theory to Guide ITS Development", Jl. of Artificial Intelligence in Education, AACE, 1996, vol. 7(3/4), pp. 277-313.

[2] A. Arruarte, I. Fernández, B. Ferrero and J. Greer, "The IRIS Shell: How to Build ITSs from Pedagogical and Design Requisites", International Jl. of Artificial Intelligence in Education, 1997, vol.8(3/4), pp. 341-381.

[3] A. Cicognani, "Concept Mapping as a Collaborative Tool for Enhanced Online Learning", Educational Technology and Society, 2000, vol. 3(3), pp. 150-158.

[4] B. Ferrero, I. Fernández-Castro and M. Urretavizcaya, "Diagnostic et évaluation dans les systèmes de «training» industriel. Diagnosis and assessment in industrial training systems", Simulation et formation professionnelle dans l'industrie, 1999, vol. 6(1) pp. 189-217.

[5] D. Jonassen, T. Reeves, N. Hong, D. Harvey, K. Peters "Concept Mapping as Cognitive Learning and Assessment Tools", Journal of Interactive Learning Research, 1997, vol. 8(3/4), pp. 289-308.

[6] E. Moen and K. Boersma, "The Significance of Concept Mapping for Education and Curriculum Development", Journal of Interactive Learning Research, 1997, vol. 8(3/4), pp. 487-502.

[7]. J.D. Novak, *A theory of education*, Ithaca, NY: Cornell University, 1977.

Acknowledgements. This work is partly supported by the Gipuzkoa Council (8/DG141.226-12934/2000) and the University of the Basque Country (1/UPV00141.263-TA-8095/2000).

Educational Web Portal based on personalized and collaborative services

Christian Martel, Laurence Vignollet
Equipe Systèmes Communicants
Université de Savoie - France
{Martel, Vignollet }@univ-savoie.fr

Abstract

The Educational Web Portal presented here has been developed in the context of an ambitious project called "Cartable Electronique®" project ("electronic schoolbag"). The objective is to design and develop an open source, dynamic and adaptive Web Portal with flexible collaborative work features. In this paper, we describe the context of this project; we detail the objectives and then we present the platform. Finally, we relate the first results of current experimentation.

1. Context

The Educational portal presented here is developed in the context of an ambitious project called "Cartable Electronique®" project ("electronic schoolbag").

Today, we are witnessing a mutation of education practices: students have quite different learning curriculums, with a mixture of distance and traditional teachings. We have to offer a unique environment able to take into account the different needs of each individual, their various backgrounds and the variety of courses they follow. Moreover, during all their scholarship, the student belongs to different communities: institutional, cultural, sports, organizational, etc. The new education tools have to address this, allowing the creation of multiple communities, giving users the power of introducing and assigning roles to participants.

The Educational Web Portal project is supported by the French National Education Ministry and by the Technology Executive, the regional council of Rhône-Alpes, the Savoie General Council and the University of Savoie.

2. Objectives

Our main objective is the design and development of an open source, dynamic and adaptive Web platform, using Web standards.
This includes the development of integrated services: unique authentication, notice board, nomad access to communication tools (Webmail, Webchat,Webforum), community creation and management, private or group organizational tools (time table, agenda, ...), course publication facilities, ...

This requires also the access to adaptive content and services from anywhere, anytime, to give users activity regulation tools and the integration of the platform in the information system of the educational institution.

3. The developed platform

The regulation concerns the social aspects of collaborative work. It can enable compromise between the interests of the group and those of the individuals, between the dependencies that stem from relationships among individuals and their autonomy.

The main contribution of our work is to take this social aspect into account in CSCW (Computer Supported Cooperative Work), and to propose a participation model [1].

3.1. Regulation

We propose two levels of regulation: the regulation at the Portal level, and the regulation into the groups.

At the Portal level, that means that a group of people could parameterize the Portal, for instance in determining for each user role the available services associated.

At the group level, an authorized user can choose the "group policy". He can, for instance, chose between *open* or *closed* group (an open group allow any user to ask for his integration), *reachable* or *non reachable* (a reachable group can receive anything from anyone in its publication space), etc.

3.3. Implementation: Zope, a very well-adapted development framework

Zope (Zope Object Publishing Environment) [2] is an open source framework for building web applications, initially developed by Digital Creation. Instead of publishing HTML files, Zope publishes objects which are able to publish themselves. Zope is multi-platform and extendable. A very active community works to improve the available components, called products.

The Zope server plays the role of a Web server, everything is an object, from document to methods. It contains an object oriented database, it runs as an ORB (Object Request Broker), a plain text search engine is proposed, and the administration is done via the Web.

One of distinguished features of this framework is the availability to define local roles associated to each object. We mainly use these local roles to organize the CSCW: it is the basis of the regulation in the shared workspaces.

Local roles give an API to the regulation of the collaborative work. We want to extend them, to allow users to define self-position to their own activity and to others group participants. We want also to propose a way to define hierarchical places in the group.

The Zope architecture is presented in [3]. We have worked on several products that we have adapted. The Portal Tool Kit (PTK) is the basis of our development. The cornerstone of security is the LDAP connection to a LDAP server, via a LDAP connector.

No external files are stored in the ZODB, but in the local servers. This reinforces the robustness of our Portal.

4. Experimentation

Today, this platform is being experimented in Université de Savoie with more than 500 students and teachers. Not all the services are proposed in the prototype under test, however, the main ones figure.

We have first of all created a "use observatory", made up of psychologists, specialists of the study of information and communication and computer scientists. In each entity where the platform is in use we have at least one pedagogical manager and one technical manager.

From the method point of view we have two strategies with regards to the different objectives. We have developed an integrated bug report tool, which captures the errors automatically and stores them with meaningful information. For the use testing part, we are defining a method based on a grid and on interviews [4].

The experiments are too recent to allow us to present final results. However, the few sets of services proposed in the experimental platform are today really and intensively used (100 000 LDAP hits for portal connection only per month). The volume of document exchange and storage is also very high. The electronic "notice board" has replaced the mural one. Many thematic groups have been created by end-users.

5. Conclusion and future work

The current success of the use of the Educational Web Portal developed reinforces the choices we have made, particularly the flexibility offered in the personalization of the environment and in letting users create communities and define the organization of the work in these communities. On the technical side, the implementation choices, using the Zope framework, lead to a Web open source platform, easy to install, maintain and extend.

The next step consists in a large deployment of this solution. This deployment requires a model of generic portal components and distributed/replicated objects.

This last point is a challenge. At this time, we are putting our efforts into ZEO (Zope Enterprise Objects). Zeo extends the Zope Object DataBase by taking into account distributed DataBases. It also manages the information replication, using intelligent caches.

Finally, even if the student can access their educational environment from anywhere, anytime, we have not yet proposed functionalities for off-line work.

6. References

[1] C. Ferraris, C. Martel, P. Brunier, "Drawing Together in the Classroom: an Application of the "cartable électronique"® Project, *ED-MEDIA 2001*, Tampere, Finland, June 25-30, 2001, to be published.

[2] Zope Web site, http://www.zope.org

[3] F. Quin "*A complete introduction to Zope in French*", http://www.zope.org/Members/ghuo/introduction_zope_fr

[4] G. Chabert, "*Les usages du cartable électronique : pour une évaluation des technologies dans l'éducation*", 2001 Bogues, Montreal, September 2001

7. Acknowledgments

We would particularly like to express our thanks for their active and efficient development collaboration to Eric Brun, from "TIC mission" of the Savoie General Council, and Steve Giraud, from "TICE mission" of the University of Savoie.

Using WebCT to Support Team Teaching

Anne Fuller, Gene Awyzio, Penney McFarlane
University of Wollongong
annef@uow.edu.au, gene@uow.edu.au, penney@uow.edu.au

Abstract

Advanced learning technologies facilitate team teaching at the college/university level. Team teaching has been shown to benefit both students and faculty. This paper explores the benefits of using a course management package in an interdisciplinary team environment. Among those benefits are: better communication, student monitoring and support, flexibility in subject delivery, and fostering a collaborative environment. The success in the early developmental stage of integrating one such learning technology (WebCT) into our course has seen the package move from a simple tool to an essential member of a teaching team.

1. Introduction

Historically, it is recognised that team teaching is beneficial for all involved parties. Although there is a plethora of research describing team teaching in elementary schools, there has been little research done in this area at the higher education level since the 1960s. For example, a 1966 study at Boston University found "team teaching strengthened and individualized relationships between teachers and students and enabled students to progress more rapidly " [1]. Despite this, team teaching continues to be regarded favourably at the tertiary level, and Davis [2] lists almost 100 examples of courses being team taught in North American colleges and universities.

Twigg, Wagner, Baker and Gloster quoted in [3] argue that higher education must move to the electronic arena rather than the face to face methodology of yore. Reasons for this movement involve multiple campuses, increased student numbers, reduced contact hours for the academic and various offshore delivery schemes. Massy and Zemsky quoted in [3] claims "that higher education cannot become more productive or hold costs down unless colleges and universities embrace technology tools for teaching and learning."

Students need to develop technological skills in order to effectively use electronic delivery systems. Bergen [4] states the development of such skills within an isolated, individualized teaching model is ineffective. Yet team teaching remains the exception in higher education[5].

Web Course Tools (WebCT) is a class management package developed at the University of British Columbia that facilitates the creation of sophisticated World Wide Web-based educational environments by non-technical users. It can be used to create entire on-line courses, or to simply publish materials that supplement existing courses. It provides tools to enhance interaction between students and faculty, and includes security, administration, facilities for backing up, etc. In addition, a broad range of course material and well-designed content from major textbook publishers is available as ready-made WebCT courses. Information regarding WebCT may be found at http://www.webct.com.

This paper discusses our experiences using WebCT to manage an introductory IT course of three hundred and eighty-five students which was team taught in Spring Session 2000. Initially we planned to use WebCT to simply streamline administration, and provide students with an online community that that could demonstrate the benefits of IT in eduction. Ultimately we conclude that WebCT and any course management package providing similar features can facilitate team teaching at the college/university level.

1. Background

Introduction to Information Technology B (CSCI102) is a mandatory first year subject for computer science and information technology students and is also available as an elective for various other disciplines. It examines a range of information and communications technologies and associated issues as well as the application and convergence of these technologies together with introducing students to a group culture.

Due to the varied nature of the content, the number of students enrolled (385) and the necessity for delivery to an external campus, the decision was made to team teach. Three staff members volunteered to be involved in this project, a software engineer, a telecommunications engineer and an IT generalist with an English background.

3. Team Teaching

Our team chose the interactive model of team teaching. Garner [6] defines this model as one in which team members understand the contributions of each member, share knowledge and experience to establish and pursue common goals. Interactive teaming is also known as trans-disciplinary [2], [6] which was suited to the needs of the subject and reflected the multi disciplinary nature of the team.

Larson and La Fasto, quoted in [2], [6] identified eight characteristics of an effective team: a clear elevating goal, results-driven structure, competent members, unified commitment, collaborative climate, standards of excellence, external support and recognition and principled leadership. Davis [2] and Garner [6] go on to examine the implications of these characteristics in the context of a teaching team.

A team exhibiting these characteristics will be better able to overcome the limitations and challenges associated with team teaching. La Fauci and Richter [1] discuss 9 factors that may combine to reduce the effectiveness of the teaching team. These are: lack of understanding and commitment, rigidity of structure, demands of a multiplicity of faculty roles, problems of scheduling. Difficulty in maintaining a balance between unity and diversity, resistance to change, necessity for pre-planning, lack of available research, difficulty of acquiring and retaining teaching faculty.

The team decided that WebCT had the potential to alleviate several administrative tasks as well as addressing the difficulty of the external campus delivery and support required. Subsequently, we also found that WebCT was able to contribute substantially to our ultimate success as a team.

4. Structure of a UOW WebCT site

All WebCT subject sites at the University of Wollongong include the following elements.

Lecture Support: The area where the lecture content is placed.

Links: This is a categorised source of useful Internet and library database links as determined by the academic. A further evolution in this area is the direct link to online databases, which is facilitated by the university's library.

Forum: Here students can post questions and comments or refer others to useful web sites or other sources of information. Communication with and between academics can take place here.

Calendar: Students can access this to double-check assessment and exam dates.

Surveys: contains a database of survey questions can be compiled by the academic and opened to the students for subject feedback. A statistical return of survey results is generated by the package in use, providing useful feedback for future course construction.

5. Team Teaching with WebCT

WebCT supports individual differences from the student and the staff perspective. On completing the course, we evaluated our success as a team by reflecting on our use of WebCT to both support Larson and Le Fasto's characteristics [2], [6], and to help overcome La Fauci and Richter's challenges[1].

As Davis [2] points out, most interdisciplinary team courses are an attempt at innovation and thus are naturally elevating. Course aims and objectives were determined by the subject, thus making the goals clear to the team. The course outline was co-written so each member was clear as to her function within the course structure.

Structure was determined by the course objectives. Each team member selected those objectives most suited to their particular skills. Once this was established, the ease of WebCT use allowed for individual loading of associated course materials. In addition, we found WebCT provided the ideal support for one very important aspect of a results-driven structure i.e. the desirability for each member to know "how the team is functioning at any given time" [2].

Competent members fortunately, were found to be an easy area for this subject. Due to the voluntary nature of their involvement with CSCI102, each team member felt comfortable collaborating, yet between them possessed the diverse technical skills and knowledge required to present the subject syllabus.

The opportunity to explore the capabilities of WebCT as a course delivery package was considered timely by each team member. All members of the team are involved in the universities offshore teaching program and recognized that more support of these programs was required. Unfortunately it is not possible for faculty to be offshore fulltime. It was felt that CSCI102 provided an ideal environment in which to examine the use of a web based course tool, such as WebCT, to support the offshore program. The large number of students and the geographical disparity of the campuses involved provided a climate that was indicative of the offshore campus whilst allowing personal contact with remote tutors on a regular basis. Thus we were unified in our commitment to the subject's success.

Having determined their relative responsibility for various sections of the subject, team communication was facilitated by emails and weekly formal and informal meetings. The private forum area of WebCT contributed to the collaborative climate, allowing personal communication between the group members during the course's duration, and was used to float ideas prior to their being considered more formally in meetings.

The team agreed that their success in teaching would be measured in terms of student grades and student IT literacy. Using WebCT as their principal contact for course information made it possible to determine an individual student's facility with the medium, as the package allows tracking of individual pathways through the system. Summaries of student activity were communicated and discussed via the private forum, ensuring each member remained clear as to what standards we were expecting. The overall average grade for this subject was slightly higher than in previous years, also a reflection of achieving our standard.

External support and recognition was realized through the support of the Head of School who encouraged and monitored the success of this implementation. Other faculty members would often ask for updates from the team members as the term progressed. As a result of CSCI102, three other staff members are mounting their subject's onto the WebCT platform. The Head of School determined at the end of the academic year to employ a part time WebCT support person to assist in the construction of sites for interested staff members following a schedule of needs as determined.

In our faculty it is customary that a single course coordinator be listed as the main contact for a subject. We departed from that norm by listing all three of us as co-coordinators. One of the team members took on the role of team leader with the responsibility of ensuring that the WebCT site was operational from the start to finish of session. The private and public fora were initiated and moderated by the team leader. Interactive learning and peer support is facilitated by this medium.

Working as an effective team clearly helps overcome La Fauci and Richter's[1] list of team teaching limitations. WebCT's private forum alleviated scheduling problems, provided a medium for collaborative brainstorming, and raised awareness of any changes to course structure. The clear division of responsibilities relating to member expertise eased any potential role conflict. As discussed previously (Section 5), our decision to participate in a team teaching environment minimized the problems in maintaining a balance between unity and diversity as well as resistance to change. The difficulty of acquiring and retaining teaching faculty is irrelevant to our situation. It should be noted that there remains a lack of available research at the tertiary level.

6. Benefits for the Students

Using a package such as WebCT provides students with a central point to collect subject resources and complete and submit assessment tasks. This establishment of an online community contributes to the collegiate atmosphere of the subject, despite its size or geographic disparity. Integrated tools for tracking student activity, allows the team to monitor student visits to pages and determine what students find useful and what they avoid. This builds a database from which to construct the following sessions' course.

Assessment tasks can be posted onto the site and be marked automatically, marked semi-automatically, or marked with full instructor intervention. In our case, we posted detailed assessment information and included links to on-line tests related to the textbook at the publisher's site. Results of the on-line tests were emailed to our tutors, while the other tasks were marked manually. Assessment results were subsequently posted to WebCT.

In addition, using one system, WebCT, for all their work, allows students to focus on learning the material rather than the system, or systems, at an earlier stage. This is particularly important for students where this is their first exposure to IT and for international students where English is not their first language.

7. Benefits for the Team

The team found that the WebCT site made contact with the students easier. Subject updates, alerts and notices could be sent easily. The forum alleviated the number of emails usually directed to the instructors, and often questions asked were answered accurately by other students.

Online marking facilitated the collation and publishing of results in a timely manner. There is further scope to ease the workload for both the academic and tutor with fully automated marking and posting of results which can then be transferred from WebCT to other administrative packages at the end of sessions.

By posting the subject via WebCT, the tutors were able, through their individual logins, to see the direction and scope of the subject, as well as future tutorials. This enabled them to lead their groups more effectively. Special administrative notices could be directed to them easily via the private forum facility.

The ability to post work from a variety of venues was found to be most beneficial in the running of this subject. The team found that they did not have to 'live in each other's pockets' in order to keep abreast of course developments. The individual nature of posting content, meant that no one academic was waiting for another to complete a body of work. Team members could work from home effectively and post their work from there, and one team member stayed in contact although teaching offshore for a period of time.

Intellectual property was not a concern as the WebCT site is password protected and only registered UOW students enrolled in the subject itself could gain access. This is proving to be a more pressing need as more people gain access to the Internet.

8. Expanding the Role of WebCT Within the Team

The success of the inaugural use of WebCT has identified a number of other areas where WebCT can assist a teaching team at the tertiary level. The areas identified include the utilization of WebCT as an examination and mark-collating tool, student monitoring, and student support.

This semester a modular approach to student learning has been introduced. Tutorials and examinations as well as subject resources are delivered via WebCT. To provide students with a greater appreciation of the adaptability and effectiveness of IT in education, students are required to complete and submit a weekly tutorial exercise from the publisher's WebCT site, and complete and submit an online exam at the completion of each module.

Previously, an individual's tutorial marks were emailed from the publisher's site to one of many tutors who then recorded the marks for later integration into the overall class result list. Whilst exams were machine marked, results were still entered manually. Under the new system, students encountering difficulty will be more easily identified. Additional tutorials can be placed on the site to provide further support. Using WebCT for online submission and marking of tutorials and exams greatly reduces administrative overheads in a class of this size and facilitates the integration of course materials to remote campuses.

WebCT's ability to support an effectively functioning team has been shown in section 5. Its use fosters good communication between team members and facilitates the sharing of ideas makes it indispensable for a teaching team in the IT area.

Although we have not used other course management packages, comparative studies suggest similar facilities to WebCT can be found. For example, Blackboard and eCollege offer similar features [7], thus these packages would also benefit team teaching. Further information regarding Blackboard and eCollege can be found at http://www.blackboard.com at http://www.ecollege.com respectively.

9. Conclusion

Previously team teaching has often been an exercise in political correctness, time management, paper chase, phone tag and sheer frustration. Our experience using a course management package shows that these problems can be alleviated if not eliminated altogether.

10. References

[1] LaFauci, Horatio M., and Peyton E. Richter, *Team Teaching at the College Level*, Pergammon Press, New York, 1970

[2] Davis, James R., *Interdisciplinary Courses and Team Teaching – New Arrangements for Learning*, Oryx Press, Arizona, 1995

[3] Stahlke, Herbert F. W and Nyce, James M. "Reengineering Higher Education: Reinventing Teaching and Learning", *CAUSE/EFFECT* Volume 19, Number 4, Boulder, 1996, pp.44-51,

[4] Bergen, Doris, "Developing the art and science of team teaching", *Childhood Education*, Volume 70, Number 4, , 1994

[5] Bakken, Linda, Clark, Frances L. and Thompson, Johnnie, "Collaborative Teaching: Many Joys, Some Surprises and a Few Worms", *College Teaching*, Gale Group, 1998, pp 154

[6] Garner, Howard G., *Teamwork Models and Experience in Education*, Allyn and Bacon, Sydney, 1995

[7] Dyrli, Odvard Egil, "The 'Webification' of College Courses: Choosing a Management System", *Matrix: the magazine for leaders in higher education*, Volume 1, Number 2, 2000

WhiteboardVCR – a Presentation Tool using Text-to-Speech Agents

Ng S. T. Chong[‡], Panrit Tosukhowong[‡], and Masao Sakauchi[†]

Institute of Advanced Studies/United Nations University[‡]
Institute of Industrial Science/University of Tokyo[†]
{chong|tosukhowong}@ias.unu.edu [‡]
sakauchi@iis.u-tokyo.ac.jp[†]

Abstract

Considerable progress has been attained in media streaming technology recently. It is now possible to combine digital images with synchronized audio and video content to create live or on-demand slide shows for the Web. In more advanced settings, the images can also be annotated for synchronized playback. However, despite advances in data compression and congestion control, media streaming is still sensitive to fluctuation of network traffic. In this paper, we propose a new web presentation tool that uses text-to-speech synthesis as an alternative to audio streaming. On playback the synchronization of slide annotations and transitions with the synthetic narration can be preserved without concerns for the individual differences in the users' environments in terms of CPU power and bandwidth availability. The implemented prototype includes a standalone production client and a web playback client.

1. Introduction

Recent years have witnessed improvements in media streaming technology. Streaming lets users view multimedia as it downloads. Many commercial software solutions now support the W3C multimedia standard SMIL [1] which allows multimedia objects to be synchronized in time and space. This has created a trend for slideshow style of presentations on the Web, where a sequence of slides is synchronized with a separate audio and/or video narration. A popular example of this type of media synchronization in streaming is Realnetworks [7]. Although streaming technology is gaining widespread use, it still faces many technical challenges. The most fundamental problem is its sensitivity to packet losses that can lead to content degradation (e.g., choppy video and audio).

In this paper, we present the architecture and a prototype of a presentation tool that supports both media streaming and text-to-speech (TTS) synthesis as an alternative to audio streaming in synchronized slideshows.

In our current implementation, we used the speech engine of Microsoft Agent [2]. With this tool, users can annotate a slide and associate any part of an annotation with a written narration, which is then converted into speech at playback time. Our approach offers a number of advantages over streaming-based presentations:

- Bandwidth requirements are lower compared to any other form of streaming;
- Synchronization is basically insensitive to network congestion as the only time-dependent data are slides that need to be downloaded conforming to a pre-determined sequence;
- There is no need for a streaming or special server other than a regular web server;
- The speech transcript is readily available and can be used as redundant media that can enhance understanding, particularly for hearing-impaired users and users who are not proficient in the language of the presentation. Similarly, using the tool, a speaker who is not proficient in the language of presentation can communicate his/her ideas easily;
- Recording a presentation amounts to preparing the speech output text, which eliminates the uneasiness of presenting in a room without a live audience and to a remote audience without any visual contact;
- Fine-grained media editing and reusability – the words and phrases of speech text and individual slides can be replaced and reused;
- TTS engines are available in many different languages;
- Storage requirements for speech text are significantly lower than those for recorded audio.

We describe the synchronization framework and preliminary results in the remaining part of the paper.

2. Synchronization Framework

We present a simple framework to deal with the specific requirements of synchronizing slide notes with a speech synthesis agent, where the notes can be interleaved or overlapped with narration. Synchronization requires that elements of the presentation be scheduled in a specified order. Unfortunately, the execution time for completing an annotation step or a speech synthesis step depends greatly on such factors as processor speed and loading conditions of the playback computer. Our framework proposes a solution to this problem.

First, we introduce media objects in our framework. Each media object consists of three main parts: the parent list, the detail of the object, and the endpoint synchronization list. The parent list points to the media objects that need to be executed before presenting the current media object. The detail part gives the characteristic of the current media object. The endpoint synchronization list is the list of media objects that are forced to end upon the completion of the current object.

We make the distinction between four media object types: drawing objects, slide-changing objects, delay objects, and TTS objects. The drawing object describes how an annotation markup (e.g., line, text, and pointer) can be shown on the screen. The slide-changing object provides information for loading background slide images. The delay object defines the wait time before starting the next object. The TTS object is a special text object, whose content will be used to control the agent's behaviors, such as move and speak.

We construct our synchronization model using a single-rooted directed acyclic graph as shown in Figure 1. Nodes in the graph represent media objects and the set of directed arcs defines the order of presentation of the media objects. A node will be executed after all parent nodes finished, and every node in the graph except the root must have at least one parent. The circle-head dashed line denotes the endpoint synchronization relationship.

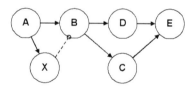

Figure 1. Media synchronization graph

It is easy to verify that our model guarantees a synchronized narration for objects on screen using only parent-child relationship. Consider node D as a drawing object on the slide, and node C a narration for node D. From the graph, the narration starts when the drawing starts, and the next object, node E waits for both C and D to finish before it starts.

However, non-time bound objects, i.e., media objects whose duration depends on system resources can take longer than desired to complete. In that case, if users want to fix the duration of a media object they can insert a delay object with an endpoint synchronization to terminate that non-time-bound object. For example, suppose node B is a slide-loading object, whose loading time depends on the network condition. If we employ node X as a two-minute delay object that imposes an endpoint synchronization on node B, then B cannot delay the overall presentation duration more than two minutes

3. Software Structure and Implementation

3.1 Production Client

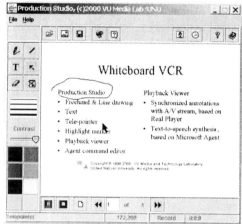

Figure 2. A screenshot of the production client

We implemented our production client as a Java application program (Figure 2). The production client maps all user actions to the described synchronization model. Then, the model information is sent to the content server to make the presentation available for the playback client over the network. Figure 3 shows the structure of the production client.

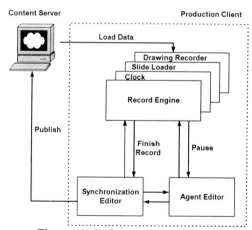

Figure 3. Software structure

The record engine records user slide annotations. It inserts appropriate delay objects in the synchronization graph based on the timing information of media objects.

The slide loader layer lets users change a background slide. Users can pause the recording at any time to add voice narration using the Agent Editor.

The synchronization editor displays the relationship between media objects. It offers visual feedback on the drawing screen when an object is selected. Users can choose to modify existing relationships, or add a completely new one. In the normal production sequence, the synchronization editor is used after users finish the creation of the presentation, but users can access it at any time during the recording process.

3.2 Playback Client

The playback client receives data from the content server. It has two view modes: RealPlayer™[7]-based and TTS based. In the first mode, the playback client uses timing information from RealPlayer™[7] for synchronization. When a slide-changing event occurs in the media, it triggers the playback client to reset its intra-slide clock and starts the display of annotations for the current page.

For the TTS mode, we define states of media objects in the algorithm as follows: *Idle*: object is idle, *Queued*: object starts waiting for its parent to finish, before start its execution, *Played*: object starts playing; *Finished*: object finished playing.

In the algorithm, we perform a breadth first search traversal on the synchronization graph. We start by marking the root node as Queued, and we repeatedly check for the Queued nodes whether they satisfy the condition that all parents are finished. If satisfied node is found, we start its execution, and change its children status to Queued. The process is repeated until the queue is empty.

Algorithm:
Initialization Phase:
- Mark the root node as *Queued*, its parent as *Finished*.
Execute Phase:
- For each node that has status *Queued*.
  If all parents of this node are *Finished* then
    * Start playing this object.
    * Mark all children of this node as *Queued*.
    * Mark this node as *Played*.
- Repeat until *Queued* node does not exist.

When an object has finished playing, we check whether the endpoint synchronization field exists. If it exists, we terminate the specified object.

```
OnFinished(){
    // perform when finished playing media object.
    if endpoint synchronization field is not empty
        for c in (endpoint sync. field)
            end c
}
```

4. Evaluation and Experimental Setting

We studied 24 presentations in our survey, of which 12 were live and 12 were electronic versions of the live presentations. Speakers delivered their talks using slide sets prepared in PowerPoint in the normal way. In making the electronic presentations, we added pointer and highlighting to identify the relevant part of the slide at the beginning of every thread of explanation. 10 people participated as the audience in the survey, of which 5 were non-native English speakers and 5 were native. The non-native presenters and audiences had their most recent TOEFL score between 500 and 600. As the participants might not be able to judge unbiasedly the degree of understandability of a recorded presentation after seeing the live version, we did not let anybody attend a same presentation more than once. Hence, we split the participants into two groups: non-native English speakers and native English speakers. We then asked each group to attend 6 distinct presentations with a 50/50 mix in presentation format (i.e., 3 live and 3 recorded) and 50/50 mix in language proficiency of the speaker (i.e., 3 native English presenters and 3 nonnative English presenters). We also asked presenters to prepare 5 quiz questions for their presentations.

5. Preliminary Results

Table 1 shows the aggregate results of the tests taken for the different combinations of presentation format and language ability. Our experimental results show that in general, native English speakers, due to their innate command of the language, could understand the presentations in all formats better than their non-native counterparts. On the other hand, for live presentations, non-native English presenters might have some difficulties presenting in English, as suggested by the lowest score for both native and non-native English audiences. For recorded presentations, which were narrated using synthesized voice, the performance gap between native and nonnative audiences narrowed. As the speaking speed of the agent was slower than that of human speech, non-native audiences were able to follow the presentation more easily. However, some native audiences found that the slow speaking rate was too slow for them. Furthermore, they felt that the flat intonation of the robotic voice was not conducive to any emotional affinity between the presenter or presentation and the audience. This partly accounted for the small performance gap between live presentations and recorded presentations. The described problem can be improved if we provide options to the presenter to express their emotions and feelings, such as when the presenter is emphasizing a point.

Table 1. Quiz results as an indicator of ease of understanding of the presentations

Presenters	QUIZ RESULTS									
	Audiences									
	Native					Non-native				
Σq, for q = 1..5 denotes q questions were answered right	Σ=1	Σ=2	Σ=3	Σ=4	Σ=5	Σ=1	Σ=2	Σ=3	Σ=4	Σ=5
Native Live	0	0	2	5	8	1	2	5	5	2
Native Recorded	0	4	5	4	2	0	0	2	6	7
Non-native Live	6	4	3	2	0	4	5	6	0	0
Non-native Recorded	0	0	3	7	5	0	1	3	4	7

6. Discussion

One feature of the presentation tool is synchronization of slide annotations with another media. That media can be streaming audio and/or video as well as synthesized audio. There are a few research and commercial systems that support synchronized slide annotations with audio and video. Common to all the systems is the requirement to capture the annotations with timing information. From the standpoint of playback strategy, they differ in terms of whether the entire content is delivered (streamed or downloaded) as a single media file or split into parts (audio and video in one media file and annotated slides in separate HTML documents) before sending.

The former has two major drawbacks. Different data types are typically stored in different tracks of the media file, but the available number of tracks is fixed. Hence, only a limited number of annotation effects can be included in the media file. Another problem is that unless the media file is streamed, users need to download the entire file even if they only want to view a part of the presentation. A recent development using this approach is Clipboard 2000 [4].

The split approach, which is the strategy we adopted, does not have any of the above limitations. The playback is based on separate players for audio/video and slides that are synchronized with each other by accessing the API of the audio/video player. Work in this area is exemplified by WLS [5] and Classroom 2000 [3]. In WLS every slide is a HTML document and the annotations are entered in a Java applet that works like a whiteboard. All media can be replayed in the same sequence as in the original presentation using a trigger from the audio player. In Classroom 2000 every slide and in-slide annotations are stored in a HTML document. Slide markings captured from a hardware whiteboard are converted automatically into static image maps.

Another media that has been used for synchronization with a narration is text. The text can be scrolled (e.g., Realpresenter [7]) and even highlighted in sync with the narration (e.g., Eloquent [6]). However, we are not aware of any existing work that applies TTS technology in the creation of synchronized slide presentations for the web. Moreover, when TTS technology is enhanced with machine translation capabilities, the same presentation can be delivered in other languages.

7. Conclusion

We have presented WhiteboardVCR, a new web presentation tool using TTS technology and it itself is a new application of TTS technology. The implementation had to deal with the scheduling problem of playing back synchronized media objects in the presence of uncertainty in the execution time of producing a speech output from text. The main benefit of our approach is lower bandwidth and storage requirements for delivering slideshow presentations with synchronized audio narration over the Internet compared with streaming audio.

Like any other offline presentation tool, the presenter can prepare a quality lecture once and will be able to reuse it many times and refine it as needed. It should be noted that, in our case, editing speech output text is much easier than editing digital audio. Although our preliminary results generally suggest a gain in speech understandability for non-native speakers, they call for extending speech output control to include gesturing and emotions, which is part of our continuing research and development.

8. References

[1] W3C SYMM Working Group, Synchronized Multimedia Integration Language (SMIL). http://www.w3.org/TR/REC-smil/

[2] Microsoft Agent: http://www.microsoft.com/workshop/imedia/agent/default.asp

[3] Gregory D. Abowd, Jason A. Brotherton, and Janak Bhalodia. Classroom 2000. A System for Capturing and Accessing Multimedia Classroom Experiences. CHI'98 Demonstration Paper, May, 1998.

[4] Severance, Charles. Clipboard 2000. Available at: http://www-personal.umich.edu/~csev/projects/cb2k

[5] Klevans, Rick and Vouk, Mladen. Web Lecture System - WLS Use and Configuration Guide, November 1999. Available at: http://renoir.csc.ncsu.edu/WLS/distribution/latest/help.html

[6] Eloquent, Inc., Presentation software. Available at: http://www.eloquent.com

[7] RealNetworks, Inc., Realpresenter. Available at: http://www.realnetworks.com/products/presenterplus/info.html

Effective Use of WebBoard for Distance Learning

Haomin Wang
Assistant Professor, College of Education
Kennedy Center 113, Dakota State University
Madison, SD 57042 USA
Haomin.Wang@dsu.edu
Phone: +1 605-256-5052
Fax: +1 605-256-7300

Abstract

This paper discusses effective use of WebBoard as an asynchronous communication tool to facilitate class interaction in distance learning. Students' perceptions of the educational functions of WebBoard were polled and analyzed. Instructors were interviewed on their views of WebBoard management. The discussion examines issues including perceived advantages and barriers in using WebBoard for class interaction in distance learning, board organization and management, instructor's direction and guidance, types of questions, and students performance evaluation.

1. Introduction

As the Internet grows in popularity as a medium of communication and information dissemination, more and more educational institutions are offering courses via the Web to learners who cannot attend on-campus classes because of geographic distance or time constraint. Such practice of distance course offering via the Web is generally defined as Web-based instruction [1].

Web-based distance education can be synchronous or asynchronous, or a combination of both. The two modes differ primarily in the presence or absence of feedback immediacy. In synchronous mode, interaction occurs in real time with immediate feedback, whereas in asynchronous mode, feedback is delayed. According to a framework proposed by Dan Coldeway and quoted by Simonson et al. [2], the purest form of distance education occurs at different times (asynchronously), with instructor and learners in different places.

Although synchronous communication (SC) is becoming more accessible to a broader population, asynchronous communication (AC) remains a primary mode of class communication in distance learning. Among the advantages of AC, time flexibility is generally a major attraction to distance learners. In addition, AC has been found to promote analytical and reflective thinking when learners have time to carefully consider and construct responses [3]. Participants also have opportunities to synthesize viewpoints and come up with integrated solutions to issues in question [4].

2. What's WebBoard?

Internet-based asynchronous communication has two primary forms: email and bulletin board system (BBS). A bulletin board system differs from email mainly in scope of audience and message organization. In BBS, messages are posted to a group, whereas email messages can be sent to either an individual or group (through LISTSERV). In addition, messages in BBS are threaded, typically with responses listed in a cascading order with some indentation. Because posted messages are relatively permanent on the board, participants can always return to review and revise their postings, or post additional messages to clarify, elaborate or modify their previous postings.

WebBoard (http://webboard.oreilly.com) is one of the many bulletin board systems currently available on the market. The server software runs on Windows and can function either as a standalone server or as a module of an external web server. WebBoard can support up to 265 boards per server.

Messages are organized under conferences and each board can host as many conferences as needed. A conference can be open or closed. An open conference allows guests to view postings, but only registered users can post and respond. A closed conference is accessible only to users assigned by the board manager.

Users can use any browser to access WebBoard. Users can self-register and set up accounts on WebBoard. A valid email address is required to receive the password for a newly created account. Alternatively, a board manager can set up new user accounts on the board. Once a user has an account, she/he can log into any board on the system.

3. Methodology

The goal of this study was to find out students' and instructors' perceptions of the use of WebBoard in distance learning. The understanding is expected to help us make better use of discussion board as an asynchronous communication tool to facilitate interaction in distance learning.

Two questions were posted on six web boards hosted by the WebBoard server at Dakota State University (http://courses.dsu.edu:8080/~admin). Four of the six boards were for undergraduate courses and two for graduate courses. The questions were posted near the end of the semester when students had had at least about a semester's experience with WebBoard. Students' responses were analyzed for convergence in perceptions and categorized accordingly.

Informant interviews were conducted with six instructors regarding their perceptions of WebBoard. Three of the instructors had used WebBoard for more than one semester.

3.1. Questions posted to students

The questions posted to students are :
- What advantages and barriers do you perceive in using WebBoard for class interaction in distance learning?
- Do you think instructors should participate in students' discussion? If yes, what kind of input would you like to have from instructors?

3.2. Questions asked of instructors

The questions asked of instructors are:
- What advantages and barriers do you perceive in using WebBoard for class interaction in distance learning?
- How do you generally organize conferences on WebBoard? Do you use closed conferences? Why?
- Do you participate in students' discussion? If yes, what kind of input do you usually contribute? If no, why?
- How do you evaluate students' performance on board?

4. Students' perceptions

Out of a total of 124 students enrolled in the courses, 103 responded to the questions. Students' responses are categorized based on convergence of perceptions and the findings are listed in descending order of frequency count in Table 1.

A primary advantage of WebBoard as identified by the majority of the students (87%) is time flexibility (shown in Table 1). Many distance students have diverse schedules with family and job commitment, and time flexibility has been a major attraction that draws them to distance learning. In discussion board, conversation on a particular topic can continue for days, and participants can come in any time to share thoughts.

However, time flexibility brings more than access convenience. It encourages reflective thinking. Without on-the-spot pressure, students have more time to contemplate and rephrase what they are writing. It allows them to link the current learning to their prior experiences through reflection. In addition, they can always come back to review, revise, or even delete their responses. Some students feel they have become a better thinker through the revising process.

Table 1. Students' perceptions

Categories	Numbers of responses	
anytime accessibility	90	87%
reflective thinking	78	75%
collaborative learning	56	54%
focus on thoughts	47	46%
less inhibition	44	43%
coherent conversation	33	32%
student-centered learning	31	30%
associative learning	22	21%
less direct conflict	19	18%
writing competency	15	15%

N = 103

Another side benefit of being allowed to delay in responding is less chance of direct conflict. Some students reflect that when they are in a live group conversation, they often accidentally hurt people's feeling without meaning any offense. In discussion board, students tend to rethink the way in which they say something before submission. But one student argues that by being cautious with what we say (type), we may get to know each other in a slower manner.

Jonassen [4] observes that while in synchronous communication, participants are more focused on establishing identity, asynchronous conferences allow participants to reflect more on issues being discussed. About half of the students (46%) argue that conversing via discussion board allows an individual to be known for her thoughts rather than looks (shown in Table 1). Without hearing or seeing each other in real-time, participants tend to focus on ideas than on appearance.

By the same token, about as many students (43%) observe that introvert or shy students would feel less inhibited and are more willing to express their views in discussion board when there is no face-to-face presence of participants or time pressure of instant response (shown in Table 1). Some find that those introvert students often make more insightful observations than their more vocal peers when they do speak up.

However, a few other students disagree and note that those who speak less may not always be shy persons. They believe that some speak less because they do not have enough knowledge of the topic and therefore do not feel comfortable enough to participate. Some respondents also admit that oftentimes they do not actively participate in discussion because they have little interest in the topics or they cannot relate the discussion to their personal experience.

Although WebBoard allows the instructor to have students' postings checked before they are posted, most instructors do not activate this function. About 30% of the students surveyed feel that discussion board allows them more control over what they can say and when they say it (shown in Table 1). When students have more control over what they can do in learning, they tend to have more to initiate. Many students feel that they contribute more messages in discussion board than in traditional classroom setting.

Student-centered learning also encourages students' collaboration. Over half of the students (54%) feel that they learn a tremendous amount by exchanging information and sharing experiences with their peers through discussion board (shown in Table 1). Some say they find it more effective to learn from peers than from instructors. Some admit that they are simply quiet listener-type students who prefer to learn by observing, though they do not contribute much.

While most of the advantages of WebBoard are associated with students' control over time and content, students also find that WebBoard has some organizational and structural attributes that can facilitate online learning. About a third of the students (30%) say that the threaded organization of messages in discussion board makes people stay more on task than in a chat, and they attribute this tendency of content coherence to the fact that respondents can always see the path of discussion.

As a hypermedia environment, WebBoard allows users to include hyperlinks in postings. It is therefore very convenient for students to cite Internet resources to support their argument, since readers can simply click the links and view the supporting resources. Many students (21%) say that Internet has now become a primary information resource for them. Linking Internet recourses to their discussion encourages them to associate what they have learned outside the course with what is covered in the course (shown in Table 1).

By posting and responding in writing, more than a dozen students feel that their writing competency is improved, though initially it may take more effort for some to put their thoughts into words. On the other hand, some other students are less confident about their writing competency, particularly grammar accuracy. One student admits that it took a lot more thought to write down something that can be easily said. When responses are put in writing, readers can analyze grammar, and grammar errors can be very embarrassing.

5. Types of input from instructor

Should instructors participate in students discussion? Most students said yes. What kind of input would be most welcome? Thought-provoking questions. The types of questions students find more likely to lead to constructive discussions are questions that have no right/wrong answers and questions that allow students to relate to their diverse individual background and experiences (shown in Table 2).

Table 2. Constructive questions

Types of questions	Numbers of responses	
no right/wrong answer	54	52%
diverse backgrounds and experiences	39	38%

N = 103

For questions with a limited range of possible answers, students find that responses tend to converge after the first few posts. In a group of 10 or more students, later respondents often do not have much to add but agree with the previous responses.

The most desired input students wish to have from instructors is direction on further resources. Students also want instructors to correct their misconceptions in a timely manner, and would be happy to get acknowledgement for their efforts. Significantly fewer students have expressed need to get encouragement from the instructor (shown in Table 3).

Table 3. Types of instructor input desired

Types of input	Numbers of responses	
direction to related resources	69	67%
correction	52	50%
acknowledgement of efforts	49	48%
encouragement	28	27%

N = 103

6. Instructors' perceptions

A major advantage of WebBoard in most instructors' view is the power to link students and build a virtual learning community. Most instructors feel that WebBoard can create an environment that helps students get to know each other and learn from each other. WebBoard has also been found to provide the instructors with opportunities to know students' meta-cognitive activities. They believe that students' responses reflect to a great extent their understanding of the course content.

Most instructors organize conferences by a combination of timeline and topics. Some instructors observe that time flexibility has to be contained within an appropriate timeframe because students may unduly postpone their responses. When these late students do respond, their responses often receive little attention.

A few instructors find closed conferences useful in preventing students from copying ideas. Some instructors set limit on the length of students' postings and responses. Others do not set limit, but encourage links to external resources.

Three of the six instructors say that they participate in students' discussion only when a question is clearly addressed to the instructor. When they do participate, they avoid giving their opinions or definitive answers to discussion questions.

All instructors evaluate students' performance in discussion by a combination of frequency of responses and depth of thought. Some also give credit to students' willingness to share experiences and thoughts with peers.

7. Discussion

The findings of this study seem to warrant the proposal of the following guidelines for using discussion board to facilitate distance learning:
- Allow time flexibility
- Give students more control
- Be around, but avoid giving opinions
- Design questions carefully
- Encourage associative learning

Since time flexibility is perceived by most distance students as a primary advantage of distance learning, we should allow students adequate time to respond to questions, both as an accommodation of diverse student time schedules and as an encouragement of reflective thinking. While giving students time flexibility, instructors may also need to train students in time management and help them become well-organized and self-responsible learners.

Collaborative learning is another major advantage of discussion board. Conferences should probably not be closed unless absolutely necessary. Students should be encouraged to share thoughts with as many peer learners as they can to maximize collaborative learning, which is generally more important in distance learning than in traditional classroom setting because collaboration can help distance learners overcome the often reported sense of isolation.

Although instructors do not need to have their presence constantly felt by students, they do need to monitor student discussions regularly and respond to students' needs in a timely manner. Students' remarkable efforts and quality responses should be acknowledged. For those less active participants, instructors may need to check with them individually through private email to find out if they need any additional help. When providing input, instructors should avoid giving their opinions because instructors' opinions tend to stifle students' active thinking and collaborative learning. As Cashing & McKnight [5] put it, the instructor's role should be that of a facilitator rather than a judge.

The best input instructors can give to students is first of all thought-provoking questions. A good question can encourage critical and reflective thinking. Constructive questions would also encourage students to elaborate by relating to personal experiences, and to critique based on different resources and diverse perspectives. Factual questions should be avoided.

Given the growing popularity of the Internet and students' familiarity with the Internet, instructors should encourage students to cite Internet-based resources to support their discussion. Instructors may also want to provide some Internet resources for students to review before each discussion session, and give necessary guidance on evaluating and filtering Internet resources.

8. Conclusion

WebBoard has now become an essential component of distance learning environment at Dakota State University. This article attempts to identify factors that are perceived to contribute to a successful discussion board in distance learning. More comprehensive and robust research effort will be needed to systematically study how discussion board can be more effectively used as a communication tool to facilitate distance learning.

9. References

[1] Khan, B.H. (Ed.), *Web-based Instruction*, Educational Technology Publications, Englewood Cliffs, NJ, 1997.
[2] Simonson, M., S. Smaldino, M. Albright, and S. Zvacek, *Teaching and Learning at a Distance: Foundations of Distance Education*, Prentice-Hall Inc., Upper Saddle River NJ, 2000.
[3] Harasim, L.M., "Online Education: An Environment for Collaboration and Intellectual Amplification," In L.M. Harasim (Ed.), *Online education: Perspectives on a New Environment*, Praeger, New York, 1990.
[4] Jonassen, D., *Computers as Mindtools for Schools: Engaging Critical Thinking*, Prentice-Hall Inc., Upper Saddle River, NJ, 2000.
[5] Cashing, W.E. and P.C. McKnight, "Improving Discussion," In R. A. Neff & M. Weimer (Eds.), *Classroom Communication: Collected Readings for Effective Discussion and Questioning*, Madison, Magna Publications, Inc., 33-40, 1989.

Decision-making resources for embedding theory into practice

Grainne Conole
Institute for Learning and Research Technology, University of Bristol, UK
g.conole@bristol.ac.uk

Martin Oliver
Higher Education Research and Development Unit, University College London, UK
martin.oliver@ucl.ac.uk

Abstract

This paper illustrates a way in which practitioners can be supported in the process of engaging with theory in order to underpin practical applications in the use of Information and Communication Technologies (ICT). This approach involves the use of decision-making resources ('toolkits'); three examples are described. The ways in which this embeds specific theoretical assumptions is discussed, and a model for toolkit specification, design and evaluation is described.

1. Introduction

Decision-making resources range from highly restrictive 'templates' or 'wizards', which provide high levels of support and step-by-step guidance but little possibility of user-adaptation, through to 'theoretical frameworks', which provide a context and scope for the work but leave the user to devise their own strategy for implementation. Between these extremes lie a range of resources, including checklists and guidelines.

A number of pedagogical frameworks have been developed to support ICT. All develop from a particular theoretical viewpoint, aiming to encourage the application of good practice according to a specific pedagogical approach. In this context, 'good practice' is taken to mean practice that is informed by and which exemplifies (or contributes to) theory. A framework for integrating ICT has been developed by Conole and Oliver [1]. It provides a structured approach to integrating learning materials into courses and is designed to support the process of 're-engineering' a course [2]. Various features of an existing course can be described and evaluated, allowing an analysis of strengths and weaknesses, the suitability of different media types and limiting factors, including resource issues and local constraints. The framework can be applied as a series of stages, starting with the review of existing provision, working through a process of shortlisting and selection of teaching techniques, and concluding with a mapping of the new course.

Another approach involves the use of structured decision-making systems: templates and wizards, which can provide structured, pre-defined layouts or structures for the user to base their document or presentation on. A wizard is a tool that makes decisions on behalf of the user, based on solicited information and drawing on pre-defined templates. In most cases, the way in which these outputs are generated is hidden from the user. As a result, they are relatively easy to use, but are restrictive in the range of outputs that can be achieved, and allow very little engagement with issues or response to the values and assumptions built into the system. There are many examples of templates and wizards that provide a generic structure that guides users through a set of options. Online shopping sites, book stores and travel centres often have 'wizards' which guide the user through a series of options or interests, helping them to focus in on topics of particular interest. It is evident that these types of semi-structured forms of support and guidance are becoming increasingly important as a way of guiding users through the plethora of online information.

These types of resource both share a common aim of supporting a users' engagement with an area. Clearly, however, they are working at very different levels and making different assumptions about the type of support that the user might need. Theoretical frameworks provide a structure and vocabulary that support the exploration of concepts and issues. Wizards provide automated processes that support the production of resources. They are predicated on the assumption that the user is primarily concerned with the efficient design and production of a resource.

These two positions can be characterised as extremes of one continuum. At one extreme there are frameworks, which are flexible and versatile, but which offer relatively little support for practitioners attempting to engage with them. At the other there are wizards and templates, which are highly restrictive, but (by virtue of the constraints that they impose) are able to offer much closer support and guidance to users. Toolkits can be viewed as a mid-point on this continuum [3]. They are more structured than frameworks; they use a model of a design or decision-making process, together with tools provided at key decision-making points, to help the user engage with a theoretical framework and apply in the context of their own practice. Each of the tools that is drawn upon as the user works through the process model is designed to help the user to access a knowledge base in order to make informed decisions. The format of toolkits means that they can be used in a standard, linear fashion, or can be "dipped into" by users whose level of expertise is stronger in some areas of the design process than others.

2. The philosophy behind developing toolkits

This section will explain the philosophy behind the concept of toolkits and will outline an approach to developing and evaluating them. Toolkits are predicated on the assumptions that they are:

- derived from an explicit theoretical framework
- easy-to-use for practitioners
- able to provide demonstrable benefit
- able to provide guidance, without being prescriptive
- adaptable to reflect the user's practice and beliefs
- able to produce outputs that reflect the local context

The process of the development of each toolkit consists of a number of steps, which can be coached in terms of the following framing questions.

1. *Assessment of need: Is there a need for a toolkit in this particular area to support practitioners?*
2. *Theoretical underpinning: What theory and models are relevant to the toolkit?*
3. *Toolkit specification: How can the range of options available at each stage be translated into a practical, but flexible form of guidance for non-experts?*
4. *Toolkit refinement: How useful and flexible is the toolkit?*
5. *Inclusion of user defined features: Is the toolkit sufficiently flexible that it can be adapted by end users to take account of local factors?*
6. *Build up of shared resources, through input of practitioner case studies: Are the completed toolkit plans produced by practitioners of any value as case studies or templates for other users?*

3. Examples

Media Advisor is a toolkit for curriculum design [1, 3]. It considers different teaching methods in terms of their relevance and value against a set of four types of teaching activity, namely – delivery, discussion, activity and feedback. The formative evaluation of an initial prototype combined with a process of iterative tailoring to meet users' needs is an important feature of this approach. The evaluation also generated a number of unexpected results. In particular, mapping teaching techniques (both traditional and new) in terms of their support for the four 'types' described above was originally a way of identifying types of teaching that were systematically emphasised or neglected in a course. However, evaluation showed this was at least as important as a way for individuals to express their own approach to teaching, or to develop ideas about new ways in which traditional resources (such as videos) could be incorporated into the course. Moreover, the simplicity of the description meant that it could be used as the starting point for discussions between practitioners about the differences in their approach to teaching.

The Evaluation Toolkit provides a structured resource to help practitioners to evaluate a range of learning resources and activities [4]. It guides them through the scoping, planning, implementation, analysis and reporting of an evaluation. It assists the practitioner in designing progressively more detailed evaluations, and allows users to access and share evaluation case studies. It consists of three sections: Planner, Advisor and Presenter, which guide the user through the evaluation process; from the initial scoping of the evaluation question(s) and associated stakeholders, through selection of data capture and analysis methods, and finally through the presentation of the findings. One of the emerging values of this toolkit is that the plan the user produces can be made available to others as a case study within the toolkit, such as assessing Web site usability or selecting resources.

The information toolkit provides a means of mapping information resources against types of information activity (gathering, processing, communicating and evaluating) [5]. This helps the user to gain an understanding of information resources to produce their own tailored information plan. It guides the user through the process of articulating their needs in the form of an information plan for a particular task.

4. Conclusion

Evaluation feedback has been positive, with many users reporting that the toolkits helped them to reflect upon and structure their thought process in a particular area by providing expert guidance and support as well as useful links to further information and support. An additional benefit was the ability to build up case studies covering common types of design, evaluation, or information maps.

Toolkits clearly represent a valuable type of resource for a range of staff, such as practitioners, educational developers, learning technologists and ICT researchers. They provide a means of articulating the use and evaluation of ICT and help users to engage with different aspects of the use of ICT towards the development of better learning and teaching practices.

5. References

[1] Conole, G. & Oliver, M. (1998) A pedagogical framework for embedding C&IT into the curriculum. *ALT-J*, 6.2, 4-16.

[2] Nikolova, I. & Collis, B. (1998) Flexible learning and design of instruction. *British Journal of Educational Technology*, 29 (1), 59 – 72.

[3] Oliver, M. & Conole, G., (2000) Assessing and enhancing quality using toolkits. *Journal of Quality Assurance in Education*, 8, 1, 32-37.

[4] Conole, G., Crewe, E., Oliver, M. & Harvey, J., (2001), A toolkit for supporting evaluation *ALT-J*. 9(1), 38-49.

Theme 11: Evaluation/Monitoring/Student Models/Information Needs

Student Models Construction by Using Information Criteria

Maomi Ueno
Nagaoka University of Technology
ueno@kjs.nagaokaut.ac.jp

Abstract

This paper proposes a construction method of Student models for Intelligent Tutoring Systems(ITSs) by using information criteria. This proposal provides a method to automatically construct the optimum Student model from data. The main problem when the traditional information criteria are employed to construct a model is that large amount of data, which are difficult to obtain in actual school situations, need to be obtained. This paper proposes a new criterion for using a smaller amount of data by utilizing a teacher's expert knowledge. Concretely, 1) the general predictive distribution is derived, and 2) the determination method of the hyper parameters by using a teacher's expert knowledge is proposed. Finally, some Monte Carlo experiments comparing some information criteria (ABIC, BIC, MDL, and the exact predictive distribution) are performed. The results show that the proposed method provides the best performance.

1. Introduction

Over the last few years, a method of reasoning using probabilities[1],[2] variously called Bayesian networks, belief networks, causal networks, and so on, has become popular within the Intelligent Tutoring System community. For example, in [3], the knowledge states diagnosis system is based on belief networks and decision theory. In [4],[5], [6],[7],[8]besides the diagnosis system, updating is concerned with the expected changes in student knowledge due to tutoring.

However, these methods subjectively constructed the student model structure without using students' response data. If the students' response data is available, then it helps our decision making for the student model construction. This paper proposes a construction method of Student models for Intelligent Tutoring Systems (ITSs) by using information criteria. The main problem when the traditional information criteria are employed to construct a model is that large amount of data, which are difficult to obtain in actual school situations, need to be obtained. This paper proposes a new criterion for using a smaller amount of data by utilizing a teacher's expert knowledge. Concretely, 1) the general predictive distribution is derived, and 2) the determination method of the hyper parameters by using a teacher's expert knowledge is proposed. Finally, some Monte Carlo experiments comparing some information criteria (ABIC, BIC, MDL, and the exact predictive distribution) are performed. The results show that the proposed method provides the best performance.

2. Representation of the student model by the Belief networks

In this paper, the student model is defined by belief networks. Let $X = \{X_1, X_2, \Lambda, X_N\}$ be a set of N variables which represent students' knowledge states; each can take r_i states in the set $\{1,\Lambda, r_i\}$. $x_i = k$ is written when it is observed that variable x_i takes k. $p(x = k \mid y = j, \xi)$ is used to denote the probability of a person with background knowledge ξ for the observation $x = k$, given the observation $y = j$. A student model is represented as a pair of knowledge structure S and a set of conditional probability parameters Θ, (S, Θ). An example of knowledge structure S in the domain "the solution of the linear equation" in a junior high school is shown in Figure.1. Here, $A \longrightarrow B$ indicates that we have to acquire knowledge A in order to acquire knowledge B. The nodes which depends on the target node are called "parent nodes" of the target node. In addition, a set of problems which correspond to the nodes has to be prepared. If a student provides a correct

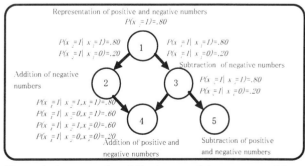

Figure.1 An example of Student model represented by belief networks

answer to i-th problem, $x_i = 1$ (he has i-th knowledge), otherwise, $x_i = 0$. A joint probability distribution over the network is shown by

$$P(X_1, X_2, \Lambda, X_N | S) = \prod_{i=1}^{N} p(x_i | \Pi_i, S), \quad (1)$$

where $\Pi_i \subseteq \{x_1, x_2, \Lambda, x_{q_i}\}$ is a set of parents nodes that renders x_i and $\{x_1, x_2, \Lambda, x_{q_i}\}$ conditionally independent. In particular, S is a directed acyclic graph such that (1) each variable corresponds to a node in S, and (2) the parents of the node corresponding to x_i are the nodes corresponding to the variables in Π_i.

The next section will propose a method to automatically construct the student model structure from the students' response data.

3. Estimation of conditional parameters and posterior distribution

This paper proposes a method to automatically construct the student model from the data. It is necessary to define the model in (1) as a statistical model in which parameters are estimated. Now, consider a database $\mathbf{X} = \{x_{sik}\}, (s = 1, \Lambda, n, i = 1, \Lambda, N, k = 0, \Lambda, r_i - 1)$, where $x_{sik} = 1$ when the s-th student takes k-th knowledge states about the i-th node, otherwise $x_{sik} = 0$. Let θ_{ijk} be a conditional probability parameter of $x_i = k$ when $\Pi_i = j$, then the following likelihood function can be obtained.

$$p(\mathbf{X} | \Theta_S, S) = \prod_{i=1}^{N} \prod_{j=1}^{q_i} \frac{n_{ijk}}{\sum_{\bar{k}=0}^{1} n_{ijk}} \prod_{k=0}^{1} \theta_{ijk}^{n_{ijk}}, \quad (2)$$

where

$\Theta_S = \{\theta_{ijk}\}, (i = 1, \Lambda, N, j = 1, \Lambda \ q_i, k = 0,1)$, and it is assumed that the prior distribution has a Dirichlet distribution, which is a conjecture distribution of (2). That is,

$$p(\Theta_S | S) = \prod_{i=1}^{N} \prod_{j=1}^{q_i} \frac{\Gamma(\sum_{\bar{k}=0}^{1} n'_{ijk})}{\sum_{\bar{k}=0}^{1} \Gamma(n'_{ijk})} \prod_{k=0}^{1} \theta_{ijk}^{n'_{ijk}-1}, \quad (3)$$

where $\Gamma()$ indicates Gamma function. Then, the following posterior distribution is obtained,

$$p(\mathbf{X}, \Theta_S | S) = \prod_{i=1}^{N} \prod_{j=1}^{q_i} \frac{\Gamma(\sum_{\bar{k}=0}^{1} n'_{ijk} + n_{ijk} - 1)}{\sum_{\bar{k}=0}^{1} \Gamma(n'_{ijk} + n_{ijk} - 1)} \prod_{k=0}^{1} \theta_{ijk}^{n'_{ijk}+n_{ijk}-1}$$

(4)

Therefore, the following maximum a posterior estimator can be derived;

$$\hat{\theta}_{ijk} = \frac{n'_{ijk} + n_{ijk}}{n'_{ij} + n_{ij}}, \quad (5)$$

where $n'_{ij} = \sum_{k=0}^{1} n'_{ijk}$, $n_{ij} = \sum_{k=0}^{1} n_{ijk}$.

Thus, the conditional probability parameters $\hat{\theta}_{ijk}$ and the posterior distribution $p(\mathbf{X}, \Theta_S | S)$ can be estimated from (4) and (5).

4. Student models construction by using information criteria

Let us consider a structure with just three nodes, then there are eight possible structures as shown in Figure 2. The problem is how to find the optimum structure of the student model. In this case, it is well known that information criteria are useful. Since Akaike's criterion[10], various criteria have been proposed. (For example, ABIC[10], BIC[11], MDL[12], and so on). The student model construction problem in this paper employs Bayesian approach, it is considered that employing Bayesian information criteria, ABIC, BIC, MDL, and so on, is valid.

ABIC is given by

ABIC(model) = $-2\log p(\mathbf{X}, \Theta_S | S) + 2K$. (6)

BIC and MDL have the same formulation by

BIC(MDL) (model) = $-2\log p(\mathbf{X}, \Theta_S | S) + 2K\log n$. (7)

Finding the structure is completed by minimizing these criteria. However, these criteria are generally derived by approximating the posterior distribution or predictive distribution, then, it is considered to expect better results by deriving directly a criterion from the belief networks model. For this motivation, Cooper [9] derived the

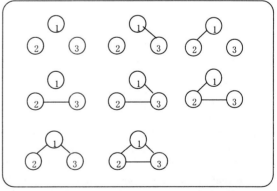

Figure 2. Possible structures of the student models with three nodes

predictive distribution of the belief networks when the prior distribution has a unique distribution. This criterion is given by

$$p(\mathbf{X}, S \mid \Theta_S) = \prod_{i=1}^{N} \prod_{j=1}^{q_i} \frac{(r_i - 1)!}{(n_{ij} + r_i - 1)!} \prod_{k=1}^{r_i} n_{ijk}!. \quad (8)$$

It should be noted that the optimum structure is obtained by maximizing this criterion. The next interest is which criterion is best for the student model construction problem. Here, some Monte Carlo studies are demonstrated. For simplicity, Figure 1 is considered as the true model. The random data is generated from the Figure 1, and the sample sizes are 50, 100, 200, 300, 400, 500, 600, 700, 800, 900, and 1000. 1000 realization for each sample size were generated. For each realization, the relative performances of the criteria mentioned before can be compared. The results, the number of times when the criteria selects the true structure among 1000 iterations for each sample size are shown in Table 1

Table 1. The results of the Monte Carlo experiments

Sample sizes	ABIC	BIC(MDL)	Cooper
50	205	75	18
100	423	347	222
200	486	641	487
300	494	786	634
400	497	908	752
500	498	951	782
600	487	972	835
700	480	983	870
800	458	988	872
900	462	988	881
1000	424	989	897

From the table, BIC or MDL shows the best performances for large sample sizes, and ABIC shows the best performance for small sample sizes. In an actual school situation, it is difficult to gather a large sample of data, in this sense, it is considered that ABIC provides the best performance. However, for a small sample (100), ABIC selects the true structure with a probability of 0.42 at most. Then, this paper proposes more effective method to construct the student model in the next section.

5. Student models construction by using teacher's expert knowledge

5.1. The general predictive distribution

All information criteria mentioned in section 4 assumes that the prior distribution, which reflects prior knowledge about the student model, has a uniform distribution. However, in education, most teachers have prior knowledge about the student model. The main idea of this paper is to develop an efficient information criterion by integrating a teacher's expert knowledge into the prior distribution. To realize this idea, an exact general predictive distribution with various prior distributions should be derived. From the assumptions in this paper, the general predictive distribution can be derived as follow: That is, from (4), the predictive distribution is given by

$$p(\mathbf{X} \mid S) = \prod_{i=1}^{N} \frac{\Gamma(n'_{ijk})}{[\Gamma(n'_{ijk})]} \frac{\prod_{j}^{q_i} \Gamma(n'_{ijk} + n_{ijk})}{\Gamma(n'_{ijk} + n'_{ijk})}. \quad (9)$$

It should be noted that the general predictive distribution has the hyper parameter n'_{ijk}. In fact, this hyper parameter acts the most important role. The predictive distribution converges to various information criteria by changing the value of the hyper parameter. When $n'_{ijk} = 1$ (the prior distribution has the unique distribution shown in Figure 3), the predictive distribution converges to Cooper's criterion, although it is natural from the definition. When $n'_{ijk} = 1/2$ (the prior distribution has the U distribution shown in Figure 3), the predictive distribution converges to BIC, or MDL. Moreover, when $n'_{ijk} = -\log \theta_{ijk} + 1/2$ (the prior distribution has the convex distribution shown in Figure 3), the predictive distribution converges to AIC. It should be noted that BIC, or MDL and Cooper's criterion assume stronger penalties for the complexity of the model, which is the number of parameters, than one of AIC. Then, AIC has a tendency to select a structure with more arcs than the true

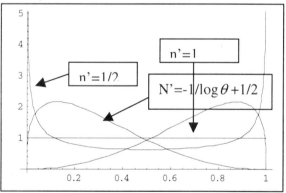

Figure 3. Prior distributions for various hyper parameters

structure, and BIC or MDL and Cooper's criterion have a tendency to select a structure with less arcs than the true structure.

5.2. Integration of teacher's knowledge

By using these properties, utilizing the teacher's prior knowledge about student model into the predictive distribution (9) can be considered. The procedure is as follows: 1) A teacher constructs a student model structure by using his or her expert knowledge, 2)based on this structure, let the value of the hyper parameter for the arcs which is considered to exist be $n'_{ijk} = -\log \theta_{ijk} + 1/2$, and let the value of the hyper parameter for the arcs which is not considered to exist be $n'_{ijk} = 1/2$. Now, consider three possible structures, as prior knowledge structures, A, B, C concerning the structure in Figure 1 as follows: A is the structure with full arcs, B is the true structure, and C is the structure with no arc. Let consider three cases of which each structure is considered as a prior knowledge about the student model. The same Monte Carlo studies as section 4 are demonstrated in Table 2. If a teacher knows the true structure, the criterion acts more exactly than the traditional criteria for small sample sizes. Moreover, if a teacher has a wrong knowledge as structures A and B, the proposed criterion acts the same as the traditional criteria.

Table 2. The results of the Monte Carlo experiments by using prior knowledge

Sample sizes	Structure A	Structure B	Structure C
50	205	982	75
100	423	972	347
200	486	1000	641
300	494	1000	786
400	497	1000	908
500	498	1000	951
600	487	1000	972
700	480	1000	983
800	458	1000	988
900	462	1000	988
1000	424	1000	989

7. Application

By using the data from 294 junior high school students data, the student model in a domain of a simple equation is estimated from (7). Expert knowledge is employed as prior knowledge, although it is omitted for want of space. It is known that the obtained structure is reasonable considering the meanings of the nodes.

8. Conclusions

This paper proposed a new information criterion for the Student model construction problem by using a teacher's expert knowledge. The Monte Carlo experiments showed the efficiency of the proposed model.

References

[1] J. Pearl, "*Probabilistic Reasoning in Intelligent System*", Morgan Kaufman Publishers, California, 1988.
[2] R.E. Neapolitan, "*Probabilistic Reasoning in Expert System*", John Willey & Sons, INC, New York, 1990.
[3] M. Ueno, "*A test theory using probabilistic network*",

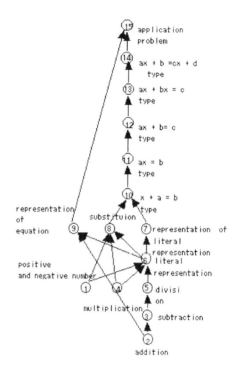

Figure 4. An example of estimated student model

Electronics and Communications in Japan, 78(5), John Wiley & Sons, Inc. ,1996.
[4] J. Reya, "Two-Phase updating of student models based on daynamic belif networks", *Proceedings of 4th Internationa Conference of Aritificial Inttelegence*, 1998, pp. 274-283.
[5] A.H. Hashimoto et al.,"A probabilistic approach for student modeling task",Proceeding of the 13th Annual Conference of JSAI, Tokyo, 1998. pp.68-69
[6] M. Ueno, "Environments of learning by using Internet, Meta knowledge navigation system", *Proceedings of ICCE 99*, IOS press, Tokyo, 1999, pp.748-751
[7]J. Martin & K.VanLehn," Student assessment using Bayesian nets. International" Journal of Human-Computer Studies, Vol. **42**.,. Academic Press.1995, pp. 575-591
[8]K.VanLehn, & J. Martin, "Evaluation on an assessment system based on Bayesian student modeling.", International Journal of Artificial Intelligence and Education, Vol.**8**.2, 1998
[9] G.F. Cooper and E. Herskovits, "A Bayesian method for the induction of probabilistic networks from data ", *Machine Learning* , **9**, 1990, pp.54-62.
[10] H. Akaike, "A Bayesian extension of the minimum AIC procedure of an autoregressive model fitting", *Biometrika*, **66**, 237-242.
[11]G. Schwarz, "Estimating the dimension of a model ", Annals of Statistics **6**, 451-454
[12] J. Rissanen, "A universal prior for integrals and estimation by minimum description length", Annals of Statistics, **11**, 416-431

The Effectiveness of an Intelligent Tutoring System for Rocket Training

Robert A. Wisher
U.S. Army Research Institute
Wisher@ari.army.mil

L. Jared Abramson
George Mason University
AbramsonJ@ari.army.mil

James J. Dees
U.S. Army TRADOC
DeesJ1@Mornroe.army.mil

Abstract

This report examines the application of an intelligent tutoring system (ITS) for use in training the complex skill of deploying a multiple launch rocket system during a reconnaissance task. The task was conventionally taught using miniature replications of vehicles and launchers on a large table of sand. An ITS version of the exercise, called the Virtual Sand Table, replicated the training with the added advantage of informative feedback and computer-base coaching. A comparison group (n=209) used the conventional sand table and the treatment group (n=105) used the Virtual Sand Table during a four-hour training session. Results, as measured by a hands on performance test, indicated superior performance by the Virtual Sand Table treatment group, with an effect size slightly more than one standard deviation unit.

1. Introduction

Distributed simulations, virtual realities, and intelligent tutors have all changed concepts for distance learning environments during the past decade, reducing the need for an in-place instructor [1]. In parallel with these advancements, the U.S. Army Training and Doctrine Command (TRADOC) is embarking on a major change to deliver standardized individual and self-development training to soldiers through the application of multiple media and networked delivery technologies.

A training technology that has matured in recent years is intelligent tutoring. Intelligent tutoring systems have evolved from an arcane art of knowledge engineering and LISP coding to development methods and delivery options that are becoming increasingly mainstream through desktop PCs. Fundamental to an intelligent tutor is a body of domain knowledge encoded as an expert system of rules [2]. This expertise is accessible to the student during a learning exercise, under the control of an instructional strategy. The goal is to have students construct a mental representation of the domain knowledge -- the expert's facts, rules, and procedures -- for later application. An ITS is usually aimed towards the training of skills and knowledge that are complex in nature. The most important aspect of an ITS is its ability to interact with learners, teaching or assisting them in processing and understanding complex information [3]. An ITS must be capable of identifying a student's strengths and weaknesses, and establishing a relevant training path. The instruction can then be tailored to the student's needs in acquiring the expertise as defined in the expert module.

This paper describes the development and evaluation of an ITS for application by the U.S. Army Field Artillery School in the Captains Career Course. Specifically, an ITS was developed for the Multiple Launch Rocket System (MLRS) sandtable exercise, a four-hour block of instruction. The MLRS is a field artillery rocket system that provides counterfire, suppression of enemy air defenses and destruction of light material and personnel targets. The MLRS delivers large volumes of firepower in a short time against critical time sensitive targets. The sandtable training is normally conducted in small groups using a conventional sand table exercise. The Virtual Sand Table (VST) is the ITS developed for conducting the same training on an individual rather than group basis.

1.1. Origin of the project

The VST stemmed from a need by the Field Artillery School to incorporate advanced training technologies to enhance the learning of complex skills. Funding for the development of the VST was a consequence of a technology transfer arrangement between Headquarters, TRADOC and the U.S. Army Research Institute (ARI): new commercial off-the-shelf technologies are assessed for their applicability to Army training needs; TRADOC was responsible for identifying criteria for selecting technologies for trial in actual Army training programs, while ARI, was charged with identifying high-potential candidates for transfer.

Of the candidates identified for a more detailed assessment, a successful training technology developed by

Sonalysts Inc., ExpertTrain™, scored highest. ExpertTrain, is an ITS that creates an intelligent learning environment which is continually modified as a student interacts with a simulation. Its success had been previously demonstrated for the Navy, as an intelligent training aid for radar system controllers. It had high applicability to teaching complex skills needed by the Army.

The published literature on the effectiveness of distributed learning is overwhelmingly anecdotal [6]. Many evaluations of DL measure only student reactions to the technology, instructor, and course rather than the outcomes of learning. Prior to conducting the experiment on the VST, previous research on the effectiveness of intelligent tutoring systems that did examine the outcomes of learning was examined. The results of these empirical studies support the implementation of intelligent tutoring technologies in place of traditional classroom training [4].

1.2. Sand table: Real and virtual

The conventional sand table exercise is a critical component of the 18-week Captains Career Course, Phase I, the purpose of which is to prepare field artillery officers for duties as fire support officers at maneuver battalion and brigade level. The conventional sand table is a low fidelity training device used to evaluate reconnaissance, selection, and occupation of position strategies (RSOP) of soldiers who have completed training on the operation of the MLRS, as well as delivery of fires, reload operations, and combat service support. The conventional sand table is an actual table of sand with molded terrain features and with MLRS assets depicted by miniature objects. One major limitation of the conventional sandtable is its sheer size. There are only a few sand tables available at the school. As a result, students are compelled to work in groups of five or six, rather than individually as is the case with the VST.

During the sand table exercise, students must review an operations order, evaluate a terrain, and strategically decide where to place firing points, ammunition hiding areas, platoon operations centers, etc. within a terrain model representing up to a 100 square kilometer operations area. The students discuss proposed plans in a group and settle on a consensus plan. The plan is then rated by a subject matter expert. Each member of the group receives the same score, even though some may have contributed little to the plan and consequently learned very little from the exercise.

The Virtual Sand Table is a computer-based intelligent tutoring system designed to replicate the conventional sand table. The MLRS VST is essentially a simulation game, where the student's actions are evaluated against a set of expectations governed by a set of operational rules. Terrain maps can be viewed on the screen or consulted off-line for greater detail. Although an actual sand table may seem a useful training aid, the individual student cannot benefit during the exercise unless there is a highly trained instructor present to critique the process through regular and informative feedback. However, considering the training throughput and instructor resources, an insufficient number of instructors is available to critique students individually during the exercise. This is another reason for the grouping of students during the conventional sand table exercise. One way to overcome such constraints is for an ITS to provide the critique and feedback throughout the exercise.

The tutoring component is designed to simulate an instructor coaching a student at a conventional sand table. The focus of the coaching is the evaluation of the student's selected positions and routes in accomplishing RSOP. The basis for the intelligent tutor is a three-step process which includes a situation assessment of the map area 3km grid, diagnosis and evaluation of the student's decisions, and generation of feedback (coaching) to the student.

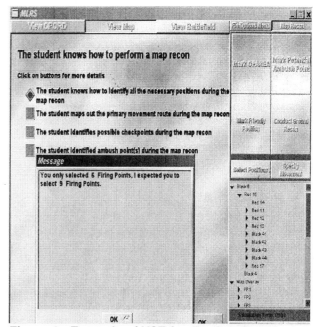

Figure 1. Example of VST feedback screen

2. Methods

2.1. Intelligent Tutor

The MLRS VST runs in a stand-alone mode on any Army Classroom XXI desktop PC. The courseware is delivered from a CD-ROM. Each student participates in a simulation-driven sand table exercise and views information about the selected scenario operations order, enters mission data, and employs assets toward mission completion from a graphical user interface. An embedded, artificial-intelligence based tutor provides coaching during the exercise and generates a student evaluation after the exercise. An example of the interface is presented in Figure 1. As the student places assets for occupation of position, a simulation component calculates the line-of-sight, mobility, and trajectories for the MLRS in real-time. The results from the simulation are displayed in the battlefield views and sent to the intelligent tutor component for evaluation and the possible triggering of coaching templates.

2.2. Participants

Soldiers (n=209) from multiple classes of the Captains Career Course constituted the comparison group. Participants were provided with a background questionnaire just before the conventional sand table exercise, and upon completion a second questionnaire was administered. The same instruments were administered to the treatment group (n=105), which was constituted from multiple classes of the same course. The background questionnaire sought demographic and experiential information on previous field artillery assignments. The post-exercise questionnaire sought feedback on perceived learning on the same topics and, in the case of the treatment group, ratings on the usability of the VST.

2.3. Procedure

For the comparison group, soldiers performed the exercise in groups of five. Instructors made a point to include at least one soldier in each group who had field experience with an MLRS unit. After being provided with an operations order, the group collaborated on a solution. Approximately three hours were required by the group to complete the plan. Afterwards, one student was selected at random (through a drawing of straw lengths) to brief the plan. The instructor scored composite performance on a ten-point scale, with ten being outstanding and one being poor. Every member of the group was assigned that score.

For the treatment group, groups of 20 students each were directed to a computer laboratory where the VST exercise was installed on personal computers. After completing the background questionnaire, an instructor explained the purpose of the VST, demonstrating procedures such as placing MLRS assets and marking routes of travel. The students reviewed an operations order displayed on their monitors. Students worked individually. They were given three hours to complete the exercise. Upon completion of the exercise, a copy of the plan (a map showing emplacements and routes) was saved electronically in jpeg format. The same instructor who scored the plans of students in the conventional sand table group applied the same criteria in scoring performance using the same ten-point scale. Scores were recorded individually.

3. Results

The experimental design for the evaluation was a two-group, multiple post-test design. The average age of the conventional group was 28.3 years compared to 28.2 years for the VST group (t=.34, ns). Five other demographic comparisons (rank, branch, MLRS experience, cannon experience, and target acquisition experience) were made, none demonstrating a significant difference between groups as measured through a chi-square test. Thus, the two groups were demographically equivalent.

3.1. Knowledge Test

A written knowledge test was administered the day after the sand table exercise. Due to a ceiling effect, the test proved to be of no practical value. Individual performance on the items corresponding to the content of the sand table exercise were extracted from an end-of-module test and analyzed for differences between groups. Both groups scored over 99% correct, so the test was not sensitive to potential differences.

3.2. Skill Performance Measure

The comparison group was trained using the conventional sand table. The approach consisted of small teams of approximately five students, and each group included at least one individual with at least one year of MLRS experience. The students were supposed to work collaboratively to solve the problems. In contrast, the Virtual Sand Table served as a one-on-one tutor, providing hints and informative feedback individually. To compare the VST students with the groups trained with the conventional approach, it was necessary to create post-hoc groups of students meeting the specifications for assembling the traditional training groups. This, in effect, would "simulate" group performance as derived from individual performance scores. The creation of these post-hoc groups required the application of resampling theory and procedures [5].

Of the 105 students who participated in the Virtual Sand Table training, 86 were appropriate for analysis. The reasons for exclusion were missing scores for either the MLRS experience variable or the after action review (AAR) score. The program Stats.exe from Resampling Stats was used to create 2,000 virtual groups, each with five students randomly selected. Each group contained at least one student with at least one year of MLRS experience, thus replicating the basis for group assignment in the conventional training procedure. The highest AAR score of each group was then assigned to the group. These scores were then compared with the group scores from the traditionally trained groups. Based on this analysis, figure 2 illustrates, a best-fit normalized curve of performance scores (AAR) computed for each group.

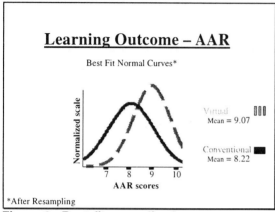

Figure 2. Best-fit normalized curves

3.3. Analysis

The hypothesis was formed that students being trained via the VST would significantly differ from students being trained on the conventional sand table on their respective learning outcomes. A two-tailed independent-sample t-test was conducted to test for significant differences. A two-tailed test was employed because there were no preconceptions about an advantage held by either group. Results from our sample indicated that students trained via the Virtual Sand Table (M= 9.07) significantly outperformed those students trained via the conventional sand table (M= 8.22), t = 11.43, p < .001.

3.4. Effect Size

The computation of effect size is a common method for determining the gain that a particular treatment has over a comparison group. In the current case, the effect size is the difference of the mean score for the VST group (9.07) and the conventional sand table group (8.22). When this difference is divided by the standard deviation of the conventional group (as a more valid estimate of the population variance when using resampling), the effect size for the VST is 1.05. This translates to a 35 percent increase in learning. The effect size found here is in line with those reported in other studies of intelligent tutors in the military and higher education, which is about 1.0 [7].

4. Discussion

The primary conclusion drawn from this experiment is that the VST intelligent tutoring system is an effective tool for training soldiers to perform the RSOP task for the Multiple Launch Rocket System. The results support the VST as a more effective training device than the conventional sand table. Two secondary implications of implementing the VST are reduced training costs and increased accessibility.

In the present study, the learning conditions were not replicated: the training was individualized, feedback during the learning process was customized, the simulation game was intrinsically motivating, and the training approach was problem based. The effect size of 1.05 reported here is indicative of the training advantage possible when intelligent tutor technology is applied in a well-designed, learner-centric training environment.

Based on the results of this research, the VST is currently in use as a replacement for the conventional sand table. It is also used as a teaching aid for the Field Artillery Officer Basic Course and the Advanced Non-Commissioned Officer Course. Modifications to the VST have been identified and funding has been provided to further improve its instructional effectiveness and extend its applicability to battery commanders.

5. References

[1] Dede, C. "The evolution of distance education: Emerging technologies and distributed learning". *The American Journal of Distance Education*, 1996, 10(2), pp. 4-36.

[2] Farr, M. J., & Psotka, J., *Intelligent instruction by computer: Theory and practice*, Taylor & Francis, Washington, DC, 1992.

[3] Kline, K. B. " Intelligent Systems for Human Resources", *Aviation, Space and Environmental Medicine*, 1998, November.

[4] Lajoi, S. P., & Lesgold, A., "Apprenticeship Training in the Workplace: Computer-Coached Practice Environment as a New Form of Apprenticeship" *Intelligent Instruction by Computer: Theory and Practice*, 1992, pp. 15-36.

[5] Simon, J. L., *Resampling: The new statistics*. Resampling Stats, Inc., Arlington, VA, 1995.

[6] Wisher, R.A. & Champagne, M.V. "Distance learning: An evaluation perspective". In S. Tobias and J. Fletcher (Eds.), *Training & retraining: A handbook for business, industry, government, and the military*, pp. 385-409, Macmillan Reference USA, New York, 2000.

[7] Woolf, B.P. & Regian, J.W. "Knowledge-based training systems and the engineering of instruction". In S. Tobias and J. Fletcher (Eds.), *Training & retraining: A handbook for business, industry, government, and the military*, pp. 339-356, Macmillan Reference USA, New York, 2000.

Evaluation of the advice generator of an intelligent learning environment

Maria Virvou
Department of Informatics,
University of Piraeus,
Piraeus 18534, Greece)
mvirvou@unipi.gr

Kabassi Katerina
Department of Informatics,
University of Piraeus,
Piraeus 18534, Greece)
kkabassi@unipi.gr

Abstract

This paper describes a formative evaluation of the advice generator of an intelligent learning environment for novice users of a GUI. 20 human tutors participated in this procedure. The human tutors had to answer theoretical questions about their way of generating advice. In addition, they were asked what they would advise learners, in real life examples, which were acquired by a previous empirical study. Their answers were compared to the responses generated by the learning environment. As a consequence, improvements were made to the system based on the evaluations' conclusions.

1. Introduction

The ultimate objective for educational software is that it should be educationally beneficial [1]. The correctness and efficiency of the advice generator of an intelligent learning environment is of vital importance for its overall usefulness. In order to ensure that the design of the advice generator was in accordance with the principles of human experts and in order to identify possible deficiencies, we conducted a formative evaluation involving human tutors. The study of the behaviour of human tutors in real educational settings may provide useful information for the development of a computer-based learning environment (e.g. [2]). A formative evaluation occurs during design and early development of a project and is oriented to the immediate needs of developers who are concerned with improving the design and behaviour of a system [3]. Its main purpose is to help the designer of an educational product during its early development stages increase the likelihood that the final product will achieve its stated goals [4]. As Dix et al. [5] point out, a design can be evaluated before any implementation work has started to minimise the cost of early design errors.

Intelligent File Manipulator (IFM) is an intelligent learning environment that can help students improve their skills while they perform their usual activities [6]. IFM is similar to a standard file manipulation program, such as Windows 98/NT Explorer but constantly reasons about every student's action. IFM is meant to operate like a human tutor who watches the student over the shoulder and provides spontaneous advice in case of an error.

The generation of hypotheses in IFM is based on a theory called Human Plausible Reasoning (HPR) [7], which has been used as a tool to generate hypotheses, about students' errors and provide a degree of certainty for advice to be proposed. HPR represents an attempt to formalise plausible inferences that occur in people's responses to different questions. The theory is grounded on an analysis of people's answers to everyday questions about the world, a set of parameters that affect the certainty of such answers and a system relating the different plausible inference patterns and certainty.

After applying the HPR theory in the advice generator of IFM we conducted a formative evaluation, to find out whether the application of this theory was successful and what improvements could take place. We also evaluated IFM's underlying rationale in relation to that usually adopted by human tutors. In this paper, we describe and discuss the evaluation methods employed and their results in connection with the design of IFM.

2. Description of the system

Intelligent File Manipulator (IFM) is an intelligent learning environment for file manipulation. IFM constantly reasons about every student's action and provides spontaneous advice in case this is considered essential. Advice is provided to students who have made an error with respect to their hypothesised intentions.

IFM's architecture follows the main line of Intelligent Tutoring Systems' architecture [8; 9; 10]. Therefore, IFM consists of four components, namely the domain representation, the advice generator, the user modeller and the user interface. As mentioned in the introduction, the generation of hypotheses is based on HPR, which is a theory about human plausible reasoning. Prior to IFM, HPR had also been successfully used for simulating the students' reasoning in another help system for a different domain [11]. However, in IFM it has been additionally used to provide a simulation of human tutors' reasoning when they form advice to be given to students.

In case, an action is considered to have been unintended by the student, IFM generates advice by applying SIM (similarity), DIS (dissimilarity), GEN (generalisation) and SPEC (specialisation) transforms in the action issued by the student. For example, in IFM if a student has clicked on a certain file having a certain name s/he could be hypothesised to have intended to click on a different file, which had a similar name. HPR is used in combination with stereotypes for student modelling.

However, there may be a variety of explanations of observed incorrect students' actions. Therefore, there is a need to attach priorities to different explanations. IFM uses the certainty parameters of HPR to determine the priority among actions belonging to the same category.

Certainty parameters of HPR have been adapted to fit IFM's requirements. The certainty parameters of HPR used in IFM are the following:
- Degree of certainty (γ)
- Degree of typicality (τ) of an action in the set of all actions issued by the student.
- Degree of similarity (σ) of a set to another set.
- Frequency (φ) of an error in the set of all actions
- Dominance (δ) of an error in the set of all errors.

In IFM, the degree of similarity (σ) is used to calculate the resemblance of two commands or two objects. The typicality (τ) of a command represents the estimated frequency of execution of the command by the particular student. The degree of frequency (φ) of an error represents how often a specific error is made by a particular student or the students of a particular stereotype. The particular student's weaknesses can be recognised by the dominance (δ) of an error in the set of all errors. Finally, all the parameters presented above are combined to calculate a probability related to every alternative command generated by IFM. This probability is called degree of certainty (γ) and represents the system's certainty that the student intended the alternative command generated.

This degree is calculated as a sum of all certainty parameters, with each parameter being multiplied to a weight, which is determined with respect to how important the particular certainty parameter is. In order to calculate the degree of certainty we proposed a formula combining all the certainty parameters of HPR. This required specifying the weights that each certainty parameter should be multiplied with. The weights that were selected are shown in the formula (1). However, these weights are to be reconsidered after the evaluation.

$$\gamma = 0.4 * \sigma + 0.2 * \tau + 0.3 * \varphi + 0.1 * \delta \qquad (1)$$

3. The experiment involving human tutors

One important aspect of an evaluation context is the reason why CAL is adopted in the first place, i.e. the underlying rationale for its development and use [1]. One of IFM's most important aims is to be able to provide plausible advice to students in a similar way as a human tutor who observes them working. As mentioned above, formula (1) represents the calculation of the degree of certainty of the system about the system's advice to be proposed to the student. In this way, HPR is used to simulate to some extent the human plausible reasoning of a human advisor. Therefore, it was among the primary aims of this evaluation to find out how close this reasoning was to human tutors' reasoning.

The experiment aimed at finding out what the human tutors' way of thinking was. For example, it was among the aims of the evaluation to find out if the human tutors were taking into account the following characteristics and how important these were to their reasoning process:
- The frequency that a user executes a particular command.
- The similarity of the action issued by the user to an alternative action that s/he might have intended.
- The frequency of execution of an error by a particular user.
- The fact that an error made by a particular user may be his/her most frequent error.

When experts are involved, there are several methods of evaluation that may be applied. Two important methods are heuristic evaluation and cognitive walkthrough. In heuristic evaluation, the expert reviewers critique to determine conformance with a short list of design heuristics whereas in cognitive walkthrough, the experts simulate users walking through the interface to carry out typical tasks [12].

In the evaluation of IFM, we have used both heuristic evaluation and cognitive walkthrough. For the heuristic evaluation, a questionnaire was given to human tutors and in cognitive walkthrough the human tutors were given real-life examples of students' use of a standard explorer to comment on their possible errors.

3.1. Questionnaires

The questionnaires given to human tutors were mainly focused on the usability heuristic about errors [13; 14]. The questions concerned the type of advice that human tutors would give to students in case of an error. More specifically, they concerned the calculation of the degree of certainty of advice given in response to an error. A sample of those questions is shown in table 1.

1	When you suggest an alternative action to a user who has just executed an action that you think as unintended, which one of the following issues do you consider and in what order of significance? a) Object or command similarity (an object is a file or a folder and a command is for example, cut or copy). b) The user's error frequency (e.g. The user has repeatedly selected an unintended command, which is neighbouring to the one s/he meant.) c) In case of error diagnosis, whether it is the most common user's mistake, related to the other types of error he commits.
2	What do you consider most probable in case a user has selected a wrong command? d) S/he has mixed up two commands with similar results. e) S/he has mixed up two commands that are either close to each other in a menu, or their shortcut buttons are close to each other.

"Table 1. A sample of questions from the questionnaire"

A lot of questions concerned what the human tutors are taking into account when generating advice so that they propose to students alternative actions to the ones they issued.

- 60% of the human tutors thought that the similarity between objects or commands was the most important aspect when generating advice.
- 55% believed that the identification of the most frequent error of an individual student was the second more important aspect that should be taken into account.
- 55% of the human tutors believed that the frequency of an observed error was to be taken into account, but not in first priority.
- Finally, 60% believed that the frequency of execution of a command was the last aspect to be taken into account.

Other question, concerned the similarity between two commands.

- 60% of the human tutors believed that the relative position of commands in the graphical representation was more important than the similarity of their result when executed. This proportion was even greater (60% - 85%) when the human tutors commented on the real-life protocols.

3.2. Protocols

After completing the first phase of the experiment, which involved the answers of the questionnaires, human tutors were asked to comment on real-life examples. All the real-life examples were acquired during an empirical study [15], which was conducted in the early stages of IFM's software lifecycle. These protocols were also given as input to IFM in order to record its responses to problematic situations of students.

For every user, the human tutors were given not only the protocols of their actions but also information about his/her user model. This was done because in order to specify what their answer would be, they should know what the characteristics of each user were. An example of an undesired user's action is the following:

Undesired action: A user issues a command for creating a Microsoft Word Document and claims s/he didn't intend to have done so.

User Model Information:
a) When this particular user creates a file, then this file is usually a sound file (80%).
b) 30% of the total errors that this user makes involve the mistaken selection of neighbouring objects or commands.
c) 40% of the total errors that this user makes involve the erroneous selection of commands in terms of their semantics.

55% of the human tutors suggested that the user intended to create a bitmap image, because this command was next to the one issued. IFM, on the other hand, suggested that the user should create a text document, because this command's result was similar to that of the command issued. However, this action was the second alternative suggested by the human experts.

IFM proposed to the user as a second alternative action to create a sound file, because this was his/her more usual action. However, this was suggested by the human tutors, as a third alternative.

4. Analysis of the results

The results of the evaluation were connected to the way that the certainty parameters were used by the advice generator. The analysis of the results from the comparison of IFM with the human tutors' comments on the protocols revealed that IFM could successfully generate the alternative actions to be suggested to the student but it was not capable of showing them in the right order, as compared to the human experts.

The categorisation of the alternative actions, which was based on the calculation of the degree of certainty, was not completely successful and had to be refined. As mentioned above, 55% of the human tutors believed that the identification of the most frequent error of a particular student was the second factor to be taken into account. Therefore, the weight of dominance was now determined to be 0.3, instead of 0.1 that it used to be. 60% of the human tutors believed that the fact that a student has repeated an error many times must have been taken into account, but not in first priority, so the weight of frequency was fixed to 0.2, instead of 0.3.

So the formula has been refined to be the following:

$$\gamma = 0.4 * \sigma + 0.3 * \delta + 0.2 * \varphi + 0.1 * \tau \quad (2)$$

Another issue of the evaluation concerned the calculation of the degree of similarity itself. The similarity of objects is calculated by taking into account the similarity of their names by 70% and their relative distance in the graphical representation. On the other hand, the similarity of commands is calculated based on their relative distance in the graphical representation by 60% and the similarity of the commands' result (40%).

These calculations of IFM were very close to those of the human experts. Therefore, the evaluation in this phase revealed that IFM was successful at reproducing the reasoning of the majority of experts concerning the calculations of similarity.

5. Conclusions

In this paper we presented and discussed the formative evaluation of the advice generator of an intelligent learning environment. The reason for this was that the advice generator plays a crucial role in the whole system and aims at rendering the advice more human-like. One important aim of this evaluation was to find out whether the reasoning from Human Plausible Reasoning theory that had been adapted to the learning environment, was successful at generating advice similar to that of human tutors. However, it was not among the goals of this evaluation to examine the overall usability of the system. Such an evaluation is to be conducted in the future.

The evaluation revealed that the learning environment had been quite successful at generating alternative commands that the students might have intended instead of the erroneous commands typed. However, it also revealed that the order that these alternative commands were proposed to students was not the same as that of the majority of human experts. The evaluation contributed to the refinement of the adaptation of HPR into the learning environment.

6. References

[1] A. Jones, E. Scanlon, C. Tosunoglu, E. Morris, S. Ross, P. Butcher & J. Greenberg, "Contexts for Evaluating Educational Software". *Interacting with Computers*, Elsevier Science, 1999, 11 (5), 499-516.

[2] Lepper, M.R. and Chabay, R.W. (1988). Socializing the intelligent tutor: bringing empathy to computer tutors. In H. Mandl and A. Lesgold (eds.), Learning Issues for Intelligent Tutoring Systems. New York: Springer-Verlag.

[3] M. A. Mark, & J. E. Greer, "Evaluation Methodologies for Intelligent Tutoring Systems". *International Journal of Artificial Intelligence in Education*, 1993, Vol. 4, pp. 129-153.

[4] C. Chou, "Developing CLUE: A Formative Evaluation System for Computer Network Learning Courseware", *Journal of Interactive Learning Research*, 1998, Vol. 10 (2).

[5] A. Dix, J. Finlay, G. Abowd, and R. Beale, *Human-Computer Interaction*, Prentice-Hall, New York, 1993.

[6] M. Virvou and K. Kabassi, "An Intelligent Learning Environment for Novice Users of a GUI. In *Lecture Notes in Computer Science (Intelligent Tutoring Systems, Proceedings of the 5th International Conference on Intelligent Tutoring Systems, ITS 2000)*, Springer, Berlin, 2000b. Vol. 1839, pp. 635-644.

[7] A. Collins and R. Michalski, "The Logic of Plausible Reasoning: A core Theory", *Cognitive Science*, Elsevier Science, The Netherlands, Vol. 13, 1989, pp. 1-49

[8] J. R. Hartley & D. H. Sleeman "Towards intelligent teaching systems". *International Journal of Man-Machine Studies*, 1973, 5, 215-236.

[9] E. Wenger, *Artificial Intelligence and Tutoring Systems*, Morgan Kaufman, Los Altos, CA, 1987.

[10] J. Self, "The Defining Characteristics of Intelligent Tutoring Systems Research: ITSs Care, Precisely", *International Journal of Artificial Intelligence in Education*, 1999, 10, 350-364.

[11] M. Virvou and B. Du Boulay, "Human Plausible Reasoning for Intelligent Help",*User Modeling and User-Adapted Interaction*, Kluwer Academic Publishers, The Netherlands, 1999, Vol. 9, pp. 321-375.

[12] B. Shneiderman (1998) *Designing the User Interface: Strategies for Effective Human-Computer Interaction*, 3rd edition, Addison-Wesley.

[13] J. Nielsen, *Usability Inspection Methods*, John Wiley, New York, 1994.

[14] D. Squires and J. Preece, "Predicting Quality in Educational Software: Evaluating for learning, usability and the synergy between them", *Interacting with Computers*, Elsevier Science, The Netherlands, 1999, Vol. 11, No. 5, pp. 467-483.

[15] M. Virvou and K. Kabassi, "An Empirical Study Concerning Graphical User Interfaces that Manipulate Files", *Proceedings of ED-MEDIA 2000. World Conference on Educational Multimedia, Hypermedia & Telecommunications*, AACE, Charlottesville VA, 2000a. pp. 1724-1726.

The Success of Advanced Learning Technologies for Instruction: Research and Evaluation of Human Factors Issues

Dan O Coldeway
Dakota State University
dan.coldeway@dsu.edu

Abstract

The development and implementation of new advanced learning technologies is an important achievement in all levels of education and in business and industry. However, the benefits of advanced learning technologies are often not fully realized until human factors issues surrounding the use of that technology by instructors, students and other users are recognized and taken into consideration. These issues are referred to as human factors issues and they can often become a major influence over the success of technological applications. This paper will identify a range of human factors issues of importance in a variety of settings. Recent evaluative and research information describing human factors in three graduate programs using advanced learning technologies will be presented and discussed. A summary discussion of these results and their impact on human factors interaction with technology will also be presented.

1. Introduction

During the mid 1980s, a group of administrators, academics and professionals at Athabasca University (a Canada wide and exclusively distance education university) produced a musical review to add humor to on-going debates about many aspects of distance education, instructional design, etc. In that review they posed the question "if technology is the answer, what was the question?". The not so hidden message in that quote suggests that technology alone is not the only answer and perhaps the question was not properly formulated in the first place.

The "not so hidden" factors within the message relate to the many human factors that influence advanced learning technologies (ALT), including human interaction, learner needs, technology use and misuse, learner support, the effect of technological change and development on human behavior. The human factors issue also extends to what has come to be known as the delivery truck debate (1). The start of this debate resulted form an article by Clark (2). Clark suggested that media of any type do not influence learning directly any more than the truck that delivers our groceries influences changes in our nutrition (2). Ten years later Kozma (3) took up the debate with this statement: "In what ways can we use the capabilities of media to influence learning for particular students, tasks, and situations?" The human factors issues reappeared and raised several questions directly related to the use of advanced learning technologies (4), (see Mellon (1) for a complete review).

Recent concerns about advanced learning technology are not dissimilar from the concerns over print, TV, and CAI. Its interesting to note that all of these "technologies" can be effective, including ALTs. The concerns reported years later appear to have more to do with human factors issues than they do with technological failure or effectiveness. It appears that if you build an instructional delivery system of any type, students and faculty will come to use it. The bigger question is how will they use it?

2. Methods

The data addressing the above themes come from graduate level programs that use a mix of delivery technologies. Although the content in all three programs differs, the mix of delivery, ALTs used and methods of instruction are similar. The goal of this case analysis methodology (4) is to identify common human factors issues of importance to the success of the programs and their respective ALT delivery systems.

The qualitative nature of this case analysis study has limitations. First, the information about each institution was obtained using a variety of methods and sources. Second, it was **not** possible to obtain a systematic sample of all student opinion regarding the range of issues under investigation. Issues concerning the use of human subjects for research and the confidentiality of evaluation data prevented a more detailed analysis of individual student responses. Finally, the comparison of information from each institution followed the modified analytic induction method (5). This method began with a formulation of potential human factors issues. As data were gathered, the issues were modified until common factors emerged. A final organization scheme was developed and information summarized for each category within that scheme.

3. Results

The methodology used in this study identified that all three categories were important in the consideration of human factors issues. The following percentages represent the relative number of comments selected for each category:

Learners: 42%
Faculty/Instructors: 38%
Technical Staff: 18%

Three raters judged the list of comments made on interviews and questionnaires in the above categories and classified them into the sub-categories.

The following opinions were chosen by raters to be human factors issues in each of the sub-categories. The top two opinions are presented.

Category One (Learners):

Needs for using ALTs:
a. Problems with interaction of ALT infrastructure requirements and home or office equipment (that could no be solved by learners).
b. The need for flexible time and alternate location for study.

Skill requirements for using ALTs:
a. Time required to learn new updates on existing ALT systems
b. Time required to learn new software applications required by ALT delivery systems.

Support for problems with ALTs:
a. Faculty lacking in support skills
b. Help desk type support not always prepared for ALT users.

Motivation to continue using ALTs:
a. Personal life issues interfering with ALT use.
b. Demand on home or office system competing with ALT needs.

Category two (Faculty/instructors):

Perception/attitude regarding ALT:
a. A view that ALT provides a second rate system compared to classroom teaching.
b. A view that time to prepare and use ALT far outweighed its benefits.

Skills in using ALT:
a. Time to mastery of skills unacceptable.
b. Changing requirements of systems require continuous learning of new skills.

Support for problems using ALT:
a. Technical support varied and non-available for some systems (i.e. PC versus Mac users, depending upon institutional policy and/or support systems.
b. Technical support availability for student related issues (compatibility of word processing systems, etc.).

Motivation to continue using ALT:
a. Changing ALT requirements and infrastructure changed motivation to continued use.
b. Frequently changing content impacted use of some systems over time.

Category three (Technical staff):

Methods of supporting students and faculty:
a. Not able to provide assistance on all software packages used by students or required in courses.
b. Difficulty with discussing problems with users while running ALT systems simultaneously.

Availability of staff to all users:
a. Limited help desk or support times (weekends and evenings)
b. In-experienced support personnel or turn-around of staff.

Range of support services:
a. Limited to institutionally supported systems not representative of all users.
b. Not often focused on individual needs.

4. Discussion

The results of this analysis indicate that human factors are inter-related to technical issues for all categories of human interaction with ALT. The direct relationship of ALT to human learning is not clear form these results. However, what is clear is that several enabling factors related to both human factor issues and ALT issues do impact learning. In other words, learning cannot take place unless certain factors are in place and under control. The delivery truck model (1, 2) may not be the best way to describe the impact of ALT, but it is also not clear that ALT alone impacts learning directly (3).

Concepts of motivation, available time, and support systems are critical to the use of ALT for all categories of users. Without these enabling factors, combined with proper equipment and user-friendly software and application programs, ALT do not meet learner or faculty needs and handicap technical assistance.

The institutional programs that provided information for this study use a wide range of ALT in course and program delivery. However, it is also clear that faculty and technical staff interact with these technologies prior to the delivery of instruction (e.g. instructional design stages, instructional preparation, and evaluation tasks). Learning is taking place at these levels, especially when new or inexperienced faculty become involved in ALT programs and courses.

Finally, an overall evaluation of the information in this study suggests that all the forms of ALT used by institutions included here work. However, there were problems in all cases, although many were overcome on an individual or small group basis. To assume any one ALT or ALT product is capable of overcoming all problems is naïve. Human factors will always play a part in the success of ALT and the more programs and institutions understand these issues the better they will be in preparing for future development and technological change.

5. References

(1) C. Mellon, Technology and the great pendulum of education, *Journal of research on Computing in Education*, 1999, 32(1), 28-36.

(2) R. Clark, Reconsidering research on learning from media. *Review of Educational Research*, 1983, 53(4), 445-459.

(3) R. Kozma, Will media influence learning? Reframing the debate. *Educational Technology Research and Development*, 1994, 42 (2), 7-19.

(4) R. Clark, Media will never influence learning, *Educational Technology Research and Development*, 1994, 42(2), 21-29.

(5) R. Stake, in Jaeger (Ed) *Methods for Research in Education*, American Educational Research Association, 1988.

(6) R. Bogdan and S. Bilken, *Qualitative Research for Education: an introduction to theory and methods*, Allyn and Bacon, 1992.

Mapping information needs

Grainne Conole

Institute for Learning and Research Technology, University of Bristol

g.conole@bristol.ac.uk

Abstract

This paper describes a way of systematising information handling in learning and research, which helps users articulate information plans within specific contexts. The rationale for this paper comes from a recognition that the increasing sophistication of online resources and virtual environments - far from making things easier for learners and researchers - can provide a distorted or biased view or filter out information that might be of relevance to the user. Furthermore, there is a need for users to think critically about their information requirements in a specific context and then assess the relevance of different online resources to meet these needs. This paper outlines a resource, which guides and support the user from the process of thinking about their information needs to map these to a specific range of online resources. The paper will describe the philosophy behind the development of this information toolkit and report an evaluation of its use to map the information needs of a researcher.

1. Introduction

Learners and researchers have a potentially vast range of resources available to them through the Internet and associated technologies. But, as the Internet increases in size and complexity, so do the associated usability and navigational issues. A range of strategies is been used to manage online information and provide different searching protocols and navigational aids and maps. Nonetheless, information overload persists. This problem is well recognised and a number of structured resources and environments attempt to address this, such as information gateways and portals, digital libraries and virtual learning environments. Furthermore there is a considerable body of research into looking at and addressing the issue of 'getting lost in hyperspace' (see for example, [1]). Similarly, Laurillard et al. have espoused the importance and role of narrative structure in supporting learning and in particular an investigation of the design features that 'afford' activities that generate learning [2]. Put simply, it is not the environments themselves that support or enhance learning, but the ways in which they are used that is important.

In principle therefore, managed environments, with tailored views for specific individuals (or types of users), are valuable, provided that they don't limit, restrict or omit information, even so it is unlikely that anyone of these tools will be sophisticated enough to meet all the needs of every individual.

In reality, user may need to adopt more of a 'mix and match' strategy, selecting individual features from relevant sources to meet specific needs. This paper outlines a resource which guides and supports the user through the process of thinking about their information needs and mapping these to a specific set of resources within a particular context.

2. Definition of toolkits

A range of resources to facilitate decision-making processes has developed to support the use and integration of Information and Communication Technologies. Within this context, the terms 'tools', 'toolkits', 'frameworks', and 'model' abound, but are very rarely used with any consistently. Indeed, there is considerable confusion and overlap within the literature on the precise nature of these types of resources. In essence, decision-making resources range from highly restrictive 'templates' or 'wizards', which provide high levels of support and step-by-step guidance but little possibility of user-adaptation, through to 'theoretical frameworks', which provide a context and scope for the work but leave the user to devise their own strategy for implementation. Between these extremes lie a range of resources, including checklists, guidelines and step-by-step tutorials.

This paper describes one example of this type of decision-making resource, in the form of a toolkit for systematising information needs. It draws on previous research work that feeds into the development of the toolkit outlined here. A detailed definition of our use of the term toolkit is defined and illustrated elsewhere [3, 4].

By way of introduction, it is worth briefly describing two related toolkits in this area. The first, Media Advisor, is a pedagogical toolkit which can be used to enable practitioners to redesign curricula and to consider how to appropriately integrate ICT alongside more traditional learning and teaching methods [5, 6]. The second, the

Evaluation Toolkit, helps practitioners to evaluate a range of learning resources and activities [4, 7].

3. The information toolkit

The information toolkit is built on a similar philosophy to Media Advisor and the Evaluation Toolkit. Toolkits are predicated on the assumptions that they will be:

- derived from an explicit theoretical framework
- easy-to-use for practitioners
- able to provide demonstrable benefit
- able to provide guidance, without being prescriptive
- adaptable to reflect the user's practice and beliefs
- able to produce outputs that reflect the local context

The information toolkit provides a means of mapping information resources against types of information activity (gathering, processing, communicating and evaluating). This gives the user a clearer view of the resources they are using, why they are using them, and allows them to form their own tailored information plan. The toolkit guides the user through the process of articulating their information needs and results in the production of an information plan for a particular task. The 'scope' of the task is one of the first stages of working through the toolkit. A task could range from considering all the information needs for a course module, a research programme or a development project.

The information toolkit consists of the following key steps:

1. Scope. Articulation of the scope of the information plan – who is this for, at what level, how long will it be used for?

2. Purpose. In this step the user considers the four information activities and specifies the relevance (or not) of each to this particular plan and the specific purpose of each activity.

3. Mapping. This final stage involves mapping the tools and methods to the information activities to produce an Information plan

1. Scoping the plan

In this section of the toolkit the user articulates the scope of the information plan (Table 1). This scoping is important because it will help the user to filter out inappropriate information sources, which may be too difficult, too rudimentary, or of secondary importance. The user may choose to describe the scope in some detail or as a rough sketch, whichever is appropriate. The scope helps to focus on what they are gathering and using the information for. Are they seeking information for a review, are they writing a report or essay, are they conducting a piece of research work?

Table 1: Defining the scope of the information plan

Scope	Illustrative examples
Description	At a minimum the scoping stage should provide a description for the focus and scale of the information plan. Questions, which might be considered here, include the following. Is this plan to be produced to support a student on a particular module? Is it information being gathering to support a research project, or relevant material for a journal review? Is it concerned with all the information needs associated with a consortium-based development project, with a view to forming a shared project resource?
Primary stakeholder	This may be the person carrying out the information plan or could be the person who has commissioned this work. Details here could include something about the person; are they a researcher, an undergraduate, a project manager. These details could be important in terms of given an indication of the quality and the level of the information plan, if a subsequent user uses it.
Secondary stakeholder(s)	These are others who might also be interested in the plan. Examples might include fellow students on a course. For example students might divide the information 'mining' for a set of modules on a course and agree to use them together as a shared resource. An example of a secondary stakeholder could be the tutor for a course or a reviewer for a research project.
Level/Type	Under/post graduate, learner/researcher, government/funding organisation, subject area?
Timescale	Is this for a particular task over the next six months or is it a general information plan for an ongoing research interest?
Resources	Who will be gathering and using the information apart from the end-user? Is there a research assistant involved in data collection for example?
Other	In this section the user can articulate any other relevant information to include in the scope of the plan, which will help them to focus on particular uses and sources of information in the next section.

2. Defining information activities

This section considers the different types of information activity that are important in the plan. The framework uses a similar classification of core information activities, comparable to the four educational activities defined in Media Advisor (delivery, discussion, feedback and activity). Of course this is a simplification. Information activities can be classified in a number of different ways and to a greater degree of granularity.

However, as with Media Advisor, the purpose here is not to provide a definite rigid classification, but to give enough guidance for the user to make informed decisions. A reasonable starting point therefore is that manipulation of information can be grouped into the following four categories:

- Seeking or gathering
- Processing or using
- Communicating or disseminating
- Monitoring or evaluating

Table 2 illustrates these four information categories in terms of their typical roles in learning and research. For simplicity these classifications will be referred to in future as gathering, using, communicating and evaluating. The user uses this table to begin to identify which of the four information activities is relevant to their own information plan. In some cases all four of activities will be involved. For example, in the production of undergraduate research project, where the user needs to gather background information on the research topic, gather and process relevant information and communicate the results in the form of a dissertation paper. The evaluation in this case will be by the tutor in terms of marking the report. In other cases, a plan might focus on only one or two activities, for example in the carrying out a literature review at the start of a research project. An information plan for a development project might focus heavily on the first three activities of gathering, using and communicating information, but be less concerned with the evaluation aspects. Therefore different plans will have different degrees of weighting across the four information activities, and it is the function of this section of the toolkit to reflect on this before beginning the more detailed mapping to information tools and methods.

Table 2: Classification of information activities

	Purpose	
	Learning	Research
Gathering	To gather information to support learning or specific activities. Information about course, university, other resources, related materials, introductory guides	To gather information for a literature review, to support a research paper, to keep abreast of current developments. Information about related research centres or individuals
Using	Manipulation of data or resources to carry out a specific learning task	Analysis of data collected using standard qualitative and quantitative research tools and methods
Communicating	Student-tutor: for checking, for support, administration, feedback, Student-student: collaboration, sharing of ideas, clarification	Communication with research peers to share or explore ideas, or with project partners, or to the wider research community to disseminate findings
Evaluating	Assessment of students, monitoring and feedback on progress	RAE, peer review through journals and conferences, success in securing funding

3. Mapping the tools and methods to information activities

This part of the toolkit gets the user to brainstorm what information will be gathered and for what purpose. The user maps a range of information sources and resources (mailing lists, email, journals, books, seminars, etc) against the four types of information activities (seeking, processing, communicating and evaluating), using the following types of questions as guidance.

- What types (or specific) information do they need to gathering?
- In the context of this work, which mailing lists are relevant, what course materials should be included, which research journals or books are important?
- What will they do with the information?
- How will they communicate their findings and with whom?
- How do they intend to monitor/evaluate or be monitored/evaluated?

This is similar to the approach adopted in the Media Advisor toolkit, where users map learning and teaching methods (including both ICT and traditional approaches) against educational activities, to give a picture of their own learning and teaching approach. Table 3 illustrates this mapping. Completed case studies of the use of the toolkit to formulate an information plan will be demonstrated at the conference.

By providing a map of the relationship between information activities and sources, the user can apply this knowledge to articulate their own individual map for specific types of activity. The map is not restrictive; the user can define additional tools or methods and can choose to adapt the information activities if they wish. The idea is to give them some structure to their thinking. It aims to guide the user through these stages and helps them to articulate different resources.

Table 3: Mapping of tools and methods with information activities

Tools and Methods	Gathering	Using	Communicating	Evaluating
Subject-gateways/digital libraries				
Online resources (course materials, tutorials, CAL-packages)				
Mailing lists, News groups, Discussion boards				
Conferences, seminars, Presentations/workshops/lectures				
Online Journals, books, etc				
Paper-based books, leaflets, etc				
Search engines				
Email				
Intranet				
Peers				
Tracking or assessment software				
Data gathering using research instruments (questionnaires, observation, video logs, focus groups, etc)				
Databases or data archives				
User defined tools and methods				

On a first iteration through this table the user is likely to identify generic sources of information, highlighting the types of tools and methods which are likely to be of most use and eliminating irrelevant ones. On subsequent iterations the user can start to build a picture of specific examples, until they are satisfied that they have a comprehensive information plan.

4. Conclusion

This paper has described an information toolkit which can be used to guide learners and researchers through a structured process of thinking about the ways in which they gather and use information to support particular activities. The toolkit has been tested with a range of users with different information needs and initial evaluation feedback has been positive. By specifying the terms of reference for the area in the form of a toolkit the user is guided through the thought processes and sequence stages which are relevant to the problem. This means that the information plan that is generated is built on a reasonable set of logical assumptions. An additional benefit is that the toolkit can be used to generate a suite of these types of information plans. These plans can be used as templates for new users which they can then adapt and update for their own needs.

Evaluation with users so far suggests that there are three potential levels of use for this toolkit. Firstly, it can be used by an individual to articulate an information plan for a particular task, however large or small. An initial plan can be developed quickly and then built over time. Secondly, information plans can be shared with peers or aggregated to form a series of templates for particular information areas, in the same way that the Evaluation Toolkit can be used to generate evaluation case studies for common types of evaluation. Thirdly, construction of these types of information plans gives an insight into the information requirements of different types of learners and researchers, and a critical analysis and evaluation of these patterns would give a fascinating insight into the cultures of different information users.

5. References

[1] Stanton, N., Correia, A.P., and Dias, P. (2000), 'Efficacy of a map on search, orientation and access behaviour in a hypermedia system', *Computers and Education*, 35, 263-279.

[2] Laurillard, D., Stratford, M., Luckin, R., Plowman, L. and Taylor, J. (2000), 'Affordances for learning in a non-linear narrative medium', *Journal of Interactive Media in Education*, 2, 1-17.

[3] Conole, 2000, 'Resources for supporting decision making', *First research seminar of the Learning Technology Theory Group*, ALT-C 2000, Manchester, September 2000, available online at www.ucl.ac.uk/~uczamao/theory/position/conole.htm.

[4] Conole, G., Crewe, E., Oliver, M. and Harvey, J., (2001, in press), A toolkit for supporting evaluation, *Association of Learning Technology Theory*, 9, 1.

[5] Conole, G. and Oliver, M. (1998), A pedagogical framework for embedding C&IT into the curriculum. *Association of Learning Technology Journal*, 6, 2, 4-16.

[6] Oliver, M. and Conole, G, (2000), 'Assessing and enhancing quality using toolkits', *Journal of Quality Assurance in Education*, 8,1, 32-37.

[7] Oliver, M., McBean, J., Conole, G. and Harvey, J., (2001, in press), 'Using a toolkit to support the evaluation of learning', *Journal of Computer Assisted Learning*.

On Workflow Enabled e-Learning Services

Joe Lin, Charley Ho, Wasim Sadiq, Maria E. Orlowska
Distributed Systems Technology Centre
School of Computer Science & Electrical Engineering
The University of Queensland, Australia
{ jlin, cho, wasim, maria }@dstc.edu.au

Abstract

Workflow technology provides a suitable platform to define and manage the coordination and allocation of business process activities. We introduce a flexible e-learning environment – called Flex-eL – that has been built upon workflow technology. The workflow functionality of Flex-eL manages the coordination of learning and assessment activities of the course process between students and teaching staff. It provides a unique environment for teachers to design and develop process-centric courses and to monitor student progress. It allows students to learn at their own pace while observing the learning guidelines and checkpoints modeled into the course process by teaching staff. We also report on the successful deployment of the concept and system for a university course and our experiences from the implementation.

1. Introduction

In the past few years, the use of information technology has increased substantially in the education domain. At the same time, expectations of students and teachers from the technology supported education systems have increased as well. More and more mature age, part time, and international students with a wide variety of education, professional and cultural backgrounds are engaging in education and training to support their career goals. They are increasingly distributed globally and have very diverse learning needs and learning styles. Flexible e-Learning solutions are required to meet their needs. The challenge is not to use new technologies to re-create traditional education systems, but rather create new learning environments, providing improvements to both teachers and students, and enhance the quality of education [1][2].

Through flexible e-learning systems students would receive education – anytime and anywhere – that matches their own pace and learning style. The idea of high quality learning experience is not to move from teacher-centered to technology-centered learning but rather to student-centered learning. Learning technologies should allow greater flexibility in supporting and enhancing learning experience.

In this paper, we will present the Flex-eL (Flexible e-Learning) system – a learning environment supported by workflow technology. We will also demonstrate how workflow technology can provide a more flexible learning solution.

2. Related Technologies

There are many research and commercial web-based educational products that have been developed and deployed all around the world. The most popular ones include Lotus LearningSpace, WebCT, BlackBoard, TopClass, etc [3][4]. Most products provide two major types of tools: Learner tools and Support tools.

Learner tools includes
- Web browsing – multimedia, security, bookmarks, etc.
- Asynchronous sharing – email, newsgroup, file exchange, etc.
- Synchronous Sharing – audio/video Chat, whiteboard, virtual space, teleconferencing, etc.
- Student tools – progress tracking, searching, motivation building, etc.

Support Tools include:
- Course – planning, managing, customizing, monitoring, etc.
- Lesson – instructional designing, presenting information, testing, etc.
- Data – marking on-line, managing records, analyzing and tracking, etc.
- Resource – curriculum managing, building knowledge, team building, etc.
- Administration – installation, authorization, registering, server security, etc.
- Help desk – student support, instructor support.

Generally, these products emphasize on learner tools such as web-based multimedia applications. Although several leading packages provide a wide range of powerful support tools for various aspects of course management, most of them are still "task-oriented" rather than "process-oriented". Some of their deficiencies can be identified as follows:

- Tools are designed to support individual learning tasks rather than the learning process.
- There is no integration of technologies that support various aspects of the study-process.
- Tools offered by educational packages are content-free resources and their adoption and integration into the study program relies on the experience of the course designer. That often results in the technology-centered learning process.
- Every educational package provides a limited set of tools and inclusion of the new tools as they become available could be very difficult.
- Generally, the educational package is used to support several individual subjects through separate accounts or workspaces and no interaction between different "accounts" is possible.
- Tracking of student learning progress is very difficult. There is very limit coordination between student's study material and time management.
- Monitoring of individual student study progress is often neglected.

We believe that integration enabled by workflow technology would provide more flexibility and a more effective learning environment. Workflows are process oriented business information systems that offer the right tasks at the right point of time to the right person along with resources needed to perform these tasks. Workflow technologies are capable of supporting control and enforcement of business processes enabling collaboration between business processes, effective time management and monitoring at various levels for various categories of users, automatic support for dynamic modification of the existing processes and relatively seamless integration of various tools and applications [5][6].

3. Flex-eL concept

The Flex-eL project is based on the concept of using workflow technology to support learning technology in order to provide an innovative, workflow-based, fully flexible learning environment to deliver education courses. In particular, Flex-eL is focusing to achieve several specific objectives:

- Relax enrolment time constraints
- Remove predefined semester duration
- Assist in enforcing academic prerequisites
- Maintain high-quality subject content
- Provide flexible learning pathways
- Support innovative learning strategies
- Allow individual time management during study
- Encourage true collaboration and work in groups
- Provide access to personal teaching assistance
- Provide effective resource management
- Provide monitoring study progress

Considering the previously mentioned deficiencies in most education packages, Flex-eL aims to support the concept of flexible learning pathways through subjects consisting of modules that, in turn, are managed by a number of learning activities. Our approach is to create student-centered learning that starts from the concept of the integrated study process that is carefully designed based on the latest educational models and supported by workflow technology. Effective integration of various learning activities is enabled by the study guide while workflow technology offers the right tasks i.e. learning activity at the right point of time to the student along with learning resources needed to perform these tasks. One of the main advantages of workflow technology, which is used as a backbone of Flex-eL, is to provide better integration of the new resources and new tools as they become available in the future. We will also demonstrate how these objectives are achieved by the workflow functionality in Flex-eL.

4. Workflow enabled learning

The idea of utilizing workflow technology to manage the learning and teaching activities came from the nature of the study process. A well-integrated study environment should include components such as learning and assessments into one fully system supported stream of activities. Workflow technology can then be used to manage these learning activities for different roles. The design of Flex-eL takes the workflow technology as the main backbone infrastructure and incorporates other technologies and tools around it to achieve a complete learning environment. Figure 1 shows the Flex-eL technology architecture.

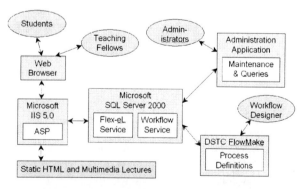

Figure 1. Flex-eL technology architecture

A process-modeling tool called FlowMake is used to capture the study process. The course activities and associated roles are identified and modeled using the tool. This predefined workflow model is then deployed in the workflow server which has been built upon Microsoft SQL server 2000. Flex-eL uses web interface to provide students and teaching fellows accessibility to the system. The study materials are presented in multimedia form. Flex-eL provides internal functionality to build study contents. However, it is also possible to link learning activities to any externally available learning material. The administration features allow setting up courses, enrolling students, and managing workflow processes.

For setting up a new course, we define the teachers in the database that are responsible for managing the new course and assign them a teaching fellow role for the new course. We then define the new course that includes creating study materials, defining tasks needed to be performed in the course, defining assessments, and scheduling assessment time slots. After that we model and export the associated process definition for coordinating the course into the workflow repository and link it with the course definition. For example, activities in the process model are associated with relevant study materials and performer roles. Exporting the process model in to a workflow repository from FlowMake also includes generating the VML code for the course process visualization. The exported process model provides a process template for the course.

Whenever a student enrolls in a course, we create his information in the database if it does not exist already. After that, we enroll the student under the requested course. Finally we start an instance of the learning process for the student based on the process template. This also means each student will have the same list of activities based on the same activity template of the same process template. It is also possible to have more than one process template for the same course. For example, one process template may have only a single assessment at the end of a study period. Another process template may have smaller assessments during the study period. The teaching fellow and student can decide between themselves which type of process template would be useful for the student.

Workflow technology offers many features that can significantly improve e-learning environments. It can automatically assign the right task to the right person. It supports individual planning of the work schedule. Students can learn at their own pace. It supports management of information and knowledge sharing. It encourages collaboration between students. It provides students, as well as teachers, an ability to monitor individual and group activities.

In Flex-eL, each course is associated with one or more workflow process templates that define the order of course activities. One of these process templates is assigned to each student when he or she enrolls in the course. This is one of the Flex-eL's unique features that offer tailored learning pathways and flexible study styles for students. Each student can learn at their own pace, without worrying about deadlines for assignments and assessments. By relaxing time constraints the flexibility for individual time management is achieved. Flex-eL's learning strategy also has the potential to speed up the studying processes. Students could complete their courses as soon as they have finalized all the required tasks assigned to them by the system. As for teaching fellows, the workload is also reduced because assessment and consultations times can be booked prior to the actual meeting. Overall, the learning and teaching effectiveness of courses is increased because of the more efficient and flexible time management.

Flex-eL offers a different learning approach than that supported by other well-known online learning management systems. Rather than making all the course material and activities available to the student at the beginning of the course, Flex-eL coordinates their availability and completion by utilizing its embedded workflow functionality. When the appropriate learning or assessment activity is completed, a new activity is assigned to the work list of the associated person.

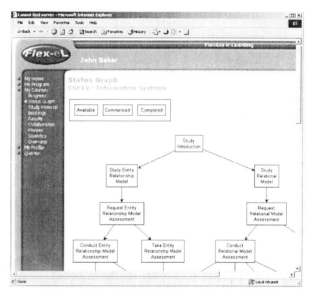

Figure 2. Flex-eL course status graph

During the study phase, the progress of each individual is captured by the workflow system. Therefore, the students have the ability to find out the information about other students who are working on the same activity. Such features encourage collaboration among students. The teaching staff is also able to monitor the progress of individual students and may provide assistance to individual students. Flex-eL provides effective collaboration between students themselves as well as students and teaching staff. In contrast to the other online learning management systems that provide chat room or discussion boards for collaboration, Flex-eL helps to identify groups of people suitable for collaboration.

One of the unique features that Flex-eL offers is the monitoring of study progress through workflow visualization. Workflow visualization plays an important part in workflow systems. It provides the information necessary to understand the interactions between workflow tasks and the decision processes. One of the ways to visualize workflow progress is to use a process model graph designed in the modeling tool as a basis for highlighting activities with appropriate status colors. Students can use this workflow functionality to visualize their current study progress and also plan for their future study pathways. Flex-eL uses VML (Vector Markup Language) to display workflow diagrams on the web interface. The diagram is dynamically generated at run time at user request and uses difference colors to represent the state of each activity. This VML workflow visualization approach has the advantage of fast accessibility and ease for understanding. It also helps teachers monitor a student's progress at a glance.

5. Observations from Flex-eL deployment

We successfully deployed Flex-eL version 1.0 at the University of Queensland for one postgraduate subject in the Master of Information Technology program. One of the positive results of the deployment was that very few students dropped out of the course offered in the Flex-eL mode in comparison to other courses offered in the traditional mode. Because of the flexibility in time management and not having to attend the lectures, students were able to manage their workload effectively.

On the basis of our experiences, we have identified several technical and design improvements that could be introduced in the new version of Flex-eL. We have also made several observations that, we believe, will help us in deploying future courses in Flex-eL environment.

Design of the study process workflow is very challenging and critical. Although we aim to provide maximum flexibility to individual study pathways, coordination between the teaching fellows and the students must be considered. For example, the assessment activity involves contributions from both parties. The definition of completing this activity should be independent for both roles so that one could not delay the other proceeding to the next activity unnecessarily.

The definition of the atomic activity is crucial. It is important to define the most appropriate size for each activity, so that the users are not repeating the same "available – commence – complete" cycle for unnecessary activities.

Another challenge is preparing students to adopt this kind of learning environment. As we provided a fully automatic self-learning system, we expect the students have basic web computer skills to use the system and it is their own responsibility to manage study time.

6. Conclusion

In this paper, we described Flex-eL - an innovative and flexible learning environment supported by workflow technology. We have identified some deficiencies of current popular e-Learning systems and proposed a new approach to overcome them by using workflow technology. We propose that a well-structured learning environment should integrate various aspects of learning by using the workflow technology. The underlying learning strategy of Flex-eL provides flexible learning pathways and possibly brings the virtual university concept closer to reality. We have also come across a number of challenges through the deployment of the Flex-eL system. These experiences have helped us identify the issues that need to be addressed when deploying workflow enabled e-learning services.

References

[1] Marjanovic, O. and Orlowska, M.E., "Making Flexible Learning More Flexible", IEEE International Workshop on Advanced Learning Technologies IWALT'2000, New Zealand, December 2000.

[2] Sadiq, W. and Orlowska, M.E., "Flex-eL – Managing e-Learning through Workflows", SAP Research and Applications Congress 2001, San Diego, USA, February 2001.

[3] "Comparison of Online Course Delivery Software Products", Centre for Instructional Technology, Marshall University, http://www.marshall.edu/it/cit/webct/compare/comparison.html

[4] "Review of leading asynchronous, web based course delivery tools", University of Tennessee, http://www.outreach.utk.edu/weblearning/reviewasych.htm

[5] Georgakopoulos, D., Hornick, M. and Sheth, A., "An Overview of Workflow Management: From Process Modeling to Workflow Automation Infrastructure", Journal on Distributed and Parallel Databases, 3(2):119-153, 1995.

[6] Sadiq, W. and Orlowska, M.E., "On Capturing Process Requirements of Workflow Based Information Systems", In Proceedings of the 3rd International Conference on Business Information Systems, Poznan, Poland, April 1999.

Acknowledgements – The work reported in this paper has been funded in part by the Cooperative Research Centres Program through the Department of the Prime Minister and Cabinet of the Commonwealth Government of Australia.

Issues and Methods for Evaluating Learner-Centered Scaffolding

Chris Quintana, Joseph Krajcik, Elliot Soloway
Center for Highly Interactive Computing in Education
University of Michigan
1101 Beal Ave., Ann Arbor, MI 48109
quintana@umich.edu

Abstract

Learner-centered design is an evolving design perspective focusing on the design of software scaffolds to support novice learners in using software to do and learn previously unknown work practices. Scaffolding evaluation is necessary to see how well learners can perform new work activity using the scaffolded software. However, scaffolding evaluation is challenging because of the need to address several issues resulting from the interplay between user-centered and learner-centered perspectives. Scaffolding evaluation methods need specific criteria to address learner-centered concerns, traditional usability measures to address user-centered concerns, and considerations of the instances when usability conflicts with the learning goals of scaffolded tools. In this paper, we contribute a discussion of these learner-centered evaluation issues and a scaffolding evaluation method to assess, not just the usability of the software, but also the effectiveness of the scaffolds in terms of the learning goals behind the software.

1. Introduction: Background and Motivation

Learner-centered design (LCD) is an evolving design perspective addressing the needs of *learners*—a specific audience trying to work in and understand new work practices in which they are novices [7]. Learner-centered design involves designing software that incorporates *scaffolding strategies* informed by learning theories [10]. Scaffolds are software features that support novices in seeing and performing previously unknown and inaccessible tasks [2]. By adopting a social constructivist "learning by mindful doing" viewpoint, scaffolds allow the doing of new tasks, which allows learners to gain new understanding about those tasks [9, 10].

Learner-centered design needs additional design methods beyond those used in traditional user-centered design (UCD) because the primary foci of the two design perspectives differ [5]. Where UCD focuses on the tool usability, LCD focuses on tools to support learning new work activity. Keeping this distinction in mind, we have previously formulated methods for analyzing work practices to design and implement necessary scaffolds [7].

Having concentrated on design methods for LCD, we are now in the next phase of our research: learner-centered software evaluation. The central question we are addressing here is: what specific LCD evaluation methods and criteria can we use to assess the effectiveness of learner-centered scaffolds in supporting learners to do and learn complex new work?

2. Issues for LCD Evaluation

LCD evaluation is a challenging endeavor. It is difficult to simply assess scaffolds in terms of a single "good/bad" rating because we need assessment methods and criteria along user-centered dimensions (e.g., are scaffolds usable?) and learner-centered dimensions (e.g., do scaffolds support "learning by doing"?). We find that there are several scaffolding evaluation issues arising from the interplay between UCD and LCD. These concerns need to be addressed to provide effective evaluation information.

2.1. Issue 1: Formulate Additional LCD Evaluation Criteria

LCD evaluation needs additional criteria to obtain evaluation information important to LCD, but not necessarily the focus of UCD. User-centered design focuses on the conceptual gaps between user and tool. So UCD assesses a tool's ease-of-use and work efficiency while using the tool [3]. Learner-centered design focuses on the conceptual gap between the learner and the expertise needed to successfully participate in a new work practice [5]. So LCD needs methods and criteria to evaluate how well learners mindfully perform and understand new work activity given the work scaffolding features in the software.

2.2. Issue 2: Integrate UCD Evaluation

LCD evaluation methods also need to integrate traditional usability evaluation because usability problems can affect the effectiveness of learner-centered scaffolds. While the focus of LCD and UCD differ, LCD should still have usability concerns. If we adopt a "learning by doing" approach, we need to insure that scaffolds make the "doing" possible. Unusable scaffolds will not allow learners to perform the underlying work supported by the scaffold.

2.3. Issue 3: Identify UCD/LCD Conflicts

LCD evaluation methods also need to consider possible conflicts between LCD and UCD because scaffolds that make work too easy to do may adversely affect LCD learning goals. Good usability in itself does not guarantee a good LCD scaffold [8]. Learner-centered scaffolds should support *mindful* completion of new work activity. Reflection on the work is necessary to understand that work. If a scaffold makes the work too easy to perform, learners may complete their tasks without the needed reflection, which can be detrimental to learning. Thus, LCD evaluation methods need to consider when ease-of-use interferes with learning.

3. Research Context

These LCD evaluation issues needed to be articulated and addressed to evaluate Symphony, a scaffolded integrated tool environment that supports the work of science inquiry [7]. Symphony integrates a collection of individual science tools with *process scaffolding* that helps learners engage in a range of science inquiry activities: planning, data collection, visualization, modeling, and argumentation.

The full set of Symphony scaffolds cannot be reviewed in this paper (see [4, 7] for more details), but here is one scaffold example to ground the discussion. Consider the work activity of planning a science investigation. Learners lack an expert's conceptual understanding of the science inquiry process and thus need information about possible inquiry activities for their plan [7]. The process wheel (figure 1) is a scaffold supporting the selection of inquiry activities by visually describing the space of possible inquiry activities. Students set up a plan by dragging inquiry activities from the process wheel into a plan they are building.

Six ninth-grade student pairs used Symphony in an after-school environmental science class to work on a variety of teacher- and student-authored questions about Michigan air quality. Each student pair worked on personal computers connected to process video kits that videotaped the computer screen as the students worked. The students also wore clip-on microphones to record their commentary. Thus, the primary data for the study consisted of a videotaped record of the students' Symphony work synchronized with their audio commentary.

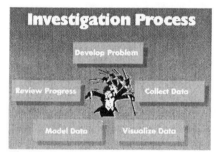

Figure 1. Symphony process wheel

4. Scaffolding Evaluation Method

The Symphony evaluation involved analyzing the student video to see how students used the individual scaffolding features to perform their different science inquiry activities. The analysis was based on a method described by Chi [1] to iteratively decompose video data of overall work activity into smaller activity units that can be interpreted in a more structured manner.

Our analysis method includes three phases to analyze the different instances when student pairs used scaffolds. Those scaffold uses are analyzed and summarized using evaluation criteria we devised. The evaluation method was applied to all the Symphony scaffolds, but our description here will refer to the earlier process wheel example. Note that this method is applied to each student pair after they completed their investigations.

4.1. Phase One: Coding by Individual Work Activity

Phase one analyzes *what* work students performed and what scaffolds students used. We isolated specific work episodes from the student video and decomposed them into smaller episodes of activity. In the Symphony trial, work episodes corresponded to specific science inquiry activities, such as planning, data collection, modeling, etc. After identifying the work episodes, we identified the scaffolds used in each episode. The result is a collection of annotated work episodes from the video. For example, a planning episode will contain uses of the process wheel (and possibly other scaffolds).

4.2. Phase Two: Scaffolding Usage Evaluation

Phase two analyzes *how* students used the software scaffolds to perform the activity in each work episode. This phase begins by applying evaluation criteria to each scaffold use instance in the different work episodes. This describes how the students used the scaffolds to perform the underlying work activity. Table 1 describes the criteria we devised to evaluate the scaffolds [6].

Table 1. Scaffolding evaluation criteria

Evaluation Criteria	Description
Accessibility	Can learners access and use a scaffold?
Use	Do learners use an accessible scaffold?
Efficiency	How fast or how easily can learners use a scaffold?
Accuracy	Do learners accurately complete the work tasks supported by a scaffold?
Progression	Do learners progress through their work tasks in a linear fashion (i.e., novice-like) or in an opportunistic, non-linear manner (i.e., expert-like) over time?
Reflectiveness	Do learners reflect on the task supported by a scaffold or are they simply completing the task without much reflection or discussion?

Each scaffold use instance in a work episode is reviewed and analyzed with respect to the evaluation criteria to create an individual scaffold evaluation report for that instance. For example, we apply the criteria in each use of the process wheel to see how well it supported the underlying task of inquiry activity selection:

- Is the process wheel *accessible* and do students *use* it?
- Can students *efficiently* select activities from the wheel?
- Do students *accurately* select activities for their plan?
- Do students *reflect* on their activity selection task?
- Do students *progress* from selecting activities in "wheel order" to selecting activities as needed?

After applying the criteria to all the individual scaffold uses, then next step involves defining specific work episode units by selecting time periods over which to group the related work episodes. By defining related sets of work episodes (or *work units*), different "slices" of scaffold use over time can be observed to gauge how scaffold usage varied and how students' work changed over time. For our study, we selected three work units corresponding to the three problem units the students performed. This was a natural decomposition to analyze how scaffold use changed as the students moved from simpler to more complex investigations.

The final step in this evaluation phase involves *summarizing* the scaffold evaluation reports for each scaffold to see how each scaffold was used in each work unit. For example, each work unit's collection of individual process wheel evaluation reports is summarized to create a process wheel *scaffold usage summary* for that work unit. The result is a collection of scaffold usage summaries for each scaffold. In the Symphony study, since there were three work units, each scaffold will have three scaffold usage summaries—one describing how the scaffold was used in each work unit.

4.3. Phase Three: Scaffolding Meta-Summarization

Phase three summarizes the individual scaffold usage summaries from phase two to describe how scaffold usage changed over time. First, a scaffold's individual scaffold usage summaries for each student pair are consolidated to create a *group summary* describing how *those learners* used a scaffold over time. For example, our study had three student pairs, so each scaffold has three group summaries. All the group summaries for each scaffold are then consolidated into a final overall group summary describing how all the students used that scaffold.

Similarly, the second step consolidates each scaffold's individual scaffold usage summaries to create a *unit summary* describing how *all* the student pairs used a scaffold in *each* work unit. For example, our study defined three work units, so each scaffold has three unit summaries. All the unit summaries for each scaffold are then consolidated into a final overall unit summary describing how all the student pairs used that scaffold over time to do their work.

The distinction between unit and group summaries is important. Group summaries describe how a particular student pair used a scaffold throughout the study. Unit summaries compare how different students used a scaffold in the same unit. The final overall group and unit summaries should be identical, serving as a "check" to make sure that the evaluation proceeded correctly.

The final step of the evaluation involves reviewing the different scaffold summaries to write an assessment overview describing the "lessons learned" from the scaffold evaluation. The assessment overview should discuss the pros and cons of the scaffold in the given work context and how the scaffold could be improved.

5. Addressing the Scaffolding Evaluation Issues

Our scaffolding evaluation method addresses the LCD evaluation issues to help assess our scaffold designs in terms of usability and how scaffolds impact work performance and learning. While the full set of Symphony evaluation findings is beyond the scope of this paper (see the complete findings in [4]), we can briefly review how the evaluation method addresses the LCD evaluation issues that we posed earlier.

5.1. Addressing Issue 1: Formulate Additional LCD Evaluation Criteria

We address this issue by adding the learner-centered evaluation criteria (i.e., accuracy doing the work, progression, reflectiveness). The learner-centered criteria helped us focus on how students engaged in different aspects of the work practice they are learning. For example, our evaluation showed that while students could use the process wheel, the "wheel" representation led them to initially select activities in a linear, clockwise manner that led to inaccurate activity selection and lower reflection on the selection task. Students improved their activity selections as they proceeded through their investigations, but we saw that we should modify the process wheel to support better reflection and activity selection.

5.2. Addressing Issue 2: Integrate UCD Evaluation

We address this issue by also integrating user-centered criteria in the evaluation method (i.e., tool accessibility, use, efficiency). As we observe students using different scaffolds, we considered usability problems to see if the scaffold's functionality is accessible or if learners can use the scaffold with reasonable efficiency. By also considering usability issues in our LCD evaluation criteria, we found some examples where poor usability interfered with proper use of some scaffolds. For example, some scaffold triggers were confusing and thus students could not access the scaffold. This usability problem interfered with appropriate performance of the underlying work activity that we wanted learners to engage in.

5.3. Addressing Issue 3: Identify UCD/LCD Conflicts

We address this issue by simultaneously correlating user-centered and learner-centered information as we look at scaffold use over time. By considering both user-centered and learner-centered criteria, we note the interplay between tool use and work understanding in our evaluation. For example, we found that process map scaffolds meant to structure some complex tasks actually made the tasks so easy and efficient to perform that task reflection suffered, which is not advantageous for LCD despite the high usability of the scaffold. Others have also noted that making a tool too easy to use can interfere with the learning goals of the software [8].

6. Concluding Remarks

Evaluation determines the effectiveness of software. For LCD, "effectiveness" is defined by how well software scaffolds support learners in doing and learning new work practices. Here we outlined issues to consider in LCD evaluation. We also described the LCD evaluation method and criteria that we have used to assess the effectiveness of learner-centered scaffolds.

The method and criteria presented here address one aspect of LCD evaluation. LCD also requires more traditional pre- and post-testing evaluation to see how well learners understand a new work practice after using the software for some time period [6]. We also want to categorize scaffold types to distill the scaffold characteristics that are more and less supportive for different types of learners (e.g., considering learners' ages, learning styles, etc.). Our future work involves such categorization along with continual development of our evaluation method as we use it to evaluate LCD software projects developed by us and by other researchers.

7. Acknowledgements

This material is based on work supported by the National Science Foundation under Grant No. REC 99-80055.

8. References

[1] M. Chi, "Quantifying Qualitative Analyses of Verbal Data: A Practical Guide", *Journal of the Learning Sciences*, 6(3), 1997, pp. 271-315.

[2] Metcalf, S.J, *The Design of Guided Learner-Adaptable Scaffolding in Interactive Learning Environments*, Unpublished dissertation, University of Michigan, 1999.

[3] D. Norman, "Cognitive Engineering", In D.A. Norman & S.W. Draper (Eds.), *User Centered System Design*, Lawrence Erlbaum Associates, 1986.

[4] Quintana, C., *Symphony: A Case Study for Exploring and Describing Methods and Guidelines for Learner-Centered Design*, Unpublished dissertation, University of Michigan, 2001.

[5] C. Quintana, J. Krajcik, & E. Soloway, "Exploring a Structured Definition for Learner-Centered Design", In B. Fishman & S. O'Conner-Divilbiss (Eds.) *Proceedings of ICLS 2000* (Ann Arbor, June), Lawrence Erlbaum Associates, 2000.

[6] C. Quintana, E. Fretz, J. Krajcik, & E. Soloway, "Assessment Strategies for Learner-Centered Tools", In B. Fishman & S. O'Conner-Divilbiss (Eds.) *Proceedings of ICLS 2000* (Ann Arbor, June), Lawrence Erlbaum, 2000.

[7] C. Quintana, J. Eng, A. Carra, H. Wu, & E. Soloway, "Symphony: A Case Study in Extending Learner-Centered Design Through Process Space Analysis", *Proceedings of CHI '99* (Pittsburgh, May), ACM Press, 1999.

[8] K. Sedighian, M. Klawe, & M. Westrom,, "Role of Interface Manipulation Style and Scaffolding in Cognition and Concept Learning in Learnware", *ACM Transactions in Computer-Human Interaction*, 8(1), 2001, pp. 34-59.

[9] J. Singer, R.W. Marx, J. Krajcik, & J. Clay Chambers, "Constructing Extended Inquiry Projects: Curriculum Materials for Science Education Reform", *Educational Psychologist*, 35(3), 2000, pp. 165-178.

[10] E. Soloway, S.L. Jackson, J. Klein, C. Quintana, J. Reed, J. Spitulnik, S.J. Stratford, S. Studer, J. Eng, & N. Scala, "Learning Theory In Practice: Case Studies of Learner-Centered Design" *Proceedings of CHI '96* (Vancouver, Canada, April), ACM Press, 1996.

[11] D. Squires, & J. Preece, "Predicting Quality in Educational Software: Evaluating for Learning, Usability, and the Synergy Between Them", *Interacting with Computers*, 11, 1999, pp. 467-483.

Towards Evaluating Learners' Behaviour in a Web-Based Distance Learning Environment

Osmar R. Zaïane and Jun Luo
Department of Computing Science, University of Alberta
{zaiane, jun}@cs.ualberta.ca

Abstract

The accessibility of the World-Wide Web and the ease of use of the tools to browse the resources on the Web have made this technology extremely popular and the means of choice for distance education. Many sophisticated web-based learning environments have been developed and are in use around the world. Educators, using these environments and tools, however, have very little support to evaluate learners' activities and discriminate between different learner's on-line behaviours. In this paper, we exploit the existence of web access logs and advanced data mining techniques to extract useful patterns that can help educators and web masters evaluate and interpret on-line course activities in order to assess the learning process, track students actions and measure web course structure effectiveness.

1. Introduction

The World-Wide Web is becoming the most important media for collecting, sharing and distributing information. Web-based applications and environments for electronic commerce, distance education, on-line collaboration, news broadcasts, etc., are becoming common practice and widespread. Distance education is a field where web-based technology was very quickly adopted and used for course delivery and knowledge sharing. Typical web-based learning environments such as Virtual-U [5] and Web-CT [1] include course content delivery tools, synchronous and asynchronous conferencing systems, polling and quiz modules, virtual workspaces for sharing resources, white boards, grade reporting systems, logbooks, assignment submission components, etc. In a virtual classroom, educators provide resources such as text, multimedia and simulations, and moderate and animate discussions. Remote learners are encouraged to peruse the resources and participate in activities. However, it is very difficult and time consuming for educators to thoroughly track and assess all the activities performed by all learners on all these tools. Moreover, it is hard to evaluate the structure of the course content and its effectiveness on the learning process. Resource providers do their best to structure the content assuming its efficacy. Educators, using Web-based learning environments, are in desperate need for non-intrusive and automatic ways to get objective feedback from learners in order to better follow the learning process and appraise the on-line course structure effectiveness.

Web-based course delivery systems rely on web servers to provide access to resources and applications. Every single request that a Web server receives is recorded in an access log mainly registering the origin of the request, a time stamp and the resource requested, whether the request is for a web page containing an article from a course chapter, the answer to an on-line exam question, or a participation in an on-line conference discussion. The web log provides a raw trace of the learners' navigation and activities on the site. While web logs are relatively information poor, present mixed accesses of different users, contain erroneous and irrelevant entries, and are extremely large, there are techniques for web log cleansing and transformation as well as advanced approaches for discovery of hidden and useful patterns from these access logs. Web usage mining refers to non-trivial extraction of potentially useful patterns and trends from large web access logs. In the context of web-based learning environments, the discovery of patterns from navigation history by web usage mining can shed light on learners' navigation behaviour and the efficiency of the models used in the on-line learning process. The patterns discovered can be used to evaluate learners' activities, but can also be used in adapting and customizing resource delivery, providing automatic recommenders for activities, etc. These patterns, however, cannot be extracted with simple statistical analysis.

Currently there is a variety of web log analysis tools available. Most of them, like NetTracker, webtrends, analog and SurfAid, etc., provide limited statistical analysis of web log data [8]. For example, a typical report has entries of the form: "during this time period t, there where *n* clicks occurring for this particular web page p". However, the results provided by these tools are limited in theirs abilities to help understand the implicit usage information and hidden trends. What is needed is summarization of these trends that can be interpreted by educators delivering their courses on-line.

There are more sophisticated tools that use data mining techniques and go beyond these rudimentary statistical analyses. Due to the importance of e-commerce and the lucrative opportunities behind understanding on-line customer purchasing behaviours, there is tremendous research effort in developing data mining algorithms and systems tailored for e-business related web usage data mining [4]. For example, WebSIFT [3] is a comprehensive web usage tools that is able to perform many data mining tasks. WUM [6] is special web sequence analyser for improving web pages layout and structure. A versatile system, WebLogMiner [8], uses data warehousing technology for pattern discovery and trend summarization from web logs.

Although these web usage-mining tools have been successfully applied to some degree in e-commerce applications, few of them are flexible enough to adapt to an on-line learning environment. Moreover, while the nature of the patterns to be discovered can be the same in both domain applications, the identification of users, hits and sessions as well as the interpretation of activities, and thus the needs of the application are significantly different. We suggest a flexible framework for web usage mining in the context of on-line learning systems where the users can express constraints at the data gathering and transformation stage, as well as at the patterns discovery and analysis steps. This way the users (i.e. educators) can tailor the data mining process to their needs and tasks at hand. The dilemma is that educators are already overwhelmed with complicated tasks pertaining to delivering courses on-line and should not be burdened with additional intricate data mining tasks, yet they need to iteratively interact with the data mining system in order to extract meaningful and useful patterns form learners' activity history. We have designed our system taking this into consideration. The complex algorithms are transparent to the users, but the needs can be simply expressed by constraining the system at different levels using plain filters and a straightforward query language to sift through the patterns discovered.

In the next section we describe the three-tier architecture for our open-structure and interactive web usage mining system. We briefly present in the third section some of our algorithms used and portray in the subsequent section, some of the experiments we conducted on real log files.

2. System framework

Data mining from web access logs is a process consisting of three consecutive steps: **data gathering and pre-processing** for filtering and formatting the log entries, **pattern discovery** which consists of the use of a variety of algorithms such as association rule mining, sequential pattern analysis, clustering and classification on the transformed data in order to discover relevant and potentially useful patterns, and finally **pattern analysis** during which the user retrieves and interprets the patterns discovered [7]. The pre-processing stage is arguably the most important step and certainly the most time consuming. Web usage data often contains irrelevant and misleading entries that need to be eliminated. Moreover, since hits of all users are combined and in impractical format in the web log, it is necessary to transform the entries into a format viable for data mining algorithms after identifying individual users, sessions and transactions.

Our web usage mining system also adopts this three-tier architecture, although we added the possibility to express specific constraints at the different levels of the system. In the context of an e-learning environment with a data-mining-based evaluation system, the users are often educators who are not necessarily savvy in data mining techniques. The constraint-based approach we suggest allows the user (i.e. educator) to simply express needs by specifying restrictions and filters during the pre-processing phase, the patterns discovery phase, or the patterns evaluation phase. Indeed, two educators using the same web server for their courses may have different requirements for learner behaviour evaluation. Even the same user evaluating different course activities at different times for different learners can have diverse requirements with regard to the data sources, the relevant attributes or the types of patterns sought for. Furthermore, defining filters during the pre-processing phase considerably reduces the search space, pushing constraints during the mining not only accelerates the process but also controls the patterns discovered, and expressing constraints at the evaluation phase helps sifting through the large set of patterns extracted. The ability to add limitation and control at all stages allows interactive data mining with ad-hoc constraint specification leading to the discovery of relevant and restrained patterns, pertinent to the evaluation task at hand. For instance, in our implementation, the user can pick filters in the pre-processing phase to select desired student or student group, the desired time period and/or the relevant subset of web pages in order to zero-in the learning tasks and activities to evaluate. In addition, educators can define their interpretation of "session" and sequence of student's clicks, concepts important in the web log data transformation. For example, a session can be defined as the sequence of clicks of one student, which happen each time from "log in" and "log out" the web environment. Also, educators can define a session as a series of clicks of one student happening in the specified period after the certain specified action. Most data mining algorithms, thereafter, use these sessions as the basic units for searching patterns.

For the pattern discovery, several algorithms, including association rule mining, inter-session frequent pattern mining, intra-session frequent pattern mining, etc., have been chosen to discover the strong trends and relationships from web usage data. The constraints that are provided to the educators to state are mainly related to these algorithms. These constraints can be used to conduct the knowledge discovery process and limit the search space. Stating the constraints is the only (optional) interaction with the data mining modules. Knowledge of the intricate algorithms is not necessary. Another noteworthy point is that the architecture of the system allows a plug-and-play of new data mining modules without significant change in the system, allowing addition of new pattern discovery functionalities.

In the last stage of pattern analysis, the objectives are to make the discovered patterns easy to interpret for the decision makers. We have implemented intuitive graphic charts and tables for pattern visualization and understanding. We intend to add an ad-hoc query language that would allow the weeding-out of irrelevant patterns and the focus on knowledge discovered to use for the evaluation of learners' on-line.

3. Algorithms

The modular design of our system allows us to add as many new data mining algorithms as necessary without compromising the effectiveness of the pattern discovery and evaluation process. We have implemented a variety of algorithms with intuitive interfaces. For example we put into practice association rule mining for discovering correlations between on-line learning activities, two variants of sequential pattern mining for studying the sequences of on-line activities within a learning session or between sessions, and clustering to group learners with similar access behaviours. In this paper, we take the association rule discovery as an example.

Association rule discovery is a classical data mining problem [2]. It shows the correlations among items within transactions. Given an item set I (in our case a set of pages or URLs), and a transaction data set T where each transaction $t \subset I$, X, Y are two different item-sets, $X \subset I$, $Y \subset I$, $X \cap Y = \phi$, then, $X \Rightarrow Y$ is an association rule with two measures: support and confidence, where support is the percentage of the transactions containing $X \cup Y$ and confidence is the percentage of transactions containing Y on the condition of containing X. For example, an association rule looks like: 30.5% of the students who successfully finished Exercise 3 also accessed Section 4 of Chapter 2. Depending upon the support and confidence thresholds, a large number of rules can be discovered and sifting through them can be tedious. A constraint-based association rule can be more useful and interesting for the educators trying to evaluate with different requirements.

For the algorithms-related constraints, the educators can set the requirements like strong support threshold, strong confidence threshold as in other association rule discovery applications. However, in addition to those two constraints, the educators can also specify constraints on the item-sets X and Y. For instance, the educators can direct the algorithms to search for the rules that answers "How often the students check out on-line resources when they read the Section 1 of Chapter 1 in one transaction."

4. Experiment

Currently we are experimenting our system on web logs from two systems: an in-house built system at the Technical University of British Columbia (TechBC), a university that delivers most its courses on-line, and Virtual-U, a web-based learning environment built in the context of the TeleLearning Canadian Centres of Excellence. The example we use for illustration in this paper comes from a TechBC web log with records of 100 students' on-line activities in two courses, TECH 142 and TECH150 from September 14th, 1999 to December 17th, 1999. There are 200,433 entries in this web log file of a size of 109 Megabytes.

One typical entry in the original log file looks as follows:
1.1999-09-14 22:02:13,200,
"/TECH150.1/Unit.2/Presentation.1/FAQ/index.html","-"

This entry shows that the user with ID "1" successfully visited at "1999-09-14 22:02:13" the web page "/TECH150.1/Unit.2/Presentation.1/FAQ/index.html".

Since the URL syntax of this web site encodes the structure of the site, when pre-processing the web log, we provide a way to generalize the log entries. For example, if a student visits these following web pages successively:
"/TECH142.1/Unit.1/LearningPath.1/ActivitySequence1/External1.html",
"/TECH142.1/Unit.1/LearningPath.1/ActivitySequence1/favicon.ico",
"/TECH142.1/Unit.1/LearningPath.1/ActivitySequence1/index.html",
"/TECH142.1/Unit.1/LearningPath.1/ActivitySequence1/tooltip.htc",
we can generalize these four clicks into one action, say, "Tech142.1, unit 1, Learning Path 1, and Activity Sequence 1". We might even generalize it in a higher-level like "tech142.1, unit 1". These drill-down and roll-up functionalities are provided to the decision makers to manipulate the data set and impose a concept level during the constraint-based mining process.

We assume that the educators use the "log-in" and "log-out" as the starting point and ending point of each transaction. We are taking 15 minutes as the upper limit of the time interval between two successive inter-transaction clicks to break the sequence of one student's click stream into the transactions.

Two experiments are presented in this paper to demonstrate the advantages of using constraint-based web usage mining in the context of e-learning. The first experiment is to find associations between visited pages using the whole web log and without use of interactive constraint specification. The second experiment takes advantage of constraint specification in particular at the data pre-processing phase. Both experiments aim at finding association rules of the same significant level with support=0.3 (supported by at least 30% of the sessions) and confidence=0.4 (the rule discovered is at least 40% confident). In the second experiment, we were interested at the students 1 to 12 and the web pages relevant to the course "TECH142". This could be the case were the educator would want to understand the on-line behaviour of students 1 to 12 who outperformed other students.

Although the second experiment deals with a subset of the filtered web log, it still finds 193 association rules with 17 frequently visited web pages, compared to 23 association rules with 4 frequent web pages found in the first experiment due to the support being 30% of the whole dataset mined. Moreover, rather than only showing the correlations among the 4 entry pages in the first experiment, the second experiment gives the educators a better idea about relationships with respect to "TECH142" web pages. For example, the educator can discover that 83% of the students who worked on "TECH142.3 Unit.1 LearningPath.1 ActivitySequence1" also visited the "PriorKnowledgeAssessment" of same Learning Path. The educator could act on this by either recommending activities or pages to students to improved their learning accomplishments, or change the structure of the on-line course towards a structure that helps the learners perform as sought for by the educator.

In summary, our system provides a powerful mechanism that makes it much easier for the educators to find the interesting rules that could be used for student access behaviour evaluation.

5. Conclusion and future work

Web usage mining has proven very useful in many e-Commerce web log analysis applications. However, the current web usage mining systems are limited in their ways to support interactive data mining and therefore they are limited in their ways to be applied in the field of web-based learning evaluation. We have implemented a system that takes advantage of the latest data mining techniques and pushes constraint specification at all stages of the web usage mining to help the educators control and guide the knowledge discovery, and effectively and efficiently understand the students' behaviours in e-learning sites.

We are in the process of enhancing the user interface of our system with the help of practitioners and educators using web-based learning environments in order to develop a more intuitive interface for constraint-based data mining and pattern visualization for the specific purpose of evaluating on-line learning.

7. References

[1] WebCT: http://www.webct.com/
[2] R. Agrawal, G. Srikant, Fast algorithms for mining association rules, Proceedings of the 20th VLDB conference, pp. 478-499, Santiago, Chile, 1994.
[3] R. Cooley, B. Mobasher, J. Srivastava, Web Mining: Information and Pattern Discovery on the World Wide Web, Proceedings of the ninth IEEE international conference on Tools with AI, 1997.
[4] M. N. Garofalakis, R. Rastogi, S. Seshadri, K. Shim, Data Mining and the Web: Past, Present and Future, Proceedings of WIDM99, Kansas City, U.S.A., 1999.
[5] C. Groeneboer, D. Stockley, T. Calvert, Virtual-U: A collaborative model for online learning environments, Proceedings Second International Conference on Computer Support for Collaborative Learning, Toronto, Ontario, December, 1997.
[6] M. Spiliopoulou, L. C. Faulstich, K. Winkler, A Data Miner analyzing the Navigational Behaviour of Web Users, Proceedings of workshop on Machine Learning in User Modeling of the ACAI'99, Creta, Greece, July, 1999.
[7] J. Srivastava , R. Cooley, M. Deshpande, P. Tan, Web Usage Mining: Discovery and Applications of Usage Patterns form Web Data, SIGKDD Explorations, Vol.1, No.2, Jan. 2000.
[8] O. R. Zaïane, M. Xin, J. Han, Discovering Web Access Patterns and Trends by Applying OLAP and Data Mining Technology on Web Logs, Proceedings from the ADL'98 - Advances in Digital Libraries, Santa Barbara, 1998.

On monitoring study progress with time-based course planning

Isaac Pak-Wah Fung
Institute of Information Sciences & Technology
Massey University
Palmerston North, New Zealand
P.W.Fung@massey.ac.nz

Abstract

This paper extends the notion of course chart by incorporating time-based information. As more and more learning materials are accumulating on the web, the burden imposed on the student in managing his own study also becomes heavier. Without suitable guidance, the ideal of providing learning opportunities to any person, anywhere, anytime would be virtually impossible to achieve if the students are getting lost in the cyber-campus. To avoid prolonged and unnecessary browsing and thereby affecting the progress of the student's entire study plan, a mechanism to monitor the progress of the students in terms of the time they spent on the topics has been developed. By adapting project management techniques into the context of study management, a course advisory system is proposed that monitors the study progress of student and alerts the instructor if the situation warrants.

1. Introduction – The Nature of Study Management

When students decided to pursue a course, they would have committed themselves to an activity with a specific goal occupying a specific period of time. Studying is a finite activity, not only in time, but also in the use of resources. Study management (SM) therefore, is concerned with the pursuit of a specific learning goal, using given resources over a defined time period. This will often require the planning and establishment of a study environment, the acquisition of resources, the scheduling of relevant activities and evaluation/review of the completed activity. Project management has always been an important component in commerce and industry but not so in educational institutions. As both the students and education providers are facing huge demand of upgrading their knowledge and the respective delivering mechanisms, the need of looking into the issue of SM becomes pressing. Figure 1 proposes a seven phase life-cycle to summarise the key stages of the endeavour of undertaking a course of study.

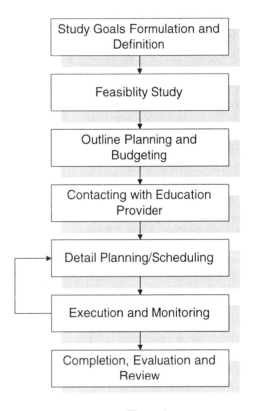

Figure 1

1.1 Study Goals Formulation and Definition.

In managing learning activities, decisions will be required on the selection of suitable courses. The student may possibly consider gaining another qualification; or merely wants to refresh his knowledge for his profession; or simply for fun. All study endeavours involve time and money so subject selection, in whatever circumstance, is of crucial importance.

1.2 Feasibility Study

Different people have different needs. Whether self-initiated or guided by a mentor, potential students would have to do some initial research on the feasibility of achieving their study goal. Is a suitable course available locally? Is the tuition fee affordable and the time required available? If no local provider can be found, does he feel comfortable with distance learning? For adult learners, would the study affect his work and family?

1.3 Outline Planning and Budgeting

Unlike commercial projects that tend to be money-sensitive, the issue of planning and budgeting in studying is more focused on the availability of time. How long is the course? How many days/hours are students expected to spend on the subject? Development of an outline plan is essential for a successful study.

1.4 Contacting with the Education Provider

In the case of the learning project is provided by formal institutions such as a university, the students are inevitably bounded by the established procedures. Enrolment must be completed before the term starts; the necessary course prerequisites must be possessed; there must still be vacancies, ... etc.

1.5 Detail Planning and Scheduling

For students enrolling in more than one course or those having other commitments such as work and family, good planning and efficient scheduling of their limited resources is extremely critical. Quite often some students not start doing their assignments until the last day. Some even went for holiday at the middle of the term and then asked for last minute help from the instructor one day before the examination.

1.6 Execution and Monitoring

In the phase of conducting educational activities, there is distinctive qualitative difference between active and passive learners. The former monitors their progress against the time-line and the learning objectives, and will be benefited greatly from the flexible online teaching. For the latter, however, removing human instructors and replacing them with online teaching might do them harm rather than good.

1.7 Completing, Evaluation and Review

On completion, the final outcome will be weighted against the student's objectives. Besides achieving a new qualification, the effectiveness of the endeavour will be assessed. Was the tuition fee worth spending? Was the learning approach suitable? Could a better result be achieved? To the providers, was the course well-designed? Was it delivered properly?

2. The Role of Study Management

In commercial projects, it is normally the role of the project manager to take responsibility for the whole project life-cycle. The situations in SM however are significantly different. Unlike commercial projects which normally operate with a matrix-type of organisations and its success often subjected to other environmental factors, delivering courses are relatively more structured and simple. Finding an instructor; preparing the materials; advertising the course; run the course; possibly mark examinations. If a course is being offered, it is very unlikely that its providers still uncertain if a suitable instructor can be found or there may not be a lecture room available. Students are mainly responsible for their own studies such as turning up at the lectures on time, studying relevant materials and after all passing the examinations. Having said that, this does not insulate the instructors from any charge of irresponsibility and in fact the course provider has to ensure:

- the instructor has the attitude of helping the students get through the course instead of letting them survive by themselves;
- the course materials are well-prepared, both content and presentation;
- the course objectives are clearly identified and ensure the enrolled students possess the necessary prerequisite knowledge;
- the learning environment is supportive;
- adequate and appropriate resources are available.

From traditional education providers such as universities to the newer private training institutes, the trend of web-based teaching is unstoppable. The implication of the roles of campus-based learning settings with human instructors becomes less prominent is the students will feel being left with a screen, keyboard and a mouse. No matter how high the quality of the course material, there are bound to be negative sentiments from the students failing the course. It is therefore essential for the course developers to monitor progress and offer supportive advice whenever appropriate. It is for this reason that time-based feature is proposed as an extension of the course chart metaphor [2].

3. Course Chart – A Metaphor for Structuring Course Topic

In [2], a compact visual notation has been proposed to organise course topics. Figure 2 is an example of a statistics course represented using this notation.

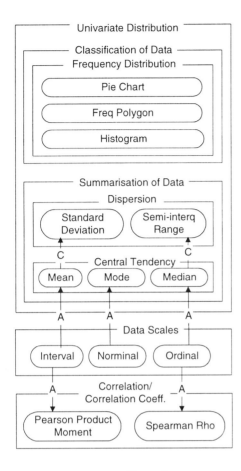

Figure 2

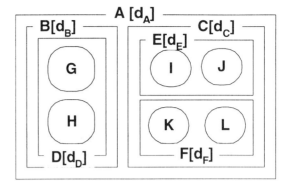

Figure 3

Figure 3 shows a sample time-based course chart which includes the expected duration of each node specified by the author. If the duration of the sub-blocs are not provided, say "I" and "J", students can spend arbitrary time on these two topics provided their total spending time not exceeding the duration (i.e. d_E) set on "E". Internal validating mechanism can also developed. Suppose the author specified two weeks for "E" and another two weeks for "F" but only three weeks for "C", the mechanism will be invoked to warn the author. This extension forcibly reminds the author to pay attention on the time resource demanded of students studying the material. With these pieces of time-based information, the system would be able to determine:

- earliest (and latest) start time of each topic; and
- earliest (and latest) finish time of each topic.

A topic cannot be started before its prerequisite topics have been completed. On the other hand, starting too late may delay the forthcoming topics. In case the student has passed the latest start (or finish) time of a topic, the system would alert the instructor. Supposing a student wants to have a break in the middle of the course, say having a holiday or visiting his sick relatives, these time-based information would be invaluable in advising the student if he can comfortably complete the course before the examination.

Activity diagrams has been widely used in operations management [3] where nodes represents activities with arcs denoting the precedence relationships among them. In Figure 4, the activities of "stripping old paper off walls" and "obtaining new paper" can be conducted independently but they must be completed before the activity of "paper the walls". This notion is equally applicable in course planning and control such as studying "mean" and "median" can be conducted independently but they must be completed before the topic of "dispersion".

4. Time-related issues on course planning

While the course chart provides an expressive facility for specifying the semantic relationships between topics, it lacks the capability of leading the student to the end within an acceptable time frame. The chart shows important information on *how* should the topics be pursued but could not tell the student *when*. When some topics are kept being browsed repeatedly by students, what is the reason? It could be that topic is particularly difficult and it is a sign to prompt the instructor for giving extra attention to the students. It could be the material was badly prepared and the students have no other choice but to repeatedly read the same material again and again in the hope of comprehending the topics. As time-based information plays a crucial role in monitoring the students, the chart metaphor is extended which allows the instructors to specify the expected time to be spent (i.e. ***duration***) on a topic. This is sensible and very useful to those students taking more than one subject in the same time.

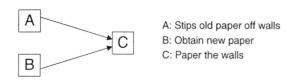

A: Stips old paper off walls
B: Obtain new paper
C: Paper the walls

Figure 4: A Network Diagram

One important characteristic of net-based project control is the identification of critical path. When the earliest finish time of an activity is the same as the latest finish time, it is on the critical path. The activities on the critical path indicate their completion on time plays a crucial role in the success or failure of a project. Any delay on the critical path would delay the entire project. Figure 5 shows a typical study network which consists of ten nodes. Nodes A, B and C can be studied independently. However, A and B must be studied prior to D as the former forms the pre-requisite knowledge for a good understanding of the latter. As node A is on the critical path, it must be finished on time otherwise the whole plan will be delayed. But for node B, there is room for a late start as long as it is finished on or earlier than its latest finishing time.

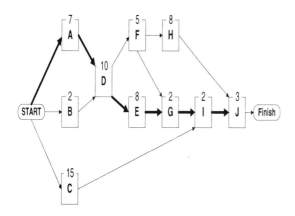

Figure 5: Sample course network with critical path highlighted.

5. Reasoning on the Temporal Relations

In the situations of two studying activities not running according to the planned network, the system has to reason about the possible implication and signal the course instructor for attention. We have adapted the 13 possible temporal relationships between two activities proposed by Allen [1]. The study nodes are modelled as finite intervals whereas the system responds according to a instant-based time marker which signals the current time point.

Possible Scenario	Actions taken by the course advisory system
A before B	If A is the prerequisite of B, no action is necessary
A meets B	
A overlaps B	Ask the instructor whether this is acceptable if A is the prerequisite B
A finished by B	Not allowed if A is the prerequisite of B
A contains B	
A starts B	These are allowed if A and B are co-requisites or they are independent topics
A equals B	
A started by B	
A during B	
A finishes B	
A overlapped by B	Ask the instructor if this is fine if A is the prerequisite B
A met by B	If B is the prerequisite of A, no action is necessary.
A after B	

Figure 6

6. Concluding Summary

In this paper, we advocate the importance of study management and put emphasis on the issue of timing. When developing a course in terms of a course network, the course developer specifies the expected time to be spent on each topic. By exploiting the inter-nodal relationships, a course advisory system works out the critical path on the network. Time to be spent on the nodes within the critical path must be monitored closely to ensure the entire study plan can be finished on time. For the non-critical nodes, however, there are rooms for potential delays and the system can advice the students accordingly.

References

[1] Allen, J. (1984). *Towards a General Theory of Action and Time.* Artificial Intelligence 23 (2), 123-144.

[2] Fung, P.W. (2000) *A hybrid approach for representing and delivery curriculum contents.* Proceedings of IWALT 2000, Palmerston North, New Zealand, IEEE Computer Press.

[3] Wild, R. (1995) *Production and Operations Management.* Cassell Educational Ltd

Acknowledgement The research reported in this paper is funded by the New Economy Research Fund of the New Zealand Government.

What's in a Prerequisite

Roland Hübscher
Auburn University
Department of Computer Science and Software Engineering
107 Dunstan Hall, Auburn, AL 36849-5347, U.S.A.
roland@eng.auburn.edu

Abstract

It is important that all pedagogical design decisions in educational learning environments are made explicit to the educator. However, implicit assumptions affecting or even contradicting the "official" pedagogical method may unbeknownst slip into the design. We discuss a concrete problem in the area of adaptive hypermedia systems related to the notion of a prerequisite. Prerequisites are frequently used to describe the many different educationally effective paths on which learners may traverse a hypermedia system. However, closer look reveals that several different concepts are called prerequisite and confusing them may lead to flawed designs. We will make recommendations as how to avoid misusing the different notions of prerequisites in adaptive hypermedia systems.

1. Introduction

There will always be some implicit design decisions in a complex system, now matter how hard designers try to avoid them. Some of these decisions may be inconsistent with the pedagogy employed by the learning environment potentially hampering its effectiveness. We will discuss this issue in the area of adaptive hypermedia systems (AHS). Although a explicit pedagogical model is part of the general architecture, assumptions inconsistent with the pedagogical model may sometimes slip in through the backdoor. One of these backdoors is the concept of a "prerequisite," or, more correctly, the concepts *(sic!)* referred to as prerequisite.

Hypermedia has been recognized as having great potential in providing content to learners. Relationships between concepts can be made explicit with hyperlinks, and the same material can be organized along different dimensions presenting the material to be learned from different views [1].

Hyperlinks allow each individual learner to traverse and explore the content in a way that fits his or her interests and learning goals the best. However, this added flexibility, compared to books which are often read in a more or less linear fashion, can also cause problems seriously impacting the pedagogical benefits of a hypermedia system. It is quite easy for the learner to lose orientation that manifests itself in the learner not knowing how the current page fits into the big picture and what hyperlinked path to follow.

This is one of the problems that adaptive hypermedia attempts to solve. Adaptive navigation support and content organization is used to present the learner with the most effective paths through the hypermedia system given the learner's characteristics, skills, learning goals, and behavior.

This approach is also known as curriculum sequencing. For most systems using curriculum sequencing, this means that the pages are organized in a partial order which depends on the pedagogical model underlying the system and the content to be presented to the learner.

Often, the concept of a prerequisite is used to define this partial order. Although, different systems vary in the details of how prerequisites are implemented and used, we can assume that if page p_a is a prerequisite of page p_b then the content a described on page p_a must be learned before p_b describing content b is allowed to be visited. Without loss of generality we will assume that each concept x is described on exactly one page p_x.

As a common result of this approach, the pages (or concepts) are partially ordered such that if concept a is required to understand concept b, then p_a is a prerequisite of p_b resulting in the constraint that p_a must be visited before p_b.

In this paper, we will show that this decision already includes some strong implicit assumptions about the underlying pedagogy, even in systems with explicit pedagogical models. We will suggest how this problem can be solved resulting in systems whose adaptive mechanism and the pedagogical model are clearly separated.

2. General Architecture

We base the discussion of the general architecture of adaptive hypermedia systems on the work of De Bra and colleagues whose model is called AHAM [2, 3]. Although other architectures could be used, AHAM is used because it explicitly refers to the pedagogical model.

The main elements of AHAM are the domain model, the user model, the teaching model, and the adaptive engine.

The domain model describes the content to be learned, that is, the concepts and how they are structured and related to each other.

The user model describes characteristics of an individual learner using the hypermedia system. For each user of the system, a specific model is built based on information gained through pretests, observed behavior, online quizzes, etc.

The teaching model (or, as we call it, the pedagogical model), describes how the user model and domain model are used to do the adaptation which is then actually executed by the adaptation engine.

The purpose of the pedagogical model is to make explicit how the system is adapted to the learner's needs. Thus, with respect to navigation, it describes what the pedagogically effective paths are for each individual learner. This is not a trivial task. It only takes 13 pages to provide a different path for every person of the earth's current population of six billion people.

3. Prerequisites

As mentioned earlier, prerequisites are often used in adaptive web sites to partially specify in what order the web pages need to be visited. The definition of a prerequisite in this context is as follows. If page p_a is a prerequisite of page p_b, then page p_a must be visited before page p_b may be visited. This definition of a prerequisite specifies a mechanism to partially order pages in an adaptive web site.

Switching over to an educational context, the concept of prerequisite is used in a pedagogical sense. For instance, if concept a is required to understand concept b then a is called a prerequisite of b. Alternatively, at a more coarse-grained level, if the material of course a is required for the material in course b to be mastered, a is declared a prerequisite of b.

Let's assume that a is required to understand b, that is, a is a prerequisite (in the pedagogical sense) of b. By definition of a prerequisite in adaptive web sites it follows that the page describing concept a must be visited before the page describing concept b may be accessed. In short, if a is required to understand b then a must be visited before b.

However, this conclusion is not correct for the general case and may contradict the specifications in the pedagogical model of the AHS.

3.1 A Prerequisite is Not a Prerequisite

Although many books present concepts in such an order that a concept a required to understand concept b is presented before b is presented, there are important situations where this approach is inappropriate. Also, some important teaching methodologies may require different kinds of orderings. Next, we briefly discuss a few examples of such methodologies.

Problem-Based Learning (PBL) is a teaching/learning method developed for medical students [4]. One of the fundamental principles is that the students need to discover on their own that they are missing a certain understanding that they have to acquire before they can solve the case at hand. This approach is not restricted to highly specialized medical students. The same method has been used in middle school science curricula where the students are expected to discover that, in order to find the solution, they need to learn more about a certain concept [5]. This may result in a trip to the library, checking a web site (which is not adaptive yet), or talk to some expert, e.g., a student's parent. Learning by Design (LBD) is another approach based on the design process that requires the students to discover the need to learn certain concepts [6].

Of course, this may not be the approach of choice if you have to teach mathematics or programming. It would be hard to imagine the students discover all those theorems. However, a final example from teaching user interface design at the college level shows that some instructors may quite often use approaches consistent with LBD.

Assume you are teaching the user interface design process and you are talking about the need for not just one but alternative approaches or prototypes during the earlier design iterations. One way to go from here is to ask the students which one of the approaches is the best. Quite soon, they realize that they need something like a set of criteria, actually, a set of quantifiable and measurable criteria. If, instead, the students are simply told in advance that measurable criteria are needed, they may not be clear about why they are needed.

In terms of prerequisites, these examples can be described as follows. A concept a is required to understand concept b. Before concept a is discussed, concept b is presented to the student who needs to discover the need to learn about a. Therefore, the student will learn about a and then be able to fully understand b. This approach, where applicable, has the advantage that the need of a for b is more clear to the student.

It is now easy to see that we talk about two different concepts when referring to prerequisites.

- A prerequisite is a mechanism to partially order web pages in an adaptive web site.

- A prerequisite is a pedagogical relationship between two concepts.

The next two sections will investigate both prerequisite concepts in more detail.

3.2 Prerequisites as Ordering Mechanism

If we consider the definition of a prerequisite as an ordering mechanism and assume that p_a is a prerequisite of p_x and also p_b is a prerequisite of p_x, it follows that the conjunction of p_a and p_b is a prerequisite of p_x. Although this approach covers many situations, it imposes an unnecessary constraint on what kinds of prerequisites can be defined.

However, extending the expressiveness of prerequisites comes at a cost. For instance, adding disjunctive prerequisites of the form "the disjunction 'p_a or p_b' is the prerequisite of p_x", where at least one of p_a and p_b need to be visited before p_x may be visited, requires computationally expensive algorithms [7]. These extensions can be useful, though. For example, it can be used to make sure that a student sees the use of a certain concept or feature, say of a programming language, in at least two examples. This constraint can be easily expressed as a disjunction of conjunctions. Another example where disjunctions come in handy is where the instructor has several examples to explain an issue in the lecture notes. He wants to select just one example and make sure he is not going over other, redundant examples. Again, conjunctions alone cannot express such a relationship easily, but a disjunction of conjunctions can. For a more elaborate discussion of disjunctions, see [7].

Adding full prepositional logic, that is, adding negation can cause even bigger problems with non-terminating updating algorithms resulting in difficult-to-define semantics [2]. In addition, replacing the visited/not-visited dichotomy of pages with continuous variables and some more or less arbitrary thresholds may make it more difficult to interpret the meaning of these variables and thus, of the prerequisites.

It is therefore of utmost importance to clearly define the semantics of the mechanisms and formalisms used.

3.3 Pedagogical Prerequisites

A pedagogical prerequisite states the relationship between two concepts with respect to a learner. If a concept a is required to understand concept b, then a is a prerequisite for b. Again, we claim that this is an incorrect overgeneralization. Let's look at an example.

Imagine you are a programmer and need a good random number generator to drive your simulation. You look up your web-based manual and find an entry "random number generator" in the index.

In the first scenario, we assume that the web-based system is *not* adaptive, thus all the hyperlinks are available to all users all the time. Clicking on the link in the index gets you to the page with the title "Random Number Generator" and the page mentions that the generator, Mersenne Twister, is fast and has a period of $2^{19937-1}$. Knowing what a period of a pseudo-random number generator is you use it in your program without any problems and the resulting simulation runs well.

In the second scenario, we use an adaptive version of the same manual. You will only see links to concepts you are ready to learn. Again, you click on the link in the index and the system understands that you want to learn the concept on that page, namely about the Mersenne Twister random number generator. So it provides you with the following list of recommended links: primitivity of characteristic polynomials, inversive-decimation method, resolution-wise lattice method, etc. Probably, you are not ready to learn about the intricate details of the random number generator called Mersenne Twister. If you study hard enough, you may be able to get to the actual random number generator page in a few weeks and run the simulation soon thereafter.

Obviously, we don't want to learn every pedagogical prerequisite even though it exists. Whether we want to learn a prerequisite a of b depends on why and for what purpose we need to learn about b. Do we need a deep understanding of b, do we need to be able to use b in non-standard situations, or are we just going to apply it in the most common cases?

This problem cannot be solved satisfactorily by just including those concepts that the user is required to learn. This would only work if every learner had to learn all the concepts in the web site. This is not a very interesting case for adaptive web sites that are supposed to cater to each learner individually. Consider an adaptive web-based manual. Most users will read parts of it only on demand and the random number generator example shows that the intended use of the concept influences what prerequisites are to be learned and which are not.

By definition of a prerequisite, a prerequisite that is not required is not a prerequisite. A potential approach may be using something like levels of understanding as proposed by Merrill's Component Display Theory [8]. In the context of adaptive hypermedia systems, this idea has also been mentioned by Murray [9]. Loosely based on Merrill's work, a concept could be used in various ways: remember, use, and understand. Remember: no need to learn the concept in any way, just being aware of its existence. Use, meaning applying the concept in normal situations: no need to understand the underlying principles of the concept. Understand: required to understand the underlying principles of the concept, thus, also being able to apply it in exceptional situations.

Of course, adding this kind of differentiation may complicate things and it may become less and less clear what the status of a specific prerequisite is. Furthermore, it is not clear who would specify the prerequisites for all the different situations, for instance, for remember, use, and understand. And finally, who would know whether the learner cares deeply about a concept or not? Maybe he really wants to know more about the Mersenne Twister, just because of its cool name.

3.4 Preconditions and Prerequisites

We have discussed two main problems with the use of prerequisites.

First, these are really two different concepts that happen to have the same name and look similar at their surface. As an anonymous reviewer has once suggested, we could call the prerequisites used as ordering mechanisms "preconditions." The pedagogical prerequisites should possibly be called exactly that, pedagogical prerequisites, to emphasize their purpose and origin. After all, in non-educational applications of AHS we still would need preconditions but probably no pedagogical prerequisites.

Second, a pedagogical prerequisite does not necessarily have to be learnt at all, which is not a newly discovered issue. However, it should be taken more seriously in many AHS.

It will be impossible to completely avoid having implicit design decisions slip in that contradict the underlying pedagogical model. We propose to clearly mark the separation between those parts (the set *P* for *Pedagogy*) in the AHS that are related to pedagogy and those that are not (the set *NP*). Clearly, the intersection of *P* and *NP* must be empty. Furthermore, the parts in *NP* must *implement* those in *P*, that is, the pedagogical model of the AHS must be fully described by the elements in *NP*.

4. Conclusions

An educational system must make its underlying pedagogical method explicit. The notion of prerequisites shows that there is always the danger that some implicit assumptions inconsistent with the "official" approach may slip in.

With respect to AHS, keeping a separate pedagogical model is important. However, every design decision needs to be understood in the light of the pedagogical method employed. In practice, this can be very difficult if not impossible. Almost every design decision will have some impact on the learner. However, if we try to keep pedagogical parts and those implementing them as strictly apart as possible, we may reduce the number of harmful, implicit design decisions.

5. References

[1] R. J. Spiro, P. J. Feltovich, M. J. Jacobson, and R. L. Coulson, "Cognitive Flexibility, Constructivism, and Hypertext: Random Access Instruction for Advanced Knowledge Acquisition in Ill-Structured Domains," *Educational Technology*, vol. May, pp. 24-33, 1991.

[2] H. Wu, P. De Bra, and G.-J. Houben, "Adaptation Control in Adaptive Hypermedia Systems," presented at Adaptive Hypermedia and Adaptive Web-Based Systems, Trento, Italy, 2000.

[3] P. De Bra, G.-J. Houben, and H. Wu, "AHAM: a Dexter-based reference model for adaptive hypermedia," presented at Proceedings of the 10th ACM Conference on Hypertext and Hypermedia, 1999.

[4] H. S. Barrows, *How to Design a Problem Based Curriculum for the Preclinical Years*. New York, NY: Springer Verlag, 1985.

[5] R. Hübscher, C. E. Hmelo, N. H. Narayanan, M. Guzdial, and J. L. Kolodner, "McBAGEL: A Shared and Structured Electronic Workspace for Problem-Based Learning," presented at Second International Conference on the Learning Sciences, Evanston, IL, 1996.

[6] R. Hübscher, S. Puntambekar, and M. Guzdial, "A Scaffolded Learning Environment Supporting Learning and Design Activities," presented at AERA, Chicago, 1997.

[7] R. Hübscher, "Logically Optimal Curriculum Sequences for Adaptive Hypermedia Systems," presented at Adaptive Hypermedia and Adaptive Web-Based Systems, Trento, Italy, 2000.

[8] M. D. Merrill, "Component Display Theory," in *Instructional-Design Theories and Models: An Overview of their Current Status*, C. M. Reigeluth, Ed. Hillsdale, NJ: Erlbaum, 1983, pp. 279-333.

[9] T. Murray, "A Model for Distributed Curriculum: From Tutor-Centered to Topic-Centered Representations," presented at AIED-99 Workshop: Ontologies for Intelligent Educational System, Le Mans, France, 1999.

Exploring Learning Problems of Cyber University

Lin K.M.
Ling-Tung College
linmin@mail.ltc.edu.tw

Chen N.S.
National Sun Yat-sen University
nschen@mail.cc.nsysu.edu.tw

Abstract

The aims of this research are to deeply exam learning problems behind cyber universities and try to draw out some concrete suggestions. Hope the effort could help people more understanding in this field and pay more attention to these issues, so that they can completely and efficiently set up and develop cyber universities. This study adopts literature analysis and qualitative exploration methods to analyze the problems in many aspects from a well-recognized cyber university in Taiwan. The results showed that problems of cyber universities could be classified into six categories, infrastructure, system, administration, curriculum, interaction and leaner. The concrete suggestions for each category problems are also proposed.

1. Introduction

In the face of the coming of information technology new era of 21st century, online e-learning has become the major trend of future teaching and learning model.

However, some scholars said that the ingrained traditional face-to-face instruction and thinking style has always been a string obstruction for the newly arousing long distance education scholars. The old thoughts claimed that the learning quality of Cyber University could never be compared to the traditional face-to-face instruction. They held doubting attitude to its instruction effect. On the other hand, Li C. J also indicated that today's online e-learning is facing the trial of transferring ideal. The orientation of online e-learning, the role it plays will be discussible issue continuously.

Some of literatures only asked questions but didn't have concrete suggestions or just proposed suggestions but didn't really find out the questions. Are there only these problems on online learning? The degree of complication on online learning is not less than the traditional face-to-face learning. How to deal and solve these problems are very essential while doing e-learning?

Our research aims focus on deeply realizing the overall learning problems of Cyber University and offering concrete suggestions, expecting to arouse attention and understanding from every field to improve the construction and development of cyber university into perfection.

2. Method

This research used qualitative exploration method. The information are gathered from all messages of 85 students taking computer networks course in 2000 academic year (2/25-6/30) on NSYSU Cyber University (http://cu.nsysu.edu.tw), including qualitative information of learning problem and questionnaire filled in both midterm and final exams, their final learning reflections, every thought members expressed on the web discussion board and also opinions from course instructor.

3. Result

Through literature exploration, analysis of members' learning problems and reflections, interviews of curriculum designers and information from other related researchers and after qualitative analysis, our research found out that the learning problems and suggestions in Cyber University can be classified into six aspects, including infrastructure, system, administration, curriculum, interaction and members. These six aspects are linked with each other. In addition to describing each problem in every aspect, we also propose active and positive suggestions to every question. Synthesized all the problems and suggestions from literature and this study into the following table.

4. Conclusions

There are still some disadvantages in online learning through Cyber University. However, if the leaders of Cyber Universities and educators can understand problems of six aspects in our study, and take a reference on the concrete suggestions proposed in our research, and cope with the improvement of native Internet infrastructure and decrease of capital of computer communication facilities, we believe that under the struggles from every aspect, the construction of Cyber

University will be more and more perfect, and learning problems of Cyber University will be less and less. Online learning in Cyber University will eventually become the best way for life-long learning in the future.

Table 1. The learning problems and suggestions in Cyber University

	Problem type	L	S	Suggestions	L	S
Infrastructure	1. The internet bandwidth	v	v	1. Offer stable, high quality and low capital Internet environment		v
	2. The Internet quality	v	v			
	3. High Internet fee		v	2. Cettification of degree and related law regulations in Ministry of Education		v
	4. Manpower capital	v		3. Increase working people in Cyber University		v
	5. Technique support	v		4. Offer standard system instruction platform		v
	6. Certification of credits and degrees	v				
Curriculum	1. Courses are not attractive.	v		1. Blackboards, animation and hyperlink explanation		v
	2. Teaching materials are restricted.		v	2. Offer practical course content	v	
	3. Unproper instruction design			3. Offer individual learning course		v
	4. Problem of link between courses		v	4. Regulate prepared courses		v
	5. Disputes on evaluation way.		v	5. Apply encouraging and rewarding measures	v	
				6. Apply forming evaluation	v	
Interaction	1. Lack of human interaction	v	v	1. Hold activities: Internet little angel, Online study circle		v
	2. Need of synchronous communication	v	v	2. Apply strategies to promote interaction	v	
	3. Problem of word communication	v	v	3. Use diversified communication methods	v	
	4. Misunderstanding of online communication	v	v	4. Reference principles of division into groups	v	
	5. Hard to achieve group consensus	v	v	5. Consider individual and group contribution	v	
	6. How to divide into groups		v	6. Teach cooperative skills and offer promoters	v	
	7. How to evaluate groups		v			
Learner	1. Burden from work and family	v	v	1. Make sure need and practice thoroughly		v
	2. Basic computer ability	v	v	2. Understand and get ready for online learning		v
	3. Not positive and active	v	v	3. Adjust attitude and learning style		v
	4. Don't ask questions	v	v	4. Form active learning attitude		v
	5. Only study but no contribution	v	v			
	6. Self laziness	v	v			
Administration	1. Problem of School Orientation		v	1. Previous planning courses and its orientation		v
	2. Lack of previous preparation		v	2. Plan proper conditions of entering school and evaluation system		v
	3. Lack of manpower support		v	3. Choose proper online learners	v	
	4. Problem of service quality		v	4. Reply messages in 24 hours		v
				5. Set up places to give exams flexibly		v
System	1. System operation and usage		v	1. Offer step-by-step operation instruction		v
	2. Immedaite discussion tools		v	2. Offer intellectual individual learning environment	v	
	3. Category of discussion boards		v	3. Offer intellectual instruction guidance and diagnosis	v	
	4. Related software usage		v	4. Offer instruction area of popular questions and related software		v
				5. Offer cooperative environment and mechanism	v	
				6. Offer alarm progress function		v
				7. Develop voice communication interface		v
				8. Classify discussion board according to topics		v

v: Where is from   L: from literature   S: from this study

Acknowledgement: This study was supported by National Science Council of R.O.C. (NSC89-2623-7-110-006).

Evaluating the Learning Object Metadata for K-12 Educational Resources

Daniel D. Suthers
Department of Information and Computer Sciences
University of Hawai`i at Manoa
suthers@hawaii.edu

Abstract

The Learning Object Metadata (LOM) is an emerging standard for organizing descriptions of digital or nondigital entities used to support learning. The descriptions include educational, legal, and technical characteristics of these resources. In this paper we describe an application of the LOM to the construction of a database of resources available to public schools in Hawai`i and report on issues encountered, focusing on structural issues such as dependencies between elements. The paper illustrates why development of metadata formats cannot be divorced from an understanding of educational context.

1. Introduction

Internet technology for learning has the potential to bring teachers and students together with a greater diversity of human, natural and technological resources than was previously possible. Additionally, the current emphasis on systemic reform in public school education in the United States is encouraging a greater diversity of stakeholders to collaborate in supporting students' achievement of high standards. These forces require that educators and their partners be aware of the resources that are potentially available to them and to understand the utility of these resources with respect to educational objectives. Already pressed for time, how will educators sort through this cornucopia of information and misinformation and find the resources appropriate for the educational needs of their students? Clearly, educators will need help. This paper is concerned with one form of help: databases of *metadata* [5] or information that describes the relevant characteristics of educational resources sometimes called *learning objects*. Properly constructed metadata should enable educators to find relevant learning objects more quickly. Examples of metadata databases include ARIADNE [1], NEEDS [6], and PEN-DOR [2].

Resource databases should adequately describe a diverse variety of resources yet relate them to educational objectives, describe the resources in terms understandable to educators, and interoperate with other major repositories. In this paper I report on our evaluation of an emerging standard, the Learning Object Metadata (LOM) with respect to its suitability for describing K-12 resources as part of a systemic initiative known as Hawai`i Networked Learning Communities. Specifically I discuss limitations and extensions to the LOM that were required, focusing on structural issues.

2. The LOM

The Learning Technology Standards Committee (LTSC), founded in 1996 by a group of academic, government, and industry representatives (including the author), is an umbrella organization that sponsors approximately 15 learning technology standards efforts, under the sponsorship of the IEEE (Institute of Electrical and Electronics Engineers, http://www.ieee.org/). The LOM draft standard (also known by its IEEE identifier as 1484.12) is arguably the most mature of the LTSC draft standards. According to draft 6 of the LOM [4], "The purpose of this standard is to facilitate search, evaluation, acquisition, and use of learning objects, for instance by learners or instructors. The purpose is also to facilitate the sharing and exchange of learning objects, by enabling the development of catalogs and inventories while taking into account the diversity of cultural and lingual contexts in which the learning objects and their metadata will be exploited."

The LOM standard is meant to provide a semantic model for describing properties of the learning objects themselves, rather than detailing ways in which these learning objects may be used to support learning. The LOM indicates the legal values and informal semantics of the metadata elements, their dependencies on each other, and how they are composed into a larger structure. It is intended to be extended, and in fact a structure has been provided specifically for the purpose. The LOM information structures are intended to support information exchange, and are neither specifications of an implementation nor specifications of a user interface. The LOM is agnostic concerning bindings or implementations of metadata in representations or notations.

The LOM metadata elements as of draft 6 [4] includes the following major element categories. *1:General* provides information such as title, a brief textual description, and keywords. *2:Life.Cycle* describes the development and current state of the learning object. *3:Metameta.Data* describes the metadata itself, e.g., who entered or validated this metadata instance and what language it is written in. *4:Technical* provides information on media type, size, software requirements, etc. for those learning objects to which these attributes apply. *5:Educational* is intended to provide basic information about the pedagogical characteristics of the learning object. This category includes some of the most controversial elements, to be discussed further below. *6:Rights* describes the conditions under which one may acquire and use the learning object. *7:Relation* can be used to describe the learning object in relation to other learning objects. *8:Annotation* records comments on the educational use of the learning object. *9:Classification* provides a means of extending the LOM to meet specialized needs.

3. HNLC Resource Database

The Hawai`i Networked Learning Communities (HNLC, http://lilt.ics.hawaii.edu/hnlc/) initiative is a partnership between the Hawai`i Department of Education (HDOE), the University of Hawai`i, and many other stakeholders in the quality of Hawai`i public education, such as business and nonprofit interests. HNLC's purpose is to prepare all students in Hawaii's public schools for careers in today's technological world by enabling them to attain high standards in science, math, and technology education. One component of the work of HNLC includes development of a web-accessible database with which educators can find and discuss educational resources available in Hawai`i. This paper analyzes the suitability of the LOM for this database.

The database describes resources for public school education ranging from US grades Kindergarten (K) to 12, abbreviated as K-12. A wide variety of digital and nondigital resources will be described, making this a particularly challenging test implementation of the LOM. To control the scope of our work, HNLC will prioritize the description of local resources and interface with other repositories of nationally available resources such the Gateway to Educational Materials (GEM, [7]).

Our method of evaluation was as follows. Initially we wrote informal textual descriptions capturing the important information about a representative sample of the resources that we wanted to describe. After reviewing these descriptions I presented the then-current LOM draft 4.1 [3] to the entire team. We then went through the textual descriptions and identified LOM elements in which the information expressed could be captured. Where we failed to find LOM elements for an item of information, we extended the LOM, either by expanding on the vocabulary of an existing element or by creating an entirely new element under 9:Classification. Where new elements were needed we searched other repositories to find metadata that we could use. Several iterations were required to understand the LOM structure well enough to define our instances of 9:Classification. Then our programmer created a Filemaker implementation of the resulting HNLC-LOM and provided the others with an interface for building metadata. Metadata for our sample was then created by two team members, and I reviewed the result to detect possible misunderstandings and issues. I also compiled a first draft of issues and recommendations. This draft was shared with the LTSC LOM committee, both via email and subsequently face to face in an LTSC meeting (Montreal, June 2000). Thanks to their feedback, many issues were resolved or re-understood as non-issues.

4. Vocabulary Issues

The data type of draft 4.1 LOM elements were either primitive (e.g., a string), referenced other standards (e.g., vCard), or consisted of a controlled vocabulary [3]. In the latter case, the vocabulary was either restricted, meaning that only the terms listed may be used, or open with recommended practice, meaning that one should attempt to use one of the terms listed as the recommended practice but may extend this vocabulary if needed. Draft 4.1 required that one accomplish an extension of a vocabulary with term *term* by placing a tuple of form ("See_Classification", *term*) in the data element, and defining an instance of 9:Classification that has the same 9.1:Purpose as the data element being extended. This 9:Classification instance would include one or more instances of 9.2:Taxon.Path as needed to indicate where the term falls within the taxonomic system indicated by sub-element 9.2.1:Source. See Figure 1 (draft 4.1) for an example. A taxon path can be thought of as a sequence of taxons, which begins at the root of a taxonomic hierarchy and works its way down the tree through intermediate nodes to the leaf node under which the object is being classified.

This arrangement provided a powerful general-purpose way of extending vocabularies with information about the taxonomic source of the term, and hence its semantics. However, subsequent to our evaluation, See_Classification was judged to be too complex, and removed in favor of a simpler system in which *all* vocabulary items have form (*Source, Value*). *Source* would either be "LOMv1.0" or an indication of the alternative source of the value, as illustrated in Figure 1 (draft 6.0). The example is simplified: Uniform Resource Identifiers would be used to identify the Source.

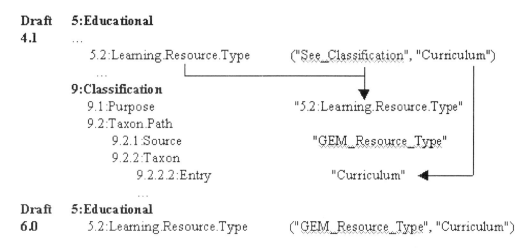

Figure 1. Extending vocabularies in LOM drafts 4.1 and 6.0.

We found several of the LOM vocabularies to be insufficient for our purposes. In one case, 5.2:Learning Resource Type, the vocabulary was open and the insufficiencies could be addressed via the extension mechanism just described. However, vocabularies for 1.9:Aggregation.Level, 5.1:Interactivity.Type (values: Active, Expositive, Mixed, or Undefined) and 5.5:Intended.End.User.Role (Teacher, Author, Learner, Manager) were restricted vocabularies, so could not be extended in this way. Subsequent to our report to the LTSC, these vocabularies and all others have been opened: there are no restricted vocabularies. This is an important recognition of the preliminary nature of vocabularies chosen by committee and the importance of allowing user communities to derive useful vocabularies.

5. Structural Issues

In some cases, including 5.1:Interactivity.Type, 5.5:Intended.End.User.Role (see next section), and 5.7:Typical.Age.Range, we felt that the vocabulary should be replaced with a structured description. Concerning 5.7:Typical.Age.Range, K-12 educational resources in the United States are often referenced by grade level rather than age range. Other applications may require other measures, for example, the military needs to access resources by rank. Therefore we recommended that 5.7:Typical.Age.Range be renamed and changed to a structured element with 5.7.1 Measure (e.g., "Chronological Age," "GEM Grade," etc.) and 5.7.2:Value (e.g., "12," "7-8," etc.).

More problematic are ways in which the value of one data element depends on another. We noted that 5.9:Typical.Learning.Time depends on the value of 5.7:Typical.Age.Range, for example, a textbook might be described as "suitable for a fast paced graduate course or a two-semester undergraduate sequence." Erik Duval (technical editor for the LOM) later pointed out that this applies to 5.4:Semantic.Density and 5.8:Difficulty as well. Hence we recommended reorganizing this information in the following manner:

5.x:Challenge Level *consisting of one or more 4-tuples:*
    5.x.1:Educational Level *(formerly 5.7):*
        5.x.1.1:Measure *(e.g., Age, US Grade)*
        5.x.1.2:Value *(e.g., 7-8)*
    5.x.2:Semantic Density *(formerly 5.4)*
    5.x.3:Difficulty *(formerly 5.8)*
    5.x.4:Learning Time *(formerly 5.9)*

Then one could create multiple instances of 5.x:Challenge.Level, with the values of 5.x.2 through 5.x.4 being dependent on the value of 5.x.1:Educational.Level. It is possible to *implicitly* achieve the same effect by replicating entire LOM metadata instances, one for each developmental level (or age); but it is far more perspicuous and efficient to acknowledge the dependency explicitly in a structure such as that above. The proposals of this section have not been adopted at this writing.

6. Our Extensions to the LOM

The following extensions were made using 9:Classification. Some will be replaced with the draft 6.0 structured vocabulary items.

We replaced 5.5:Intended.End.User.Role with the GEM *Audience*[1] a two-part classification consisting of

---

[1] See http://www.geminfo.org/Workbench/Metadata/ for all GEM Metadata described here.

ToolFor (who uses the tool) and Beneficiary (who benefits). For example, a professional development resource that helps teachers handle learning-disabled children in their classes is for the teacher but benefits the particular population of learning disabled students. Using LOM draft 6.0 we could write (GEM_Tool_For, *value*) and (GEM_Beneficiary, *value*).

Discipline describes the subject matter area covered by the resource. The LOM intentionally leaves this for 9:Classification. We are using the GEM Subject, originally designed to be a simple pointer to a subject area, with provisions for subject terms added from other controlled vocabularies [8]. This is a two-level classification system, requiring a two-step Taxon Path, for example Science/Astronomy.

Educational Objective addresses content and performance standards, specifically the Hawai`i Content and Performance Standards in our application. It is distinct from Discipline because it is more specific: it aligns the resource with the particular educational standards that the resource is intended to help achieve.

Educational Level augments LOM 5.7:Typical.Age.Range, and is structured as described in the previous section, using Source for Measure and Taxon for Value. The current 9:Classification is not sufficiently expressive to achieve the more complex 5.x:Challenge.Level we proposed.

We designed a classification called *Pedagogy* to replace 5.1:Interactivity.Type, which has an extremely deficient vocabulary of *active, expositive, mixed, undefined* and was restricted in draft 4.1 A rich description of interactivity is available in the GEM Pedagogy controlled vocabulary. This vocabulary has three facets: Teaching Methods (GEM provides a large vocabulary), Grouping (individual, small group, large group, etc.), and Assessment. Subsequently, 5.1:Interactivity.Type has been unrestricted.

7. Conclusions

In this paper I described our attempt to use the LOM for a K-12 resource database. We found that it provides a solid foundation in the form of many well thought-out data elements as well as a means for extension. We also found that the LOM does not address all the needs of our application. This is not surprising, as the LOM is being designed to serve a variety of applications in government and industry as well as public education. We were able to deal with most of the limitations through the Classification method of extension. Some of these solutions would be converted to structured vocabulary items under LOM draft 6.0. However, there remain some structural dependencies between LOM elements that are not well captured. These issues were illustrated with examples from K-12 education. It is hoped that this paper will help increase awareness within the primary/secondary education sector of the LOM standards effort, and encourage this sector's contribution to further development of the standard to be more appropriate for primary/secondary education needs. Further information may be obtained at http://ltsc.ieee.org/wg12/index.html.

8. Acknowledgements

Thanks to Beth Tillinghast and Susan Johnson for for contributing to the design of the HNLC metadata and constructing metadata, David Nickles for implementing the first prototype of our resource database, and Laura Girardeau for editorial assistance. Thanks to Erik Duval, Wayne Hodgins, Tom Murray, Brendon Towle, and Tom Wason for their "meta-comments" on my LOM commentary. This work was funded by a development grant from the National Science Foundation's Rural Systemic Initiative.

9. References

[1] Forte, E., Haenni, F., Warkentyne, K., Duval, E., Cardinaels, K., Vervaet, E., Hendrikx, K., Forte, M. W., & Simillion, F. (1999). Semantic and Pedagogic interoperability mechanism in the ARIADNE educational repository, SIGMOD Record, 28(1), 20-25.

[2] Fullerton, K, Greenberg, J., McClure, M., Rasumussen, E., & Stewart, D. (1999). A digital library for education: the PEN-DOR project. The Electronic Library, 17 (2), 75-82.

[3] IEEE Std. 1484.12 (draft). Draft 4.1 of the Learning Objects Metadata (LOM), IEEE, Piscataway, N.J., USA, February 5, 2000.

[4] IEEE Std. 1484.12 (draft). Draft 6.0 of the Learning Objects Metadata (LOM), IEEE, Piscataway, N.J., USA, February 8, 2001. Available: http://ltsc.ieee.org/wg12/index.html

[5] Milstead, J. & Feldman, S. (1999). Metadata: Cataloging by any other name. Online 23(1), 24-31.

[6] Muramatsu, B. & Agogino, A. (1999). The National Engineering Education Delivery System: A digital library for engineering education. D-Lib Magazine 5(4).

[7] Sutton, S. (1998). Gateway to Educational Materials (GEM): Metadata for networked information discovery and retrieval. Computer Networks and ISDN Systems 30(1-7), 91-93.

[8] Sutton, S. (1999). Conceptual design and deployment of a metadata framework for educational resources on the Internet. Journal of the American Society for Information Science 50(13), 1182-92.

Measuring Knowledge Transfer Skills by Using Constrained-Student Modeler Autonomous Agent

Safia Belkada, Alexandra I.Cristea, and Toshio Okamoto
AI & Knowledge Engineering Lab., Graduate School of Information Systems,
The University of Electro-Communications,
Choufugaoka 1-5-1, Chofu-shi, Tokyo 182-8585, Japan
safia@ai.is.uec.ac.jp

Abstract

In this paper we propose a methodology for the design of student modeler autonomous agent based on constraint model. The function of our autonomous agent is to interpret the student's constraint violations and draw a behavioral schema of the student's knowledge transfer skills on neural networks concepts. We aimed to build a student modeler by taking advantage of both the theoretical framework of the constraint-based model and the autonomous agent. This enables us to generate from a set of constraints desirable functions such as, task generation, hints, and domain application metaphors related to the student's learning goals. We present the fundamental issue of how to represent a student knowledge transfer skills in an autonomous agent.

1. Introduction

The task of building a student model is extremely difficult due to the huge search spaces involved [1]. Several researchers have pointed out the inherent intractability of the task if the goal was to model the student's knowledge completely and precisely [1,2]. However, a student model can be useful although it is not complete and accurate is sufficient for instruction purposes. Therefore, Ohlssen introduced a framework of student modeling for Intelligent Tutoring Systems based on constraint-based model [1]. This concept reduces the complexity of student modeling by focusing on faults only. Systems based on this approach need to have an interface and an internal representation of the domain knowledge components; each action is recorded in the internal state as a collection of constraints and pattern matcher that can determine when the constraints are satisfied or not. Our research is aimed to build a student modeler based on the theoretical framework of the constraint-based model and autonomous agent to be able to generate from a set of constrains a desirable functions

such as, help (i.e. generating dynamic hints) and measurements (i.e. measuring the student's knowledge transfer skills on neural network applications), task generation, and metaphors related to the student's learning goals (i.e., cognitive tools). Therefore, we faced the student-modeling problem within the context of student reasoning in the domain of neural network applications within our simulation-based learning environment: DiscoverNet [3]. DiscoverNet environment consists of three core components aimed at facilitating the student's knowledge transfer skills to solve problems in the domain of neural networks. The three core components are: (i) an interface that fosters a general representation of neural networks [3,4], (ii) a cognitive model for learning neural networks' concepts [3], and (iii) a student modeler knowledge. To develop the student modeler, which is the focus of the present paper, there are three stages:
- Develop an appropriate method to model the domain knowledge based on the constraint-based model framework.
- Develop cognitive tools for neural nets' concepts whose role is to enhance learning.
- Develop a model of an agent whose three specifications are: task level skills, communication skills, and problem solver skills.

We organized the paper as follows: First, we examine the modeling of structural aspects of the constraints in the present neural network domain knowledge. Section 3 describes issues in the knowledge transfer skills and its measurement. Sections 4 and 5 describe the modeling of the autonomous agent behavior and concluding remarks respectively.

2. Constraints Modeling

As mentioned previously, Ohlssen and Ress introduced a representational format that simplifies the domain knowledge as a collection of constraints. The unit of this format is called a state constraint. Each state is an ordered pair $<C_r, C_s>$, where C_r, the relevance condition, identifies the class of problem states of which the constraint is relevant, and C_s, the satisfaction condition, identifies the class of states on which the constraint is satisfied. Each member of the pair can be thought of as a conjunction of features or properties of a problem state [1]. To describe the constrained domain knowledge of neural networks, let $K = <P_a, P_p, P_b, P_i>$ be a finite set of problem states of neural network's model design. P_a, P_p, P_b, P_i are finite subsets of problem states in pre-designed learning goals (i.e., constraints associated with neural nets' architecture components usage, tuning parameters, using the training algorithms, and interpreting of the outputs respectively). Let C_r a cluster of relevance conditions over a finite set of problem states $P' \in K$ (e.g. P' $\in$ the stage of tuning parameters' learning goals). Then, the set of constraint satisfaction C_s can be defined inductively in the following way :

(i) All elements in C_r are action types,

(ii) If $O_1...O_n$ are distinct objects in P' and $A_1...A_n$ are the appropriate instructions during the design steps, then every expression which conforms to one of the following is a constraint satisfaction of the form:

(a) $[O_1:C_{r1}=>A_1.... O_n:C_{rn}=>A_n]$; sub-expressions $O_i:C_{ri}=>A_i$, $1<i<n$ are called components whose constraints should be satisfied.

(b) $\{Cr, C_s\}$ are agent's inputs.

The objects are classified into three main classes in the system: Application objects: APPL_OBJ, Configuration objects: CONFIG_OBJ, Parameter objects: PARM_OBJ. The objects of each class can be static and dynamic except for the applications objects where all objects are static. A static object does not have a constraint that should be specified. By contrast, dynamic object has an implicit constraint. This means that the dynamic objects are related to their information contents.

In the following section, we look at some relevant constraints in the context of an example.

2.1. Example

Before going into further explanation, lets us give a brief description of neural networks' problem solving. Neural network learnability is related to the ability of the learning algorithm to find a set of constraints satisfaction values (weights and parameters), which gives the accuracy needed for a good mapping approximation. The network performance depends on the network architecture and the learning algorithm as well as on the training set. For thorough discussion of neural network concepts, readers are referred to references [5] and [6].

The example consists of a simple neural network with three layers: input layer, hidden layer and output layer. We assume the hidden unit has a sigmoid activation function and the output unit has a linear activation function. The constraint to satisfy, here in this example, is to find the weight vectors of the hidden and output layers that approximate the discontinuous functions $g(x) = sin(x-2)+3$; for all $x \in R$, to any degree of accuracy. The relevant constraint clusters of the above example during the design steps are given in table 1. The student's configuration consists of a training set that denotes input vector, the desired output vector, a set of parameters, and random weight vectors. To build the corresponding model, a student uses several components in the system and may follow different solution paths (e.g. assigns the weight vectors in the layers and then affects parameters for the whole net). An equivalent cluster of constraints, witch triggers the same instructional action, is evaluated with respect to the current problem state.

Table. 1: An example of relevant constraints for backpropagation

Objects	Relevant Constraint's clusters
Parameters as learning rate	Choose not constant values
Hidden activation function, function criterion	Choose a differentiable and bounded functions
Data sets	Divide the set in training set and test set t and validation set

In the following section, We describe how to model the autonomous agent.

3. Student's Knowledge Measurements

Knowledge transfer occurs when students who have learned and mastered a complex domain of knowledge learn a second one that shares the similarities with the first one [7]. Knowledge transfer is not constrained to only the similarity of surface elements; there is also some very large positive transfer of knowledge between two skills that have the same logical structure even if they have different surface elements. We believe that students intending to learn neural nets were exposed to solve at least basic domain application problems. Although the basic elements of domain applications are different, they still share similarities in solving them. In more general terms, the present tutoring system can be described as a system to teach the concepts of neural nets and use the cognitive tools as supporting knowledge to reinforce learning. This technique offers relatively optimal tutoring mechanism in that students can develop structures of basic modeling concepts. We do not believe that a student can complete the whole modeling process. However, we believe that students with some background can quickly acquire basic skills of neural nets modeling. The student's knowledge transfer skills are recognized from the point of view of the agent, considering: the tasks performed, the knowledge needed by the student, and how the student is expected to have/acquire that knowledge. This is done by measuring: the average task difficulty for a particular activity, average latency in the activity, dependency between tasks, constraints violated, and hints taken.

4. The Autonomous Agent's Modeling

The autonomous agent needs a way to look at student's actions and a mechanism to analyze these actions and store the results. So, in this section we will show the requirements to design the autonomous agent and turn them into a list of functions that satisfy the following requirements:
- *Delegation:* perform a set of tasks on behalf of the students (i.e. activate the training process by setting some appropriate constraints if needed)
- *Communication skills:* the agent needs to interact with the student and the environment during the problem solving stage.
- *Autonomy:* the agent operates without direct intervention from the system or the student.
- *Intelligence:* the agent needs to be able to interpret the monitored events to make appropriate decisions.
- *Functions:* cognitive tools retrieval, hints' generation, and knowledge transfer skills measurements.

Now that we have specified the functions that our agent must provide, we will describe our design decisions.

4.1. Task Level/ Communication Skills

The agent accomplishes two types of tasks, the task addressing the analysis of static knowledge (knowledge on neural concepts). The agent knowledge is linked to static knowledge on the domain; it models the knowledge compulsory for the task (e.g. probable parameters setting for the learning algorithm selected for the model). The second type of agent's task addresses analysis of dynamic knowledge (knowledge on functional aspects of neural concepts). The agent holds links to the knowledge used to build new knowledge. This type of task gets the static knowledge as input. The general representation has the form:
If sensor (sensorName, rulevariable) then effector (effectorName, parameters).

The agent analyses the student's "initial model" in order to infer a consistent model. Since the task performed by the student is the main focus of this analysis, the agent identifies the clusters $C_{r1}...C_{rm}$ of student's problem states of the task at hand and where instructional $A_1...A_n$ are appropriate to retrieve the cognitive tool associated to them.

As the construction of the agent is java framework, the easiest way for us to build such functions is to have the agent run in a separate thread, but use events to communicate with the whole system. By using the observer/ observable framework, the agent would be an observer and whenever the agent was notified that an event occurred, it would be executed. This allows us to develop a special-purpose for the agent, which uses Rulebase (i.e., the Rulebase supports forward and backward processing with sensors and effectors) and knowledge- based neural network to determine what actions to take by sending signal events to the student. The event modeling consists of developing: an event library for the agent, an object event model encapsulating appropriate action, and an event for each problem state. This event model is based on event source and event listeners (see Java specifications for more details [9]).

4.2. Knowledge Acquisition: Mental State Behavior

The agent's knowledge concerns rules that the agent follows to complete its tasks. To model the agent in term of belief states and how it changes as a consequence of interactions, the mental state of the agent, the agent's own belief, and problem solving procedures are necessary. The following are the specifications of the agent during the agent learning stage:

(i) The agent is able to learn to do classification and prediction of the knowledge transfer skills.

(ii) The agent is persistent; that is, there must be a way to save the mental sate in a file and reload it whenever is needed. Java supports serialization of objects and makes the saving and loading of objects very easy. Actually, any data members that are not explicitly declared as static or transient will be saved to a file.

(iii) The agent is autonomous; that is, it solves problems and updates autonomously its mental state such as behaviors and commitments.

At a given time, the behavior of the agent is given by the function being performed. Thus, modeling the behavior of the agent requires the knowledge of each function (called also method) applied to each component and their effect on them (the components) and the corresponding pair of relevant/satisfaction constraint clusters <Cr,Cs>.

The following type of rules used to update the agent mental state are as follow:
(i) Rules defined in the Rulebase for solving specific task,
(ii) Rules for updating its working memory,
(iii) Rules to request for message from the environment that is handled by the event model.

- **Update the Agent's Mental State**

Programming the agent mental state is giving it temporal logic specifications. The agent's specifications are the process of constructing linear sequences of states. The agent is a base class, which defines a common programming interface and behavior. The agent class implements the Runnable interface, which requires that our agent subclass provides a *run()* method [9], which can serve as the body of the thread. This is the mechanism we use to give the agent autonomy. A vector of listeners holds all sequences of states that implements the behavior interface and have registered themselves using the *modifybehavior()* method. A notify method is used to send events to registered listeners and this method must be synchronized to control access to the listener' vector in multithread environment.

- **Problem Solver**

To generate a model (architecture, weight values, parameters, and pattern or error graph) of the student, the agent uses the knowledge-based neural networks as a search space for the correct network topology and finds the closest to the student-designed network. It also shows to the student the violated constraints, see for more details, the algorithm is described [8].

5. Conclusion

Ohlsson et al. described a new approach based on the idea of representing subject matter knowledge in sets of constraints. This approach promises to eliminate the need for runnable models of either the expert or the student models and to reduce the computations required for student modeling to pattern matching to fault only. In this paper, we proposed a different approach to the student modeler where we introduced the need of software computation at different steps in the design of the student modeler. The function of the student modeler is to measure the knowledge transfer skills of the student. We give more importance to the capability of the student to transfer his knowledge from one application to another or from one neural concept to another rather than traditional focus on diagnosis and assessment. We combined two concepts: the constraint-based model framework and the agent technology to measure the students' transfer of knowledge skills. A first step consisted of modeling the domain knowledge with a collection of relevant/satisfaction constraint clusters. The relevant constraints are used to be matched with the students' actions. The second step consisted of specifying the agent's requirements and its functions. The matcher is the agent's function during the design process. While this function is running the mental state of the agent is updated. Therefore, to update its mental state, it checks if all relevant constraints related to the problem are satisfied or not, and generates a new factor of belief of students' reasoning in the context of knowledge transfer.

References

[1] Ohlsson, S. & Rees, E., "Constrained-Based student Modeling", Student Modeling: the Key to Individualize Knowledge-Based Instruction, NATO ASI Series, Quebec, 1991, vol.125, pp.103-179.

[2] Holt, P., Dubs, S., Jones, M., Greer, j., "the state of student modeling" Student Modeling: the Key to Individualize Knowledge-Based Instruction, NATO ASI Series, Canada, 1991, Quebec, pp. 3-35.

[3] Belkada Safia, Cristea, A., Okamoto, T., "DiscoverNet: a Discovery Learning Support System on Neural Networks", Journal JET (Japan Society for Educational Technology), Tokyo, 2000, vol.24.

[4] Belkada, S., Cristea, A., Okamoto, T., "DiscoverNet: a Discovery Learning Support System on Neural Networks", Smart Engineering System Design, ASME press, USA, 2000, vol.10, pp.1045-1050.

[5] Haykin, s., "Neural Networks: a Comprehensive Foundation", macmillan, New York, 1994.

[6] Mohamad, H.H, "Fundamentals of Artificial Neural Networks", MIT Press Cambridge, Massachusetts London, England, 1995.

[7] Singly, M.K & Anderson, J.R., "Transfer of Knowledge Skills", Harvard University Press, Cambridge, MA, 1989.

[8] Okamoto, T., Belkada, S., "Knowledge based Neural Network: A Method to Support Learners in Building Neural networks Models", Frontiers in Artificial Intelligence and Applications, IOS press, Czech republic, 2000, pp. 352-360.

[9] Zukowski, J., " Mastering Java 1.2", SYBEX Inc., USA, 1998.

Studying the Learning Practice: Implications for the Design of a Lifelong Learning Support System

Giasemi N Vavoula and Mike Sharples
University of Birmingham
G.Vavoula@bham.ac.uk, M.Sharples@bham.ac.uk

Abstract

A phenomenological study of personal experiential learning was conducted. This informed a descriptive Framework of Lifelong Learning (FoLL) which describes four facets of lifelong learning: the learner, the organisation of learning, the process for carrying out learning projects, and the breakdowns that occur during, or because of, learning.

The FoLL was then used to derive a set of general requirements for a lifelong learning support system. System design has commenced based on these requirements. The paper discusses the FoLL and presents the system's general requirements.

1. Introduction

We report on an investigation aimed at the design of personal technologies to support adult learning over long periods of time. Sharples [1] proposed general requirements for this type of system based on a study of theories of personal and lifelong learning. In this study we take a different approach: we are seeking to ground the requirements for a system that supports the user in their everyday learning over a lifetime on a model of lifelong learning which originates in a theory-informed phenomenological study of learning.

Section 2 describes the methods used for the study. The result of the study, a descriptive Framework of Lifelong Learning (FoLL), is presented in section 3. Section 4 outlines the general requirements for a (lifelong) learning support system, which were derived from the FoLL.

2. Study methodology

The study employed the "diary: diary-interview" method [2] for data collection. Twelve diaries were distributed to adults whose occupations were learning intensive (e.g. postgraduate students), or who were involved in continuing education programmes. The participants kept the diaries for 4 days. A total of 118 learning experience descriptions were collected. This amount of data was enough to allow us to identify what a learning experience is in practical terms and what different forms it might take. Categorisation of these forms provides a means to conjecture possible configurations of lifelong learning.

In-depth interviews, focusing on the diary entries, were carried out shortly after the diary period. The data obtained with this method were then subjected to analysis using Grounded Theory (GT) techniques [3]. These techniques have been used before in the study of specific learning situations (e.g. novice mathematician's encounters with mathematical abstraction during tutorial study [4]) and as part of knowledge elicitation in knowledge-based systems design [5].

Theories of learning informed the analysis in two stages: at the beginning where a set of categories arising from the theories was superimposed on the data and at the end, during the definition and discussion of the emergent categories.

3. FoLL

The learner. At the core of the learning practice, and therefore of FoLL, is the learner: a person with certain physical characteristics, who assumes a number of social roles, and who has a number of characteristics that relate to how (s)he practices learning. All these characteristics change in time as the learner develops and changes, or as the learning topic/project changes.

Organisation of learning. Learning is organised in three operational levels. At the lowest level, the learner performs learning activities such as reading, discussing, observing, and taking notes. These activities are then grouped at the middle level into distinct learning experiences based on (learner's) criteria such as the topic of learning, the time, and the context in which the activities are performed.

At the top level, the learner organises learning experiences into learning projects based largely on purposes and outcomes: experiences which add to the achievement of a certain aim are likely to be grouped under a single project. A number of learning projects may be pursued during the same period of time.

The physical, social and organisational environment, the time, the topic, the objects used, the relevant information presented to the learner, his/her objectives, and the tree(s) in this hierarchical organisation where the specific learning episode lies, further specify the learning context.

Learning projects. As regards the procedure for carrying out learning projects, a cyclic pattern was inferred. Learning needs or opportunities are translated into objectives, which are translated into plans. Plans prepare the learner for learning action [6]. Action results in learning outcomes, which are used in other (non) learning situations, which give rise to further needs or opportunities for new projects, and so forth. Oscillations between the different phases are possible. Through the study we could see that all the phases in this cycle could be initiated either by the learner or by someone else and that some cycles may start with a phase other than the first one described above. Based on which is the starting phase for a learning project and who initiates it, we could arrive at a useful typology of learning projects (table 1).

Table 1: Typology of learning projects

Phase	Initiated by learner	Initiated by others
Identify needs	Intrinsic necessity learning	Extrinsic necessity learning
Identify opportunities	Intrinsic opportunistic learning	Extrinsic opportunistic learning
Formulate objectives and plot plans	Self-managed goal-driven learning	Institution-managed goal-driven learning
Learning action	Self-initiated experiential learning	Externally-initiated experiential learning
Evaluation of, and reflection on experience	Self-managed reflection	Externally-managed reflection

The process of learning was first described as a cyclic one by Kolb [7], with the well-known experiential learning cycle. What we present here, however, is different in that it describes the life cycle of learning projects, highlighting the way learning is intertwined with everyday life.

Breakdowns. Breakdowns may occur in all phases and levels of the learning practice. A learner may fail to formulate an objective/identify a need; plan how to satisfy it; carry out a learning activity both practically and conceptually; apply his/her existing knowledge; organise his/her learning activities into experiences and projects; and develop new abilities in order to respond to physical, social, and situational changes. Breakdowns, however, may also occur not during but because of the learning experience: some new piece of knowledge may require further exploration; or some background knowledge might be missing which does not allow learning to continue. Breakdowns form possible starting points for new, intervening learning experiences/projects, carried out as problem-solving exercises.

4. Requirements

Based on FoLL as we described it above, a set of general requirements for a lifelong learning support system was devised. Such a system should be able to:
1. Follow the user's development in life and aid them to consider new responsibilities they have undertaken, new abilities they have developed, and new learning habits and tactics they've adapted.
2. Aid the user in synthesising serendipitous learning, planning deliberate learning, and managing semi-structured learning
   2.1 Aid the organisation of learning activities into experiences or the planning of activities for an experience.
   2.2 Aid the association of experiences with learning projects or the planning of experiences to complete projects.
3. Support the user in carrying out learning projects, in:
   3.1 identifying and articulating learning needs
   3.2 creating/ grasping learning opportunities
   3.3 planning and assessing learning
   3.4 performing learning activities
   3.5 identifying and linking learning outcomes
   3.6 using outcomes in future tasks
4. Support the user in recovering from breakdowns

5. References

[1] M. Sharples, "The Design of Personal Mobile Technologies for Lifelong Learning", *Computers and Education*, 34, 2000, pp. 177-193.

[2] D.H. Zimmerman, and D. L. Wieder, "The Diary: Diary-Interview Method", *Urban Life*, 5(4), 1977, pp. 479-498.

[3] B.G. Glaser, and A.L. Strauss, *The Discovery of Grounded Theory,* Aldine, Chicago, 1967.

[4] E. Nardi, *The Novice Mathematician's Encounter With Mathematical Abstraction: Tensions in Concept-Image Construction and Formalisation,* Doctoral Thesis, Linacre College – Oxford University, 1996.

[5] N.F. Pidgeon, B.A. Turner and D.I. Blockley, "The use of Grounded Theory for conceptual analysis in knowledge elicitation", Int.J.Man-Machine Studies, 35, 1991, pp. 151-173.

[6] L.A. Suchman, *Plans and Situated Actions: the problem of human machine communication,* Cambridge University Press, UK, 1987.

[7] D. Kolb and R. Fry, "Toward an applied theory of experiential learning", in C.L. Cooper (ed), *Theories of Group Processes*, John Wiley and Sons, UK, 1975.

Courseware Accessibility: Recommendations for Inclusive Design

Robert Luke
Curriculum Coordinator
Special Needs Opportunity Windows (SNOW) Project
Adaptive Technology Resource Centre, University of Toronto
http://snow.utoronto.ca   robert.luke@utoronto.ca

Abstract

Providing educational opportunities within online environments, while beneficial, also has the potential to exclude a significant portion of the population. The learning and/or physically disabled may be prevented from accessing online learning environments due to problems both in the design of the technology itself, and with the pedagogy directing the use of this technology. This paper presents an overview of the results of research into courseware accessibility. Recommendations for inclusive design are presented within disability-specific requirements for accessible courseware.

People with learning and/or physical disabilities may be prevented from accessing online learning environments. There are problems in the design of the technology itself, as well as with the pedagogy directing its use. *Inclusion in an Electronic Classroom* [1], a study funded by the Office of Learning Technology and conducted at the University of Toronto's Adaptive Technology Resource Centre, recently examined accessibility within various courseware platforms in order to better assess the limitations faced by persons with disabilities using online learning environments. The study focused on eight individuals from the following disability groups: Blind, Vision Impaired, Mobility Impaired, and Learning Disabled, and encompassed both experienced and inexperienced users. The study is a joint initiative of the Adaptive Technology Resource Centre at the University of Toronto, The Centre for Academic Technology at the University of Toronto, the Special Needs Opportunity Windows Project at the University of Toronto, the Learning Disabilities Association of Ontario, the Canadian National Institute for the Blind and Dr. Bruce Landon. Results indicate that courseware environments need to consider problems in the design of the technology itself, as well as with the pedagogy directing its use.

Accessibility Issues by Disability Group

Blind

Java-based technologies such as chat rooms, whiteboards, and progress displays are inaccessible to users who cannot use the mouse and are reliant on the keyboard. This also extends to "Browse" buttons (for file uploads) that are not properly programmed, as well as poorly labeled form elements. Inconsistencies in layout and in language used to describe functions and features can result in confusion around how to accomplish tasks.

Users who rely on screen readers have problems with multiple frames and nested tables, given that screen readers generally read across a screen and cannot differentiate between text in columns. Frames that are either unlabelled or improperly labeled also present complications for blind users as regards the specific function of each frame (navigation, content, etc.). In addition, surprise popup windows can confuse screen readers with respect to matters of their function, as it may not be clear why the window has opened, and how the window contents relate to the rest of the site.

Low Vision

People with low vision experience problems similar to those experienced by blind users. However, due to the fact that these users can fall back on residual vision, these complications can be less confusing in some instances. As with blind users, problems with the adaptive technology itself can create accessibility issues, although users may more easily overcome these given more experience with both the adaptive technologies and online learning environments. One low vision participant in this study experienced problems resulting from (or exacerbated by) limited experience with web navigation and courseware environments in general, inexperience with the adaptive technology, and general assumptions and preconceptions of online learning. Problems experienced by the other low vision participant were inherent to the courseware itself, including the illogical display of steps required for task completion, and confusing and ambiguous use of terminology

Learning Disabled

Problems encountered by learning disabled participants were largely a result of inconsistencies in layout and in the language used to explain task requirements, the absence of alternative information formats, and the absence of instructions for multi-step activities. Learning disabled participants faced problems related to inexperience with online learning environments and difficulties related specifically to their respective disability. However, problems with absent or incomplete instructions coupled with assumptions of online skills and ability exacerbated these difficulties. The latter problems can and should be addressed by courseware manufacturers (by making adequate and complete instructions available on demand) and by course developers/instructors. Course developers should create an inclusive environment for learning that includes provisions made for people who need these instructions as well as the redundant display of key information.

Mobility Impaired

The mobility impaired study participant (a quadriplegic) experienced no difficulties in accessing each courseware function. This was largely a result of this participant's familiarity with both adaptive technology and online instructional environments. The only problems experienced by this participant related to specific courseware functions that were not operational.

Recommendations for Inclusive Design

The major obstacles to accessibility are complexities in page layouts, inconsistencies in item labelling, a lack of instructions for task completion and the absence of consistent and clear functions related to items within courseware platforms. To ensure full accessibility of information, courseware developers need to ensure their platforms conform to the current WAI guidelines [2]. Redundant information display is needed to aid those who are learning disabled. In addition, problems with courseware platforms and adaptive or assistive technology need to be acknowledged, addressed and tested to lessen the effects of incompatibilities. The courseware environments studied in *Inclusion in an Electronic Classroom* failed on several accounts to comply with the WAI guidelines, including the generation of inaccessible content by automated markup tools [3].

Separating the media used to access educational material from the content puts the emphasis on content flexibility; it must fit a variety of presentation media (text, audio, etc.). New mark-up languages (XML, XSL, CSS, DOM, XUL, Java) "separate content and structure from presentation" and "separate function from input [and output] method" [4]. This allows people who need alternative or redundant output devices to access media that may otherwise be inaccessible to them. Future courseware applications should strive to include the option of coding in these modalities. "The transformative power of computers makes most of this information available for the first time to many people with disabilities" because digital media can be adapted to multiple outputs [4]. Technical considerations aside, the most important obstacle to accessibility is effective pedagogical deployment of the technology in educational contexts.

Course instructors using online learning technology must be aware of both pedagogy and accessibility issues and how the platform being used might fail to meet these standards. Technical faults can be worked around, although this can be costly, time consuming, and frustrating for all concerned. The need for built-in accessibility in the tools facilitating online learning must be prioritized. It is also imperative to provide adequate and comprehensive instructions for the use of the actual courseware, as well as dedicated institutional support for both instructors and end users.

Just as buildings are built with accessibility factored in from the ground up, so too must WWW and Internet architecture factor in accessibility initiatives from the outset to ensure equitable access to online resources. Ensuring accessibility encourages people with physical and/or learning disabilities to become producers of information, instead of remaining passive consumers. Accessibility in online course designs will ensure that the wider population benefits from these programs: "For people without disabilities, technology makes things convenient; for people with disabilities, it makes things possible" [4]. Accessible online programs offer disabled persons an avenue to pursue educational options where none might have existed before. By making information more accessible to all, everyone benefits [5]. It is imperative to ensure the accessible design of all web materials from their first iteration.

References

1. Inclusion in an Electronic Classroom. <http://snow.utoronto.ca/initiatives/access_study/inclusion.html>
2. World Wide Web Consortium. Web Content Accessibility Guidelines 1.0 <http://www.w3.org/TR/WAI-WEBCONTENT/>
3. Harrison, Laurie. Inclusion in an Electronic Classroom - 2000: The Role of the Courseware Authoring Tool Developer. <http://snow.utoronto.ca/initiatives/access_study/ATrec.html>
4. Treviranus J. Expanding the Digital Media in More Human Directions. Towards the Digital Media Institute. University of Toronto: KMDI Lecture Series, 11 March, 2000.
5. Rose D, Meyer A, 2000, *The Future is in the Margins: The Role of Technology and Disability in Educational Reform* <http://www.air.org/forum/pdf/rose.pdf>

Design for Web-Based On-Demand Multiple Choice Exams Using XML

Raymond Lister
Faculty of Information Technology
University of Technology, Sydney
raymond@it.uts.edu.au

Peter Jerram
School of Computing and I.T.
University of Western Sydney
p.jerram@uws.edu.au

Abstract

We describe a web-based system for the delivery and marking of multiple-choice questions. While other web systems exist that deliver such exams, those systems require either the manual mark-up of static exam questions, or generate different versions of an exam with limited variability. Such approaches result in exams that should only be administered once to a specific student. Our system is more general in how it generates different versions of an exam, using XML.

1. Introduction

A popular idea in pedagogy is the concept of assessment for learning. where student learning is structured via the assessment tasks. Ideally, students should be able to work at their own pace. Another popular idea is criterion-referenced assessment, where criteria are specified for a "pass" and all higher grades. Most university assessment is norm-referenced, where students receive a percentage score, and arbitrary boundaries determine a student's grade. A small number of large assessment tasks, and final exams in particular, lead to stress and an unhealthy approach to learning, where the goal of students is to pass rather than to learn.

The above pedagogical issues led us to conceive of the following assessment regime, for a single subject with very large class numbers (typically 500), occurring early in a university degree program. Students are assessed by small regular assessment tasks. Students work at their own pace. Each task is graded as either "satisfactory" or "not yet satisfactory". Students with the latter grade need to repeat the task until the "satisfactory" grade is achieved. An overall-passing grade for the subject is awarded when a student attains a "satisfactory" grade for all tasks comprising a core set. Higher grades are awarded if a student completes further optional tasks.

Given the marking-intensive nature of the above assessment regime, each assessment task is an exam of multiple choice questions. Many existing web-based assessment systems support multiple-choice testing (e.g. Blackboard, WebCT, and TopClass). However, they do so in a way that does not easily lend itself to supporting the above assessment regime. In some existing web-based systems, an exam is a static web page. Such exams should only be administered once to each student. In other web-based system, static questions are selected at random from a pool. While an improvement on completely static exams, such exams are still limited in their usability, and require intensive invigilation to prevent students from plagiarism.

To support our above assessment regime, where it is expected that students will routinely repeat exams, we have developed a system that uses a pool of potential questions, marked-up in XML. Whenever a student requests an exam, a subset of questions are chosen at random from the pool. Furthermore, the system also randomises the order of the choices within each chosen question. Thus students may sit the exam many times, until a minimum acceptable score is achieved, without intensive invigilation or manual intervention from the examiner.

2. The XML Mark-up of an Exam

The complete XML mark-up of an exam is illustrated in Figure 1. This particular example is very small, containing only four potential questions. In practise there should be many more potential questions.

The integer between the <Size> and </Size> tags specifies the number of questions that should be drawn from this pool to create an exam page for a particular student. In Figure 1, two questions are to be chosen.

The "Instruction", "RightAnswer" and "WrongAnswer" tags are self-explanatory. Together, they define a single potential multiple choice question. For each question, five choices are presented to a student, one of which is the right answer. However, the XML script may contain a larger number of wrong answers, from which four are chosen at random.

The "MUTEX" in one of the tags is an abbreviation of "mutually exclusive". A <Question> and </Question> pair contain one or more pairs of <MUTEX> and </MUTEX>. When selecting questions from the pool, the system selects at most one mutually exclusive element within any <Question> and </Question> pair. Thus the XML script in Figure 1 can generate four different exams, each exam containing two questions.

3. The Delivery System

The student initiates the generation of an exam by pointing his/her browser at a given URL. The web server passes the request to a Java™ Servlet. The servlet reads the XML file, and selects a random subset of questions. For each question, the right answer and four randomly selected wrong answers are placed into random order. The correct answers are saved to a back-end "Session DB" and the exam is sent to the student. By clicking on a "Submit" button, the student sends his/her selected answers to another servlet, which retrieves the correct answers from the Session DB. The student's mark is recorded in another database, and is also sent back to the student. Incorrectly answered questions are indicated, with the correct answer highlighted.

3.1. Security and Other Issues

The greatest security risk is one student impersonating another student. If student A is struggling with a particular assessment item, his friend student B, who has already passed that item, might sit the exam for student A.

To avoid this security problem, the system was designed so that the exams would be conducted in a room with an invigilator (who is not required to have knowledge of the subject under examination). The invigilator may demand photo-identification. When satisfied, the invigilator enters a password into a special field on the student's exam page. The grading system will not accept the student's submission without the password. An attractive feature of this simple approach is that there is plenty of time for the invigilator to verify the identity of students, as the students work on their exam questions. There is not a rush at the beginning or end of the session to verify identities.

Our aim was to build a simple yet flexible system. The approach to security is one example of that. Another example is the task of establishing whether a student is eligible to sit a particular test at a particular time, which is left to the invigilator, who provides authorisation via their password. Students frequently have valid reasons for sitting a weekly exam outside their normal allotted time, such as illness. Policing this issue is best left to the invigilator, rather than complicating the delivery system.

Acknowledgements:

The Java™ Servlets implementation of the online system was done by Mark Burgess, Kin Keung Kwan, Shi Sui Ling, Meng Tze Liu, and David Nguon. Java is trademark of Sun Microsystems.

```
<?xml version="1.0"?>
<Test>
<Size>2</Size>
<Question>
<MUTEX>
<Instruction>The unsigned binary
number 10101110 is the base 10
number: </Instruction>
<RightAnswer>174</RightAnswer>
<WrongAnswer>172</WrongAnswer>
<WrongAnswer>168</WrongAnswer>
<WrongAnswer>166</WrongAnswer>
<WrongAnswer>142</WrongAnswer>
<WrongAnswer>178</WrongAnswer>
</MUTEX>
<MUTEX>
<Instruction>The unsigned binary
number 10100110 is the base 10
number: </Instruction>
<RightAnswer>166</RightAnswer>
<WrongAnswer>174</WrongAnswer>
<WrongAnswer>172</WrongAnswer>
<WrongAnswer>168</WrongAnswer>
<WrongAnswer>142</WrongAnswer>
</MUTEX>
</Question>
<Question>
<Instruction>The base 10 number 189
is the unsigned binary number:
</Instruction>
<RightAnswer>10111101</RightAnswer>
<WrongAnswer>10111111</WrongAnswer>
<WrongAnswer>10011111</WrongAnswer>
<WrongAnswer>10100101</WrongAnswer>
<WrongAnswer>10110001</WrongAnswer>
</Question>
<Question>
<Instruction>The base 10 number 124
is the unsigned binary number:
</Instruction>
<RightAnswer>1111100</RightAnswer>
<WrongAnswer>1101100</WrongAnswer>
<WrongAnswer>1101110</WrongAnswer>
<WrongAnswer>1110110</WrongAnswer>
<WrongAnswer>1110010</WrongAnswer>
</Question>
</Test>
```

Figure 1. An XML mark-up of a small multiple-choice exam.

Theme 12: Linkages/Networks for Linkages and Lifelong Learning/Distance Education/Internet Resources

Resource Manager for Distance Education Systems

Goran Kimovski   Vladimir Trajkovic   Danco Davcev
Faculty of Electrical Engineering and Computer Science, Skopje, Macedonia
gkimovski@yahoo.com; {trvlado; etfdav}@cerera.etf.ukim.edu.mk

Abstract

In this study, we extend the concept of distance education with adding a new service, that we call Resource Manager. The Resource Manager offers a possibility to the attendees to share different resources out of time and space boundaries. It enables geographically separated users to effectively facilitate remote access to various, presumably diverse, (real) resources.

The Resource Manager has to provide an efficient sharing of resources among distance learning students according to the student profiles. We defined several protocols in XML suitable for communication between agents.

The first experiments show that users are satisfied with the Resource Manager's usability. They found Resource Manager as a very convenient service within the Distance Education Systems.

1. Introduction

In this paper, a Resource Manager (RM) defined within the Virtual University [1] is presented. The RM system represents a collaborative software environment that enables geographically separated users to effectively facilitate remote access to various, presumably diverse, (real) resources. It represents an extension, a new service, to our concept of Virtual Classroom found in [2]. Namely, the concept of RM is to offer a possibility to the attendees to share different resources outside the time and space boundaries. It tends to provide remote users an experience that approaches the same as "being there". An example of a possible scenario can be to control remotely a robot system in a chemical laboratory from a PC connected on Internet.

The general idea of the system is to enable sharing of many different resources between as many users as possible. The system is structured in common software framework building blocks or "middleware" to enable the user-resource collaboration to succeed. The idea of the resource is not limited to some laboratory equipment. It can be some software application or some interface to data stored in a remote database for example. In this way, we are not concerned with the nature of the resources that this system should work with. On the other hand, it adds another level of complexity, since we have to take into account that the system can be used for resources that possibly have nothing in common. In order to satisfy these requirements, when designing the system, we used the object-oriented approach. Common interfaces that different types of objects should implement have been defined. Thus, the object communication is standardized and the RM system is relieved from the details of the nature of every resource that can be possibly shared through it.

In our design we use agents as entities that work on different tasks in the system. Different types of agents have different specialization, thus encapsulating different services, but they still have similar features. All agents used in the RM are of stationary type and do not expose any mobility behavior. They act independently, cooperating amongst themselves through the act of exchanging messages. We designed several protocols for the communication between the agents of different levels and different hosts in the system. Having in mind the features and increased implementations in XML (Extensible Markup Language), we decided to define all communication protocols within RM using XML.

In section 2, few resource management approaches are addressed. Section 3 describes the general RM architecture. The discussion about the implementation is given in section 4. Section 5 shows some experimental results, while section 6 concludes the paper.

2. Related work

An overview of resource discovery systems and their approaches is given in [3]. In [4], a model for type- specific, user- customizable information extraction and a system implementation called Essence is presented. It generates file summaries that can be used to improve both browsing and indexing in resource discovery systems. In [5], distributed system architecture called Harvest that supports Internet resource discovery systems is presented. For motivation of the research problems addressed by Harvest (of which Essence is one component) see [5] as well. The main objectives of these systems are about resource discovery.

In [6], agent technology is used for developing Virtual Laboratory that is also used as a tool for researchers at three collaborating universities. A dynamic repository of information on ATM topics, ATM simulators and ATM hardware is among resources being made available.

In our approach, our Resource Manager has to provide an efficient sharing of resources among distance learning students according to the student profiles. We defined several protocols in XML suitable for communication between agents. In order to provide better functionality of the Resource Manager, common software framework building blocks or "middleware" ties all services together. This approach should enable an easy cooperation among distributed teams as well.

3. Resource Manager Architecture

3.1. Logical Model

The RM service may be considered within the virtual laboratory defined as an alternative to the "classical laboratory", but it is obvious that this service is quite general. It offers universal, ubiquitous, easy access to any type of remote resources in a distributed system.

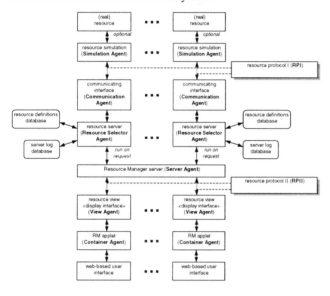

Figure 1 The architecture of the RM

Figure 1 shows the logical model of the RM system. Every user can work with the system by using a web-based interface. The Container Agent enables different resources, with different implementations of their graphical user interface (GUI), to be placed on the web-based interface. In this way every specific resource can be presented with a GUI representing the nature of the resource and can communicate with the users in a manner that is close to the "classical", e.g. having a software compiler as a resource, a source code editor and an output field could represent its GUI. The Container Agent's task is to initialize the web-based interface by gathering information about the resources available in the system and providing an effective way for the user to select a desired resource.

The View Agents are key players on the client side (the web browser that the user is using to view the RM system). They are responsible for the GUI of the particular resources.

The system can hold many definitions of different View Agents, since every different resource or a group of resources should present a GUI of its own to the user.

All of the system communication runs through two agents that run on the RM server machine, the Server Agent and Resource Selector Agent. Every View Agent in the system communicates with the Server Agent when a connection to the RM server is required in order to communicate with a particular resource. The Server Agent handles different connection requests and runs a Resource Selector Agent as a thread that handles the specific requests and passes back the resource responses to the View Agent.

The Resource Selector Agent's task is to get the request from the View Agent, find the appropriate Communication Agent to communicate with the specific Simulation Agent and the real resource and then pass back and forth the requests/responses between the View Agent and Communication Agent (with possible message translation). One Resource Selector Agent's instance handles only one View Agent (the one that had a request for a particular resource) and one Communication Agent. This means that if many View Agents make request for the same resource and this resource cannot be shared in parallel between several users, then a mechanism for waiting queues and timeout methods should be defined in the system design.

For a better communication and error tracking, Server Agent and Resource Selector Agent keep an event log database with information on the system work. Another database is used for mapping the View Agents' resource request with a matching Communication Agent. This database is used for the initialization of the Container Agent as well.

The Communication Agent's task is to enable the message translation and communication in general with the Simulation Agent and the (real) resource.

All of the agents mentioned above are stationary agents and implemented as pure Java classes, but the implementation of the set of Simulation Agents is based on DCOM (Distributed Component Object Model). This enables a distributed computing of the resource requests. The Simulation Agent's task is to wrap around a physical resource in a software entity working in the RM system. The Simulation Agent doesn't always need to be connected to a specific real resource. It can simply be a software simulation of a particular resource.

3.2. Object-oriented Design

The whole system is separated in three different modules: view, server and simulation module. Every module contains its own set of agents and other objects and communicates with the objects in the other modules via XML defined protocols (see an excerpt of a simplified protocol on Figure 2) that give a possibility to exchange several common messages.

```
      View Agent              Resource Selector Agent
<REQUEST>                     <RESPONSE>
   <COUNT>1</COUNT>              <COUNT>2</COUNT>
   <ITEM>                        <ITEM>
      <NAME>Direction</NAME>        <NAME>x</NAME>
      <VALUE>LEFT</VALUE>           <VALUE>135</VALUE>
   </ITEM>                       </ITEM>
</REQUEST>                       <ITEM>
                                    <NAME>y</NAME>
                                    <VALUE>23</VALUE>
                                 </ITEM>
                              </RESPONSE>
```

Figure 2 XML defined protocol (excerpt)

Figure 3 represents the collaboration diagram of the scenario that runs when the system is servicing a resource request.

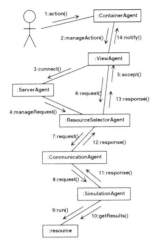

Figure 3 Collaboration diagram

User initiates the scenario by taking an action in the web interface. The Container Agent communicates the action to the View Agent, which tries to connect to the Server Agent in order to send the resource request. The Server Agent is a connections coordinator and it does not process the requests itself; instead it runs a Resource Selector Agent that takes its role in the communication between view and simulation modules. After the acceptance of the connection request by the Resource Selector Agent, all of the requests from the View Agent are passed through the Communication Agent and Simulation Agent to the (real) resource. After the resource processes the request and acts accordingly, the Simulation Agent gets the result back and passes up through the Communication Agent and Resource Selector Agent to the View Agent that shows the result to the user, by redrawing it self for example.

4. Implementation

RM is implemented in Windows 9x/NT environment. All agents are developed in Java, with exception of Resource Simulation Agents, which are based on DCOM technology. Being of stationary type, no mobility support platform is required for the RM service agents. Figure 4 shows the component layout of the RM service.

Our system runs on three hosts. The first host is the client machine of the user browsing the RM via Browser. The Container Agent and the set of View Agents reside here, presenting the different resources' GUIs to the users.

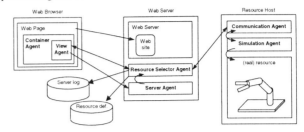

Figure 4 The main components of the RM service

The second host is the RM server. It runs the web server software, and Server and Resource Selector Agents. These agents have access to two databases, Server Log database and Resource Definitions database. The Server Log database is used as a log for every event on the RM server and for the requests history received by the Server Agent and Resource Selector Agent. The Resource Definitions database purpose is to contain information on the resources available in the system. The list of resources available in the system, required by the Container Agent at its initialization is drawn from this database, as well as the information for the correct Communication Agent to which the Resource Selector Agent has to pass the request for the particular resource.

The third host is a computer connected to the (real) resource. The specific Communication Agent and Simulation Agent for a particular resource reside there. As we mentioned before, the Simulation Agent doesn't necessarily have to work with some real resource, it can simulate the resource behavior to the rest of the system.

The next two figures show the implemented RM in action. Figure 5 shows the resource called "Robot Car" under the control of our RM. This resource implements an interface that allows users to control robot movements in six possible directions. Additionally, a web camera is following the robot movements, so it can be watched on-line while moving.

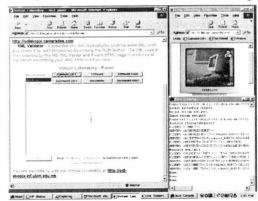

Figure 5 RM in action (resource: Robot Car)

Figure 6 shows a very different resource running in the same framework defined by the RM. This resource represents a software code syntax checker (in this case XML code). The user can type his/her code into the input provided by the resource's GUI and than it will be checked by the

Microsoft XML Parser running on a totally different machine.

Figure 6 **RM in action (resource: XML Validator)**

5. Experimental Results

The latest phase of the work on the RM system consisted of performing tests for its functionality, usability and education efficiency.

Functionality tests were performed during the design and implementation of the modules of the system. We involved end users (students) in these phases and used feedback from them to add all requested functionality.

Usability and education efficiency of the system was tested in order to show when and how RM should be used, using participants interview technique.

We involved 23 users from the University "Sv. Kiril i Metodij" – Skopje, Macedonia (students from Computer Science, Telecommunications and Electronics), 10 users from the Technical University "Sv. Kliment Ohridski" – Bitola, Macedonia, and five users from "Politecnico di Torino" – Turin, Italy (38 in total) in the analysis.

In general, users are satisfied with the RM usability, although some of them would like to see more extensive list of resources before they state their conclusion. They found RM as a very convenient service within the Distance Education Systems and suggested various possibilities: immersion, distance control of unsafe laboratory procedures, following the professor that is not in a fixed position during the lectures, inexpensive solution for sharing expensive laboratory equipment, etc.

Since the user interface is different for different resources, some of the users suggested that it could be different for different user groups as well. We found that it is a very good idea, and we will implement it in the next version of our system.

This kind of feedback will continue to contribute to our system. More extensive research and evaluation of the impact of the system on the students' learning process will be performed during the following work on the system.

6. Conclusion

In this study, we extended the concept of distance education with adding a new service, that we call Resource Manager.

The designed Resource Manager offers a possibility to the attendees to share different resources at once and work with them as if they were at the same place where (real) resources are.

Resource Manager uses agents as entities that work on different tasks in our modular system. By combining this feature with the object-oriented design and definitions of common interfaces for the different types of agents for different types of resources, we designed extensible system, in which resources are added in an easy way.

In order to provide better functionality of the Resource Manager, common software framework building blocks or "middleware" ties all services together. This approach should enable an easy cooperation among distributed teams, which may be one of the possibilities we should try to identify in the next steps.

Research is needed to identify the most useful resources for distance education students and how the Resource Manager service should be best integrated with the Virtual Classroom service.

More extensive research and evaluation of the impact of the system on the students' learning process will be performed during the following work on the system.

7. References

[1] D. Davcev, V. Cabukovski, "Agent-Based University Intranet and Information System as a Basis for Distance Education & Open Learning", In. Proc. of the 1st UICEE Annual Conf. on Engineering Education, Monash University, Clayton, Melbourne, February 1988, pp.253-261;

[2] V. Trajkovic, D. Davcev, G. Kimovski, Z. Petanceska, "Web - Based Virtual Classroom", In Proc. Of the TOOLS 34, Santa Barbara, California, USA, July 30 - August 4, 2000

[3] M.F.Schwartz et al., A Comparison of Internet resources discovery approaches, Computer Systems 5, 4 (Fall), 1992, 461-493

[4] D.Hardy, et al., Customized Information Extraction as a Basis for Resource Discovery, ACM Transactions on Computer Systems, Vol.14, No. 2, May 1996, 171-199

[5] C.Bowman et al., Scalable Internet Resource Discovery: research Problem And Approaches, Communications of the ACM 37, August 1994, 98-107

[6] P. Baxendale, et al., An Introduction to Trilogy Virtual Laboratory, IEE 15th Teletraffic Symposium, March 1998

The Tuneup and Integration of Resources in Web-Based Learning

Li Xiao   Li Jianguo   Zhang Xiaozhen

Faculty of Computer and Information Science, Southwest China Normal University
Chongqing, P.R.China
ivy@swnu.edu.cn   jgli@swnu.edu.cn   zhangxz@swnu.edu.cn

Abstract

A personalized intelligent and interactive learning environment suitable for the personal aptitude of a distance learner can give better play to the learner's special skills and abilities and help him improve the efficiency of his learning activities.

But unfortunately, the existing learning systems have some defects in the way learning resources are organized and do not support personalized learning.

This paper, which centers on how to provide the learner with a personalized intelligent and interactive learning environment, discusses the construction of learning resources, user model, and a resource agent (RAGENT for short), which has been built for the tuneup and integration of web resources to meet learner's personalized expectation on web-based learning.

1. Introduction

We have ample proof that a personalized intelligent and interactive learning environment can give better play to a distance learner's special skills and abilities and help him improve the efficiency of his learning activities. In such an environment, all the web resources are organized according to the learner's personal aptitude, with all the elements of human-computer interface, such as size and color of fonts, shape and size of illustrations, etc, displayed to the preference of the learner. Moreover, the learner can also study with other learners who share the same interest and goal so as to improve himself.

But unfortunately, the existing learning systems have some defects in the way learning resources are organized and do not support personalized learning, which manifest themselves mainly as follows:

(1) The way learning resources are organized does not effectively support the teaching/learning based on different students. What the existing systems provide are largely uniformed and formulated learning contents and learning methods, failing to consider the individual differences in aptitude and cognitive skills different learners already possess, not to say the respective characteristics of different learners' learning behaviors.

(2) The organization of resources does not support the tuneup and integration of the learner's knowledge structure and learning skills. The organization of resources in existing systems only allows the learner to learn discrete knowledge cells, making it very difficult for him to establish a global relationship among all these knowledge cells to discover the underlying rules or principles. In such an environment, it is also difficult for the learner to find satisfactory solutions to difficult problems, which may cause him some mental setbacks and affect the efficiency of his study.

(3) The way does not include any user model. On the one hand, there is no complete information to describe the user's personal traits; on the other hand, there is no mechanism to tune up the user's information.

(4) Also the organization of resources does not support collaborative learning.

Therefore, this paper discusses the construction of web resources and user model, advances a new way of organizing web resources — tuneup and integration, and, by adopting AGENT technique, constructs a resource agent (RAGENT for short) to implement the tuneup and integration, thus solving the first three problems.

Through the tuneup of resources, the learner can discover for himself the interrelationship among the discrete knowledge cells and the underlying principles to form a complete system of knowledge. Moreover, this tuneup can also contribute to the development of the learner's learning skills and other abilities.

By integration we mean: (1) the organization of learning materials, the knowledge cell display and learning methods can be self-adaptive to the aptitude of the learner's learning behavior; (2) the processing of data that are not interrelated or the regularities of which are not yet established, such data are temporarily integrated together to be tuned up in due course.

This idea discussed in this paper meets learners' personalized expectations on web learning and provides learners with an intelligent and interactive learning environment to teach students according to their aptitude.

2. Construction of Web Resources

Based on the Object-Oriented, web resource (*WebR*)

consists of learning resource room (*LSRM*) and learner room (*LRM*), namely, *WebR* ::= { *LSRM, LRM* }

2.1. Elements of LSRM

LSRM consists of learning resources, test bank and other resources. Here we will focus on the construction of learning resources (*LSR*).

LSR is one of the basic elements of *LSRM*. It is made up of knowledge cells of different granularities in a bottom-up stratum structure, which can be tuned up with other resources.

Definition 1: knowledge cell

A knowledge cell (*K* for memory) is the smallest and most basic knowledge unit the learner learns in the learning process to obtain knowledge and attain his desired goal. It is also the smallest element in the LSR on the web. It has nodes of the following properties:

$K ::= <W, U, F, S, E>$

In which, the value of *W* determines the properties of knowledge cell *K*;

U is the corresponding URL set for knowledge cell *K*, and $n(U)=1$, and there exists function $g: K \rightarrow U$,

for any $k_1, k_2 \in dom\ g$, when $k_1 \neq k_2$, $g(k_1) \neq g(k_2)$;

F—the URL set for preexisting knowledge cell *K*;

S—the URL set for subsequent knowledge cell *K*;

E—the URL set for expression of knowledge cell *K*.

Definition 2: Knowledge Cell Class

The knowledge cell set, obtained by categorizing knowledge cells according to their interrelationship, aim, and granularity, is defined as knowledge cell class (*KC*), namely,

$KC ::= \{ K_i | K_i \in KC,\ i\ \text{is a natural number and}\ i \leq 6 \}$.

Definition 3: Course

A course (*SU*) is a set of interrelated knowledge cell classes (*KC*), namely,

$SU ::= \{ KC_i | i\ \text{is a natural number} \}$.

Definition 4: Major

A major (*MG*) is a set of interrelated courses (*SU*), namely,

$MG ::= \{ SU_i | i\ \text{is a natural number} \}$.

Definition 5: Domain Knowledge

A set of majors of the same nature constitute the domain knowledge (*DK*), namely,

$DK ::= \{ MG_i | i\ \text{is a natural number} \}$.

Definition 6: Learning Resource

Learning resource is defined as the set of various domain knowledge (*DK*), namely,

$LSR ::= \{ DK_i | i\ \text{is a natural number} \}$.

Definition 7: Overall Set of Knowledge Cells

The overall set of knowledge cells is the set of all the knowledge cells in web learning, denoted as $\cup K$.

The overall set of knowledge cells constitutes the web learning resource (*LSR*), namely, $\cup K = LSR$.

The relationship among *KC, SU, MG, DK*, and *LSR* is as follows: $KC \subset SU \subset MG \subset DK \subset \cup K = LSR$.

Furthermore, similar to conceptual graph [1], a knowledge-learning relationship should be added so as to present the relationship between them in learning process.

Figure 1 represents the hierarchy of learning resources, in which the knowledge-learning relationship is marked.

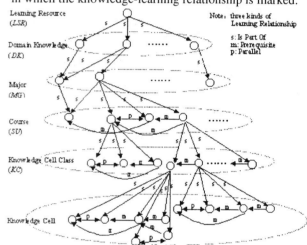

Figure 1. Hierarchy of Learning Resources

The arc following different tags (*s, m* or *p*) describes exiting three kinds of knowledge-learning relationship (arc *s* for 'Is Part Of', *m* for 'Prerequisite', and *p* for 'Parallel').

2.2. User Model

The learner is the main object of the web learning system and the object to which the web resources, after being tuned up, will be presented. A personalized intelligent learning system should suit the individual learning behavior of the learner. A user model [2] is used to describe the individual characteristics of the learner.

Definition 8: User Model

User model (*UBHR*) represents the aptitude or characteristics of the distance learner's learning behavior and is defined as:

$UBHR ::= \{ UB,\ UH,\ UT,\ UA,\ UP,\ UC \}$,

Where,

(1) *UB* describes the user's basic information including the following five units:

<user identification, password, name, e-mail, learning type>;

(2) *UH* records the user's learning history, which is a knowledge tree (just like figure1) made up by tuneup in a bottom-up fashion of the knowledge cells the user has already mastered, reflecting the user's knowledge structure;

(3) *UT* records the user's learning characteristics and learning methods, initiated by a aptitude-processing program upon log-in as the processed result of the user's learning aptitude and, in the course of learning, to be dynamically updated based on an analysis of the learner's progress;

(4) *UA* is the result of estimate of the user's ability, initiated as the result of a user-ability-affirmation program, to be renewed in the course of learning;

(5) *UP* describes the user's preference for human-computer interface to organize specific user interfaces, initiated as the result of a visualized interface-processing program;

(6) *UC* describes the necessary information for collaborative control between the users.

It can be noted from the above definition that the construction of the user model is a dynamic process that will approximate the actual behavior of the learner (if given enough time). Now, let's define learner's room to support collaborative learning to solve the fourth problem in chapter 1.

Definition 9: Learner's Room

Learner's room (*LRM*) is the set of user models describing various learners' behavior room, namely,

$LRM ::= \{ UBHR_i \mid i \text{ is a natural number} \}$.

3. Architecture of RAGENT

By adopting the maturing AGENT technique [3], we have constructed a web resource agent (RAGENT for short) to tune up and integrate web resources.

Figure 2 is the system architecture of a RAGENT, composed of client-side RAGENT and server-side RAGENT.

Its functions at the client side in the management of web resources are as follows:

(1) Monitoring the learner's learning behavior;

(2) Constructing the user model and providing the mechanism to tune up the user's personal information;

(3) Dynamically organizing learning materials and building the learner's learning environment according to the user's personal preference.

Its functions at the server side are as follows:

(1) Tuning up and integrating web learning resources;

(2) Serving as the manager and coordinator of web resource room;

(3) Serving as the manager of the system's operation and the manager of the server.

The RAGENT is made up of the external inlet and the control module, including five modules in two strata..

The first stratum is the external inlet. It is the RAGENT's inlet with the external environment (such as the server, the internet, the user's inlet, the learner, etc), which, by detecting changes in the environment, produces behaviors that act on the environment (to send a message for example) and communicates with other Agents. It is made up of three modules: Module Detect (responsible for detecting and obtaining the learner's learning behavior and transmitting the data to learner behavior control chunk); Module Messge (responsible for transmitting learning materials); Module Communi (responsible for communication with other Agents in the learning system).

The second stratum is the control module. This is the core of the RAGENT, mainly functioning to tune up and integrate web-learning resources and provide the learner with a learning environment self-adaptable to the characteristics of his own learning behavior. It consists of two modules:

(1) Learning Behavior Processing Module (Module BhrP), which is connected with the Detect Module to receive all the behaviors the learner demonstrates in the learning process and process them to obtain the learner's learning characteristics, learning method, personal tendency, and habitual way of thinking, and then, while modifying user model (*UBHR*), to send the message to the Module Plan;

(2) Learning Plan Module (Module Plan), which receives the learner's aptitude from the Module BhrP, then searches the resource bank (*LSRM*) for proper materials according to the learner's aim and behavioral characteristics before organically integrating these materials to generate a learning plan suitable for the learner's individual aptitude, and finally activates the Module Messge to deliver materials.

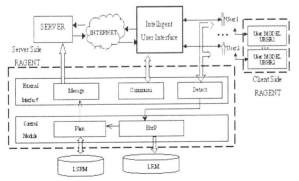

Figure 2. Architecture of RAGENT

4. The Tuneup and Integration of Resources

The tuneup and integration of resources are implemented through RAGENT. The whole process is illustrated by an example of a learner requesting to learn a new knowledge cell.

Step1 Detecting a learner's request, RAGENT (Module Detect) delivers the message to Module BhrP for user model update and to Module Plan for adaptive learning materials;

Step2 RAGENT gets the user model of the learner's from Client-Side;

Step3 Module Plan in RAGENT judges what the learner should study is by seeking *UH* of *UBHR*. The current learning content should be decided by learner's learning history.

Step4 Module Plan makes suggestion on how to study by hunting *UT* and *UA* in user model (*UBHR*);

Step5 Module Plan starts to search data from LSRM to get appropriate learning content in accordance with the learner's aptitude;

Step6 Module Plan generates the plan of learning;

Step7 Module Plan activates Module Message for delivering learning content.

If necessary, Intelligent User Interface will re-construct the interface according to the learner's preference for human-computer interface from *UP* of *UBHR*.

5. Conclusions and Future Work

This paper represented a new way of organizing web resources to provide the learner with a personalized learning environment. Using Java [4] we have succeeded in constructing a prototype and obtained anticipated results.

This paper has failed to cover the collaboration among learners. Actually, collaboration is also an important aspect of organizing web resources and will great conduce to the progress of the individual. This is what we are going to study next.

References

[1] Olivier Corby, Rose Dieng, and Cédric Hérbert, "A Conceptual Graph Model for W3C Resource Description Framework", http://www.lirmm.fr/Ecogito/papers.html, 2000

[2] Edward Ross, "Intelligent User Interfaces: Survey and Research Directions", http://www.cs.bris.ac.uk/tools, 2000

[3] *Intelligent Agent Technology, System, Methodologies, and Tools*, by World Scientific Publishing Co., Pte.Lt, Singapore, 1999

[4] Simon Roberts, Philip Heller and Michael Ernest, Complete *Java 2 Certification Study Guide*, Publishing House of Electronics Industry, Beijing, 2000

Interactivity and Integration in Virtual Courses

Claus Pahl
Dublin City University, Dublin 9, Ireland
cpahl@compapp.dcu.ie

Abstract

Web-based virtual courses focussing on content delivery only have turned out not to be as successful as expected. We will investigate reasons and suggest some remedies – using our own virtual Database course for illustration. Interactivity and integration are the two main remedies, which shall lead towards improved, learning-oriented virtual course environments. We will introduce a formalism to support their development and analysis.

1. Interactivity and Integration

In virtual courses traditional educational activities or services such as lectures, tutorials and lab sessions are replaced by Web-based counterparts. In this paper we will discuss the design and evaluation of interactive integrated multi-service virtual courses. Despite potential advantages of Web-based delivery [1], there are also risks involved. In order to support active learning virtual course systems need to provide more than a static representation of knowledge. Interactivity is a means for engaging and encouraging a student. Observing student behaviour, we can see that students typically use different activities with different regularity, e.g., a 'just-in-time'-learning in labs close to tests/exams vs. a more regular lecture attendance. Virtual courses need to be made more *interactive* or engaging. The realisation of educational services such as lectures or tutorials should be fully *integrated* in order to exploit the potentials of the new technology such as availability and integrated access. Both Web-based systems and study habits might lead us away from a successful learning experience, i.e., the usability of the system, success in exams, and user satisfaction. In order to overcome these shortcomings, we propose *integrated interactive courses*. Integrating interactivity allows students to learn actively – a necessity if practical skills have to be taught in a course – integrated with more conceptual parts of a course. Our hypothesis is that students learn by doing – which is true for most science or engineering subjects – and by reflection. To teach computer programming skills, as in our own virtual course [2], requires in particular interactivity support. *Interactive* elements are typically accompanied by theory. Interactivity might be supported by guided tours where topics are demonstrated and by elements where the student can freely train skills. These forms - corresponding to lectures, tutorials and labs - should be integrated, e.g. supported by an adequate navigation structure allowing the student to access complementary material, integrating learning activities around course topics.

We shall focus on techniques to support the design and evaluation in a coherent framework. In order to formulate the structure of a multi-service course and the behaviour of students within this structure, we suggest a reference architecture for the integration of educational services. A notation based on this model will be introduced that allows us to describe behaviour based on this architecture.

2. A Model for Integrated Interactive Services

When a multi-service virtual course has to be developed an in-depth understanding of the course structure and the expected dynamics within the system is essential. We will present a model of a layered reference infrastructure and a notation to express the dynamics in such a course. It allows the developer to formulate expected behaviour in the design phase, but also to analyse student behaviour during and after delivery. A suitable formal approach is sought to capture the static dependency issues and the navigation infrastructure, which requires means to express the sequential but also concurrent use of services. A second criterion is a notion of state in order to remember the usage history (which pages have been visited). We use Petri-nets in combination with a notation called path-expressions that can express user behaviour over time.

In a system with fully integrated educational services, the student could interrupt using the lecture service, practice relevant skills using the lab service, and then resume the lecture. In a virtual system, changing between modes of learning can easily be supported by appropriate navigation support connecting the services. The services can be represented in layers. The first layer is the lecture service, the second layer represents interactive services:

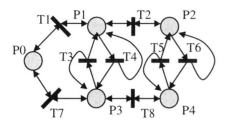

Figure 1. Virtual course topology

*Usage pattern*s defined by the developer are based on causal dependencies and can describe the intended use. *Path expressions* formulate the patterns. The expression P1;(P1∥P3);P2 means that P1 shall be accessed first, then P1 and P3 are looked at in parallel (link T4, Figure 1), and finally page P2 is visited. Besides the sequence operator ';' and the parallel composition '∥', other operators such as choice P1|P2 (choose between P1 and P2) and iteration P1* (visit P1 any number of times) are available. These path expressions, formulated on the defined course topology, describe the expected usage and should be implemented by a respective navigation infrastructure. The expression P1;(P1∥P3);P2 can be interpreted as a guideline for using the system: look at lecture page P1 first, then open tutorial/lab page P3 in parallel and practice, then go to the next lecture page P2. The formal model also allows the teacher to capture actual usage of the system for the course evaluation in terms of the notation provided, see [3].

Our own virtual database course [2] replaces traditional *lectures* with a lecturer's speech and overheads by audio and synchronised visuals. The audio presentation controls the presentation of visuals. Our course system improves the classical delivery by making it more personalised and more interactive through individualised delivery. The student can interact freely with the lecture service by controlling the audio stream. A *tutorial service* illustrates the lecture material dynamically. It guides the student through a series of examples with increasing complexity. The tutorial service allows the student to practice material learned in the lectures. The system offers also individual feedback in form of a check-function. A *laboratory service* provides practical material beyond the examples used in the lectures. The integration of these services is essential. An infrastructure allowing students to navigate between corresponding parts of different services needs to be provided. The topology model allows us to formulate this infrastructure. Using the path expressions, a developer can specify behaviour which is expected and which shall be supported in the system by the navigation infrastructure.

3. Evaluation and Conclusions

We will address evaluation here only in order to validate and justify the ideas and concepts that have been presented – more can be found in [3]. We have been looking at *student behaviour* in two dimensions: usage in time (frequency/regularity of usage, number of accesses) and usage in space (usage patterns based on the topology). These evaluations are based on the topology model, which forms the backbone of our approach. A look at usage patterns shows that the lecture service is typically used without accessing the interactive parts, but when lab or tutorial services are used, students do look up definitions, concepts and examples in lectures. Here, a close integration of static and interactive services is useful. The analysis of frequency and regularity of system usage shows that lectures are used on a more regular basis than tutorials and labs. The majority of students leave exercising until close to tests and exams. An improved integration of interactive elements into the lectures might have lead to a more regular use of tutorials and labs, but the integration has certainly improved the usability of the interactive services themselves. We have monitored the *student success* in exams over a period of three years. There is no difference between traditional and virtual delivery with respect to performance in questions about topics supported by interactive services – our last delivery with improved interactivity showed even better results. The same is true for an overall comparison between traditional and virtual delivery [4]. Two usability criteria emerging from *student surveys* are the need for interactive elements that provide good feedback to student input and the need for a good integration and navigation structure. Active learning and feedback are important elements of a course, see e.g. [5].

Our main objective has been to support the development and delivery of virtual courses, providing educational activities equivalent to those in traditional courses. Interactivity is a desirable feature in virtual courses; the integration of interactive and other parts is essential. Our formal model and the notation allow us to describe infrastructure and behaviour in integrated interactive virtual courses. Model and notation provide a tool for teachers and developers to design and analyse advanced virtual Web-based courses.

References

[1] J.A. Lennon. *Hypermedia Systems and Applications*. Springer-Verlag, 1997.

[2] C. Pahl. CA218 Databases (Course Web Site), *2000/2001*, URL for Web page: http://www.compapp.dcu.ie/~cpahl/.

[3] C. Pahl. The Evaluation of Educational Service Integration in Integrated Virtual Courses. *IEEE SAINT Workshop on Internet Supported Education*. IEEE Press. 2001.

[4] A.S. Smeaton and G. Keogh. An Analysis of the Use of Virtual Delivery of Undergraduate Lectures. *Computers & Education*, 32(1):83-94, 1999.

[5] American Association for Higher Education. *Seven Principles for Good Practice in Undergraduate Education*. http://www.aahe.org. 1997.

Networked Facilitated Open and Distance Learning in Continuing Engineering Education

Sabine Payr
Austrian Research Institute for Artificial Intelligence
sabine@ai.univie.ac.at

Franz Reichl
Vienna University of Technology
reichl@ai.tuwien.ac.at

Gottfried S. Csanyi
Austrian Association for Didactics in Higher Education
gottfried.csanyi@univie.ac.at

Ursula E. Vierlinger
Vienna University of Technology
ursula.vierlinger@tuwien.ac.at

Abstract

Open and Distance Learning in Continuing Engineering Education brings about a set of specific challenges: production and delivery have to take place in a network of providers in order to be cost-effective, and the demands of an audience of professionals and practitioners have to be met. The project FACILE had as a goal to find solutions to these challenges. Active facilitation turned out to be a key factor. The experiences allow us to discuss the role and relevance of different ODL elements and technologies for the specific requirements of CEE.

1. Background

Coping with an increasing pace of world-wide information exchange, traditional Universities of Technology are challenged with an increased need to update and upgrade their graduates' knowledge, increasing competition with market leaders operating in a global market, and increasing specialization.

For Universities of Technology, it will sooner or later not be sufficient any more to provide locally available knowledge to local customers. One way to adapt towards the challenges of a global knowledge market is to apply Open Distance Learning (ODL). Universities of Technology have to team up and build strategic networks to complement their Continuing Engineering Education (CEE) programs by Open Distance Learning. But the Centers for CEE are still at the beginning of this process; they still have to gain experience.

The SOCRATES/ODL Project FACILE (**FACI**litated open distance **L**earning for continuing **E**ngineering education) was a starting point to complement CEE provisions by close collaboration within a network of Centers for Continuing Engineering Education (CCEEs). It provided European Centers for Continuing Engineering Education with the opportunity to enhance the knowledge and skills of their staff on the application of facilitated ODL to their daily work [1].

2. The FACILE project

The SOCRATES/ODL project FACILE (1998-2000) added a dimension of collaborative CEE provisions by facilitated ODL to a network of CCEEs. By exchange of information, exchange of educational material and training modules, and by dissemination and transfer of evaluated and validated training material as well as of the evaluation results, the project made expertise on facilitated ODL widely accessible.

2.1 Joint curriculum development

The project partners (among them 7 CCEEs from different countries) attempted to collect experiences about applying facilitated ODL in different learning situations, especially related to continuing education, and to collect, analyze and adapt existing learning material about facilitated ODL. This attempt did not work out as expected. Thus, the partnership jointly designed a curriculum ("FACILE Staff Training") for managers and teachers at CCEEs. Based on this curriculum, 10 modules covering different areas of facilitated ODL (educational, organizational, financial, and technological aspects of facilitated open distance learning) and its application to CEE have been developed and, after two pilot runs, been offered to learners during spring 2000.

These modules were delivered to the learners by means of facilitated ODL. A web-based learning environment developed by the project partner HCI was used mainly for dissemination of the learning materials and other resources. Together with other communication technologies such as e-mail, mailing-lists, a chat-room and audio-conferences, the learning environment also provided a framework for interaction through group workspaces and discussion fora.

2.2 Networking among Providers

Since creating online learning material is very time consuming, dissemination of (perhaps adapted) material of other providers will be a key issue for future online learning provisions. The project FACILE implemented the nucleus for an international network for the joint organization and for the support of distance learners in CEE. CCEE's personnel had the opportunity to gain practice and experiences from applying their new knowledge in FACILE pilot courses delivered to a geographically dispersed audience via this network.

The FACILE project developed transferable products based on the application of generally available technology (e.g. telecommunication through the Internet, video conferencing with ISDN). FACILE concentrated on best practice for the support of distance learners by facilitation of the learning process.

Nevertheless, this could only be a first step in the networking process, and many additional experiences have yet to be gained, e.g. on establishing an organizational, logistic, legal and financial framework for such network activities and implementing successful procedures for mutual transfer activities.

2.3 ODL for Practitioners

The target groups for both FACILE training events were practitioners and professionals in their field (further engineering education for the staff training, engineering for the pilot courses). Many learners came to the course with already a high level of expertise. Motivation was generally high, but did not regard completion of the course itself (as would be the case e.g. with students for whom a certificate is relevant), but the benefit of the course for practical problem solving and development. In short, one could define participation in the course as "need-driven", where, however, the pressure and imminence of the "need" (to develop ODL in the home institution in the case of the staff training) varied widely. Thus, some participants were already involved in ODL, some others only wanted to be prepared for ODL introduction in the near future. Similarly, the familiarity with ICT and the level of mastery of the skills necessary for telematics-supported ODL varied widely.

3. Three types of knowledge and learning

The consequences of this setting and target group are far-reaching. We will outline them here on the basis of three types of knowledge, each connected to a different type of learning and teaching (cf. [2], [3]).

3.1 "know that" (factual knowledge)

This type of knowledge was represented in the modules which, in general, were informative texts. Especially the first pilot course showed that the written online material was, at best, scanned or read once and did not initiate group learning with questions and answers, let alone discussions. The reason is related to the specific target group: practitioners refuse to accumulate factual knowledge for some later purpose. While students are trained to do just that - to put facts "on stock" for some later application -, practitioners have successfully un-learned this way of knowledge acquisition and are interested in facts and information only when relevant for their work. What they put on stock, however, are clues to where and how to find this information once it is needed.

3.2 "know how" (procedural knowledge, skills)

Hands-on experience turned out to be the most valued aspect of the FACILE staff training. In the domain of experimenting with and evaluating of technologies used in ODL, we can safely assume that most of the actual learning process took place. Of course, this kind of learning is highly interactive and requires intensive support and active facilitation.

3.3 "can do" (mastery, practice)

This final step of the learning process, where participants transfer their knowledge and skills to their own practice, supported by a coach or mentor, can only take place when the course coincides in time with a practical problem to be solved by the learner. "Constructed" project work is no sufficient substitute for this kind of learning. In this stage, individual support by experienced practitioners is necessary. Most ODL courses cannot, for organizational reasons, be "open" to such a degree that they would allow for this kind of "just-in-time"-learning, however desirable it would be from an educational point of view.

4. The role of facilitation in networked ODL

Both the pedagogic requirements of CEE and the situation of networked production and delivery of ODL led to a concept of facilitation adapted to this situation.

4.1 Facilitation in ODL

Typically, "facilitation" in ODL is understood as the support of learner activities at a remote learning site [4]. The facilitator, in such a setting, is the first or only person involved in a course with whom the learners are in face-to-face contact. The role of a facilitator is to ensure communication between faculty and students, to ensure smooth running of technical equipment, to respond to students' questions concerning course administration, learning facilities, but rarely on the subject itself.

In the FACILE project, on-site facilitation was replaced by distance facilitation. While in most ODL courses the role of support, tutoring, coordination and contact at a distance is taken over by the teacher/trainer him/herself, this model did not work out satisfactorily for this new type of networked ODL. The first approach that had been taken in the project was to make the author of each module responsible for its delivery. Different contents areas would be facilitated by different persons. Learners experienced problems with this approach, probably for several reasons at once:

- Each author/teacher brought his/her own concepts and methods of ODL to the undertaking.
- Learners did not know any of the teachers personally and did not have time to get to know (and to adapt to) their differing methods and pedagogic backgrounds.
- The individual authors of the modules were "bound" to the specific contents they wanted to deliver, so that no continuous course activities could develop.

4.2 The role of online material

Having stated that "know-how" is the most important type of knowledge that can be developed in an online course for practitioners, it would result that the contents - being mostly informative texts - of the course modules would be almost irrelevant. Indeed, the participants only referred to the texts marginally in their interactions with the facilitator and the authors of the modules. Online materials were, at best, incentives for discussion. However, it would be wrong to conclude that it does not matter what contents and in what form is put online. Written materials can fulfill two important functions:

If online courses are to *prepare practitioners for "just-in-time"-learning*, the materials should provide the learner with the clues for later information retrieval in a very clear and concise form. They should, for example, lay out what the aspects are that the learner has to take into account when starting out into practice, and give hints to resources where information on these aspects can be found (not in the form of an exhaustive bibliography, as would be the case in an academic paper, but in the form of very well reflected and commented recommendations).

Another function of online materials can be seen in *laying the ground for the mentoring network*. This could be achieved by building contacts between learners and authors/experts. To this aim, the "experts" have to be aware of the fact that they are only a little more experienced than the learners, and that only in very specific fields. In other words: they have to leave behind their role as "teachers" and should not try to "lecture" participants on how things should be done. Instead, their role would be to share their experience in the form of case studies. We suppose that case studies are the more beneficial to learners the less they are "finished" or "polished". It is the unsolved problem, the lack of success, the open question that motivates participation and discussion. At this point, the roles of teachers and learners become indeed interchangeable, as the problems to be discussed are real and not artificially constructed by a seemingly omniscient authority for the learners to solve. Case studies give the learners the opportunity to get to know the author, her competencies and problems, and could facilitate future contact at the time when the learner needs exchange and support in her practical work.

4.3 The facilitator as the "spider in the web"

The (tele-)presence of a single facilitator in the staff training course as it was delivered in its final form turned out to be a substantial factor in providing stability and continuity in this networked ODL environment. The facilitator did not only coordinate students' activities, but also became a mediator between the students and the different authors. While the modules still were authored by different people from different countries, backgrounds, etc., student activities were designed and monitored centrally and over the whole duration of the course. The facilitator's role could best be compared to a "spider in the web", tying together the most diverse persons, interests and activities.

Facilitation has turned out to be the key to success in the FACILE courses. This result is connected to the insight that the most important type of learning that can take place in an ODL course for practitioners is of the "know how"-type. In this type of learning, intensive assistance is necessary. But even more: While motivation for the first type of learning ("know that") is mostly extrinsic (e.g. a certificate) and almost entirely intrinsic in the third type (own practice, professional networking), it can be considered "mixed" in the second type ("know how"). Thus, the facilitator also has to take "motivating" action, not in the sense of reinforcing extrinsic motivation (e.g. by sanctions), but by trying to build a learning group that develops its own social dynamics and commitment.

For the final launch of the FACILE Staff Training, the roles of facilitation and consequences for the facilitator profile have been defined, e.g.:

- the facilitator is the primary contact person for all the participants and contacts authors and experts;
- the facilitator has to take the initiative e.g. in soliciting questions, contributions or personal information from participants; (s)he elicits questions from participants and acts as moderator between the learning group and the authors/experts;
- the facilitator monitors each participants' activities, e.g. by reminding them of their tasks;
- the facilitator creates a social learning environment - a virtual learning group, using the available technology; (s)he thus has to be technically and socially competent in communication technologies.

In short, the facilitator has to react timely to every input by a participant, triggers self help and group support instead of answers by facilitator or expert, and keeps the team learning going by setting tasks and deadlines, by ensuring information flow among participants, by being supportive and communicative.

5. ICT for networked facilitated ODL

Several ICT were used in the FACILE courses, as mentioned before: besides an electronic learning environment (ELE), these were e-mail, mailing-lists, and a chat-room. The web-based ELE, specifically developed for the FACILE project on Lotus Notes Domino basis, also served as a joint workspace for the project partners.

The ELE offered a whole range of functions for course delivery and communication. The modules (text, graphics and links) were disseminated via this environment, as were contact information about authors and participants. The ELE also offers discussion fora and feedback facilities for online tutors. Mailing-lists and chats were external to the ELE, although dissemination of e-mail to different groups was supported.

The experiences showed that participants accepted the learning environment for distribution of materials, but preferred e-mail and mailing-lists for communication. The ELE's discussion forum function was hardly used, although such a forum offers much more comfort to users, such as access to past discussions and structuring of contributions in threads. We suppose that participants prefer communication via e-mail because this is their everyday working and communication environment. It is not necessary to log into the learning environment each time they want to read or write messages to the course group. While it would be desirable to collect messages exchanged in the framework of an online course in the ELE together with the material they (possibly) refer to, it would be the wrong approach to force participants to use the ELE. Learning environments should adapt to users' needs and habits, and not the other way around.

Technically, it would therefore be desirable to integrate mailing-list communication into the ELE in such a way that postings and replies are distributed via a list-server and simultaneously stored in the discussion forum of the ELE, thus enabling both modes of access (much the same way as in free discussion servers such as yahoogroups.com). In addition, automatically generated reports about activities in the learning environment, distributed via e-mail (as implemented in the project partners' workspace), are necessary as a service to participants: it is a reminder to log into the learning environment and, at the same time, allows them to reduce logins to those cases where it is worth while. In training busy professionals, every chance has to be taken to help them save time. It is known that online courses, especially when communication is intensive, take much more time to complete than foreseen by either teachers or students.

Networked ODL as in the FACILE project, where contents originate from different institutions and authors, requires a learning environment that makes it easy to import and adapt learning material from different sources.

Information about participants of online courses is often limited to the presentation they give of themselves at the beginning of the course. A mode of representation of online discussions where contributions are grouped by their authors - together with personal information - could be a means to create a "dynamic profile" of the participant which would be of interest not only for the facilitator, but also for other participants in view of future contacts and networking.

Conclusions

The FACILE project has shown, especially in the second staff training course, that the double role of the facilitator - both (technical, organizational, personal) assistance and motivation - requires a high level of skills of technical, organizational and, maybe first of all, social nature from the person who is in charge of facilitation. Facilitators have to be professionals in their specific field, and in any ODL development attention has to be paid to developing facilitation competencies in the persons who are to play this role.

References

[1] F. Reichl, "Joint European Continuing Education Courses with Facilitated Open Distance Learning", paper presented at 2001 CIEC, ASEE, San Diego, 2001.
[2] P. Baumgartner, S. Payr, "Lernen mit Software", 2nd ed., Studienverlag, Innsbruck, 1999.
[3] S. Payr, "Tele-Training on the Job. Experiments and Experiences in Media Integration", paper presented at EdMedia/EdTelecom, Seattle, 1999.
[4] L. Sherry, "Issues in Distance Learning", *International Journal of Educational Telecommunications*, 1 (4), 1996, pp. 337-365.

Linking Experiences: Issues Raised Developing Linkservices for Resource Based Learning and Teaching

Hugh C. Davis and Su White
IAM: Learning Technologies, The University of Southampton, UK.
hcd@ecs.soton.ac.uk, saw@ecs.soton.ac.uk

Abstract

Microcosm is an open hypertext system designed to facilitate a resource-based approach to learning and teaching. In spite of winning accolades the system never established itself as a mainstream educational product. This paper describes the technical and pedagogical features of Microcosm, and attempts to explain the reason for its limited take up. The paper is in the form of a case-study of both good and questionable practice.

1. Introduction

1988 was a good year for educational technology. The Apple Macintosh was well established, and came bundled with Hypercard; OWL had produced a hypertext system called Guide, which was available, on both the Macintosh and the Windows 2 platform, which had recently established itself as the GUI front-end of choice for DOS based PCs. Videodisc technology was advancing rapidly and genlock cards were enabling video output to appear within Windows on a computer screen. PC's were becoming affordable, massive storage was possible on CD-ROMs and local area networking was a reality.

At this point in time a number of researchers at Southampton sat down to design a new hypertext system. The Microcosm system was first described in 1990 [1]. In 1994 the system became the platform for a number of projects within the Teaching and Learning Technology Programme (TLTP) in the UK. At about the same time a spin-off company, Multicosm Ltd, was established to exploit the intellectual property. In 1995 Microcosm won a European Software (ITEA) Award. In 1996 it won the British Computer Society's Software Innovation Prize and StoMP, an application of Microcosm for teaching of Physics, won a European Academic Software Award.

In spite of all its successes, Microcosm never became a big player in the educational technology market, and the company that currently licenses the product is no longer actively marketing it. In this paper we examine the good points that Microcosm embodied, both from a technological and pedagogical point of view, and then we analyze the reasons why such apparently notable features have failed to impress the purchasing public

2. The Pedagogical Perspective

The fundamental premise of the Microcosm design team was that learning is enhanced by access to an information system that allows a user to explore a set of resources related to a specific domain. Students who actively query a resource-base will build a better model of the topic than those who passively receive information [2]. An advanced hypertext presentation was seen as being a good way to provide such a learning environment; links may be used to associate a phrase representing a concept to related information. However, simple browsing and clicking on blue links does not necessarily encourage deep learning [3], so the team designed an environment which encourages the user to actively engage with the materials. This encouragement is provided in a number of ways:

1. The user has access to a document explorer window which maintains a hierarchy of the corpus of available documents. This is not rocket science, but many hypertext systems fail to provide this obvious parallel to the table of contents in a book.
2. The author may give the documents attributes (e.g. keywords) for the user to find appropriate documents.
3. Links may be hidden - rather than standing out in blue underlined text. To activate such links, the user must select an enclosing chunk of text and choose "Show Links" from a menu, as shown in figure 1. The theory is that this makes users ask the question "what would I like to be linked?" rather than waiting to find links in the text [4]. Of course it is important that there are lots of links, so that they are generally successful in finding links. If not they will quickly give up.
4. Users may make queries, using much the same approach as a modern web based search-engine, by typing in words they might expect to find in the document they are looking for. Similarly a user can select a chunk of text and choose "Compute Links" from a menu. In both cases the system will produce a ranked list of other documents (from within the corpus) that contain the similar vocabulary.

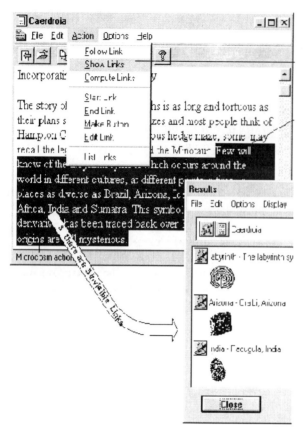

Figure 1: Finding invisible links. The user selected text and asked to show links. Three were found.

5. The user may create their own links and trails, and even add their own documents to the corpus.
6. The user may add annotations to existing documents.

Another important concern of the designers was to provide a system in which it would be possible to provide different links depending on *context*. So it is possible, for a given corpus of documents, to have a different set of links depending, for example, on whether the user is a beginner to the subject or an expert, or perhaps reflecting the user's intended purpose for the document. This is now referred to as "link level adapted hypertext" [5].

The designers wanted it to be possible for authors to include the full range of media types within their corpus of documents. This meant that it must be possible to create links into and out of any type of file, including sound (e.g. on a CD), video (e.g. on a videodisc), word processed documents, spreadsheets, bitmapped pictures, vector drawings etc.

Building a resource-base is a time consuming task. If you add the requirement that the resource base is hypertext linked then the task grows exponentially with the number of documents. Early experiments building small hypertexts using packages such as Hypercard had encouraging results with students, but were prohibitively expensive to build [6] and that effort did not scale well to producing large corpuses with thousands of documents and tens of thousands of links. Consequently the Microcosm designers considered it very important to produce a system which enabled rapid development of learning materials; the goal was to produce a system which would allow the author to take existing documents (of any media type) and to import them into the document hierarchy. The index for querying would be built automatically, and existing links would be automatically re-purposed for use within the new document, thus making the task of linking much faster.

3. The Technological Perspective

The most significant feature of Microcosm is that the links and anchors [7], which are the fundamental components of any hypertext system, are not held in the document (as they are, for example, as HREF's in html), but in separate databases, called *linkbases*. When a user loads a document, the viewer queries all currently selected databases for any links that should be shown in this document, and displays them appropriately. The links have enough information stored with them so that the viewer can tell where in the document the link belongs, how to display the link, and the anchor at the other end of the link. This separated link approach has the immediate benefits that:

1. Authors, when making links can choose in which linkbase to store their links and users (or some agent acting on the user's behalf) may select which linkbases to activate. It is thus possible, for example, to keep links suitable for a beginner in a linkbase separate from that holding links suitable for an advanced student, and links suitable for an engineering student in a linkbase separate from that holding links intended for a business studies student.
2. Users may keep their own linkbases, in which they keep their private links and trails, without altering the set of links delivered by the author.
3. Documents do not need to be altered in order to add a link to the document. This is important if users are to be able to make their own links (it would not be acceptable to allow users to alter the author's documents). It also means that links can be made in read-only documents such as those on CD-ROMs, videodiscs and those to which we do not have write access.
4. Data formats do not need to be adapted in order to allow linking. Microsoft have recently extended the data formats of their office products to allow embedded HREFs but, for example, if you wish to put a link in a bitmap the only way this can be done is to hold the link externally.

5. The final advantage is the facility to make *generic links*. These are links which are associated with some word or phrase, rather than the position that they occur within a document. So for example an author might create a link on the phrase "insertion sort" which might link into a document about sorting methods, to a section about insertion sort. Now, wherever *any* document includes the phrase "insertion sort", this phrase will automatically be made into a link at runtime. The power of generic links is significant for two reasons. First, it is possible to automatically generate links to documents using their keywords (e.g. glossaries). Secondly, a generic link, once made, applies to all documents added to the corpus, so link authoring grows linearly with the number of documents, rather than growing exponentially.

A second important feature of the Microcosm architecture was the ability to display links in a very large number of data formats. In earlier versions of Microcosm this was achieved in two ways. a) by writing specialized viewers (e.g. for bitmaps, GIFs, MPEGs etc.) that understood the protocol to talk to the linkservice and to save and retrieve links. b) by adapting existing packages (such as MS-Word and emacs) using their built-in macro programming facility. Later versions of Microcosm used a third party viewer that had been adapted at source code level to talk to the linkservice; this viewer could present the vast majority of known data formats.

4. Analysis

From the point of view of the research group at Southampton, Microcosm was an enormous success [8]. The architecture allowed plug in modules to be inserted into the system which in turn enabled us to experiment with new ideas in hypertext. The linkservice approach turned out to be very timely, and became one of the reference systems for the work of the Open Hypertext Systems Working Group [9]. The approach to resource-based learning was seen as important [10], [11] and lead to funding from the UK University funding councils including the SCOLAR project [12] based in Southampton. Microcosm and the ideas it embodied formed the basis of the Multimedia Research Group which consisted of 6 people in 1990 and grew to a team of about 60 by 2000. Microcosm has been available as a commercially supported and maintained product since 1996, and is sold into the training, educational and technical documentation markets.

So the question which must be asked is "Why has the take up of this product been limited, and what lessons can be learned?"

4.1. The World Wide Web *was* mind narrowing

At about the same time that Microcosm was being introduced, the World Wide Web was also becoming available. Microcosm was implemented initially on Windows, and conversion to work across the Internet happened too late. The World Wide Web is without doubt one of the killer applications of the age; it certainly killed hypertext research and development for a few years. The problem was that it introduced a whole generation of users to the idea of hypertext as embedded unidirectional binary links that were used almost exclusively in a structural manner (contents to sections, this page to next page etc.). It is rare to see "associative hypertext" which links a concept to related information, and web browsers are only now becoming capable of hypertext features that were commonplace in research lab systems 10 years ago.

However, an important corollary to this lesson is that global access to shared information from heterogeneous platforms is a more fundamental requirement than the clever navigation and seamless information retrieval offered by Microcosm.

4.2 Systems must pass the elevator test

The elevator test [13] requires that you must be able to explain what your system does in the time that it takes to complete a ride in an elevator. You might for example describe hypertext as the ability to put buttons as signposts in your documents so that users could elect to follow links to further information. Much more than that and you lose the attention of the average decision maker.

The problem for systems like Microcosm is that simple descriptions as above do not distinguish them from everyday web browsers. So why would anyone want the product? The answer to that question requires you to read sections two and three of this paper, or at least whatever parts might be relevant to your needs.

The fact is that no-one wants to buy a general purpose hypertext widget; all major sales of Microcosm have been in response to user-defined needs and specifications.

4.3 Users give conflicting feedback

The people who first authored in Microcosm were enthusiastic "early adopters" [13]; they had used other systems such as CBT or the web and had found limitations. We received a great deal of advice and requests for enhancements. Being academics the design team always took the wide view; any individual enhancement is an instance of a class of enhancement, and the whole class must be solved.

The resultant system was enormously feature rich, and when users with less background knowledge tried to use it they were baffled by the array of features and didn't know where to start. So the designers tried to hide the less important features and then the experienced users (and the pundits who have a view on everything but knowledge of nothing) complained again. This leads us to the final lesson.

4.4 A user-friendly system is a simple system

Some of the important design features of Microcosm were hard to implement in a simple and user-friendly way.

For example, integration with third party applications (using DDE/OLE) was fraught with problems. What worked on one machine did not work on another, and versions of Microsoft software became impossible to track.

Microcosm requires users to let the system know something about context (e.g. ability and topic of interest) in order to decide which linkbases to use to retrieve and store links. But users find this too great a cognitive overhead; they want the system to "just know" what linkbases to turn on and where to store personalizations.

When designing Microcosm we set out to produce a system that was simple enough for a non-technical computer user to easily put together a corpus. We overestimated the level of computer literacy of the average academic. We had assumed familiarity with the file system.

The bottom line is that research teams are not necessarily the best people to specify systems as they are more interested in the solution to tricky problems than the interface which allows an ordinary user to understand what they are doing.

5. Conclusion

Many of the ideas in Microcosm were visionary. Separating the links from content and associating rich resource bases are vital prerequisite to individual learning environments. Systems that provide such functionality are now emerging in the Web and internet environment. These include GENTLE [14] (based on Hyper-G), the DLS [15] which is the Southampton research group's offering and Portal Maximizer [16] which is produced by the company that originally commercialized Microcosm. The question remains whether these successor systems will be sufficiently straightforward for academic authors to achieve widespread use.

6. References

[1] Fountain, A., Hall, W., Heath, I. and Davis, H.C.. MICROCOSM: An Open Model for Hypermedia with Dynamic Linking. In: A. Rizk, N. Streitz and J. Andre. Eds. *Hypertext: Concepts, Systems and Applications. Proc European Conference on Hypertext*, INRIA, France, November 1990, Cambridge University Press, 1990

[2] Jonasson, D., Mayes, J.T. and McAleese, R. A manifesto for a constructivist approach to uses of technology in higher education. In: Duffy, T.M., Lowyck, J. and Jonassen, D.H. Eds. *Designing Environments for Constructivist Learning*, p 231-247, Springer Verlag, 1993

[3] Hammond N. Tailoring Hypertext for the Learner. In: Kommers, P.A.M., Jonassen, D,H. and Mayes, J.T. Eds. Cognitive Tools for Learning. Springer Verlag, 1991

[4] Hall, W. Ending the Tyranny of the Button. *IEEE Multimedia* 1(1). 1994

[5] Brusilovsky, P., Kobsa, A. and Vassileva, J. *Adaptive Hypertext and Hypermedia*. Kluwer. 1998

[6] Lechtenberg, S and Joubert, G.R. Effort Estimation for Multimedia Information Systems Development. In Roger, J-Y., Stanford-Smith, B and Kidd, P.T. *Technologies for the Information Society: Developments and Opportunities*. pp 615-622. IOS Press. 1996

[7] Halasz, F. & Schwartz, M. The Dexter Hypertext Reference Model. In: *Proc Hypertext Standardization Workshop*. pp 95-133, Gaithersburg. US Government Printing Office. Jan. 1990

[8] Hall, W, Davis, H.C., & Hutchings, G.A. *Rethinking Hypermedia: The Microcosm Approach*. Kluwer.1996

[9] Reich, S, Wiil, U.K., Nürnberg, P.J., Davis, H.C., Grønbæk, K., Anderson, K.M., Millard, D.E. & Haake, J.M., Addressing interoperability in open hypermedia: The design of the open hypermedia protocol. *New Review of Hypermedia and Multimedia*, Vol 5. pp 207-248. 1999

[10] Davis, H.C., Hutchings G.A. and Hall, W. A Framework for Delivering large-scale Hypermedia Learning Material. In: Hermann Maurer. ed. *Educational Multimedia and Hypermedia Annual. Proceedings of ED-MEDIA '93, Orlando, Florida, USA*, pp 115-122. AACE. 1993

[11] Hall, W,. Hutchings, G.A. and White, S.A. Breaking Down the Barriers: An Architecture for Developing and Delivering Resource Based Learning Materials. In: *The Proceedings of The World Conference on Computers in Education*, Birmingham, England. July 1995

[12] White S.A. SCOLAR - A campus wide structure for Multimedia Learning. In: *The Proceedings of the AETT Annual Conference: Designing for Learning*, Glasgow, Kogan Page. 1995

[13] Moore, G. Crossing the Chasm. Harper Collins, 1995

[14] Dietinger T and Maurer H.: GENTLE – General Network Training and Learning Environment, Proc. of ED-MEDIA98/ED-TELECOM98, Freiburg, Germany, AACE, June 1998

[15] Carr, L.A., DeRoure, D., Hall, W and Hill. G.J. The Distributed Link Service: A Tool for Publishers, Authors and Readers. *Proc Fourth International World Wide Web Conference: The Web Revolution*, Boston, USA. 1995

[16] Portal Maximizer. http://www.activenav.com/ . 2001

Distance Learning Technologies and an Interactive Multimedia Educational System

Guorui Jiang
jiangg@missouri.edu
Department of Computer Science
Hebei Normal University, China
Junqiang Lan
lanjq@meru.cecs.missouri.edu
Xinhua Zhuang
zhuang@cecs.missouri.edu
Computer Engineering and Computer Science Department
University of Missouri-Columbia, USA

Abstract

The innovation and development brought by the new technologies in our information era have started to appear in education. Distance learning has become a core educational strategy around the world in recent years, with extending to a broad cross-section of online universities and virtual universities [1-3]. The Internet is providing mechanisms for fundamental changes in the way people learn. Video conferencing, Web conferencing and Collaboration provide interactive, real-time multimedia communications for distance learning. In this paper, we review the evolution of technologies and tools in distance learning, and discuss the current state of distance learning, with a focus on the tools or technologies on the Internet available for distance learning. Then we present a newly developed interactive multimedia educational system. The system is based on new compression algorithm with low bit-rate, software-implemented video codec to achieve real-time, multicast transport. Finally we provide experiment results of this system.

1. Introduction

In recent years, personal computers are widely used and most of them have been connected to network, more and more people can surf on the Internet. This provides an advantageous environment in web-based learning for all computer mediated communication learners. Since information and knowledge are increasing so rapidly, most companies need to facilitate the distribution of this "wealth" to their employees anywhere and at any time quickly to protect current knowledge assets or increase their value. Most citizens need to learn certain new knowledge and practical skills beyond secondary school in order to be a full participant in the workforce[4]. This creates a great need for distance learning. A variety of businesses regard Web-based learning as an important emerging market and therefore as a business opportunity, and have started engaging in activities in distance learning. Some new technologies of communications, networks, and multimedia have begun to be used in distance learning [5-11]. For examples, the merger of computer and video into multimedia desktop terminals as well as videoconferencing. Though there are over 20,000 videoconferencing systems installed worldwide, and the growth rate is around 40% per year. Yet most of them are unicast. Multicast video conferencing systems are not used widely so far. Two major limitations are relatively low network bandwidth and the lack of multicast routers. Truly interactive, multimedia courseware and digital learning environments require a greater level of network capacity and quality-of-service guarantees than the current public Internet can provide. However, one should witness a dramatic increase in bandwidth capacity over the next few years. For example, Internet2 provides advanced IP networking services to support an evolving set of new advanced networked applications in the US [12]. While in China, China Net Communication Corporation has built broadband high speed CNCnet last year. CNCnet has connected 17 China main cities. The transmission bandwidth of its backbone network is up to 40Gbps so far. Meanwhile because MBONE (multicast backbone) programs are today's implementation of IP multicast protocol, the design and development of multicast video conferencing for distance learning is very significant. In this paper, we first describe three historic phases in the evolution of technologies and tools in distance learning, and discuss the current state of distance learning, with a focus on the tools on the Internet available for distance learning. Then we present a newly developed interactive multimedia educational system. The system is based on new compression algorithm with low bit-rate, software-implemented video codec, IP multicast transport. Finally we provide experiment results of this system.

2. Technologies and tools in distance learning

2.1 Evolution of technologies in distance learning

Though distance learning has developed very fast and has become a core educational strategy in the world only in recent years, the phenomenal growth in avenues of information distribution and dissemination in human history has occurred in the past centuries [13]. We could see evolution of technologies available for distance education by *three different historical phases*. Before human used radio, manuscripts, printing materials and books with correspondence courses were main tools for distance education. Correspondence schoolteachers monitored study by responding to mailed-in assignments. This is *the correspondence phase* of distance education. After radio was used, the barrier of physical distance was greatly reduced. Radio, television, telephone and broadcast became the main technologies for distance education during *the analog signal times*. Since we entered *the information age*, information superhighway has started to provide fundamental mechanisms for distance learning. The merge of computer, multimedia, communication and network has driven the development of distance learning. The technologies of Web-based learning and computer-based learning have appeared.

2.2 Current technologies in distance learning

We can divide current technologies in distance learning into three general categories.

Telecourses and videotapes Telecourses are tape or live television programs provided by broadcast or cable television stations. These telecourses are used at large in distance education.
Videotapes are videos recorded during a class period or tapes reproduced especially for a distance learning classes. They can be viewed later by a student via his TV and videocassette recorder (VCR) at home.

Computer-based training and computer aided instruction Computer-based training courses are developed for multimedia delivery. Interactive multimedia instructions allow learners to read text, view image or video, and hear information. These courses are offered on CD-ROM or diskettes. They usually are supplemented with on-line materials.
Computer aided instruction (CAI) provides complement to traditional learning methods. It usually provides tutorials delivered in computer laboratories to supplement lecture-based instruction and to provide hands on exercises to learners. Recently, the Internet has started playing an increasing role in CAI by providing "on-line" tools on the Web to complement classroom instruction, or to provide additional exercises for educational purposes.

Web-based training and teleconferencing Web-based training uses personal computer and the Internet (or the Intranet) as its primary medium for delivery of instruction. The front-ends are most often designed in HTML, enriched by Java, Javascript, or Dynamic HTML. Teleconferencing has mainly three types: video, audio, and web.
Video and Audio conferencings are performed using computer multimedia facilities such as cameras, appropriate multimedia monitors, microphone, speakers. Their signal is compressed using protocol such as MPEG, motion JPEG, H.26x, G.72x standards and etc.
Webconferencing is any time, any place conferencing over the Internet. Participants only need a web browser and an Internet connection, and interact in virtual meetings with anyone in the world. It has functions as shared visuals, webcasting, recording etc.

2.3 Tools on the Internet for distance learning

There are many tools on the Internet that can be used for distance learning. The commonly used tools are summarily as follows three kinds.

Tools for information disseminating and group discussions *Bulletin board, newsgroup, listserv, USENET and BITNET*. They provide a mechanism for concurrent information dissemination, organizing topical mailing and facilitating discussion among learners.
Electronic Mail. It provides a mechanism for rapid communication between individuals and groups.

Tools for offering resources and remote access *World Wide Web(WWW)*. It provides a collection of internet sites that offer text and graphics and sound and animation resources through the hypertext transfer protocol.
File Transfer Protocol (FTP). It provides a mechanism for retrieving and storing files.

Telnet. It provides interactive connections with remote computer system.

Tools for interactive and real-time communication
Teleconferencing, Whiteboard, Network Text Editor and Chat facility etc. They provide interactive and real-time multimedia communication, where some are multicast, and some include sharing documents.

As the communication technologies are developing, and the bandwidths of networks are increasing, multimedia conferencing will play an important role in distance learning in the future.

3. An interactive multimedia educational system

Recently, we have developed an interactive multimedia educational system. It comprises multicast video conferencing (MVC) and multifunctional whiteboard (MW). Design and implementation of video conferencing over Internet is a complicated project. Its key technologies include compression algorithm and implementation, video / audio real-time transport constrained by delay, video / audio synchronization, collision free, video / audio capture and play on PC, multicast on Ethernet based on UDP. Whiteboard is a shared workspace, a widely-used remote collaboration application based on multicast. Its key technologies include H.323 and T.120 conferencing standards, scalable reliable multicast (SRM).

3.1 New algorithms

Fast H.263+codec H.263+ is the traditional DCT+Motion compensation video compression standard[14], it has very good performance in terms of compression ratio. But in packet loss environment, it is not robust though. To improve the performance of the H.263 codec, key modules such as DCT and motion estimation are optimized and implemented using MMX technology.

Motion block JPEG codec (MJPEG) It includes DCT decomposition of motion blocks and JPEG encoding of motion block.

Innovative implementation On video compression and associated techniques, we consider the following innovative implementations:

1). High quality, low bit-rate, software-implemented video codecs used for interactive multimedia communication;

2). Joint source / channel coding for optimizing performance;

3). Application cooperated and coordinated codecs with protocols to achieve real-time transport.

On graphics representation of the whiteboard, we adopt structured-object model, each page in WB acts as a container for a set of graphical items. The item types supported include lines, rectangles, ovals, freehand sketches, text segments, and images. All participants can draw or manipulate different items simultaneously. The items are ordered by causal-broadcast protocol (CBP).

3.2 Hardware and software environment

Networks: 10 / 100 base-T Ethernet; TCP/IP support

Computers: PC Pentium II 350 MHz, 64 MB Memory; PC Pentium III 700 MHz, 384 MB Memory

Camera: 3Com Bigpicture camera (analog video + capturing card), which captures full color NTSC video under 320×240 pixels at up to 30 frames per second; US Robotics camera (analog video + capturing card), which supports up to 30 fps continuous capture in CIF full color.

Audio card: sound blaster compatible 16 bit audio card, full duplex.

Software: Windows 98 platform; Video capturing driver; Audio driver; TCP/IP socket support; MS Visual C++ Studio 6.0.

System Testbed The system testbed (prototype) for video conferencing was built on Local Area Network in Multimedia Communication and Visualization Laboratory (MCVL), Computer Engineering and Computer Science Department, University of Missouri-Columbia.

4. Experimental Results and Evaluations

4.1 Experiment 1

First we introduce several concepts that are used in the following tables.
ms/f stands -- milli-second per frame.
Low Motion – Head-shoulder Image;
Middle Motion – Hand Gesture;
High Motion – Full Motion Image.

QCIF 176×144 resolution
Pentium II 350 MHz, 64MB.

Algorithm: H.263+				
Motion	Bit rate (kbps)	Frame rate (fps)	Encode (ms/f)	Decode (ms/f)
Static	21-28	30	2	1
Low	50-60	30	2	1
Mid	160-180	30	3	1
High	300-320	30	4	1

QCIF 176×144 resolution
Pentium II 350 MHz, 64MB.

Algorithm: MJPEG				
Motion	Bit rate (kbps)	Frame rate (fps)	Encode (ms/f)	Decode (ms/f)
Static	21-28	15	2	3
Low	50-60	15	3	5
Mid	160-180	15	4	8
High	300-320	15	8	21

Note: Codec time does not include end-to-end delay and overhead time.

320 × 240 resolution
Pentium II 350 MHz, 64MB.

Algorithm: H.263+				
Motion	Bit rate (kbps)	Frame rate (fps)	Encode (ms)	Decode (ms)
Static	21-28	30	12	2
Low	50-60	30	15	2
Mid	160-180	30	18	2
High	300-320	30	20	2

Note: Codec time does not include end-to-end delay and overhead time.

H.263+ codec was quite time consuming, but after optimization with MMX technology, the codec speed was improved more than 4 times.

4.2 Experiment 2

Time: December 9 2000.
Place: 301 and 350 Lab, in West Engineering Building, University of Missouri-Columbia.
We gave presentation in classroom 301, used camera and microphone to get the video and audio. We connected to three personal computers using video conferencing by local area network. Three students sat in front of computers to watch the remote lecture.
Experiment result: In any bandwidth, we guaranteed the quality of voice, but the quality of video was dependent on transmittal speed. When the bandwidth was not restricted, we set the resolution 320×240, frame rate as 15, and the resulting video quality was very good. When the bandwidth was restricted to lower than 56 kbps, we took GCIF resolution, and decreased frame rate, and the resulting video quality was acceptable.

5. Conclusion

Since the new instructional model of distance education was introduced in the society, it has changed the way we do teaching and learning. Many new technologies of communication, network, and multimedia computing can be used in this field. Interactive real-time multimedia communication system, advanced video conferencing, web conferencing, as well as collaboration will play important role in distance learning in the future.

References

[1] T. Pison. *Distance learning is an opportunity*, IEEE Circuits & Devices Magazine. v 13 n 2 Mar 1997. p 41-43.
[2] L. K. Lau. *Distance learning technologies: Issues, trends and opportunities*, Idea Group Publishing, Hershey USA, 2000.
[3] F. Belanger, and D. H. Jordan. *Evaluation and implementation of distance learning: technologies, tools and techniques*, Idea Group Publishing, Hershey USA, 2000.
[4] A. G. Chute, M. M. Thompson, and B. W. Hancock. *The McGraw-Hill handbook of distance learning*, McGraw-Hill Companies,Inc, New York, 1999.
[5] B. Liu. *A high-performance ISDN/ATM multimedia integrated access system for real-time interactive distance learning*, High Technology Letters. v 5 n 1 1999. p1-6.
[6] Y. Takefuji, N.Takahashi, H. Tsuchida, Y. Fukuhara, and R. Neff, *ATM and wireless experiments for remote lectures*, IEEE Communications Magazine. v 37 n 3 Mar 1999. p 98-101.
[7] W. Doube. *Browser-based system to support & deliver DE*, Proceedings - Frontiers in Education Conference. v 1 1998. IEEE, Piscataway, NJ, USA,98CB36214. p 479-484.
[8] J. Carrasquel, W. P.Dann etc. *Using the web to support distance learning of computer science*, Proceedings - Frontiers in Education Conference. v 1 1998. IEEE, Piscataway,NJ, USA,98CB36214. p 477-478.
[9] S.G. Deshpande, J. N. Hwang. *An interactive virtual classroom-multimedia distance learning system*. 1999 IEEE Third Workshop on Multimedia Signal Processing, IEEE. 1999, p.575-80.
[10] M. Kassouf, S. Pierre, C. Levert, and J. Conan, *Modeling a telecommunication platform for remote access to virtual laboratories*. 1999 IEEE Canadian Conference on Electrical and Computer Engineering, IEEE. Part vol.1, 1999, p.127-32.
[11] L.Guan, S. Kung, J. Larsen. *Multimedia image and video processing*, CRC Press. 2000, p487-505.
[12] B. Teitelbaum, S. Hares, L. Dunn, R. Neilson, V. Narayan, F. Reichmeyer, *Internet2 QBone: Building a testbed for differentiated services*, IEEE Network. v 13 n 5 1999. p 8-16.
[13] P. Wiesner. *Education and careers: Distance Education: Rebottling or a new brew?* Proceedings of the IEEE, vol. 88, No. 7 2000. p1124-1130.
[14] ITU-T Rec.H.263 "*Video coding for Low Bit Rate Communication.*" 1998-01.

The Continuous Education Solution for a Country Wide Telecommunication Company

Marcelo Leifheit
CRT-Brasil Telecom
Mleifheit@crt.net.br

Juarez Sagebin Correa
CRT-Brasil Telecom
Jsagebin@crt.net.br

Daniel Fink
CRT-Brasil Telecom
Dfink@crt.net.br

Abstract

Competition among companies has put up a challenge regarding training and development of human resources. In a country-wide company this challenge takes into account constant trainee traveling back and forth, so, generating costs. This paper presents a telecommunication development program that uses top of the available technology to broadcast synchronous classes, over a system with full motion images in real time. It uses a transmission rate as low as 2 Mbps, easily available almost everywhere, allowing training at remote sites with all of its advantages.

1. Introduction

Actually there are many education methodologies based upon new technologies, but a lot of them without satisfactory results evidences. The lack of knowledge about new applied education technologies and methodologies is one of the main problems among professionals who are not being prepared to use those tools to be competitive in this new globalized world.

Considering new education methods, associated to new technologies, one of the greatest challenges posted to instructors is keeping people motivated and using distance learning tools for actualization, focusing in the company business.

A decade ago, there were the first attempts in distance learning as a corporate tool [5], which means using various methods to reach and teach people who are widely dispersed geographically.

In practice, how is it possible ? What is the technology that can actually be counted on ? How is it implemented ? What methodology could be used for a Telecommunication Company environment ?

2. Methodology

This work refers to, as it is seen, the best-implemented solution until today in a Brazilian Telecommunication Company, that is seeking for continuous education for all of its employees scattered all over Brazilian ten states.

The methodology used for this system, counts on WEBTV resources, chat-room, a Teleducation system with full audio and video interactivity, which allows the participant a total interactive environment with the instructor, minimizing the discomfort caused by the non presental situation.

As a solution for those challenges it was sought one that could put together management instruments, top of technology, new distance learning methodologies, allowing organizing culture dissemination and valorization of the enterprise human capital.

PEN BrT, which is Brasil Telecom´s Extension Program is a technical and institutional knowledge-recycling program, focused in training employees on technical and administrative areas, wherever the professionals are, and they are spread out all over the Country.

The program issues were selected to provide a basic formation oriented to the Company's core business and the main departments internal processes, searching the improvement of quality service benchmarks.

At the end of the program, the participant is supposed to have a clearer idea of each activity importance in the company's general context, reinforcing the value of cooperative work among involved areas.

The employees qualifying matrix, which was developed under training consultants orientation and their managers, showed training necessity.

According to the philosophy that the best Brasil Telecom´s company business specialists are their own employees [6], they themselves conduct the classes. This methodology increased competitive advantage that derived from having a well-trained workforce that is up-to-date on the entire latest trends, collaborating with one another and sharing information throughout the company.

The classes are filmed, recorded and broadcasted through the auditorium's television system.

3. Technology

The Teleducation system implemented at Brasil Telecom uses an audio and video network exclusively for training [1]. It is based on compressed video using a MPEG-2 algorithm that allows showing VHS quality images, in real time and using a bandwidth equivalent to an E1 channel at 2Mbps. This low transmission rate allows using the public telecommunication network for this task. This bit rate is nowadays easily available and the backbone access can be made through a variety of means. It can be pointed out, as examples, coaxial cable, HDSL modems and twisted pair.

4. Distance Learning System

For those students who do not have a TV reception point nearby, it is possible to participate in classes through the WebTV system, since the web connection has a bandwidth larger than 300 KBPS with the main server so having an excellent audio and video quality. Through the WebTV class, students can watch the instructor, presentation slides, interacting with others participants and making questions through a chat-room

For distance learning classes, the interactivity with the instructor happens by e-mail, phone call and a chat-room provided in the course home page [2]. An instructor assistant receives questions and forward them to the instructor during classes.

At the beginning of each class, presential participants receive the training material which was made by the instructor in advance. For distance online learning participants, the slides can be downloaded directly through the PEN home page – http://www.ade.crt.net.br/pen , for watching, printing, or online consulting during the class as well as afterwards.

Recorded classes remain stored at the Company's library for future references. Participants who could not watch a class can request the video tape through an on-line library service. The video will reach the participant desk in less than three days, for home watching. After this, he proceeds to the PEN home page to fill out the on-line evaluation and reaction forms for attendance registration in a training database [3].

Questions will be answered by e-mail or phone call directly by the instructor, who is a company employee too, during work time.

Distance learners fill out the registration participation form directly on the PEN site. There is a suggestion box, where the students can leave messages for the organizing team who manages the program.

5. Results

Until now 2.653 students participated in this program that began in September 10, 2000 and is supposed to end by June 6, 2001.

The results obtained up to this date can be considered satisfactory; some questions have been pointed out such as new technology equipment handling and a little delay pertaining to the information flow in this system. With some practice instructor and students were able to overcome these characteristics of communication.

Through the reaction and opinion evaluation and interviews with students and their managers, it was indicated that this teaching methodology has reached an efficiency of 82 percent considered through the question tasks applied under the PEN web page after classes and acceptance of 98 percent measured by a reaction evaluation .

The low cost associated to commuting and also, the fact that the employee does not leave his post to go to classes, contribute to validate the investment in this methodology.

6. Conclusions

Actually, the on-line learners can, not only access their studies from remote locations using multimedia, but they can also build online communities [4], swapping questions and answers with their tutors, and fellow students via e-mail and a chat-room.

This online learning methodology can add up to presential learning to build a successful knowledge management system and using technology to leverage the intellectual capital of the entire company, which in turn, leads to increased productivity, shorter time to access new knowledge to the company, and a superior competitive advantage.

7. References

[1] Leifheit M., Correa J.;"A Corporate Distance Learning Experiment"; Expocomm´99-Brazil; 1999.

[2] Hall, Brandon; "Web-Based Training Cookbook"; Wiley Computer Publishing, 1996

[3] Demo; Pedro; Questões para Teleducação; Editora Vozes; Petropolis, RJ; 1998.

[4] Tiffin, John and Rajasingham, Lalita; "In Search of the Virtual Class"; Routledge, 1995

[5] J.M. Wilson, W. C. Jennings, "Studio Courses: How Information Technology Is Changing the Way We teach, On Campus and Off"; Proceedings of the IEEE, vl. 88, no.1, January 2000, pg 72-79

[6] Drucker, Peter F.; "Aprendizado Organizacional"; Harvard Business Review Book, 1999, Editora Campus

The Practice in the Web-Based Teaching and Learning for Three Years

Fuzong Lin
State Key Laboratory of Intelligent Technology and Systems, Computer Science & Technology Department, Tsinghua University, Beijing, China
Email: linfz@mail.tsinghua.edu.cn

Xiaoyan Xie
Department of Communication, Tsinghua University, Beijing, China
Email: wendy@s1000e.cs.tsinghua.edu.cn

Abstract

The WWW has created major innovation in the way knowledge is transferred. Since 1998, we have offered Web-based multimedia technology course for on-campus and off-campus graduates in our country. In this paper, we describe the teaching and learning system we have developed, the model for teaching and learning, and report the results of it in the past three years.

1. Introduction

The rapid growth of the Internet is leading many educators to experiment with distance learning [3][4][5][6]. Colleges and universities are rapidly moving courses and even entire degree programs onto the Internet at a staggering rate [2]. This paper reports our first-hand experience in the Web-based teaching and learning for three years, including Web-based course named Multimedia Fundamentals and Applications, the teaching and learning system, which were developed only for our courses, the model for teaching and learning, and the results of Web-based teaching and learning.

2. The teaching and learning system

In the teaching and learning system, both teachers and students use Web browsers as the interface for reading and writing, and to simply publish or download the materials that supplement existing courses. Both teachers and students can work at home or office via the Internet.
During the developing, we have paid a great attention for the interaction between teachers and students. The interaction is non-real time and also text-based. We also encourage students to communicate student-to-teacher and student-to-student by telephone, for it is a real time interaction in speech. A powerful editor like Microsoft Word Processor and a simple email program are embedded in the system, which students have enjoyed very much.

3. The course materials

The content of the course is consisted of 4 parts, which are media (speech, audio, image, and video) computing, Web Programming Languages, such as HTML (Hypertext Makeup Language), XML (Extensible Markup Language), JavaScript and VBScripts, storage for multimedia and multimedia networking. Besides, there are supplementary articles and papers related to state-of-the-art technologies. These course materials are mainly text-based and organized in non-linear fashion, learners can conveniently use them. Since the speed of network is very slow and the fee paid for using the Internet is high in China, we have not used audio/video stream for delivering the course during the past three years. Meanwhile, our students strongly ask me to publish traditional textbook.

4. The model for teaching and learning

There is no doubt that nothing will replace synchronous learning through face-to-face interaction, but it is sometimes not feasible for students to attend conventional classes due to distance or time constraints [1]. That is why we use the hybrid model of face-to-face plus distance learning.
We arrange face-to-face discussion for three or four times in a semester. At the first time, teacher introduce students how to teach and study the Web-based course, in the second and third sections, teacher instruct students again and answer the question occurred in their study. In the end, there is exam for students in classroom, since there is no good means to prevent cheating until now. The teaching and learning through face-to-face is arranged in

campus for on-campus students, and delivered to off-campus students through a satellite, the Internet or CD-ROM.

Students learn the course materials under teacher's guidance, put their questions to our Web site (http://166.111.68.180/). We strongly encourage students to answer their questions by themselves, and to use search engine, such as Google (http://www.google.com), for discovering answers and new technology.

In fall 2000, we had assigned two tasks, from which students could select one. One is to understand wavelet and write a overview on its applications, another is to understand text to speech for mandarin and write a paper how to improve the naturalness of output synthesized speech. Many students were very active to find the materials via the Internet, put their opinions and what they found to our Web site, and wrote their papers.

5. The facts and experience

Since 1998, we have offered the Web-based degree course, Multimedia Fundamentals and Applications, for conventional on-campus students and off-campus students. The off-campus students are distributed in our country whose circumstances require that they be asynchronous in time or space. It is the first Web-based course for graduate students in our university. Since then, a total of 1463 students have attended this course.

In the end of fall 2000, we arranged the exam for these students. Since all of examination questions are the same, it is very convenient to compare between the on-campus and off-campus students.

There are one third of off-campus students who selected the course only for updating their knowledge, and they did not take part in the exam. The others are pursuing for master degree.

The average of on-campus score is higher than the average of off-campus score. There are mainly two reasons for explaining the result. One is that on-campus students were selected from many students by exam, and off-campus students were not. The other is that most off-campus students have to work during the daytime.

6. Conclusion

The course, Multimedia Fundamentals and Applications, is the first formal Web-based course offered in Tsinghua University, maybe the first one in China. Since 1998, we have noticed, several courses in our university have been following the teaching and learning model, that is face-to-face plus distance learning via the Internet, text-based web pages, and developing the almost same teaching and learning system as we have done. It is true that this kind of Web-based course is more suitable for the situation of our country, most Chinese are unable to get high education and the network environment is very poor. However, the situation in China is rapidly changing today, what we are doing is to get the help of companies for developing the teaching and learning system with audio/video streaming technology, and let the teacher concentrate on developing course contents.

7. References

[1] H.A. Latchman, C. Salzmann, D. Gillet, and H. Bouzekri, "Information Technology Enhanced Learning in Distance and Conventional Education", *IEEE Transactions on Education*, vol. 42, 1999, pp. 247-253.

[2] B. Mehlenbacher, C.R. Miller, D. Covington, and J.S. Larsen, "Active and Interactive Learning Online: A Comparison of Web-Based and Conventional Writing Classes", *IEEE Transactions On Professional Communication*, vol. 43, 2000, pp. 166-194.

[3] G. Greenberg, "Distance education technologies: Best Practices for K-12 Students", *IEEE Technology and Society Magazine*, winter, 1998/1999, pp. 36-40.

[4] S. Palmer, "On- and off-campus computer usage in engineering education", *Computers & Education*, 34, 2000, pp. 141-154.

[5] K. Bolding, "hybrid distance-learning OR face-to-face class experiment", *29th ASEE/IEEE Frontiers in Education Conference*, pp. 12b2-2.

[6] F. J. Doyle III, "Use of the World Wide Web for distance education in a process control course", *Proceedings of the American Control Conference*, Chicago, Illinois, June 2000, pp. 3454-3457.

Running A European Internet School – OTIS at Work

*Martin Beer (1), Gillian Armitt (2), Johanna van Bruggen (3), Ramon Daniels (3), Ludo Ghyselen (4),
Sharon Green (5), Jan Sandqvist (6) & Andrew Sixsmith (5)*

*Computing and Management Sciences (1), Sheffield Hallam University, United Kingdom
Computer Science (2), and Primary Care (5), University of Liverpool, United Kingdom
Hogeschool van Amsterdam (3), Amsterdam, Netherlands, Hogeschool West-Vlaanderen (4), Belgium
Linköpings Universitet (6), Linköping, Sweden*

This paper discusses the OTIS experience in providing Internet based courses to student and professional occupational therapists in four centres across Europe. It uses a problem based learning approach to promote collaboration between students based in several different European countries, who study and evaluate the different strategies for assessment and treatment of patients across Europe.

Some of the issues arising from the development of this pilot course are discussed, together with the experiences of the project team in bringing it through to a successful pilot. These have included not only technical matters, but also managerial, pedagogic and classroom control issues that are often neglected. Only if all these are addressed can large-scale distance-learning activities be developed.

1. Introduction

This paper discusses the OTIS (*Occupational Therapy Internet School*) experience in providing Internet based courses to student and professional occupational therapists in four centres in the UK, Belgium the Netherlands and Sweden. The OTIS consortium is currently running a full award-bearing pilot module in "*High Level Assistive Technology in European Occupational Therapy*". The pilot course started with twenty-three students registered from all four participating Occupational Therapy Schools, who are also all providing tutorial input on a course-wide basis. The course uses a problem based learning approach [1] to promote collaboration between students. All students complete:

1. a reflective account of their learning progress in relation to assistive technology.
2. one completed case study, concerning either a younger or an elderly client
3. one critical review of a second case study completed by one of their European counterparts, and concerning the alternate age group to that above

A similar approach has been followed elsewhere, for example at Roskilde [2].

2. The OTIS System

The system has been developed specifically to facilitate the educational and collaborative philosophy, while being sympathetic to a non-technical user base. Above all, the technical design has to facilitate a high degree of synchronous (real time) communication [3]. This permits groups of people to meet both formally in weekly tutorials and in an ad hoc manner, to discuss different matters separately from other groups who might be meeting at the same time.

The course materials interface has been developed using a Virtual Campus metaphor [4]. The Virtual Campus comprises a number of virtual rooms such as the library, student work area, etc. This facilitates navigation around the OTIS site, as students readily identify the purpose of each room, and know intuitively what activities would take place in that room.

The virtual rooms also provide the means by which groups of users can congregate to discuss specific issues with every occupant of that room. This provides an effective way in which a group can discuss for example Case Study 1, at the same time as another group is discussing Case Study 2 in another room. Both students and tutors can book rooms dynamically in the Room Booking area, and invite the other members of their group to attend. Once understood, this has proved very popular and is used extensively for both formal and informal meetings. There was initially a problem with participants finding the rest of the group, since they can move around the virtual campus. A Page facility allows direct communication with one or more individuals, wherever they are on the system, and can be used to locate and shepherd wayward members of the group, or to conduct personal conversations.

The organization of the OTIS library, within the problem based learning model, generated much debate. The problem based learning approach requires that students should not be told the answers, but encouraged to find solutions for themselves. It was important that items in the OTIS library should not be described in too much detail in a cataloguing system, but some guidance for students was felt to be desirable. Students have only a limited time to complete their coursework, and much time can be wasted in checking out websites. Early thoughts were to categorize the library into broad topics, such as physical disability, learning difficulties and ageing. On examining the library materials, it quickly became apparent that many materials spanned several topics. For example, an assistive technology device might be equally appropriate to a physically disabled person or an old person with difficulties with mobility. Establishing a list of keywords, and attaching keywords to each library material, has solved this problem. A library search facility allows all materials matching specific keyword combinations to be identified.

3. Evaluation of OTIS Progress

It is clear that OTIS is seen as an educational environment, rather than purely as a technical system. Students mostly join the course in order to share experiences with students and tutors in other countries. Students in the first pilot found that their interactions with people in other countries were an enjoyable and highly motivating part of the course. Students did, however, feel that they would like more guidance on how to prepare for online tutorials. This feedback was extremely important developing the final course structures. Students viewed the case studies as giving either too much or too little information, and the problem-based learning approach does not allow tutors to give information to the students. In the main pilot experts (e.g. representatives of companies marketing assistive technology devices) are available in communications sessions to answer questions about the products. This has proved very popular with students, as has the opportunity to conduct online consultations with patients (or in practice, tutors role-playing patients).

Facilities are provided through the "Staff Room" to allow tutors to meet in private, and regular Staff Meetings have ensued. This has proved very important for the cohesiveness of the course, as immediate issues have been dealt with effectively as they occur.

Classroom management has been an issue with all tutors as the environment has been very new. New techniques have had to evolve to ensure that tutorials take place at the right times, and everyone is present. The synchronous nature of the online tutorials has at times put a considerable strain on the computer networks, and individual group members have missed parts of the discussion. The facility to record online discussions in meeting rooms has been of great help here, as the transcript can be used to fill in any missed comments, as well as acting as a highly informative record of the tutorial.

4 Conclusions

The OTIS project shows that problem-based learning techniques can be used effectively in a distance-learning environment to bring members of a highly practical profession together across many different countries. This allows experiences to be shared across a wide range of problems and best practices to be disseminated rapidly across many national boundaries. An important feature of the OTIS course structure is the use of multi-national tutorial groups and tutorials. The relatively high level of synchronous communication encouraged by the system means that mock consultations are practical, and the use of role playing characters means that they are easy to arrange, as different tutors (or indeed students) can act the patient whenever necessary. This leads to an extremely dynamic and flexible learning environment, in which students, staff and others interact much more closely than with electronic mail based tutorial systems.

The regular liaison and negotiation between all parties closely mirrors the advanced communication necessary in a team of health professionals who are aiming to deliver a best possible integrated health care package to a consumer. In the case of OTIS, the student as the consumer needs to both experience the results of such integration, and be trained in the methods to achieve such ends.

As with all projects of this type, much more needs to be done. The advantages of synchronous communication are considerable, but meetings need to be arranged, and diaries coordinated. This is done at present through the use of timetables and electronic mail, very much as one would within a regular University Department, but as the groups grow larger and more diverse, this becomes increasingly difficult. An online diary management and booking facility would be a considerable advantage for all concerned. Meetings could then be arranged fully within the system.

Acknowledgements

The OTIS project is funded by the European Union through the TEN-Telecom programme. The system is based on CoMentor, developed by the University of Huddersfield, whose help in setting up the system is gratefully acknowledged.

References

1. Davis M, Harden R (1999) AMEE Medical Education Guide No. 15: Problem-based learning: a practical guide. *Medical Teacher* 21 (2) 130-139.

2. Robin Cheeseman, Simon Heilesen (1999). Supporting Problem-based Learning in Groups in a Net Environment. In C. M. Hoadley and J. Roschelle (Eds.), *Proceedings of the Computer Support for Collaborative Learning (CSCL) 1999 Conference* (pp. 94-100). Palo Alto, CA: Stanford University [Available from Lawrence Erlbaum Associates, Mahwah, NJ].

3. Armitt, G. M., Green, S., & Beer, M. D (2001) "Building a European Internet School: Developing the OTIS Learning Environment", in 'European Perspectives on Computer-Supported Collaborative Learning', Proceedings of Euro-CSCL, March 2001, Maastricht, Netherlands, pp 67-74

4. Ginsberg A, Hodge P, Lindstrom T, Sampieri B & Shiau D (1998) "The Little Web Schoolhouse" Using Virtual Rooms to Create a Multimedia Distance Learning Environment, *ACM Multimedia 98*, September 13-16, 1998, Bristol, UK, 89-98.

Considering automatic educational validation of computerized educational systems

Alexandra Cristea and Toshio Okamoto
University of Electro- Communications Graduate School of Information Systems
{alex, okamoto}@ai.is.uec.ac.jp

Abstract
Evaluation of educational systems is difficult and often subjective. Many computerized educational systems are proposed, find their ways into classrooms and are already forming new generations of pupils. However, it is risky to leave the actual validation of these new systems to a trial-and-error technique, especially when the involved target implies humans. We propose with this paper automatic, virtual students for a more unitary and consistent testing of computerized educational systems, as an alternative to the traditional trial-and-error method.

1. Introduction
Critics warned about a "dehumanization of the learning process" [5, 8], but computers in education are an evident success [2]. New systems emerge all the time, and vary in teaching methods, expression ways, technical qualities, etc. This makes it difficult for educators to choose, as well as for individual learners to find the best one for their learning goals. Today's state of the art researches propose a new system, build it, test it on a class of real students or more, and then publish their results. This method says little of the actual usability of the proposed system or about results' reproducibility. Attempts to systematize evaluation of computerized educational systems are mostly of technical nature [3]. Some sites gather information to help teachers decide on products, via criteria such as transfer format, service, platforms, instructor or student tools, etc. This information is useful, but says little of the educational values of the product. As the main purpose of computerized education systems is education, moreover, as authors usually claim that their systems are improvements over classical education, this lack of educational evaluations is extremely dangerous. The field obviously lacks benchmarks with which each researcher can compare his/her own application. We propose here an alternative method of evaluation of computerized educational systems, based on automatic students.

2. User modeling
In educational applications, user modeling means to find out *how the student learns best*, and to offer him/her the suitable form of education. Student analysis is limited by the human-computer interaction: questionnaires (single/ multiple choice, etc.), tests, learner's steps tracing, etc. The latest student models contain a layered evaluation of the student: *knowledge & cognitive, learning profile* and finally, *believability & emotional layer* [1]. Student modeling has the following steps: *1) gathering, 2) filtering,* and *3) interpreting information for selecting the appropriate pedagogical strategy.* Although difficult, this process has been thoroughly researched by psychologists, and draws back on mankind's hundreds of years of teaching and learning practice. Talented teachers select instinctively the best tutoring strategy for each student. Also, many educational systems lean on IQ, personality or various other cognitive style tests, to classify students and offer them tailor made educational programs [6].

3. Automatic students
As testing with real students is often time- and resource-consuming some researchers have proposed systems based on simulated tests. They used what is called *simulated*, or *automatic students*. Of course, many of the previous critics for the current evaluation methods, such as subjectivity issues, etc., can apply in such cases as well. I.e., it is questionable if a simulated student, built entirely according to the predicted student model, will not simply always generate good results, which, translated into real life, might result in completely different real evolutions of the human students. Therefore, to validate the usage of automatic students, their specifications must be more general, and not decided by the system designer, but should be accepted by an international community.

4. Learning Styles
The literature provides various definitions of *cognitive styles* (Allport, 1937) and *learning styles* (Thelan, 1954); often the two terms are used interchangeably. We focus on learning style, as the specific individual approach to acquiring new knowledge of each student. According to the student's learning style, the student is able to receive knowledge easier or not via a certain teaching style. Learning style is independent from other abilities, which have direct sequels [6], whereas styles control and define the internal preferences and value system. Below are some of the most important cognitive/ learning styles.

4.1. Hill's cognitive style mapping
Hill has built a cognitive style coefficient as a function of symbols and meanings (i.e., the preferred form in which an individual encodes information), cultural determinants (i.e., family, colleagues, etc.), modalities of inference (reasoning style, i.e., inductive, deductive, etc.) and a memory function. Please note that the cultural determinants, in the form of the influence of the country and cultural background on ones information processing style (learning style) have only recently been proposed for

studying in the adaptive hypermedia community.

4.2. Kolb's learning styles

Kolb (1984) defined a 2-dimensional scale of learning styles, with 4 extreme cases: 1) converger (abstract, active): abstract conceptualization and active experimentation; advantage in traditional IQ tests, decision making, problem solving, practical applications of theories; knowledge organizing: hypothetical-deductive; question: "How?". 2) diverger (concrete, reflective): concrete experience and reflective observation; advantage in imaginative abilities, awareness of meanings and values, generating alternative hypotheses and ideas; question: "Why?". 3) assimilator (abstract, reflective): abstract conceptualization and reflective observation; advantage in inductive reasoning, theoretical models creation; focus on logical soundness and preciseness of ideas; question: "What?". 4) accomodator (concrete, active): concrete experience and active experimentation; focus on risk taking, opportunity seeking, action; solve problems in trial-and-error manner; question: "What if?".

4.3. Dunn and Dunn's Learning styles

Rita and Kenneth Dunn developed in 1974 a comprehensive learning style model on the following axes: 1) environmental factors (sound/noise level, light level, temperature, design setting). 2) emotional factors (motivation, persistence, responsibility, structure). 3) sociological factors (self-orientation, colleague orientation, authority orientation, pair orientation, team orientation). 4) physical factors (perception, intake, time, mobility). Although this model deals very little with the cognitive factor, this model is currently used in schools for pupils of grades 3-12 and a version has been developed for adults.

4.4. Herman brain dominance model

Ned Herman classified in 1976 thinking styles according to the brain quadrants: 1) Quadrant A (left brain, cerebral): analytical, logical, factual, critical and quantitative. 2) Quadrant B (left brain, limbic): sequential, structured, organized, planned, conservative and detailed. 3) Quadrant C (right brain, limbic): interpersonal, emotional, sensory, kinesthetic, symbolic and spiritual. 4) Quadrant D (right brain, cerebral): visual, holistic, innovative, conceptual, imaginative, artistic. His model classifies people according to their preferences, determining a dominant style.

4.5. Other models

Among other models we mention the classification: field dependent, field independent. Field independence means the extent to which a person can perceive analytically, and can distinguish the study object from the surroundings. Field dependent people are dependent on external cues, and can, for instance, learn better if they have graphical support.

4.6. Considerations on learning styles

It is obvious that, although the correlation between learning styles and teaching styles has not yet been clearly defined, the existing findings already can provide guidelines. Many of the above classifications are actively used in evaluating students in the regular school system, as well as in higher education or life-long learning. E.g., it is good to consider before introducing an on-line course with all the new multi-media technology gimmicks, that it might not improve learning for field independent students. Or, before taking over collaborative learning as the latest in computer based learning developments, to consider that self-oriented students might actually be perturbed by the added interaction. Moreover, it is obvious that there exists an information exchange deficit between psychological researches and the computerized educational systems community, which results often in a poor reflection of the findings of one domain into the other, to the detriment of both.

5. Proposing an automatic student pool

Educational systems evaluation is difficult. Mathematically, the function: $O = f(I)$ is not well defined. The input, I, which are the students, can vary a lot from case to case. The (learning) output, O, can be measured from many different angles, so is also poorly defined. In-between there is the computerized educational system, given by our function f. A good result $O1$ for input set $I1$ cannot guarantee a good result $O2$ with input set $I2$. So, if students are good, the result will be probably good, disregarding the appropriateness of f. Most current tests of educational systems presume an even distribution of the cognitive styles among their students. This is often not the case.

To solve these problems, we propose building an automatic student pool for computerized educational system testing and validation. By designing specifications for simulated students representing different cognitive styles, such a pool of students can be created. A computerized educational system designer can refer to this student pool when testing his/her system, and report the results that s/he achieved by using the virtual students from the student pool.

Although simulated students represent only simplified versions of real learners, using the same simulated students for different systems can be the common denominator, and provide a uniform measurement system and benchmark. Moreover, the pool of students can be permanently improved, as more information is gathered about the particularities of the different learning styles, and the process can feedback. Results from new researches can be, e.g., "our system performed well with the convergent students with strong field independence", etc. In this way, concrete and clear information about the educational usage of the respective system can be expressed. Furthermore, comparison of performances of real students and simulated student pool students can validate these evaluations and suggest improvements for the virtual. The virtual students representing different cognitive styles can be exposed to different teaching strategies provided by the system, in order to validate the system. By showing which students performed best, the system designers can show concretely for which type of students their system is applicable. Beside of the obvious difficulties of implementing such virtual students, real students have another characteristic that is difficult to implement: the tendency to sometimes change their cognitive style, i.e., their approach to new information and its processing. This change is dependent or independent on the teaching strategy used. Some teaching strategies are specifically aimed at changing the cognitive style. Such simulation as in the latter situation could also be designed, and percentile results of the students who could be "converted" by the teaching strategy can be obtained. However, modifications in cognitive style are long-term processes, and it cannot be reasonably presumed that such changes could take place by, for instance, attending one single on-line course.

The final evaluation and validation results should show that a system is efficient for students representing some specific cognitive styles.

So, we propose to test existent systems with more than the available real students, and more often than is possible if following the available curriculum and semester structure. In this way, more of the learning styles population can be covered, and the justification of a system can advance from "it suited my class so it might suit yours" to more exact statements as to precisely what kind of students the system is targeted at, what type of students can profit more from such a system, and what type of students could just as well use some different learning method, etc. This type of testing should not replace completely testing with real students, at least for as long as this method is not sufficiently verified.

6. Conclusion

This paper attempts to bring some structuring ideas in a field that is inherently difficult to structure: computerized educational systems evaluation and validation. We propose a new uniform evaluation method via virtual, automatic students, derived from a public student pool. This pool should work as a benchmark for the field, and can grow together with the field itself. Moreover, the paper attempts to unify "old fashioned" psychological theories and the new developments in computerized educational systems. Findings from the usage of the student pool could feedback into the psychological domain, bringing clarifications about learning styles and cognitive processes.

References

[1] Abou-Jaoude, S., and Frasson, C., Integrating a Believable Layer into Traditional ITS, AI-ED99, Le Mans, France, 1999.

[2] Blair, J., Colleges, Education on the Web, http://www.edweek.org/ew/ , Jan. 19 2000.

[3] Marshall's University's Comparis. of Online Delivery Soft. Products, http://multimedia.marshall.edu/cit/webct/compare/comparison.html

[4] Nkambou, R., IsaBelle, C., and Frassoun, C., Supporting some pedag. issues in a Web-based distance learning envir., NTICF'98, 1998.

[5] Oppenheimer, T., The Computer Disillusion, The Atlantic Monthly, http://www.theatlantic.com/issues/97jul/computer.htm 1997.

[6] Pham,P., http://www.payson.tulane.edu/ppham/Learning/lstlyses.html

[7] Rich, E., User Modeling via Stereotypes, Cog.Sc.,3(4),329-354, 1979.

[8] Self, J., The Develop. of the Comp. Based Learning Unit: A Discussion Doc., http://cblslca.leeds.ac.uk/~jas/discussion.html, 1996.

The Distance Ecological Model to Support Self/Collaborative-Learning in the Internet Environment

Toshio OKAMOTO, Mizue KAYAMA, Alexandra CRISTEA and Kazuya SEKI
University of Electro- Communications Graduate School of Information Systems
{ okamoto, kayama, seki}@ai.is.uec.ac.jp

Abstract

With the rapid development of information technology, it is widely accepted that computer and information communication literacy has become extremely important, and is the main new ability required from teachers everywhere. Therefore, for enhancing their teaching skills and information literacy about the Internet environment with multimedia, a new teachers' education framework is necessary. The purpose of this study is to propose and develop a distance educational system-RAPSODY, which is a School-based curriculum development and training-system. In this environment, a teacher can learn subject contents, teaching ways, and evaluation methods of the students' learning activities, related to the new subject called "Information", via an Internet based self-training system. In this paper, we describe the structure, function and mechanism of the distance educational system based on Distance Ecological Model, in order to realize the above-mentioned goal, and then describe the educational meaning of this model in consideration of the new learning ecology, which is based on multi-modality and new learning situations and forms.

1. Introduction

Recently, with the development of information and communication technologies, various teaching methods using Internet, multimedia, and so on, are being introduced. Most of these methods emphasize, in particular, the aspect of collaborative communication between students and teacher during interactive teaching/learning activities. Therefore, now-a-day it is extremely important for a teacher to acquire computer communication literacy [10].

So far, there were many studies concerning system development, which aim at fostering and expanding teachers' practical abilities and comprehensive teaching skills, by using new technologies, such as computers, Internet, multimedia, and so on. In Japan, systems using communication satellites such as SCS (Space Collaboration System) are developed and used as distance education systems between Japanese national universities. In the near future, a teacher's role will change from text based teaching, to facilitating, advising, consulting, and his/her role will be more that of a designer of the learning environment. Therefore, a teacher has to constantly acquire/learn new knowledge and methodologies. We have to build a free and flexible self-learning environment for them under the concept of "continuous education" [11]. At the same time, we need to build a collaborative communication environment to support mutual deep and effective understanding among teachers. Here, the word of "self" means "autonomous behavior" for either individual or group learning, including the collaborative attitude.

In this paper, we propose a Distance Educational Model, which is based on the concept of School Based Curriculum Development and Training System, advocated by UNESCO [15] and OECD/CERI (Center for Educational Research and Innovation), and describe the structure, function, mechanism and finally the educational meaning of this model. Based on such a background, it is necessary to construct an individual, as well as a collaborative learning environment, which supports teachers' self-learning/training, by using Internet distributed environments and multimedia technologies. A teacher can choose the most convenient learning media (learning form) to learn the contents (subject units) that s/he desires.

The new information and communication technologies bring with them rich and useful opportunities for the self-development of people. Knowledge concerning the teaching contents and methods can be stored as multimedia information, in the form of pictures, videos and sound tracks. Moreover, by using the network environment, it is possible to make use of all resources over the net, without any constraints or restrictions of time and/or geographical location. With this goal in mind, we are developing the integrative distance education/training system for supporting teachers' self-learning/training, called RAPSODY (Remote and AdaPtive educational System Offering DYnamic communicative environment).

2. Distance Educational System- RAPSODY based on Ecological Model

Until now, when a teacher wanted to take a class on "IT-education", s/he had usually to leave the office or school. However, now it became possible to learn various kinds of subject contents by building a virtual school on the Internet environment.

2.1 RAPSODY Distance Ecological Model Rationale

RAPSODY is an integrated guide system that can

logically connect individual learning units, called CELLs. The CELL corresponds to the Learning Object Metadata proposed by IEEE-LTSC[5], and is intentionally focused on three primary aspects in order to represent educational meaning within the distance learning environment: learning goals, learning contents and learning media We call this conceptual scheme the Distance Ecological Model. Each of the CELLs has also the other several attributes (slots) such as features of the material, available tools, a related CELL, Guide-Script, and so on, besides those three primary aspects. From a user's (learner) point of view, this model seems to be quite transparent in order to identify/select his/her leaning conditions, and the system can easily guide towards an adequate Learning Object, according to his/her requirements. The word of "Ecological Model" in distance learning means multi-modal "learning gestalt" reflecting learning goals, learning contents and learning media/environment including any situation of individual/self and group/collaborative learning. We use the ecological model in the wider sense of ecology, as a closed, perpetuum-mobile system, which functions without interference from the outside, once the actors (such as designer, author, and learner) and their interactions are defined. In this research (system), the word "designer" means a person who designs/describes each value of a Learning Object Metadata (CELL) and a Guide-Script. On the other hand, the word "author" means a person who produces digital course materials such as Web-based contents, movies and sound of VOD, etc, by means of any authoring tools. A Learning Object Metadata for any learning course-material would be defined, modified and registered from far sites by designers or authors according to a certain educational goal. At the same time, authors such as schoolteachers, university professors, etc., would produce and store their digital learning contents in their local server machines individually. Of course, each of contributors may play any of the two parts. This system can logically link the CELLs based on the Distance Ecological Model of RAPSODY, which provides the learning guide environment by taking into consideration each user's individual learning needs/conditions. Our Distance Ecological Model is built on three aspects. The first one represents the learning goals, which are 1) the study of subject –contents, 2) the study of teaching ways (knowledge and skills), and 3) the study of evaluation methods of the students' learning activities. The second one stands for the aspect of the curriculum (subject-contents) of "information", which represents what the teachers want to learn. From the third aspect, the favorite learning media (form) can be chosen, e.g., VOD, CBR, etc. By selecting a position on each of the three aspects, a certain CELL is determined. A CELL consists of several slots, which represent the features/characteristics of the Learning Object. Especially, the most important slot in the CELL is "Script", which describes the instruction guidelines of the learning contents, the self-learning procedure, and so on. In the following, we will explain the meaning of each aspect in more details.

2.2 Learning media and environments

This aspect represents five different learning environments, as follows: 1) Distance teaching environment (Tele-Teaching) based on the one-to-multi-sites telecommunications. 2) Distance individual learning environment (Web-CAI) based on CAI (Computer Assisted Instruction) using WWW facilities. 3) Information-exploring and retrieving environment using VOD, CBR (Case Based Reasoning). 4) Supporting environment for problem solving, by providing various effective learning tools. 5) Supporting environment for distributed collaborative working/learning based on the multi-multi-sites telecommunications. Brief explanations for each environment are given in the following.

(1) Distance teaching environment (Tele-Teaching): This environment delivers the instructor's lecture image and voice information through the Internet, by using the real-time information dispatching function via VOD (Video On Demand).
(2) Distance individual learning environment (Web-CAI): This environment provides CAI (Computer Assisted Instruction) courseware with WWW facilities on the Internet.
(3) Information-exploring and retrieving environment: This environment delivers, according to the teacher's demand, the instructor's lecture image and voice information, which was previously stored on the VOD server. For delivery, the function of dispatching information accumulated on the VOD server is used. In addition to it, this environment provides a CBR system with short movies about classroom teaching practices.
(4) Supporting environment for problem solving: This environment provides a tool library for performance support, based on CAD, Modeling tools, Spreadsheets, Authoring tools, and so on.
(5) Supporting environment for distributed collaborative working and learning: This environment provides a groupware with a shared memory window, using text, voice and image information for the trainees.

2.3 CELL definition

The concept of the CELL in Distance Ecological Model is quite important, because it generates the training scenario, including the information to satisfy the teacher's needs, the learning-flow of subject materials and the guidelines for self-learning navigation. The frame representation of a CELL is omitted for lack of space. The frame slots are referred when RAPSODY guides the process of the teacher's self-learning.

3. The System Configuration of RAPSODY

The teacher's training environment is composed of two subsystems based on the Distance Ecological Model. One of the subsystems is the training system, where a trainee can select and learn the subject adequate for him/her guided by the script in the CELL. Another subsystem is an authoring system with creating and editing functions for CELL description. The users of the second environment are the designers, e.g., IT-coordinators or IT-consultants, who can design lecture-plans in this environment.

3.1 Training system in RAPSODY

The training system aims to support teachers' self-training. The role of this system is first to identify a CELL in the model, according to the teachers' needs. Then, the system tries to set up an effective learning environment, by retrieving the proper materials for the teacher, along with the Guide-Script defined in the corresponding CELL. So, the system offers programs for both Retrieving and Interpreting.
STEP 1: Record the teacher's needs.
STEP 2: Select a CELL in the Distance Ecological Model according to the teacher's needs.
STEP 3: Interpret the CELL in the WM (Working Memory).
STEP 4: Develop the interactive training with the teacher according to the Guide-Script in the guide WM.
STEP 5: Store the log-data of the dialog. (log-data collects info on the learning histories and teachers' needs and behaviors)
STEP 6: Provide the needed and useful applications for the user's learning activities and set up an effective training environment.
STEP 7: Give guidance-info, according to the CELL script guidelines, decide on the proper CELL for the next learning step.

Here, it is necessary to explain the dialog mechanism (algorithm) between user and system. The interpreter controls and develops the dialog process between user and machine according to the information defined in our Guide-Script description language. This Guide-Script description language (GSDL) consists of some tags and a simple grammar for interpreting a document, similar to the HTML (Hypertext Markup Language) on the WWW. The interpreter understands the meanings of the tags, and interprets the contents. A GSDL example is shown below.
(1) <free> *Definition*: description of the text (instruction)
(2) <slot (num.)> *Definition*: a link to a slot value in the CELL
(3) <question> *Definition*: questions to a trainee
(4) <choice>: branching control according to a trainee's response
(5) <exe> *Call*: to relevant CELLs
(6) <app> applications for training activities (e.g.,Tele-Teaching)

Firstly, a user inputs his/her needs or requests for self-training. Then, RAPSODY computes the appropriate training course based on the user's request, and gives the user some consulting information in order to confirm the contents of his/her training course.

3.2 Authoring system (creating/editing a CELL)

The system provides an authoring module to create and edit the information in the CELL. This module also offers the function of adding new CELLs, in order to allow a designer (supervisor/experienced teacher) to design the teachers' training program. The tasks that can be performed by this system are: adding new CELLs, editing the existing CELLs, receiving calls for Tele-teaching lectures and managing the lectures schedule. This system is composed of the CELL frame creating module, and the Guide-Script creating module. A CELL design can be performed as shown in the following.
STEP 1: Get slot-values of "learning goals", "subject-contents/teaching ways/evaluating methods", and "useful tools" from the CELL.
STEP 2: Substitute the return value of the slot of the prepared media with the training-contents according to the user's needs.
STEP 3: Substitute the slot-value in the CELL for the corresponding tag in the Guide-Script template.
STEP 4: For "Tele-teaching", get some info about the lecture, by referring the lecture-DB and VOD short movie-DB.
STEP 5: Add a new CELL to the Distance Ecological Model.

The lecture-database consists of "lesson managing files" containing user-profile data, lecture schedules, trainees learning records, lecture abstracts, and so on. The Guide-Script template file contains tag-information, written in the Guide-Script description language (GSDL), for all subject-contents items in the Distance Ecological Model. If a certain designer/author wants to register a new training course into RAPSODY, s/he has to describe the values of its course characteristics, which are: a learning objectives, an abstract of the course, a recommended learning environment (learning tool), guide-scripts of this course, and so on. This Guide-Script is a kind of short scenario about the lessons in the course. A designer/author inputs the subject-contents, the student learning objective and the learning media (form) as the lesson attributes.

4. Technical assessment

Here, we focus on traffic problem on transmitting/receiving data of real movie and sound via the Tele-teaching mode. We found that the system can transmit image data of 10 frames per a second under the situation of 100kbps (bit rate). This value depends on the capability of the computer that controls the VOD server, and the network paths bandwidth (rooting). In the case of receiving such data via a PPP connection (Point-to-Point Protocol) by using a public telephone line, the transmission rate is 1 frame per 20 –30 seconds. Furthermore, in the case of receiving such data via a Provider, the value becomes 1 frame per 10-20 seconds. As a result, we are compelled to set up a high-speed line for far receiving sites in order to conduct real learning efficiently. Figure 1 shows the experimental results in details about time-characteristics for each of Movie-frame rate, Movie-bit rate, and Sound-bit rate. As for transmitting sound data, it is much more stable in comparison with transmitting movies. However, we must set up the condition of echo-canceling, noise filtering and so on. In a real situation, we have used two personal computers, one for transmitting/receiving movies and sounds from a digital camera and a microphone, and another for looking at Web-materials such as Power-Point manuscripts. From far sites, users can ask the lecturer some questions via a convenient dialogue window developed by us. In this environment, a lecturer can give collaborative calls to each of the participants and adjust the teaching/learning process interactively.

5. Conclusions

This paper proposed the Distance Ecological Model for building the integrated distance learning environment. This model stands for the networked virtual leaning environment based on a 3 aspects-representation, which has on the axes 1) learning goals (on the study of the subject–matter, teaching knowledge/skills, and evaluation methods), 2) subject-contents in the designated subject-matter, and 3) learning media (forms). This represents a new framework for teachers' education in the coming networked age. We have mentioned the rationale of our system and explained the architecture of the training system via the Distance Ecological Model. Furthermore, we have described a Guide-Script language. The aim of our system is to support teachers' self-learning, provided as in-service training. At the same time, we need to build rich databases by accumulating various kinds of teaching expertise. In such a way, the concept of " knowledge-sharing" and "knowledge-reusing" will be implemented. As a result, we trust that a new learning ecology scheme will emerge from our environment.

RAPSODY is a platform that provides various kind of learning places based on the Distance Ecological Model according to the learner's need in consideration of the relationship among Learning Objects. In this system, the function of a Tele-conference with real video/ sound and the shared window of chatting/application software are provided. So, if a learner wants to change the mode of the learning media from the individual (self) one to the collaborative (group) for some reason after he/she has finished with a certain Learning Object, then the system asks a group manager for permission to attend this discussion group to satisfy lecturer's request according to the Distance Ecological Model. As another example, the system may recommend a selection of the collaborative mode for encouraging deeper recognition/understanding for that learner, if such a navigation message is described in the Guide-Script. In this way, RAPSODY provides the integrated/free environment for distance learning that can adaptively change between self/collaborative mode according to a user's needs based on the Guide-Script in the Distance Ecological Model. In this sense, the Guide-Script contains the core information about the stream developed between Learning Objects by reflecting both the learner's needs and curriculum relationships.

With this system, we can construct various kinds of learning forms and design interactive and collaborative activities among learners. Such an interactive learning environment can provide a modality of externalized knowledge-acquisition and knowledge-sharing, via the communication process, and support learning methods such as "Learning by asking", "Learning by showing", "Learning by Observing, "Learning by Exploring" and "Learning by Teaching/Explaining'". Among the learning effects expected from this system, we also aim at meta-cognition and distributed cognition, such as reflective thinking, self-monitoring, and so on. Therefore, we expect to build a new learning ecology, as mentioned above, through this system. Finally, we will apply this system to the real world and evaluate its effectiveness and usability from experimental and practical point of view.

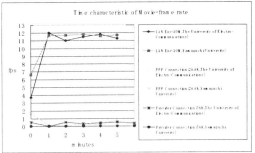

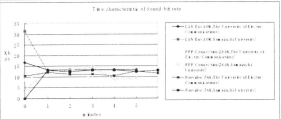

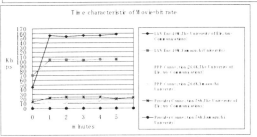

Figure 1: Movie-frame &bit rate, Sound-bit rate

References

[1] Aoki, Y. and Nakajima, A. (1999): User-Side Web Page Customiz., 8th Int. Conf. on Human-Comp. Interact., 1, 580-584.
[2] Kaye, A. R. (1994): Computer Supported Collaborative Learning in a Multi-Media Distance Education Environment, O'Malley, C. (Ed) Comp. Sup. Collab. Learn., 125-143, Springer.
[3] Collis, B. (1999): Design, Development and Implementation of a WWW-Based Course-Support System, ICCE99, 11-18.
[4] Colazzo, L. and Molinari, A. (1996): Using Hypertext Projection to Increase Teaching Effectiveness, J. of Educ. Multimedia and Hypermedia, AACE.
[5] IEEE-LTSC (2000): Draft Standard for Learning Object Metadata,IEEE P1484.12/D4.0
[6] Jones Univ. home-page: http: //www.jonesinternational. edu/
[7] Kobayashi, et al. (1998): Collaborative Customer Services Using Synchronous Web Browser Sharing, CSCW 98, 99-108.
[8] Lewis, J. and Shaw, E. (1997): Using Agents to Overcome Deficiencies in Web-Based Courseware, Proc. of the workshop "Intelligent Educational Systems on the World Wide Web" at the 8th World Conference of the AIED Society, IV-2.
[9] Nishinosono, H. (1998): A Teacher Education System for IE High School Teachers, Information Society., JET, 523-524.
[10] Salvador, L. G. (1999): Continuing Education Through Distance Training. Proc. of ICCE99, 512-515.
[11] Seki, K., Takaoka, R., Matsui, T. & Okamoto, T. (1998): A Teaching Model in Distance Education based on Teacher's Needs. IEICE Technical Report, ET98-12, 49-56.
[12] Shimizu, Y. (1999): Toward a New Distance Education System, Proc. of ICCE99, 69-75.
[13] The UK Government (1997): Connecting the Learning Society. United Kingdom Government's Consultation paper.
[14] UNESCO Report (1998): World Declaration on Higher Educ. for the 21st Century Vision and Action, UNESCO Paris.
[15] Phoenix Univ. home-page: http://www.uophx.edu/

A Cooperative Linkage between University and Industry via an Internet Distance Education System

Toshio Okamoto, Hisayoshi Inoue, Alexandra Cristea,
Mizue Kayama, Tatsunori Matsui and Kazuya Seki
University of Electro- Communications Graduate School of Information Systems
{okamoto, ino, alex, kayama, matsui-t, seki}ai.is.uec.ac.jp

Abstract

The Industry and University cooperation program is a project of the Japanese Ministry of Education (Mombusho). The target objective of this program is the harmonization of university level educational research, society practice and practical business. The University of Electro-Communications has been appointed by Mombusho to fulfill some specific parts of this program. In this paper we report our experience with the introduction of this program, its framework, settings, actual implementation and preliminary findings. The problems encountered are discussed together with a number of solutions to assist good practice in the building of such an applied educational program.

1. Introduction

The objective of the Industry and University cooperation program is a learning environment for upstanding members of the society, from business, industry or university background. The program aims to offer the university student access to practical business, and the industry and company employee an opportunity to improve and an access to high-level education. In this way, we are promoting an educational collaboration program that binds university and industry. The implementation was done not via a new, large-scale curriculum, but by expanding and enriching the existing graduate school level one, by adding professional knowledge, to enable practical, efficient re-training and re-education. The Japanese Ministry of Education appointed the University of Electro-Communications (UEC) with the enacting of this cooperation program between Industry and University. Target learners are industry students and regular students of the university graduate courses, who wish to learn about new, advanced Information Technology (IT) issues. Putting it into practice involves curriculum extension towards more flexibility of contents and form, in order to also allow learning conditions for industry persons. The implementation method is via long-distance Internet courses, which allow business and industry people to take courses from distant sites, without leaving their working places, as well as flexible hours. Other three universities were appointed to start parallel projects in complementary fields. Here we report about our experiences with the program initiation, framework, settings, actual implementation and first results. We analyze results and problems we encountered and offer constructive solutions.

2. The Program

Career development, via acquiring of experience and knowledge in the field of information communication technology, is an important subject for the future information communication society. Our program's starting courses involve this discipline. The lectures were focused on high information technology, large-scale information system planning and application, and network technology. The first year courses involved the following topics: *Multimedia Communication Technology* and *Information Security*. The appointed lecturers were not only researchers and professors from our graduate school, but also company researchers and implementers as well. This is due to the fact that it is important for students to acquire knowledge about not only the theoretical side of information systems, but also the practical side. The lectures are held as collaborative lectures of "*omnibus*" type (each lecturer presents only one lecture). Although this new curriculum is flexible, with many evening hours (to ensure easy participation from distant sites) and a relatively concentrated information contents given over a short time period (again, to ensure that company workers don't loose too many hours with this program), the UEC graduate school has established a regular credit system for certification purposes. The level of certification is Master level.

3. Experimental situation

Here we report an example situation of a lecture entitled "multimedia communication science/ technology and application". The syllabus of this lecture includes: multimedia and distributed cooperation, CSCL (collaborative system collaborative learning) and collaborative memory, multimedia communication technology, ATM networks, new Internet technology, media representation form and application, data mining, multimedia and distributed cooperation learning support systems, knowledge management, standardization and new business models. The total of lecture attendees was 63, among who 13 were curriculum students. The distance company sites that cooperated were located near Tokyo, in Kanagawa prefecture. The distance companies were linked via Internet and ISDN circuit, therefore establishing a real time bi-directional information transmitting & receiving environment. The round trip of the closest distance company site to the University of Electro-Communications is of 2 hours by public transportation, and for the furthermost of about 6 hours. Therefore, considering the hours of time saved and the convenience of the distance education method, the system presents evident merit for both learners from far company sites as well as company managers.

4. Distance education system configuration

The lecture environment should guarantee lecture movie, sound and lecture materials presentation distribution for attending students, as well as a dialogue-communication function between lecturer and students. We have implemented these functions by using a VOD (Video On Demand) server and a WWW (World Wide Web) server. Figure 1 shows the system configuration of the distance learning system.

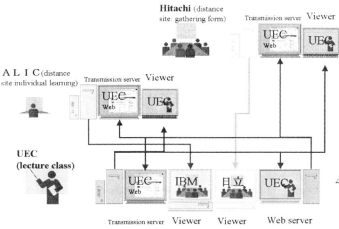

Figure 1. Distance learning system configuration

Each site has to configure the transmission server and the viewer. Each distance lecture site establishes two dedicated channels: one for *receiving only* and one for *sending only*. By connecting respectively the lecture classroom transmission with the distance site reception and the distance site transmission with the lecture classroom reception, bi-directional communication is implemented. Below, each function is outlined.

1) *Transmission of lecture:* movie and sound of the lecture are distributed to the attending sites via the VOD server real time transmission function. Moreover, in the case of some companies, their Internet access and therefore, the transmission/ reception of lecture movie and sound is not free. To cope with such a network environment, we had to ensure a dedicated Proxy server that relays the distribution of lecture movie and sound between Internet sites, to ensure the data reception. Students at distant sites attend lectures by viewing the lecture movie and sound data distributed by the VOD server.

2) *Lecture material presentation function:* the lecture materials are formatted as HTML (Hyper Text Markup Language) sources and offered to students via the WWW server. Students use the WWW browser to access the lecture materials and to follow the lecture progress by referring the appropriate page.

3) *Dialogue/chat function*: communication tool between lecturer and student: ensured by a CGI (Common Gateway Interface) program written in Perl. Lecturer and students access the CGI program via the WWW browser and perform a questions and answers session via the chat terminal screen.

One problem appearing is that the lecture movie and sound data sending and reception conditions depend on the network traffic conditions. Namely, if the reception data has a delay, distant site students have difficulty in viewing the lecture, therefore failing to understand the lecture contents. Therefore we have setup a bi-directional communication channel between lecture site and distant sites. In this way the communication channel between lecturer and students is guaranteed, independent of the Internet network conditions. Figure 2 shows the distance lecture environment structure (lecture site -> distant sites) as well as the interface of both lecturer and students. As can be seen in the figure, the lecture manager transforms the lecture materials into HTML form and performs the registration into the lecture materials database. Figure 3 displays the main menu lecture site and attending sites registration window. Concretely, data recorded are machine name (organization name), VOD server IP address and port number and VOD contents stream identifier. Students attend lectures (figure 3) via our distance lecture environment interface. Firstly, they select the lecture movie "reception" button from the main menu. The system than enquires for the movie to display.

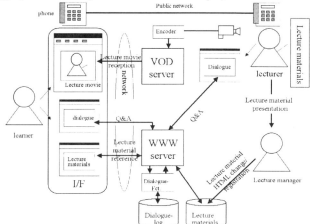

Figure 2. Distance lecture environment architecture

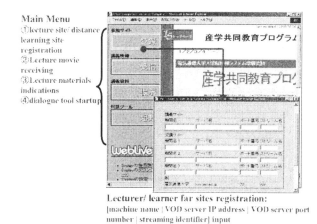

Figure 3. lecture / distance learner site registration

After pressing the button "lecture site ->lecture movie", the system pops up the "lecture movie playback" window and starts the playback (figure 4). Moreover, when clicking the main menu's "lecture materials presentation" button, the system presents the lecture materials, too. Students can browse the lecture material via the buttons "advance" and "return". To enable the questions and answers session, the lecturer has to select from the main menu the dialogue tool "startup", action that will generate a "chat" window. Lecturers attend lectures via a similar procedure to the students. Firstly, they select the lecture movie "reception" button from the main menu (figure 3). If the students' distant sites are equipped with VOD server, video camera, microphone and movie encoder, it is possible for lecturers to receive the image (and sound) of the students attending the lectures. Similarly to the students, the lecturer can view the movie and sound and use the playback function. Moreover, in the case of multiple distant sites, the lecturer can switch between them. Questions and answers sessions are possible as full movie and sound exchange, as chat (in text format) or via the regular telephone line.

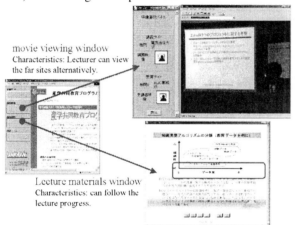

Figure 4. movie playback/ lecture material indication

5. Program Evaluation

The evaluation of the program focuses on how many of the objectives stated in the introduction were actually achieved, and was performed from the 3 points of view enumerated below:
1) *Science-technology aspect:* Here we analyze the operations, functions and the lecture movie and sound data transmission and reception.
2) *Educational aspect:* Here we examine the meaning and significance of the distance education lecture within the industry and university collaboration project, via questionnaires filled in by the students.
3) *Application and organizational aspect:* Here we collect information from administrators participating in the same program; the focus is on possible improvements of the distance education system, and the data collection is done via work sheets analysis and interviews.

In this section we report the results of points (1) and (2) above.
(1a) Science and technology aspects; evaluation 1: movie and sound data transmission and reception aspect:

We have measured the delays, time-lapses and evolution of the lecture movie and sound data transmission. For this purpose, we have analyzed a situation with 3 sites exchanging VOD data. Each site was respectively connected to the Internet via either a high speed -, low speed dedicated circuit (maximum rate 128 kbps), or via a business provider dial up IP link. For the analysis of the transmission, we have collected data from each site on the movie bit rate, movie frame rate and sound bit rate. As can be seen in the figures, the maximum real transmission values are, respectively, 48 kbps for the maximum real movie bit rate, 10 fps for the greatest real movie frame rate and 13.2 kbps for the highest real sound bit rate. We firstly discuss the movie transmission. For movie bit rates of 40 kbps and above, the lecture transmission is stable, with little interruptions, and the frame transmission is of about 5 frames per second. However, for the dial up connections, reception at the students' distant sites can be of about 1 frame per 10 to 20 seconds. At this rate, the lecture movie playback window often freezes into a still image for several seconds. Students attending lectures from such sites complained therefore that movie and sound reception is often asynchronous. Compared with the movie data reception, the sound reception was relatively stable for all participating sites. The above experimental result shows clearly that for such distance lectures the limitation is given by the movie reception, and that it is recommended to implement distance lecture environments on network circuits allowing maximum transmission capacity of 128 kbps or above. If a provider IP connection cannot be avoided, the distance lecture environment can be improved with a number of adjustments, as follows. The first adjustment is related to the lecturer's actions. The VOD system used in our distance lecture environment complies with the H.261 standard, which bases the movie distribution on information compression via a frame prediction mechanism. Namely, the transmitted/ received information weight changes according to the movie data, movie complexity and details, movement intensity, etc. If the lecturer's actions are various, the transmission data weight grows. However, if the lecturer's actions during the lecture are moderate, the movie changes are reduced and therefore the lecture movie distribution can become a little more stable. As another adjustment, the creation of a dedicated low speed circuit for the VOD server is necessary. If these adjustments are made, all sites can reach similar bit rate, and the distribution occurs according to the different frame rates. I.e., according to the network bandwidth of the access points, the distribution to high speed circuit students' distant sites should be high rate, whereas the low speed circuit distant sites should receive low rate movie and sound data.

(1b) Science and technology aspects; evaluation 2: distance education system operativity and functions

We identified possible improvement points for lecturing via a distance education system, with focus on application result, operativity and function. Table 1 shows the recommended improvements for the distance lecture environment. Some items in the table point to proposed solutions, others to counter-measures, and again others just present situational examinations.

Table 1: Problems and solutions of distance learning

Encountered problems	Solutions/ Results
Web lecture materials are not synchronous with lecturer indications. If lecturer browses through the material (e.g., in the second half of the lecture), distance learners cannot keep up with the current page.	By using applications with share function, the synchronous display in both lecture room and distant sites is possible. One drawback is that the lecturer's personal browsing through the material is not possible anymore.
The circuit speeds of the distance lecture sites differ. Reception is therefore difficult for some sites.	Multiple preparations are necessary for correct movie & sound transmission. The server setup should follow the various requirements of the distance sites.
During the Q&A session, the communication quality is reduced by the existent delay between lecture	There is no fundamental replacement scheme at present. New infrastructure and communication hardware is

room and distant sites.	necessary. Rehearsal prior to the actual Q&A session is useful (or session guidance by a chair person).
The phone line Q&A sound problems: 1) Noise, howling; 2) Speaker & receiver volume differ.	1) Noise, howling are caused by the large combination of equipment 2) Volume difference can be cancelled by circuit resistance value.

(2) Educational aspects evaluation: questionnaire survey

After each industry and university cooperation program lecture we have asked the learners to fill in a questionnaire. The questionnaire contains 21 questions that are a combination of free description form questions and 5 steps assessment scale questions (5: I strongly think so, 4:I incline to think so, 3: Neither Nor, 2: I incline not to think so, 1: I definitely don't think so). As the lectures cycle currently not yet finished, we cannot establish conclusive results yet. However, as at present we have finished the first round of lectures, it is possible to derive some suggestions and hints. Figure 5 shows the average answers of the first lecture. Questionnaires were anonymous and 38 questionnaires were returned from the 63 distributed. As the questionnaire (table 2) average is in general around 3-4, the result can be called satisfactory. Especially, questions Q4-6 (about students interest in the subject, the motivating effect of the lecture and about the knowledge they acquired) showed a high score. Students had an average of 2.8 preliminary knowledge on the subject (Question Q1). This result possibly points to the fact that learners with no prior knowledge, up to learners with some prior specialty knowledge were interested in the lecture. As the main object of the program is acquisition of high-level specialty knowledge, this is an extremely important pointer. A low result was noticeable especially for questions Q10 and Q11 (regarding the distant site transmission situation). Therefore, an important sub-goal and improvement point for the next lectures is the harmonization of the transmission with the distant sites. As these were the results of the first lesson in the series, such problems were perhaps unavoidable, and the management and synchronization of lecturer, lecture manager, technical supporters, and teaching assistants was difficult to establish. Some of these problems have already been solved in the following lectures of the same cycle, but as the results were not yet analyzed, we present here our first experiences only, to be of use and serve to guide other educators worldwide.

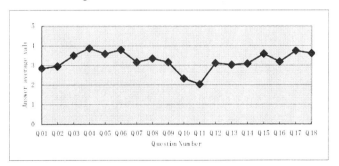

Figure 5. Curriculum questionnaire result average

Table 2: Collaboration Program Questionnaire

Q1. Did you have previous knowledge on the contents of this lecture?
Q2. Is there an important connection between today's lecture and your study field?
Q3. Do you think you can use today's lecture contents in your future research/ job?
Q4. Are you interested in the topics presented by today's lecture?
Q5. Do you feel that today's lecture has opened new perspectives?
Q6. Did today's lecture deepen your knowledge?
Q7. Was today's lecture according to your expectations?
Q8. Was today's lecture level appropriate?
Q9. Was today's presented subject of expertise of importance?
Q10. Was the cooperation with the distant sites smooth?
Q11. Did you feel the distant sites as integrated in the lecture environment?
Q12. Did the lecture materials help you in understanding the lecture?
Q13. Did lecturing and lecture material presentation run smoothly?
Q14. Was question asking made easy by the offered environment?
Q15. Were today's sound and movie transmission clear?
Q16. Was today's lecture complete and sufficient for you?
Q17. Was the time distribution appropriate for today's lecture?
Q18. Was the time distribution appropriate for distant sites in the Q&A session?

6. Conclusion

In this paper we have presented our experience with the introducing of the industry and university education cooperation project at our graduate school to promote the future high level information communication society based on life-long education. Especially, we have highlighted the following points.

1) The implementation method as well as possible problems of a distance lecture environment offering synchronous study at the lecture site as well as distant sites, based on bi-directional communication. In this sense we have presented the lecture form, the representation and presentation technique, as well as the implementation of the questions and answers session.

2) The issue of curriculum integration and certification of both on-site and far-site learners. Students in the graduate school as well as distant site learners achieve credits according to the Master course credit system. Credits can be obtained course-based or program-based. We have also pointed to the merits of continuous learning and self-improvement for company employees, within our program.

3) Moreover, we have discussed and presented our efforts in the direction of building the infra-structure necessary to support the lecturing within the industry and university education cooperation program. Specifically, we referred to the prior collecting of information, which is to be sent together with the lecture movie and sound transmission. This service packet contains advice and assistance information concerning the lecture contents, the collection and classification of lecture materials, text forwarding, etc.

In this way we hope to respond to the forecasted growing future learning demand, especially on advanced IT topics, and to create the environment for a wide area high-level education.

Acknowledgements

Special thanks to Mr. Akihiko Koga (HITACHI), Mr. Yuji Tokiwa (IBM) and Mr. Kenji Ito (ALIC) for cooperating in this program.

References

[1] Okamoto, T. Adv.Info.Soc;synthetic research on contents, system &form of teachers'educ-final report",2000, (A)(1)09308004 stage report.
[2] http://www.mext.go.jp
[3] http://www.open.ac.uk

Internet Based Course Delivery: Technology and Implementation

M. Sun

School of Construction and Property Management
University of Salford, Salford, M7 9NU, UK
m.sun@salford.ac.uk

Abstract

This paper describes the implementation experience of an Internet based distance-learning course currently offered at University of Salford in the UK. The course uses the Blackboard system which provides a virtual teaching and learning environment. It facilitates the interactions between the tutors and students to be conducted over the Internet, which were traditionally only possible in classrooms. This paper seeks to offer some observations on distance learning practice by reflecting the experience of the course delivery. As the Course Director, the author is well positioned to gather extensively qualitative data of feedback from students and tutors. These data together quantitative data captured by the Blackboard system form the basis for this paper.

1. Background

The University of Salford is a leading UK university in the field of IT and Construction Management. Its School of Construction and Property Management runs an on-line MSc course in IT Management in Construction, one of the first exclusively Internet delivered courses in the UK. The course was originally developed jointly by leading industrialists and academics with UK government funding as a part-time course for UK construction companies. Its Internet distance learning delivery started in 1998. After a review of the existing distance learning systems, CourseInfo supplied by Blackboard was chosen as the platform to deliver this course. Using this system, students can access course materials, engage in on-line discussions with fellow students, interact with course tutors, and undertake on-line assignments and assessments.

2. The on-line learning environment

CourseInfo is an Internet based teaching and learning platform. It integrates Web technology with database technology and provides a user-friendly teaching and learning environment for tutors and students (Figure 1). It consists of a course server where the CourseInfo software is installed, and many client PCs from which tutors and students interact with the server using WWW as the interface. The learning environment offers the following essential services.

Course documents dissemination: CourseInfo allows files in multiple formats, e.g., acrobat, Microsoft Word and others, to be uploaded to the course server and organized in hierarchical folders. A student will be granted access to these materials once he or she registers for the course. To save student's downloading time and cost, course materials including workbooks, presentations and additional reading materials are also sent to the students on CD-ROM disks.

Tutor/learner interaction: The system supports both synchronous and asynchronous communication between tutors and learners. Virtual Classroom is a synchronous chat room for student and group communications. It can be used to hold "live" classroom discussions, tutorial sessions, and office hour type question/answer forums. However, due to many students live in different time zones, we find it very difficult to co-ordinate a specific time to deliver a live lecture. The learning environment provides a discussion board as an asynchronous communication tool. Conversations are grouped into forums that contain threads and all related replies.

On-line assessments: The learning environment supports a variety of assessment methods, written tests, multiple choice questions, essays, projects, etc. The tutor can use a combination of these instruments when determining a course assessment strategy. Students can either complete these assignments on-line or they can complete them off-line and submit their answers using the digital dropbox. The learning environment provides a Gradebook which integrates the students' submission to the tutor's marking. It gives the tutor a clear picture of who has submitted coursework. It also informs the student about the assessment results once the tutor has marked an assignment. This is one of the most liked features of the system by the tutors.

© 2001 British Crown Copyright

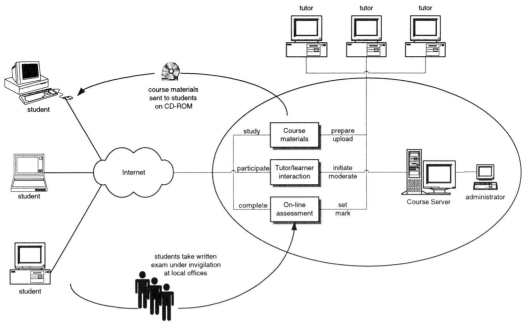

Figure 1. Distance learning environment

Course management and statistics: The learning environment provides a suite of functions to help the tutor gathering information on the student's interaction with the course server. This feature provides instant access to course statistics including page usage, link usage, and access times. The system can generate report on individual student's access pattern to the server. This information will help the tutor to identify students who might need extra support.

3. Discussions and Conclusions

The course described above has been running for more than three years. At present, over 60 students from several countries, including the UK, South Africa, Australia, Singapore, Hong Kong, etc., are studying the course from their own countries. It no doubt achieved one of its main objectives in widening the recruitment base. However, there are a number of issues need to be addressed in the future including:

Tutor skill training: Tutors play a vital role in the successful delivery of distance learning programs. Implementing an Internet based course delivery requires tutor to acquire skills in using the course delivery technologies. A good teacher in the conventional face-to-face classroom setting may not be well-prepared to teach in a distance learning environment because it requires very different skills.

Tutor commitment: It is equally important for tutors to be committed to distance learning course delivery. Otherwise, they will very conveniently ignore the existence of the "virtual" students.

Student support: Most distance learning students study the course in their spare time usually alone. After a long day's hard work, to study late at night can be very stressing. The distress and frustration suffered by distance learning student have been widely acknowledged. The most complaints we got from students were about the lack of prompt response to their enquiries. It is therefore very important to provide effective student support to make them feel they belong to a learning community.

Harmonization with University administration: One of the major advantages of distance learning is its flexibility. It allows individual student to progress at his or her own pace. Unfortunately, this often cause tension with the existing university administration system which is more geared for the traditional full-time taught courses.

4. Conclusions

On a whole, the described course has been a success. It enables students in many countries to benefit from the course with regardless physical distances. However, it also revealed a number of issues need to be addressed in its future development.

The adoption of Internet based teaching and learning technology allows universities to share teaching resources and expertise. With experts from different universities, even different continents can easily teach on the same course, the students will have a richer learning experience. Salford University is in a process of forming collaboration links in relation to distance learning courses with a number of international universities.

Posters

ITS Design Technology for the Broad Class of Domains

Ildar Kn. Galeev, Sergey A. Sosnovsky, Vadim I. Chepegin
Kazan State Technological University
monap@kstu.ru, chepegin@bancorp.ru

Abstract

The elements of Intelligent Tutoring Systems (ITS) design technology for a broad class of domains are described. Characteristic features of proposed technology are considered. Restrictions on the area of its application are posed. Describing technology has been realized in the authoring tools for ITS design, which are the part of program complex MONAP-II. For a number of domains MONAP-II provides full automation of ITS design.

1. Introduction

Development of authoring tools for ITS design and creation on its basis useful systems, which realize didactically-valid algorithms of education and control the leaning process adequately, is traditionally in the center of attention for many researchers of computer-based learning (see, for example, [1,2]). However, at present these researches have experimental nature. To move ITS from the level of laboratory study to the level of commercial product researchers have to decide a number of serious problems. On of them is high labor-intensity of ITS creation. Unfortunately, we cannot say that authoring tools of ITS design meant for deciding of this problem are completely satisfy these requirements. The most often authoring tools help to create ITS relatively easy in very close domain. The main reason of this is the fact, that to obtain the maximal didactic effect in chosen domain developers of authoring tools frequently create systems possessing subject-dependent architecture and based on the subject-dependent algorithms and models.

Authoring tools, which are the part of program complex MONAP-II, provide the automation of ITS design, realizing the algorithms of adaptive management of the learning process in the chosen domain. Algorithms of educational problem solving are developed on the basis of domain analysis by a teacher and described by means of an aggregate of rules (operations) of type: IF (condition), THEN (action). Authoring tools of ITS design of MONAP series have gone through the several stages of program implementation: MONAP-MICRO, MONAP-PLUS, MONAP'99 (see, for example, [3]). At present time this tools are the main component of program complex MONAP-II. These authoring tools are far less exacting to the nature of domain. However, some set of restrictions on the area of application is imposed.

2. Restrictions on the area of application

The problem of learning personalization in *MONAP-II* involves in the assessment of student's knowledge on the each step of learning process taking into account the history of learning process and providing student with problem of optimal difficulty. The difficulty rate of the task $T(k)$ is an average portion of errors anticipated in the task of accomplishing it. This parameter of MONAP model is described in detail in [4] with all computations and bonds with another parameters. The optimal difficulty value is determined by the teacher during the tuning of ITS.

Note that stabilization of difficulty value is possible only in the case when the domain characterized by following features.

A set of vectors $\{L_{rq}\}$ describes the properties of tasks, which can be present for student:

$$L_{rq} = [L_1, L_2, \ldots L_j, \ldots L_J], \quad (1)$$

where r – identifies the class of learning task $(r = 1, 2, \ldots R)$;

q – identifies the subset of synonymous tasks inside the r class $(q = 1, 2, \ldots Q_r)$;

L_j – is the number of operations of j type necessary to be used when solving a task with the identifier rq.

Learning tasks are isomorphic regarding the complication rate if their associated vectors L_{rq_1} and L_{rq_2} are interconnected by the following relationship which is true for each j:

$$\frac{L_j(rq_1)}{L_j(rq_2)} = 1(q_1/q_2) = \text{const}, \quad (2)$$

where $L_j(rq_1)$ – is the j component of the vector L_{rq_1};

$L_j(rq_2)$ – is the j component of the vector L_{rq_2};

$l(q_1/q_2)$ – a coefficient characterizing the complexity of the vector L_{rq_1} relative to the vector L_{rq_2} complexity.

Thus, if condition (2) is met, learning task, their properties being described by the corresponding vectors L_{rq}, will possess the same complication rate for each particular learner at the k step of learning. It is intuitively clear, as they will have one and the same ratio of "badly" and "well" mastered operations.

Consequently, the adaptation to the student as far as the complication is concerned is fundamentally impossible. That is, the less tasks of the learning domain are connected by relationship (2), the higher is the potential possibility to find a learning task (among the available ones) with the complication optimum for the student.

3. Automation of ITS design

Generally accepted method of ITS design inevitably conducts us to the inflexible dependence on domain. That is ITS being developed for example, for learning correct writing of German adjective endings, can not be used for learning correct writing of Russian adjective endings (though in this case the grammar rules are similar). The same is true for a more strategic change of domain, for instance, from one natural language grammar component to another. This is stipulated by the need of developing subject-dependent subsystems of ITS, executing the educational problem formation, its solving and diagnostics of student's answers.

In MONAP-II we propose another approach, which is based on the rejection of designing of the subject-dependent components of ITS, such as the generator. MONAP-II uses the base of educational problems (PROBLEMS base) with tools of its creation and maintenance as a subsystem of educational problem formation. Teacher forming the PROBLEMS base has to indicate for each educational problem not only correct using of all operations but also identifiers of every operation. In this case, if the student makes an error, the diagnostics subsystem will be able not only to fix its presence, but also to determine the place where wrong operation has been applied.

Thus, authoring tools MONAP give the teacher possibility of rapid designing all ITS subsystems. The main advantage of such ITS designing process organization is the possibility for the teacher, who is not a specialist in programming, to create automatically an ITS from the beginning to the end without any help for a number of domain. The proposed approach does not negate an alternative method of ITS design with using the educational problem generator, the educational problem solver and the subject-dependent components, containing information about domain. MONAP technology allows to develop automatically an ITS learning control subsystem, and to develop other ITS components independently or to use an existing developments, as it was made, for instance, in [4]. In this case a calling sequence has to be applied for compatibility of all subsystem interfaces.

4. Conclusion

Thus, the elements of ITS design technology for a broad class of domains were considered in this paper. Restrictions on the area of application and some characteristic features of proposed technology are posed. Described technology is implemented in the program complex MONAP-II. Authoring tools of ITS design, which are part of MONAP-II provide full automation of ITS creation for a number of domains. At present intelligent CALL-systems for German and Russian languages in the part of adjectives declensions were developed by the means of these tools. In the future we are going to approve the proposed technology in another domains. Besides, the work on transferring of elements of described technology of ITS design to Internet is conducted.

5. References

[1] L.H. Wong, C. Quek, and C. K.Looi, TAP-2: A Framework for an Inquiry Dialogue Based Tutoring System, *International Journal of Artificial Intelligence in Education*, 1998, 9, pp. 88-110

[2] M. Virvou, M. Moundridou, An authoring tool for algebra-related domains, *Proceedings of HCI'99, V. 2*, edited by H.-J. Bullinger and J. Ziegler, Lawrence Erlbaum Associate, Publishers, Munich, Aug., 22–26, 1999, pp. 642-646.

[3] I.Kh. Galeev, V.I. and Chepegin, S.A. Sosnovsky, MONAP: Models, Methods and Applications, *Proceedings of the KBCS 2000*, Mumbai, India, 2000, pp. 217-228.

[4] J.C. Yang, K. Akahori, Development of Computer Assisted Language Learning System for Japanese Writing Using Natural Language Processing Techniques: A Study on Passive Voice, *Proceedings of AI-ED'97*, Kobe, Japan, 1997, pp. 263-270.

Constructivist Learning Systems: A New Paradigm

Weiqi Li
Michigan Technological University
weli@mtu.edu

Abstract

The educational enterprise is in the midst of a philosophical shift from a behaviorist to a constructivist paradigm. If constructivism is to be taken seriously as a new educational paradigm in the information age, learning technology R&D must pay more attention to learners instead of teachers.

1. New Times Give New Challenges

Advances in information technology are creating a new infrastructure for business, scientific research, social interaction, and education. Future workforce is required to effectively use information technology to remain competitive, employ creativity and critical thinking to solve problems, possess the ability to communicate and collaborate with others, and have the capacity to readily acquire new knowledge and skills. Given this trend, we must emphasize the learning environment that enable individuals to understand the changing world, create new knowledge and shape their own destinies. We must respond to new challenges by promoting learning in all aspects of life, through all institutions of society, in effect, creating environments in which living is learning.

Building an education enterprise suited to the new times requires developing new education strategies, designing new teaching and learning modes, and creating learning environments that enhance learners' proficiency in understanding, thinking, reasoning, and problem solving. The real promise of technology in education lies in its potential to enhance the learning experience for learners. Better learning will not come from finding better ways for teachers to instruct classes but from giving learners better environments to construct knowledge.

The increasing understanding of knowledge, learning process, and learning environments enables us to have a better perspective of how learning technology should be used to create "authentic" learning environment that correspond to the real world. During the past decade, constructivism has become an important intellectual movement in education as well as in many other fields. The challenge is how to cast this new learning theory and vigorous information technology advances as an opportunity for creative and innovative learning paradigm that takes the whole learning environment into account.

2. Constructivism

Constructivism is a philosophical view about knowledge, understanding and learning. Constructivism holds that learning is a process of building up structures of experience. By contrast with the traditional view of education as a process involving the transmission of knowledge from teachers to students, a constructivist view believes that learning occurs through a process in which learners play active roles in constructing the set of conceptual structures that constitute their own knowledge base.

Constructivism focuses on the learner's control of learning processes. Learners are viewed as active constructors of knowledge. They develop understanding through observation, reflection, experimentation, and interactions with the surrounding environment that continually confirm, challenge, or extend ongoing theories or beliefs. In summary, constructivism holds several general assumptions and beliefs about learning [1]:

- Knowledge is constructed, not transmitted.
- Knowledge construction is embedded in learner's interests and personally meaningful activities.
- Learners take active roles in developing their learning environment.
- Social interaction is an essential factor in the construction of knowledge.

3. Constructivist Learning Environment

Constructivism provides both theoretical foundation and practical opportunity to move towards building constructivist learning environments. A constructivist learning environment (CLE) is a technology-rich, open place where a learner can use a variety of tools and information resources in his pursuit of learning goals and problem-solving activities. Wherein the learner can draw upon information resources and tools to actively construct knowledge, generate a diverse array of ideas, develop multiple modes of representation, engage in social interaction, and solve authentic problems.

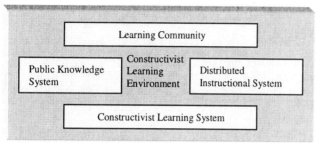

Figure 1 Elements of constructivist learning environment

Four major elements are bound together by their roles in the establishment of a CLE (Figure 1):

- **Public knowledge system**: It consists of all kinds of knowledge management systems that facilitate the creation, capture, storage, manipulation, and dissemination of public knowledge.
- **Distributed instructional system**: It offers educational programs, produces instructional materials, and is entitled to give out legally recognized certificates.
- **Learning community**: It is a group of individuals who are interested in a common topic or area and engage in knowledge-related transactions as well as transformations within it.
- **Constructivist learning system**: Constructivist learning system is a tool with which learners combine the most appropriate learning information and tools for a certain kind of learning situation. This system emphasizes the active and purposeful role of learners in configuring learning environments to resonate with their own needs, echoing the notions of learning with technology through mindful engagement [2].

4. Constructivist Learning System

The fundamental principle of the constructivist theory is that learning is a constructive activity that the learners themselves have to carry out. The insights gained from constructivist perspective can be instrumental in forming our views about the nature of learning system and the purpose of learning technology.

A constructivist learning system (CLS) is a technology-based knowledge-construction tool with which a learner develops his CLE and constructs his knowledge base. Since CLEs are constructed from the perspective of learners, sensitive to their learning needs, styles, paces, local cultures, interests, and aspirations, learning system developers provide learners with only scaffoldings that contain tools, strategies, and guides, which enable learners to interact with construction tools in ways that best enable them to build the learning systems at different levels of knowledge structure and technological sophistication.

As shown in Figure 2, in general, a CLS contains six components [1, 2]:

- **Information bank:** It serves as a source of explicit information about topics through which a learner can access databases of information, including textbooks, dictionaries, encyclopedias, journals, digital library, music, films, and other electronic documents.
- **Notebook**: It is a highly organized set of note-taking tools for a learner to construct his own artifacts in the form of presentations, written documents, reports, models, pictures, etc.
- **Learning tool kit**: It provides a learner with access to the tools to understand and solve a problem. Learning tools include visualization tools, modeling tools, simulation tools, remote instrumentation, remote engineering tools, remote workstations, etc.
- **Microworld**: It presents scientific and social phenomena and makes them accessible to scrutiny and manipulation. It may contain virtual realities, scientific simulations and emulations, virtual laboratories, virtual museums, online field trips, online special events, and other miniature scientific and social world.
- **Conferencing Room**: It provides conversation and collaboration tools, using a variety of computer-mediated communication methods, to facilitate communication among communities of learners and with teachers and experts.
- **Task manager**: It provides a learner with self-service and self-administrative mechanism that enables the leaner to plan his study, create learning maps, manage and track learning process, update records, register for courses and event, and schedule activities.

Figure 2 The components of a constructivist learning system

References

[1] D. H. Jonassen, K. Peck, & B. G. Wilson, *Learning with Technology: A Constructivist Perspective*, Columbus, OH: Prentice-Hall, 1999.

[2] D. N. Perkins, "Technology Meets Constructivism: Do They Make Marriage?" *Educational Technology*, vol. 31, no. 5, 1991, pp. 18-23.

Promoting the Introduction of Lifelong Learning Related Concepts in the Description of Information Resources Using Metadata Technology

Rocío García Robles
Institute for Prospective Technological Studies
rocio.garcia@jrc.es

Abstract

The rise of Internet has created an urgent need to define widely agreed-upon methods and vocabularies for describing its contents in a consistent and orderly manner. This necessity can be satisfied by means of the use of Metadata Schemas. In achieving this aim in relation to the Educational, Vocational and Training sector, several initiatives have emerged at an international level, but there is a need for integrating core skills and competencies issues, which are very important in relation to the new paradigm of Lifelong Learning (LLL) and especially regarding with informal education and training, into the metadata solutions that are arisen in the framework of those initiatives.

1. Introduction

In order to improve the retrieval of information from electronic data sources, since a few years ago a number of Metadata initiatives have arisen, developing metadata schemas and specifications to describe all kind of information sources. But **how can Metadata be defined?**. Metadata is a broad term that covers many types of **"structured data about data"** - and it can be applied to any kind of information resource from traditional ones such as library catalogues to new forms of technical and descriptive data for Web resources. Therefore Metadata is used to describe an information resource, but Metadata can serve a variety of purposes, not only for localisation aims identifying a resource that meets a particular information need, but also evaluating their suitability for use or tracking the characteristics of resources for maintenance or usage over time.

In order to allow a common platform for applying Metadata Schemas ensuring interoperability between all information resources, it is important to get consensus among all related initiatives. This process culminates with the arising of worldwide standards on metadata schemas, as in the development of global specifications to be used freely by any person, company or institution. An area where Metadata have been successfully agreed-upon is on the Vocational, Education and Training (VET) sector.

According to one of the Metadata Watch Reports delivered by the SCHEMAS Project[1], there are four important Metadata Initiatives on the Education and Training domain:

1. First of these initiatives in the IEEE LTSC LOM[2], The IEEE (The Institute of Electrical and Electronics Engineers, Inc) LTSC (Learning Technology Standardization Committee)

2. The second initiative is the DCMI[3]: (Dublin Core Metadata Initiative).

3. The third initiative is the CEN (European Committee for Standardization) ISSS (Information Society Standardization System) LTWS[4] (Learning Technology Workshops).

4. The fourth initiative is the ISO (ISO: International Standardization Office) IEC JTC1[5] (Joint Technical Committee) SC36 (SubCommittee number 36).

2. Rationale

In an innovation-driven economy, and a world characterised by the emergence of a mosaic of lifestyles and intermixing of cultures, a commitment to continuous learning and the generation of new knowledge has become vital to sustain economic, social and cultural development. This applies equally to individuals, organisations, whole communities, regions and countries in a worldwide scope [6].

Therefore, there is a new paradigm on the Vocational Education and Training (VET) field, which is called "Lifelong Learning". This term describes a goal set by the most developed countries, whereby all citizens would participate in learning activities over the complete course of their lives. This includes structured learning ranging from initial formal education, to job related training courses provided by employers, public labour market training schemes, and adult education programmes. But it also includes the informal learning which takes place in all varieties of life-situations, outside educational institutions and structured learning activities, such as early childhood learning, learning-by-doing, self-motivated study and learning, learning via social interaction, learning from the mass-media, and so on. In this sense, the term Life-Wide Learning is sometimes used to describe the breadth and variety of formal and informal learning that takes place in each life phase, and includes an implicit message that the informal aspects be left less to chance than before[6].

But Lifelong learning is still far from a reality and must continue to be pursued vigorously to provide all groups in society with proper learning opportunities, regardless of age, gender, ethnicity and basic formal qualifications[6].

In both cases, formal and informal learning, (core) skills[7] and competencies are key issues that must be tackled as "drivers" for learning[6]. They are essential components of modern VET systems, becoming important means by which innovative teaching and learning practices are implemented into educational institutions, leading to an stronger emphasis on learning by doing (which is especially important in relation to informal learning)[8].

Although main Metadata initiatives on the VET sector are supposed to be aware of the new paradigm of Lifelong Learning, there seems to be a gap in introducing those key issues, skills and competencies, into the current approved versions of Metadata Schemas. But, as a previous stage for integrating skills and competency concepts into Metadata Schemas and Specifications, there is a need to get agreement on the definition of all these related concepts and also on the possible ways of categorization[10].

3. Conclussion: Advantages of considering Core Skills discourse in relation to Metadata Initiatives

As it was previously explained, skills are the "drivers" for formal and especially informal learning in the framework of LLL. In this sense, and in relation to metadata initiatives, it is true that there are some national approaches to produce taxonomies on competencies, (NVQs), but there seems to be little concern, if any, in trying to introduce all these important concepts into metadata schemas which will be used to describe information resources on VET.

Main advantages of integrating skills and competencies discourse in relation to the Metadata Initiatives would be: to offer a new way for localisation of information, up-to-the-minute curricula development and connection between VET and the labour market.

Therefore, core skills can guide the development of new up-to-date curricula, because to include skills information into metadata schemas would serve to give a tool to different targeted user groups (teacher, learner,) to be able to choose learning materials (for example, a course) using as searching criteria the type of skills the user group wants to acquire, and not only focusing the search on content details of the learning material. This is particularly relevant to informal learning, where skills are the "drivers" for learning activities.

Moreover, as it was mentioned above, the ideal aim would be to be able to specify skills with the same terminology used in the labour market in relation, e.g., to the profile of the employees who are already working or who are demanded in the workplace. Therefore, to include skills information as part of the metadata structure could improve the process of searching for information using the terms of reference of the labour market.

After the previous analysis we can conclude that metadata schemas that, according to the original definition, are used to describe information resources, should be linked to core skills and competencies concepts.

Notes & References

[1] Document Number: SCHEMAS-PwC-WP2-D22-Final-20000602. http://www.schemas-forum.org/

[2] IEEE LTSC Initiative: http://ltsc.ieee.org/

[3] Dublin Core Metadata Initiative (DCMI): http://purl.org/DC/

[4] CEN/ISSS LTWS Initiative: http://www.cenorm.be/isss/

[5] ISO/IEC JTC1 SC36: http://jtc1sc36.org/

[6] The IPTS Futures Project Report Series N. 14: "Knowledge and Learning: Towards a Learning Europe"

[7] For the purposes of this study, the terms skills and competencies are considered equivalent. According to Neville Bennett's definition (Bennett N., Dunne E., Carré C., (1999), "Patterns of Core and Generic Skill Provision in Higher Education", Higher Education 37), the word "core" refers to disciplinary skills, that are generic enough to represent the skills which can support study in any discipline, and which can potentially transferred to a range of contexts, in formal and informal education or in the workplace. There are many ways of classifying skills. Those skills go from skills for learning and developing competence to skills for looking for opportunities to start or improve businesses and, generally speaking, skills for developing effective performance across a wide range of settings at work but also in social and economic life.

[8] Report on "Development of Core Skills Training in the Partner Countries", June 1998, European Training Foundation (ETF). http://www.etf.eu.int/etfweb.nsf/pages/downloadtheme1

[10] García Robles, R. April 2001, "Metadata on Education and Training promises Lifelong Learning", IPTS Report.

A Framework for Constructing Adaptive Web-Based Educational Systems

Marek Obitko
Dept. of Cybernetics, FEE, Czech
Technical University, Prague,
Czech Republic
obitko@labe.felk.cvut.cz

Lubomír Kurz
Neuralgen s.r.o., Brno,
Czech Republic
http://www.neuralgen.com/
lubomir@kurz.cz

Igor Glücksmann
Faculty of Mathematics and
Physics, Charles University,
Prague, Czech Republic
igluck@atlas.cz

Abstract

Adaptive educational systems are often built from scratch with no general methodology that would enable the system to be easily reused in another domain [5]. We propose a general methodology to structure the taught domain into so-called "knowledge units" that can be treated in a unified way. Individualization of the learning process for a particular student consists of two layers - a cognitive and a presentation layer.

We will briefly describe our framework for the construction of web-based educational systems. A course in elementary mathematical functions is presented as an example of our approach.

1. Goals of Learning

In a typical educational learning process, a student is expected to acquire knowledge and, more importantly, to apply the new knowledge to problem solving.

Since we cannot measure the student's knowledge directly, we have to build a student model by observing outputs (i.e. answers) resulting from inputs (i.e. questions) in the learning process. The goal of learning is to make the student's answers correspond with the required outputs. A flexible question-answer mode, which individualizes the presented information based on feedback from the student, can be employed here very efficiently.

To summarize, learning can be understood as the adaptation of the student's ability to react correctly to presented tasks. This adaptation can be reached and measured by a controlled dialogue with feedback.

2. Learning System Knowledge Base

The most visible result of the learning process is how correct the student's answers are. Based on this, the setting up of an appropriate question-answer structure is a basic part of defining the knowledge base. In fact, a knowledge base can always be represented by a question-answer set and any knowledge item can be presented in a question-answer mode. We can structure the whole course in units built around the questions expected to be answered correctly as a result of learning.

2.1. Knowledge Units

In our view, a knowledge unit is a compact set of related knowledge items that can be taught and examined independently. Knowledge units can be defined with various granularity (e.g. one unit can correspond to one paragraph or to one chapter in a book) for different groups of students. The granularity of a knowledge unit can be changed dynamically.

Considerable attention has been paid to various relations between the knowledge units, which can result in quite a complicated structure of the knowledge domain [5]. The actual effect of such structures is often not worth the effort. In our approach, we use just one type of relation, prerequisite or predecessor.

2.2 Knowledge Unit Functions

A knowledge unit teaches the embedded knowledge by presenting to the student various "screens", to which the student can react. The simplest unit could use only a question and an answer while a more complicated unit may contain a deeper explanation, hints, examples etc. at several levels (see below) and/or a generator of sample questions (exercises) with corresponding answers.

A unit presents embedded knowledge to the user, registers his reactions, updates the mastery level, and produces another presentation. When the required mastery level is reached, the unit transfers control to another unit.

3. Individualization for a Particular Student

In order to achieve maximal efficiency in a learning process, each student needs his own personalized treatment. It has been shown (e.g. [2]) that we cannot expect the student to search proactively for the right information needed to answer questions from an unknown domain. Moreover, the ability to take in the information presented to the student depends on his or her background. It means there is no general presentation pattern meeting the needs of every student.

The individualization for a particular student can be obtained by a combination of changes at the cognitive and the presentation layer.

3.1. Cognitive Individualization

The cognitive state indicates how the student has mastered the domain [4]. The cognitive student model is expressed by the set of student's mastery levels of knowledge units. This model is used to control the flow between units. A possible algorithm for cognitive individualization follows:
1. Select the unit that has not been mastered yet and that has all prerequisite units mastered.
2. Let the unit deliver a presentation. Evaluate the user's answer/reaction and update the cognitive student model (note that the student's answer may also influence the mastery level of other units).
3. Repeat from step 1 until all units have been mastered.

Such cognitive individualization together with presentation individualization enables us to repeat the presentation at different levels and in different styles (as discussed below) until the required mastery level is achieved. It also enables students to learn at their own pace without any unnecessary presentations.

3.2. Presentation Individualization

Every item of knowledge can be presented in several ways with differing levels of comprehension, i.e. the complexity and amount of textual and other explanation, and it can also be presented in various forms. We assume that the expected comprehensibility can be parameterized and all possible presentations generated by the unit can be ordered by these parameters. Ordered presentations create presentation levels; each knowledge unit can have several levels. The goal is to lead each student through the levels that are most appropriate for the individual.

4. Case Study - Elementary Mathematical Functions

Using the methodology described above, we have created several particular teaching systems implemented in the WWW environment (reasons for this environment are summarized e.g. in [3]). In this section, we will describe a case study that uses our domain-independent Java framework.

The core of the framework is an invisible Java applet that generates presentations for the user as dynamic HTML pages. These pages use their own embedded applets from [1,6] for displaying graphical information.

The goal of the course is to teach the students to recognize elementary mathematical functions, especially relations between a formula and the corresponding graph. Each knowledge unit corresponds to one type of function. For each function, the student is expected to know its name, general formula, graph, and to recognize the values of the parameters of the function from the graph and vice-versa. Both knowledge units and their presentation levels are placed in order of expected difficulty to enable cognitive and presentation individualization as described above.

This system will be used at the Mendel University in Brno to help first-year students prepare for a part of their Mathematics I exam.

5. Concluding Remarks

We have presented a methodology for constructing an intelligent adaptive tutoring system together with a fully functional, flexible, web-based framework based on this methodology. This framework was tested on several domains. One domain, an elementary functions course, was described briefly. The advantage of this methodology is the simplicity in authoring the system while keeping to the main goal of the teaching process - to teach the student to use the acquired knowledge on practical examples. We do not force authors to develop unreal, complicated structures of the system control; the system uses data collected from students' responses to adapt a learning sequence to particular student's needs. Currently we are planning a careful evaluation of the system based on real data obtained from students.

Acknowledgements

This research was supported by Neuralgen s. r. o., Czech Republic. The first author was supported by the MSMT Grant No. 212300013.

References

[1] "HotEqn – The IMGless Equation Viewer", http://www.esr.ruhr-uni-bochum.de/VCLab/software/HotEqn/HotEqn.html
[2] Vincent Aleven and Kenneth Koedinger. Limitations of Student Control: Do Students Know When they Need Help?, In [5], pages 292-303
[3] Peter Brusilovsky. Adaptive Hypermedia: From Intelligent Tutoring Systems to Web-Based Education. In [6], pages 1-7.
[4] Nicola Capuano, Marco Marsella and Saverio Salerno. "ABITS: An Agent Based Intelligent Tutoring System for Distance Learning", in C. Peylo (ed.), *Proceedings of the International Workshop on Adaptive and Intelligent Web-based Educational Systems*, Montréal, Canada. Pages 17-28.
[5] Gilles Gauthier, Claude Frasson and Kurs VanLehn (Eds.): *Intelligent Tutoring Systems, ITS2000*, Montréal, 2000. Springer-Verlag Lecture Notes in Computer Science Vol. 1839.
[6] Marek Obitko and Pavel Slavík. Visualization of Genetic Algorithms in a Learning Environment. In J. Žára (ed.), *Spring Conference on Computer Graphics, SCCG'99*, pages 101-106, Bratislava, 1999. Comenius University.

Domain Instruction Server (DIS)

Juan E. Gilbert, Ph.D. [1]
Auburn University
Computer Science and Software Engineering
107 Dunstan Hall
Auburn, AL 36849 USA
+1 334 844 6316
gilbert@eng.auburn.edu

Dale-Marie Wilson [1]
Auburn University
Computer Science and Software Engineering
107 Dunstan Hall
Auburn, AL 36849 USA
wilsodc@eng.auburn.edu

Abstract

Web based instruction is growing at an incredible rate. Teachers, instructors, trainers and several others are putting their instructional content online. The format, style and media types vary from instructor to instructor. In essence web based instruction is being done by an unknown number of people in an unknown number ways using different media types and platforms. In this paper, an instruction repository will be introduced that has the ability to unite all web based instruction under one umbrella. The media types, platforms, instructors and formats will remain independent. The repository will create a new instructional model for Web based instruction.

1. Introduction

In the traditional classroom, there exists a one-to-many instructor-student ratio. There is one instructor teaching many students. In a tutoring environment, there exists a one-to-one instructor-student ratio. In this paper, the instructional model will be changed to a many-to-one instructor-student ratio [3]. Creating a distributed repository of instruction will provide each student with many instructors facilitating a many-to-one instructional model. In this instructional model, the major task is matching students to instructors that accommodate their learning style [1]. In the sections that follow, the architecture for a distributed instruction repository will be introduced.

2. Repository Model

The internet consists of several domain name servers. Each domain name server contains a list of domain names, i.e.: www.eng.auburn.edu , and a corresponding IP address. When a browser connects to the Web, it will ask a domain name server for an IP address before it is connected to the requested web site. This model can be adopted to create an instruction repository that consists of domain names and other instructional meta data.

In the instruction repository, users can submit queries that will yield a domain name. The domain name will correspond to an instruction unit on the web. This function is similar to the domain name service, but it also models credit card processing on the web. For example, most electronic commerce sites buy or lease merchant services from a merchant service provider. The merchant service provider gives the store owner access to their credit card processing server. The store owner submits a query to the credit card processing server and the server returns a response code. The response code informs the store owner of an acceptance or rejection of the credit card transaction. This model of credit card processing fits a model for the creation of an instruction repository called DIS.

3. Domain Instruction Server (DIS)

The Domain Instruction Server (DIS) is a repository consisting of web deliverable instruction units. A single unit of instruction is defined as an instruction unit. In traditional terms a single unit of instruction can be viewed as a lesson found on a syllabus. The sequential organization of a collection of instruction units defines a course. Instruction units are created by an instructor and placed on a web server. The web server may belong to their college, department, company, internet service provider or it may be their own personal machine. In any case, the web server is world accessible and it contains their instruction units. Each instruction unit has several common attributes. These attributes can be viewed as instructional meta data. By collecting several instruction units from several instructors the repository creates a high level view for each course as seen in figure 1.

In figure 1, there are three instructors teaching the same course. This course is composed of five lessons and fifteen different instruction units. Each row in figure 1 illustrates that each lesson is taught three different ways using three different instructors. The columns under each instructor illustrate the use of different instructional methods used by the individual instructor. Each instructor teaches the same course using a different media type or style. Each rectangle in figure 1 is an instruction unit consisting of instructional meta data. Figure 1 is a high level view of the DIS organization. The lower level architecture is discussed next.

Figure 1. High level course view.

3.1. DIS Architecture

The architecture that supports the DIS environment is very flexible. The primary objective in defining the architecture is to obtain total flexibility with respect to varying platforms across various implementations. This architecture must be platform independent, universally accessible and easy to use. With these requirements in mind, the architecture in figure 2 was defined.

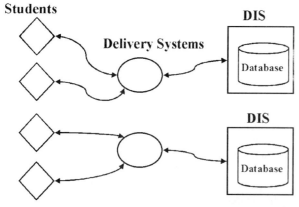

Figure 2. DIS architecture.

In figure 2, there are three layers. The first layer consists of the students. These are designated by the diamonds. The students are using a web browser to connect to the second layer, delivery systems.

Delivery systems are the middleware that provide instructor flexibility for their instruction unit selection process. Within the DIS architecture, there exists the concept of instruction method selection. As described by figure 1, the DIS contains several instruction units. These instruction units have to be selected for use. The process of selecting an instruction unit for use is called "instruction method selection" [2] The final layer is the server layer.

The server layer is composed of domain instruction servers. Each server contains the same repository information. The delivery systems request instruction units from the domain instruction servers. Upon receipt of the instruction unit, the delivery systems will deliver instruction to the student's browser.

4. Conclusion

The DIS environment is being created to unify web based instruction. This environment also serves the purpose of changing the instructional model from the traditional one-to-many or one-to-one models to a many-to-one instructor-student relationship model. This new model of instruction will link students to instructors that before now, may have been impossible to learn from. This environment also creates a repository of instruction that is modeled after existing, yet proven technologies, i.e. domain name server. The future design and implementation will consists of a synchronous component in order to facilitate group participation.

5. References

[1] Dunn, K. and Dunn, R., *Teaching Students Through Their Individual Learning Styles: A Practical Approach*, Reston Publishing, Reston VA, 1978.

[2] J.E. Gilbert, "Case Based Reasoning Applied to Instruction Method Selection for Intelligent Tutoring Systems", *In Workshop Proceedings of ITS'2000: Fifth International Conference on Intelligent Tutoring Systems*, Montreal, CA, 2000, pp. 11-15.

[3] J.E. Gilbert and C.Y. Han, "Adapting Instruction in Search of 'A Significant Difference'", *Journal of Network and Computing Applications*, Academic Press, 22, 3, 1999, pp. 149-160.

[1] The research of this author was supported under the National Science Foundation grant #EIA-0085.

A Web-Enabled Exam Preparation and Evaluation Service: Providing Real-Time Personalized Tests for Academic Enhancement

Syed Sibte Raza Abidi
School of Computer Sciences
Universiti Sains Malaysia
11800 Penang, Malaysia.
Email: sraza@cs.usm.my

Alwyn Goh
OpenSys (M) Sdn Bhd
KLCC Tower 2, Level 23
50088 Kuala Lumpur, Malaysia.
Email: agoh@opensys.com.my

Abstract

We present a technology-enriched, Web-enabled, value-added Distance Exam Preparation and Evaluation Service that provides (a) offline execution of fully-featured preparatory exercises and evaluation tests in a real-life simulated examination environment; (b) content personalization to address scholastic weakness and (c) the use of data mining techniques to ensure content effectiveness and the pro-active identification of the academic needs of various student segments. The solution is designed as a client-server architecture featuring Java technology and XML-mediated information exchange over the Internet.

1. Introduction

In Malaysia, secondary-level academic evaluation is government mandated and centrally administered. The evaluation exams comprise multiple-choice questions. Traditionally, students whilst preparing for such exams tend to refer to a vast pool of past examination questions and in-house tests. However, for more thorough exam preparation students prefer to seek multiple perspectives of a topic/subject vis-à-vis evaluation material prepared by educators different than their own teachers [1].

To meet the above requirement, in this paper we present a technology-enriched, web-enabled, value-added *Distance Exam Preparation and Evaluation (DEPE) Service* that focuses to support secondary-level students in the undertaking of real-time tests, with particular emphasis on long-term scholastic support and corrective measures. [2, 3]. The DEPE system (as shown in Figure 1)—a web-based client-server application incorporating XML and Java technological components—exhibits the following technical and functional features:

The featured DEPE service exhibits four functional components: (a) *Content Compilation* vis-à-vis the population of a test bank comprising past exams together with school-specific tests prepared by an ensemble of teachers; (b) *Real-time Test Administrator* allowing for WWW-mediated generation of 'personalized' tests based on an individual's longitudinal evaluation record, offline real-time execution of the tests, followed by automatic test evaluation and reporting. Built-in regulation mechanisms ensure 'non-alterable' timekeeping, policy-enforcement and on-request hints; (c) *Solution Constructor and Performance Monitor* to track the student's overall performance on a longitudinal basis and provide the necessary guidance; and (d) *Student Response Analyzer* leveraging cohort student responses for *content review*—i.e. to gauge the relevance, quality and impact of the test questions—and *student profiling*.

2. Functional Description

2.1. Content compilation

A collaborative content representation scheme distinguishes each singular test/exercise in terms of its origin (i.e. school at which the test was administrated); date of administration; subject; class level; type (exercise, test or final exam), coverage (i.e. single/multiple chapters); topic(s) covered; duration and no. of questions.

We have developed a generic Windows-based client-side content compilation application that features: (a) electronic forms for providing test-identification information; (b) a question specification GUI—enabled with multimedia and mathematical formulae inclusion facilities—to provide the question text, multiple solutions, the correct answer, hints, difficulty level and some relatively difficult answers for enhanced testing options; and (c) an Internet-based content upload mechanism to directly send the compiled content to the DEPE service provider.

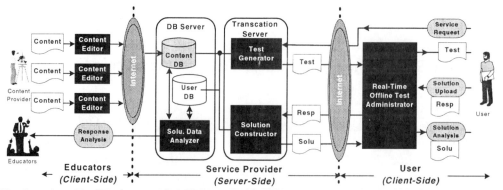

Figure 1. The functional architecture of the DEPE service. Information exchange is realized via XML document objects, whereas the functional modules are implemented using Java technology.

2.2. Real-time test administration

An operational session is initiated by a service request by a registered user which triggers the following server-side functionalities: (a) *Personalized Test Construction* (PTC) and (b) *Remote Test Management* (RTM). The DEPE service offers three options for PTC:

1) *Pre-designed Tests* that the user can undertake.
2) *Customized Test* that the user can dynamically design based on the following parameters: topic(s) to be included, single or multiple origin, test currency (a date range), difficulty level and test duration.
3) *Guided Test* whereby the DEPE system advices the user, based on observed test performances, designs tests aiming to rectify the academic shortcomings.

The dynamically designed 'test document is transmitted to the user who can subsequently undertake the test in a simulated real-life exam setting. For operational efficiency as soon as the test document is downloaded the Internet connection is automatically terminated and subsequent test administration is carried out in an offline mode at the client-side. The downloaded test document is initially 'sealed' and is activated—i.e. the test content becomes visible and the test timer starts—when the user agrees to start the test. The test activity initiates client-side RTM processes that involve:

a) Test time regulation via a timer embedded in the test document—the timer is non-alterable even if the system clock is modified.
b) User assistance in terms of hints and partial elimination of the solution options, all at the expense of reduced credit.
c) Capture of the user's responses, as an encoded response string, at the expiry of the allocated test duration and automatic up-load to the DEPE service server once an Internet connection is established.

2.3. Solution construction and response analysis

Upon receipt of the response string the DEPE system:

a) autonomously evaluates the user's test response as recorded in the response string.
b) generates a solution document comprising test evaluation, explanations, performance indicators and future preparatory suggestions.
c) pro-actively 'pushes' the solution document to the user over an active Web channel.

2.4. Student response analysis

We have implemented a data-mining agent to 'mine' population-wide student responses to effectuate detailed analysis of: (a) Population-wide student profiles as the basis for formulation of individually focused programs; (b) Difficulty level based inductive grouping of questions into different levels; (c) Effectiveness of a set of questions if presented together; (d) User response patterns i.e. whenever a user correctly responds to question x then he/she also correctly responds to question y; (e) Students performance across different schools, regions or states; and (f) Performance comparison between fundamental, problem-solving and analytical type questions.

3. Concluding Remarks

We believe that the DEPE service can provide ubiquitous access to exam preparatory material of relatively high quality and in relatively substantive quantities to students disadvantaged by their physical location. DEPE is accessible at www.eschoolplus.com.my

References

[1] Dede, C., "The Evolution of Distance Education: Emerging Technologies and Distributed Learning", *The American Journal of Distance Education*, 12(2), 1996, pp. 4-36.
[2] Yoshino T., "Application of Distance Learning Support System Segodon to Exercise-type Classes", *Journal of Information Processing Society Japan*, 40(11), 1999.
[3] Okada, A. and Tarumi, H., "Real-Time Quiz Functions for Dynamic Group Guidance in Distance Learning Systems", In *Proc. of 1st Intl. Conf. on Web Information Systems Engineering*, 2000, Hong Kong.

An evolving instructional design model for designing Web-based courses

Siew-Woei Ling
Faculty of Creative Multimedia, Multimedia University, Malaysia
e-mail: swling@mmu.edu.my

Chee-Weng Khong
Faculty of Creative Multimedia, Multimedia University, Malaysia
e-mail: cwkhong@mmu.edu.my

Chien-Sing Lee
Faculty of Information Technology, Multimedia University, Malaysia
e-mail: cslee@mmu.edu.my

Abstract

The complexity of design for Web-based course requires the incorporation of instructional design principles. However, majority of the existing models were not adequate to address the needs of Web-based learning. This project therefore stems from the fact that a specific Instructional Design model is needed to guide designers in developing Web-based courses. A study was conducted in a multimedia course offered by Multimedia University, Malaysia to investigate the scaffoldings for Web-based learning. The finding from the study conforms to the components in which the SWLing instructional design model was developed.

1. Introduction

Web-based courses can be explained as online courses that utilize the World Wide Web technologies as a scaffold to facilitate teaching and learning process. It involves four application phases [1] that the level of interactivity increases incrementally corresponding to the forms of interaction and the functions of these applications. As the dimensions for interactivity is augmented, the complexity of design increases. Thus requires the guide of an instructional design model to how information and knowledge can be presented on the World Wide Web for learning interests.

Instructional design is a systematic process of designing instruction that aims to ensure the quality of knowledge transferred from an instructor to a learner. The evolvement of ID_2 [2], "The Attack on ISD" [3] and concerns from the annual New Media Instructional Design Symposium [4] indicated that the conventional instructional design method is unable to respond to the demands of Web-based courses.

2. Project

This project aims to develop a specific instructional design model to guide designers regarding the design and development of Web-based courses in the context of higher education. The model is built upon the observation of the researcher, feedback and suggestions of experienced Web-based course learners. Qualitative methods were used to investigate the scaffoldings for Web-based learning.

The study took place within the Faculty of Creative Multimedia. A Web-based course on Computer Graphics I was designed and implemented on the fundamental year students who were computer literate. Throughout the 8 weeks, the participants attended one-hour lecture and application demonstration and tutorial sessions on Web-based course setting. Discussion and learning activities were carried out to increase the learning retention and interest on the Web.

3. Measurement

The qualitative data collected at the end of the courses identifies the scaffolds required by learners e.g. interaction, multimedia, support and motivation. The quantitative data determines the importance of each scaffolds for Web-based learning. Frequency counts, median and mode at p 0.5 are taken into account for quantitative data.

Figure 1 shows the importance of five features namely, Support, Learners Engagement, Learners Participation, Multimedia Integration and Learner Interaction. Support, Participation, Multimedia Integration and Interaction were viewed as very important elements whereas Engagement was viewed as of moderate importance.

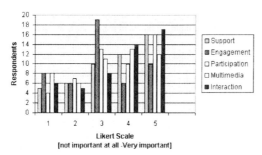

Figure 1: Students feedback on the importance of five features to learning.

The observation showed the participants interest in learning via the Web. Not only students could learn self-paced, they were motivated to further their interests in the subject to the World Wide Web as long as explicit guidance are provided to them.

4. SWLing instructional design model

The literature reviews on existing instructional design models and data finding in this study contributes to the development of a specific instructional design model called the SWLing model. Similarly, SWLing model comprises of five iterative phases, Situate, Analyze, Develop, Formative Evaluation and Implementation.

The Situate phase determines the content and learning goals. This stage similar to the phases from *Identify instructional goal* till *Develop criterion-referenced test items* available *in* Dick and Carey model [5]. It ensures the correctness in course content.

The major challenge in developing Web-based courses is presenting the information in an interactive, multimedia and navigational form for individual learners. The Analyze phase involves well-designed instructional plan. It identifies the attitude of learners, context, resources and Web characteristics to determine the instructional strategies and learner activities. Due to the rapid changes in Web technologies, the components within this phase indicated flexibility and can be modified according to media availability.

The development phase challenges the designer's intelligence and knowledge in identifying the suitable tools or format for Web-based course delivery.

Formative evaluation phase gathers feedback from learners about the system and improves on the Web-based course.

Finally, the implementation phase sees the result of the prior learning and makes necessary refinement.

5. Conclusion

A good instructional design model would be one that acknowledges the relationship among learners, instructors, learning materials and learning environment. Due to the dynamic changes with Information Technology, multimedia capabilities and content, it will be necessary for one to review the model often or as when required to ensure quality instruction.

Acknowledgement:

The authors are deeply grateful to Dr. Philip Duchastel, CastelNet for the development of SWLing model.

References:

[1] R. Ells, *Effective User of the Web for Education: Design Principles and Pedagogy*, Available: http://Weber.u.washington.edu/~rells/workshops/design/ Retrieved October 25, 1998.

[2] M. D. Merrill, L. Zhongmin, & M. K. Jones, Second Generation Instructional Design (ID_2), Educational Technology, 30(1), 7-11 and 30(2), 1991, pp. 7-14. Winner of AECT/DID 1991 Outstanding Journal Article in Instructional Development Available: http://www.coe.usu.edu/it/id2/id1&id2.htm

[3] J. Gordon and R. Zemke, The Attack on ISD, *TRAINING magazine*, v37 n04 p42, April 2000. Available: http://www.trainingsupersite.com/publications/archive/training/2000/004/004cv.htm Retrieved February 24, 2001

[4] V. Hisey, Fourth annual New Media Instructional Design Symposium, 2000. Available: http://www.influent.com/nmid2000/menu.html

[5] W. Dick and L. Carey, The systematic design of instruction, Third Edition, Library of Congress Cataloging in Publication Data, USA, 1990.

Web-Based Educational Tools in Buraydah College of Technology

Jasir Al-Herbish, Amar Arbaoui
*Computer Technology Department, Buraydah College of Technology, POBox2663 Buraydah,
Saudi Arabia*
jherbish@bct.edu.sa, AArbaoui@bct.edu.sa

Abstract

The World Wide Web has been widely recognized as a powerful medium for distributing course based information and has seen a tremendous expansion over the last three years. This paper outlines the importance of the Web in enhancing student learning and describes the applications of Web Course Development Tools for large campus wide deployment. The paper presents a general look at WebCT as a powerful tool that facilitates the creation of sophisticated Web-based educational environments. Finally, the paper presents a model for improving study effectiveness based on our student's way of thinking.

Introduction

The World Wide Web (www) has become the tool of the century. It has been revolutionizing the way we do things in general from studying and learning to any sort of operation. Using the www technology, in enhancing our education and training policies with the current information explosion, will shape our very near future.

In this paper an attempt is given for educational institutions to improve their methodology of teaching and learning. We will describe in this paper the role of the Internet to enhance student learning, what is online learning and finally the importance of the www technology to produce tools for course management.

We will conclude our discussion with a brief summary concerning distance education and learning experienced by department of Computer Technology in our institution.

What do we expect from our graduates?

The graduates are expected to be able to (Figure 1) [1]:
- Communicate effectively
- Stay up-to-date of what is changing in their field
- Have formed ethical principles which will guide them in their carrier
- Work effectively in a team environment

The rich resources available on the web lead to more active, independent student learning if student is held responsible for the information posted [1].

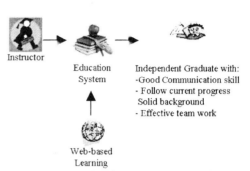

Figure 1: The Role of the Web in Enhancing Student Learning

Online Education

Analysis showed that the three main obstacles for providing effective online materials and learning environment are [2,3]:
1. Lack of support for the collaborative and dynamic nature of learning.
2. Lack of standards for locating and operating interactive platform-independent materials.
3. Lack of incentives and structure for developing and sharing content.

Based on these shortcomings, a better approach has to be found. The solution [2,3] is to develop web course tools and provide a collaborative environment and standard interface for creating and distributing course contents.

Course Tools Based on the Web Technology

The integrated web-based tool provide a three way interaction:
1. Student with contents
2. Student with Instructor
3. Student with other students.

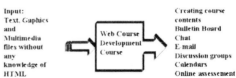

Figure 2: Showing features of Web course development tools.

A comparative study [2] of the following products of web course management tools that conform IMS (Instructional Management Systems) specification:
- BlackBoard CourseInfo

- Course-in-a-box
- Lotus LearningSpace
- TopClass
- WebCT
- WebMentor

showed that WebCT is the recommended development tool. This study was based on committee discussions, online evaluation, vendor demonstration, review literature, individual contacts, testimonials from existing users and institution, scalability, integration with current infrastructure, and rating comparison. Other studies also showed the same result [4,5].

Figure 3 shows the designer view of a sample course using WebCT.

Figure 3: Designer view of a sample course

Buraydah College Infrastructure

With the current infrastructure the college [6] has all the requirements needed to implement an environment for educational purposes that facilitates the creation of on-line courses for students. From the above discussion, the college has chosen the course tool WebCT for the following reasons: 1) free to download, install, and create courses, 2) low cost comparing to other product, 3) ease of use.

The college has now around 1700 students registered in 9 different specialties. The college will focus in the first stage on developing on-line courses for the Computer Technology students (20 students). During this study we will provide all the necessary teaching support for students and follow the progress of student learning through assessments and student progress tracking provided by WebCT. The results are expected to improve study effectiveness.

The new Course model format

To improve study effectiveness, the following course model format has been adopted beside the text book:
- The introduction of interactive visualisation tools for each course. CBT courses from Smart Force is one of example.
- The creation of Web-based study support system where the student can find all information needed such as delivering study material and communiation tools necessary for exchanging electronic messages such as E-mail, chating and discussion groups.
- Offering interactive progress to master the course contents. This include self-test, exercices and quizzes.
- Offering simulation packages to help student solving problems and finding solutions to their specific case.

Conclusion and Discussion

A Web-based study support environment was designed and developped based on the proposed model. The benefits expected are important as far as student motivation during the entire course and final marks improvement. The development cost for the new course model requires a lot of initial investment in both labor and money.

The use of WebCT Tools has given teachers a powerfull mean to control the course as well as tracking students over the entire course.

One of the main concern in the next study will focus on the problem associated with the arabic support with the WebCT environment.

The knowledge and technical experience gained from this project are expected to help educational institutions plan and restructure their programs of training in the region.

References

[1] Mary V. Connory: "The role of the Web in Enhancing Student Learning", 1999, ASCUE Proceedings.
[2] Sunil Hazari, Ed.D.: Evaluation and Selection of Web Course Management Tools, University of Maryland, College Park, 1998.
[3] Linda Harashim: A framework for Online Learning: The Virtual-U, IEEE Computer, Sept 1999, p44.
[4] Murray W. Goldberg, Editor: Badrul H Khan, Publisher: Educational Technologies Press. July, 1996.
[5] Carolyn Gard, Greg Ashley: Web Course Tools (WebCT) at the University of Goergia, 1998.
[6] A. Arbaoui, D. Hinz: "A Low Cost Intranet Solution for Educational purpose", Internet Gulf98 International Conference, June 1998.

The University of Bristol DataHub - a prerequisite for an integrated learning environment?

Paul Browning & Grainne Conole
University of Bristol
paul.browning@bristol.ac.uk, g.conole@bristol.ac.uk

Abstract

Virtual or Managed Learning Environments (VMLEs) cannot be fully effective unless well-integrated with existing campus information systems.

Institutional planning for the UK Research Assessment Exercise (RAE) has provided an unexpected impetus at the University of Bristol. The DataHub, a reporting database created initially to back-end a locally developed Web application (IRIS) to support the RAE submission, has been extended to include undergraduate and curriculum information drawn from the corporate student information system.

The same authentication framework used by IRIS has also been used in developing a faculty-based staff and student intranet which is itself underpinned by information drawn from the DataHub. The university now wishes to develop this model further to provide a generalised learning and teaching portal that could be offered to all academic departments.

1. Problems

The problem of hardened information siloes is familiar to large organisations and universities are no exception. A student's term-time address is recorded in at least six places but no-one can be sure which one, if any, is right. The student, who knows the answer, can't see the information, let alone correct it. The "Joined-Up Web" was the theme of a recent conference and nowhere is the joining-up question of more relevance than in the context of VMLEs.

2. Incentives

Whilst national initiatives (such as Information Strategies and the Teaching Quality Enhancement Fund) have acted to raise the profile of information and communications technology (ICT) and learning and teaching strategies, they are nothing compared to the effect of the UK's Research Assessment Exercise (RAE).

With the logistical nightmare of the 1996 exercise still fresh in the institutional mind, and serious doubts about the quality of the software being offered to universities to collate information for the 2001 exercise, it proved possible to secure internal resources to build an Integrated Research Information System (IRIS).

The short-term goal of IRIS was to help deliver a successful RAE return in 2001; the longer term goal is to provide Web access to information on our research and enterprise activities.

3. Solution

The IRIS project involved building a Web-based application (using Java Server Pages) that would allow academic departments to view and, where appropriate, manipulate the information that would go to provide their RAE submission.

An important part of the project was the construction of a reporting database (Figure 1) to hold a partial snapshot of the relevant parts of our personnel, student and finance systems which could be regularly refreshed. Departmental users could then add local value in the shape of publications and free-text statements.

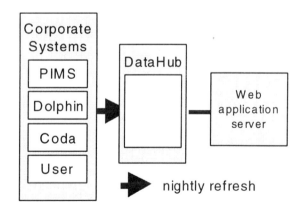

Figure 1: The DataHub. PIMS = personnel, Dolphin = student, Coda = finance, User = user registration.

Users of IRIS authenticate (via Samba) against the existing campus-wide NT ("UOB") domain which controls access to network services for 4000+ staff and 12000+ students - crucially, in terms of administrative overheads, no new usernames and passwords were required.

On logging in a role is assigned to each user which then determines what information can be seen and the operations that can be performed on it. The IRIS project has therefore provided a Web-based "research portal" for the university.

Importantly IRIS exposed data held in the existing corporate systems that had been rarely seen and, unsuprisingly, was therefore out of date. A feedback loop was consequentially established and the quality of the information in the back-end systems has been improved.

4. Plans

It has been a relatively small step to extend the range of information being stored in the reporting database beyond that required for the purposes of the RAE. For example, selected information about undergraduates as well as postgraduate students is now exported nightly, as is curriculum information.

In this way the IRIS reporting database has evolved into the "University DataHub" - a general vehicle for re-distributing corporate information around the institution. The DataHub venture draws inspiration from the MIT Data Warehouse project.

Figure 2 summarises the initial subset of data that is being considered - well stuctured ("catalogue") information in the category of "read-by-all" at the university.

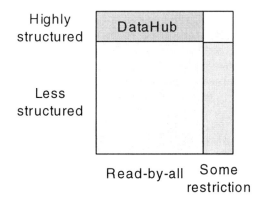

Figure 2: The nature of institutional information.

We also view a Content Management System (CMS) as having an important role to play in promoting less well structured ("narrative") information into the structured domain so that it can be better managed.

The availability of DataHub information has allowed an intranet to be constructed for the Faculty of Law. The use of Zope has permitted the rapid development of a modest but effective integrated learning environment. This provides a prototype for a future "learning & teaching portal" that is complementary to the "research & enterprise" portal being delivered by IRIS.

Users of the Law Intranet also authenticate against the UOB domain and then, according to their role (student, staff or portal manager), are assigned appropriate levels of access.

At present the mode of use is prosaic but labour saving - the Law Intranet is being used to provide an alternative way to distribute information that previously arrived on paper. Class lists are now dynamic, being built from the DataHub, but departmental information (such as tutorial groups) is maintained locally (but still resides in a centrally-hosted database).

A news and events section (with automatic switch on and expiry of items) and bulletin boards (available on a course by course basis) are offering the Faculty of Law an opportunity to experiment with novel (for the staff concerned) approaches to learning and teaching.

5. Conclusions

This paper has outlined an opportunist approach to building the information architecture required if integrated learning environments are to be realised.

It is inevitable that for the present integrated learning environments will be delivered as Web portals. Such portals will be content-starved (or at least prohibitively resource intensive to maintain) unless they can can be underpinned by data currently held in corporate student information systems. Mechanisms such as the DataHub provide a vehicle for exposing such information to a variety of end-uses.

Commercial-off-the-shelf VLEs may be attractive in terms if presentation but expensive or hard to integrate with extant campus systems. The Faculty of Law experience suggests that using open source toolkits such as Zope then 75% of a well-integrated VMLE may be built for 25% of the cost of a commerical offering.

As institutions struggle to manage both their structured and less structured information assets better, it may be that a generic CMS solution is a better investment that a specific VLE.

Displaying Mental Powers: The Case of VirtualMente

María del Carmen Malbrán and Claudia Mariela Villar
National Scientific and Technical Research Council (CONICET)
National University of La Plata (UNLP). Argentina
malbranserdarevich@infovia.com.ar, cvillar@netverk.com.ar

Abstract

VirtualMente is a web based scenario for displaying cognitive processes in university graduates. The name VirtualMente was coined combining two words virtual and mente (mind). The name stresses its virtual nature, a web scenario targeted at cognitive competences. It is based on selected parts of the Triarchic Theory of Human Intelligence [1]. It focuses on metacomponents and insight processes. The structure consists of three dimensions: minds in action (problem solving, illustrations, self- assessment and collaborative glossary); minds in communication (horizontal interaction between participants through e-mail, chats and forums) and aids for the mind (suggestions for searching and selecting information and frequently asked questions). It includes materials coming from the local context and culture (songs, icons, words, cartoons) proposed by designers and participants. It promotes a collaborative atmosphere and demands continuous action from the participants. Content validity controls involve face validity and ecological validity. Procedures are content analysis and expert judgement. Available data come from a graduate level course on cognitive processes developed at the University of La Plata, Argentina. A demo version of VirtualMente and some data will be presented.

"Introduction"

The development of high level cognitive abilities such as critical thinking, reasoning, and problem solving are considered to be important aspects of university education. To meet this goal, an online graduate level course on cognitive processing was developed and offered in Spanish at the National University of La Plata, Argentina. The use of the Web is rather frequent in our University context for searching and exchanging information. Its use as a mediator for displaying cognitive processes is much less frequent. Strictly speaking, VirtualMente is not a course for teaching how to learn or how to think. It assumes that users have previously developed the implied cognitive abilities. It intends to offer a chance to set cognitive processes in motion, both individually and cooperatively. It tries to build a relatively spontaneous atmosphere to foster personal initiative and interaction.

The main features of VirtualMente include the following:

- the conceptual framework is based on the Triarchic Theory of Human Intelligence developed by R. J. Sternberg [2] [3]. The Triarchic Theory draws on the paradigm of human intelligence as a complex way of processing information, which is particularly pertinent : a virtual environment designed for putting into practice cognitive processes (mainly metacomponents and insights). Sternberg [2] states that if you are interested in fostering cognitive processes in others, you have to previously develop your own. VirtualMente is an attempt to crystallize this principle. Mental self-government and autonomy are reflected in VirtualMente through the emphasis on the participants´ activity.

- the program is a case of computer-mediated communication where problems are posed and solved in a virtual setting. The discussion list and the communication via e-mail provide a chance for horizontal and continued interaction. Participants ´contributions are judged on the basis of not strictly academic parameters. Criteria used to assess the contributions are : frequency, by the number of contributions and the number of contributions referred to by others; quality, by relevance, novelty, clarity and coherence; character of the contribution, by discussion, opinions, self-assessment, questioning, reformulating peer contributions, formulating new ideas and soundness, by quality of the sources and documentation. The forum allows more experienced peers to support others and may lessen eventual tensions produced during the performance of cognitive tasks.

- the language used is Spanish. Its constitutes an advantage over other Webs programs which require fluency in English.

- some materials come from local traditions and modes. The data coming from the graduate level course shows that the challenge of finding elements from social experience and the local culture may have motivational power in computer-mediated communication.

- as the program demands some knowledge of the Triarchic Theory, an online assessment instrument called Monitor Triarchic Test (MTT) was developed. Completing the online test and deciding the time of taking is optional.

"VirtualMente Structure"

The structure of VirtualMente includes three dimensions: minds in action, minds in communication and aids for the mind. Minds in action contains: problem solving, examples and illustrations, self-assessment, collaborative glossary, and other contributions from the users. Minds in communication implies: horizontal interaction between the participants through e-mail, chats and forums (discussion list). Aids for the mind gives: cues for navigating, suggestions for searching and selecting information and frequently asked questions.

The program is flexible, the screens are not arranged in a linear sequence; interactive, allowing participants to enter data or commands, and minimalist, controlling the amount of the information given.

It was considered desirable that the icon selected to give identity to VirtualMente should be culturally relevant. After an extensive search in magazines, comics, and advertisements, it was decided to adapt the icon used to advertise a popular analgesic called Geniol. The original icon consisted of a big head with pins and nails that symbolized headache. Only the head was kept. A survey of 100 university students on the semantic charge of words associated with "thinking" was done. The data indicated the Spanish word "mente" (mind) as the best equivalent of thinking. This link supported the decision to include "mente" in the coined name for the program. An interesting finding was that a colloquial word "mate", was located on the top of the ranking. "Mate" refers to a typical beverage of Argentina and other South American countries. Saying that someone has "mate" means that he/she is clever or witty.

Due to the cultural significance of the "mate", its use as the icon of the program is currently considered .The multimedia environment of VirtualMente includes two extracts of modern tango composed by a leading Argentine musician, Astor Piazolla. This composer is considered as a reformer of tango Music. He introduced a different way of conceiving tango. It is hoped that this special kind of tango may encourage people to think in a non routine way. Whenever VirtualMente is presented, people's first reaction show astonishment to hear tango in this scenario. The Music tries to focus attention at the onset of VirtualMente creating a climate with a local taste.

"Validity Controls"

A pilot study of VirtualMente was conducted using a sample of 15 university teachers and researchers from different disciplines. They were enrolled in studies of Computing Technology Applied to Education (Faculty of Computer Sciences. National University of La Plata). Content analysis was performed to determine the participant´s reactions to the Triarchic Theory. It was considered relevant for inducing and displaying cognitive processes. Also as a guide to select material for illustrating different kinds of intelligence. The metacomponents are seen useful as strategies for navigating through VirtualMente. As it is previously said, an online assessment instrument on the Triarchic Theory was developed (MTT). Comments of the participants indicate that assigning responsibility for discussing the content and proposing changes, increase interest in the online format. Moreover, changes in presentation, ways of going through the test, structure and content, will be taking into consideration for preparing the complete online version bearing in mind the answering styles of the audience. Participants advised to avoid a linear, rigid sequence. A free transit through the test is considered more coherent with the nature of VirtualMente. They added that the beep following the wrong choice may affect the careful consideration of the text. Related to the appropriateness of the icon, participants said that the Geniol head, in spite of not having pins or nails, is usually linked with headache. Therefore, this association may be transferred to VirtualMente. It was suggested that another icon to represent VirtualMente has to be considered.

"Final Remarks"

It seems that it is possible to include cultural materials into a course about human intelligence and cognitive processing through a virtual environment. In this case words, icons, habits, music and comics are used to put mental powers into action. From the standpoint of the Triarchic Theory the contributions made by the participants showed analytical and creative competences (metacomponents and insights). Metacomponential abilities were displayed in the application of constructs and principles of the Triarchic Theory to the specific purpose of the program, in monitoring self-progress and in making profit from the Internet resources. Creative abilities were shown in the searching, selection and adaptation of cultural materials relevant to VirtualMente.

"References"

[1] R. Sternberg. *Beyond IQ. A triarchic theory of human intelligence,* Cambridge University Press, USA, 1985.

[2] R. Sternberg, *Intelligence Applied. Understanding and increasing your intelectual skills.* Harcourt Brace Jovanovich, USA, 1986.

[3] R. Sternberg, *Successful Intelligence. How practical and creative intelligence determine success in life.* Simon & Shuster, USA, 1996.

[4] R Denning & Philip, Smith. *A case study in the development of an interactive learning environment to teach problem-solving skills.* Journal of Interactive Learning Research.9(1),,pp3-36.1998.

An Evolutionary Distribution System for Web-Based Teaching Materials

Takashi Ishikawa, Hiroshi Matsuda, and Hiroshi Takase
Nippon Institute of Technology
{tisikawa, hiroshi, takase}@nit.ac.jp

Abstract

The paper describes an evolutionary distribution system for web-based teaching materials that can sophisticate teaching materials stored in the system spontaneously. The system consists of web server, assisting tools and repository of teaching materials represented as HTML/XML data. The layered structure of the information model of teaching materials enables to treat content of teaching materials as scenarios, subjects and elements independently. For this purpose XML language is used to represent teaching materials. The proposed system is being built at Nippon Institute of Technology to evaluate its usefulness to open for the Internet within the year 2001.

1. Introduction

Web-based teaching (WBT) is spreading to wide range of schools through increasing of personal computers connected to the Internet in the schools (ex. [1]). The state of art in WBT is using web pages as teaching materials created with HTML by teacher's self. The excellent feature of WBT is its interactive property to bring up active learning of students.

The paper will describe a distribution system for WBT materials, of which evolution mechanism can sophisticate *spontaneously* WBT materials stored in the system.

2. Evolutionary distribution system

The objective of the research is to promote WBT especially for teachers with less training of computer-based teaching. To attain this objective, we have set up two goals:
  (1) to distribute WBT materials freely via the Internet
  (2) to provide assisting tools for modifying WBT materials by less trained teachers.
In this research, the system not only distributes WBT materials but also sophisticates the materials spontaneously by itself. A key element of the system is XML representation that encodes features of teaching materials like genes of lives.

The proposed distribution system will be built based on an architecture comprised of a web server and repository of teaching materials (Fig. 1). The web server provides web pages with links to teaching materials and assisting tools via servlet and XML technology. The repository stores teaching materials as HTML and XML data to be accessed by any web browsing software via the Internet.

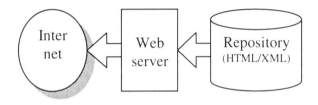

Figure 1. Architecture of the system

3. Evolution mechanism

The idea of the evolutionary distribution system is rooted in evolution mechanism of biological system. Darwin's theory of natural selection assert that "*genes* that fit to the environment would survive", where genes means a suite of DNA representing features of lives. Genes are mixed by reproduction of lives conveying the genes (so called "crossover"), or are changed by mutation. The digital feature of genes enables these mixing and changing of genes in biological evolution process (Fig. 2).

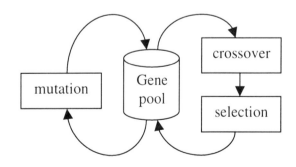

Figure 2. Evolution process

4. Information model of WBT materials

Preliminary observation for existing WBT materials on the Internet reveals that three types of information exist:
(1) *scenario* information describing lesson plans, teaching strategy, steps of presentation, and so on,
(2) *subject* information describing content of lessons, fundamental knowledge, and so on,
(3) *element* information specifying types of data such as texts, images, scripts, and so on.

The need to reuse WBT materials on the Internet requests the three types of information can be treated independently. This requirement leads to the layered structure of information model of WBT materials.

5. XML representation

XML (*the eXtensible Markup Language*) has emerged as a standard for data representation and exchange on the Internet [3]. The basic elements of XML are *tags* and *relationship*:
(1) tags on data element identify the meaning of the data,
(2) relationship between data elements is provided via simple nesting and references.

The following XML data shows an example teaching material for a mathematics lesson with layered structure.

```
<?xml version="1.0" encoding="UTF16"?>
<SCENARIO item="Lesson plan">
    <GOAL> To understand that fundamental
           figures are sets of points
    </GOAL>
    <STRATEGY> Solve an every problem by
           applying the procedure of drawing
           figures
    </STRATEGY>
</SCENARIO>
<SCENARIO item="Problem 1">
    <SUBJECT item=
       "equi-distance points from two points">
        <ELEMENT text="There are ...">
        </ELEMENT>
        <ELEMENT image="image1.gif">
        </ELEMENT>
    </SUBJECT>
    ...
</SCENARIO>
```

6. Discussion

The outstanding features of the proposed system are summarized as follows:

(1) allowing less trained teachers to use WBT materials by tailoring to fit for their own lesson plans
(2) spontaneous sophistication of WBT materials in the repository by reusing materials as the evolution mechanism.

7. Related work

Using XML to create and use WBT materials is getting familiar in recent few years (ex. [4]). But most of those activities are merely language change from HTML to XML to apply the extensibility of XML to WBT materials. On the contrary, the proposed system in this paper enables not only to create and use WBT materials but also to sophisticate in spontaneous way.

The search tool for WBT materials is proposed in several papers in Japan (ex. [5]). These systems use information filtering mechanism with user profiles and evaluation by users. The filtering mechanism is fundamental in search tools for WBT materials, so the proposed system in this paper use more sophisticated filtering mechanism such as *collaborative filtering* [6] than the previous systems.

8. Conclusion

The paper has described the design and implementation of an evolutionary distribution system for WBT materials. The system allows less trained teachers to use WBT materials that match for their own lesson plans. Furthermore the system enables to sophisticate WBT materials in the repository by the evolution mechanism of the system. The system is implementing with XML and agent technology to open for teachers in Japan within the year 2001.

References

[1] http://www.webct.com/
[2] Caglayan, A. and Harrison, C., *Agent Sourcebook*, Wiley Computer Publishing, 1997.
[3] http://www.w3.org/
[4] Kasai, S., "Creation and Use of Multimedia Teaching Materials by using XML" (in Japanese), *In Proceedings of 6th JCET*, Naruto, Oct. 7-9, 2000, pp. 351-354.
[5] Fukami, K., Ochi, Y., Yano, Y., and Wakita, R., "Educational-Resources Sharing System on Web-Customizing Environment" (in Japanese), *In Proceedings of 6th JCET*, Naruto, Oct. 7-9, 2000, pp. 355-358.
[6] Shardanand, U. and Maes, P., "Social Information Filtering Algorithms for Automating 'Word of Mouth'", *In Proceedings of CHI-95*, ACM Press, 1995.

ACME Project, Internet-based Systems that Advocate Academic Credit for Military Experience and Analyze Options For Veterans in Career Transition

Denise A. Wenger, Ph.D.
Educational Consultant, ACME Project; President, Instructional Materials Depot, Inc.
Pewaukee, WI, USA   email: imdepot@execpc.com

Mike Rufflo, M.S.
Instructional Technology Consultant, ACME Project; President, Metavue Corporation, Inc.
Sun Prairie, WI, USA   email: rufflo@metavue.com

F. Joe Bertalan
Director, ACME Project, *Supervisor, Employment and Training Section*
Wisconsin Department of Veterans Affairs (WDVA), Madison, WI
email: joe.bertalan@dva.state.wi.us

Abstract

The ACME Project provides Internet-based systems that promote equity in the awarding of Academic Credit for Military Experience (ACME) and training in college and technical college programs in Wisconsin, USA. ACME Project analysis systems provide program credit and cost analyses for academic programs on a college-by-college basis. A unique Databank of Equivalencies documents credit awards for military courses using the military ACE Guide as a standard. All branches of the United States military are served by these systems. This paper focuses on Phase I piloting results and implications of systems utilization for solving transcript analysis problems for college admissions officers, military personnel, and active-duty military personnel and veterans in career transition in Wisconsin and elsewhere. A demonstration of Internet-based system functions is included in the Poster Presentation.

1. Introduction

The ACME Project answers problems related to delays in training teachers for the Troops to Teachers Program (TTT) and advocates for equity and increased academic credit awards for military experience (ACME) and training that is equivalent to college coursework at private and public colleges in Wisconsin and elsewhere.

The project anticipated development of a dynamic, user-friendly, Internet-based system with information, analysis, and record keeping functions. This Internet-based system would make ACME services available to veterans around the world. It would work as a central record-keeping system, a two-way communications system, and a tracking system for measuring progress of veterans in retraining from distant locations. Project plans proposed development of proactive systems to help military personnel achieve maximum academic credit for military experience (ACME), thereby reducing time and costs for achieving teacher certification or completion of college programs. Statistics would compute costs and benefits of programs where ACME is awarded, record completion rates for candidates, report on years of service, and support a 5-year longitudinal study of ACME awards (reporting on retraining success, service benefits, and tuition reimbursement savings to taxpayers). Databanks and systems developed during the project would serve as models across the nation. The project would establish the ACE Guide (American College of Education Guide for all military courses, in online format, ACEnet) as a standard reference for building course equivalencies between military and college courses. ACME systems would serve as ÒvirtualÓ personnel and would perform credit and cost analysis not previously available to veterans, running 24-hours a day, 365 days a year.

2. Design Strategy and Project Goals

The ACME Project focused on creating internet-based systems that respond to real, field-based needs of Troops to Teachers and WDVA personnel, veterans, and college admissions personnel, including registrars. To develop understanding of complex issues related to the project, the ACME team conducted field interviews and surveys.

The ACME Project focused on solving three real problems involved with career transition. First, a lack of Academic Credit for Military Experience (ACME) often means that few veterans enter professions where extensive retraining is required for licensure. Because no history of awards exists, ACME provided to one veteran is not always awarded with ÒequityÓ to others. A lack of credit awards means the cost of retraining is too high for many veterans who served as instructors at the service academies and other professionals, such as computer and medical technicians, military police, etc.

Second, because no history of awards exists, the possibility of ACME awards is unknown to veterans.

Third, the absence of a record keeping system means that benefits and values of ACME awards to a retraining program go unassessed and unreported.

3. The Work Plan

The ACME team conducted field surveys and interviews to investigate service needs and to develop technical system design and content requirements. They used qualitative methods to conduct structured interviews and quantitative survey methods to find solution paths to problems listed above.

The team mailed an ACME survey to 2100 Troops to Teachers (TTT) Program participants across the nation. Surveys returned by over 1500 veterans involved in TTT training programs provided a profile of veteranÕs needs.

The team interviewed college admissions personnel, deans of academic affairs, credit decision makers, registrars, and college veteransÕ representatives who serve as credit evaluators in the 61 colleges and technical colleges in Wisconsin. They conducted interviews with career transition program administrators for the Military Career Transition Program (MCTP in Norfolk, VA), deans, and registrars at Old Dominion University (Norfolk, VA) and military credit analysts at Thomas Edison State College (Trenton, NJ). These interviews provided behind-the-scenes views of how transferable credit and advanced standing issues and problems are addressed in colleges and universities across the nation.

4. Designing Internet-based Systems

Research indicated that users and stakeholders would benefit from the design and development of specific data management and analysis systems. First, there was need for an account and transcript management system that veterans could use in preparing multiple applications.

Second, there was need for a credit tracking program for an individual with an account in the system. This program would need to track credits granted over time. It needed to track ACE-based equivalency credit awards, CLEP, DANTES, AP exam, test out credit awards, and others types of credit awards, including portfolio awards.

Third, there was need for a transcript analysis program that could compare schools in terms of what credit may be granted to the applicant, given an individual transcript of awards and the course requirements for a selected program. During credit analysis, the program would identify how much credit might be applied against the graduation requirement and how much credit might be awarded for a specific program. Note: The ACME systems do not guarantee credit awards for any college program. They report on the history of awards entered by college and technical college registrars.

To support these functional systems, four information management systems were developed. The Curriculum Catalog system provides a description of college programs on a program-by-program basis. The End-User Information system manages information about the end user, including demographic and security information. The Transcript Subsystem supports the recording of ACE transcripts for military personnel, including contents of the ACE transcript and a record of granted continuing education credits. The Course Equivalencies system contains information on equivalencies formalized between ACE and college courses. This system is used by registrars and serves as the heart of the historical system of awards called the Databank of Equivalencies.

5. Piloting ACME Systems

To test ACME systems, two focus groups (18 veterans from the University of Wisconsin, Madison, WI) and 49 veterans (from WDVA Online databases) were recruited to Òtest driveÓ ACME systems. The focus groups asked veterans to use ACME systems, then rate the experience, assess how systems worked, and judge the value of ACME systems for themselves and others.

Results indicate that ACME systems all worked with excellence. Veterans gave the highest ratings to system designs and reported that all systems worked very well. They rated the ACME system as having high value to themselves and others.

6. Implications of System Utilization

The development of the Databank of Equivalencies, using the ACE Guide as a standard, produces new transfer credit maps for colleges in Wisconsin (and across the nation) and offers new credit analysis services to veterans. For colleges seeking to attract students, these maps represent Òmilitary-friendlyÓ career paths that promote equitable delivery of ACME awards, save taxpayer dollars, and provide a valuable Òleg upÓ for veterans in career transition.

Today, ACME Project databases and information systems are serving as powerful advocates for increasing awards of academic credit for military experience and training in many fields of study. Unexpected applause for ACME systems is coming from college registrars who see how ACME systems can save admissions personnel hours of credit evaluation time and streamline college admissions processes for veterans.

Acknowledgements: The ACME Project was funded by the WDVA, Madison, Wisconsin, USA.

Panels

Panel 1: Agents, Believability, and Embodiment in Advanced Learning Environments

Agents, Believability and Embodiment in Advanced Learning Environments
Introduction to a Panel Discussion

Anton Nijholt

University of Twente, PO Box 217
7500 AE Enschede, the Netherlands
anijholt@cs.utwente.nl

Abstract

On the world-wide-web we see a growing number of general HCI interfaces, interfaces to educational or entertainment systems, interfaces to professional environments, etc., where an animated face, a cartoon character or a human-like virtual agent has the task to assist the user, to engage the user into a conversation or to educate the user. What to say about the effects a human-like agent has on a student's performance? We discuss agents, their intelligence, embodiment and interaction modalities. In particular we introduce viewpoints and questions about roles embodied agents can play in educational environments.

1. Introduction

This short paper is meant to introduce the issues that underlie the introduction of embodied agents in learning environments. Embodied agents appear in different forms. We can just have a simple 2D talking face or a cartoon-like human figure on a web page or in a separate window making suggestions to the user, a desktop virtual reality environment where we have 3D avatars representing tutors or other learners or we can have an immersive Cave-like virtual reality environment where we can really experience interaction with a tutor, with objects and with other learners.

2. Systems, Agents and Intelligence

Before zooming in on some examples of embodied agents in learning environments it is useful to say something about the impact of computer systems in general, the impact of intelligent agent-like systems, the impact of believability, trustworthiness, emotion and personality modeling, and the impact of animated and life-like characters on the behavior of a human user of the interface or system.

Systems as Social Actors: Experiments have shown that when users engage with computer systems they attribute human characteristics to these systems. Not much intelligence has to be included in order to see this effect. Humans engage in social behavior toward computers. Studies and experiments show that users apply politeness norms to computers, they respond to computer personalities in the same way they respond to human personalities, they are susceptible to flattery and they apply gender stereotypes to computers (see e.g. Reeves & Nass [13]).

Intelligent Software Agents: When we really aim at making a system more intelligent, as, for example, in intelligent tutoring systems, we may expect that apart from influencing the social behavior of the student toward the system, we have of course possibilities to steer a student's learning behavior, but also the student's cooperative, or motivational attitude can be influenced. This is even more true when we present the system or the interface to the system as some kind of actor (tutor) that knows, that reasons, that communicates and that displays consistent behavior in its environment. Agent technology is a research field that emerged in the 1990's and that can be considered as a field in which exactly such actors have to be developed, although not necessarily in the context of human-computer interaction. There have been a lot of discussions about what is exactly an agent and is not every computer program an agent. Some researchers explain that the answer is no (see e.g. Franklin & Graesser [4]), other researchers have a pragmatic view: does the agent point of view helps us to develop ideas, helps us to become aware of possibilities and does it help us to communicate ideas. We don't think it is wise to underestimate the value of a good metaphor. Without going into details and especially controversial details, we want to mention properties of software modules that are generally assumed to be present before being allowed to talk about them as agents: autonomy, reactive and proactive behavior and the ability to interact with other agents (or humans). For an agent to act appropriately in a domain it has been useful to have an internal model in which we distinguish beliefs (what the agent regards to be true, this may change in time), desires (the goals the agent has committed himself to) and intentions (short-term plans that it tries to execute).

Interacting Personalities: Software shows itself to the (human) user in the interface. This interface, whatever its form, may aggressively push information, it may try to pull information from a user, it may try to sell, to cheat, to seduce, to persuade, to flatter, etc. We need to mention the notions of believability, trustworthiness and emotions. Believability is an important notion that has been emphasized by Joseph Bates in the early 1990's. An agent is called believable, if some version of a personality shows in the interaction with a human. It does not necessarily mean that the agent is embodied although it is certainly true that in designing believable agents much can be learned from character-based artists that develop animate characters. In (Loyall [8]) requirements for believable agents have been investigated and attempts are given to fulfill these requirements. The main requirements are: personality, emotion, self-motivation, change, social relationships, consistency of expression and, finally, a list of properties that help to create the illusion of life in an agent (reactive and responsive, situatedness, appearance of goals, etc.). Trustworthiness is an other issue. How does a system show its good will and does it build credibility? In a text-based system face-to-face interaction cues (facial expressions, gestures, intonation, posture and gaze) are not available.

Embodied Agents: Now that we have discussed social, intelligent and believable behavior, it is time to consider the role of embodiment. Do we need embodiment to display the previously mentioned kinds of behaviors and when we assume embodiment of an agent, what is the extra impact of this behavior, how does this show in the agent's activities, and not less important, how can we use the embodiment as a multimedia modality to show information (e.g., the sequence of actions to handle complex machinery), to support verbal communication, and to display nonverbal behavior of the agent? Several authors have investigated nonverbal behavior among humans and the role and use of nonverbal behavior to support human-computer interaction. See e.g. (Cassell [2]) for properties and impact of embodied conversational agents (with an emphasis on coherent facial expressions, gestures, intonation, posture and gaze in communication).

3. Agents, Embodiment and Learning

In the previous sections we surveyed developments in computer science (artificial intelligence, agent technology and graphics) that make it possible to talk about software modules and use them in application domains as agents and as embodied agents that can take the form of a 2D or 3D talking face or an animated human-like body. Such agents are finding their way in learning environments. Are they pushed by the technology, are learners – having become accustomed to them in computer games – asking for them or do we have careful considerations about their use and careful experiments that evaluate their effectiveness in learning environments? And when we agree they can be effective, where and how to use them in a continuum between a constructivist and an instructionist approach? How should be their relation with teaching strategies such as tutoring, coaching, cognitive apprenticeship or Socratic dialogue? These questions need to be asked and answered. When we look at the current literature and survey the systems that have been designed and implemented in such a way that they allow experiments, two observations can be made First of all, several impressive research systems employing animated pedagogical agents have been built (see section 4). Secondly, and not surprisingly, we must observe that the abundance of ideas and technological possibilities, the multi-disciplinarity that is required and the lack of resources to have really comprehensive research programs that involve both advanced technology and large-scale empirical study, have not made it possible to give text-book-like decisive answers on how to use animated pedagogical agents, in what situations, and to achieve what goals. Nevertheless, with the observations on the abilities of animated agents in the previous section it is not difficult to predict that researchers will employ these agents in their systems.

Animated pedagogical agents have particular competence. As a real teacher they can show how to manipulate objects, they can demonstrate tasks and they can employ gesture to focus attention. As such they can give more customized advice in a rich learning environment, probably leading to improved problem solving by the student. Lester et al. [7] use the term deictic believability for agents that are situated in a world that they co-inhabit with students and in which they use their knowledge of the world, their relative location and their previous actions to create natural deictic gestures, motions, and utterances. There are more possibilities using animated agents to broaden the bandwidth of tutorial communication. When the agents are sufficiently expressive they can increase the student's enjoyment of the learning experience and the student's motivation. An agent can be designed for emotive believability, showing contextually appropriate facial expressions and expressive movements, not only to support and enhance the communication but also emotion (appreciation, enthusiasm, concern, disagreement, etc.) appropriate to the context. Encouragement, avoiding a student's frustration, conveying enthusiasm and making learning more fun are benefits that are mentioned when discussing the possibility to endow agents with emotive behavior and hence making it an interacting personality. As a result, students may spend more time using the (constructivist) learning environment, but also, as has been reported, there is a positive effect on student's perception of their learning experience. Such animated agents stimulate reflection and self-explanation and have a strong motivational effect. In Moreno et al. [9] a detailed report, including results on retention (recall of factual knowledge), problem-solving transfer (the ability to solve new problems based on similar principles) and motivation and interest, obtained by comparing learning in an animated agent-based environment with learning in a computer-based text environment, can be found.

4. Embodied Agents: Learning Environments

We mention some projects that we think are illustrative for the work on embodied agents in educational environments. We would like to mention the Soar Training Expert for Virtual Environments (STEVE, see Johnson et al. [6]) as an example of an advanced immersive 3-D learning environment with a virtual animated agent. In STEVE an animated, 3D, pedagogical agent gives instruction in procedural tasks in an immersive virtual environment. STEVE is able to demonstrate and explain a sequence of actions, monitor the movements and manipulations of the user, comment on them and suggest possible continuations to complete a task. In the JACOB project [3] a 3D agent walks and grasps objects in a particular order to help students how to solve the problem of the Towers of Hanoi. The student interacts by performing actions as well as by using natural language. The 'Design-A-Plant' project [7] is an interactive learning environment in which Herman the Bug acts as an agent that helps student to learn about plants and their environment. Especially this project has been subject of careful experiments concerning constructivist learning yielding very interesting results. AutoTutor [5,12] is another tutoring system that uses NL dialogues for tutoring. The dialogue is delivered using an animated agent. Intonation and facial expressions of the talking head have been incorporated in order to present affective responses.

5. Conclusions and Discussion

We surveyed developments in computer science (artificial intelligence, agent technology and graphics) that make it possible to talk about software modules and use them in application domains as agents and even embodied agents that can take the form of a 2D or 3D talking face or an animated human-like body. Such agents are finding their way in learning environments. Are they pushed by the technology, are learners – having become accustomed to them in computer games – asking for them or do we have careful considerations about their use and careful experiments that evaluate their effectiveness in learning environments? And when we agree that they can be effective, where and how to use them in a continuum between a constructivist and an instructionist approach?

How should be their relation with teaching strategies such as tutoring, coaching, cognitive apprenticeship or Socratic dialogue? These questions need to be asked and answered. When is it worth the trouble? Human-like agents raise expectations. The learner expects human-like concern, social and competent behavior whatever he or she as learner is doing, etc. Isn't possible to increase the effect of computer-based learning environments without getting involved with creating models of emotion and personality of artificial embodied agents? Enough topics and approaches have been mentioned here to make a fruitful discussion possible.

6. References

[1] A.L. Baylor. Beyond butlers: Intelligent agents as mentors. *Journal of Educational Computing & Research*. To appear.

[2] J. Cassell et al. (eds.). *Embodied Conversational Agents*. MIT Press, Cambridge, 2000.

[3] M. Evers & A. Nijholt. Jacob, an agent for instruction in VR environments. *Education and Information Technologies*, T.A. Mikropoulos & I.D. Selwood (eds.), to appear.

[4] S. Franklin & A. Graesser. Is it an agent, or just a program? 3rd *Intern. Workshop on Agent Theories, Architectures, and Languages*, Springer, Berlin, 1996.

[5] A.C. Graesser et al. AutoTutor: A simulation of a human tutor. *J. of Cognitive Systems Research*, 1: 35-51, 1999.

[6] W. L. Johnson et al. Animated Pedagogical Agents: Face-to-Face Interaction in Interactive Learning Environments. *Intern. J. of Artificial Intelligence in Education* (2000) 11, 47-78.

[7] J.C. Lester et al. Deictic and emotive communication in animated pedagogical agents. In: Cassell et al., 2000.

[8] A.B. Loyall. *Believable Agents: Building interactive Personalities*. CMU-CS-97-123, Carnegie Mellon University.

[9] R. Moreno. Life-like pedagogical agents in constructivist multi-media environments. *EDMEDIA* 2000, 741-746.

[10] A. Nijholt & H. Hondorp. Towards communicating agents and avatars in virtual worlds. Proc. *EUROGRAPHICS 2000*, A. de Sousa & J.C. Torres (eds.), August 2000, Interlaken, 91-95

[11] A. Paiva & C. Martinho. A Cognitive Approach to Affective User Modeling. Proc. *Affect in Interactions*, Siena, 1999.

[12] N. Person et al. The integration of affective responses into AutoTutor. In [11].

13] B. Reeves & C. Nass. *The Media Equation*. New York, Cambridge University Press, 1996.

Teaching with the Help of Talking Heads

Arthur C. Graesser, Xiangen Hu
Department of Psychology, 202 Psychology Building, University of Memphis, Memphis, TN
38152-3230, a-graesser@memphis.edu, xhu@memphis.edu

Natalie Person
Department of Psychology, Rhodes College, 2000 N. Parkway, Memphis, TN 38112,
person@rhodes.edu

Abstract

Talking heads were integrated with two learning systems. In AutoTutor, students learn about computer literacy by holding a conversation with a student. AutoTutor is an animated pedagogical agent that asks deep reasoning questions and engages in a mixed initiative dialog as answers emerge. Students type in information via keyboard whereas AutoTutor delivers discourse sensitive contributions with facial expressions, synthesized speech, and gestures. In the Human Use Regulatory Affairs (HURA) Advisor, high ranking officers in the military learn about the ethical use of human subjects on a web site with a conversational navigational agent.

1. Introduction

Researchers have recently developed computer-generated, animated, talking heads that have facial features synchronized with speech and in some cases appropriate gestures [1, 6]. These agents are pedagogical agents if they are designed to promote learning. Conversational pedagogical agents have the potential to help learning in two fundamental ways. The first is to serve as a conversation partner that conveys subject matter knowledge and that produces substantive utterances in a turn-by-turn conversation. The second is to serve as a navigational aid that points out what actions users might take while they interact with the human-computer interface in pursuit of goals. We have developed systems that illustrate these two uses of conversational agents: AutoTutor and the HURA Advisor.

2. AutoTutor

AutoTutor simulates the discourse patterns and pedagogical strategies of a typical human tutor [2, 5, 9]. The dialog tactics are based on a previous project that dissected 100 hours of naturalistic tutoring sessions [3]. AutoTutor is currently targeted for college students in a introductory computer literacy course, who learn the fundamentals of hardware, operating systems, and the Internet. Evaluations of AutoTutor have shown that the tutoring system improves learning and memory of the lessons by .5 to .6 standard deviation units compared with an experience of rereading a chapter. The AutoTutor architecture is currently being incorporated in a conceptual physics tutor that resides on the web [4].

Instead of being a mere information delivery system, AutoTutor serves as a collaborative scaffold that assists the student in actively constructing knowledge. A dialog manager coordinates the conversation that occurs between a learner and a pedagogical agent, whereas lesson content and world knowledge are represented in a curriculum script and in latent semantic analysis, i.e., a statistical representation in high dimensional space [7]. The tutor presents dialog moves that are both responsive to the student and that facilitate the student in actively constructing knowledge. The tutor dialog move categories include feedback on the student information (positive, negative, neutral), pumps ("What else?"), prompts for specific information ("The primary memories of the CPU are ROM and _____"), hints ("What about the hard disk?"), assertions ("CD ROM is another storage medium."), corrections, and summaries. AutoTutor's dialog moves are delivered by a talking head that synchronizes synthesized speech, facial expressions, and some rudimentary gestures. Microsoft Agent is currently being used as the talking head with synthesized speech, with parameters of the facial expressions and intonation being generated by production rules.

3. HURA Advisor

The HURA Advisor is being developed to help users learn the policies and regulations that are relevant to the use of human subjects in research that is funded by the U.S. Department of Defense [8]. The web site has several modules, including a historical overview, cases, ethical issues, lessons, decision support, mechanisms for querying documents, lessons, a glossary, a library of documents and links to other web sites. However, the most salient components, from the present standpoint, are the talking head and the dialog advancer network (DAN). The talking head guides the user in navigating through the system by giving short messages (typically 2 words or less) that guide the user on what to do next. The navigational agent is particularly helpful for new users of a site who are overloaded with display content and who have difficulty knowing how to achieve their goals by taking effective action. The DAN manages the interaction between the user and the HURA Advisor by considering alternative pathways in a state transition network.

4. Acknowledgements

Research on AutoTutor was supported on grants from the National Science Foundation (SBR 9720314) and the Office of Naval Research (N00014-00-1-0600). The HURA Advisor was supported by the Institute for Defense Analyses (AK-2-1801), under Robert Foster and Dexter Fletcher, on a contract to Thoughtware Corporation.

5. References

[1] J. Cassell, and K.R. Thorisson, "The power of a nod and a glance: Envelope vs. emotional feedback in animated conversational agents", *Applied Artificial Intelligence*, 1999, pp. 519-538.

[2] A.C., Graesser, N.K. Person, D. Harter, and the Tutoring Research Group, "Teaching tactics and dialog in AutoTutor", *International Journal of Artificial Intelligence in Education*, in press.

[3] A.C. Graesser, N.K. Person, and J.P. Magliano, "Collaborative dialog patterns in naturalistic one-on-one tutoring", *Applied Cognitive Psychology*, 1995, pp. 359-387.

[4] A.C. Graesser, K. VanLehn, C. Rose, P. Jordan, and D. Harter, "Intelligent tutoring systems with conversational dialogue", *AI Magazine*, in press.

[5] A.C. Graesser, K. Wiemer-Hastings, P. Wiemer-Hastings, R. Kreuz, and the Tutoring Research Group, "AutoTutor: A simulation of a human tutor", *Journal of Cognitive Systems Research*, 1999, pp. 35-51.

[6] W.L. Johnson, J.W. Rickel, and J.C. Lester, "Animated pedagogical agents: Face-to-face interaction in interactive learning environments", *International Journal of Artificial Intelligence in Education*, 2000, pp. 47-78.

[7] T.K. Landauer, P.W. Foltz, D. Laham, "An introduction to latent semantic analysis", *Discourse Processes*, 1998, pp. 259-284.

[8] N.K. Person, B. Gholson, B., S. Craig, X. Hu, C. Stewart, and A.C. Graesser, "HURA Advisor: An interactive web-based agent that optimizes information retrieval in a multimedia environment", *Proceedings of ED MEDIA 2001*, in press.

[9] N.K. Person, A.C. Graesser, R.J. Kreuz, V. Pomeroy, and the Tutoring Research Group, "Simulating human tutor dialog moves in AutoTutor", *International Journal of Artificial Intelligence in Education*, in press.

Cognitive Requirements for Agent-Based Learning Environments

Amy L. Baylor
Instructional Systems Program
Department of Educational Research
Florida State University
baylor@coe.fsu.edu

Abstract

While there has been a significant amount of research on technical issues regarding the development of agent based learning environments, there is less information regarding cognitive requirements for these environments. The management of control is a prime issue with agent-based computer environments given the relative independence and autonomy of the agent from other system components. I discuss four dimensions of control that should be considered in designing agent-based learning environments. The first dimension of control involves instantiating the instructional purpose of the environment on a constructivist (high learner control) to instructivist (high program/agent control) continuum. The second dimension entails managing feedback, and several issues need to be considered: type, timing, amount, explicitness, and learner control of agent feedback. Third, agent vs learner control is further defined through the desired relationship of the learner to agent(s) (e.g., agent as learning companion, agent as mentor, multiple pedagogical agents, agent as personal assistant, or agent as resource). Fourth, to be instructionally effective, the agent(s) must assert enough control so that the learner develops confidence in the agent(s) in terms of believability, competence, and trust. Overall, an array of possible permutations of system versus learner control must be carefully considered.

1. Introduction

Part of the value of intelligent agents as a computing paradigm is that they act independently from each other and the systems in which they operate; this potential for autonomous agent control and reactivity is also a possible liability if not properly planned. Consequently, a prime cognitive consideration is the management of control within an agent-based learning environment (see [1]). In other words, who has the figurative "ball," the agent(s) or the learner? How much and to what extent and in what capacities? These are important issues to consider because they impact learning and instruction.

2. Determining instructional purpose of environment

The first dimension of control involves instantiating the instructional purpose of the environment on a constructivist (high learner control) to instructivist (high program/agent control) continuum. A critical issue from a constructivist approach to agent-based learning environments is in moderating between the agent taking over thinking for the student with the agent training the student to think more effectively [2]. [3] refers to this as the difference between the effects "of" and "with" technology, with effects "with" technology being more desirable. In this sense, the computer technology in and of itself is of little interest whereas what activities it affords is of interest. Further, in terms of the amount of artificial intelligence that should be used by an agent, it is important to consider that more intelligence is not necessarily better from a pedagogical perspective [4]. Intelligent agents for learning can come in both varieties: constructivist and instructivist. Overall, in the instructivist approach, the role of the agent is to teach the student knowledge, similar to the role taken traditionally by human teachers. In the constructivist approach the agent would be a medium that does not teach the student directly.

3. Managing agent-learner feedback

The second dimension of control entails managing feedback, and several issues need to be considered: type, timing, amount, explicitness, and learner control of agent feedback. An important consideration in terms of feedback is that the pedagogical agent should not provide too many insights and thereby annoy the student. As [5] suggests, the human act of winking can connote a lot of information to others simply in the lack of information. This sort of familiarity is needed for the pedagogical agent to avoid relentless explicitness. To address this issue, part of the pedagogical task should include the monitoring of the timing and implementation of the advisements. With the principle of minimal help as the default, there could also be the possibility for the student to select a feedback option depending on the amount of structure, interaction, and feedback s/he desires when problem-solving. In this way the learner-agent relationship becomes mutually collaborative as each provide feedback for each other.

A related issue is in terms of how active the agents should be to provide explanations of their pedagogical behavior. Assuming that the agents do have some planning role in the instructional environment, does the learner needs understanding of what happened pedagogically and why? One advantage of providing explicit teaching strategy differences from the mentoring agents to the learner (as opposed to being built-in to the system and invisible to the learner) is that it can facilitate reflective thinking [6].

4. Defining the relationship of learner and agent(s)

Agent vs. learner control is further defined through the desired relationship of the learner to agent(s). Some examples of instantiating the learner-agent relationship include the agent as learning companion, agent as mentor, multiple pedagogical agents, agent as personal assistant, or agent as resource.

5. Establishing confidence

To be instructionally effective, the agent(s) must assert enough control so that the learner develops confidence in the agent(s) in terms of believability, competence, and trust. The Guides [7] project (as discussed by [8]) is an anecdotal study that investigated the issue of believability for agent-like computer programs. The project involved the design of an interface to a CD ROM encyclopedia (focusing on early American history) with a set of travel guides, each of which was biased towards a particular type of information (settler woman, Indian, inventor). They found that students tended to assume that the guides, which were presented as stock characters, embodied particular characters. For example, since many of the articles in the encyclopedia were biographies, learners would assume that the first biography suggested by a guide was its own! Students also wondered if they were seeing the article from the guide's point of view (they weren't). Further, they sometimes assumed that guides had specific reasons for suggesting each story and wanted to know what they were (in line with learners' general wish to understand what adaptive functionality is actually doing). Some of the students got emotionally engaged with the guides; one student getting angry that the guide had betrayed her; in another case the guide inadvertently disappeared and the student interpreted this as "...the guide got mad, he disappeared." As [8] comments, while no controlled experiment was involved in these findings, rather these findings are anecdotal, it is hard to believe that the learner would have made such an inference if the suggested articles had been presented in a floating window that had vanished.

6. Conclusion

Overall, an array of possible permutations of system versus learner control must be carefully considered: 1) instantiating the instructional purpose of the environment on a constructivist (high learner control) to instructivist (high program/agent control) continuum; 2) managing feedback (in terms of type, timing, amount, explicitness); 3) defining the desired relationship of the learner to agent(s) (e.g., agent as learning companion, agent as mentor, multiple pedagogical agents, agent as personal assistant, or agent as resource); and 4) establishing learner confidence in the agent(s) in terms of believability, competence, and trust [1].

[1] A. L. Baylor, "Permutations of control: Cognitive guidelines for agent-based learning environments," *Journal of Interactive Learning Research.*, in press.

[2] A. Baylor, "Beyond butlers: Intelligent agents as mentors," *Journal of Educational Computing Research*, vol. 22, pp. 373-382, 2000.

[3] G. Salomon, *Distributed cognitions: Psychological and educational considerations*. Cambridge: Cambridge University Press, 1993.

[4] G. Salomon, D. N. Perkins, and T. Globerson, "Partners in cognition: extending human intelligence with intelligent technlogies.," *Educational Research*, vol. 20, pp. 2-9, 1991.

[5] N. Negroponte, "Agents: From direct manipulation to delegation," in *Software Agents*, J. M. Bradshaw, Ed. Menlo Park, CA: MIT Press, 1997, pp. 57-66.

[6] A. L. Baylor and B. Kozbe, "A personal intelligent mentor for promoting metacognition in solving logic word puzzles.," presented at Workshop "Current Trends and Applications of Artificial Intelligence in Education" at The Fourth World Congress on Expert Systems, Mexico City, Mexico, 1998.

[7] T. Oren, G. Salomon, K. Kreitman, and A. Don, "Guides: Characterizing the Interface.," in *The art of human-computer interface design*, B. Laurel, Ed. Reading, Mass.: Addison Wesley, 1990, pp. 367-381.

[8] T. Erickson, "Designing agents as if people mattered," in *Software Agents*, J. M. Bradshaw, Ed. Menlo Park, CA: MIT Press, 1997, pp. 79-96.

Contributions to Learning in an Agent-Based Multimedia Environment: A Methods-Media Distinction

Roxana Moreno
University of New Mexico
moreno@unm.edu

Abstract

How can we foster the process of knowledge construction in learners using an agent-based multimedia environment? The present paper reviews a set of studies where students learn in various agent-based multimedia environments designed to promote the understanding of an environmental science lesson. The role of media is examined by comparing how students learn from the same microworld delivered via a desktop display, head mounted display without walking, and head mounted display with walking. The role of method is examined by comparing how students learn when the agent communicates with the student via spoken or written words, the agent's image is present or not, and students actively participate in the process of knowledge construction or receive direct instruction.

1. Introduction

The dissemination of highly visible software agents in instructional design brings up the need to investigate the cognitive and motivational effects of agent-based multimedia environments [4]. The central feature of a social agency environment is an animated pedagogical agent--a likable character who talks to the learner and who responds to the learner's input. In the reviewed studies, the major aspects of the social agency environment include: (a) presenting a visual image of the agent, (b) presenting the agent's voice, and (c) allowing the learner to interact with the agent by providing input and receiving feedback. These three aspects are intended to help the learner accept his or her relation with the computer as a social one, in which communicating and participating are as natural as possible within the limits of available technology [4]. In addition, a fourth aspect that might promote learners' sense of presence in the agent-based environment consists of the level of immersion induced by virtual reality technologies.

The purpose of the reviewed set of studies was to manipulate the described features of the social agency environment to determine which of the attributes are most important in the promotion of meaningful learning [4].

In all studies college students learned the relation between the physical characteristics of a plant and its ability to function in various environments by interacting with a multimedia program called "Design-a-Plant" [1]. The program uses a fictional pedagogic agent who offers individualized advice concerning the relation between plant features and environmental features by providing students with feedback on the choices that they make in the process of designing plants. After interacting with the respective computer program, all participants were given a retention test, a problem-solving transfer test, and a program-rating sheet. It was determined for each study, whether the groups differed on measures of retention, transfer, and program-ratings.

2. Studies 1 & 2: The role of agents' visual and auditory presence

The first two studies examined the role of agents' visual and auditory presence. In the first study, the learning of four groups of students was compared: the first group learned by interacting with the image of a fictional agent who gave narrated explanations to them; the second group learned by interacting without the image of a fictional agent who spoke to them; the third group learned by interacting with the image of a fictional agent who gave explanations as on-screen text; and the fourth group learned by interacting without the image of a fictional agent who gave explanations as on-screen text. The second study included identical treatment conditions but the image and voice of the fictional agent were replaced by the video and voice of a human agent.

Past research has demonstrated a modality effect in multimedia learning with animations: Students perform better on tests of retention and problem-solving transfer when words are presented as speech rather than on-screen text [2,3]. These findings are consistent with a dual-processing theory of multimedia learning [3]. Congruent with this theory, it was predicted that students who learned with the voice of a pedagogical agent would outperform students who learned the same verbal materials as on-screen text. Respect to the visual presence of the agent, two alternative hypotheses were tested. First, according to an interference hypothesis, the use of visual characters in a computer lesson may impair learning by adding irrelevant visual materials that compete with the processing of the relevant visual materials of the lesson [4]. Conversely, according to interest theory of learning, the visual presence of the agent may increase students' motivation [4]. More motivated students try harder to make sense of the presented material than less motivated students and are more likely to form a coherent mental model of the to-be-learned system [3,4].

The first two studies demonstrated a modality effect: Students who learned with the voice of the agent recalled more, were better able to use what they have learned to solve new problems, and rated the lesson more favorably than students who learned the same verbal materials as on-screen text. However, for both studies, neither interest theory nor the interference hypotheses were supported.

3. Study 3: The role of students' interaction

The third study tested the hypothesis that one of the main attributes in promoting meaningful learning in an agent-based lesson is students' active interaction and participation [4]. The learning outcomes of students who learned by receiving direct instruction from the agent was compared to those of students who were given the opportunity to participate in the design of plants before receiving the instruction from the pedagogical agent.

The results of the third study gave evidence in favor of using interactive agent-based environments to promote deeper learning. Students who learn by participating in the learning task with a pedagogical agent learn more deeply than students who learn in a non-participating agent-based environment. However, participation did not have an effect on students' program-ratings.

4. Study 4: The role of media

The goal of the fourth study was to examine whether a more immersive agent-based environment would result in different cognitive or motivational outcomes. The performance of students learning with three alternative delivery mediums was compared: a desktop display, a head mounted display where the learner was seated at a computer station, and a head mounted display where the learner could walk. Some students learned with the voice of the agent and some learned by reading the same messages as on-screen text. Two hypotheses were tested.

First, a "media affects learning" hypothesis. According to this approach, immersive environments have the potential of making computer-based learning feel more real by promoting a sense of presence, which in turn can promote deeper learning. Second, a "method affects learning" hypothesis according to which it is not technology per se that promotes learning but rather how technology is used. The findings from the fourth study revealed that students in more immersive conditions felt a higher level of presence, rated the program as being friendlier, and were significantly more eager to interact with the program again. However, despite the difference in motivation, groups did not differ in their learning outcomes. Similar to the first two studies, students who learned with narrated explanations outperformed in tests of retention, transfer and gave higher program-ratings than those who learned with on-screen text explanations.

5. Discussion

The goal of the present review was to pinpoint some of the features of an agent-based environment that contribute to the meaningful learning of a computer lesson. The agency features studied included students' immersion in the environment, students' participation, agents' visual presence, and agents' auditory presence. The results support the use of agent-based learning by demonstrating enhanced recall and transfer when two social cues are present in the interface: the agent's voice and students' participation. Consistent with the classic "method affects learning" hypothesis, it was demonstrated that it is the instructional method (i.e., auditory or visual verbal information) and not the delivery medium (i.e., desktop or virtual reality displays) which impacts how students learn.

6. References

[1] Lester, J. C. and B. A. Stone. "Increasing believability in animated pedagogical agents" *Proceedings of The First International Conference on Autonomous Agents*, ACM Press, New York, NY, 1997, pp. 16-21.

[2] Mayer, R. E. and R. Moreno, "A split-attention effect in multimedia learning: Evidence for dual processing systems in working memory", *Journal of Educational Psychology*, 1998, 90, 312-320.

[3] Moreno, R. and R.E. Mayer, "Cognitive principles of multimedia learning: The role of modality and contiguity", *Journal of Educational Psychology*, 1999, 91, 1-11.

[4] Moreno, R., R. E. Mayer, and J. C. Lester, "Life-like pedagogical agents in constructivist multimedia environments: Cognitive consequences of their interaction", *ED-MEDIA 2000 Proceedings*, AACE Press, Charlottesville, VA, 2000, pp. 741-746.

Panel 2: The Integration of Television and the Internet

Integration of TV and the Internet: Design Implications and Issues

Mohamed Ally, Ph.D.
Associate Professor, Centre for Distance Education
Athabasca University
mohameda@athabascau.ca

Abstract

The integration of the Internet and TV (InternetTV) is a powerful medium for delivery of instruction to learners. However, the instruction has to be designed to cater to the needs and characteristics of the learners. Proper design and analysis strategies must be followed when designing instruction for InternetTV. More research is needed on the integration of the different messages on InternetTV and the most appropriate design and support strategies for delivery.

1. Introduction

In recent years there has been considerable development in telecommunication technology to link businesses and individuals together. These include satellite, ADSL (Asymetric Digital Subscriber Line), and Cable. These communication technologies allow businesses to send data, audio, video, and graphics through the same line. The speed of the communication also allows the communication to be synchronous using videoconferencing, audio conferencing, and interactive chat. Also, the speed of the communication allows the information, video, audio to be transmitted digitally. However, most of the telecommunication is being used by business and industry. Education and training need to conduct research and develop methods for using the merging of the Internet and TV (InternetTV} to delivery instruction to students. This paper will identify how education and training can design and deliver training using InternetTV to teach students and suggest research should be conducted to make the merging of the two technologies more effective.

2. Expertise Required to Develop InternetTV Instructional Materials

A team of experts is required to develop instructional materials for InternetTV delivery. These experts include television production experts, Internet experts, content experts, instructional designers and a project coordinator to coordinate the activities to develop the instructional materials. Figure 1 shows the expertise required to develop InternetTV instructional materials.

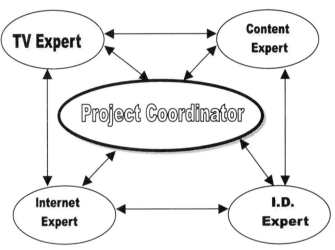

Figure 1: Expertise required to develop InternetTV

3. Design for the Learner

When designing instruction for InternetTV, the learner must be analyzed to determine the characteristics of the learner. Different learners may have different learning styles and preferred modes of learning [1]. Some learners like to perceive and absorb the information in the environment. This can range from concrete experience to abstract conceptualization. Concrete experience relates to one's preference to learn things that have personal meaning in life. The second dimension of perceiving is reflective observation. Students who prefer this style like to take the time to think and reflect on the learning materials. The second component, processing, is related to how one understands and processes the information that is absorbed after perceiving the information. Processing ranges from abstract conceptualization to active experimentation. Learners who have a preference for abstract conceptualization like to learn facts and figures and like to research new information on different topics. Learners who have a preference for active experimentation prefer to apply what they learn to real life situations. They like to try things and learn from their experience.

Learning materials should include activities for the different styles to allow learners to experience all of the activities. Refer to Figure 2 for the dimensions of learning.

Concrete experience learners prefer specific examples in which they can be involved, and they relate to peers and not to people in authority. They like group work and peer feedback and they see the instructor as coach or helper. These learners prefer support methods that allow them to interact with peers and obtain coaching from the instructor. For learners who prefer concrete experience, support mechanisms should be in place to allow these learners to interact with other learners. Interactive video or chat will be an appropriate approach for concrete experience learners.

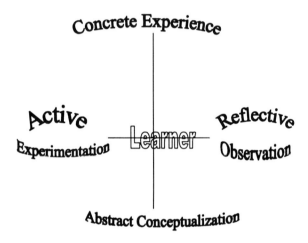

Figure 2: Dimensions of learning

Reflective observation learners like to observe carefully before taking any action. They prefer the lecture method and see the instructor as the expert. They tend to avoid interaction with others. The reflection observation learners like to obtain instruction on demand where they can access the instruction at any time. This could take the form of a recorded lecture, video, or online materials. All of the instruction must be available to the reflective observation learners and they need the time to reflect on the learning materials.

Abstract conceptualization learners like to work with things and symbols and less with people. They like to work with theory and to conduct systematic analysis. In the InternetTV mode, the abstract conceptualization learners like to be presented with the theory and suggestions on how do hands-on activities to apply the theory.

Active experimentation learners prefer to learn by doing practical projects and group discussions. They like active learning methods and prefer to interact with peers for feedback and information. They tend to establish their own criteria to evaluate situation and things. The active experimentation learners need the opportunity to apply what they learn to real life situations.

The challenge for educators is how to develop instruction to meet the needs of different types of learners. If designed properly, InternetTV, perhaps uniquely among delivery methods, can customize the instruction to meet learner needs.

Research is needed to determine the most effective layout of the information when delivering instruction on InternetTV. More research is needed to determine the learning style of students and how to develop learning materials for InternetTV based on the learning style of students. There has to be research on how to sequence the instruction for InternetTV and the instructional strategies to use to promote learning and transfer

In summary, because of the increasing bandwidth for transmission of information, the use of InternetTV to deliver instruction is inevitable. If done properly, this could be an effective medium for teaching students. The interactive nature of InternetTV will allow learners to interact with each other and will present the materials in a high quality format

4. Research and issues to be addressed

- Can everyone learn from InternetTV and how can educators design instruction to meet the needs of all learners?
- What is the role of the instructor when teaching with InternetTV?
- How can we make InternetTV portable so that learners can learn anywhere?
- How can educators convert existing educational materials for InternetTV?
- What strategies that can be used to make InternetTV instruction interactive?
- What are the social implications of using InternetTV for instructional purposes?

5. References

[1] Kolb, D. A. (1984). Experiential learning: Experience as the source of learning and development. Englewood Cliffs, NJ: Prentice-Hall.

Genres, User Attitude and Prospects for Learning through Video on the WWW

Dr. Piet Kommers

Assistant Professor, Faculty of Educational Science and Technology, U. of Twente, Netherlands
kommers@edte.utwente.nl

Abstract

Building upon the notion that TV and video programs will be accessible via the web, there is the question if and how it can bring added value for education. How will these two genres (working on internet versus relaxing on the sofa) go together. From a post-modern point of view, the mass media like TV, contribute to the supposed mechanism of 'social constructionism'. However the program makers have a large control over the underlying message that is articulated, whereas Constructionism for learning is the more conscious process by which students 'build' new knowledge through experimentation and conceptualization.

1. Introduction

The idea of 'one-click' learning following the e-shopping model is not likely to be successful; Learning demands longer-term interactions (or better: negotiations) with teachers and co-students about meanings, hidden mechanisms and strategies. The question is then how interactive video on demand will work out for the more conceptually complex learning domains. Research on the effects of simulation-based studying http://citeseer.nj.nec.com/30314.html reveals that simulation program as a learning tool has limited transfer to the average situation that needs this particular knowledge. Schär, Schluep, Schierz and Krueger demonstrate that the interaction mode is the critical factor in the success/failure of learning systems http://imej.wfu.edu/articles/2000/1/03/index.asp#4. Its lesson for the use of video on the Web is that visualizations and interactions do not guarantee learning.

2. Goal of the Discussion

We might say that learning is exactly the growing capacity to generate what cannot be seen anyway. The continuous need for ever more flexible, quicker and less cannot be explored enough is the perceived discrepancy between entertainment and study that might obstruct users to learn from video in web applications. A classical view expensive media and networks for learning communities brings us to the added value of full video via the www: Bollozos (1997). In order to make a fair prediction in this direction, a number of orientations should be made.

1) The technology to enable and disseminate the needed infrastructure. Just like the early pioneering in educational computing, here we initially rely on the larger forces in the consumer market generating new opportunities for commercials and marketing; The higher its general economical impact: the more any educational side effects will come through: Hollar (1997) and Sagan (1997). .

2) The content providers like publishers, universities, branch organizations (for industrial and on-the-job training) need to manifest a new mutual relation. They realize that content in itself like books, video and even interactive programs do not necessarily bring its recipients to learning. Learning communities as cultural and social groups with a common learning need get more impact. Even universities understand that their future role depends on maintaining their alumni and playing a sustaining role in the life-long learning of former students; the certificate is no longer the single passport for expertise and qualified jobs for life.

3) The attitudes of intellectuals and employees towards information 'services' like broadcast programs and conferences are as modes of 'learning experiences' rather than 'information access'. This trend is in line with the recent development of learning paradigms such as 'collaborative', 'project-oriented', 'constructivistic' and 'cognitive apprenticeship'.

4) The status of 'learning' in itself undergoes an evolution as learning is almost integrated with working, entertainment and leisure time; Citizens who are in the position to travel, specialize in hobbies and cultural activities will spend a large part of their useage to 'change themselves' and to experience immersive emotions as can be seen in reality TV, risky sports and travels through fictitious landscapes created by VR.

It is this more intertwined combination of factors that will decide upon the speed and the amount of penetration of video on the web. One element that

is to see the ultimate learning attitude as one of optimal attention, analytic and highly active. In this respect the "TV meets the Web" would have a detrimental effect on the students' learning if it generates the typical mood of 'divertissement', which might be called 'learning by forgetting'. Interaction as a highly desired property of new media like hypermedia, simulations and VR seems to have an inherent value for those who already realized in the early eighties that receiving information whether on a TV and through a www interface is only a tiny step in the facilitation of meaningful learning. Strom (1997) and Miles (1998). The main question has migrated from "Who has access and who doesn't?" into "Who has interest and who hasn't?" Feinleib (1999).

3. Directions for Solutions

Given the new opportunities for learning, it becomes a key question if and how persons are willing to make learning existential; to what extent do I want to change myself in order to accommodate the wider horizon in tolerance, multiculturalism and participation in ecology (both in the moral, ethnic and aesthetic senses). How "Video on the Web" will finally serve this wider goal is hard to answer. We are just in the stage of early experiments and still highly conceptual. Videoconference via the WWW will bring us a more realistic understanding of how body language extends communication over the highly cryptic email correspondence. The moment that we can flounder through virtual streets, and theatres it will become more clear what can still be offered by normal broadcast video on demand. Such ultimate visuals could provoke similar feelings for conventional media as browsing through a photo or video album of yesterday.

4. Directions for Solutions

The time that project-oriented learning was a relief from one's individual responsibility is far behind; Teachers may start to show the more successful cases of previous years, and the learning group is often seen as an enterprise, competing with other groups. But will students suffer from over ambition and time stress.

Two directions for solutions will be brought forward: a) On-site learning communities should keep a high degree of informality with only a mature amount of competition and sufficient super-ordinate goals. b)If this is not affordable because of an over-stressed grading system, the student should find and relax in external WWW-based communication with students who belong to a different institute.

Though the first solution is inherently better, it might be inevitable to fall back on the second alternative. Here the communication serves a more emotional and personal function like consolation and regaining self-confidence. It is the more overall reflection on the link between studying and life itself that counts, and that is hardly reflected in the institutional context. Besides the direct learning community with students from the same institute, it seems beneficial to join the larger www-based learning communities for the sake of existential support without any threat of repercussions in one's own institute.

An adjacent aspect is the teachers' learning network that should be apart from their employer or supervisor. It should lead to real partnerships between teachers. Here the importance is to have a low-threshold consultancy mechanism for external peer review and sharing experiences. Teachnet and the 21st Century Teachers Network (21CT) are examples. However, these still have a large emphasis on the provision of information. The community aspects still have to be worked out. The student and teacher network may have a lot in common than realised.

5. Conclusion

The integration of TV and the Internet will pose many social questions and pedagogical problems while it rolls out into education. How we answer the former will be critical to how we solve the latter.

6. References

1. Bollozos, s. & Co. (1997). *Webcasting. Push technology.* http://userwww.sfsu.edu/~sbollozo/
2. Feinleib, D. (1999). *The inside story of Interactive TV and Microsoft WebTV for Windows.* CA: Academic Press, San Diego.
3. Hollar, M.J. (1997). Webcasting on the Internet. http://members.aol.com/mikehollar/Webcasting.html
4. Miles, P. (1998). *Guide to Webcasting. The Complete Guide to Broadcasting on the Web.* New York: John Wiley & Sons.
5. Sagan, P. (1997). *Webcasting. Managing all media means mastering new media.* http://www.medialink.com/webcast/foreword.html
6. Strom, D. (1997). *Push media. Examining the latest crop of personalised web publishing techniques.* http://www.webreview.com/96/12/13/feature/index.html

Prefetch Agent: Virtual Internet Based on CATV

Lu Haiming, Lu Zengxiang, Li Yanda
Department of Automation, Tsinghua University, Beijing, China
luhm@webinfo.au.tsinghua.edu.cn

Abstract

Internet provides an interactive mode for information retrieval. But its bandwidth is narrow. TV can provide broadband service. Usually, it is not interactive. In China, the CATV (CAble TeleVision) is very popular and is mainly running in one-way mode. It is expensive to rebuild the CATV from one-way to two-way. Using Prefetch System, we can achieve the virtual Internet and virtual two-way communication through CATV.

1. Introduction

The Prefetch System can provide Internet information through CATV. The information has the format of HTML. All the information resource is packed and delivered through CATV. In the client, the STB (set-top-box) connects the Cable with the user's computer. It can receive the analog signals and translate them to digital signals. Then the user can read the information on his computer. When the user is watching one page, the Prefetch Agent (PA) will save the content of the hyperlinks in this page. The PA plays an important role in the Prefetch System. When the user clicks the hyperlinks, which has been saved by the PA, the corresponding content will be displayed, as if the information is downloaded through Internet. Since all the information is broadcasted repeated. You must save the information when it is being broadcasted. Otherwise, you may get it during the next round. So the PA should save multi-level hyperlinks and satisfy the user's multi-step click during one round. It looks like that the user is browsing the Internet. We call this as Virtual Internet. The Prefetch system is in experimental stage. Figure 1 is the architecture of the system.

2. Character of the PA

The PA runs in the user's personal computer and is a kind of personal information agent [1]. It accompanies you from page to page as you browse the information and highlights hyperlinks that it believes will be of interest.

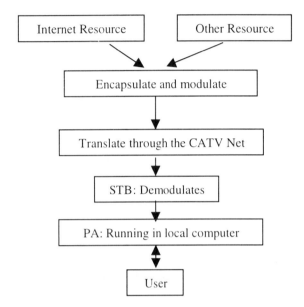

Figure 1. Architecture of Prefetch System

Its strategy for giving advice is learned from feedback of earlier tours. This strategy is inspired by WebWatcher [2].

The CATV is running in the one-way mode. In the Prefetch System, all the information is broadcasted repeated. You must save the information when it is being broadcasted. Otherwise, you may get it during the next round.

The PA has the following characteristics:

(1) Multi-level prefetch with limited scope.
Since there may be too many hyperlinks in a page and such as the next level pages. So the total pages increase with exponential law. The user is only interested in part of them. It can save the storage space by selecting out the interesting pages. Or under the same storage space, we can save more levels.

(2) Prefetch the information-on-demand (IOD) during the next round.
The PA highlights the hyperlinks that have been prefetched. It believes the user will be interested in them.

The user can click these hyperlinks saved by the PA. If the user click the hyperlink not saved by the PA. The PA will remember it and trigger an IOD event. Since all the information is broadcasted repeated. The PA will get it during the next round.

(3) Filter the incoming information and recommend them to the user.

The PA monitors all the incoming information and compares them with the user's profile. When finds the interesting information, it will recommend them to the user. The information can pop-up when user opens the PA's interface, or when user is waiting for the IOD.

(4) Set up and update the user's profile through "watches over the users shoulder".

The PA avoids involving the user in its learning process. It solely records the addresses of pages requested by the user and highlights interesting hyperlinks. In the learning phase (typically during the night), clicked pages and requested pages are analyzed, a model of user interests is generated/updated. This model is used to prefetch information.

3. Model of the PA

We use the Vector Space Model [3] to represent the PA's model. The PA's model is the user's profile. In VSM, both profile and documents are represented using vectors of term weights.

Let p be the user's profile vector, d be the document vector. The terms (words or phrases) are after stop-word elimination and stemming. For Chinese, there are no spaces between words and expressions, so word cutting is necessary. p_i and d_i are the term weights, usually computed by term frequency inverted document frequency (TFIDF).

The keyword vector is in high dimensions which slower the system, so a threshold q is set to simplify the word frequency vector. Each component in p that is smaller than q is omitted. And a keyword vector is formed with a smaller dimension m.

For any information with vector d, if the similarity between p and d is high enough, it will be prefetched and recommended to the user, i.e. the corresponding hyperlink will be highlighted and the corresponding page will be saved.

4. Provide other interactive programs

The Prefetch System provides a broadband platform. We can develop many different applications on this platform.

Distance education: During the video-courseware is being broadcasted, we may have some questions. Many general questions are broadcasted together with the courseware. Since the courseware is being broadcasted, we can not stop the courseware. But we can save the questions/answers and read them when the courseware is over. We can take some examinations. The PA prefetches the answers. Then the user's answers are compared with the prefetched answers. We can get the score.

Interactive sports: We can guess the game's result. The PA saves the guess. When the game is over, the result is broadcasted and the PA can judge whether the guess is right. The system can save the reward in the card, which is plugged in the STB. The user can get his reward by providing his card to the operator or running the special program. The program is developed by the operator. It can extract the reward from the card and send it to the operator through Internet. A problem is that the user can save the past information and repeat them. Then the user can give the answer after the game is over. So the user can cheat the system and always get the award. The timestamp and sequence identification can prevent this problem. We can play the interactive entertainment by the same way.

5. Conclusion

The Prefetch System can provide Internet information through CATV. Using Prefetch System, we can achieve the virtual Internet and virtual two-way communication through CATV. The model of the PA and the corresponding learning algorithm are presented.

The Prefetch System provides a broadband platform. Multimedia information can be translated with high speed. Many interactive applications can be developed under this platform, such as Distance education, Interactive news, and interactive sport. Of course, we should develop the corresponding PA for each new application.

References

[1] H.M. Lu, Z.X. Lu, Y.D. Li, "Notify the Interesting Web Change-Based on Information Agent", *Computer Science*, 27(9), 2000, pp. 75-77

[2] D. Mladenic, "Machine learning used by Personal WebWatcher", *Proceedings of ACAI-99 Workshop on Machine Learning and Intelligent Agents*, Chania, Crete, July 5-16, 1999.

[3] G. Salton, *Automatic Text Processing: The Transformation, Analysis, and Retrieval of Information by Computer*, Addison-Wesley, Reading, Pennsylvania, 1989.

The Integration of Television and the Internet

Chris O'Hagan

*Dean of Learning Development and Professor of Educational Development,
University of Derby, UK*

c.m.ohagan@derby.ac.uk

Abstract

The ongoing development of digital television to provide hundreds of channels, plus the provision of Internet services through domestic televisions with interactivity via the subscribers' telephone line (or very soon, return satellite signal) is but the beginning of a process of integration between television and Internet services that heralds new dimensions to entertainment and e-commerce. The advent of mass home storage technology – personal video recorders (PVRs) and video-on-demand (VOD) - also heralds a related revolution. (See www.durlacher.com, 'Digital Local Storage – PVRs, Home Media Servers and the future of broadcasting'.) In most countries, mass culture is still driven by television rather than the Internet, and thus changes in access to the former remain much more influential. Even if all this only heralds a further dumbing-down of the mass media – **fifty-seven thousand channels and still nothing on!** *(to adapt Bruce Springsteen's ironic lament) – what could it mean for forms of e-learning such as educational broadcasting or online learning? This sequence of six papers reports on some of the leading-edge developments and seeks to provide insights into the impact that the integration of television and video with Internet protocols will have on the provision of educational opportunities, and how educators might prepare themselves and their institutions for the revolution ahead.*

1. Interactive Television

Is there an educational role as opposed to shopping, betting, and game shows? Stephen Skelton believes there is, as he outlines in the second of the papers. However, there remains a real difficulty in that beyond a certain number of active participants, from an educational perspective, the interaction can be almost meaningless. There is perhaps greater potential in narrowcast, rather than broadcast, interactive satellite TV. The University of Derby has demonstrated this, technically and commercially, with its five-studio Telepresence Teaching Centre for Interactive Distance Learning, transmitting in IP via satellite to 22 classrooms located in five different university extension centres. In the first ten weeks of operation it delivered 7,000 classroom hours - about 150,000 hours of student earning. (See end of paper for a definition of telepresence.)

In the classrooms every student has a telephone handset through which they log in, answer questions and converse with the teacher. One example of use will help elucidate the process: a teacher connected to five classrooms puts up a multiple choice question in Powerpoint or under the document camera, which the students read on the large projection screen in each classroom. They input their answers into the handsets and within seconds a pie-chart graphic of them is displayed on the teacher's 'instructional management' screen. She mouse-clicks on one of the segments and the names of every student giving that answer, and their location, are displayed. She clicks a name and the student's telephone rings. Simultaneously, the return video switches to that location and classroom. The student picks up the handset and the teacher now selects the 'talking head' output from her studio to ask him why he gave that particular answer – students in all other classrooms hear his reply.

A student can also electronically 'raise' a hand to flag-up a question to the teacher, but only the teacher can activate the phone to initiate a conversation, at a time of her choosing: the teaching flow is thus not interrupted, and the student knows he has been 'seen'. Sometimes the question is answered before the teacher is ready to take it and the student can 'lower' his hand.

At Derby, we believe this approach constitutes a significant examplar for *real*, synchronous interactive television, and are currently using it to teach over 6,000 registered University of Derby students in Israel, with our partners, Inter College. Some highly distributed companies are using similar systems for in-company training: the Royal Bank of Scotland trains staff in 400 branches around the UK from a single centre in Edinburgh.

The difference between such multipoint *telepresence teaching* and multipoint videoconferencing is the use of sophisticated instructional management computer systems, which allow the teacher to control the level of interactivity in sound, vision and data, regardless of the number of students on line. Satellite point-multipoint delivery provides a reliable, even-quality IP broadband path, regardless of the number of receiving sites. Adrian Vranch, in the third of these papers, outlines another interesting case study of satellite-delivered teaching, in the field of medicine.

Broadcasters are also experimenting with TV/WWW combinations. In the UK, Channel 4 television (a national terrestrial channel, which has just launched a new digital-only channel, E4) is seeking ways of enhancing its educational broadcasting by using the Internet. It has set up an innovative interactive website, Homework High, www.homeworkhigh.com, to support its educational broadcasting.

2. Digital Video Storage

PVRs and VOD will expand the potential for downloading learning packages to the home. Interactivity will be both contained within the package, and through dedicated websites. Thus very high quality resources can be integrated with relatively narrow online bandwidth. Broadband may never be broad enough, and so whatever the case, mass home digital storage will ease the pressure. TiVo and ReplayTV are but the start of this process.

Some argue that the rapidly expanding DVD market can offer an educational solution to high-quality volume storage, more simply and at lower charges. The Ohana Learning company in Hawaii www.ohanalearning.com believes it has a solution applicable to all levels of education – it has combined a DVD player with a Web interface, which can be connected to the Internet via a modem or network point. Ohana anticipates the development of software combining high quality video and graphics with dedicated web sites as well as general web access. The box would replace the VHS players in fixed and rollabout video playback systems used currently in education, providing an attractive tool for classroom, group or individual use. However, the relative failure of the CDROM in education might advise some caution that DVD can succeed in its place; but the Ohana box will retail at around a third to a quarter of the price of a PC – and price is one of the major determinants of the eventual ubiquity of technology, whether domestically or educationally.

3. The 'One-box' Screen

The practices, projects and potentials described so far are all forms of the integration of TV and Internet protocols which are leading towards a fully integrated entertainment/communications environment for the home - with interactivity via telephone lines or return transmission via domestic satellite dishes. What will this mean for the social context of learning - at home, at work, or at an educational centre? Will it accelerate the process of the de-institutionalisation of learning and knowledge generation? What does it mean for our schools and universities? Can the public and private sectors work synergistically to develop a more student-centred use of broadcast and narrowcast media? Or are we facing a step-too-far in the commodification and marketisation of learning?

Gary Chapman, director of the 21st Century Project at the University of Texas, and pundit on new technologies for the Los Angeles Times, has observed, "It's obvious that a battle is shaping up about whether the Internet will quickly become dominated by giant companies that will mimic the programming and advertising models of TV today, or an explosion of creative and diverse content gradually will replace mass-market programming. Whichever model wins will have an immense effect on society for years to come." Already there has been at least one failure to run on the Web scheduled broadcast programming such as on TV. Either the level of integration is insufficient at the moment to support such a model, or it is anathema to the whole philosophy of the Web – and some would argue that the latter is the case because the Internet is a lean-forward technology of active engagement, which is incompatible with the lean-back, mind-switched-off nature of TV. As education is essentially lean-forward, the crucial question is which of the two characteristics will finally dominate the integrated medium, or can it support both, which broadcast TV on its own has largely failed to do, despite examples like the UK and Chinese open universities. Piet Kommers, in the fourth paper, explores some of these issues.

Of course, one barrier to the one-screen concept is the poor quality look of Web pages on a standard television, regardless of the 'box' delivering it, whether Microsoft's new Ultimate TV or AOLTV, for example. Some would argue that this is irrelevant in an educational context, as material can easily be authored for the Web which works well on a conventional TV screen. Meanwhile a company called Ch.1, www.ch1.com, is working with TV set producers to provide high definition TVs which connect directly to the Internet. Haiming Lu and colleagues, in the fifth paper here, describe a research project at Tsinghua University in Beijing to provide a 'virtual' Internet through cable television.

4. Broadband

"Open access broadband is almost here" is a frequently heard refrain, but the evidence for imminent arrival is somewhat thin. There are experiments with the so-called 'next generation' Internet2. The University of Pennsylvania and the University of Grenoble have used this technology for videoconferencing, with insignificant delays in interaction, regardless of distance. However, we do not have to wait for the next generation for this. Dedicated e-lines or satellite bandwidth can provide the same. A London-based company, Global Intercasting, www.globalintercast.com, is offering corporate communications using a "unique fusion of state-of-the-art interactive broadcast television, Internet platforms and delivery systems" and one of its customers, TELERIS: Interactive Telepresence, www.teleris.com, is currently providing boardroom multipoint conferencing technology with extraordinary high levels of telepresence realism (without noticeable signal delays) at 4.5Mbps. But none of this will be cheap enough for ubiquitous use in public education for some time to come.

National university networks around the world are being beefed-up to support high levels of use of full-motion, full-screen video, whether for conferencing or VOD. In the UK the Distributed National Electronic Resource (DNER), www.jisc.ac.uk, is being developed, with a number of trial projects investigating how it can be effectively accessed and used. One such project is Click and Go Video, www.clickandgovideo.ac.uk/, which is investigating and reporting on best practice in developing a video-enriched learning environment through the integration of archived moving images, locally produced video, web resources and asynchronous and synchronous communication tools. The "prime object of this project is to make use of…IP-based conferencing and streaming tools that are available freely for PCs to remove the barriers of cost and platform." Could it be that the specialist commercial hardware and software providers are going to be sidestepped by an education sector able to develop highly cost-effective solutions for itself? There are growing signs such as this one that higher education is capable of developing the integration of video and the Web, and also creating VLEs and MLEs in much more flexible and distinctive ways than the corporations rushing to exploit the speculated multi-billion dollar education market would like. Will the lean-forward democracy of the Web prevail in the end over commercially driven interactive television?

As the Massachusetts Institute of Technology (MIT) moves to place all its course materials on the Web, Mohamed Ally, in the last of these papers, brings us back down to earth by reminding us that, in the end, however intuitive and versatile the technology, or abundant and accessible the content, ensuring good instructional design and faculty training remains a major obstacle to effective pedagogical use. Without effective support for students, the rest is of little use, particularly for the millions who have been unable to access education in the past, and lack the skills of independent learning.

5. Knowledge and Learning

MIT's apparently altruistic act should remind us that knowledge – even education itself – is date-marked. What I know today can become quickly irrelevant: with every day that passes the fraction of what I know divided by what I don't know gets ever smaller, despite my best endeavours. The effect is exacerbated by mass communications, and the Web in particular.

Ironically, in the knowledge-based society the value of knowledge per se is in a sense diminishing. On the other hand the skills of searching out new knowledge and knowing how to use it are ever-increasing in importance. Thus within higher education courses, knowledge could be viewed more as a vehicle for the development of intellectual and communication skills than as an asset in itself. 'Content' is relatively unimportant as such because it is temporally and socially situated, while the skills of learning and using that learning last a lifetime, and are transferable to other contexts.

Are both TV and the Internet so strongly linked to the ephemeral nature of their content dissemination that they will remain largely peripheral to real education, even when integrated? Are we just seeing another phase, which began with radio, in the continuing failure of each new electronic technology to live up to the educational dreams of the proselytisers?

I hope this and the following five papers here will provide some clues to both the future potential and the possible limits of integration, and also provide some provocative ideas for further discussion.

6. References

Chapman G (2001), Will interactive televsion turn into a two-headed monster? Los Angeles Times, March 22.

O'Hagan CM (2001), Telepresence teaching for interactive distance learning – an introduction. Video produced by the Centre for Educational Development and Media, University of Derby, England: approx 15 mins.

Palmer, Desnos, Babanoury, Corbett (2001), Un enseignement sous forme de visioconferences entre Philadelphie et Grenoble, paper at EUNIS 2001 conference, Berlin.

Note: *Telepresence* can be defined as the experience of presence in another environment by means of a communication medium. Someone experiencing telepresence can work and receive stimuli in a similar way to being present at the remote site; and those at the remote site can likewise interact with the telepresent person.

Edutainment – The Integration of Education and Interactive Television

Stephen Skelton
UNITEC, Auckland, New Zealand
skeltand@xtra.co.nz

Abstract

This paper describes the process of an educational institution implementing an interactive television solution. The fundamental issue is the educational role within a TV environment as research continues to show that viewers expect entertainment on television. Edutainment, the successful integration of education into the entertainment environment of television, is not going to be a trivialisation of real education. Instead, the new paradigm of interactivity for education in broadcast TV formats offers sophisticated, personalised, exciting and innovative ways to present traditional academic courses.

1. Achievements in Commercial ITV

In the ever-changing world of commercial TV broadcasting, everyone is looking for gold. Analysts claim that there is a US$20 billion interactive television (ITV) market, that 30 million households in the US will be wired for ITV by 2004 and that 625 million people around the world will have access to online services on their TV sets. The question remains, what will all those people really be doing and will it benefit education?

Consider the current situation. In the UK 97% of UK homes have at least one TV set! What has commercial interactive television achieved in this market? So far over 5 million people on BSkyB UK have access to interactive services. Woolworth's interactive TV site in the UK (on Open Interactive, a company which is currently valued at over US$3 billion) generates over 5000 orders a week and is the 3rd largest store in the UK out of 800 in entertainment sales. The Pantene Study, trialed in 78,000 homes, found that 60,000 went into an interactive ad space and 30,000 worked their way through 10 screens to request a product sample! Interactivity provides an engaging platform for people to connect directly with the broadcasters.

2. Implications for Education

How can this commercial model be applied to education in a way that enhances educational services? In today's business environment universities and other educational institutions are competing for the diverse needs of students worldwide across a wide range of subjects and material. It is commonplace for a student in New Zealand, for example, to study anywhere in the world and new media opportunities like interactive television allow institutions to reach those students directly in their homes wherever they are.

The familiarity of television makes ITV a stronger B2C model, enabling educational institutions to reach critical market segments, such as those who don't have the Internet or are not computer literate. The security problems of the Web are almost eliminated. An Oftel Study reported only 2% of ITV users were concerned with security thus enabling a great relationship with students in terms of safety of their fees. All analysts agree that t-commerce will surpass e-commerce heralding excellent potentials for any commercial transactions such as fees conducted through ITV. "By 2004 ITV will generate US$11 billion in advertising, US$7 billion in t-commerce, US$2 billion in subscriptions" (Forrester Research).

Interactive television enables, most significantly, students, faculty and administrators to interact face-to-face over live sound and video with colleagues anywhere in the world. Many Universities are already using this communication technology, such as Illinois State University, USA. The potential for distribution efficiencies and fantastic export opportunities, without the "clunky" video streaming challenges of the Internet, makes early mover status for educational institutions in this new industry critical.

3. Strategic Model for Education in ITV

Step 1: The Approach

An educational institution's approach to interactive television must cover the following key areas:
(Example using AUSTAR, Australia)
a) Service Creation – the creation of interactive templates, media, advertising, courses and tests (Massive Interactive)

b) Service Platform – an end to end solution for creating, scheduling and operating interactive services between institution/teacher and student over multiple networks (Oracle Interactive Services Solution)

c) Middleware – the software that enables the student to interact with the television, through the set top box, by pushing the remote control (Open TV)

d) Deployed Digital Boxes – the hardware boxes that sit on top of your television set e.g. Motorola

e) Merchandising and Fulfilment – the group responsible for the successful delivery of any services and t-commerce e.g. collection of student fees (TV Shopping Network)

Step 2: The Phases

Any approach to interactive television must evolve in phases as technology allows. The rollout of interactive services would take place in two stages:

Stage 1- available at launch. Digital channels can allow viewers to access back-up text while still watching the video stream at a reduced size. It will be accessed via the viewer's remote control. Some of the text may be enhanced with pictures and could include:

a) Interactive tools designed to help you identify your educational strengths and weaknesses e.g. subject tests. Also meet and chat with other students, opinions/comments on past courses

b) Self-diagnosis tools, text-based back-up information for the institution's subjects and programmes

c) Guides to local resources including courses online, local events

d) Links to a host of other sources, for example to official bodies in education or student clubs

d) Regularly updated news, keeping students in touch with the latest courses

The e-shop would contain courses at launch as well as supportive products including text books etc

Stage 2 - available within 2 to 3 years. Viewers would be able to programme 'smart' STBs to select which programme areas or lectures to record for viewing, and then eventually call up content direct from central servers in the combinations that best suit them. Video on demand (VOD) has significant bandwidth requirements so the costs must be considered carefully. WAP software would be added to the set-top boxes (STBs, for provision of the same services to mobiles).

There are already many educational websites – education is said to be the one of the largest subject areas on the WorldWideWeb – but none of them are integrated with authoritative, impartial, empowering interactive television channels. The total NZ tertiary educational market is worth some $500 million and growing.

Research among consumers and professionals indicates widespread support for an interactive educational channel and associated website.

A key element in the success of edutainment is the explosion of new media and the potential for personalization. Viwers will expect to be able to accss content and information where and when they want – thus 'smart boxes' will download and store programmes which suits the consumers unique identity and user profile.

All content created and broadcast or published will be tagged and stored digitally, ready for on-demand access in the interactive and broadband world, when new revenue streams will come from on-demand, personalized educational services.

There are enormous opportunities for revenue potential for educational institutions. Apart from charging fees for interactive lectures and forums run through interactive television, such as interactive advertisements, sponsorship, infomercials and content sales to other networks.

4. Conclusion

The traditional view of "the place" of education in television as being more static and informational could be replaced by a new paradigm, that of education as a dynamic, entrepreneurial and innovative medium. Channels like Discovery and National Geographic point the way to the possibilities of engaging the viewer in a sophisticated way. The interactive TV department, like the Internet in universities, could become a whole new growth area, providing revenue and resources to students, teachers and administrators alike.

Interactive TV is not an enhancement. It is a fundamental paradigm shift in the evolution of television, how it is used and what opportunities and financial models it will generate. A successful model needs to encompass the advantages of using interactive advertising, innovative multimedia and personalised services in a way that truly makes use of interactivity and the medium of television. The opportunities for educators are enormous…are you going to be there?

5. References

[1] G. O'Driscoll, *The Essential Guide to Digital Set Top Boxes and Interactive TV*, Prentice Hall, New York, 2000

[2] www.opentv.com (industry statistics and research)

Surgeon training via video conferencing and the Internet

A. T. Vranch
Academic Developments Manager, Information and Learning Services, University of Plymouth, UK
a.vranch@plymouth.ac.uk

A. Kingsnorth
Professor of Surgery, Postgraduate Medical School. University of Plymouth, UK
andrew.kingsnorth@phnt.swest.nhs.uk

Abstract

SANTTSUR is a pilot project which was set up to evaluate the benefit of, live, interactive, digital TV broadcasting, combined with video conferencing and access to the Internet, as an integrated distance learning medium for doctors in basic surgical training. A series of thirty 90-minute TV broadcasts has been developed, aimed at doctors in the working in 26 hospitals across the UK and Ireland. Overall, the response from trainees on the general approach was favourable. Looking to the future, it is envisaged that the TV broadcasts will be enhanced by further integration of web resources and e-mail communications, both by enhancing the learning environment and for developing a viable business plan

1. Introduction

SANTTSUR (SAtellite Network: Telematics Training for SURgeons) is a pilot project at the University of Plymouth, in collaboration with the Royal College of Surgeons of England, which was set up to evaluate the benefit of, live, interactive, digital TV broadcasting as a distance learning medium for doctors in basic surgical training. A need for reviewing the methods used in basic surgeon training has been identified in the context of new opportunities arising from developments in learning technologies [1,2]

A series of thirty 90-minute TV broadcasts has been developed, aimed at doctors in the workplace who are preparing for the MRCS (Member of the Royal College of Surgeons) examinations. Broadcast via satellite between October 2000 and June 2001 to 26 centres across the UK and Ireland, SANTTSUR provides a combination of presentations, case studies, expert panel discussions, live surgery, anatomy demonstrations and current affairs to a potential audience of approximately 400 trainee doctors. Interaction is provided using ISDN2 video conferencing, e-mail and telephone for discussion between trainee doctors at the receiving centres and the studio presenters. Live surgery from participating hospitals is incorporated into broadcasts via ISDN6 video conferencing links (Figure 1). A web site provides a means for on-line chat and enables access to presentations and other content from the broadcasts.

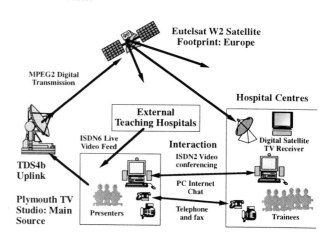

Figure 1. Overview of system architecture [3]

2. Broadcast content and approach

The presenters for the SANTTSUR broadcasts are consultants and specialists in specific areas of surgery related to the module topic covered in the transmission. In order to maintain a consistent, high standard of production, all presenters attend a special "training the presenters" course. The whole day course is run at the University of Plymouth TV studio and covers aspects of studio presentation techniques, practical work in using chromakey, preparation of content for broadcasts and aims of the SANTTSUR project. Broadcasts were transmitted in three sets of ten programmes, during Autumn 2001, Spring 2001 and Summer 2001. All modes of operation proposed in the project plan were

implemented in these Phase 1 broadcasts and these modes provided the various styles of delivery during each transmission, including:
- studio panel discussion;
- live anatomy from The Royal College of Surgeons of England via ISDN6 video conferencing;
- topical input from the editor (and other staff) of Hospital Doctor magazine via ISDN6 video conferencing;
- VT inserts for "Case of the Week";
- ISDN2 video conferencing inserts for " Case of the Week";
- ISDN2 video conferencing for surgeon trainee questions;
- telephone dial-in for surgeon trainee questions;
- e-mail submission of surgeon trainee questions.

In order to streamline communications and encourage feedback, a special e-mail address was set up, studio@plymouth.ac.uk. The mailing list was publicised during the live TV broadcasts so that anyone involved in the project (as a trainee, centre manager or surgical tutor) could send messages to the management team easily.

A web site was set up to provide support for the SANTTSUR series of broadcasts. (http://www.fac.plymouth.ac.uk/studio/index.html) The site provided information on the broadcast schedules, copies of Powerpoint presentations for download, access to a group discussion mailing list and links to other relevant web sites.

3. Evaluation

Feedback from trainees was collected for each of the 30 broadcasts, in collaboration with the Royal College of Surgeons of England. This took the form of a weekly questionnaire on a single sheet of paper, which was distributed at each receiving centre to each trainee. The completed questionnaires provided valuable feedback of the trainees' responses and preferences in terms of programme content, style and quality. These included:
- holding attention/stimulating interest
- thematic coherence
- variety of presentation methods
- pace
- technical quality
- relevance of content

In addition, information on pattern of attendance and suggestions for improvement was also gathered. The weekly nature of this process meant that this feedback could be used to fine-tune the style and content of broadcasts very quickly.

4. Discussion

Overall, the response from trainees on the general approach was favourable. The six detailed content criteria and overall rating (on a scale of 6) showed peaks in the 4 and 5 categories, which was encouraging. Individual programmes showed differences, reflecting the variation in content, subject matter and presenter style. The trainees were very positive about the "mini-lecture" PowerPoint/chromakey style of presentation and stressed the importance of having a good presenter. This confirmed the need for the "training the presenters" courses. One disappointing aspect at the beginning of the series was the lack of interaction with questions from trainees. Although questions were submitted via e-mail, phone and through the web site, the level of activity was lower than expected. It appears that, at the beginning, there was a cultural issue in terms of trainees getting used to the different approach offered by SANTTSUR.

Looking to the future, it is envisaged that the TV broadcasts will be enhanced by further integration of web resources and e-mail communications. It is hoped to encourage trainees to make more extensive use of web sites by featuring reviews and live demonstrations of relevant web sites during the TV programmes, with prior notice so that trainees can explore each site in advance of its review. The development of the web site, to include downloadable video, in parallel with the live TV programme content and approach are seen as key factors for enhancing the learning environment and for developing a viable business plan to ensure that the SANTTSUR programme continues.

5. References

[1] Dawson J. L.. Are basic surgical trainees being "short-changed"? *Annals of the Royal College of Surgeons of England*, London, 1998, 80, 163-165.

[2] Bunch G. A., Bahrami J. and MacDonald R. Basic surgical training: how good is it? *Annals of the Royal College of Surgeons of England* London, 1998, 80, 219-222.

[3] Kingsnorth, A., Vranch, A and Campbell, J. Training for surgeons using digital satellite television and video conferencing. Journal of Telemedicine and Telecare, London, 2000, 6, S1-29-S1-31.

SANTTSUR is partly funded by the European Space Agency under the ARTES 3 Programme.

Panel 3: Collaborative Context-Mediated Experiential Learning through Asynchronous Learning

Collaborative Learning Through Versatile Representations in Asynchronous Learning Transactions via the WWW

Elspeth McKay
Lecturer, School of Business Information Technology
RMIT University, Melbourne, Australia
elspeth@rmit.edu.au

Abstract

Asynchronous learning environments currently support individual training needs, reflecting the limitations of online instructional materials ([1]; [4]; [10]; [18]; [19]; [21]). Contemporary approaches to instructional design are hampered by the failure to recognise and accommodate the cognitive processes necessary for online learning, specifically, the interactive effects between cognitive style and instructional format, and the need to adapt the instructional format dynamically. It may be concluded that a mechanism to achieve such dynamics lies in the concurrent acquisition of the meta-knowledge about the learner's cognitive performance within a contextual framework defined by a knowledge-level-analysis of task difficulty [9]).

1. Discussion Context

The topic chosen for this panel follows on from the contributions which arose during the International Workshop on Advanced Learning Technologies (IWALT 2000). The term asynchronous learning networks (ALN) describes a people network for learning that is largely asynchronous [3]. This type of learning paradigm requires non-traditional instructional/learning techniques whereby there is an expectation for a combination of self-study mixed with the need for self-expression in an online and therefore public forum. Moreover the ALN forum requires the often widely distributed participants, to access common learning resources without the need for concurrent communication.

The panel members were selected to provide an interesting view that ranges from a pure theoretical stance to current projects which implement asynchronous learning frameworks. Ass.Prof. Kommers offers the position that capturing meta-knowledge is complicated because of social and emotional reasons. Prof. Garner identifies the need to discover new instructional strategies that activate context-mediated reasoning processes within a given cultural metaphor, and provides an innovative approach for future research, which deals with the external knowledge contexts. However, Prof Okamoto proposes that the decision making process of both learners and learning mediators needs to be managed effectively before reuse in an educational context. Dr Wiley proposes that a new paradigm of peer-to-peer, learning object-based systems provide a viable opportunity for online collaborative learning to be appropriately architected.

2. Background

The collaborative learning process when implemented online, offers innovative solutions to the requirement for interactive knowledge acquisition and knowledge-sharing in problem solving. Collaborative modeling in E-learning environments (see [5], within a web-enabled meta-knowledge architecture[12];[13], is deemed to constitute the new educational paradigm for novice-learners [15]. In a sense, in the online learning environment we are now able to promote a type of distributed cognition. Because of this researchers must develop knowledge-mediation strategies that permit both the facilitators and the learners to reflect and revise understandings. To this end, the term collaborative memory refers to the use of a kind of an intelligent chat-type window, which has a knowledge-sharing capacity to manage and edit information entered by learners and instructors [20]. Consequently, each learner can reflect on their collaborative activities, which involve other learners and facilitators, to see such interactive-knowledge asynchronously.

The challenge for courseware designers is therefore to be engaged with information technologies in finding practical solutions to the dilemma multi-media platform generates. The range of media related offerings include: text, graphics, and the emerging streaming technologies for voice and video. However while multi-sensory instruction is known to improve a learner's capacity to learn effectively, the overarching role of knowledge-mediated, human-computer interaction (HCI) has been poorly understood.

This is particularly so, in the design of instructional strategies, which integrate contextual E-learning components in asynchronous mode frameworks. Contemporary learning environments often fail to address the need for efficient and effective strategies for capturing vital meta-knowledge [16]. Once we understand the HCI phenomenon, learn how to manage the ALN environment successfully, manage media in an efficient and effective manner, conquer collaborative online communication and knowledge-sharing; we may be able to claim that context-mediated learning has arrived.

3. Contextual Effects

The relationships between instruction and learning are dynamic. In the past research has isolated these two important variables, maintaining that the former induces learners to use learning strategies for particular tasks, while the latter involves the control of learning processes. Moreover, mental skills were thought of as those processes that learners use productively in learning.

However it can be seen in many online learning interfaces that imagery must become one of the key mental skills an individual should utilize for E-learning, and although most people use imagery as a cognitive process, only some of them are skilled in its use [17].

More research is needed to guide courseware authors to focus on addressing the meta-knowledge processing issues, which involve the interactive relationship between instructional strategies and E-learning contexts [14].

3.1 The Question of Spatial Ability

E-learning throws open the question of how people interact with computerized instructional strategies. Novice-learners are immediately aware that digital instruction draws on a wider range of experiential knowledge and skills. In the main, they must be multi-skilled to cope with the duality of being highly computer literate to cope with most E-learning resources, while at the same time entering a new knowledge (learning) domain, as a complete novice. In the past, verbal (or analytic) ability was taken to be measure of crystallized intelligence, or the ability to apply cognitive strategies to new problems and manage a large volume of information in working memory [7], while the non-verbal (or imagery) ability was expressed as fluid intelligence [8].

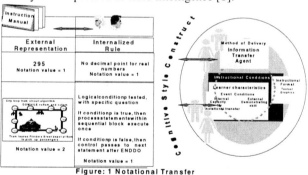

Figure: 1 Notational Transfer

However, as electronic courseware lends itself to integrating verbal (textual) and non-verbal (graphical) instructional conditions that generate novel (or fluid) intellectual problems, more research needs to be carried out to provide instructional designers with prescriptive models that predict measurable instructional outcomes for the broader range of cognitive abilities.

It was speculated that the method of instructional delivery will affect cognitive performance of highly-verbal/low-spatial learners, because they need a direct notational transfer agent (Figure:1); or whether the instructional conditions will disadvantage high-spatial/low-verbal learners, because they will be less able to pick out the unstated assumptions [11].

Picking out these important instructional variables for some types of instructional outcomes provides appropriate instructional environments for a broader range of novice-learners by means of the: *information-transfer-agent, thereby controlling the choice of instructional format and instructional event conditions.* To this end, Figure:1 shows how isolating the key components of the instructional conditions provides the means to manipulate the method-of-delivery, which in turn may bring about a choice of information-transfer-agent.

On the one hand, the external-representation of the instructional material may require a direct notational transfer of the symbol-system used for the instructional strategy (from the external representation of the instructional material to an internalized form in an individual's memory) [6]. For instance: the graphical details in a road map directly relate to the physical environment (in a 1:1 or direct notation ratio, like the explicit representation of basic data-type rules in computer programming). Therefore, in a programming environment, another example would be that a real number must not contain a decimal point (Figure:1).

On the other hand, the embedded details in an abstract metaphor are said to require a non-notational transfer process.

For instance, the programming loop shown as a graphical metaphor in Figure:1 requires a 2:1 transfer for the non-notational characteristics of the external representation to a single internal notational representation.

Researchers have been pursuing the question of what happens when humans process abstract information [2]. While there have been a number of research studies conducted using younger participants, studies on the interaction of instructional strategies with adult-learners are not as common. For instance, asynchronous learning transactions may require substantial conceptual awareness. However, as Ass.Prof Kommers points out, this higher cognitive load may in the end stimulate learners to integrate information between a number of variable learning contexts into more meaningful knowledge domains, thus becoming more conscious about the structure of their own developing knowledge.

The meta-knowledge processing model [11] accommodates the novice-learner, regardless of age. It is now possible to evaluate whether cognitive style in adults is more entrenched.

4. Summary

There can be no doubt that ALN represents an approach to learning that requires a substantial research effort. Panel members agree that we are faced with a complex environment. However as we rush to implement new systems, let us not forget, to deal with the social and emotional context.

5. References

[1] Beiers, H. (2000). Information Architecture's Potential Contribution to an Asynchronous Learning Environment. Kinshuk, C. Jesshope, & T. Okamoto (Eds.), *International Workshop on Advanced Learning Technologies: Advanced Learning Technology: Design and Development Issues*, Palmerston North, NZ, pp.253-254.: IEEE.

[2] Biggs, J.B., & Collis, K.F. (1982). *Evaluating the Quality of Learning: The SOLO Taxonomy (Structure of the Observed Learning Outcome)*. New York: Academic Press.

[3] Campbell, J. (1997). Evaluating ALN: What Works, Who'w Learning? *ALN Magazine, 1* (2), August, http://www.aln.org/alnweb/magazine/issue2/campbell_alntalk.htm.

[4] Garner, B.J. (2000). Meta-Knowledge Acquisition Strategies in Asynchronous Learning Frameworks. Kinshuk, C. Jesshope, & T. Okamoto (Eds.), *International Workshop on Advanced Learning Technologies (IWALT 2000): Advanced Learning Technology: Design and Development Issues*, Palmerston North, NZ, pp.247-249.: IEEE.

[5] Geertshuis, S.A., Bristol, A., Holmes, M.E.A., Clancy, D.M., & Sambrook, S. (2000). Learning and Business: Supporting Lifelong Learning and the Knowledge Worker through the Design of Quality Learning Systems. Kinshuk, C. Jesshope, & T. Okamoto (Eds.), *International Workshop on Advanced Learning Technologies (IWALT 2000): Advanced Learning Technology: Design and Development Issues*, Palmerston North, NZ, pp.186-187.: IEEE.

[6] Goodman, N. (1968). *The Languages of Art: An approach to a theory of symbols*. New York: Bobbs-Merrill.

[7] Hunt, E. (1997). The status of the concept of intelligence. *Japanese Psychological Research, 39* (1 March), 1-11.

[8] Kline, P. (1991). *Intelligence: The psychometric view*. United Kingdom: Routledge.

[9] McKay, E. (2000). Measurement of cognitive performance in computer programming concept acquisition: Interactive effects of visual metaphors and the cognitive style construct. *Journal of Applied Measurement, 1* (3), 293-327.

[10] McKay, E. (2000). Asynchronous Learning Process Dynamics. Kinshuk, C. Jesshope, & T. Okamoto (Eds.), *International Workshop on Advanced Learning Technologies (IWALT 2000): Advanced Learning Technology: Design and Development Issues*, Palmerston North, NZ pp.239-243.: IEEE.

[11] McKay, E. (2000). Instructional Strategies Integrating the Cognitive Style Construct: A Meta-Knowledge Processing Model (Contextual components that facilitate spatial/logical task performance): Doctoral Dissertation (Computer Science and Information Systems): An investigation of instructional strategies that facilitate the learning of complex abstract programming concepts through visual representation, Deakin University, Australia. In Deakin (Ed.). Geelong.

[12] McKay, E. (2000). Towards a Meta-Knowledge Agent: Creating the context for thoughtful instructional systems. S.S.-C. Young, J. Greer, H. Maurer, & Y.S. Chee (Eds.), *Paper presented at the 8th International Conference on Computers in Education/International Conference on Computer-Assisted Instruction (ICCE/ICCAI 2000): New human abilities for the networked society, held 21-24 November*, Vol. 1 (pp. 200-204), Taipei: National Tsing Hua University, Taiwan.

[13] McKay, E. (2000). The Thoughtful Instructional System. S.S.-C. Young, J. Greer, H. Maurer, & Y.S. Chee (Eds.), *Poster Session Paper presented at the 8th International Conference on Computers in Education/International Conference on Computer-Assisted Instruction (ICCE/ICCAI 2000): New human abilities for the networked society, held 21-24 November*, Vol. 2 (pp. 1654-1655), Taipei: National Tsing Hua University, Taiwan.

[14] McKay, E. (2001a). (In referee process). Cognitive Skill Acquisition Through a Meta-Knowledge Processing Model. *Interactive Learning Environments, Cognitive Skills Acquisition in Life-long Learning*.

[15] McKay, E. (2001b). (In referee process). Prepare for the Onset of Asynchronous Learning Platforms: Courseware designers put on notice. *Educational Technology & Society* (Special Issue: Integrating Technology into Learning and Working).

[16] McKay, E., & Garner, B.J. (1999). The complexities of visual learning: Measuring cognitive skills performance. G. Cumming, T. Okamoto, & L. Gomez (Eds.), *Paper presented at the 7th International Conference on Computers in Education: New human abilities for the networked society, held 4-7 November*, Vol. 1 (pp. 208-215), Japan: IOS Press.

[17] McNamara, S.E. (1988). *Designing Visual Analysis Training for the Individual Learner: An Examination of Individual Learner Differences and Training Content and Procedures*. Doctor of Philosophy, Monash.

[18] Merrill, M.D. (2000). Knowledge Objects and Mental Models. Kinshuk, C. Jesshope, & T. Okamoto (Eds.), *International Workshop on Advanced Learning Technologies (IWALT 2000): Advanced Learning Technology: Design and Development Issues*, Palmerston North, NZ, pp.244-246.: IEEE.

[19] Okamoto, T. (2000). RAPSODY: Distance Ecological Model for Self- and Collaborative-Learning. Kinshuk, C. Jesshope, & T. Okamoto (Eds.), *International Workshop on Advanced Learning Technologies: Advanced Learning Technology: Design and Development Issues*, Palmerston North, NZ, pp.249-250.: IEEE.

[20] Okamoto, T., Cristea, A., Matsui, T., & Miwata, T. (2000). Development and Evaluation of a Mental Model Forming Support ITS - the Qualitative Diagnosis Simulator for the SCS Operation Activity. Kinshuk, C. Jesshope, & T. Okamoto (Eds.), *International Workshop on Advanced Learning Technologies (IWALT 2000): Advanced Learning Technology: Design and Development Issues*, Palmerston North, NZ, pp.137-140. IEEE.

[21] Steeples, C. (2000). Reflecting on Group Discussions for Professional Learning: Annotating Videoclips with Voice Annotations. Kinshuk, C. Jesshope, & T. Okamoto (Eds.), *International Workshop on Advanced Learning Technologies: Advanced Learning Technology: Design and Development Issues*, Palmerston North, NZ, pp.251-252. IEEE.

Collaborative Knowledge Management Requirements for Experiential Learning (CKM)

Brian J. Garner
Professor of Computing, School of Computing & Mathematics
Deakin University, Geelong, VIC 3127
Email: brian@deakin.edu.au

Abstract

Exploratory studies in Collaborative Knowledge Management (CKM) across four domains have identified significantly expanded research requirements for experiential learning. This paper reports preliminary conclusions/propositions. The quality of collaborative (group) learning, particularly in experiential processes such as problem solving and professional practice requires innovative support of knowledge-mediated, human interaction requirements and the associated sharing of knowledge between participants.

1. Background

Our research investigates the technical framework for CKM processes that support:
1. Learner activation of, and access to, stored knowledge.
2. Knowledge as Contextual forms and reference models.
3. Study of spreading activation theories in experiential learning
4. Context mining algorithms that accelerate the learners' acquisition of problem solving skills and professional practice effectiveness.

The significance of future research, in association with Dr. McKay and Professor Okamoto, lies both in the development of novel evaluation instruments for measuring cognitive performance improvements derived from external knowledge (contexts), and in the discovery of new instructional strategies that activate context-mediated reasoning processes within a given cultural metaphor.

2. Role of Collaboration Frameworks in Knowledge Management

The study of collaboration frameworks in a number of domains has identified six (6) significant roles germane to the requirements of experiential learners in an Asynchronous Learning Networks (ALN) environment:
1. Basis for Issues Management Determination (eg Global Tax Harmonisation initiatives)
2. Common focus on Ontological engineering
3. Source of explicit Process Knowledge (eg in Professional Practice)
4. Structural Knowledge for Business Modelling (e.g Scenario Definitions)
5. Elaboration of Behavioural Patterns:
   - Role <--> Actor relationships
   - Activity definitions
   - Contextual Forms
6. Specification of Knowledge Integration Requirements:
   - Common notation for knowledge transfer.
   - Formalisation of the model dynamics
   - Specifications for Knowledge Domains may be formalised
     (Garner and Lawrence, 1998)

2. Human aspects of Knowledge Management

Davenport & Prusak (1998) define knowledge as

a fluid mix of framed experience, values, contextual information, and expert insight that provides a framework for evaluating and incorporating new experiences and information. It originates and is applied in the mind of knowers. In organisations it often becomes embedded not only in documents or repositories but also in organisational routines, processes practices and norms.

The effectiveness of CKM frameworks is thus seen to lie in the motivational value of context-mediated, spreading activation mechanisms for experiential learning, and in the associated meta-knowledge acquisition strategies (Garner, 2000), given that there are two classes of knowledge involved in professional practice;
1. Formal Knowledge
2. Informal (tacit) Knowledge (cf Nonaka-Takeuchi)

4. Preliminary Conclusions/Propositions

1. Thematic interpretation of human discourse patterns and group interactions during knowledge transfer processes requires a personalised knowledge dictionary for each participant (refer Figure 1)

2. Novel activation mechanisms may be discovered in CKM frameworks, due to the

fact that tacit knowledge provides the process frameworks for our understanding of human behaviour, and the consequential demand (inferred) for new contextual forms of human interaction.

3. Dynamic contexts, requiring dynamic knowledge fusion (Garner & Lukose, 1992), justifies the paradigmatic approach to CKM, in contrast to contemporary implementations of knowledge management, which are largely based on static data/knowledge structures.

4. Context mining (ie the discovery of new contextual forms) is the essential tool for evaluating the effectiveness of collaborative knowledge management through the identification of thematic behaviour patterns.

5. References

(1) Garner, B.J. and Lawrence, E.(1998):Knowledge Domains in Scoping Educational and Support Issues for Internet Commerce; in "Proceedings of Eleventh International BLED Electronic Commerce Conference", pp109-117; BLED, Slovenia

(2) Davenport, T., & Prusak, L. (1998): "Working Knowledge: how organisations manage what they know."; Harvard Business School Press, Boston, MA, USA.

(3) Garner, B. J. (2000): "Meta-Knowledge Acquisition Strategies in Asynchronous Learning Networks"; proceedings of IWALT 2000, pp 247-248. Published by IEEE Computer Society: ISBN 0-7695-0655-0

(4) Garner, B. J., and Lukose, D. (1992): "Knowledge Fusion"; Proceedings of the 7th Annual Conference on Conceptual Graphs, New Mexico State university.

(5) Garner, B. J., and Lukose, D. (1991): "Expert Advisor utilising Goal Interpretation Methods", in Proceedings of the First World Congress on Expert Systems, Florida, USA.

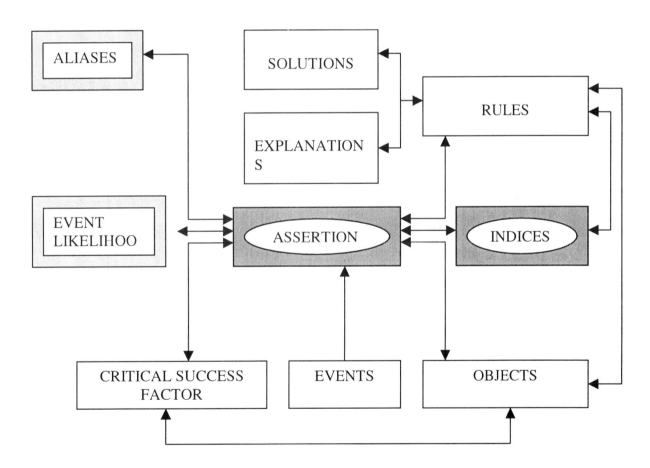

Figure 1:Knowledge Dictionary Structure (Garner & Lukose, 1991)

Collaborative Learning Support Knowledge Management for Asynchronous Learning Networks

Toshio OKAMOTO, Mizue KAYAMA and Alexandra CRISTEA
University of Electro-Communications Graduate School of Information Systems
{okamoto, kayama, alex}@ai.is.uec.ac.jp

Abstract

The large amount of information available on the distributed network environment is an excellent source basis for remote learning environments. However, to support the decision making process of both learners and learning mediators in such a huge information space, some prior arrangement and integration of the learning information is necessary. If managed accordingly, various information can be referred and reused in educational context. At the Laboratory of AI & Knowledge Computing, University of Electro-Communications, Japan, we are constructing a digital portfolio database, which is rapidly increasing, as well as building a learning infrastructure. The remote learning support environment called RAPSODY-EX (Remote and Adaptive Educational Environment: A Dynamic Communicative System for Collaborative Learning) can effectively carry out the collaborative learning support in asynchronous/synchronous learning mode, based on the following two functions:
1) *learning control function for individual learner and group.*
2) *learning information managing function for mediation.*

1. Management of learning information

The managing mechanism consists of two modules: the *learning environment* and the *collaborative memory*, which controls information and data produced in the learning environment.

The learning environment is characterized by two functions: *learning progress monitoring* and *collaborative learning tools and applications*. The first function controls the learning record of individual learners and the progress of the collaborative group learning. The learning information created in such a learning environment is used by the collaborative memory.

The collaborative memory is based on two functions: the *knowledge processing* (input learning information is re-shaped into a standard form) and the *knowledge storage* (attributes related to content are added).

2. Knowledge management in RAPSODY-EX

Knowledge management can be expressed as a conversion cycle between tacit knowledge and expressive knowledge [8]. Tacit knowledge has a non-linguistic representation form, while expressive knowledge results when transforming tacit knowledge into a linguistic form, which enables sharing. In the SECI model [8], socialization (S), externalization (E), combination (C) and internalization (I) of knowledge are expressed.

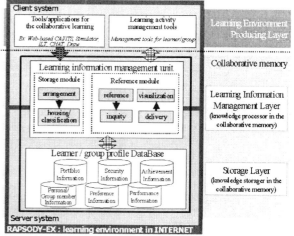

Figure 1: The mechanism schema of the RAPSODY-EX

Knowledge management in educational context is *"the systematic process of finding, selecting, organizing, distilling and presenting information in a way that improves a learner's comprehension and/or ability to fulfill his/her current learning objectives."* (from [14])

RAPSODY-EX, supports item C, as well as the knowledge conversions: $C \to I$ and $E \to C$. The learning information is expressed learners' knowledge, as a result of transformation of tacit knowledge via a language. The questions to consider are:
- *What are the knowledge resources in the learning group?*
- *What is the expected gain for the learning group?*
- *How can the knowledge resources be controlled, to guarantee a maximum gain for the learning group?*

3. The collaborative memory

In the collaborative memory, expressive knowledge is managed via information generation, arrangement, housing, reference and visualization. In the RAPSODY-EX learning space, the following take place:
1) *learning process review,*
2) *summarization of the problem solving process* and
3) *reference of problem solving methods of the other learners.*

The main objects of the knowledge management are generation and management of the learning performance information and the construction of the learner and group portfolio. Learning information results from the collaborative learning application tools, from ones own learning record, from the reference log of other learners as well as the log of problem solving and learning progress.

The collaborative memory consists of two units, for *information storage* and for *information management*.

The information storage unit mainly processes four kinds of information:
1) *Learning information,*
2) *Learner information,*
3) *Learning environment settings information* and
4) *Learning result information.*

The information management unit deals with the reference, arrangement and integration of the learning information. The individual learner profile information is built as required by the IEEE Profile information guidelines [6]. Group information is built from the cumulated individual learner profiles. This process requires conversion from learning log data to learning information. Learning information includes:

- *Information on the learning context and situation*
- *information about sender and sendee of information*
- *significance in educational context*
- *Information on the relational structure of the learning information*
- reference pointer to individual learner(s) and/ or group(s) who produced or linked the information

By adding the above attributes, the learning information can be arranged into a unique, standard form, and can be retrieved for the learner as meaningful information.

4. Conclusion

Our contribution to the present panel discussion is to present a working project, as real example implementation of a collaborative learning support environment based on the new concept of knowledge management for asynchronous learning networks.

5. References

[1] ADLNet, "*Shareable Courseware Object Reference Model: SCORM*", Ver.1.0, http://www.adlnet/org/ , 2000.

[2] A. R. Kaye, "Computer Supported Collaborative Learning in a Multi-Media Distance Education Environment", in Claire O'Malley (Ed.) *Computer Supported Collaborative Learning*, pp.125-143, Springer-Verlag, 1994.

[3] B. Collis, "Design, Development and Implementation of a WWW-Based Course-Support System, *Proceedings of ICCE99*, pp.11-18, 1999.

[4] G. Cumming, T. Okamoto and L. Gomes, "*Advanced Research in Computers in Education*", IOS press , 1998.

[5] J. Elliott,: "What have we Learned from Action Research in School-based Evaluation", *Educational Action Research* ,Vol.1, No.1, pp175-186 , 1993.

[6] IEEE, "*Draft Standard for Learning Technology -Public and Private Information (PAPI) for Learner*", IEEE P1484.2/D6, http://ltsc.ieee.org/, 2000.

[7] IMS, "*Learning Resource Metadata : Information Model, Best Practice and Implementation Guide*", IMS Ver1.0, http://www.imsproject.org/ , 1998.

[8] I. Nonoka, "*The Knowledge-Creating Company*", Oxford University Press, 1995.

[9] L. Colazzo, and A. Molinari, "Using Hypertext Projection to Increase Teaching Effectiveness", *International Journal of Educational Multimedia and Hypermedia*, AACE, 1996.

[10] M. Kobayashi, M. Shinozaki, T. Sakairi, M. Touma, S. Daijavad, and C. Wolf, "Collaborative Customer Services Using Synchronous Web Browser Sharing", *Proceedings of CSCW 98*, pp.99-108, 1998.

[11] The United Kingdum Government, "Connecting the Learning Society", The United Kingdom Government's Consultation paper, 1997.

[12] S. McNeil et al., "*Technology and Teacher Education Annual*", AACE ,1998.

[13] T. Chan et al., "*Global Education ON the Net*", Springer-Verlag , 1997.

[14] T. Davenport, "*Working Knowledge*", Harvard Business School Press, 1997.

[15] T. Kuhn, "*The structure of scientific revolutions*", University of Chicago Press ,1962.

[16] T. Okamoto, A.I.Cristea and M. Kayama, "Towards Intelligent Media-Oriented Distance Learning and Education Environments", *Proceedings of ICCE2000*, 2000.

[17] University of Phoenix Home Page: http://www. uophx.edu/

[18] Jones International University Home Page http://www.jonesinternational.edu/

[19] W. L. Johnson, and E. Shaw, "Using Agents to Overcome Deficiencies in Web-Based Courseware", *Proceedings of the workshop "Intelligent Educational Systems on the World Wide Web"* at the 8th World Conference of the AIED Society, IV-2, 1997.

[20] Y. Aoki and A. Nakajima, "User-Side Web Page Customization," *Proceedings of 8th International Conference on HCI*, Vol.1, pp.580-584, 1999.

[21] Y. Shimizu , "Toward a New Distance Education System", *Proceedings of ICCE99*, pp.69-75, 1999.

Sharing Learning Experiences through Correspondence on the WWW

Piet Kommers
Ass. Professor University of Twente, Faculty of Educational Science and Technology
Division of Educational Instrumentation; Box 217,
7500 AE Enschede, The Netherlands
Tel. +31 53 4893576 or 3611 Fax +31 53 4894580
kommers@edte.utwente.nl
http://www.to.utwente.nl/user/ism/kommers/personal.htm

Abstract

Asynchronous Learning Networks are the facilities and procedures to allow members of the learning communities to be more effective and efficient in their learning. One approach is to see how the 'sharing' of knowledge can be augmented through metadata descriptions attached to portfolios and project work. Another approach is to facilitate the reflection upon individual or collaborative learning experiences (Okamoto, Cristea, Matsui, & Miwata, 2000). The position to be defended in my panel contribution is that both the metadata approach and the attempts to capture the students' meta-knowledge are rather complicated because of social and emotional reasons.

1. Introduction

The general awareness nowadays is that both broad- and narrow casting media provide significant support in the more tailored distribution of insight and understanding within and between learning communities, like in the statement below:

> "All available instruments and channels of information, communications and social action could be used to help convey essential knowledge and inform and educate people on social issues. In addition to the traditional means, libraries, television, radio and other media can be mobilized to realize their potential towards meeting basic education need of all".
>
> *From: Final Report World Conference for All: Meeting Basic Learning Needs, Jomtien, 1990.*

It may be clear that the communicative quality of WWW-based media is superior to one-way media like radio and TV. See also Girard (1990) and Burke (1990); many countries nowadays can still benefit a lot more if we tune radio program structures to the way students actually learn.

2. Goal of the Discussion

"Asynchronous Learning" implies that the sending and receiving co-students reside in a different learning phase and thus have an asymmetric added value like the tutor/tutee mechanism in collaborative learning. The problem in this situation is however that most of the students have the persistent idea that being a tutor means 'giving' and being a tutee means 'receiving'. This preconception paralyzes the synergy between potentially fruitful learning partners. That is why I will plead here for a less conventional mechanism in which the students informally support each other by sharing emotional and existential experiences. Some heterogeneity in course domain, prior knowledge, learning style or ethnic background may even be beneficial. So far the theories on cooperative- and collaborative learning have articulated its value for cognitive synergy between students with a certain complementary cognitive style or different prior knowledge. This is a definition that still fits in the instructional paradigm; expecting that students' learning is affected by the way they are confronted with information, the feedback and the further reconciliation with prior knowledge.

The underlying contribution to this ICALT panel is to claim that beyond the technology of asynchronous learning networks we need to articulate again the notions from social psychology that have been applied to face-to-face education already the years before. For instance the humanistic-oriented definition of learning like: "Learning is the process whereby knowledge is created through the transformation of experience"; David Kolb (1984). One step further will take into account the situational and existential factors that determine the student as a real living person and the way this learning matches / interferes with his/her emotional and affective stage. Here the term "ecology of learning" fits well; It means that the cognitive, attitudinal, social and communicative factors should work together in order to provoke the right mentality for learning.

It is this time scale and granularity that demands for the right support from a "learning community". We may expect that students have significant study skills and knowledge consults to learn in an operational way; they have also acquired the needed memory strategies like for instance the LeiText method:

- Students learn autonomously and motivated
- They plan and actively integrate the learning outcomes in their habits and problem solving approach
- Teachers tend to coach the group process rather than the individual learning steps
- Both intellectual and competence progress is at stake
- Students are supposed to learn quicker and ever more flexible

The price for this shift into student autonomy is a higher moral- and existential pressure on the student as both the teacher and co-students have high expectations from the individual student; As in the more traditional settings the student could complain with the co-students about the unrealistic speed and the severe tests criteria, it is now the student him/her self who has invented the expedition. And it is even more problematic in learning projects where also the co-students are dependent from individual contributions; not to speak about the effects when students are finally grading each other's contributions.

3. Directions for Solutions

The time that project-oriented learning was a relief for one's individual responsibility is far behind; Teachers may start to show the more successful cases of previous years, and the learning group is often seen as an enterprise, competing with other groups. Through realistic cases learning is close to real labor organizations. The question is now how students can still recover from over ambitions and time stress.

Two directions for solutions will be brought forward:

1. On-site learning communities should keep a high degree of informality with only a mature amount of competition and sufficient super-ordinate goals.
2. If this is not affordable because of an over-stressed grading system, the student should find and relax in external WWW-based communication with students who belong to a different institute.

Though the first solution is inherently better, it might be inevitable to fall back on the second alternative. Here the communication serves a more emotional and personal function like consolation and regaining self-confidence. It is the more overall reflection on the link between studying and life itself that counts, and that is hardly reflected in the institutional context. Besides the direct learning community with students from the same institute it seems beneficial to join the larger www-based learning communities for the sake of existential support without any threat for repercussions in ones own institute.

An adjacent aspect is the teachers' learning network that should be apart from their employer or supervisor. It should lead to real partnerships between teachers who have doubts in progressing their skills. Here the importance is to have a low-threshold consultancy mechanism for external peer review and sharing experiences. Teachnet and the 21st Century Teachers Network (21CT) are examples. However they have still a large accent on the provision of information like lesson models. The community aspects still have to be worked out. In fact the student and teacher network may have a lot in common.

4. References

- Kolb, D., *Experiential Learning*, Prentice Hall, Englewood Cliffs, NJ, 1984.
- Jomtien, 1990, http://www.col.org/events/0006commradio.htm
- Burke, A. *Communications & Development: a practical guide*. London: Department for International Development, 1999.
- Girard, B. *Radio Broadcasting and the Internet: Converging for Development and Democracy*. Voices, Journal on Communication and Development, December 1999. 3
- Okamoto, T., Cristea, A., Matsui, T., & Miwata, T. (2000). Development and Evaluation of a Mental Model Forming Support ITS - the Qualitative Diagnosis Simulator for the SCS Operation Activity. Kinshuk, C. Jesshope, & T. Okamoto (Eds.), *International Workshop on Advanced Learning Technologies (IWALT 2000): Advanced Learning Technology: Design and Development Issues*, Palmerston North, New Zealand, pp.137-140.: IEEE Computer Society.

Peer-to-Peer and Learning Objects: The New Potential for Collaborative Constructivist Learning Online

David Wiley
Postdoctoral Fellow, Dept. of Instructional Technology
Utah State University, Logan, Utah
wiley@cc.usu.edu

Abstract

Despite the fact that collaboration is as well studied as any instructional strategy, online collaborative learning efforts have been slow to succeed. Client-server computing models are currently giving way to newer peer-to-peer models. At the same time, traditional conceptions of online learning as electronic page turning are falling to the learning object model. The new paradigm of peer-to-peer, learning objects-based systems provides an opportunity for online collaborative learning to be appropriately architected, and therefore, successful. Specifically, learning objects systems force their designers to consider the important role of context mediation in learning, and design accordingly. Peer-to-peer systems make real time collaboration, a critical component for successful collaborative learning, possible.

1. The Problem

Despite the fact that, as Slavin [1] has observed, "cooperative learning is one of the most thoroughly researched of all instructional strategies," the adoption of cooperative learning strategies to the Internet has been both slow and poor. This is most likely due to the fact that first generation Internet applications, such as e-mail and the World Wide Web, are architected in ways antithetical to collaboration. In this first generation of applications, each of the clients (such as a web browser) was reliant on a central authority (such as a web server) to dispense the desired information. This closely resembles the traditional classroom in which students are dependent on a single authority, in this case their teacher, to "dispense learning." In practice, mangling client-server applications to support collaboration has frequently been about as graceful and successful as convincing lecture-ingrained professors to adopt collaborative instructional strategies. And just as many traditional teachers have slandered collaboration as "academic dishonesty," many traditional client-server applications programmers have been equally reluctant to leave the world they know in order to architect technology that truly supports collaboration.

2. Changing Paradigms

But the times have changed, as they are wont to do. The second generation of Internet applications is emerging, built around a peer-to-peer (P2P) computing paradigm. In a P2P paradigm no machine is privileged over any other, and each is theoretically capable of contributing to the network equally. Just as there are different models of cooperative or collaborative learning, there are different models of P2P computing. In distributed file sharing systems (such as Napster, Gnutella, and Freenet), each machine contributes to the system by providing access to its spare storage space to store files. In distributed computing or cycle sharing systems (such as SETI@home, GENOME@home, and distributed.net), each machine contributes to the system by providing access to its spare computational resources. Advances in our understanding of the characteristics of certain types of groups (e.g., Watts and Strogatz's "small world" or "six degrees of separation" model [2]), combined with P2P computing architectures, open new doors for online collaborative learning.

Simultaneously, our conception of online learning is changing. What was once an exercise in "converting" existing material into web pages has given way to the "learning object" model. The learning object model, in which educational content is stored in small chunks that are reusable in many different learning environments, has gained such broad acceptance that the IEEE has formed the Learning Technology Standards Committee (with 20 working groups) to pursue the creation of standards for the description, interchange, and management of learning objects [3]. The authors of learning object systems praise their ability to reuse content in multiple learning environments, allowing for the creation of scalable learning environments with the potential to be significantly more efficient and adaptive than previous computer-based instructional systems [4], [5], [6] Gibbons, Nelson, & Richards, 2000; Wiley, 2000a). Within the academic field of instructional technology, learning objects have become a topic of intense research and publication, including the appearance of dissertations [6] and books [7] dedicated to the topic, as well as numerous journal articles and book chapters.

3. Context Mediation

Context-mediation assumes the preeminent role in learning object systems. As small chunks of content cannot facilitate learning, these chunks must be placed within instructionally functioning contexts before they can begin to do so. For example, a digital image of a Van Gogh does not teach about art history. Neither does the label, "Starry Night." Neither does a brief exposition of impressionism. But when these elements are juxtaposed, either physically (on a screen or page) or temporally (sequenced in a presentation or lecture), they lend a context to each other that cultivates *meaning*.

The assignment of this context provisioning responsibility, namely, "Who will perform this juxtaposition of elements?," has been the topic of some debate. Some have claimed that automated systems should be given this responsibility, while others have argued that teachers should simply use learning object repositories as digital libraries, assembling their instruction themselves much as they have always done. However, there is another point of view: if we were to charge students with this responsibility given certain guidelines and conditions, the opportunity for a new type of constructivist, collaborative learning may emerge.

Constructive hypertexts provide a conceptual model of where this new technological and pedagogical hybrid could take us. As individual users of a P2P system share learning objects of their own creation with each other in a distributed file sharing system, these resources could replace centrally managed nodes in a CSILE or CFH. Students could then collaborate to elaborate relationships between objects in the system. Current client-server-based systems like Everything2 and Perl Monks show that it is possible for distributed peer review to scale to support this type of distributed knowledge creation activity. Additionally, students could draw on resources in the system and collaborate to build goal-based scenarios or other resource-dependent constructivist learning activities for other students.

4. Collaborative Synchronicity

The Internet, especially the Open Source or Free Software movement, has shown that some types of asynchronous collaboration can be extraordinarily successful. John "Mad Dog" Hall summed up the success of the Linux operating system with the phrase "World Domination through World Cooperation." Open Source lore is filled with stories of distributed programming teams with two or more volunteers on each continent, each team finishing their work day just as the next team awakes and picks up where the first left off. This 24 hour per day asynchronous collaboration schedule (or more realistic variants of it) has been extremely successful at producing high quality software. However, it is very likely that the same model would *not* be equally successful for distributed learning teams.

Distributed software projects are carefully designed from the beginning to be extremely modular so that each team can work in relative isolation, bringing their near finished products together for a final, furious integration period. The overwhelming majority of the time is spent with individuals crafting their individual contribution, while integration and troubleshooting is designed to take as little time as possible.

In a somewhat analogous distributed learning Jigsaw, this would translate into students spending the vast majority of their time in their initial groups, gaining some portion of the expertise their team needs to complete their collaborative project. Accordingly, the team members would spend very little time in their collaborative learning team, engaging in the collaborative part of the learning experience. But this is exactly the *opposite* of the way teachers would hope students working in a Jigsaw format would spend their time. We want the students to spend the majority of their time working together, experiencing the give, take, and joint discovery of team problem solving.

Asynchronous collaboration tools such as those generally available today (e-mail, web-boards, CVS) amply facilitate distributed individuals' modularized work, which can be later be quickly integrated into a whole, as in the case of Open Source Software development. However, individuals' meaning making frequently cannot be meaningfully modularized ahead of time. Rather than asynchronous tools for "divide and conquer" approaches, online collaborative learners need "united we stand" tools that allow them to work together in real time, similar to the structure of their face-to-face meetings. Peer-to-peer systems such as Groove offer learners the hope of finally getting a real time collaborative learning platform.

5. Conclusion

The learning object and peer-to-peer computing paradigms have the potential to overcome the obstacles collaborative learners have traditionally faced online. As they do, new types of collaborative learning strategies are likely to emerge. For example, significantly larger learning group sizes can be employed relying on the small world phenomenon and the distributed peer review facilitated by P2P. Better connections among members of larger groups will increase the complexity of problems collaborative learning groups are able to undertake. By allowing larger group sizes and removing other technology-imposed limitations to online collaborative learning, learning objects and peer-to-peer may both speed the adoption and improve the quality of online collaborative constructivist learning.

6. References

[1] Slavin, R. E. (1989/1990). Research on cooperative learning: Consensus and controversy. *Educational Leadership, 47* (4).
[2] Duncan J. Watts and Steven H Strogatz, "Collective dynamics of 'small-world' networks", Nature 393, 1998.
[3] Learning Technology Standards Committee. http://ltsc.ieee.org/
[4] Advanced Distributed Learning Network; http://adlnet.org/
[5] Gibbons, A. S., Nelson, J. & Richards, R. (2000). The nature and origin of instructional objects. In D. A. Wiley (Ed.), The Instructional Use of Learning Objects: Online Version. http://reusability.org/read/chapters/gibbons.doc
[6] Wiley, D. A. (2000). Connecting learning objects to instructional design theory: A definition, a metaphor, and a taxonomy. In D. A. Wiley (Ed.), The Instructional Use of Learning Objects: Online Version. http://reusability.org/read/chapters/wiley.doc
[7] D. A. Wiley (Ed.), The Instructional Use of Learning Objects: Online Version. http://reusability.org/read/

Tutorials

Learner-Centered Design: Developing Software that Scaffolds Learning

Chris Quintana and Elliot Soloway
*Center for Highly Interactive Computing
in Education
University of Michigan*
quintana@umich.edu

Cathleen Norris
*Department of Technology and Cognition
University of North Texas*

Abstract

Learner-centered design (LCD) is an evolving design approach for designing tools that support learners trying to engage in and understand complex work practices in which they are novices. Learner-centered design focuses on developing tools that incorporate support (or "scaffolding") to support the novice in seeing and doing complex, unknown work so that the learner can begin developing an understanding of that work in a "learning by doing" fashion. In this tutorial, we will give participants a more in-depth treatment of learner-centered design by describing a definition for LCD, theoretical background, and the LCD process and methods to give participants experience with the mindset needed for the LCD approach.

1. Introduction

Learner-centered design (LCD) is an evolving design approach addressing the needs of learners—a specific audience trying to work in and understand a work practice in which they are novices [6]. The central tenet of learner-centered design involves the design of *scaffolding* into software to make complex work activity accessible to novices [6]. Scaffolding allows learners to engage in unknown work activities that would normally be out of their reach to begin developing an understanding of the work [7].

Many researchers are exploring learner-centered design as evidenced, for example, by two special issues of *Communications of the ACM* focusing on LCD, a workshop at the ACM CHI 99 human-computer interaction conference, and previous tutorials at CHI and other university settings. Here, we continue this work with a more in-depth treatment of learner-centered design by describing a LCD definition, theoretical background, and the LCD process and methods to give participants experience with the mindset needed for LCD.

There are two basic, high-level learning objectives for this tutorial:
• *Objective 1*: Learning what learner-centered design is. Participants will learn background about LCD: LCD definitions and descriptions, theoretical background, and examples of learner-centered tools.
• *Objective 2*: Learning how to "do" learner-centered design. Participants will learn design approaches for doing learner-centered design. Through examples presented in the tutorial, participants will see design methods for designing learner-centered software.

2. Objective 1: Defining Learner-Centered Design

We will describe several high-level themes about LCD:
• "Learning by doing" approach. Learner-centered design is informed by a "learning by doing" approach that says learners gain expertise in new work by mindfully performing authentic activities from the work practice.
• Scaffolding in software. Learner-centered software incorporates "scaffolding" to support learners as they do new work. Scaffolds can be considered to be software features that support work practice novices in performing previously inaccessible work activity. As these novices develop expertise in their work, the scaffolds are no longer needed and can fade away from the software.

We will also give a more structured definition of learner-centered design by comparing LCD to user-centered design (UCD) [1]. Specifically, we will compare UCD and LCD along three dimensions [3]:
• The audience being addressed by each design perspective.
• The central design problem being addressed by each design perspective.
• The underlying theory used by each design perspective to address the design problem.

Audience. We will discuss the audience addressed by each design perspective by taking into consideration the audience's level of work expertise, growth, diversity, and motivation. In UCD, the audience of users is considered knowledgeable and motivated about their work. The user audience is less diverse because designers design around their tasks, not the users themselves. Finally, users grow less in their work expertise.

In LCD, the audience of learners—primarily K-16—has little and inaccurate work expertise and wavering motivation to engage in the work. Learner diversity is a key issue because of the differences in gender, learning styles, etc. Finally, learners should gain work expertise, so the growth of the learner must be considered in the design of the tool.

Central Design Problem. We will discuss the central design problem being addressed by each perspective. In UCD, the central design challenge is to design tools that are easy to use and understand and that allow the easy and efficient completion of work activities. Because tool use is the central challenge, designers must address the conceptual gap between user and tool (i.e., the gulfs of execution and evaluation).

In LCD, the central design challenge is to design tools to help learn new work practices. Because learning new work practices is the central challenge, designers must address the conceptual gap between learner and work of expertise (i.e., a gulf of expertise).

Underlying Theory. We will discuss the underlying theory used by each design perspective to address the central design problem. In UCD, the underlying theory is a "theory of action" that describes how people use tools. In LCD, the underlying theory includes social constructivist learning theories that give a perspective on how people learn.

3. Objective 2: Describing the Learner-Centered Design Process

Given a description of learner-centered design, we will describe some design methods and techniques that can be used to develop and evaluate learner-centered software. The traditional approach to software design involves high-level phases such as task/work analysis, requirements specification, design, and evaluation.

We will describe how we should modify the traditional software design cycle to develop specific methods within each phase for learner-centered design. Our primary example for doing so focuses on the design of Symphony, a tool to support the work of science inquiry [4]. Specifically, we will describe:

• Work analysis: How can designers study the work practice to fully understand, especially understanding the *tacit knowledge* that experts use to perform the work? We will discuss the process-space analysis method for analyzing work practices to isolate different components of the work that need to be made explicit and accessible to learners [4].

• Requirements specification: What are *learner-centered* software requirements? Identifying learner-centered requirements involves identifying learner support needs that isolate the areas where learners will need support to engage in the new work practice.

• Design: What kinds of conceptual and physical user interface and software design decisions are needed for LCD? Designers first need to identify the scaffolding strategies that will address the learner support needs identified earlier. Second, designers need to determine how those scaffold strategies will be implemented in software to help learner mindfully engage in the new work practice. We ground these ideas with software examples from University of Michigan, Northwestern University, University of California at Berkeley, and Vanderbilt University

• Evaluation: How can designers evaluate the effectiveness of learner-centered software? Designers need to evaluate scaffolded software to analyze the "effects of" and "effects with" software [5]. Designers need to assess the impact that the individual scaffolds have on the learner's new work understanding (i.e., evaluate the "effects with" software). Designers also need to assess whether learners learned the new work (i.e., evaluate the "effects of" software) [2].

5. Acknowledgements

This material is based on work supported by the National Science Foundation under Grant No. REC 99-80055.

6. References

[1] D. Norman, "Cognitive Engineering", In D.A. Norman & S.W. Draper (Eds.), *User Centered System Design*, Lawrence Erlbaum Associates, 1986.

[2] C. Quintana, E. Fretz, J. Krajcik, & E. Soloway, "Assessment Strategies for Learner-Centered Tools", In B. Fishman & S. O'Conner-Divilbiss (Eds.) *Proceedings of ICLS 2000* (Ann Arbor, June), Lawrence Erlbaum, 2000.

[3] C. Quintana, J. Krajcik, & E. Soloway, "Exploring a Structured Definition for Learner-Centered Design", In B. Fishman & S. O'Conner-Divilbiss (Eds.) *Proceedings of ICLS 2000* (Ann Arbor, June), Lawrence Erlbaum Associates, 2000.

[4] C. Quintana, J. Eng, A. Carra, H. Wu, & E. Soloway, "Symphony: A Case Study in Extending Learner-Centered Design Through Process Space Analysis", *Proceedings of CHI '99* (Pittsburgh, May), ACM Press, 1999.

[5] G. Salomon, D.N. Perkins, & T. Globerson. "Partners in Cognition: Extending Human Intelligence with Intelligent Technologies". *Educational Researcher*, April, 1991.

[6] E. Soloway, S.L. Jackson, J. Klein, C. Quintana, J. Reed, J. Spitulnik, S.J. Stratford, S. Studer, J. Eng, & N. Scala, "Learning Theory In Practice: Case Studies of Learner-Centered Design" *Proceedings of CHI '96* (Vancouver, Canada, April), ACM Press, 1996.

[7] D. Wood, J.S. Bruner, & G. Ross. "The Role of Tutoring in Problem-Solving". *Journal of Child Psychology and Psychiatry*, 17, 1975, pp. 89-100.

Animated pedagogical agents for education training and edutainment

W. Lewis Johnson
USC / Information Sciences Institute
4676 Admiralty Way
Marina del Rey
CA 90292 USA
johnson@isi.edu
Tel: +1 310 448 8210
Fax: +1 310 822 0751

Abstract

Animated pedagogical agents are animated characters that can interact with learners in computer-based environments, in order to stimulate and encourage learning. They improve the effectiveness of education and training applications, and can also be employed in many other interactive applications to assist novice users. They draw on techniques from interactive entertainment, making them well suited for entertainment applications with an educational purpose ("edutainment"). This tutorial provides an introduction to animated pedagogical agents, their use in education, training, and entertainment applications, and methods and technologies used to develop them. The important capabilities of APA's will be described and illustrated in the context of example pedagogical agent systems. Common methods and architectures for building and controlling the behavior of such agents will be described and contrasted. Empirical evaluations of effectiveness of APA's will also be presented. Particular emphasis will be placed on assessing the current state of the art in APA research, highlighting recent accomplishments and identifying areas of current and future research.

Adaptive Educational Environments for Cognitive Skills Acquisition

Ashok Patel
Director of CAL Research
School of Business, De Montfort University
Leicester LE1 9BH, United Kingdom
Email: apatel@dmu.ac.uk

Kinshuk
Information Systems Department
Massey University, Private Bag 11-222
Palmerston North, New Zealand
Email: kinshuk@massey.ac.nz

Abstract

This tutorial deals with the cognitive skills based adaptive educational environments - an approach to designing tutoring systems that has been very effective, especially within applied domains where the learning is predominantly concerned with acquiring operational knowledge. Since such environments can accommodate both the 'instruction' and 'construction' of knowledge and actively engage the learner, they have been successful. This is evident from the popularity and wide acceptance of simulation based tutoring systems. This tutorial aims to demonstrate and discuss various aspects of actually implemented systems, with a view to provide intelligent educational system developers and implementers with the knowledge they need in order to make informed decisions about such environments.

1. Introduction

Competence in any domain requires both the subject specific knowledge and skills, besides the general skills that apply across all the domains. The skill based competence, in turn, consists of physical skills - which refer to physical expertise needed to perform the procedural tasks efficiently; and cognitive skills - which refer to cognitive expertise to successfully perform the procedural tasks, involving situational analysis, interpretation of information, orderly execution of sub-tasks and decision making. The acquisition of physical skills is greatly facilitated by the visibility of the processes and tasks involved in a procedural task, as these can be observed and imitated. The cognitive skills, however, require a much more sophisticated learning process as the cognitive processes run inside a human mind, invisible to an observer. For example, learning to enter data for computer processing requires physical skills that may be obtained by observing and imitating an experienced operator's eye-hand co-ordination and other physical movements, but the consistency verification of the data being entered for the purpose of isolating invalid data requires cognitive skills and very experienced operators would frequently spot the invalid data before consciously rationalising about its invalidity. This tutorial is mainly concerned with applied domains where cognitive skills assume perhaps greater significance for the practitioners than the conceptual knowledge.

2. Cognitive Skills Acquisition in Academy

A major aspect of learning in applied domains is the acquisition of subject specific knowledge and skills to add to the general knowledge and skills. The subject specific skills include cognitive skills required in applying the domain's conceptual knowledge to solve problems. The learners need to acquire a sense of judgement and decision-making ability within different current and foreseen contexts and to be successful, they require skills pertaining to analysis, methodical approach and systematic build up of the full solution through identification and solution of intermediate steps.

In case of some applied domains such as those within science and engineering, academic learning does provide practical 'hands-on' learning sessions where the learners can obtain physical skills and also cognitive skills surrounding the practical work. However, academic learning of some other applied domains, for instance accounting, is quite deficient in terms of opportunities for acquiring cognitive skills. Such skills are regarded so important by the professional bodies that they require adequate work experience before granting membership to those who pass their examinations and thus emphasise the apprenticeship mode of learning.

For academic institutions, the only feasible way of providing a similar learning experience is through cognitive apprenticeship [1] based adaptive educational environments. The tutorial aims to demonstrate an entry level practical application of these ideas in the form of intelligent tutoring software developed under the Byzantium project (for more details please see [2]) with substantial funding from the Higher Education Funding

Councils of the United Kingdom under their Teaching and Learning Technology Programme (TLTP). Many institutions of Further and Higher Education in U.K. and abroad have used the software in teaching more than 10,000 students till-date. It has been tested independently and the researchers found statistically significant improvement in student performance [3].

The Byzantium intelligent tutoring environment (ITE) is primarily aimed at numeric disciplines, which are task oriented and require the learners to achieve cognitive skills [4] as a major constituent of domain competence besides the domain concepts. The tutorial will discuss current implementation and plans for further development, including the need for redesign to achieve a web-based open and flexible learning environment that is increasingly being favoured by many authors (for instance, see [5]).

To contrast with Byzantium, the tutorial will also look at the InterSim project, which provides an educational multimedia system that can be used in and outside educational institutions. The system facilitates acquisition of competence not only in domain knowledge but also in related cognitive skills. The system is activity oriented and supports acquisition of competence in both knowledge and task oriented performances. The InterSim system provides a risk free simulated environment for explorations by a learner without the fear of harming a fellow human and includes real world scenarios based on actual patients. Such systems are applicable to a wide geographic area, since multimedia techniques facilitate the adoption of multilingual support with little effort.

3. Conclusion

The tutorial is intended for all those who are engaged either in research, development or implementation of adaptive and intelligent educational environments. While discussing the two implementations of cognitive apprenticeship based educational environments, it aims to consider five questions that can be asked about any such cognitive skills based adaptive educational environments: 1. What are the various aspects of learning facilitated by such environments; 2. What are the theoretical issues that underlie development of such environments; 3. What functions is served by adaptation and how they are twinned with assessment issues; 4. What pedagogical issues are important in the development of such environments; and 5. How one can go about practically developing such environments. The purpose is to provide intelligent educational system developers and implementers the knowledge for assessing and making informed decisions about such environments.

4. References

[1] Collins A., Brown J. S. & Newman S. E. (1989). Cognitive Apprenticeship : Teaching the crafts of reading, writing and mathematics. In Lauren B. Resnick (Ed.) *Knowing, Learning and Instruction*, Hillsdale, N. J.: Lawrence Erlbaum Associates, 453-494.

[2] Patel A. & Kinshuk (1997). Intelligent Tutoring Tools in a Computer Integrated Learning Environment for introductory numeric disciplines. *Innovations in Education and Training International Journal*, 34 (3), 200-207.

[3] Stoner G. & Harvey J., (1999) Integrating learning technology in a foundation level management accounting course: an e(in)volving evaluation. CTI-AFM Annual Conference, Brighton, U.K., April.

[4] Patel A., Kinshuk & Russell D., (2000) Intelligent Tutoring Tools for Cognitive Skill Acquisition in Life Long learning, Educational Technology & Society, (Ed. Sinitsa K.), Journal of IFETS and IEEE-LTTF, Available: http://ifets.ieee.org/periodical/

[5] Khan B. & Ealy D. (2001). A Framework for a Comprehensive Web-based Authoring System (Ed Khan B.), Educational Technology Publications Inc., New Jersey (ISBN 0-87776-303-9) pp.355-364

5. Biography

Ashok Patel is the Director of CAL Research, De Montfort University, Leicester, United Kingdom. He has been designing educational software for more than ten years. By profession, he is an accountant and teaches Management Accounting and Business Systems. He is a co-initiator of the International Forum of Educational Technology & Society, and co-editor of the Educational Technology & Society journal. He is also an Executive Committee Member of IEEE Learning Technology Task Force.

Kinshuk is a Senior Lecturer at the Massey University, New Zealand. He has been involved in large scale research and development projects for cognitive skills based adaptive educational environments and has published over 30 research papers in international refereed journals, conferences and book chapters. He is an active researcher in learning technologies and human computer interactions. He is currently chairing IEEE Learning Technology Task Force, New Zealand chapter of ACM SIGCHI, and the International Forum of Educational Technology & Society. He is also editor of 'Educational Technology & Society' journal (ISSN 1436-4522) and 'Learning Technology Newsletter (ISSN 1438-0625).

Author Index

Abbas, J. 107
Abidi, S. 441
Abramson, L. 335
Ainsworth, S. 189
Albalooshi, F. 231
Albano, G. 241
Aldridge, G. 11
Al-Herbish, J. 445
Alkhalifa, E. 231
Allison, C. 29
Ally, M. 469
Aluísio, S. 257
Anastasiades, P. 5
Andreas, P. 17
Angehrn, A. 174, 225
Anido, L. 3
Arbaoui, A. 445
Armitt, G. 413
Arruarte, A. 309
Aurum, A. 160
Austin, J. 7
Awyzio, G. 315
Axmann, M. 105
Baird, R. 123
Barcelos, I. 257
Batrum, P. 153
Baylor, A. 462
Beer, M. 413
Belkada, S. 135, 375
Bennett, C. 285
Bertalan, F. 453
Bieliková, M. 193
Blauvelt, G. 301
Bouras, Ch. 13
Boyer, B. 35
Brantner, S. 215
Browning, P. 447
Caeiro, M. 3
Callear, D. 139
Cao, J. 78
Carchiolo, V. 96
Cardinali, F. 21
Cerri, S. 109, 203
Chan, A. 78
Chan, S. 78
Chang, M. 289
Chen, J. 135
Chen, N. 369
Chen, N.-S. 94
Chepegin, V. 111, 431
Chien, F.-Y. 53, 127
Chiu, C.-H. 57
Chong, N. 319
Chu, U. 291
Chun, H. 180
Clark, J. 37
Conole, G. 327, 345, 447
Correa, J. 305, 409
Crawford, L. 180
Cristea, A. 135, 267, 375, 415, 418, 422, 490
Cross, J. 160
Csanyi, G. 397
Cupp, E. 67
da Nóbrega, G. 109
Dagdilelis, V. 166
Danchak, M. 67
Daniels, R. 413
Davcev, D. 387
Davis, H. 401
Dees, J. 335
del Carmen Malbrán, M. 449
Dingley, A. 199
do Socorro da Silva, A. 243
Duffin, J. 61
Efopoulos, V. 166
Eisenberg, M. 301
Elorriaga, J. 309
Enzi, T. 215
Erkens, G. 269
Evangelidis, G. 166
Fernández, M. 3
Fernández-Caballero, A. 283

Fink, D.	305, 409
Formato, F.	241
Foster, K.	67
Fuller, A.	315
Fung, I.	361
Galeev, I.	111, 431
Gamper, J.	197
García Robles, R.	435
Garner, B.	488
Georgantis, N.	41
George, I.	291
Ghyselen, L.	413
Gibbons, A.	61
Gilbert, J.	439
Glücksmann, I.	437
Goh, A.	441
González, P.	283
Gopinathan, S.	180
Graesser, A.	460
Green, S.	413
Guth, S.	215
Han, B.	297
Handzic, M.	160
Hartley, R.	145
Hashimoto, M.	221
Hatzilygeroudis, I.	239
Hawkes, M.	247
Heh, J.-S.	289
Hernández-Domínguez, A.	243
Ho, C.	349
Hong, H.	227, 297
Hornig, G.	13
Hoskins, J.	291
Hu, X.	460
Hübscher, R.	365
Hubscher-Younger, T.	113
Illmann, T.	90
Inoue, H.	135, 422
Ishikawa, T.	451
Jaspers, J.	269
Jerram, P.	383
Jerrams-Smith, J.	139
Jiang, G.	405
Jianguo, L.	391
Johnson, A.	67
Johnson, W.	501
Juell, P.	37
Kale, I.	279
Kanselaar, G.	269
Karadimitriou, P.	209
Karagiannidis, C.	21, 209
Kasyanov, V.	307
Katerina, K.	339
Kayama, M.	267, 273, 418, 422, 490
Keegan, H.	149
Kemp, R.	153
Khine, M.	180
Khong, C.-W.	443
Kim, C.	67
Kimovski, G.	387
King, F.	157
Kingsnorth, A.	480
Kinshuk	227, 297, 502
Klett, F.	63
Kommers, P.	471, 492
Kort, B.	43
Koutsojannis, C.	239
Krajcik, J.	353
Krukowski, A.	279
Kubota, Y.	213
Kuittinen, M.	178
Kuminek, P.	261
Kuo, C.-H.	53, 127
Kuo, R.	289
Kurz, L.	437
Lan, J.	405
Lawrence, C.	123
Lee, C.-S.	235, 443
Leifheit, M.	409
Lepouras, G.	125
Li, W.	433
Li, Y.	473
Lin, F.	411
Lin, J.	349
Lin, K.	369
Ling, S.-W.	443
Lister, R.	383
Llamas, M.	3
Longheu, A.	96
López, V.	283

Lu, H.	473
Lu, Z.	473
Luchini, K.	55
Luke, R.	381
Luo, J.	357
Malgeri, M.	96
Manolis, S.	17
Maraschi, D.	203
Mariela Villar, C.	449
Martel, C.	313
Martens, A.	90
Martinengo, G.	203
Mastrangelo, M.	291
Matheson, M.	74
Matsuda, H.	70, 164, 451
Matsui, T.	422
Mayall, H.	157
McClean, P.	37
McFarlane, P.	315
McKay, E.	485
McKechan, D.	29
Metcalfe, A.	7
Michaelson, R.	29
Milrad, M.	141
Montañés, J.	283
Montero, F.	283
Mora, M.	82
Moraes, C.	305
Moreno, R.	464
Moriyón, R.	82
Moundridou, M.	185
Nabeth, T.	174, 225
Nakabayashi, K.	213
Narayanan, N.	113
Nejdl, W.	197
Neumann, G.	215
Ngoh, M.	180
Nijholt, A.	457
Norris, C.	107, 499
O Coldeway, D.	343
O'Donoghue, J.	263
O'Hagan, C.	475
Obitko, M.	437
Oehler, P.	55
Okamoto, M.	221
Okamoto, T.	135, 267, 273, 375, 415, 418, 422, 490
Okui, Y.	221
Oliveira, Jr., O.	257
Oliver, M.	327
Orlowska, M.	349
Ou, S.-C.	121
Pahl, C.	395
Paine, C.	170
Papageorgiou, A.	209
Papaioannou, V.	209
Papaterpos, C.	41
Papatheodorou, T.	41
Paris, A.	17
Park, A.	291
Patel, A.	227, 297, 502
Payr, S.	397
Person, N.	460
Petrushin, V.	129
Picard, R.	43
Pilkington, R.	253, 261, 285
Pöntinen, S.	178
Prentzas, J.	239
Qu, C.	197
Quintana, C.	55, 353, 499
Ravenscroft, A.	74
Raybourn, E.	151
Razmerita, L.	225
Reichl, F.	397
Reilly, R.	43
Roda, C.	225
Rodríguez, J.	3
Ruddle, A.	29
Rueda, U.	309
Rufflo, M.	453
Sadiq, W.	349
Saini-Eidukat, B.	37
Sakauchi, M.	319
Sala, N.	103
Sallantin, J.	109
Sampaio, J.	257
Sampson, D.	21, 209
Samsuri, P.	119
Sánchez, T.	283
Sandqvist, J.	413
Santos, J.	3

Sarlin, D.	67
Satratzemi, M.	166
Schwert, D.	37
Seitz, A.	90
Seki, K.	418, 422
Semrau, P.	35
Shabajee, P.	199
Sharpe, L.	180
Sharples, M.	379
Shih, Y.-C.	94
Shindo, Y.	70, 164
Shinohara, M.	221
Shinohara, T.	213
Simon, B.	215
Simos, R.	17
Singh, G.	263
Singh, Y.	235
Sixsmith, A.	413
Skelton, S.	478
Slator, B.	37
Snitzer, M.	7
Soh, V.	139
Soine, R.	49
Soloway, E.	55, 107, 353, 499
Sonntag, A.	305
Sosnovsky, S.	111, 431
Spore, M.	137
Stewart, T.	153
Sun, M.	426
Sung, W.-T.	121
Suthers, D.	25, 371
Sutinen, E.	178
Takase, H.	451
Taso, N.	127
Terashima, S.	221
Thomas, P.	170
Tosukhowong, P.	319
Trajkovic, V.	387
Tretiakov, A.	227
Triantafillou, V.	13
Tsiatsos, Th.	13
Tsiriga, V.	131
Ueno, M.	331
van Bruggen, J.	413
Van Toorn, C.	160
Varley, G.	145
Vavoula, G.	379
Vierlinger, U.	397
Vignollet, L.	313
Virvou, M.	131, 185, 339
Vranch, A.	480
Walker, D.	251
Walker, S.	253
Wang, C.-C.	53
Wang, H.	323
Weber, M.	90
Weir, G.	125
Wenger, D.	453
White, A.	37
White, S.	401
Wible, D.	53, 127
Wijekumar, K.	86
Wiley, D.	494
Wilkinson, J.	205
Williams, B.	189
Wilson, D.-M.	439
Wisher, R.	335
Witzke, D.	291
Witzke, W.	291
Wong, A.	180
Wood, D.	189
Xiao, L.	391
Xiaozhen, Z.	391
Xie, X.	411
Yoshida, H.	213
Yusof, N.	119
Zaïane, O.	357
Zhuang, X.	405

Notes

Press Operating Committee

Chair
Mark J. Christensen
Independent Consultant

Editor-in-Chief
Mike Williams
Department of Computer Science, University of Calgary

Board Members

Roger U. Fujii, *Vice President, Logicon Technology Solutions*
Richard Thayer, *Professor Emeritus, California State University, Sacramento*
Sallie Sheppard, *Professor Emeritus, Texas A&M University*
Deborah Plummer, *Group Managing Editor, Press*

IEEE Computer Society Executive Staff
Anne Marie Kelly, *Acting Executive Director*
Angela Burgess, *Publisher*

IEEE Computer Society Publications

The world-renowned IEEE Computer Society publishes, promotes, and distributes a wide variety of authoritative computer science and engineering texts. These books are available from most retail outlets. Visit the CS Store at *http://computer.org* for a list of products.

IEEE Computer Society Proceedings

The IEEE Computer Society also produces and actively promotes the proceedings of more than 160 acclaimed international conferences each year in multimedia formats that include hard and softcover books, CD-ROMs, videos, and on-line publications.

For information on the IEEE Computer Society proceedings, please e-mail to csbooks@computer.org or write to Proceedings, IEEE Computer Society, P.O. Box 3014, 10662 Los Vaqueros Circle, Los Alamitos, CA 90720-1314. Telephone +1-714-821-8380. Fax +1-714-761-1784.

Additional information regarding the Computer Society, conferences and proceedings, CD-ROMs, videos, and books can also be accessed from our web site at *http://computer.org/cspress*

Revised April 20, 2001